Physical and Chemical Hydrogeology

Patrick A. Domenico
David B. Harris Professor of Geology
Texas A&M University

Franklin W. Schwartz
Ohio Eminent Scholar in Hydrogeology
The Ohio State University

John Wiley & Sons
New York   Chicester   Brisbane   Toronto   Singapore

Library of Congress Cataloging in Publication Data:

Domenico, P. A. (Patrick A.)
    Physical and chemical hydrogeology / Patrick A. Domenico, Franklin
  W. Schwartz.
      p.   cm.
    Includes bibliographical references and index.
    ISBN 0-471-50744-X
    1. Hydrogeology.   I. Schwartz, F. W. (Franklin W.)   II. Title.
  GB1003.2.D66   1990
  551.46—dc20   90-39772
                                                              CIP

Printed in Singapore

10 9 8 7 6 5 4 3 2 1

*To Diane and Cynthia
and
Lucy and the Memory of Phil*

Preface

We view *Physical and Chemical Hydrogeology* as a textbook rather than a reference volume or collection of facts and formulas with little underlying organization. The approach taken is process-oriented, and one of the goals is to provide an intuitive feeling for the total science so that the subject can be seen as a whole rather than a collection of unconnected pieces. However, we have not overlooked the value of reference material, and original and other sources are included in a rather extensive bibliography. We recognize further that there is great interest in the practice of hydrogeology. To this end we include chapters or sections that deal exclusively with state-of-the-art applications, including a significant number of worked examples of computational procedures and a problem set for most of the chapters. We also recognize that in the absence of theory there can be no practice, only chaos. As some of the mathematical ideas in hydrogeology are among the most abstract that beginning students in the physical sciences will ever encounter, we have attempted a reader-friendly style of exposition so that the mathematics can at least be understood conceptually. The price one pays for this style of exposition is increased length and a fair amount of redundancy, some of which may hopefully reinforce the more difficult concepts.

Physical and Chemical Hydrogeology is divided into three sections. After presenting preliminary information on the hydrologic cycle and the porosity and permeability of porous material, the sections deal with fluid, energy, and mass transport in porous media. We believe these sections will serve a diversity of interests, including those of geologists, practicing hydrogeologists and engineers, geochemists, and geophysicists interested in fluid dynamics. We claim little originality in the basic content of the book, but we are pleased with the manner in which it is organized and treated. Whoever the reader, the book is intended as an introduction to hydrogeology, either at the advanced undergraduate or beginning graduate level. We have assumed the reader knows something about calculus, basic physics, chemistry, and geology.

The consideration of notation and units is always a central one in textbooks. Because of the integration of ideas from a variety of disciplines, it is difficult to employ a consistently familiar system of symbols. Whenever one symbol is generally accepted for two different quantities, choices have to be made. The question of units is equally perplexing. The SI system is used to some degree within the United States and more or less universally everywhere else. In the United States, some units are derived from the English system, some from the metric system, and some are best regarded as "field" units because their use is so common in field practice. In this text, we generally use SI metric units but will also provide an equivalent measure in field units where the use of such units is common practice.

For us the work was truly a collaborative effort with each of us contributing equally to the book. We would like to express our thanks to others who contributed directly and indirectly, in particular Alan Fryar, who prepared Section 15.4, and Vic Palciauskas, who commented on most of the theoretical developments through Chapter 9. Large parts of the manuscript were reviewed by Dick Jackson, Darrell Leap, Gary Robbins, and Don Siegel. Jim Hendry, Les Smith, Rob Schincariol, John Tinker, Jr., and Hans-Olaf Pfannkuch reviewed sections of the manuscript. Their helpful comments and suggestions helped us shape the final draft. We are also indebted to other colleagues and friends, and students who have influenced us through the years.

July 1990   PATRICK A. DOMENICO
                                                  FRANKLIN W. SCHWARTZ

Contents

Chapter 1
Introduction / 1

1.1 What Is Hydrogeology? / 2

Physical Hydrogeology Before the Early 1940s / 3
Chemical Hydrogeology Before the Early 1960s / 5
Post 1960 Hydrogeology / 5

1.2 The Relationship Between Hydrogeology and Other Fields of Geology / 7

1.3 How to Use This Book / 8

1.4 The Hydrologic Cycle / 9

Components Of the Hydrologic Cycle / 9
Evapotranspiration and Potential Evapotranspiration / 12
Infiltration and Recharge / 14
Baseflow / 14
Hydrologic Equation / 18

Chapter 2
The Origin of Porosity and Permeability / 23

2.1 **Porosity and Permability** / 24

Porosity and Effective Porosity / 24
Permeability / 27

2.2 **Continental Environments** / 28

Weathering / 28
Erosion, Transportation, and Deposition / 31
    Fluvial Deposits / 32
        Alluvial Valleys / 33
        Alluvial Basins / 33
    Eolian Deposits / 35
    Lacustrine Deposits / 35
    Glacial Deposits / 35

2.3 **The Boundary Between Continental and Marine Environments** / 36

2.4 **Marine Environments** / 38

Lateral and Vertical Succession of Strata / 38
Ancestral Seas and Their Deposits / 39
    The Paleozoic Rock Group / 39
    The Mesozoic Rock Group / 41
    The Cenozoic Rock Group / 41
Diagenesis in Marine Environments / 42
    Porosity Reduction: Compaction and Pressure Solution / 42
    Chemical Rock-Water Interactions: Secondary Porosity in Sandstones / 44

2.5 **Uplift, Diagenesis, and Erosion** / 46

The Style of Formations Associated with Uplift / 47
Secondary Porosity Enhancement in Carbonate Rocks / 49

2.6 **Tectonism and the Formation of Fractures** / 50

Style of Fracturing / 51
Fluid Pressure and Porosity / 53
Connectivity / 53

Chapter 3
Ground Water Movement / 55

3.1 **Darcy's Experimental Law and Field Extensions** / 56

The Nature of Darcy's Velocity / 57
Hydraulic Head: Hubbert's Force Potential / 58

The Gradient: An Introduction to Field Theory   / 60
Physical Interpretation of Darcy's Proportionality Constant   / 61
Units and Dimensions   / 63

**3.2   Hydraulic Conductivity and Permeability
of Geologic Materials   / 63**

Observed Range in Hydraulic Conductivity Values   / 63
Character of Hydraulic Conductivity Distribution   / 66
Anisotropicity and Heterogeneity: Origin and Manifestations   / 67
    Heterogeneity and the Classification of Aquifers   / 70
Darcy's Law for Anisotropic Material   / 71
Measurement of Hydraulic Conductivity   / 74
    Laboratory Testing   / 75
    The Search for Empirical Correlations   / 75

**3.3   Darcy's Law and the Field Mapping
of Hydraulic Head   / 77**

3.4   Flow in Fractured Rocks   / 83

Continuum Approach to Fluid Flow   / 83
    Intergranular Porous Rocks   / 83
    Fractured Rocks   / 84
The Cubic Law   / 87

3.5   Flow in the Unsaturated Zone   / 88

Unsaturated Flow in Fractured Rocks   / 90

Chapter 4
Elastic Properties and Main Equations of Flow   / 99

4.1   Conservation of Fluid Mass   / 100

Main Equations of Flow   / 102

4.2   The Storage Properties of Porous Media   / 104

Compressibility of Water and Its Relation to Specific Storage   / 105
Compressibility of the Rock Matrix: Effective Stress Concept   / 107
Matrix Compressibility and Its Relation to Specific Storage   / 109
Equation for Confined Flow in an Aquifer   / 115
Specific Yield of Aquifers   / 116

4.3   Boundary Conditions and Flow Nets   / 118

4.4   Deformable Porous Media   / 124

One-Dimensional Consolidation   / 124
    Development of the Flow Equation   / 124
    The Undrained Response of Water Levels to Natural Loading Events   / 126
    The Drained Response of Water Levels to Natural Loading Events   / 131

Three-Dimensional Consolidation / *132*
    Elastic Properties in Deformational Problems / *132*
    Flow Equations for Deformable Media / *135*

4.5 Dimensional Analysis / *136*

Chapter 5
Hydraulic Testing: Models, Methods, and Applications / *141*

5.1 Prototype Geologic Models in Hydraulic Testing / *142*

5.2 Conventional Hydraulic Test Procedures and Analysis / *144*

The Theis Nonequilibrium Pumping Test Method / *144*
    The Curve Matching Procedure / *148*
    Assumptions and Interpretations / *150*
Modifications of the Nonequilibrium Equation / *151*
    Time-Drawdown Method / *152*
    Distance-Drawdown Method / *153*
Steady State Behavior as a Terminal Case of the Transient Response / *155*
The Hantush-Jacob Leaky Aquifer Method / *155*
Water Table Aquifers / *159*

5.3 Single Borehole Tests / *161*

Recovery in a Pumped Well / *162*
The Drill Stem Test / *163*
Slug Injection or Withdrawal Tests / *164*
Response at the Pumped Well: Specific Capacity / *168*
Direct Determination of Ground Water Velocity: Borehole Dilution Tests / *171*

5.4 Partial Penetration, Superposition,
and Bounded Aquifers / *172*

Partial Penetration / *172*
Principle of Superposition / *173*
Bounded Aquifers / *175*

5.5 Hydraulic Testing in Fractured or Low
Permeability Rocks / *181*

Single Borehole Tests / *182*
Multiple Borehole Tests / *182*

5.6 Some Applications to Hydraulic Problems / *185*

Planning a Pumping Test / *185*
    Screen Diameter and Pumping Rates / *186*
    Well Yield: The Step-Drawdown Test / *186*
    Distance to the Observation Well(s) / *187*
Planning a Dewatering Operation / *188*
A Problem in Water Supply / *190*

Chapter 6
Ground Water as a Resource   / 199

6.1   Development of Ground Water Resources   / 200

The Response of Aquifers to Pumping   / 200
Yield Analysis   / 202
Water Law   / 203
Artificial Recharge and Conjunctive Use   / 205
    Artificial Recharge   / 206
    Conjunctive Use   / 209

6.2   Simulation of Aquifer Response to Pumping   / 210

Numerical Simulation   / 211
    The Method of Finite Differences   / 211
        Computational Techniques, Steady Flow   / 214
        Computational Techniques, Unsteady Flow   / 216
    The Method of Finite Elements   / 216
    On the Use of Numerical Models   / 216

6.3   Land Subsidence and Sea Water Intrusion   / 217

Land Subsidence   / 217
    Physical Properties of Sediments   / 219
    Mathematical Treatment of Land Subsidence   / 224
        Vertical Compression   / 224
        The Time Rate of Subsidence   / 228
        Simulation of Subsidence   / 230
Salt Water Intrusion in Coastal Aquifers   / 230
    The Fresh Water–Salt Water Interface in Coastal Regions   / 231
        The Ghyben-Herzberg Relation   / 232
        The Shape of the Interface with a Submerged Seepage Surface   / 233
    Upconing of the Interface Caused by Pumping Wells   / 235

6.4   Simulation-Optimization Concepts   / 236

Chapter 7
Ground Water in the Basin Hydrologic Cycle   / 241

7.1   Topographic Driving Forces   / 242

The Early Field Studies   / 242
Conceptual, Graphical, and Mathematical Models of Unconfined Flow   / 243
    Effect of Basin Geometry on Ground Water Flow   / 247
    Effect of Basin Geology on Ground Water Flow   / 249
Ground Water in Mountainous Terrain   / 255

7.2   Surface Features of Ground Water Flow   / 260

Recharge-Discharge Relations   / 260

Ground Water-Lake Interactions / 264
Ground Water-Surface Water Interactions / 266

**7.3 Some Engineering and Geologic Implications
of Topographic Drive Systems** / 268

Large Reservoir Impoundments / 268
Excavations and Tunnels: Inflows and Stability / 269
    The Sea Level Canal / 269
    Ground Water Inflow into Excavations / 271
    The Stability of Excavations in Ground Water Discharge Areas / 273
    Ground Water Inflows into Tunnels / 274
Landslides and Slope Stability / 278

Chapter 8
Ground Water in the Earth's Crust / 283

**8.1 Abnormal Fluid Pressures in Active
Depositional Environments** / 284

Origin and Distribution / 284
Mathematical Formulation of the Problem / 288
Isothermal Basin Loading and Tectonic Strain / 290
    One-Dimensional Basin Loading / 291
    Extensions of the One-Dimensional Loading Model / 294
        Vertical Compression with Compressible Components / 295
        Horizontal Extension and Compression / 295
        Tectonic Strain as a Pressure-Producing Mechanism / 297
Thermal Expansion of Fluids / 298
Fluid Pressures and Rock Fracture / 301
Phase Transformations / 304
Subnormal Pressure / 306
Irreversible Processes / 306

8.2 Pore Fluids in Tectonic Processes / 307

Fluid Pressures and Thrust Faulting / 308
Seismicity Induced by Fluid Injection / 309
Seismicity Induced in the Vicinity of Reservoirs / 312
Seismicity and Pore Fluids at Midcrustal Depths / 312
The Phreatic Seismograph: Earthquakes and Dilatancy Models / 313

Chapter 9
Heat Transport in Ground Water Flow / 317

9.1 Conduction, Convection, and Equations of Heat Transport / 319

Fourier's Law / 319
Convective Transport / 321
Equations of Energy Transport / 323
    The Heat Conduction Equation / 324
    The Conductive-Convection Equation / 325
    Dimensionless Groups / 326

9.2 Forced Convection / 327

Temperature Profiles and Ground Water Velocity / 328
Heat Transport in Regional Ground Water Flow / 332
Heat Transport in Active Depositional Environments / 335
Heat Transport in Mountainous Terrain / 341

9.3 Free Convection / 343

The Onset of Free Convection / 344
Sloping Layers / 345
Geologic Implications / 346

9.4 Energy Resources / 347

Geothermal Energy / 347
Energy Storage in Aquifers / 348

9.5 Heat Transport and Geologic Repositories for Nuclear Waste Storage / 349

The Nuclear Waste Program / 349
The Rock Types / 350
Thermohydrochemical Effects / 352
Thermomechanical Effects / 354

Chapter 10
Solute and Particle Transport / 357

10.1 Advection / 358

10.2 Particle Transport / 360

10.3 Basic Concepts of Dispersion / 362

Diffusion / 366
Mechanical Dispersion / 369

10.4　Character of the Dispersion Coefficient / 371

Studies at the Microscopic Scale　/ 371
Dispersivity as a Medium Property　/ 372
Studies at Macroscopic and Larger Scales　/ 372

10.5　A Geostatistical Model of Dispersion　/ 377

Mean and Variance　/ 378
Autocovariance and Autocorrelation Functions　/ 378
Estimation of Dispersivity　/ 379

10.6　Mixing in Fractured Media　/ 381

Chapter 11
Principles of Aqueous Geochemistry　/ 387

11.1　Introduction to Aqueous Systems　/ 388

Aqueous Solution Phase　/ 389
Gas and Solid Phases　/ 391

**11.2　Structure of Water and the Occurrence
of Mass in Water**　/ 391

**11.3　Equilibrium Versus Kinetic Descriptions
of Reactions**　/ 392

Reaction Rates　/ 393

11.4　Equilibrium Models of Reaction　/ 394

Activity Models　/ 395

11.5　Deviations from Equilibrium　/ 398

11.6　Kinetic Reactions　/ 399

11.7　Organic Compounds　/ 402

11.8　Ground Water Composition　/ 408

The Routine Water Analysis　/ 408
Specialized Analyses　/ 410

11.9　Describing Chemical Data　/ 411

Abundance or Relative Abundance　/ 411
Abundance and Patterns of Change　/ 414

Chapter 12
Chemical Reactions　/ 421

12.1　Acid-Base Reactions　/ 422

Natural Weak Acid-Base Systems　/ 423
CO_2-Water Systems　/ 424

Alkalinity / *425*

12.2 Solution, Exsolution, Volatilization, and Precipitation / *427*

Gas Solution and Exsolution / *427*
Solution of Organic Solutes in Water / *428*
Volatilization / *430*
Dissolution and Precipitation of Solids / *432*
Solid Solubility / *432*

12.3 Complexation Reactions / *434*

Stability of Complexes and Speciation Modeling / *434*
Major Ion Complexation and Equilibrium Calculations / *437*
Enhancing the Mobility of Metals / *438*
Organic Complexation / *438*

12.4 Reactions on Surfaces / *440*

Sorption Isotherms / *440*
Hydrophobic Sorption of Organic Compounds / *443*
Multiparameter Equilibrium Models / *446*

12.5 Oxidation-Reduction Reactions / *450*

Oxidation Numbers, Half-Reactions, Electron Activity, and Redox Potential / *450*
Kinetics and Dominant Couples / *455*
Control on the Mobility of Metals / *455*
Biotransformation of Organic Compounds / *457*
   Rates Limited by Substate Availability / *458*
   Rates Limited by Availability of Electron Acceptors / *459*

12.6 Hydrolysis / *460*

12.7 Isotopic Processes / *462*

Radioactive Decay / *462*
Isotopic Reactions / *464*
Deuterium and Oxygen-18 / *465*
Carbon-13 and Sulfur-34 / *467*

Chapter 13
The Mathematics of Mass Transport / *471*

13.1 Mass Transport Equations / *472*

The Diffusion Equation / *472*
The Advection-Diffusion Equation / *473*
The Advection-Dispersion Equation / *475*

13.2   Mass Transport with Reaction / 475

First-Order Kinetic Reactions / 476

Equilibrium Sorption Reactions / 476

Heterogeneous Kinetic Reactions / 478

13.3   Boundry and Initial Conditions / 480

Chapter 14
Mass Transport in Ground Water Flow:
Aqueous Systems / 483

14.1   Mixing as an Agent for Chemical Change / 484

The Mixing of Meteoric and Original Formation Waters / 485

Diffusion in Deep Sedimentary Environments / 487

14.2   Inorganic Reactions in the Unsaturated Zone / 492

(1) Gas Dissolution and Redistribution / 492

(2) Weak Acid-Strong Base Reactions / 494

(3) Sulfide Oxidation / 497

(4) Gypsum Precipitation and Dissolution / 497

(5) Cation Exchange / 497

14.3   Organic Reactions in the Unsaturated Zone / 498

(1) Dissolution of Organic Litter / 499

(2) Complexation of Fe and Al / 499

(3) Sorption of Organic-Metal Complexes / 499

(4) Oxidation of Organic Compounds / 499

14.4   Inorganic Reactions in the Saturated Zone / 500

(1) Weak Acid-Strong Base Reactions / 500

(2) Dissolution of Soluble Salts / 504

(3) Redox Reactions / 504

(4) Cation Exchange / 509

14.5   Case Study of the Milk River Aquifer / 510

**14.6   Quantitative Approaches for Evaluating
Chemical Patterns** / 515

Homogeneous and Heterogeneous Equilibrium Models / 515

Mass Balance Models / 520

Reaction Path Models / 523

14.7   Age Dating of Ground Water / 525

Tritium / 526

Carbon-14 / 527

Chlorine-36 / 530

Chapter 15
Mass Transport in Ground Water Flow:
Geologic Systems / 533

15.1 Mass Transport in Carbonate Rocks / 534

The Approach Toward Chemical Equilibrium in Carbonate Sediments / 536
The Problem of Undersaturation / 539
Dolomitization / 542

15.2 Economic Mineralization / 544

Origin of Ore Deposits / 545
    Roll Front Uranium Deposits / 545
    Mississippi Valley-Type Lead-Zinc Deposits / 548
        Water from Compaction / 549
        Gravity Flow Origins / 551
    The Expulsion of Fluids from Orogenic Belts
    and Continental Collisions / 551
Noncommercial Mineralization: Saline Soils and Evaporites / 554

15.3 Migration and Entrapment of Hydrocarbons / 556

Displacement and Entrapment / 556
Basin Migration Models / 558

15.4 Self-Organization in Hydrogeologic Systems / 561

Patterning Associated with Dissolution / 561
Patterning Associated with Precipitation and Mixed Phenomena / 562

15.5 Coupled Phenomena / 563

Chemical Osmosis / 564
Electro and Thermo-Osmosis / 566
Experimental, Theoretical, and Field Studies / 566
Generalized Treatment / 567

Chapter 16
Introduction to Contaminant Hydrogeology / 573

16.1 Sources of Ground Water Contamination / 574

Radioactive Contaminants / 578
Trace Metals / 579
Nutrients / 579
Other Inorganic Species / 580
Organic Contaminants / 580
Biological Contaminants / 583

16.2 Solute Plumes as a Manifestation of Processes / 583

Fractured and Karst Systems / 588
Babylon, New York Case Study / 591
Alkali Lake, Oregon Case Study / 594

16.3 Multifluid Contaminant Problems / 595

Saturation and Wettability / 597
Imbibition and Drainage / 597
Relative Permeability / 598
Features of Nonaqueous Contaminant Spreading / 601
Secondary Contamination Due to NAPLs / 604
A Case Study of Gasoline Leakage / 607
Hyde Park Landfill Case Study / 608

**16.4 Design and Quality Assurance Issues
in Water Sampling** / 611

Design of Sampling Networks / 611
Assuring the Quality of Chemical Data / 612

16.5 Sampling Methods / 616

Solutes in the Zone of Saturation / 616
Nonaqueous Phase Liquids in the Zone of Saturation / 621
Dissolved Contaminants and NAPLs in the Unsaturated Zone / 623
Solid and Fluid Sampling / 623

16.6 Indirect Methods for Detecting Contamination / 624

Soil Gas Characterization / 624
Geophysical Methods / 627

Chapter 17
Modeling Contaminant Transport / 633

17.1 Analytical Approaches / 634

Advection and Longitudinal Dispersion / 634
The Retardation Equation / 639
Radioactive Decay, Biodegradation, and Hydrolysis / 642
Transverse Dispersion / 643
Models for Multidimensional Transport / 645
Continuous Sources / 645
Instantaneous Point Source Model / 649

**17.2 Programming the Analytical Solutions
for Computers** / 651

17.3 Semianalytical Approaches / 656

17.4 Numerical Approaches / 660

    A Generalized Modeling Approach / 661
    The Common Solution Techniques / 662
    Adding Chemical Reactions / 664

17.5 Case Study in the Application of a Numerical Model / 665

Chapter 18
Process and Parameter Identification / 677

18.1 Tracers and Tracer Tests / 678

    Field Tracer Experiments / 679
        Natural Gradient Test / 679
        Single-Well Pulse Test / 679
        Two-Well Tracer Test / 680
        Single-Well Injection or Withdrawal with Multiple-Observation Wells / 681

18.2 The Diffusional Model of Dispersion / 681

    Borden Tracer Experiment / 683

18.3 Dispersivities Estimated Using the Advection-Dispersion Equation / 686

    Interpreting Data from a Two-well Tracer Test / 687
    Domenico and Robbins (1985b) Graphical Procedure / 689
    Trial and Error Fit of a Numerical Model / 693

18.4 New Approaches in Estimating Dispersivity / 693

18.5 Identification of Geochemical Processes / 695

    Discovering Systematics in Plume Configurations / 695
    Complete Characterization of Possible Reactions / 695
    Interpreting Reaction Pathways in the Laboratory / 698

18.6 Quantification of Geochemical Processes / 700

    Field Approaches / 701
    Laboratory Techniques / 703

Chapter 19
Remediation / 711

19.1 Overview of Corrective Action Alternatives / 712

    Containment / 712
    Contaminant Withdrawal / 716
    In Situ Treatment of Contaminants / 718
    Management Options / 718

**19.2   Technical Considerations with Injection
Withdrawal Systems** / *719*

**19.3   Methods for Designing Injection/
Withdrawal Systems** / *724*

Expanding Pilot Scale Systems / *724*
Capture-Zone-Type Curves / *725*
Trial and Error Simulations with Models / *731*
Simulation-Optimization Techniques / *731*

19.4   Interceptor Systems for LNAPL Recovery / *732*

19.5   Bioremediation / *735*

Biochemical Reactions / *735*
Oxygen and Nutrient Delivery / *736*
Comments on System Design / *739*

19.6   Steps for Dealing with Problems of Contamination / *739*

19.7   Case Studies in Site Remediation / *743*

Gasoline Spill / *744*
Oil Spill: Calgary, Alberta / *744*
Gilson Road: Nashua, New Hampshire / *746*
Hyde Park Landfill: Niagara Falls, New York / *748*

Answer to Problems / *751*

References / *757*

Index / *807*

Chapter One

Introduction

1.1 What is Hydrogeology?

1.2 Relationship Between
     Hydrogeology and Other
     Fields of Geology

1.3 How to Use This Book

1.4 The Hydrologic Cycle

This book was written from the perspective that the reader is interested in becoming a hydrogeologist. We recognize that this is not likely to be the case generally. However, for us to take a different perspective would require a different kind of book, one that is perhaps not as detailed or as rigorous. And this would be unfair to all students in several different ways. First, the student interested in hydrogeology as a career is entitled to know the breadth of the field and the fact that there is much to study, so much in fact that it cannot all be learned in one course. Additional courses will be required, not only in hydrogeology but in supporting sciences such as soil physics, soil mechanics, geochemistry, and in numerical methods. Graduate training is an essential requirement. Additionally, all students are entitled to know that hydrogeology contains not only geology but a heavy dose of physics and chemistry as well, and its main language is mathematics. Thus, the entry level to the field is high, but so are the intellectual and practical rewards. For earth scientists not interested in hydrogeology as a career, let us state outright that virtually every activity in the earth sciences requires some knowledge of subsurface fluids, and rock-water interactions in particular, and that is what hydrogeology and this book is about.

Last, from our own biased perspective, we think it would be difficult to find a more rewarding introduction to the concepts of science than those offered by the study of hydrogeology. The discipline blends field, experimental, and theoretical activities. Sometimes the experimental activities take place in the laboratory; other times they take place in the field where the comfort of laboratory control is lost. Additionally, the field and experimental activities often play a major role in the formulation of reliable theoretical models of processes and events. In increasingly more cases, the methods of hydrogeology give us the wherewithal to study and perhaps quantify some of nature's experiments. These same methods help us study, quantify, and sometimes rectify some unfortunate "experiments" of an industrialized society. All of this, and much more, is hydrogeology.

1.1 What Is Hydrogeology?

We have chosen to call this book "Physical and Chemical Hydrogeology" for two reasons. First, this is what the book and the field is about. Second, most other potential candidate titles have been preempted. The simple title "Groundwater" is appealing, and was used by Tolman in 1937 and by Freeze and Cherry in 1979. The books are quite different in scope because during the period 1935 to the present, the field of knowledge was undergoing tremendous expansion and this expansion is expertly captured by Freeze and Cherry (1979). Lamarck wrote a book entitled "Hydrogeology" in 1802, as did Davis and DeWiest in 1966. Obviously, due to the times, there is no similarity between the subject matter of these books. A book entitled "Geohydrology" was prepared by DeWiest in 1965 and one entitled "Groundwater Hydrology" was written by Todd, first in 1959 with a later edition appearing in 1980.

A few people remain uncomfortable with the situation where the subject matter of subsurface fluids can be organized under a variety of titles and taught under a variety of disciplines. The meaning of the terms hydrogeology and geohydrology in particular have caused some debate. Frequently it is stated that the former deals with the geologic aspects of ground water whereas the latter places more emphasis on hydraulics and fluid flow. This arbitrary division is no longer taken seriously by most people in the field. In fact, the term hydrogeology was defined long before the modern era in hydrogeology, which differs markedly from its early beginnings, and both definitions likely reflect the special interest of their promulgators. In 1919, Mead published a book on hydrology where he defined hydrogeology as the study of the laws of occurrence and movement of subterranean water. There is nothing wrong with this definition. However, Mead stressed the importance of ground water as a geologic agent, especially as it contributes to an understanding of rivers and drainage systems. As a hydrologist interested largely in surface phenomena, this emphasis was well suited to Mead's interests. In 1942, Meinzer edited a book called "Hydrology," which he defined within the context of the hydrologic cycle, that is, the march of events marking the progress of a particle of water from the ocean basins to the atmosphere and land masses and back to the ocean basins. He divided this science into surface hydrology and subterranean hydrology, or "geohydrology." Meinzer had an illustrious career devoted almost exclusively to the study of ground water as a water supply. In 1923, he published his famous volume on the occurrence of ground water in the United States, which essentially brought to a close the exploration period that started before the turn of the century. Indeed, because of this volume and some modern supplements, we are no longer exploring for ground water in North America and have not been for several decades. Shortly after this publication, Meinzer turned to inventorying the resource, for example, measuring or estimating the inflows and outflows and the changes in storage over time within ground water bodies (Meinzer and Stearns, 1928; Meinzer, 1932). Such studies require detailed information on the interrelationships between subterranean water and other components of the hydrologic cycle to which it is connected. Thus, Meinzer's definition of geohydrology as the subsurface component of the hydrologic cycle suited his interests well. It still remains a good definition, but it does not go far enough. We will offer a definition after we develop an understanding of what the field is today and how it got that way, and who were the major players. Only then will we become aware of the scope of hydrogeology.

Physical Hydrogeology Before the Early 1940s

The turn of the century was an exciting time for hydrogeologists, especially those inclined to the rigors of fieldwork. Their main tools were rock hammers, compasses, and some crude water-level or fluid-pressure measuring devices. These hydrogeologists were likely aware of two important findings of the previous century. First was the experimental work of Henry Darcy in 1856 providing a law that described the motion of ground water, and second was some work by T. C. Chamberlin in 1885 that described water occurrence and flow under "artesian" conditions. Armed with these details, these workers were busy delineating the major water-yielding formations in North America and making important measurements of the distribution of hydraulic head within them. It was a good time to be a field hydrogeologist, with the emphasis being on exploration and understanding the occurrence of ground water and its interrelationship with other components of the hydrologic cycle.

In addition, there was some abstract thinking occurring at this time, but it had little or no impact over the following three or four decades. Slichter (1899) in particular did original theoretical work on the flow of ground water, but he was several decades ahead of his time. King (1899), too, attempted to provide some calculations to support his field findings.

The culmination of this era probably occurred in 1923 when Meinzer published his book on the occurrence of ground water in the United States. The major water-yielding formations were described, and water supply was the order of the day. As we shall learn later, there are generally four stages in the ordered utilization of ground water for water supply: exploration, development, inventory, and management. With the exploration stage completed and the resource already undergoing development, much of the effort of the U.S. Geological Survey turned toward inventory. However, at least one major theoretical finding resulted from this work, and this was provided by Meinzer in 1928 as a result of his study of the Dakota sandstone (Meinzer and Hard, 1925). In his inventory it seemed that more water was pumped from a region than could be accounted for; that is, the inventory (which is really a water balance) could not be closed. Meinzer concluded that the water-bearing formation possessed some elastic behavior and that this elastic behavior played an important role in the manner in which water is removed from storage. Although nothing was made of these ideas for another seven years, this was the start of something that would change the complexion of hydrogeology for at least two decades, if not forever.

That "something" occurred in 1935 when C. V. Theis, with the help of a mathematician named C. I. Lubin, recognized an analogy between the flow of heat and the flow of water (this analogy was also recognized earlier by Slichter as well as by others). The significance of this finding was that heat flow was already mathematically sophisticated whereas hydrogeology was not. However, by analogy, a solution to a heat flow problem, of which there were several, could be used to provide a solution to a fluid flow analogy. Thus, Theis presented a solution that described the transient behavior of water levels in the vicinity of a pumping well. Imagine the significance of such a finding to a group of scientists that were totally committed to studying the response of ground water basins to pumping for water supply. For this work, Theis was awarded the treasured Horton medal almost 50 years later (1984).

Two additional major contributions came about in 1940. The first was by Hubbert (1940), who published his detailed work on the theory of ground water flow. In this work Hubbert was not interested in small-scale induced transients in the vicinity of pumping wells, but in the natural flow of ground water in large geologic basins. The hydrogeologic community at that time was far too preoccupied with transient flow to wells, so this contribution did not become part of the mainstream of hydrogeology until the early 1960s.

The second finding, and we think the most important of the first half of the twentieth century, was provided by Jacob (1940). Now let us fully appreciate the situation at the time. The Theis solution was provided by analogy from some heat flow solution that was rigorously derived from some equation describing heat flow. Is there not some equation describing the flow of fluid from which it (as well as other solutions) could be derived directly? In 1940, Jacob derived this equation, and, significantly, it incorporated the elastic behavior of porous rocks described by Meinzer some 12 years earlier. Now, why all the fuss about something as abstract as a differential equation? A differential equation establishes a relation between the increments in certain quantities and the quantities themselves. This property allows us to say something about the relation between one state of nature and a neighboring state, both in time and in space. As such, it is an expression for the principle of causality, or the relation between cause and effect, which is the cornerstone of natural science. Thus, the differential equation is an expression of a law of nature.

Following Theis, Jacob, Hantush (one of Jacob's students), and several of Hantush's students, the transient flow of water to wells occupied center stage for about two decades. Training in well hydraulics today is as necessary to a hydrogeologist as training in the proper use of a Brunton compass is to a field geologist. Today, Meinzer, Theis, Jacob, and Hantush are no longer with us. However, of all the hydrogeologists who ever lived, over 95% are still alive and still working. In spite of this enormous pool, it is unlikely that such an era as occurred from 1935 to about 1960 will ever repeat itself.

Chemical Hydrogeology Before the Early 1960s

It would be pleasing indeed if we could trace a bit of historical development in chemical hydrogeology that parallels the common interests demonstrated on the physical side of things. However, no such organized effort materialized, largely because there was no comparable early guidance such as that provided by Darcy, Chamberlin, and the early fieldworkers. Back and Freeze (1983) recognize an "evolutionary" phase in chemical hydrogeology that started near the turn of the century and ended sometime during the late 1950s. By evolutionary, we mean a conglomeration of ideas by different workers interested in a wide variety of hydrochemical aspects. Several good ideas came out of this period, including certain graphical procedures that are still used to interpret water analyses (Piper, 1944; Stiff, 1951) and a particularly good treatment of sodium bicarbonate water by Foster (1950) that is still highly quoted today. In addition to these studies, two others stand out from the perspective of the chemical evolution of ground water with position in ground water flow (Chebotarev, 1955; Back, 1960). For the most part, however, the early effort on the chemical side of things was directed at determining water quality and fitness for use for municipal and agricultural purposes.

The first extensive guidance to the working chemical hydrogeologist was provided by Hem's (1959) treatment on the study and interpretation of the chemical character of natural waters. This work demonstrated, among other things, that most of the important reactions were known. The majority of the contributions to this knowledge come from chemists, sanitary engineers, biologists, and limnologists. Hem's volume served as the bible for the working hydrogeologist for several years.

The early parts of the 1960 decade marked some unification of the divergent interests of the previous 50 years. Garrels' (1960) book in particular focused on the equilibrium approach in chemical thermodynamics. From this point on, a main focus of much of the work in chemical hydrogeology shifted to understanding regional geochemical processes in carefully conducted field investigations. Interestingly, at about that same time, hydrogeologists became interested in regional ground water flow as described by Hubbert (1940) some 20 years earlier. Thus, the timing was perfect for a serious bonding of regional hydrogeology and the chemical evolution of formation waters. Today, some 30 years later, training in regional ground water flow and the chemical evolution of ground water basins is an essential part of hydrogeology.

Post-1960 Hydrogeology

We have already given the reader a glimpse of post-1960 hydrogeology by making reference to the marriage of physical and chemical hydrogeology by our contemporaries well aware of the pioneering work of Hubbert (1940) and Garrels (1960).

Basically, we may isolate three contributing factors that are more or less responsible for additional developments beyond the early 1960s, one technological and two institutionally motivated. The technological factor was the development and eventual accessibility of high-speed computers. Problems that could be solved tediously by analytical mathematics (or those that could not be solved at all) are readily handled by numerical methods requiring computer devices. Thus, numerical methods in hydrogeology is a core course in any serious curriculum in hydrogeology.

Two institutionally motivating factors initiated the interest of hydrogeologists in transport processes, or processes where some physical entity such as heat or chemical mass is moved from one point to another in ground water flow. Interest in heat flow and geothermal energy in particular was at its peak in the early 1970s in response to an effort to develop alternative energy sources due to the oil embargo. With a return to pre-1970 prices for oil and gas, the interest in geothermal resources diminished considerably. However, the groundwork was laid for a continued interest in heat transport by ground water flow, especially with regard to the thermal evolution of sedimentary basins, low-temperature mineralization, the dissipation of frictional heat following earthquake generation, thermal pollution, and the fluid and thermomechanical response of rocks used for repositories to store high-level nuclear waste.

A second institutionally motivating factor occurred in the mid-1970s in the form of a federal environmental law. Environmental laws were quite common during the 1960s and 1970s: the Clean Water Act, the Clean Air Act, the Clean Drinking Water Act, the Surface Mining Act. Many of these dealt largely with surface water and were relatively simple to initiate and enforce. The Clean Water Act, for example, made funds available for treatment of sewage disposed of in surface waters, thereby contributing to the cleanup of polluted rivers. In 1976, the environmental loop was closed with the introduction of the Resource Conservation and Recovery Act (RCRA), which is fundamentally a ground water protection act. The purpose of RCRA is to manage solid hazardous waste from the time it is produced to its ultimate disposal. The act had provisions for forced monitoring of waste disposal facilities, which eventually revealed the extent of ground water contamination throughout the land. This led to the Comprehensive Environmental Response, Compensation, and Liability Act (CERCLA) in 1980, more commonly known as the "superfund" for cleanup. RCRA does not address radioactive or mining wastes. Nor until recently did it address leaky storage tanks. All wastes will eventually be regulated. Hydrogeologists will be involved with the fate and transport of contaminants in the subsurface for many years to come.

Contaminant transport is merely a special application of mass transport in ground water flow. The transport processes are essentially physical phenomena, for example, the movement of dissolved mass by a moving ground water. This movement is frequently accompanied by the transfer of mass from one phase (liquid or solid) to another. Mass transfer from one phase to another proceeds at a decreasing rate until the two phases come to equilibrium with each other. Thermodynamic information regarding this equilibrium is basic to an understanding of mass transfer, but a more immediate concern is the rate at which the transfer takes place. This is a kinetic, as opposed to an equilibrium concern, which in many cases is influenced by the rate of ground water flow. Here again the timing is perfect for a serious bonding of physical and chemical hydrogeology. Training in mass transport and reactions in ground water flow is an essential part of modern hydrogeology.

Given this brief description of what hydrogeology is and how it got that way, we are now in a better position to attempt a partial description of the field. We should still like to hedge a bit here by making a distinction between the principles of the science and their various applications. The principles of the science reside in the principles of fluid, mass, and energy transport in geologic formations. The applications of the science are not so easily enunciated, but so far we have emphasized problems in human affairs such as water supply, contamination, and energy resources. There is, however, a historical connotation to hydrogeology, and this is taken up in the next section.

1.2   *The Relationship Between Hydrogeology and Other Fields of Geology*

The cornerstone of geology resides in four areas: mineralogy, petrology, stratigraphy, and structure. The role of hydrogeology in geologic studies can best be appreciated if, when making observations, we ask ourselves, "How did it get this way?" This is an historical question that is fair in this case because geology is an historical science studied largely from the perspective of observation. Let us remember further that the rock mass we are examining consists of a stationary phase and a mobile phase. The terms "stationary" and "mobile" refer to the solids making up the rock body and the fluids contained within the rock, respectively. This is essentially a long-term "through-flow system," with the fluids continually replaced over geologic time frames. Thus, the ability of a moving ground water to dissolve rocks and minerals and to redistribute large quantities of dissolved mass has important implications in chemical diagenesis, economic mineralization, and geologic work in general. Additionally, it has long been understood that the fluids need not reside in a passive porous solid, but instead there frequently exists a complex coupling among stress, strain, and the pore fluids themselves. This coupling is often complicated by heat sources in crustal portions of the earth. The net effect of this coupling is fully demonstrated in several modern environments throughout the world where "abnormally" high fluid pressures reside in deforming rocks. Ultimately, these pressures can lead to rock fracture and further deformations. Is it possible that such environments were equally widespread throughout geologic time, and what consequences can be attributed to them?

Even metamorphic processes cannot be fully examined in the absence of a fluid phase. According to Yoder (1955), water is an essential constituent of many minerals common to metamorphic rocks, it is the primary catalyst in reaction and recrystallization of existing minerals, and it is one of the chief transport agents of material and possibly heat in metasomatic processes. As early as 1909, Munn recognized the role of ground water in hydrocarbon migration and entrapment. To discuss these aspects further would preempt some of the main considerations in this book. However, it should be clear from the perspective of processes that one cannot ignore the role of ground water in performing geologic work.

Bearing in mind all the factors discussed in the preceding paragraphs, we venture the following definition of hydrogeology. Hydrogeology is the study of the laws governing the movement of subterranean water, the mechanical, chemical, and thermal interaction of this water with the porous solid, and the transport of energy and chemical constituents by the flow.

1.3 How to Use This Book

"Physical and Chemical Hydrogeology" deliberately contains much more material than can be covered in a single course. This provides the instructor with a choice of topics suited to the class objectives. Because there are aspects of hydrogeology that bear on human affairs as well as paleohydrologic aspects of interest to earth scientists, perhaps some guidance may be in order. There are three introductory courses that may be served by this book, one being a survey-type course with an emphasis on hydrogeology and human affairs, one dealing with the preeminent problem of contaminant transport in various geologic environments, and one concerned with the geologic aspects of ground water of interest to earth scientists. Whichever of these options is chosen, parts of seven chapters contribute to all:

Chapter 2:   The Origin of Porosity and Permeability

Chapter 3:   Ground Water Movement

Chapter 4:   Elastic Properties and Main Equations of Flow

Chapter 10:   Solute and Particle Transport

Chapter 11:   Principles of Aqueous Geochemistry

Chapter 12:   Chemical Reactions

Chapter 13:   The Mathematics of Mass Transport

By and large, these chapters contain the main principles of hydrogeology. The practice of hydrogeology rests on theory, and this "practice" is provided in the remaining chapters.

For the survey course treating hydrogeology and human affairs, the seven chapters just cited should be supplemented with parts of the following:

Chapter 5:   Hydraulic Testing: Models, Methods, and Applications

Chapter 6:   Ground Water as a Resource

Chapter 7:   Ground Water in the Basin Hydrologic Cycle

Chapter 16:   Introduction to Contaminant Hydrogeology

The heart of any course in contaminant hydrogeology will include parts of the four chapters dealing with contaminants:

Chapter 16:   Introduction to Contaminant Hydrogeology

Chapter 17:   Modeling Contaminant Transport

Chapter 18:   Process and Parameter Identification

Chapter 19:   Remediation

In addition, some elements of hydraulic testing (Chapter 5) and ground water in the basin hydrologic cycle (Chapter 7) are necessary if the students are unfamiliar with these topics.

The geologic aspects of groundwater are reasonably well served by the following chapter supplements to the original seven cited:

Chapter 7:   Ground Water in the Basin Hydrologic Cycle

Chapter 8:   Ground Water in the Earth's Crust

Chapter 9:   Heat Transport in Ground Water Flow

Chapter 14:   Mass Transport in Ground Water Flow: Aqueous Systems

Chapter 15:   Mass Transport in Ground Water Flow: Geologic Systems

Alternatively, if the instructor has the luxury of providing a two-semester course in hydrogeology, parts of Chapters 1 through 10 provide information for the first course, whereas parts of Chapters 11 through 19 seem appropriate for the second course.

1.4 The Hydrologic Cycle

This book is concerned solely with ground water. Because ground water is one component of the hydrologic cycle, some preliminary information on the hydrologic cycle must be introduced to set the proper stage for things to come.

Components of the Hydrologic Cycle

Schematic presentations of the hydrologic cycle such as Figure 1.1 often lump the ocean, atmosphere, and land areas into single components. Yet another presentation of the hydrologic cycle is one that portrays the various moisture inputs and outputs on a basin scale. This is shown in Figure 1.2, where precipitation is taken as input and evaporation and transpiration (referred to as evapotranspiration)

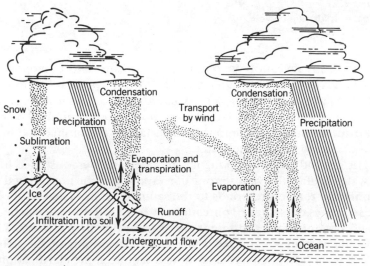

Figure 1.1
Schematic representation of the hydrologic cycle.

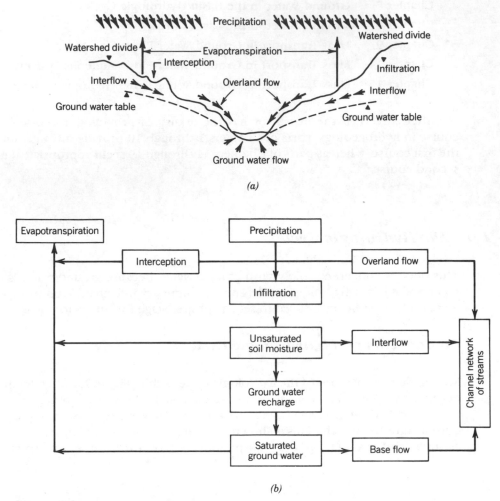

Figure 1.2
The basin hydrologic cycle.

along with stream runoff are outputs. The stream runoff component, referred to as overland flow, can be augmented by interflow, a process that operates below the surface but above the zone where rocks are saturated with water, and by base flow, a direct component of discharge to streams from the saturated portion of the system. Infiltration of water into the subsurface is the ultimate source of interflow and recharge to the ground water.

The ground water component is more clearly demonstrated in Figure 1.3. In this water profile, the vadose zone corresponds to the unsaturated zone, whereas the phreatic zone corresponds to the saturated zone. The so-called intermediate zone separates the saturated phreatic zone from soil water. It can be absent in areas of high precipitation, and hundreds of feet thick in arid areas. The water table marks the bottom of capillary water and the beginning of the saturated zone.

The terms saturated and unsaturated require some clarification. Given a unit volume of soil or rock material, designated as V_T, the total volume consists of both solids (V_s) and voids (V_v). Only the voids are capable of containing a fluid, either air or water. The degree of saturation is defined as the ratio of the water

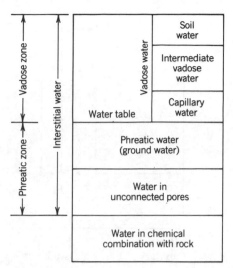

Figure 1.3
The water profile.

volume to the void volume, V_w/V_v expressed as a percentage. For a fully saturated medium, the ratio is one (or 100%). A degree of saturation less than 100% indicates that air occupies some of the voids. Another term in common usage is moisture content, θ, which is defined as the volume of water divided by the total volume (V_w/V_T), expressed as a percentage.

At the contact between the lower part of a dry porous material and a saturated material, the water rises to a certain height above the top of the saturated material. This gives rise to the capillary water zone of Figure 1.3. The driving force responsible for the rise is termed surface tension, a force acting parallel to the surface of the water in all directions because of an unbalanced molecular attraction of the water at the boundary. The tensional nature of these forces can be compared with those set up in a stretched membrane.

Capillarity results from a combination of the surface tension of a liquid and the ability of certain liquids to wet the surfaces with which they come in contact. This wetting (or wetability) causes a curvature of the liquid surface, giving a contact angle between liquid and solid different from 90°. The idealized system commonly utilized to examine the phenomenon is a water-containing vessel and a capillary tube. When the tube is inserted in the water, the water rises to a height h_c (Figure 1.4). The meniscus, or curved surface at the top of the tube, is in contact with the walls of the tube at some angle α, the value of the contact angle depending on the wall material and the liquid. For a water-glass system, α is taken as zero.

Points A and C in Figure 1.4 are at atmospheric pressure (1.013×10^5 newtons per meter squared (N/m²) or 14.7 lb per square inch (psi) or approximately 1 bar). A fundamental law of hydrostatics states that the pressure intensity of a fluid at rest can vary in the vertical direction only. Hence, point B is also at atmospheric pressure, and the pressure at point D must be less than atmospheric by an amount dictated by the pressure exerted by the length of the water column above B, or $h_c\gamma_w$, where γ_w is the unit weight of water. At a height of 10.3m, the value of $h_c\gamma_w$ is just equal to atmospheric pressure. Hence, 10.3m is a theoretic upper bound for the height of capillary rise, at least in glass tubes. For pure water at 20°C, the height of capillary rise in a glass tube is expressed as

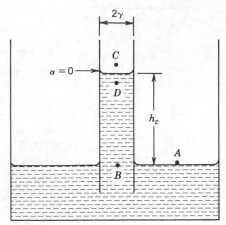

Figure 1.4
Capillary rise in a glass tube.

$$b_c = \frac{0.153}{r} \tag{1.1}$$

where b_c and r are expressed in centimeters and r is taken as the radius of the tube. In real systems, r is taken as the radius of passage.

Of further significance to this discussion is now an absolute definition of the water table. From Figure 1.4 we recognize that water above the water table is at a pressure below atmospheric while water below the water table is at a pressure greater than atmospheric. The water table is the underground water surface at which the pressure is exactly equal to atmospheric.

The intermediate zone lies above the capillary fringe and consists of water in the form of thin films adhering to the pore linings. This water is free to drain downward under the forces of gravity. The soil, intermediate, and capillary zones are collectively taken as the unsaturated or vadose zone (Figure 1.3).

Evapotranspiration and Potential Evapotranspiration

Evaporation can take place from both soil and free water surfaces. Evaporation from plants is called transpiration, and the combined process is often referred to as evapotranspiration.

The concept of potential evapotranspiration (or the potential for evapotranspiration) and methods for calculating it were introduced by Thornthwaite (1948). Potential evapotranspiration is defined as "the amount of water that would evaporate or transpire from a surface if water was available to that surface in unlimited supply." There is thus a clear distinction between what is actual and what is potential evapotranspiration. For example, the actual evapotranspiration off the surface of Hoover Reservoir very likely equals the potential rate. However, during most of the year, the actual evapotranspiration in regions adjacent to the reservoir is generally less than the potential rate, simply because of the lack of unlimited water. The actual evaporation off the Sahara desert is nil whereas the potential evapotranspiration (or potential for evapotranspiration) is quite high. In fact, if we wished to irrigate parts of the Sahara, we would have to supply irrigation water at about the potential rate.

From this definition and the few examples stated earlier, potential evapotranspiration is a maximum water loss (or upper limit of actual evapotranspiration), is a temperature dependent quantity, and is a measure of the moisture demand for a region. The last point stated is better demonstrated by the ratio of precipitation to potential evapotranspiration. In a desert or tundra region, this ratio may be less than 0.1. For these regions, precipitation is not sufficient to meet the water demand so that very little grows naturally. For a successful irrigation scheme in such a region, water must be supplied at the potential rate. The ratio of precipitation to potential evapotranspiration in parts of Montana and North Dakota ranges from 0.2 to 0.6, again indicating a need for irrigation water in crop production. The range in eastern United States is from 0.8 to 1.6, indicating a rather well-balanced situation and, in some cases, a water surplus. The vegetation (or lack thereof) in a region is thus a reflection of the precipitation–potential evapotranspiration ratio. This is obvious when comparing the Sahara desert with the Amazon basin, both of which have excessive potential evapotranspiration, but only one of which is characterized by an equally excessive amount of precipitation.

Figure 1.5 is a qualitative demonstration of the natural water demand and water supply in a region as measured by the potential evapotranspiration and the precipitation, respectively. From these curves, we note three important time periods:

1. Precipitation equals potential evapotranspiration.

2. Precipitation is less than potential evapotranspiration.

3. Precipitation is greater than potential evapotranspiration.

In formulating some rules (laws) that govern the behavior of water in the hydrologic cycle, we first stipulate that the demands of potential evapotranspiration must be met (if at all possible) before water is permanently allocated to other parts of the cycle. Thus, when precipitation is equal to potential evapotranspiration, the surplus is theoretically zero, and all the rainfall for that period is available to satisfy the evaporation needs. During such time periods the actual evapotranspiration equals the potential evapotranspiration. This does not mean that infiltration cannot take place. It means simply that any infiltrated water will be available in the soil moisture for use by plants in the transpiration process. In addition, any time the rate of precipitation exceeds the rate of infiltration, there will be some water available for overland runoff so that this suggested balance of precipitation and potential evapotranspiration is not exactly true. At any rate, the surplus after all demands of potential evapotranspiration are met will be zero.

When precipitation is less than potential evapotranspiration, all the precipitation is available to partially satisfy potential evapotranspiration. During such time periods, the actual evapotranspiration would appear to be less than the potential amount. Such is not the case, however, as the part of the demand not met by precipitation may be met by drawing on whatever moisture is in the soil zone. So here again the actual rate can equal the potential rate. If this situation continues over a prolonged period, the soil becomes depleted of its moisture (Figure 1.5), and the actual rate of evapotranspiration will fall below the potential rate. Such periods are normally labeled droughts.

Finally, when temperature drops sufficiently so that precipitation exceeds potential evapotranspiration, a water surplus is realized. This surplus immediately goes into rebuilding the soil moisture component. This rate of infiltration will

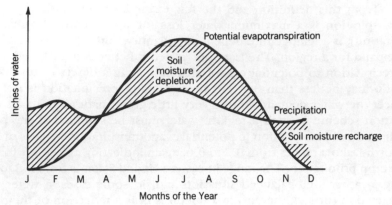

Figure 1.5
Relationship between precipitation and potential evapotranspiration.

generally decrease with time as the soil zone becomes more saturated with water. It is during these periods that the underlying ground water becomes recharged and, in response to a decreased infiltration rate, overland flows become more pronounced. It is thus seen that the soil zone is a buffer in the workings of the basin hydrologic cycle, taking in moisture during surplus periods and releasing moisture during deficit periods.

Infiltration and Recharge

If water is applied to a soil surface at an increasing rate, eventually the rate of supply has to exceed the rate of entry, and excess water will accumulate. Not so obvious is the fact that if water is supplied to a soil surface at a constant rate, the water may at first enter the soil quite readily, but eventually will infiltrate at a rate that decreases with time. Horton (1933, 1940) was the first to point out that the maximum permissible infiltration rate decreases with increasing time. This limiting rate is called the infiltration capacity of the soil, an unfortunate choice of words in that capacity generally refers to a volume or amount whereas the infiltration capacity of Horton (1933) is actually a rate.

The concept of field capacity is another useful idea in soil science to denote an upper limit of moisture content in soils. At field capacity the soil is holding all the water it can under the pull of gravity. In a conceptual sense, both interflow and ground water recharge can commence when the moisture content exceeds field capacity. The proportioning of the available moisture to interflow or ground water recharge is dependent to a large extent on the nature of the soil zone. If the soil zone is thin and underlain by rather impervious rock, the interflow component may be the dominant one. For thicker pervious soils, downward migration may dominate, with most of the moisture being allocated to ground water recharge.

Base Flow

The term recession refers to the decline of natural output in the absence of input and is assumed or known from experience to follow an exponential decay law. The base flow component of streams represents the withdrawal of ground water from storage and is termed a ground water recession. As this recession is generally determined from stream hydrographs, the hydrograph must be separated into its component parts, which normally consist of overland flow, interflow, and base flow. Frequently, the interflow component is ignored so that the hydrologist generally deals with an assumed two-component system.

Whatever the assumptions, methods employed in hydrograph separation are wrought with difficulties. One of the most acceptable techniques is illustrated in Figure 1.6, which portrays an ideal stream hydrograph that plots the discharge of a river, usually in cubic feet per second (cfs) at a single point in the watershed as a function of time. According to Linsley and others (1958), the direct runoff component is terminated after some fixed time T^* after the peak of the hydrograph, where this time T^* in days is expressed

$$T^* = A^{0.2} \tag{1.2}$$

where A is the drainage area in square miles and 0.2 is some empirical constant. Normally, the recession existing prior to the storm is extended to a point directly under the peak (A-B, Figure 1.6) and then extended again to the point T^* after the peak (B-C, Figure 1.6). The depressed base flow component under the peak is justified because high water levels in the stream have the tendency to cut off or retard the ground water discharge component.

Consider now the point on the hydrograph T^* days after the peak. From this point onward, the total flow is the base flow component derived from groundwater discharge into the stream, at least until the next storm. This recession starting at time T^* can be described as

$$Q = Q_o e^{-kt} \tag{1.3}$$

where Q_o is the discharge at time T^*, Q is the discharge at any time, k is a recession constant, and time t is the time since the recession began. In this equation, the time t varies from zero, which corresponds to the time (T^*), to infinity. Thus, at the beginning of the recession $t = 0$ and $Q = Q_o$. For later times, Q decreases and follows the exponential decay law of Eq. 1.3.

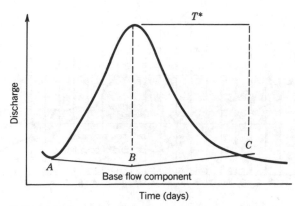

Figure 1.6
Determining base flow component from stream hydrograph.

From Eq. 1.3, the recession constant can be expressed

$$k = \left(\frac{1}{t}\right) \ln \left(\frac{Q}{Q_o}\right) \tag{1.4}$$

where, as described, Q_o is the base flow at time $t = 0$ and Q is the base flow t time units later. This provides a means for determining the recession constant. Clearly this constant will be some number less than one, and will be large (approach one) for flat recessions and small (approach zero) for steep recessions.

The influence of geology on the shape of the recession curve has been discussed and demonstrated by Farvolden (1964). Streams in limestone regions are characterized by flat recessions, a reflection of the fact that most of the drainage takes place in the subsurface. Granitic and other low-permeability regions typically have rather steep base flow recessions.

A plot of discharge versus time on semilogarithmic paper will yield a straight line, the slope of which defines the recession constant (Figure 1.7). For this case the recession expression becomes

$$Q = \frac{Q_o}{10^{t/t_1}} \tag{1.5}$$

where Q_o is the discharge at time $t = 0$, t_1 is the time 1 log cycle later, and t equals any time of interest for which the value of Q is desired. For example, for $t = t_1$

$$Q = \frac{Q_o}{10} \tag{1.6}$$

The total volume of base flow discharge corresponding to a given recession is found by integrating Eq. 1.5 over the times of interest

$$\text{Vol} = \int_{t_0}^{t} Q \, dt = - \left. \frac{Q_o t_1/2.3}{10^{t/t_1}} \right|_{t_0}^{t} \tag{1.7}$$

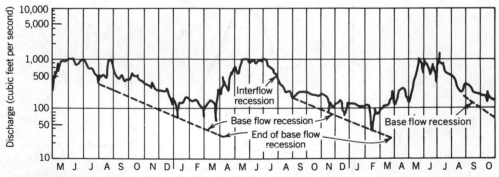

Figure 1.7
Stream hydrograph showing base flow recession (from Meyboom, J. Geophys. Res., v. 66, p. 1203–1214, 1961. Copyright by Amer. Geophys. Union).

where t_o is the starting time of interest. Meyboom (1961) has employed Eq. 1.7 to determine ground water recharge between recessions. For example, if t_o equals zero and t equals infinity

$$\text{Vol} = \frac{Q_o\, t_1}{2.3} \qquad (1.8)$$

which is termed the total potential ground water discharge. This volume is defined as the total volume of ground water that would be discharged by base flow during an entire recession if complete depletion takes place uninterruptedly. It follows that this volume describes the total volume of water in storage at the beginning of the recession. The difference between the remaining potential ground water discharge at the end of a given recession and the total potential groundwater discharge at the beginning of the next recession is a measure of the recharge that takes place between recessions (Meyboom, 1961).

Example 1.1

(From Domenico, 1972) Determine the approximate recharge volume between the first two recessions of Figure 1.7.

The first recession has an initial value of 500 ft³/sec, and takes about 7.5 months to complete a log cycle of discharge. Total potential discharge is calculated from Eq. 1.8:

$$Q_{tp} = \frac{Q_o t_1}{2.3}$$

$$= \frac{500 \text{ ft}^3/\text{sec} \times 7.5 \text{ months} \times 30 \text{ days/month} \times 1440 \text{ min/day} \times 60 \text{ sec/min}}{2.3}$$

$$= 4222 \times 10^6 \text{ ft}^3$$

The ground water volume discharged through the total recession lasting approximately 8 months is determined by evaluating Eq. 1.7 over the limits t equals zero to t equals 8 months

$$\frac{Q_o t_1}{2.3} - \frac{Q_o t_1/2.3}{10^{t/t_1}} = 4222 \times 10^6 \text{ ft}^3 - \frac{4222 \times 10^6 \text{ ft}^3}{10^{8/7.5}}$$

or about 3800 ft³. Base flow storage still remaining at the end of the recession can be determined by evaluting Eq. 1.7 from t equals 8 to t equals infinity, or by merely subtracting actual ground water discharge from total potential discharge, which gives $422 \times 10^6/\text{ft}^3$.

The second recession has an initial value of about 200 ft³/sec and takes about 7.5 months to complete a log cycle of discharge. Total potential discharge is calculated

$$Q_{tp} = \frac{Q_o t_1}{2.3}$$

$$= \frac{200 \text{ ft}^3/\text{sec} \times 7.5 \text{ months} \times 30 \text{ days/month} \times 1440 \text{ min/day} \times 60 \text{ sec/min}}{2.3}$$

or about 1400×10^6 ft³. The recharge that takes place between recessions is the difference between this value and remaining ground water potential of the previous recession, or 978×10^6 ft³.

Hydrologic Equation

The hydrologic cycle as shown in Figure 1.2 and described throughout this chapter is a network of inflows and outflows that may be conveniently expressed as

$$\text{input} - \text{output} = \text{change in storage} \tag{1.9}$$

The word equation given as Eq. 1.9 is a conservation statement and assures us that all the water is accounted for; that is, we can neither gain nor lose water. The interconnections between the components of the hydrologic cycle can be demonstrated on a global scale (Figure 1.8) or on a basin scale (Figure 1.9). On a global scale, the atmosphere gains moisture from the oceans and land areas E and releases it back in the form of precipitation P. Precipitation is disposed of by evaporation to the atmosphere E, overland flow to the channel network of streams Q_o, and infiltration through the soil F. Water in the soil is subject to transpiration T, outflow to the channel network (the interflow component) Q_o, and recharge to the ground water R_N. The ground water reservoir may receive water Q_i and release water Q_o to the channel network of streams and the atmosphere.

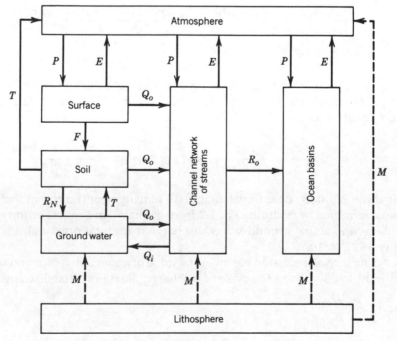

Figure 1.8
Elements of the global hydrologic cycle.

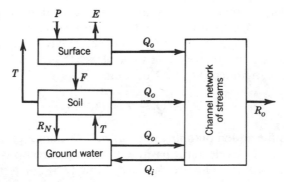

Figure 1.9
Elements of the basin hydrologic cycle.

Streams that receive water from the ground water reservoir by the base flow component Q_o are termed effluent or gaining streams. This occurs when the ground water table in the vicinity of the stream is above the base of the stream. Streams that lose water to the ground water reservoir as designated by the Q_i component are termed influent, or losing, streams. This occurs when the ground water table is well below the base of the stream. Also noted on Figure 1.8 is the potential contribution of water of volcanic or magmatic origin.

Figure 1.9 illustrates that isolation of the basin scale hydrologic subsystem cannot exclude the lines of moisture transport connecting this subsystem to the global cycle. This connection is accomplished by accounting for precipitation derived from the global system and total runoff R_o to the global system. From this diagram we reformulate our word equation given earlier as

$$P - E - T - R_o = \Delta S \tag{1.10}$$

where ΔS is the lumped change in all subsurface water. Each item in the equation has the units of a discharge, or volume per unit time.

The hydrologic equation (Eq. 1.10) must balance for every time period to which it is applied. This balancing will occur provided the accounting of the inflow and the outflow is done over a common period of time. An exact balance is an unreasonable goal due to poor instrumentation, lack of data, and the assumptions that are generally applied.

Equation 1.10 may be expanded or abbreviated, depending on what part of the cycle the hydrologist is interested in and, of course, depending on the available data base over the period of record. For example, for the ground water component, another form of Eq. 1.10, is arrived at from Figure 1.9:

$$R_N + Q_i - T - Q_o = \Delta S \tag{1.11}$$

Over long periods of time, providing the basin is in its natural state and no ground water pumping is taking place, the natural inputs R_N and Q_i will be balanced by the natural outputs T and Q_o so that the change in storage will be zero. This gives

$$R_N + Q_i = T + Q_o \tag{1.12}$$

or, stated simply, input equals output. This means that the ground water component in the basin is hydrologically in a steady state, and the variables (or, more properly, the averaged values of the variables) have not changed over the time period over which the averaging took place. On the other hand, if a pumpage withdrawal is included, Eq. 1.12 becomes

$$R_N + Q_i - T - Q_o - Q_p = \Delta S \qquad (1.13)$$

where Q_p is the added pumping withdrawal. As pumpage is new output from the system, the water levels in the basin will decline in response to withdrawals from ground water storage. The stream will eventually be converted to a totally effluent one. In addition, with lower water levels, the transpiration component will start to decline and eventually approach zero. If potential recharge to the system was formerly rejected due to a water table at or near land surface, the drop in water levels will permit R_N to increase over its steady value. Thus, at some time after pumping starts, Eq. 1.13 becomes

$$R_N + Q_i - Q_p = \Delta S \qquad (1.14)$$

A new steady state can, at least theoretically, be achieved if the pumping withdrawal does not exceed the inputs R_N and Q_i. If pumping continually exceeds these input values, water is continually removed from storage and water levels will continue to fall over time. Here, the steady state has been replaced by a transient, or unsteady state, where some parameter (in this case, ground water storage) continually changes over time. Not only is ground water storage being depleted, but some of the surface flow has been lost from the stream.

The term inventory is generally reserved for investigations in which a detailed accounting of inflow, outflow, and changes in storage is attempted for time intervals, such as years or other units of time, during a period of observation (Meinzer, 1932). In the Pomperaug Basin, Connecticut, for example, available data were organized on a monthly basis for a three-year period of study (Meinzer and Stearns, 1928). The hydrologic equation is the basis for such studies.

Last, our hydrologic equation may be expressed in the form of an ordinary differential equation

$$I(t) - O(t) = \frac{dS}{dt} \qquad (1.15)$$

where I is input and O is output, both taken as a function of time. Equation 1.15 is also referred to as a lumped parameter equation, a term that is reserved for any equation or analysis wherein spatial variations in the parameters are not considered.

Example 1.2

The following is a simplified inventory based on the hydrologic equation

{recharge from direct precipitation plus recharge from stream flow}

{minus discharge by pumping minus discharge by evapotranspiration}

{equals change in ground water storage}

With regard to this inventory, the following are worth noting:

1. Previous to pumping outputs, natural recharge was equal to natural discharge and the ground water basin was in a steady state.

2. With the addition of pumpage and in the course of withdrawals from storage, net recharge from stream flow tends to increase and reaches some maximum value, whereas discharge by evapotranspiration seems to decrease and approach some minimum value.

3. During the course of these withdrawals, the basin is in a transient state where water is continually being withdrawn from storage. Although not shown, this results in a continual decline in water levels. A new steady state can be achieved by reducing pumping to about 38×10^6 m³/yr.

Time Year	Recharge from Direct Precipitation m³	Net Recharge from Stream Flow m³	Discharge by Pumping m³	Discharge by Evapotranspiration m³	Change in Ground Water Storage m³
1	3×10^7	0	0	3×10^7	0
2	3×10^7	6×10^5	1×10^7	3×10^7	94×10^5
⋮	⋮	⋮	⋮	⋮	⋮
7	3×10^7	1×10^6	3×10^7	9×10^6	8×10^6
8	2.8×10^7	2×10^6	3.5×10^7	5×10^6	10×10^6
9	2.5×10^7	3×10^6	3.5×10^7	3×10^6	10×10^6
10	3.5×10^7	4×10^6	4×10^7	1×10^6	2×10^6
11	3.5×10^7	4×10^6	4.2×10^7	1×10^6	3×10^6
12	3.5×10^7	4×10^6	4×10^7	1×10^6	2×10^6

Chapter Two

The Origin of Porosity
and Permeability

2.1 **Porosity and Permeability**

2.2 **Continental Environments**

2.3 **The Boundary Between Continental and Marine Environments**

2.4 **Marine Environments**

2.5 **Uplift, Diagenesis, and Erosion**

2.6 **Tectonism and the Formation of Fractures**

A porous body is a solid that contains holes. All rocks are considered porous to some extent, with the pores containing one or more fluids—air, water, or some minority fluid such as a hydrocarbon. The holes may be connected or disconnected, normally or randomly distributed, interstitial or planar cracklike features. The degree of connectivity of the pores dictates the permeability of the rock, that is, the ease with which fluid can move through the rock body. There is nothing of more vital interest to those concerned with the subsurface movement of fluids than permeability, its creation and destruction, and its distribution throughout a rock body. Hence, this introductory chapter is intended to be as complete as possible—given our limited state of knowledge—in describing those processes that result in the formation of porous bodies along with other processes that ultimately modify the rock body, the connectivity of its pore space, and its contained fluids. To help us along in these tasks we call upon simple concepts known to most students in the earth sciences. Thus, we approach this topic from the simple rock cycle to describe the origin and formation of porous bodies, and from tectonic and chemical processes to ascertain whatever mechanical or chemical alterations rocks have been subjected to at some point in their history.

2.1 Porosity and Permeability

Porosity and Effective Porosity

Total porosity is defined as the part of rock that is void space, expressed as a percentage

$$n = \frac{V_v}{V_T} \tag{2.1}$$

where V_v is the void volume and V_T is the total volume. A related parameter is termed the void ratio, designated as e, and stated as

$$e = \frac{V_v}{V_s} \tag{2.2}$$

expressed as a fraction, where V_s is the solid volume. As total volume is the sum of the void and solid volume, the following relationships can be derived:

$$e = \frac{n}{1 - n} \quad or \quad n = \frac{e}{1 + e} \tag{2.3}$$

Figure 2.1 shows some typical kinds of porosity associated with various rocks. The term primary porosity is reserved for interstitial porosity (Figures 2.1*a* through *d*), and the term secondary is used for fracture or solution porosity (Figures 2.1*e* through *f*). Interstitial porosity has been investigated by Graton and Fraser (1935), who demonstrated that its value can range from about 26% to 47% through different arrangements and packing of ideal spheres. In actuality, the porosity of a sedimentary rock will depend not only on particle shape and arrangement, but on a host of diagenetic features that have affected the rock since deposition.

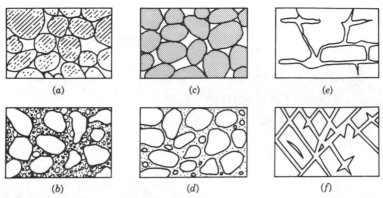

Figure 2.1
Relation between texture and porosity. (*a*) Well-sorted sedimentary
deposit having high porosity; (*b*) poorly sorted sedimentary deposit
having low porosity; (*c*) well-sorted sedimentary deposit consisting of
pebbles that are themselves porous, so that the deposit as a whole has
a very high porosity; (*d*) well-sorted sedimentary deposit whose
porosity has been diminished by the deposition of mineral matter in
the interstices; (*e*) rock rendered porous by solution; (*f*) rock rendered
porous by fracturing (from Meinzer, 1923).

Porosity can range from zero or near zero to more than 60% (Table 2.1). The
latter value is reflective of recently deposited sediments whereas the former value
is for dense crystalline rocks or highly compacted soft rocks such as shales. In
general, for sedimentary materials, the smaller the particle size, the higher the
porosity. This is best demonstrated by comparing the porosity of coarse gravels
with fines, and the total gravel assemblage with silts and clays.

An important distinction is the difference between total porosity, which does
not require pore connections, and effective porosity, which is defined as the
percentage of interconnected pore space. Many rocks, crystallines in particular,
have a high total porosity, most of which may be unconnected. Effective porosity
implies some connectivity through the solid medium, and is more closely related
to permeability than is total porosity. Some data on effective porosity are shown in
Table 2.2. As noted, effective porosity can be over one order of magnitude smaller
than total porosity, with the greatest difference occurring for fractured rocks.

Heath (1982) recognizes five types of porosity in dominant water-bearing
bodies at or near the Earth's surface in the United States and attempts to map their
distribution (Figure 2.2). There are some difficulties with this map because of the
necessity of mapping a single type of opening in areas where two or more types
are present. However, this is a useful presentation and one to which we will refer
frequently in this chapter. Each pattern on Figure 2.2 is associated with one or
more major water-yielding formations in the United States. Thus, the solution-
enlarged openings in carbonate rocks that make up the Florida peninsula are
known as the Floridian system; the sands and gravels stretching from New Jersey
into Texas are sediments of the Atlantic and Gulf coastal plains; the sand and gravel
in the Midwest represents glacial deposits; the sandstones in the northern midcontinent represent several formations, including the Dakota sandstone and the
Cambrian-Ordovician system; the sands and gravels in the western part of the

Table 2.1

Range in values of porosity (in part from Davis, 1969, and Johnson and Morris, 1962)

Material	Porosity (%)
SEDIMENTARY	
Gravel, coarse	24–36
Gravel, fine	25–38
Sand, coarse	31–46
Sand, fine	26–53
Silt	34–61
Clay	34–60
SEDIMENTARY ROCKS	
Sandstone	5–30
Siltstone	21–41
Limestone, dolomite	0–20
Karst limestone	5–50
Shale	0–10
CRYSTALLINE ROCKS	
Fractured crystalline rocks	0–10
Dense crystalline rocks	0–5
Basalt	3–35
Weathered granite	34–57
Weathered gabbro	42–45

Table 2.2

Range in values of total porosity and effective porosity

	Total porosity (%)	Effective porosity (%)
Anhydrite[1]	0.5–5	0.05–0.5
Chalk[1]	5–20	0.05–0.5
Limestone, dolomite[1]	5–15	0.1–5
Sandstone[1]	5–15	0.5–10
Shale[1]	1–10	0.5–5
Salt[1]	0.5	0.1
Granite[2]	0.1	0.0005
Fracture crystalline rock[2]	—	0.00005–0.01

[1]Data from Croff and others (1985).
[2]Data from Norton and Knapp (1977).

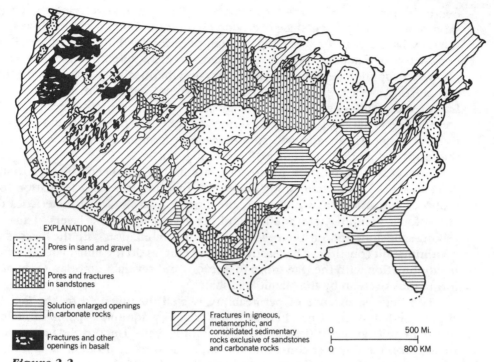

EXPLANATION

Pores in sand and gravel

Pores and fractures in sandstones

Solution enlarged openings in carbonate rocks

Fractures and other openings in basalt

Fractures in igneous, metamorphic, and consolidated sedimentary rocks exclusive of sandstones and carbonate rocks

0   500 Mi.

0   800 KM

Figure 2.2
Map of conterminous United States showing types of water-bearing openings in the dominant aquifers (from Heath, 1982). Reprinted by permission of Ground Water. © 1982. All rights reserved.

nation occupy alluvial basins, whereas the occurrence of these same sediments in central United States represents the remnant of a giant alluvial apron that formed on the eastern slope of the Rocky Mountains; and the basalt in western United States occupies the Columbia River Plateau.

Permeability

Permeability may be described in qualitative terms as the ease with which fluid can move through a porous rock and is measured by the rate of flow in suitable units. This qualitative definition will suffice for this section, although later the topic is taken up in quantitative terms.

Contrasts in permeability from one rock to another and, less frequently, within a given rock unit have given rise to a variety of terms and definitions in hydrogeology, most of which pertain to water supply. A rock unit that is sufficiently permeable so as to supply water to wells is termed an aquifer. Major aquifers are referred to by their stratigraphic names, such as the Patuxent formation in Maryland, the Ogallala formation throughout most of the High Plains, and the Dakota sandstone. These aquifers may be considered ''commercial'' in the sense that they supply sufficient quantities of water for large-scale irrigation or municipal usage.

Aquitards have been defined as beds of lower permeability in the stratigraphic sequence that contain water but do not readily yield water to pumping wells. Major aquitards are generally considered to be low-permeability formations that overlie major aquifers. Examples include the Pierre shale overlying the Dakota sandstone, the Maquoketa shale overlying high-permeability rocks in Illinois, and the Hawthorn formation overlying the high-permeability Floridian aquifer. Rocks considered to be aquitards in one region may serve as aquifers in others. For example, water supplies are obtained from many low-permeability materials throughout North America, such as glacial tills in the Midwest and in Canada, and fractured crystallines and shales in many parts of the continent. These materials are not able to supply sufficient quantities of water for municipal and irrigation use, but frequently are adequate for domestic or farm usage in rural areas. Thus, the terms "aquifer" and "aquitard" are ambiguous.

More recently, the term hydrostratigraphic unit has been employed rather extensively. A hydrostratigraphic unit is a formation, part of a formation, or a group of formations in which there are similar hydrologic characteristics that allow for a grouping into aquifers and associated confining layers. Thus, the Dakota sandstone and various other aquifers in combination with the Pierre shale confining unit constitute a hydrostratigraphic unit, as do the Floridian carbonates in combination with the Hawthorn formation, and several Cambrian-Ordovician formations overlain by the Maquoketa shale.

The field occurrence of permeability is well documented by the fact that about one-half of the United States is underlain by aquifers capable of yielding moderate to large quantities of water to pumping wells. Unconsolidated deposits such as sand and gravel constitute the largest and most productive aquifers in terms of the volumes of water produced annually. Of the consolidated rocks, sandstones are the most important from the perspective of annual withdrawals, followed by carbonate and volcanic rocks. Sandstones and carbonates are virtually ubiquitous in sedimentary terranes. Volcanics, as the fourth major aquifer in the United States, depend on fractures for the transmission of water.

2.2   Continental Environments

Weathering

Rocks at the Earth's surface are subjected to either physical disintegration or chemical decomposition. The weathering products can accumulate in place to form a soil, or can be transported by wind or water to accumulate elsewhere as sedimentary materials. Weathering is also responsible for much of the secondary porosity illustrated in Figure 2.1. This is especially so with the chemical decomposition of carbonate rocks where sinkholes and crevices can readily form due to dissolution along preexisting fractures or other pathways of fluid movement (Figure 2.3). A similar type of process occurs in less soluble igneous and metamorphic rocks, where chemical weathering tends to open preexisting fractures, contributing to more previous pathways. An excellent example of this occurs in the Columbia Plateau, which is characterized by numerous individual basaltic flows, along with occasional sand or gravel units that mark periods of flooding (Figure 2.4). After a particular flow is extruded, it undergoes a cooling period accompanied by the formation of cooling joints, fractures, and vesicles. These rocks then become subjected to weathering, a process that is immediately terminated when another flow is extruded. However, the weathering features are retained in the rock and constitute major pathways for fluid movement. The path-

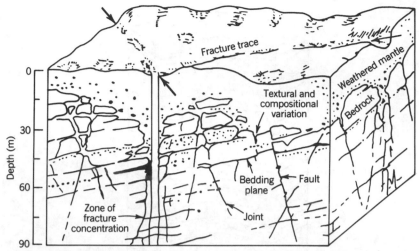

Figure 2.3
Occurrence of permeability zones in fractured carbonate rock. Highest well yields occur in fracture intersection zones (from Lattman and Parizek, 1964). Reprinted with permission of Elsevier Science Publishers, from J. Hydrol., v. 2, p. 73–91.

ways are largely horizontal, reflecting a widespread surficial weathering process, with the depth of weathering restricted to a few tens of feet, depending on the length of time before the volcanic rock becomes inundated in another sea of lava. Below the weathering zone, the rocks remain rather dense and are characterized by poorly connected cooling joints. Hence, this weathering pattern produces a predictable layercake type of hydrology, with some of the permeable flow tops extending over great distance in the plateau.

Recognizing that rocks are exposed at the Earth's surface for long periods of time, chemical weathering can operate for millions of years, contributing to pervious pathways in even the most insoluble of rocks. Figure 2.5 is a conceptual presentation of the occurrence of fractures in crystalline rocks in the Piedmont region of the Appalachian Mountains. The higher permeability resides where the fractures are connected. In these rocks, the enhanced fracture or joint system frequently becomes tighter with depth. The data of Figure 2.6 were collected from this region and show that the average yield of wells per foot of depth below the position where the rocks are saturated with water decreases with increasing depth. Thus, either the connectivity is decreasing with depth, or the fractures are getting tighter due to limitations on the depth of weathering, or both. It is noted that even the most productive wells in this region have very low yields, commonly less than 25 gallons per minute (1.58 liters/sec). However, in the Piedmont, much of the rural domestic supply is furnished by these fractured low-permeability rocks. Once more, the weathering pattern produces a predictable type of hydrology.

Weathering can take place both above and below the water table, but is generally slower in the saturated environment. In addition, a primary mineral subjected to weathering may simply dissolve or a portion of it may reprecipitate to form secondary minerals. The weatherability of igneous silicate minerals is directly proportional to their temperature of formation from molten materials, that is, the high-temperature-phase olivine weathers faster than do the various plagioclase

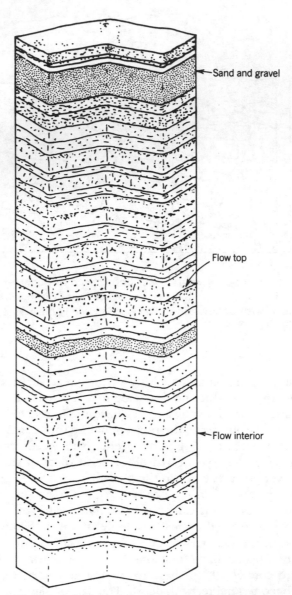

Figure 2.4
Flow tops and flow interiors, basalt, Hanford
Reservation (after Department of Energy, 1986).

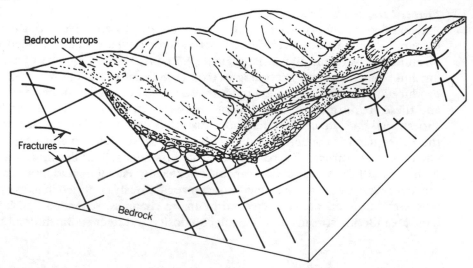

Figure 2.5
Topographic and geologic features of the Piedmont and Blue Ridge region showing fracture interconnections (from Heath, 1984).

families. One of the more common minerals, quartz, weathers the slowest. On the other hand, halite, gypsum, pyrite, calcite, dolomite, and volcanic glass weather faster than do the common silicate minerals.

Erosion, Transportation, and Deposition

The main surficial process in continental environments is erosion of the land-scape. There are several agents of erosion, but rivers are the most dominant. They carve their own valleys, transporting the eroded material downstream to lakes, rivers, or oceans. Rivers also receive the products of weathering, frequently by an overland flow process.

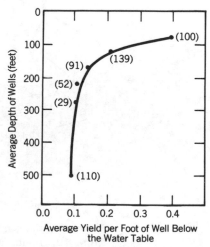

Figure 2.6
Decrease in well yields (gpm/ft. of well below the water table) with depth in crystalline rocks of the Statesville area, North Carolina. Numbers near points indicate the number of wells used to obtain the average values that define the curve (from LeGrand, 1954).

Fluvial Deposits

Fluvial deposits are generated by the action of streams and rivers. Boggs (1987) recognizes three broad environmental settings for fluvial systems: braided rivers, meandering rivers, and alluvial fans. Figure 2.7 contrasts the geometry of meandering and braided streams. According to Cant (1982), meandering streams generally produce linear shoestring sand bodies that are aligned parallel to the river course, and these are normally bounded below and on both sides by finer materials. The shoestring sands are many times wider than they are thick. The braided river, on the other hand, frequently produces sheetlike sands that contain beds of clay enclosed within them. The reason postulated by Cant (1982) lies in the meander width, with the braided streams capable of extensive lateral migration whereas the meandering streams are considered to be more rigorously confined in narrow channels. Braided rivers are characterized by many channels, separated by islands or bars. Meandering streams have a greater sinuosity and finer sediment load.

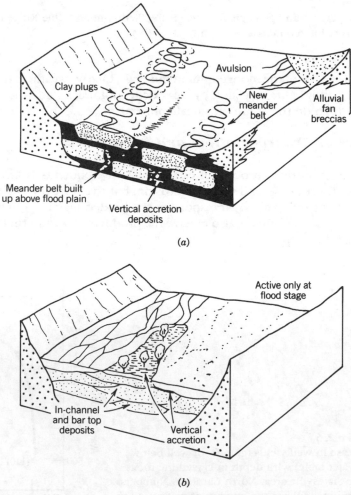

(a)

(b)

Figure 2.7
Contrasting the geometry of meandering (*a*) and braided (*b*) rivers (from Walker and Cant, 1984). Reprinted with permission of the Geol. Assoc. of Canada.

Alluvial fans form at the base of mountains where erosion provides a supply of sediment (Figure 2.8). Fans can occur in arid areas, such as those in Death Valley, California, and in humid areas as well. The upper part of the alluvial fan is characterized by coarse sediment due to confinement of flow to one or a few channels. The toe of the fan has the finest sediment, where more than one channel persisted over long periods of time.

Alluvial Valleys   The most important alluvial valleys in the United States are shown in Figure 2.9. The sand and gravel deposits are in hydraulic communication with their stream systems, which provides for their continual replenishment. The more permeable material occurs in clearly defined deposits that normally do not extend beyond the flood plain. Figure 2.9 does not differentiate between those channels cut by glacial melt water, such as those in the Midwest, and those stream channels not affected by glacial melt water, such as those in the southern Appalachians. Nor is there any distinction between meandering and braided patterns. Alluvial valleys are frequently underlain by silt and clay deposits and are among the most productive aquifers in the United States. Heath (1982), in his modern classification of ground water regions in the United States, established the alluvial valley system as a separate entity, independent of their geographic occurrence.

As rivers become rejuvenated from time to time, they cut through their own deposits, with the margins of the original valley floor left as terraces (Figure 2.10). These terraces also contain permeable materials and are considered part of the alluvial valley system. In Figure 2.10, the current flood plain alluvium is the youngest deposit and the terrace occupying the highest ground is the oldest, and was part of a flood plain prior to rejuvenation.

Alluvial Basins   Alluvial basins in the United States occupy a discontinuous region in excess of 1×10^6 km² extending from the state of Washington to the western tip of Texas. They are demonstrated on Figure 2.2 as the patterns of pores in sand and gravel. The material filling most of the western basins was derived by erosion of the adjacent mountains, with alluvial fan development extending to the

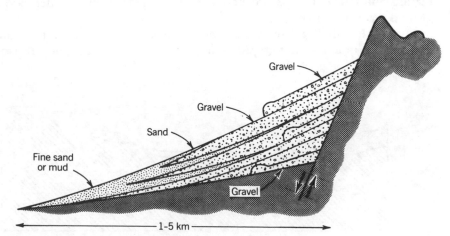

Figure 2.8
Diagrammatic cross section of an alluvial fan (from Rust and Koster, 1984). Reprinted with permission of the Geol. Assoc. of Canada.

Figure 2.9
Alluvial valleys (from Heath, 1984).

basin bottoms. The basins themselves are characterized by block faulting, volcanism and intrusion, high average elevation, and high average heat flow.

Also noted on Figure 2.2 is an extensive sand and gravel region in central United States extending from the southern part of South Dakota into Texas. This is part of an extensive alluvial apron deposited on the sloping bedrock surface extending from the foot of the Rocky Mountains eastward. Although much of the apron has been removed by erosion or otherwise dissected, the remaining part is known today as the Ogallala formation. Pumping from the Ogallala for irrigation purposes is the one factor accounting for the agricultural prosperity of the High Plains. Unfortunately, this prosperity is threatened because of depletion of the ground water resource.

Other than the Ogallala, which was deposited during Late Tertiary time, alluvial fans are noted in stratigraphic sequences of earlier ages. The Van Horn sand

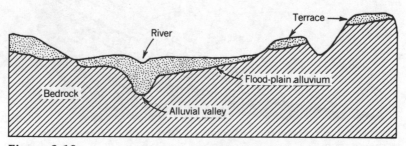

Figure 2.10
Cross section showing associations between alluvial valley and terrace development.

stone of West Texas is an example of a Precambrian fan system, consisting of massive conglomerates that grade into thinner pebbly sandstones, and finally into mudstones (Figure 2.11). Triassic sediments on the margins of the Newark Basin are also made up of the remnants of alluvial fans and were noted by Meinzer (1923) for their water-yielding characteristics.

Eolian Deposits

Eolian is a term used in reference to wind erosion or deposition. There are two types of wind deposits: loess, which is an unstratified deposit composed of uniform grains in the silt size range, and dunes or drifts composed of sand. Loess deposits are frequently associated with glacial terrane, whereas dunes and drifts are associated with deserts. The porosity of loess is very high, 40% to 50%, but they are not good transmitters of water because of the poor connectivity of the pore space. The best examples of ancient eolian deposits are in the Colorado Plateau and include the Jurassic Navajo sandstone and the Entrada, Wingate, White Rim, Coconino, and Lyons sandstones. Both the Navajo and the Wingate form what is known as the La Plata group, noted by Meinzer (1923) for its water supply. Other eolian deposits made up of dunes and drifts are expected to have a reasonably high permeability.

Lacustrine Deposits

The term lacustrine pertains to lakes, which currently form about 1% of the Earth's surface. Ancient lake deposits may include rock types ranging from conglomerates to mudstones, as well as carbonates and evaporites. Lacustrine deposits formed during the glacial period are generally characterized by low-permeability silts and clays, with an occasional high-permeability delta associated with the terminus of rivers. Ancient lacustrine deposits include the Morrison shale in the Colorado Plateau, the Chugwater Series of Wyoming, and the Green River formation in Utah, Wyoming, and Colorado. These are all low-permeability rocks.

Glacial Deposits

The glacial environment is a composite one and incorporates fluvial, lacustrine, and eolian environments (Figure 2.12). The high permeability resides in the glaciofluvial depositional environments, including the alluvial valleys, the alluvial deposits in buried bedrock valleys, and the well-sorted sand and gravel resulting

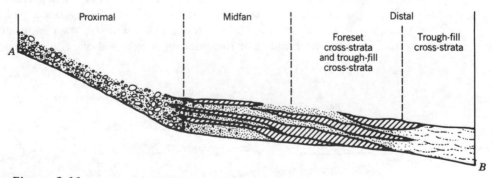

Figure 2.11
Facies and sedimentary structures in Van Horn sandstone (from McGowen and Groat, 1971).

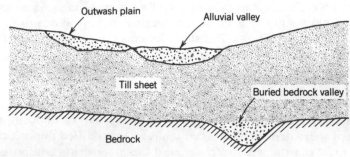

Figure 2.12
Glaciofluvial deposits in glaciated terrain.

from glacial melt water (outwash plains). Buried bedrock valleys are valleys that are no longer occupied by the streams that cut them. The till sheets are the most extensive deposit and consist of an assortment of grain sizes ranging from boulders to clay and are normally of low permeability. Frequently, the till is fractured and contains a higher permeability than was present originally.

The advance and retreat of continental glaciers generally results in a complex vertical stratigraphy. This is demonstrated on Figure 2.13 for a hypothetical case. The basal till sheet marks the advance of a glacier across a region. The terminal moraine marks the furthest advance, with subsequent melting forming the glaciofluvial deposits downstream from the moraine. We also note the presence of lacustrine deposits associated with a glacial lake that was trapped between the retreating glacier and the end moraine complex. The recessional moraine marks the position where the retreating glacier was temporarily halted, with its melt water forming the lake and associated delta deposits. This is only one simple example of a complex stratigraphy.

2.3   *The Boundary Between Continental and Marine Environments*

The environments that occupy the boundary between the continents and the ocean have been referred to as marginal marine (Boggs, 1987). Within this boundary, the following environments may be identified: deltas or deltaic systems, beach and barrier island systems, estuarine and lagoonal systems, and tidal flats. A delta is a fluvial deposit that is built into a standing body of water. Deltas consist of an upper deltaic plain, with characteristic braided or meandering patterns above the influence of

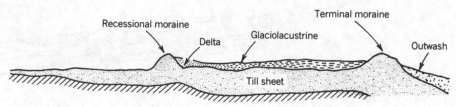

Figure 2.13
Deposits associated with a hypothetical advance and retreat of a glacier.

tides; the lower deltaic plain, which extends from the low-tide shoreline landward to the uppermost tidal influence; and the subaqueous delta plain, which is submerged (Coleman and Prior, 1982). The finer-grained material is normally transported seaward in the subaqueous region. Ancient deltaic systems have been identified in the Illinois Basin and in Gulf Coast sediments. In the Gulf Coast region, both the Frio and the Wilcox are well-known deltaic aquifers. The features and occurrence of ancient deltas are discussed by Eliott (1978).

The beach and barrier island environment consists of offshore bars or barrier islands with shallow lagoons between the islands and the mainland (Figure 2.14). Such coasts normally form where the offshore terrain is smooth and the seas shallow. The beach and barrier island system results in a narrow body of sediments parallel to the shoreline. Numerous sandstones have been recognized as beach and barrier island systems, including the Gallup and La Ventana sandstone in New Mexico, the Muddy sandstone in Wyoming, and the Mission Canyon sandstone in the Williston Basin. These are permeable formations, with the Mission Canyon being part of the Madison group, an important aquifer system in the Williston Basin.

Estuarine and lagoonal environments consist of the estuary, which is the lower part of a river open to the sea, and lagoons, which are shallow sea water bodies such as sounds or bays. Tidal flats are the regions affected by the rise and fall of tides. One of the better documented tidal flat environments is the upper Dakota group along the Front range in Colorado. The dominant stratigraphy consists of fine-grained tidal flat deposits overlying thicker coarse-grained channel deposits. The depositional environment for the Dakota group in general ranged from fluvial to marine in a marginal marine environment, with the greatest amount of sediment deposited as deltas in southward spreading seas.

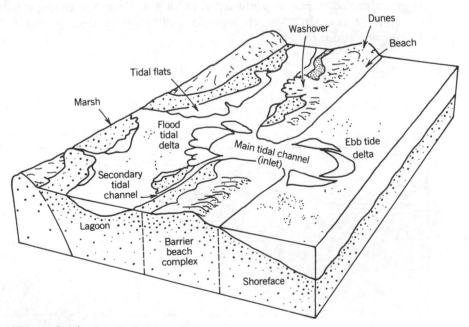

Figure 2.14
Subenvironments of a barrier island system (from Reinson, 1984). Reprinted with permission of the Geol. Assoc. of Canada.

2.4   Marine Environments

The marine environment extends from the continental shelf to the oceanic ridges. Shelf sediments normally consist of sands or, where tropical to subtropical climates persist, carbonates. Ancient shelf deposits are well known in the geologic record. Examples of modern carbonate platforms include the Yucatan shelf and the eastern Gulf of Mexico off Florida. The lower part of the Edwards formation in Texas has been cited as a possible carbonate shelf deposit (Wilson and Jordan, 1983). Deep-water sediments can be of various kinds, but generally are muds of one sort or another, along with carbonates.

Lateral and Vertical Succession of Strata

The term facies is defined as any arealy restricted part of a designated stratigraphic unit that exhibits a character significantly different from those of other parts of the unit. Thus, a lithofacies is a distinct unit, that is, sand, that may grade into finer material (silt and clay) as one moves from the shoreline toward the deeper parts of the ocean. The concept of a facies is one that relates to the nature of the depositional environment. As shorelines shift over geologic time, the deposits of one environment will eventually underlie the deposits of yet a younger and different environment. With a transgressing sea, near-shore deposits become progressively overlain by deeper-water environments (Figure 2.15). With a regressing sea, the order is reversed. Figure 2.16 shows a complete transgressive-regressive sequence. Such sequences are common in many parts of central and southern United States (Driscoll, 1986). Frequently, periods of erosion following the regression will remove one or more of the layers. Shifts in the shoreline bring different major environments in contact. A seaward shift of the shoreline (regression) can result in the establishment of lagoons and beach-barrier systems on older finer-grained deeper-water sediments; a landward shift of the shoreline (transgression) can result in the formation of beach deposits on finer-grained sediments associated with lagoons and marshes.

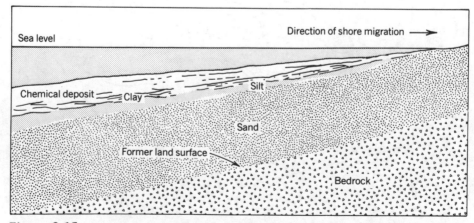

Figure 2.15
Cross section of sediments associated with a transgressing sea (from Driscoll, 1986). Reprinted with permission from Groundwater and Wells, 2nd Ed., 1986.

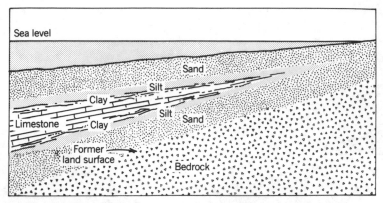

Figure 2.16
Cross section of transgressive-regressive sedimentary sequences
(from Driscoll, 1986). Reprinted with permission from Groundwater
and Wells, 2nd Ed., 1986.

Ancestral Seas and Their Deposits

We have already seen Figure 2.2, which was presented as a map showing the major
types of openings in dominant aquifers across the United States. Let us supplement Figure 2.2 with yet another map that gives the ages of the major rock groups
associated with the porosity patterns (Figure 2.17). Excluding the environments
previously discussed, three major groups remain:

1. The Paleozoic consolidated sedimentary group occupying a large south-
    west-northeast trend across the United States, with a few areally restricted
    occurrences elsewhere.

2. A rather dispersed pattern of unconsolidated and semiconsolidated
    Mesozoic rocks.

3. A continuous pattern of unconsolidated and semiconsolidated sedi-
    mentary rocks of Cenozoic age extending along the eastern coastline
    from New Jersey to Texas along with a few areally restricted occurrences
    elsewhere.

Virtually all these rocks are marine or marginal marine in origin and are related to
ancestral seas that inundated parts of the continent during periods of geologic
time. We cannot discuss all the formations in this grouping, but we can discuss
some of the major ones.

The Paleozoic Rock Group

The paleogeographic maps of Dunbar and Waage (1969) show that seas occupied
virtually all the southwest-northeast region of the United States from the Cambrian period through the Carboniferous (sometimes divided into the Mississip-
pian and Pennsylvanian periods). The resulting deposits consist chiefly of
sandstone, shale, limestone, and dolomite, with an aggregate thickness of several
thousand feet. During this period of time there were several regressions and trans-
gressions with a substantial amount of erosion. Except for those sediments along
the eastern margins (the Appalachian Geosyncline region), the formations are

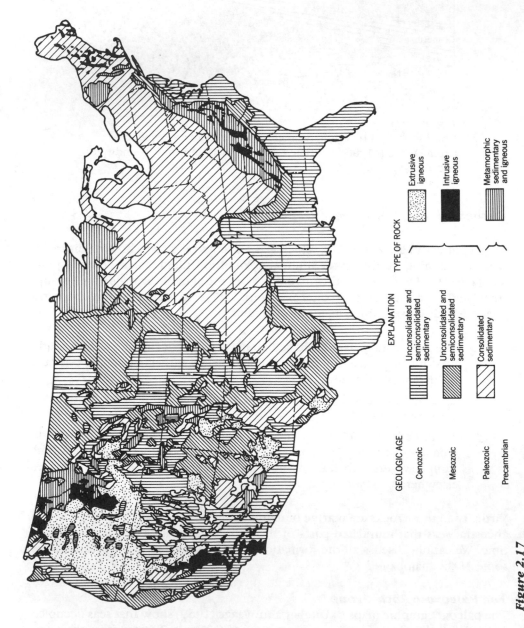

Figure 2.17
Geologic age of major rock groups in the United States (after U.S. Geological Survey, 1970).

nearly horizontal. It was during this period that the Cambrian-Ordovician aquifer was deposited, extending from the Northern Great Plains (Williston Basin) eastward. This formation occurs near the land surface in the Appalachian Mountains and in the Midwest. Cambrian sandstones in the Midwest include the Mount Simon and Eau Claire, whereas Ordovician sandstones include the St. Peter. Important carbonate rocks of Ordovician age include the Plattville and Galena formations. The Maquoketa shale marks the top of the Ordovician in Wisconsin and Illinois. Thus, the Cambrian-Ordovician-Maquoketa shale sequence constitutes a well-defined hydrostratigraphic unit. Other important aquifers above the Cambrian-Ordovician group include the Silurian age Monroe (restricted to Ohio) and the Niagara groups, both of which are carbonates. There are relatively few good water-producing rocks in the Devonian and Carboniferous. Extensive deposits of Permian age occur only in the southwest part of this region and are poor water producers. Thus, from the perspective of water supply, the Cambrian-Ordovician and alluvial valley systems in the Midwest are of considerable importance.

The Mesozoic Rock Group

The paleogeographic maps of Dunbar and Waage (1969) for the Mesozoic era show a rather dispersed seaway pattern, which accounts for the dispersed pattern of Mesozoic rocks shown on Figure 2.17. Seaways were very restricted from the Triassic period through the Early Jurassic, except for what is now the west coast of the continent. During the Late Jurassic period, the region that now constitutes the Rocky Mountains was inundated along with some parts of Texas and Louisiana. During Early Cretaceous a major seaway inundated most of Texas and Louisiana and the northern Rocky Mountain states. During Late Cretaceous a major seaway extended from Mexico through the entire Rocky Mountain region, both in the United States and Canada.

The marine formations associated with the Triassic seas in western United States produced few aquifers worth noting. Some of the Jurassic rocks in the Colorado Plateau are good water producers, for example, the La Plata group, but these have already been described as eolian deposits. By far, it is the Cretaceous period with its extensive seaway that gave rise to the major water producers in western United States. The Early Cretaceous produced the Trinity group in Texas, the Dakota sandstone in the Black Hills, the Cloverly sandstone in Wyoming, and the Kootenai in Montana. The Upper Cretaceous produced the Woodbine in Texas and the Dakota group in the Northern Great Plains, both of which are noted on Figure 2.17, and the Milk River sandstone in Alberta, Canada.

The Cenozoic Rock Group

The Cenozoic era is noted for the rise of the Rocky Mountains, the extrusion of lava beds in the Columbia River Plateau, block faulting in the basin and range region, and the advance of glaciers across the northern part of the continent. Our interest here, however, is restricted to the Cenozoic band of rocks that bound the eastern coast of the United States, extending around most of the Gulf of Mexico. These are marine deposits and, once again, the result of continental inundation by seas.

Seas covered the eastern continental shelf from Newfoundland to the Gulf Coast of Mexico throughout the Cenozoic and, on occasion, spread over the coastal plains. Deposits from these seas have a minimal thickness at their westernmost outcrop and thicken toward the Atlantic. The deposits were progressively tilted seaward. Carbonate facies are not abundant in the northern parts of the region and are restricted to Georgia and Florida. The Mississippi River was formed by Miocene time, carrying large volumes of sediments to the Gulf of Mexico, building the Mississippi Delta. Thus, Cenozoic sediments in the Gulf of Mexico region are largely of the clastic variety.

Within this geologic setting, numerous sandstones were deposited in the Atlantic coastal plain region, giving way largely to carbonate groups to the south. This fact motivated Heath (1984) to break from precedent in his modern classification of ground water regions to establish the Atlantic and Gulf coastal plain region, which he describes as interbedded sand, silt, and clay, and a southeast coastal plain region, which he describes as layers of sand and silt over carbonate rocks.

Diagenesis in Marine Environments

The progressive burial of sediments in depositional environments is accompanied by physical and chemical changes that affect the sediments. These physical and chemical changes occur because of increases in the overburden pressure, increased temperature, and the chemical interactions between minerals and migrating pore water. From a hydrologic perspective, diagenesis is important because of its effect on pore space. Some processes act to reduce the existing pore space and others act to enhance it. An end result of these processes is the lithification of the sedimentary pile, that is, the conversion of sediments to consolidated sedimentary rocks.

Porosity Reduction: Compaction and Pressure Solution

The term compaction refers to a diagenetic process where the weight of the overburden contributes to porosity reduction from some initial higher value. Compaction can take place by changing either the arrangement of the grains, or their shape. In a fluid-saturated environment, a decrease in porosity requires an expulsion of the pore fluids from the sediments, with the total volume of fluid expelled balanced by the total volume of porosity loss. The term normal compaction is reserved for the condition where the fluid expulsion takes place more or less concurrently with the deformation such that there is no apparent increase in the pressure of the fluid. Disequilibrium compaction implies a time lag between geologic loading of the sediments and the expulsion of their pore waters, meaning that some of the load is carried by the pore water, resulting in fluid pressures higher than would normally be expected in a hydrostatic (oceanic) environment. As a general rule, the finer grained the sediment, the greater its initial porosity, and the greater the potential porosity decrease. Hence, clays will be affected by compaction to a greater extent than sands, and carbonate muds may not be materially affected if crystallization occurs early in the depositional history.

Measured porosity versus depth curves for shales and sandstones are shown in Figure 2.18. The curve for shale was produced by Athy (1930) from a study of Paleozoic shales in Oklahoma. Athy proposed the relationship

$$n = n_o e^{-az} \tag{2.4}$$

where n is porosity, n_o is the average porosity of near surface clays, z is the depth below surface, and a is an empirical constant, equal to 1.42×10^{-3} m^{-1} for his data. The data of Athy suggest that given a porosity of surface clays on the order of 0.4 to 0.5, the porosity of compacted clay (shale) is reduced to about 0.03 at 2000m. The curve for sandstone (Blatt, 1979) is based on some 17,000 measurements of

Figure 2.18
Porosity versus depth curves. Curve *A* from Athy (1930) for shales; curve *B* from Blatt (1979) for sandstones. Data for Blatt's curve represent 1000-ft averages of 17,367 porosity measurements (from an unpublished manuscript by Atwater and Miller).

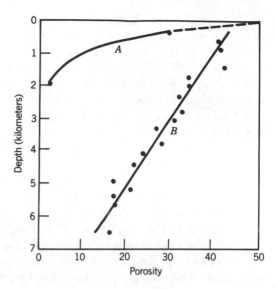

Late Tertiary sandstones in Louisiana. Below 350m, the relationship is a linear one, with porosity decreasing from 38% to about 18% at 6000m.

The decrease in porosity versus depth for sandstones shown in Figure 2.18 does not necessarily prove that grain rearrangement in closer packings is the responsible mechanism for porosity reduction. Another candidate is pressure solution, where grains dissolve at grain-to-grain contacts where the stress is greatest. If the material so dissolved is precipitated locally on the unstressed surface of the solid exposed to the pore space, the bulk volume is reduced at the expense of the pore volume. A conceptual visualization of this process has been given by Weyl (1959) and demonstrates that the reduction in bulk volume is compensated by a commensurate decrease in pore volume (Figure 2.19).

The deformation resulting from pressure solution is directly proportional to the grain-to-grain stress established in the rock. Thus, we expect it accompanies basin loading in depositional environments and is driven by the same stress as that

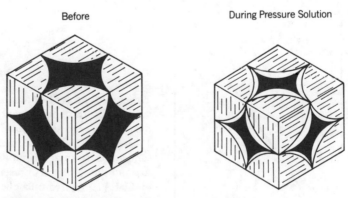

Figure 2.19
Pressure solution of identical spheres of simple cubic packing (from Weyl, J. Geophys. Res., v. 64, p. 2001–2025, 1959. Copyright by Amer. Geophys. Union).

responsible for a closer packing of the grains. There is an abundance of literature on pressure solution in rocks, including overviews by Rutter (1983) and DeBoer (1977). However, there still remains some controversy regarding the driving forces. Bosworth (1981) argues for a plastic deformation mechanism, whereas Green (1984) suggests that volume transfer creep during phase transformations has many characteristics similar to that of pressure solution. Whatever the mechanism, the reduction of porosity causes a rather large reduction in permeability. Hsu (1977), for example, determined that permeability of sandstones in the Ventura Basin in California decreased by three orders of magnitude over 3000 meters of depth.

Palciauskas and Domenico (1989) present a simple constitutive law for porosity reduction based on interpenetration of spheres and the melting heat at points of contact. Figure 2.20 shows their theoretical curve for porosity reduction with depth for sandstone. Based on this model, they concluded that the pore volume change for sandstone subjected to this inelastic deformation is substantially larger than the pore volume changes associated with grain rearrangement. Also shown on this figure are the data of Blatt (1979) for sandstones. Since the sandstone cores contain unspecified amounts of pore filling clays and cements, the measured porosity should be lower than the theoretical curve, but ideally should have a similar slope with depth. The theoretical curve given in Figure 2.20 is in good agreement with observations by Maxwell (1964) who stated that porosity of sandstones diminished with depth to less than 5% porosity below 10 km.

Chemical Rock-Water Interactions: Secondary Porosity in Sandstones

From a strictly chemical perspective, porosity can be reduced by cementation, replacement of one mineral by another, and recrystallization. Conversely, porosity can be enhanced by dissolution of grains and cements. A very large percentage of existing porosity in sandstones is not original intergranular porosity, but a secondary or solution-type porosity created by some dissolution process. The following statements by Hayes (1977) sum up much of modern thinking on porosity development in sandstones:

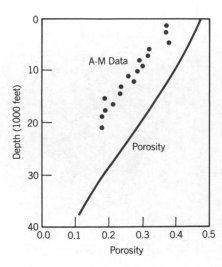

Figure 2.20
Theoretical upper bounds of porosity variation with depth in sandstone. The 17 data points represent 1000-foot averages of 17,367 porosity measurements of Late Tertiary sandstones from an unpublished manuscript by Atwater and Miller, as presented in the work by Blatt (1979) (from Palciauskas and Domenico, Water Resources Res., v. 25, p. 203–213, 1989. Copyright by Amer. Geophys. Union).

1. Primary intergranular porosity is subject to almost total destruction by several mechanical and chemical processes early in the diagenetic period.

2. Later in diagenesis, secondary porosity can be produced by dissolution.

3. Chemical diagenesis of sandstones is a kinetic process involving reactions between minerals and a moving aqueous solution. The main source of water in the depositional environment is from dewatering of shales interbedded with sandstone.

The first two statements appear to be amply supported by thin section and field studies (see, for example, Schmidt and McDonald, 1977). The third statement is a hypothesis, albeit a reasonable one. Before accepting any of these statements, there are several questions that must be addressed, including the responsible driving forces for water movement from shale into sandstone in an oceanic (hydrostatic) environment, the mass transport mechanisms, and the reactions that contribute to the diagenesis. These are major topics that will be covered elsewhere in this text. However, at this time a short statement on the driving forces for fluid movement may be in order in that these driving forces relate to the equilibrium and disequilibrium concepts introduced earlier, as well as to the process of pressure solution.

Simply, fluid moves from the shales into the sands because the shales are in a greater state of disequilibrium compaction than the sands. Indeed, if sands are continuous to the overlying oceanic waters, they are likely undergoing equilibrium compaction. Given this statement, there are two questions that may be pursued further: (1) For a given thickness of overburden and in the absence of fluid flow, will the fluid in all rocks be equally pressurized? (2) What is responsible for maintaining high fluid pressures in certain rocks and not in others over long periods of geologic time; that is, what contributes to disequilibrium compaction?

The first of these can be addressed immediately with a statement describing the fluid pressure production per interval of time in the absence of fluid flow

$$dP/dt = \chi\gamma'\omega \qquad (2.5)$$

where P is fluid pressure, t is time, γ' is the unit weight of the submerged sediment, which is the weight of the saturated sediment minus the weight of water, ω is a constant rate of sedimentation, and the factor χ is called the pore pressure coefficient. The product $\gamma'\omega$ expresses the time rate of change of the overburden pressure, so that if $\chi = 0$, $dP/dt = 0$, and if $\chi = 1$, the change in fluid pressure with time equals the change in overburden pressure with respect to time. The pore pressure coefficient expresses the change in fluid pressure with changes in stress in the absence of fluid flow; that is,

$$\chi = \left(\frac{dP}{d\sigma}\right)_M \qquad (2.6)$$

where the subscript M means at constant fluid mass, that is, no loss or gain of mass due to fluid flow, and σ is the stress. If the rocks and the water contained within them are totally incompressible, the pore pressure coefficient is zero, and the fluid pressure would not be affected by the stress changes. Here the total load is carried by the rigid matrix. On the other hand, if virtually all the overburden is carried by the fluid and none by the matrix, the pore pressure coefficient is close to one, and the fluid pressure change with time keeps up with the overburden pressure change with time. As the individual grains, the pores, and the fluids within the pores all exhibit compressibility to some extent, the pore pressure coefficient will be greater than zero but less than one. Work by Palciauskas and Domenico (1989) suggests that the pore pressure coefficient is on the order of 0.99 for clays, 0.83 for mudstones, 0.67 for sandstones, and 0.25 for limestones. Thus we note that the more rigid rocks (that is, limestones and, to a lesser degree, sandstones) are not capable of being excessively pressurized even in the absence of fluid flow. On the other hand, the fluid pressure in clays and mudstones can approach the overburden pressure because most of the incremental load is carried by the water in the pores. Hence, depending on the magnitude of the pore pressure coefficient, it may be difficult to generate high fluid pressures in some rocks relative to others, irrespective of the rate of loading and limitations on the rate of fluid flow.

Low-permeability sediments, such as shales, are those in which the fluid pressures will be maintained the longest because they do not drain efficiently. This statement is only partially correct, but it is sufficient for our purposes here. It follows that the driving force for fluid movement is the pressure differences between the fluid in the various rocks in the depositional environment, and this pressure difference is established because all rocks cannot be equally pressurized due to compressibility factors. Those that are pressurized the most maintain their pressures longer due to permeability factors.

Boggs (1987) cites seven major diagenetic processes other than compaction, at least four of which have the potential for effecting porosity and permeability in depositional environments. These are replacement, where one mineral is replaced by another; inversion, which is the replacement of a mineral by its polymorph (polymorphs are minerals having the same chemical composition but different crystal forms); recrystallization, or a change in crystal size and shape, generally resulting in an increase in grain size; and dissolution, a selective process that removes the most soluble minerals for a given set of environmental conditions.

2.5   Uplift, Diagenesis, and Erosion

With uplift and erosion, the sedimentary sequence we have been following has gone through one complete cycle. Over all geologic time, weathering and erosion act to modify any elevated landscape, with numerous rock bodies entering the sedimentary cycle time and again. From the perspective of permeability development, there are several important factors associated with the uplift of sedimentary rocks from their marine or marginal marine environment. First, the rocks are thrust into a meteoric regime so that fresh water can continually enter their outcrop areas. The entering meteoric water mixes with and displaces the original formation waters of the depositional environment. In some places, the displacement has been quite effective, with the original formation waters being completely displaced from the formation, at least in their current continental positions. In other places, the displacement process has not been effective. At any rate, the rocks apparently undergo yet another stage of diagenesis, rather unevenly perhaps, this time meteorically driven. With regard to porosity development, which is our main concern here, sandstone diagenesis merely means yet another generation of secondary porosity development by dissolution of those mineral phases that are not stable in the meteoric regime. For carbonates, again from the perspective of porosity development, our main interest here is the dissolution of mature carbonate terrain uplifted into a ground water circulation system.

The Style of Formations Associated with Uplift

Figure 2.21 shows three styles of development for formations associated with uplift. Figure 2.21*a* shows the geologic features associated with the Rocky Mountain development throughout the Cenozoic era. This style of development gives rise to exposed outcrops in the uplifted area, with the Dakota group referred to earlier providing a classic example. Several of the beds pinch out with depth and distance eastward, whereas others continue across the basin and outcrop or subcrop in structurally undisturbed areas. The Dakota sandstone, for example, outcrops along the Black Hills in South Dakota and underlies about 90% of Nebraska, and both Iowa and Minnesota to some extent. The Dakota also occurs in the Williston Basin in North Dakota and in the Denver Basin, where it contains salt water.

Figure 2.21*b* shows those features associated with the uplift of the Colorado Plateau, which was raised vertically along with the Front Range of the Rocky Mountains. The classic basin-type structure is evident, with outcrops or subcrops circumscribing the outer rim of the basins. It is noted that water can enter this type of basin along its entire circumscribed outcrop area.

Yet a third type of association of formations with mountains is shown in Figure 2.21*c*, but the relationship here is a passive one. This setting is supposed diagrammatically to represent the Atlantic coastal plain, where the sediments were derived largely from the Appalachians, and were not "uplifted" with them. As mentioned previously, the sediments thicken toward the sea, and, although there is evidence of some eastward tilting, the association of these sediments with tectonics is quite different from that mentioned earlier for the western United States. However, because of their topographic relationships, water can enter the formations in the outcrop areas and move eastward. Relatively fresh water zones extending several miles offshore into the Atlantic Ocean are not uncommon.

Following Heath (1984), we have attempted to show the occurrence of salty or brackish waters in association with the uplift patterns of Figure 2.21. In these diagrams, the transition from fresh water to salt water is portrayed as an abrupt interface. As noted, the transition is a progressive one that can occur over many miles. However, the diagrams do provide a background for a general description of our observations, several of which can be made here.

One notable feature of Figure 2.21 is that the younger permeable units can be more or less completely flushed of their original waters, while the deeper (older) ones are only partly flushed, with the degree of flushing getting poorer with depth. One possible reason for this is that the younger beds may have had more recent regressions and the formations subjected to long-term meteoric flushing in the depositional environment prior to uplift. Further, they are closer to the surface and may be replenished from meteoric sources other than those available at the outcrop. A second feature, as shown in Figure 2.21*b*, is the effect of basin perimeter recharge. Down-dip flushing in this case has the effect of concentrating the original formation water and can only be effective for short distances from the intake areas. As the formation fluids become more concentrated with depth, there is a marked density difference between the incoming meteoric waters and the original fluids, and density stratification occurs. For these cases, the lighter fluid

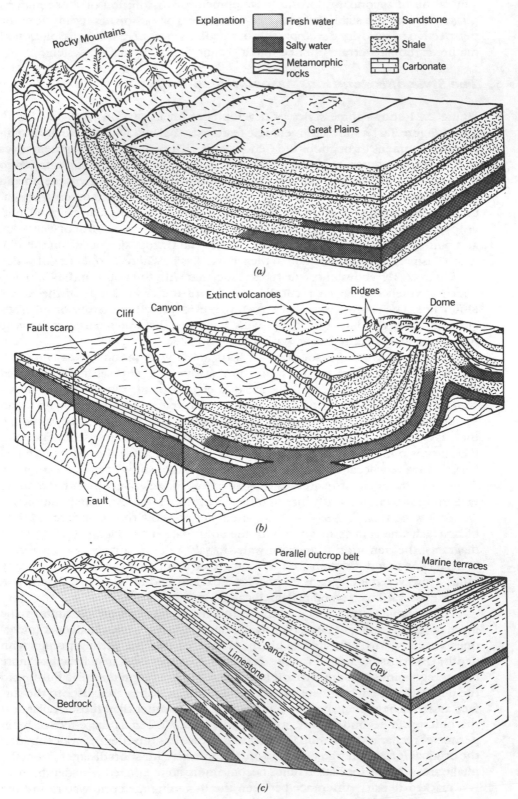

Figure 2.21
Style of uplift (from Heath, 1984).

will actually attempt to move above or around the more dense one if at all possible, or actually discharge cross formationally where further movement is impeded by the more dense fluid. It is difficult to conceive of any manner in which flushing can be effective in such environments.

Secondary Porosity Enhancement in Carbonate Rocks

According to Stringfield and LeGrand (1966), if the deposits overlying a carbonate terrain are of low permeability, and are thick, and if the carbonate was never elevated into a ground water circulation system, little secondary porosity will develop. On the other hand, early elevation into a ground water circulation system will lead to development of secondary porosity and permeability and to partial or complete removal of the carbonate. Hence, if a carbonate rock is unconformably overlain by a low-permeability rock such as shale, there is little question that the carbonate rock contains permeability. The same can be stated for carbonates in direct contact with glacial till laid down during the Pleistocene, for such carbonates represent long-term erosional surfaces that were characterized by paleohydrologic drainage systems. On the other hand, secondary permeability development cannot be assured when limestones are in conformable contact with overlying marine shales. Thus, the significance of the unconformity should not go unnoticed in the geologic record, especially when the concern is with secondary porosity and permeability development in carbonates. Along these same lines and of equal importance is the emergence of mature carbonate terrain relative to sea level when sea level drops by several hundreds of meters, as during the Pleistocene. This lowers the base level for ground water discharge, permitting meteorically derived fluids to penetrate the "elevated" terrain more deeply, contributing to secondary porosity and the formation of deep-seated karst. This was very likely the case for most of the porosity development in major carbonate aquifers in the Southeast coastal plain region of the United States.

Sinkholes are the primary landform associated with a mature karst terrain. They can be classified generally into four types: (1) the classical solution sink, which involves a closed depression in the bedrock with or without a soil cover, (2) a cavern collapse sink that develops when the roof falls in on a near surface conduit, (3) subsidence sinks caused by upward stopping of a collapsing solution cavity through substantial thicknesses of bedrock, and (4) soil piping sinkholes, which represent the abrupt collapse of a soil arch on mechanically sound bedrock through soil loss into solution cavities (White and White, 1987). Sinkholes are a serious geologic hazard. Considerable damage has resulted from their formation beneath highways, railroads, dams, reservoirs, pipelines, and vehicles (Newton, 1984).

Karstification is an evolutionary feature in which the dissolution of rocks over time results in the gradual development of an integrated conduit system (Quinlan and Ewers, 1985). Initially, diffuse flow moves through a large number of small fractures (Figure 2.22a). With time, a network of conduits begins to develop that carries an increasing proportion of the flow (Figure 2.22b). The development of sinkholes also contributes to the development of the conduit system because surface water will begin to flow directly into the conduits. In a mature karst system (Figure 2.22c), there is essentially no diffuse flow. Runoff proceeds directly into tributary conduits and moves through the rock at velocities as high as 400 m/hr (Quinlan and Ewers, 1985).

The key to the development of a conduit system is the circulation of ground water. Thus, the capability must exist for (1) chemically aggressive ground water to recharge the system, (2) the rock unit to transmit water through the fractures, and

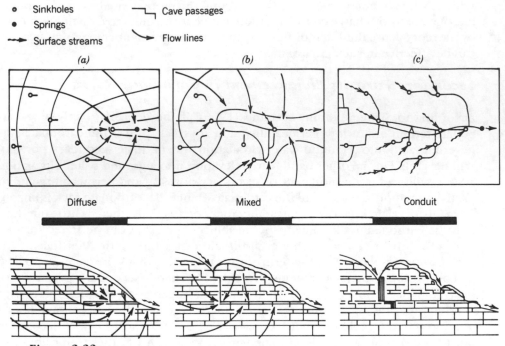

○ Sinkholes   ⌐_ Cave passages

● Springs

↝ Surface streams   ↘ Flow lines

(a)   (b)   (c)

Diffuse   Mixed   Conduit

Figure 2.22

Diffuse, mixed, and conduit flow in a hypothetical ground water basin and the sequence of its evolution. Flow lines are shown for the diffuse and mixed cases (from Quinlan and Ewers, 1985). Reprinted with permission of Ground water. © 1985. All rights reserved.

(3) ground water to drain from the system. If any of these conditions is not met, karst will not develop (Stringfield and LeGrand, 1966). It is for this reason that not every limestone or evaporite has large conduits in it and that carbonate aquifers in particular end up on a continuum between diffuse and conduit-type systems.

2.6   *Tectonism and the Formation of Fractures*

A fractured rock may be regarded as intact rock bodies separated by discontinuities. The rock bodies are generally considered to be impermeable, but frequently some permeability has to be attributed to these intact rock blocks. From the perspective of ground water flow, a fluid-conducting medium can be described as fractured if the majority of the flow takes place through discrete fracture channels that in some fashion form an integrated interconnected system.

Van Golf-Racht (1982) defines a fracture as the surface along which a loss in cohesion takes place. A fault is a fracture where displacement has occurred, whereas a joint is a fracture along which no notable displacement has occurred. From a purely geologic perspective, these definitions are adequate; from a hydrologic perspective, the most important factors are fracture properties irrespective of displacements, including orientation, density, aperture opening, smoothness of fracture walls, and—possibly above all else—the degree of connectivity. For example, given a set of fractures in a rock, none of which extend over the scale of the formation and none of which are interconnected, the rock cannot transmit water via the fracture network.

Style of Fracturing

Fracture style is closely related to stress history and rock type. Brittle rocks of low porosity are most susceptible to fractures, given the proper tectonic setting. Van Golf-Racht (1982) cites three cases where stress-related fractures may occur:

1. In response to folding and faulting.
2. Deep erosion of the overburden, which will produce differential stresses that can cause fractures.
3. Rock volume shrinkage (shrinkage cracks) where water is lost, say, for example, in shales or shaley sands.

Frequently, the terms microfracture and macrofracture are applied to describe the magnitude of fracture. These terms are imprecise, but a microfracture is on the order of the grain size or smaller and is not readily detected by the naked eye. Microfractures are frequently caused by differential thermal expansion of individual minerals. A macrofracture is a fracture that is readily detected without a microscope and can range from a simple joint to a throughgoing discontinuity.

Stearns and Friedman (1972) use the following descriptions in the study of fractures:

1. Conjugate fractures, or fractures related to a common origin that form under a single state of stress. Conjugate fractures are intersecting fractures, with the angle of intersection commonly found to be about 60°.
2. Orthogonal fractures, which are normal to each other and are not related to a single stress state.
3. Regional fractures, which extend over large areas and apparently are unrelated to local structures.

Orthogonal fractures are independent of structure and are generally well defined, even in flat lying rocks. Structurally related fractures are associated with specific structures such as faults and folds. Fractures associated with faults are thought to be caused by the same stresses that caused the faulting, and their strike will in general parallel that of the fault. Stearns (1964, 1967) reported five types of fracture patterns commonly associated with folding, only two of which produced a significant fracture density. As noted in Figure 2.23, both patterns consist of two conjugate shear fractures and an extension fracture. The pattern designated type I is thought to be earlier in origin than type II, and is common even on low

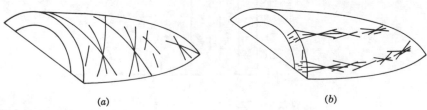

(a)   (b)

Figure 2.23
Schematic illustration of most common fractures associated with a fold: (*a*) type I, (*b*) type II. Both types maintain a consistent relation to bedding, but not to folding (from Stearns and Friedman, 1972). Reprinted by permission.

dipping folds where type II is absent. The two patterns occur in the same bed and represent two different stress states. Pattern type I has been described as single zones of parallel fractures across the entire structure whereas type II fractures may range in length from a few inches to several tens of feet. Type II fractures, however, consist of a fracture zone with three different orientations, whereas type I fractures appear to have a single orientation and are not an assemblance of all three orientations. Hence, Stearns and Friedman (1972) postulate that the type II fractures, although smaller in length and aperture, have a greater degree of connectivity and may be the better fluid conductors. On the other hand, because type I fractures are continuous over large regions, they may be an important regional pathway for fluid movement.

In comparing rock deformation under the same environmental influences, Stearns (1967) notes that the density of fractures is dependent on the lithology of the rock (Figure 2.24). This is in agreement with DeSitter (1956).

Several others studies attempt to relate fractures to lithology or to regional and local structures. Harris and others (1960) conclude that fractures associated with compressional deformation have certain features in common, such as repetition and continuity of trend. The influence of lithology is again pointed out along with the fact that thinner sedimentary rocks are more susceptible to fracturing than are thicker ones. Regan and Hughes (1949) discuss the productivity of fractured shales in California, where the fractures occur in chert zones and in zones of calcareous and platy siliceous shale.

Other types of fractures include those produced by volume shrinkage, caused by cooling in igneous rocks such as basalts, and desiccation in sedimentary rocks. Landes and others (1960) conclude that water-bearing fractures are a general rule in crystalline rocks between the depths of 600 and 1000 feet below the surface.

Weathered till is yet another material that shows a high fracture density. According to Barari and Hedges (1985), studies in the Midwest and Canada show that the upper part of most glacial tills has been weathered and contain a dense fracture pattern. Few if any fractures persist in the lower unweathered portions of till sheets. The permeability for weathered tills is three to four orders of magnitude larger than that for the unweathered tills.

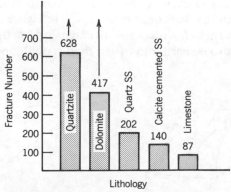

Figure 2.24
Average fracture density for several common rock types naturally deformed in the same physical environment (from Stearns, 1967).

Fluid Pressure and Porosity

The in situ porosity and permeability of fractured rocks are generally considered to be a function of fluid pressure. This statement holds for intergranular material as well, but to a lesser extent. In discussing these concepts, Snow (1968) provides abundant evidence that fractures "breathe," or open and close in response to changes in fluid pressure. As there is a direct relationship between fluid pressure and aperture opening (and, consequently, permeability), there follows some relationship between fracture strength, frictional resistance to sliding, and fluid pressure. Brace and others (1968) discovered an increase in permeability in fractured granites due to an increase in fluid pressure, which caused a decrease in the stress acting on the rocks. Serafin and del Campo (1965) suggested that the percolation of reservoir water beneath dams caused a widening of the fracture openings with a consequent weakening of the structure. A series of field experiments by Jouanna (1972) revealed that as the normal stress increases, the permeability decreases, and this process is irreversible. According to Sharp and Maini (1972) the relationship between stress and permeability is nonlinear because the normal stiffness of the fracture increases with decreasing aperture width. Thus, decreases in stress in fractured rock caused by increases in fluid pressure have a significant effect on rock permeability. Witherspoon and Gale (1977) review the relationship between fracture permeability and fluid pressure. Quantitative evaluations have been provided by Gangi (1978) and Walsh (1978, 1981).

Connectivity

De Marsily (1985) demonstrates the significance of connectivity of fractures with a few examples shown in Figure 2.25. Figure 2.25a shows some fracture connections common to two different areas in otherwise similar crystalline rock. Wells drilled in set *A* were quite successful; wells drilled in set *B* were not, even when fractures were encountered. De Marsily (1985) postulates that the density of fractures in set *B* is so low that they do not form an interconnected network whereas when a fracture is encountered in set *A*, it is likely that it is connected with all or most of the others in the set because of a greater fracture density.

    In Example 2, two fracture sets at one site occur at different depths. Injection into or pumpage from one of the sets does not affect the water levels in the other. The sets are obviously not connected.

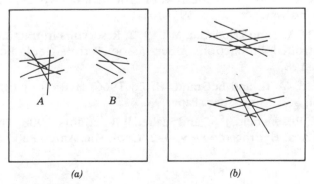

(a)   *(b)*

Figure 2.25
Schematic illustration showing fracture sets (from de Marsily, 1985).

Percolation theory deals with the idea of connectivity (Shante and Kirk-patrick, 1971; Dienes 1982). A percolation threshold is defined as the density of fractures that intersect sufficiently to promote flow. A finite set of fracture clusters can be interconnected, but the connections between various finite clusters may not exist. Above the percolation threshold, one cluster becomes "infinite" so that flow can take place through the medium. However, many other clusters may not be connected to the infinite one. As the fracture density increases, isolated clusters become more rare. De Marsily (1985) gives several examples where the percolation threshold has been related to the average number of intersections of a single fracture with all other fractures, as well as to the density and length of the fractures. Robinson's (1982) results are based on the latter and indicate that when the product of the fracture density and the square of their average length equals 1.5, the percolation threshold is achieved and flow commences. Below this value, the fractures are insufficiently connected to promote flow.

These results at this stage of this research are the consequence of numerical experiments wherein fracture patterns are randomly generated with a computer and the individual intersections noted. Key questions and difficulties remain in identifying the percolation threshold and obtaining field data on actual fracture densities. Three-dimensional data on fracture density and volumes (Charlarx and others, 1984) are obviously even more difficult to obtain. Further, from our discussion of fracture style, it is clear that there exists in nature different classes of fractures and that connectivity studies for the individual classes as well as any possible assemblage of classes would give results that might be less abstract and more pertinent to the problems of flow in real rocks. That is to say, if classes of fractures can be identified from field mapping, perhaps something can be said of a qualitative or semiquantitative nature about the potential for connectivity and the degree of connectivity that might exist on the basis of theoretical studies.

Suggested Readings

Hayes, J. B., 1979, Sandstone diagenesis, the hole truth, in Aspects of Diagenesis, ed. Scholle, P. A., and P. R. Schluger: Soc. Econ. Paleont. and Min. Spec. Pub. 26, p. 127–139.

Heath, R. C., 1982, Classification of ground water systems of the United States: Ground water, v. 20, p. 393–401.

Stearns, D. W., and Friedman, M., 1972, Reservoirs in fractured rock, in Fracture Controlled Production: Amer. Assoc. Petrol. Geol. Reprint Series 21, p. 174–198.

Stringfield, V. T., and LeGrand, H. E., 1966, Hydrology of limestone terranes: Geol. Soc. Amer. Spec. Paper 93.

Back, W., Rosenshein, J. S., and Seaber, P. R., editors. 1988, The Geology of North America, Hydrogeology: v. 0–2, Geol. Soc. Amer. Pub., p. 15–263.

Chapter Three

Ground Water Movement

3.1 Darcy's Experimental Law
     and Field Extensions

3.2 Hydraulic Conductivity and
     Permeability of Geologic
     Materials

3.3 Darcy's Law and the Field
     Mapping of Hydraulic Head

3.4 Flow in Fractured Rocks

3.5 Flow in the Unsaturated
     Zone

Considering the fact that the flow of fluid can be observed in "real time" in laboratory experiments, it is not surprising that laws have been developed to describe some of the macroscopic details of the motion. Because "motion" involves measurements of time and distance, the experimental observations are quantitative ones, so that the laws take on a mathematical form. Our main law may be stated outright as

The velocity, or distance traveled over some time interval, is proportional to some driving force.

The surprising feature of this statement is that—provided certain conditions are met—it holds irrespective of the complex details of the connected pore space described in Chapter 2.

3.1   Darcy's Experimental Law and Field Extensions

Henry Darcy was a civil engineer concerned with the public water supply of Dijon, France, in particular the acquisition of data that would improve the design of filter sands for water purification. In search of this information, Darcy set out "to determine the laws of flow of water through sand." His method was experimental, the results of which were published in 1856.

The experimental apparatus employed by Darcy is shown in Figure 3.1. In these experiments, water was passed through the sand column and the volumetric flow rate Q was measured at the outlet. The cross-sectional area of the sand column was known, as was the length of the sand in the column. During the experiment, Darcy measured the distance between the water levels in the two manometers at various flow rates. The pertinent measurements included Q, a volumetric flow rate, L^3/T; A,

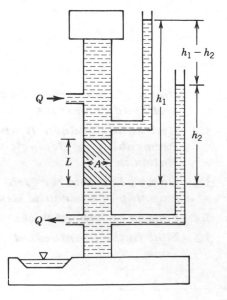

Figure 3.1
Darcy's laboratory apparatus.

the cross-sectional area, L^2; l, the length of sand column, L; and $(h_1 - h_2)$, the difference in elevation of manometer readings, L. From these measurements, the pertinent calculations were Q/A, a velocity L/T; and $(h_1 - h_2)/l$, a dimensionless quantity. The quantity Q/A is a volumetric flow rate per unit surface area and is termed specific discharge. The dimensionless quantity $(h_1 - h_2)/l$ represents the change in water-level elevation in the manometer tubes over the distance through which the change takes place. This term is referred to as the gradient, or hydraulic gradient, and is often expressed with the calculus as $\partial h/\partial l$.

In the terms of the measurements and calculations just given, Darcy's law is expressed

$$\frac{Q}{A} = q = \frac{-K(h_1 - h_2)}{l} = -K\frac{\partial h}{\partial l} \qquad (3.1)$$

where q is the volumetric flow rate per unit surface area, with units of velocity, and K is a constant of proportionality. Because the gradient is a dimensionless quantity, the proportionality constant K has units of velocity. The minus sign is a convention that is employed because flow is in the direction of a decrease in water levels in the manometers, that is, from where h is high to where h is low (Figure 3.1). Darcy's equation may be stated in words as

The velocity of flow is proportional to the hydraulic gradient.

Darcy's law is valid for flow through most granular material. The law suggests a linear relationship between the specific discharge and the hydraulic gradient. This relationship holds as long as the flow is laminar. Under conditions of turbulent flow, the water particles take more circuitous paths. At the other extreme, for very-low-permeability materials, a minimum threshold gradient may be required before flow takes place (Bolt and Groenevelt, 1969).

Four aspects of Darcy's law require clarification from a field perspective: the specific discharge q, the water-level measurements, the gradient, and the proportionality constant K.

The Nature of Darcy's Velocity

As defined, the specific discharge q is a volumetric flow rate per unit surface area of sample. Because water only moves through the pore openings making up the surface area, Darcy's q is a "superficial" velocity. We thus define a more realistic velocity v that is a volumetric flow rate per unit area of connected pore space. The expression for v comes directly from Eq. 3.1

$$\frac{Q}{n_e A} = \frac{q}{n_e} = v = -\left(\frac{K}{n_e}\right)\frac{\partial h}{\partial l} \qquad (3.2)$$

where $n_e A$ is the effective area of flow and n_e has been introduced as the effective porosity. Here, the quantity v is referred to as the linear or pore velocity of ground water. The linear velocity v will always be larger than the superficial velocity (specific discharge) and increases with decreasing effective porosity. Finally, it is convenient to designate the gradient $\partial h/\partial l$ as i so that Darcy's law may be expressed in the following forms:

$$q = Ki \tag{3.3a}$$
$$Q = KiA \tag{3.3b}$$
$$v = Ki/n_e \tag{3.3c}$$
$$v = q/n_e \tag{3.3d}$$

The quantity q (as well as v) has both a magnitude and a direction, the later being toward decreasing elevation of water in the manometers. This makes q and v vector quantities, a fact that shall be exploited in later chapters.

Hydraulic Head: Hubbert's Force Potential

In Figure 3.1, the water-level elevations in the manometers are measured with reference to a common datum, taken arbitrarily at the base of the sample. Thus, the absolute value of the water-level elevations were of no concern to Darcy, only the differences between them. In this section, we are not interested in these differences but in what the actual water-level measurements mean. For this purpose, it is convenient to introduce a counterpart of the laboratory manometer, known as a piezometer.

The piezometer is a tube or pipe used to measure water-level elevations in field situations. It is open at the top where measurements are taken and open at the bottom to facilitate the entrance of water. Such a device is shown in Figure 3.2, where the common datum is taken as sea level (zero elevation). As demonstrated by Hubbert (1940), the terms elevation, pressure, and total head on Figure 3.2 can be explained in terms of the conventional Bernoulli equation. This equation states that under conditions of steady flow, the total energy of an incompressible fluid is constant at all positions along a flow path in a closed system. This may be written as

$$gz + \frac{P}{\rho_w} + \frac{v^2}{2} = \text{constant} \tag{3.4}$$

where g is the acceleration due to gravity, z is the elevation of the base of the piezometer, P is the pressure exerted by the water column, ρ_w is the fluid density, and v is velocity. Dividing through by g, Eq. 3.4 becomes

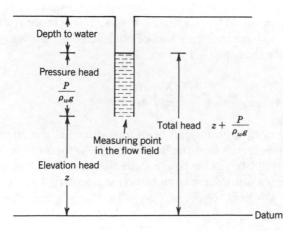

Figure 3.2
Diagram showing elevation, pressure, and total head for a point in the flow field.

$$z + \frac{P}{\rho_w g} + \frac{v^2}{2g} = \text{constant} \tag{3.5}$$

where the quantity $\rho_w g$ equals the unit weight of water. Equations 3.4 and 3.5 describe the total energy contained by the fluid, where the first term is the energy of position, the second term is the energy due to sustained fluid pressure, and the third term is the energy due to fluid movement. The dimensions of Eq. 3.4 are L^2/T^2, or energy per unit mass. Equation 3.5 is expressed as energy per unit weight, say, foot lb per pound, which reduces simply to feet. In SI units, the dimensions are newton-meters per newton, or simply meters.

The three terms of Bernoulli's equation, when expressed as energy per unit weight, are referred to as elevation head, pressure head, and velocity head, respectively. The term z in Eq. 3.5 is the elevation head and represents the elevation at the base of the piezometer. In a theoretical sense it represents the work required to increase the elevation of a unit weight of water from datum to height z (Figure 3.2). Stated another way, every body at the surface of the earth has a gravitational attraction toward the earth's center. To raise this body counter to this attraction requires work, and this work is stored in the form of a potential energy. The quantity $P/\rho_w g$ represents the length of the water column in the piezometer. It represents the work that a fluid is capable of doing because of its sustained pressure. The sum of these terms is called the potential energy of the fluid. The third term, $v^2/2g$, is the kinetic energy, or energy due to the fluid motion. As the velocity of ground water is slow, this term is ignored. Thus, the sum of the elevation head z and the pressure head $P/\rho_w g$ represents the total head h in the system; that is,

$$h = z + \frac{P}{\rho_w g} \tag{3.6}$$

Stated, simply, the total head h is the sum of the elevation of the base of the piezometer plus the length of the water column in the piezometer. Hence, the total head at a point is found by measuring the elevation of the water level in a piezometer (Figure 3.2). However, the point to which this head refers to is not the water level, but the point at the terminus of the piezometer at elevation z. The total head is sometimes referred to as the energy of position, and is a scalar quantity.

An additional expression for the energy of position of a fluid can be obtained from the Bernoulli expression of Eq. 3.4 where, ignoring the velocity term

$$\phi = gz + \frac{P}{\rho_w} = gh \tag{3.7}$$

where ϕ is referred to as the hydraulic potential, in units of energy/mass. Defining potential in this way provides yet another form of Darcy's law commonly found in the literature. From the statement $\phi = gh$, it follows that

$$\frac{\partial h}{\partial l} = \left(\frac{1}{g}\right)\frac{\partial \phi}{\partial l} \tag{3.8}$$

Substituting this result in Eq. 3.1 gives

$$q = -\left(\frac{K}{g}\right)\frac{\partial\phi}{\partial l} \qquad (3.9)$$

where $\partial\phi/\partial l$ is the potential gradient (Hubbert, 1940).

The Gradient: An Introduction to Field Theory

To obtain a clear idea of the gradient from a field perspective requires an introduction into field theory. A field represents a continuous distribution of scalars, vectors, or tensors described in terms of a spatial coordinate system and time. For example, a scalar field is defined as a property that takes on a value (number) at every point of a specific region of space and time. There are numerous quantities of interest that satisfy the definition, including temperature $T(x, y, z, t)$, concentration of some substance $C(x, y, z, t)$, and energies of position such as hydraulic head $h(x, y, z, t)$. Here, the Cartesian coordinate system is employed to identify the value of the stated scalar quantity at a point of interest at a given time.

Given this brief introduction to scalar fields (which are really scalar functions) suppose now that several piezometers are placed in some region. Starting at any one of the piezometers, it is likely that the head will increase in some directions and decrease in others. The gradient in x can be written as $\partial h/\partial x$, the partial derivative indicating how h changes in x irrespective of changes in y and z. That is, $\partial h/\partial x$ is the change in head as we move along the x axis. Further, the gradient in y becomes $\partial h/\partial y$, which describes the change in head along the y axis, and the gradient in z is $\partial h/\partial z$. Thus, Darcy's law expressed as

$$q_x = -K_x \frac{\partial h}{\partial x} \qquad (3.10a)$$

$$q_y = -K_y \frac{\partial h}{\partial y} \qquad (3.10b)$$

$$q_z = -K_z \frac{\partial h}{\partial z} \qquad (3.10c)$$

describes the flow of fluid along the x, y, and z axes, where the material properties in x, y, and z are different; that is, $K_x \neq K_y \neq K_z$.

As it is rather limiting to restrict ourselves to one of three directions, we may ask for the rate of change in head in any direction and, most important, the direction of maximum change. For this purpose, the gradient of a scalar field h can be defined as

$$\text{grad } h = \nabla h = \mathbf{i}\frac{\partial h}{\partial x} + \mathbf{j}\frac{\partial h}{\partial y} + \mathbf{k}\frac{\partial h}{\partial z} \qquad (3.11)$$

where the vectors $\mathbf{i}$, $\mathbf{j}$, $\mathbf{k}$ are unit vectors in the x, y, and z directions, respectively, and ∇h (pronounced del h and is an upside-down "delta" to remind us of differentials) is the abbreviation for grad h. Thus, in vectorial notation, for $K_x = K_y = K_z$, Darcy's law becomes

$$\mathbf{q} = -K \text{ grad } h = -K\nabla h \qquad (3.12)$$

where grad *h* can be read "gradient of *h*." Equation 3.12 is a vectorial equation where the *x*, *y*, and *z* components are given by Eq. 3.11.

The gradient of *h* or ∇*h* is a vector that represents the spatial rate of change of hydraulic head (Eq. 3.11). It consists of three components *x*, *y*, and *z*, each of which represents how fast the head changes in that respective direction. The direction of ∇*h* is that which coincides with the direction in which the head changes the fastest. For the condition $K_x = K_y = K_z$, this direction is perpendicular to lines of equal head. This direction is of particular interest in that it coincides with the direction of ground water flow.

Example 3.1

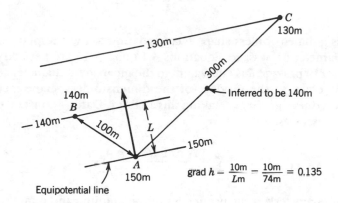

$$\text{grad } h - \frac{10m}{Lm} - \frac{10m}{74m} = 0.135$$

This example concerns the determination of the direction of flow and the hydraulic gradient from three piezometers in a horizontal plane with heads of 150m, 140m, and 130m at points *A, B*, and *C*, respectively. Note the analogy to the three-point problem in structural geology.

Physical Interpretation of Darcy's Proportionality Constant

By experimentally varying fluid density, viscosity, and the geometrical properties of sands, Hubbert (1956) reported that Darcy's proportionality constant *K* varied in the following manner

$$K \propto \rho_w$$
$$K \propto 1/\mu$$
$$K \propto d^2$$

where μ is viscosity and *d* is the mean grain diameter of the sand. This can be expressed as

$$K = K^* \frac{\rho_w \, d^2}{\mu} \tag{3.13}$$

where K^* is yet another constant of proportionality containing variables not yet evaluated. It is obvious at this point that Darcy's K is a function of both properties of the medium and properties of the fluid.

Insight into the parameter K^* may be obtained by comparing Darcy's law with the Hagen-Poiseuille equation long known to govern laminar flow through small-diameter passages. Incorporating Eq. 3.13 into Darcy's law gives

$$\mathbf{q} = \frac{-K^* \, \rho_w \, d^2}{\mu} \, \text{grad} \, h \tag{3.14}$$

whereas the Hagen-Poiseuille equation is

$$\mathbf{q} = \frac{-N \, \rho_w \, g \, R^2}{\mu} \, \text{grad} \, h \tag{3.15}$$

where N is a dimensionless shape factor relating to the geometry of passage and R is the diameter of passage. Equations 3.14 and 3.15 are perfectly equivalent if the diameter of passage R is equivalent to the mean grain diameter so that the constant K^* is equal to the product of the dimensionless shape factor N and the acceleration due to gravity g. Making this analogy, Darcy's constant of proportionality is expressed as

$$K = \frac{N \, \rho_w g \, d^2}{\mu} = \frac{k \, \rho_w \, g}{\mu} \tag{3.16}$$

where Nd^2 characterizes the properties of the medium and ρ_w/μ the properties of the fluid. The parameter K is referred to as the hydraulic conductivity and contains properties of both the medium and the fluid, and the parameter k is referred to as the intrinsic permeability (equal to Nd^2) and contains properties of the medium only. The hydraulic conductivity with units of velocity characterizes the capacity of a medium to transmit water, whereas the permeability with units L^2 characterizes the capacity of the medium to transmit any fluid.

The term "hydraulic conductivity" is used most frequently in ground water literature when dealing with the single water phase. The term "permeability" is used in the petroleum industry where the fluids of interest are oil, gas, and water. Typical values of permeability in square centimeters or square meters are quite small so that the "darcy" is commonly defined as a unit of permeability. For a material of 1-darcy permeability, a pressure differential of one atmosphere will produce a flow rate of 1 cc (cubic centimeter) per second for a fluid with 1-centipoise viscosity through a cube having sides 1 cm in length. One darcy is approximately equal to 10^{-8} cm^2 or, for water of normal density and viscosity, 10^{-3} cm/sec. For tight materials, the millidarcy (md) is used, where 1 md equals 0.001 darcys, or approximately 10^{-6} cm/sec.

It is now possible to state yet another form of Darcy's law commonly used in the petroleum industry. Incorporating our definition of the constant of proportionality K, the hydraulic head h, and the gradient of the head h, Darcy's law becomes

$$\mathbf{q} = \frac{N \, \rho_w g \, d^2}{\mu} \, \text{grad} \left(z + \frac{P}{\rho_w g} \right) \tag{3.17}$$

For those cases where all piezometers are bottomed at the same elevation in a flat lying bed, z is a constant so its derivative goes to zero. Taking both the fluid density and the acceleration of gravity as constants, Eq. 3.17 becomes

$$\mathbf{q} = -\frac{Nd^2}{\mu} \operatorname{grad} P = -\frac{k}{\mu} \operatorname{grad} P \qquad (3.18)$$

This expression is quite convenient when dealing with multiphase fluid systems where the permeability k by definition is invariant with respect to whatever fluid is being considered.

Units and Dimensions

Various units and dimensions are given in Table 3.1 in both the foot-pound-second system with its FLT base (force, length, and time) and the International System (SI) with its MLT base (mass, length, and time). In this text we will attempt to utilize SI metric units but will also provide an equivalent measure in "field" units where the use of such units is common in field practice.

3.2 Hydraulic Conductivity and Permeability of Geologic Materials

Observed Range in Hydraulic Conductivity Values

Virtually thousands of measurements of hydraulic conductivity and permeability have been obtained in both the laboratory and in the field. Davis (1969) provides the best summary of these data. Table 3.2 is taken largely from this review along with additional input from Johnson and Morris (1962) and Croff and others (1985). The values cited are given in meters per second, but conversions can be made to centimeters per second, feet per second, feet per year, or gallons per day per square foot, the last referred to as a meinzer unit in honor of O. E. Meinzer. The meinzer unit is defined as the flow of water in gallons per day through an opening of 1 ft^2, under a unit hydraulic gradient at a temperature of 60°F. Conversion to permeability (square centimeters, square feet, or darcys) is readily accomplished with the cited conversion factors.

As noted in Table 3.2, hydraulic conductivity can range in value over about 12 orders of magnitude, with the lowest values for unfractured igneous and metamorphic rocks and the highest values for gravels and some karstic or reef limestones and permeable basalts. The range in hydraulic conductivity within a given rock type is greatest for the crystalline rocks and smallest for the sedimentary material. In general, a hydraulic conductivity approaching 10^{-9} m/sec and smaller can be characterized as low permeability material. Clay, shale, chalk, and unfractured igneous and metamorphic rocks fall within this category. However, if these rocks or sedimentary accumulations are fractured, the conductivity can easily exceed this limiting value by two or three orders of magnitude.

Table 3.1
Dimensions and common units for flow parameters

Parameter	Symbol	ft-lb-sec system		Conversion factor	SI	
		Dimension	Units	Multiply by	Dimension	Units
Hydraulic head	h	L	ft	3.048×10^{-1}	L	m
Elevation head	z	L	ft		L	m
Pressure head	—	L	ft		L	m
Fluid pressure	P	F/L^2	lb/ft^2	4.788×10	M/LT^2	N/m^2 or P_a
Fluid potential	ϕ	L^2/T^2	ft^2/sec^2	9.29×10^{-2}	L^2/T^2	m^2/sec^2
Density of water	ρ_w	—	—		M/L^3	kg/m^3
Gravitational constant	g	L/T^2	ft/sec^2	3.048×10^{-1}	L/T^2	m/sec^2
Unit weight of water	$\gamma_w = \rho_w g$	F/L^3	lb/ft^3		—	—
Volumetric flow rate	Q	L^3/T	ft^3/sec	2.832×10^{-2}	L^3/T	m^3/sec
Specific discharge	q	L/T	ft/sec	3.048×10^{-1}	L/T	m/sec
Hydraulic conductivity	K	L/T	ft/sec	3.048×10^{-1}	L/T	m/sec

Table 3.2
Representative values of hydraulic conductivity for various rock types

Material	Hydraulic conductivity (m/sec)
SEDIMENTARY	
Gravel	3×10^{-4} – 3×10^{-2}
Coarse sand	9×10^{-7} – 6×10^{-3}
Medium sand	9×10^{-7} – 5×10^{-4}
Fine sand	2×10^{-7} – 2×10^{-4}
Silt, loess	1×10^{-9} – 2×10^{-5}
Till	1×10^{-12} – 2×10^{-6}
Clay	1×10^{-11} – 4.7×10^{-9}
Unweathered marine clay	8×10^{-13} – 2×10^{-9}
SEDIMENTARY ROCKS	
Karst and reef limestone	1×10^{-6} – 2×10^{-2}
Limestone, dolomite	1×10^{-9} – 6×10^{-6}
Sandstone	3×10^{-10} – 6×10^{-6}
Siltstone	1×10^{-11} – 1.4×10^{-8}
Salt	1×10^{-12} – 1×10^{-10}
Anhydrite	4×10^{-13} – 2×10^{-8}
Shale	1×10^{-13} – 2×10^{-9}
CRYSTALLINE ROCKS	
Permeable basalt	4×10^{-7} – 2×10^{-2}
Fractured igneous and metamorphic rock	8×10^{-9} – 3×10^{-4}
Weathered granite	3.3×10^{-6} – 5.2×10^{-5}
Weathered gabbro	5.5×10^{-7} – 3.8×10^{-6}
Basalt	2×10^{-11} – 4.2×10^{-7}
Unfractured igneous and metamorphic rocks	3×10^{-14} – 2×10^{-10}

To convert meters per second to	Multiply by
cm/sec	10^2
(gal/day)/ft²	2.12×10^6
ft/sec	3.28
ft/yr	1×10^8
darcy	1.04×10^5
ft²	1.1×10^{-6}
cm²	1×10^{-3}
To convert any of the above to meters per second	Divide by the appropriate number above

Character of Hydraulic Conductivity Distribution

Given a set of conductivity data from a particular formation or unit, it is a simple task to obtain the mean or average value. If the conductivity data fit some normal distribution, the mean value would be the one that occurs most frequently. On the basis of thousands of histograms constructed from conductivity data (Law, 1944; Bennion and Griffiths, 1966), a definite positive skewness in the conductivity distribution has been ascertained. Data from histograms skewed to the right will generally plot as a straight line if plotted as a cumulative curve on a log normal probability scale (Figure 3.3). Given this straight-line plot, the probability density curve for hydraulic conductivity is log normal. Csallany and Walton (1963) and Davis (1969) have shown that this log normal tendency is reflected further in histograms of water well yield in specific geologic formations. This result is not totally unexpected in that well yield depends to a large extent on the hydraulic conductivity of the material.

According to Collins (1961), the most frequently occurring value in a log normal distribution is closer to the harmonic mean than to the arithmetic mean. The harmonic mean of a set of numbers is determined by

$$H = \frac{N}{\Sigma \, (1/X)} \tag{3.19}$$

where H is the harmonic mean, N is the number of values in the sample, and X is the individual values. The harmonic mean of the numbers 2, 4, 8 is 3.43 whereas the arithmetic mean for the same set is 4.66. Detailed work on actual distributions by Cardwell and Parsons (1945), Warren and Price (1961), and Bennion and Griffiths (1966) has determined that the "average" conductivity lies between the

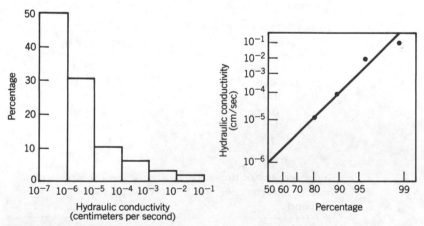

Figure 3.3
Typical frequency distribution for hydraulic conductivity and corresponding plot as cumulative curve on log normal probability paper. Notice in this plot that 80% of the samples are equal to or less than 10^{-5} cm/sec, 90% of the samples are equal to or less than 10^{-4} cm/sec, and 99% of the samples are equal to or less than 10^{-1} cm/sec.

harmonic and arithmetic mean and is better described by the geometric mean, given as

$$G = \sqrt[N]{X_1 \, X_2 \, X_3 \ldots X_N} \qquad (3.20)$$

which is the Nth root of a product of N numbers. The geometric mean of the numbers 2, 4, 8 is 4. Hence, the geometric mean is intermediate between the harmonic and arithmetic mean.

In contrast to hydraulic conductivity, porosity values from a single formation have normal rather than log normal distributions. Thus, porosity will plot as a straight line on a normal probability scale, whereas permeability plots are straight lines on the log normal probability scale. Davis (1969) demonstrates these facts for some sandstone data.

Anisotropicity and Heterogeneity: Origin and Manifestations

Most rocks have a directional quality to their overall structure. Metamorphic rocks are noted for their schistosity, most sediments for their horizontal stratification, and basalts for their preferred orientation of shrinkage cracks when the cooling period is rather short (and a scarcity of such cracks when the cooling history is long). For such materials, the hydraulic conductivity as measured from some representative sample will not be equal in all directions. The term anisotropic is used to describe materials where the permeability or conductivity is different in different directions. When permeability is the same in all directions, the material is isotropic. Davis (1969) cites several cases for bedded sediments in particular where the permeability is greatest in the direction of the stratification and smallest perpendicular to the stratification. Although supporting data are sparse, similar statements can be made for maximum directional permeability associated with metamorphic structures such as schistosity. Table 3.3 provides a summary of information on the anisotropic nature of some sedimentary materials as determined from core samples. In these tabulations, the horizontal conductivity is taken in the direction of the structural features, such as stratification, and the vertical conductivity is taken at right angles to the stratification.

A porous material is homogeneous if the permeability is the same from point to point. Materials that do not conform with this condition are heterogeneous. Freeze and Cherry (1979) draw attention to three broad classes of heterogeneity.

Table 3.3
The anisotropic character of some rocks

Material	Horizontal conductivity (m/sec)	Vertical conductivity (m/sec)
Anhydrite	$10^{14}–10^{-12}$	$10^{-15}–10^{-13}$
Chalk	$10^{-10}–10^{-8}$	$5 \times 10^{-11}–5 \times 10^{-9}$
Limestone, dolomite	$10^{-9}–10^{-7}$	$5 \times 10^{-10}–5 \times 10^{-8}$
Sandstone	$5 \times 10^{-13}–10^{-10}$	$2.5 \times 10^{-13}–5 \times 10^{-11}$
Shale	$10^{-14}–10^{-12}$	$10^{-15}–10^{-13}$
Salt	10^{-14}	10^{-14}

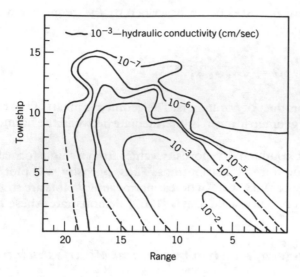

Figure 3.4
Map of the Milk River sandstone showing hydraulic conductivity distribution (from Schwartz and others, 1982). Reprinted with permission of Elsevier Scientific Publishing Company.

The first is referred to as trending heterogeneity, which refers to a progressive increase or decrease in hydraulic conductivity within one single unit or formation. Figure 3.4 shows an example of trending heterogeneity for the Milk River sandstone in Alberta, Canada. The Milk River sandstone has been interpreted as the seaward margin of a littoral environment, some of the sand having been deposited by streams or currents running parallel to the coastline. This depositional environment has resulted in a northwest-southeast linear trend of sand lenses (Meyboom, 1960). The trend observed in Figure 3.4 indicates highest conductivity values associated with thick clean sands in the southern and central parts of the formation, with a progressive decrease northward and to the west and east as well in response to greater proportions of fine-grained material in the formation. Trending heterogeneity in sedimentary accumulations is often controlled by the early geologic history as established within the depositional environment.

Yet another class of heterogeneity identified by Freeze and Cherry (1979) is termed layered heterogeneity. This kind of heterogeneity is demonstrated in Figure 3.5 where several beds of high and low permeability make up the stratigraphic framework. If the conductivity variation within a bed is much smaller than the conductivity contrasts between beds, as might be the case for interlayered sands and shales, it may be assumed that each layer is homogeneous and isotropic. Each layer, however, is characterized by a different hydraulic conductivity, rendering the sequence as a whole heterogeneous. The assumption of

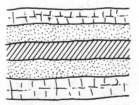

Figure 3.5
Layered heterogeneity.

homogeneity and isotropicity for each layer provides us with some useful information between layered heterogeneity and anisotropy. According to Leonards (1962), an equivalent horizontal hydraulic conductivity for the vertical section can be expressed as

$$K_x = \frac{\Sigma(m_i K_i)}{\Sigma m_i} \tag{3.21}$$

where K_x is the equivalent horizontal conductivity, K_i is the homogeneous conductivity of an individual layer, and m_i is the thickness of the layer. For the direction at right angles to the stratification

$$K_z = \frac{\Sigma m_i}{\Sigma (m_i/K_i)} \tag{3.22}$$

where K_z is an equivalent vertical conductivity for the layered system. These equations give the hydraulic conductivity values parallel and vertical to the stratification for a single homogeneous, anisotropic rock that is the equivalent of the layered system shown in Figure 3.5. In general $K_x/K_z \geq 1$. The ratio K_x/K_z is referred to as the anisotropy ratio.

Example 3.2

To illustrate the use of Eqs. 3.21 and 3.22, consider a 1000-ft sequence of interbedded sandstone and shale that is 75% sandstone. The sandstone has a horizontal and vertical permeability of about 1 darcy ($K = 10^{-3}$ cm/sec), whereas the shale has a horizontal and vertical permeability of about 2×10^{-7} darcys ($K = 1.92 \times 10^{-10}$ cm/sec). From Eq. 3.21,

$$K_x = \frac{750 \text{ ft} \times 1 \times 10^{-3} \text{ cm/sec} + 250 \text{ ft} \times 1.92 \times 10^{-10} \text{ cm/sec}}{1000 \text{ ft}}$$

$$= 7.5 \times 10^{-4} \text{ cm/sec}$$

From Eq. 3.22,

$$K_z = \frac{1000 \text{ ft}}{(750 \text{ ft}/1 \times 10^{-3} \text{ cm/sec}) + (250 \text{ ft}/1.92 \times 10^{-10} \text{ cm/sec})}$$

$$= 7.7 \times 10^{-10} \text{ cm/sec}$$

Thus, for horizontal flow the most permeable units dominate the system; for vertical flow the least permeable units dominate the system. Under the same hydraulic gradient, horizontal flow is on the order of six orders of magnitude faster than vertical flow. For this example, horizontal flow through 1000 ft of sediments with a uniform horizontal hydraulic conductivity of 7.5×10^{-4} cm/sec would be the same as the sum of the flows through each of the individual layers in the example.

Discontinuous heterogeneity has been recognized as a third broad class. This spatial structure is characterized by large contrasts in hydraulic conductivity, as might be expected in fractured rocks where the fractures are the major pathways for fluid movement. Discontinuous heterogeneity in fractured rocks is often associated with major structural features that are responsible for the establishment of fluid pathways, say, along the axes of anticlines where failure of the rock material has occurred in response to folding (Figure 3.6). In other cases, the pattern of the heterogeneity is not so easily recognized. Figure 3.7, taken from the environmental assessment of the Columbia River basalts at Hanford, Washington (1986), presents alternative concepts for ground water movement based on anisotropy and heterogeneity. Figures 3.7*a* and 3.7*c* are examples of layered heterogeneity, where the dense basalt interior is depicted as having either a low (Figure 3.7*a*) or a high permeability (Figure 3.7*c*), but in either case considerably lower than the flow top permeability. Figures 3.7*b* and 3.7*d* are examples of discontinuous heterogeneity in an anisotropic medium. In Figure 3.7*b*, most of the vertical transfer of fluids takes place through major structural discontinuities. In Figure 3.7*d*, both the major and minor fractures cutting the interior account for the vertical transport of water.

Heterogeneity and the Classification of Aquifers

Layered heterogeneity is of some importance in categorizing the occurrence of ground water in the hydrologic cycle. If a homogeneous rock of high permeability exists continuously from land surface to some great depth, the water in this aquifer will occur exclusively under unconfined conditions. An unconfined aquifer is one in which the water table forms an upper boundary (Figure 3.8). In some instances, where low-permeability materials are interbedded with higher-permeability units, downward percolating water in the unsaturated zone may become "perched" on the low-permeability units. Thus, a localized zone of saturation could form above the low-permeability unit. This condition is referred to as a "perched water table," which is a local zone of saturation completely surrounded by unsaturated conditions (Figure 3.9).

Given the layered heterogeneity demonstrated in Figure 3.8, a confined aquifer can occur as a high-permeability unit between two low-permeability units, or aquitards. The aquitards in this case are referred to as confining layers. The confined high-permeability unit frequently contains water under pressure due to the elevation of the intake area in dipping beds. The water level in wells registered in wells tapping such a confined aquifer can be above or below the regional water table in the overlying unconfined aquifer (Figure 3.8). In some

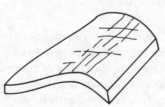

Figure 3.6
Discontinuous heterogeneity.

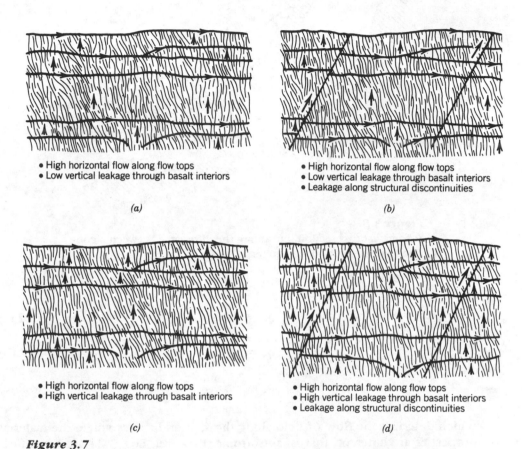

> Ground water flow path,
> no horizontal scale intended
>
> Note: Assume hydraulic heads
> increase with depth

- High horizontal flow along flow tops
- Low vertical leakage through basalt interiors

(a)

- High horizontal flow along flow tops
- Low vertical leakage through basalt interiors
- Leakage along structural discontinuities

(b)

- High horizontal flow along flow tops
- High vertical leakage through basalt interiors

(c)

- High horizontal flow along flow tops
- High vertical leakage through basalt interiors
- Leakage along structural discontinuities

(d)

Figure 3.7
Conceptualization of ground water occurrence and movement in Columbia River
basalts based on anisotropy contrasts and hypothetical structures (from Department
of Energy, 1986).

cases, the wells may even flow at the surface, in which case they are referred to as
artesian wells.

Darcy's Law for Anisotropic Material

In most of the discussions of Darcy's law, it has been tacitly assumed that conductivity was independent of direction. For this assumption, Darcy's law was
expressed as

$$\mathbf{q} = -K \operatorname{grad} h \tag{3.23}$$

which is correct only when $K_x = K_y = K_z$, that is, the material is isotropic. For
anisotropic material, the following forms of Darcy's law were given as

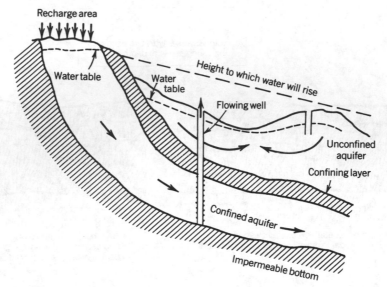

Figure 3.8
Schematic cross section illustrating the difference between a confined and an unconfined aquifer.

$$q_x = -K_x \frac{\partial b}{\partial x} \tag{3.24a}$$

$$q_y = -K_y \frac{\partial b}{\partial y} \tag{3.24b}$$

$$q_z = -K_z \frac{\partial b}{\partial z} \tag{3.24c}$$

which describes the flow of fluid along the x, y, and z axes where the material properties are different, that is, anisotropic. However, Eqs. 3.24 are simplifications of yet a more complex form of Darcy's law for anisotropic material. This more complex form can be examined by considering the fact that the velocity in each of the expressions of Eqs. 3.24 is a vector and can be resolved into components parallel to the x, y, and z axes. Consider that these x, y, and z axes form the

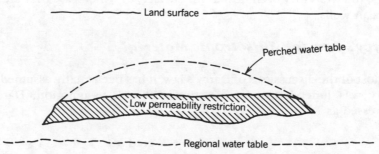

Figure 3.9
Schematic presentation of a perched water table.

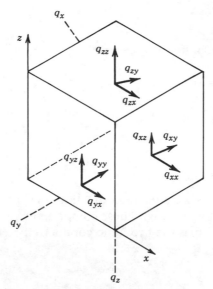

Figure 3.10
Array of nine components of three
velocity vectors acting at a point on
the three faces of a cube.

edges of a small cube through which the fluid is flowing (Figure 3.10). We may now condense and express the information on Figure 3.10 in terms of q_{ij}, where the first subscript indicates the direction perpendicular to the plane upon which the velocity vector acts and the second subscript indicates the direction of the velocity vector in that plane. That is to say, q_x given in Eq. 3.24a describes the flow of fluid along the x axis. Thus, the components of the velocity vector q_x on the plane of the cube normal to the x axes become q_{xx}, q_{xy}, q_{xz}. Here q_{xx} is a normal velocity component in that it acts along the normal to the designated plane, whereas q_{xy} and q_{xz} act tangential to the plane in question. Thus, Eq. 3.24a can be expressed

$$\text{fluid flow in the plane normal to } x = q_x = -K_{xx}\frac{\partial h}{\partial x} - K_{xy}\frac{\partial h}{\partial y} - K_{xz}\frac{\partial h}{\partial z} \qquad (3.25)$$

$$\text{in the direction of} \quad\quad\underset{x \text{ axis}}{\uparrow} \quad \underset{y \text{ axis}}{\uparrow} \quad \underset{z \text{ axis}}{\uparrow}$$

Thus q_x depends not only on gradients in x, but on gradients in y and z as well.

In a similar fashion, the velocity vector q_y of Eq. 3.24 can be resolved into three components, all of which act on a plane normal to the y axis (Figure 3.10). One of these will be a normal velocity component (q_{yy}) in that its acts in the direction normal to the designated plane whereas the others will be tangential. The same line of reasoning is followed for the resolution of q_z into three components. Hence, Eqs. 3.24 can be expressed as

fluid flow

in the plane normal to $x = q_x = - K_{xx} \dfrac{\partial h}{\partial x} - K_{xy} \dfrac{\partial h}{\partial y} - K_{xz} \dfrac{\partial h}{\partial z}$

in the plane normal to $y = q_y = - K_{yx} \dfrac{\partial h}{\partial x} - K_{yy} \dfrac{\partial h}{\partial y} - K_{yz} \dfrac{\partial h}{\partial z}$

in the plane normal to $z = q_z = - K_{zx} \dfrac{\partial h}{\partial x} - K_{zy} \dfrac{\partial h}{\partial y} - K_{zz} \dfrac{\partial h}{\partial z}$

$\qquad\qquad\qquad\qquad\qquad\uparrow\qquad\quad\uparrow\qquad\quad\uparrow$

in the direction of $\qquad\qquad\qquad x$ axis $\quad y$ axis $\quad z$ axis

In this form it is seen that there are nine components of the hydraulic conductivity in anisotropic material. These components can be placed in matrix form to give what is known as the hydraulic conductivity tensor

$$
\begin{matrix}
K_{xx} & K_{xy} & K_{xz} \\
K_{yx} & K_{yy} & K_{yz} \\
K_{zx} & K_{zy} & K_{zz}
\end{matrix}
$$

This is a second-order symmetric tensor that has the property $K_{ij} = K_{ji}$; that is, $K_{xy} = K_{yx}$, $K_{xz} = K_{zx}$, and so on. For the special case where the principal directions of anisotropy coincide with x, y, and z directions of the coordinate axes, the six components K_{xy}, K_{xz}, K_{yx}, K_{yz}, K_{zx}, and K_{zy} are all equal to zero. In this case, the x, y, and z axes are called the principal axes of the porous medium. The conductivity tensor for the principal axes then becomes

$$
\begin{matrix}
K_{xx} & 0 & 0 \\
0 & K_{yy} & 0 \\
0 & 0 & K_{zz}
\end{matrix}
$$

Stated another way, when the coordinate axes are oriented parallel to the principal axes of the porous medium, we recover the original expressions for Darcy's law for anisotropic material (Eqs. 3.24) when x, y, and z are principal axes and $K_x \neq K_y \neq K_z$. For the very special condition where the material properties do not differ with direction, we obtain the isotropic form of Darcy's law (Eq. 3.23), where the hydraulic conductivity is taken as a scalar quantity.

Measurement of Hydraulic Conductivity

Hydraulic conductivity can be measured in a variety of ways. In practice, there are only three common methods that are employed: field tests, laboratory tests, and empirical or semiempirical methods based on grain diameter and grain size distributions, or simple hydraulic models. Field tests are by far the most reliable for they permit the testing of large volumes of rock with one pumping well and one or more observation wells. Pumping-in tests have a considerably smaller area of influence but are important in low-permeability rocks that do not readily yield water to pumping wells. Field testing is an extensive topic and will be covered in Chapter 5.

Laboratory Testing

Permeameters for measuring conductivity are shown in Figure 3.11. In the constant head test, a valve at the base of the sample is opened and the water starts to flow. After a sufficient volume of water is collected over the time of the test, the volumetric flow rate Q is ascertained. Hydraulic conductivity is then determined with Darcy's law of the form

$$K = \frac{QL}{Ah} \qquad (3.26)$$

where L is the length of the sample, A is the cross-sectional area of the sample, and h is the constant head shown on Figure 3.11a.

In the falling head test, the head is measured in the standpipe of Figure 3.11b, along with the time of measurement. For a sample of length L and a cross-sectional area A, the conductivity is determined by

$$K = 2.3 \frac{aL}{A(t_1 - t_0)} \log_{10} \frac{h_o}{h_1} \qquad (3.27)$$

where a is the cross-sectional area of the stand pipe and $(t_1 - t_0)$ is the elapsed time required for the head to fall from h_o to h_1. Some permeameters are designed so the sample can be brought to the original stress condition at the depth it was collected. This is important for deep samples where the measured conductivity changes as a function of the applied stress.

The Search for Empirical Correlations

As hydraulic conductivity can be readily measured in the laboratory, there have been numerous attempts to relate the measured values to various properties of a

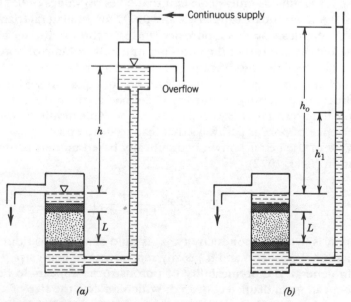

(a)   *(b)*

Figure 3.11
Constant head (*a*) and falling head (*b*) permeameters.

porous medium. One commonly accepted relationship has been proposed by Hazen (1911)

$$K = C\, d_{10}^2 \tag{3.28}$$

where K is the hydraulic conductivity in centimeters per second, d_{10} is the effective grain size in centimeters, and C varies from 100 to 150 (cm sec)$^{-1}$ for loose sands. The effective grain size is defined as the value where 10% of the particles are finer and 90% coarser.

Another formula of the form given by Eq. 3.28 has been proposed by Harleman and others (1963), stated as

$$k = (6.54 \times 10^{-4})\, d_{10}^2 \tag{3.29}$$

where k is the permeability in square centimeters and d_{10} is again the effective grain size. Masch and Denny (1966) incorporate the spread of the grain size curve by using the median grain size d_{50} in formulations in an attempt to correlate permeability with grain size. Krumbein and Monk (1943) include both particle size and sorting in their search for an empirical correlation and provided the formula

$$k = 760\, d^2\, e^{-1.31\sigma} \tag{3.30}$$

where k is the permeability in darcys, d is the geometric mean diameter, in millimeters, σ is the log standard deviation of the size distribution, and 760 is a constant.

In addition to these empirical approaches, there are other more hydraulically based attempts to relate permeability to porous medium properties. Kozeny (1927) considered the porous medium to be a bundle of capillary tubes and demonstrated that permeability must have the form

$$k = C\, n^3/S^{*2} \tag{3.31}$$

where C is a dimensionless constant that takes on values of 0.5 for circular capillaries, 0.562 for square capillaries, and 0.597 for equilateral triangles; k is permeability in length squared; n is porosity; and S^* is the specific surface, defined as the interstitial surface area of the pores per unit bulk volume of porous material. Collins (1961) reports a modification of the Kozeny relationship where, for capillary tubes of unequal length, the right-hand side of equation 3.31 is multiplied by the reciprocal of tortuosity. Tortuosity is defined as the ratio of flow path length to sample length and is always greater than one. This modification accounts for the occurrence of porous pathways that lead to dead ends.

One of the better known hydraulically based models is the Kozeny-Carmen equation (Bear, 1972)

$$K = \left(\frac{\rho_w g}{\mu}\right) \frac{n^3}{(1-n)^2} \left(\frac{d_m^2}{180}\right) \tag{3.32}$$

where K is hydraulic conductivity, ρ_w is fluid density, μ is fluid viscosity, g is the gravitational constant, and d_m is any representative grain size.

In general, the permeability of porous rocks appears to be proportional to some mean grain diameter squared, which reflects the size of a pore, along with the spread or distribution of the grain sizes (pores). Collins (1961) points out that the grain (pore) size distribution also influences the magnitude of specific surface, which increases with decreasing grain size. Consequently, pore size distributions are indirectly incorporated in the Kozeny-type formulations.

Hydraulic arguments have also been applied to fractured rocks. For a parallel array of planar joints of aperture width b, with N joints per unit distance across the rock face, the permeability may be described by (Snow, 1968)

$$k = Cnb^2 = CNbb^2 \tag{3.33}$$

where k is the permeability with dimensions of lengths squared, C is a dimensionless constant, in this case (1/12), and the porosity n is taken to be a planar type of porosity equal to Nb. Here, N has the units L^{-1}.

3.3 Darcy's Law and the Field Mapping of Hydraulic Head

Measurements of hydraulic head obtained from water wells in a confined aquifer can be contoured on a map, each individual data point representing a measure of the potential energy of the water in a rock unit in which the measurement is made. Lines connecting points of equal hydraulic head are called equipotential lines. The resulting hypsometric surface is referred to as a potentiometric map. Another term in common usage is piezometric surface, defined by Meinzer (1923) as an imaginary surface that everywhere coincides with the level of water in the aquifer. If the aquifer is unconfined, the contoured surface is referred to as a map of the water table.

Figure 3.12 shows the potentiometric surface for the Dakota sandstone. This artesian aquifer crops out along the eastern flanks of the Black Hills in South

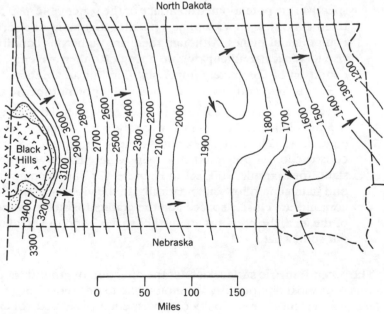

Figure 3.12
Potentiometric surface of the Dakota sandstone, contour interval 100 ft (from Darton, 1909).

Dakota and dips eastward. Water enters this unit at its elevated intake areas and moves downdip in an easterly direction. The water moves from where the head is high to where the head is low, with each of the lines presumably connecting points of equal head.

Regarding the use of such surfaces the following should be noted:

1. A potentiometric map must be related to a single aquifer. Other aquifers deeper or shallower in the section will have different potentiometric surfaces that may exhibit heads that are higher or lower than the one of immediate concern. For example, the Madison formation underlies the Dakota sandstone throughout much of South Dakota and has its own potentiometric surface showing more or less the same slope and same direction of flow, but higher heads than those encountered in the Dakota. Head measurements from more than one aquifer are not interchangeable in the construction of the potentiometric surface. If a potentiometric map is drawn from data obtained from different aquifers, the surface obtained is a composite of potential measurements.

2. It is assumed that flow in the aquifer is horizontal, that is, parallel to the upper and lower confining layers. Thus, there is an assumed absence of vertical gradients in the aquifer. If a piezometer is placed in such an aquifer, and a hydraulic head is noted, the hydraulic head is presumed not to change with increasing depth of piezometer penetration. Hence, the potentiometric surface is a projection of vertical equipotential planes into the horizontal plane. This arrangement of equipotential lines is demonstrated on Figure 3.13 which shows a cross-sectional (x-z) and a horizontal (x-y) projection of lines connecting points of equal head. The surface depicted in the x-y plane is a potentiometric surface and is merely a projection of vertical equipotentials in the x-y plane. From another perspective, the potentiometric surface or the water table map for the unconfined condition appears to be a parallel alignment of a series of vertical equipotential planes. Although there are numerous conditions that give rise to vertical gradients within a single aquifer, we emphasize that the condition of completely horizontal flow (or an absence of vertical gradients) is frequently assumed when using potentiometric surface maps.

3. Head losses between adjacent pairs of equipotential lines are equal, and the hydraulic gradient varies inversely with distance between lines of equal head. The first statement is simply a necessary convention in the construction of potentiometric maps where the contour interval is constant, for example, 100 feet in Figure 3.12. This accomplished, the second statement follows automatically. In the vicinity of the Black Hills, the contours are closely spaced and the gradient is steep. In the eastern part of the area, the contours are further spaced, and the gradient is flattened out somewhat.

The potentiometric surface defines the direction of ground water movement, and this is its most obvious use. We should like to do something more with such surfaces, say, calculate the velocity of movement, or the total volumetric flow in the system, or even obtain some idea of the spatial distribution of hydraulic conductivity, at least in a relative sense. Before these things can be done, something further must be said about hydraulic conductivity and some of the properties that

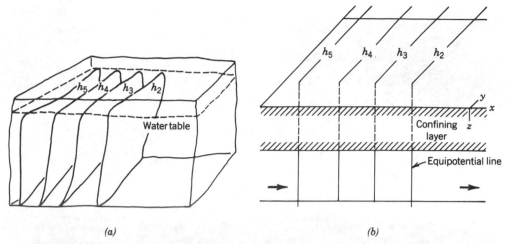

Figure 3.13
Cross section and horizontal projection of lines connecting points of equal head for an unconfined (*a*) and confined (*b*) aquifer.

real ground water systems possess. For example, it has been stated that in an isotropic medium, the flow is at right angles to the equipotential lines. This condition is not satisfied in anisotropic media. The condition of isotropicity is thus another assumption in the interpretation of potentiometric maps.

Just how much more information can be obtained from the potentiometric surface depends on the flow condition it is presumed to depict. The simplest type of flow condition is shown in Figure 3.14*a*, which portrays an aquifer of uniform thickness that neither loses nor gains water in the direction of flow. The volume of flow entering in a unit time at the outcrop area is equal to the volume of flow leaving the aquifer at its discharge end. Figure 3.15 is presumed to be a potentiometric map of part of this aquifer. By construction, the head losses between adjacent equipotential lines are equal, and the hydraulic gradient varies inversely with distance between the equipotential lines. As long as we assume the system is isotropic, the flow takes place at right angles to the equipotential lines. Further, as there are no gains or losses in the total flow, the volumetric flow rate passing one equipotential line must equal the flow rate passing the adjacent or downgradient equipotential line. In other words, there can be no gaps in the fluid. This means that

$$Q_1 = K_1 \frac{\Delta h}{L_1} A_1 = Q_2 = K_2 \frac{\Delta h}{L_2} A_2 \qquad (3.34)$$

Recognizing that $A_1 = A_2$ for a uniformly thick aquifer, and $\Delta h_1 = \Delta h_2$ by construction

$$\frac{K_1}{K_2} = \frac{L_1}{L_2} \qquad (3.35)$$

or the ratio of the conductivities equals the ratios of the lengths between the equipotential lines.

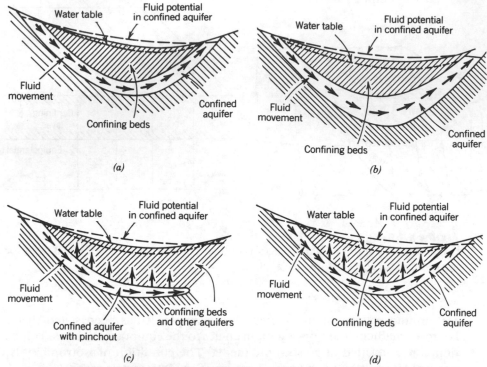

Figure 3.14
Four examples of the occurrence of confined aquifers.

Equation 3.35 provides information on the relative conductivity of isotropic, nonhomogeneous aquifers. In Figure 3.15, K_1 is greater than K_2 (about twice as much) and K_3 is greater than K_1 (again, about twice as much). The explanation of this is straightforward and should already be clear from Darcy's law. The region characterized by the low-permeability zone K_2 requires a steeper hydraulic gradient to drive the common Q through it. Once this Q passes this low-permeability region and enters the K_3 region, the higher conductivity requires that it be accommodated with a flat gradient lest there be discontinuities in the fluid. Thus, the hydraulic gradient adjusts to variations in hydraulic conductivity. The Snake River basalts have been depicted as such a system (Skibitzke and da Costa, 1962).

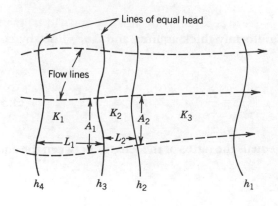

Figure 3.15
Potentiometric map and intersecting flow lines for a nonhomogeneous aquifer.

If the equipotential lines are equally spaced, Eq. 3.35 becomes

$$K_1 = K_2 \tag{3.36}$$

This equation applies for the isotropic, homogeneous case where the hydraulic conductivity is constant throughout the formation. The Patuxent formation in the Baltimore area has been depicted as such a formation (Bennett and Meyer, 1952).

Figure 3.14*b, c,* and *d* present more complex flow conditions. In Figure 3.14*b,* the aquifer gets thicker in the direction of flow. For a homogeneous aquifer that neither gains nor loses flow, Eq. 3.34 becomes

$$\frac{A_1}{A_2} = \frac{L_1}{L_2} \tag{3.37}$$

where $A_2 > A_1$ so that $L_2 > L_1$. Hence, the equipotential lines become further spaced in the direction of flow in response to a progressive decrease in ground water velocity. The potentiometric surface in this case no longer reflects conductivity contrasts. Indeed, if conductivity increases in the flow direction, the wider spacing becomes more pronounced. The Madison limestone in and near the Williston Basin may be a good example of this behavior. At its outcrop areas in Wyoming, flow rates as high as 15 to 23 m/yr have been reported (Downey, 1984a). These are reduced to 3 to 6 m/yr and ultimately to less than 0.6 m/yr in the Williston Basin proper. This formation thickens considerably from outcrop to the depocenter of the Williston Basin.

As noted in Figure 3.14*c,* cross-formational discharge is a necessary condition for aquifers that pinch out in the subsurface. The equality given as Eq. 3.34 is no longer valid here as $Q_1 > Q_2$. If the aquifer is homogeneous, $L_2 > L_1$, and the equipotential lines become further spaced in the flow direction. The gradient then flattens and the velocity decreases with distance of flow. Similar conditions are noted in Figure 3.14*d.* Changes in aquifer thickness or hydraulic conductivity in combination with progressive cross-formational discharge will give rise to complex potentiometric surfaces that reflect these combination of events.

From these discussions, the spacing of equipotential lines as portrayed on a potentiometric map can reflect a variety of conditions. The gradient can either steepen or become quite flat in response to spatial variations in hydraulic conductivity, a loss of flow to cross-formational discharge, or changes in aquifer thickness. Last, the potentiometric surface depicts the energy levels at one point in time. For the Dakota sandstone (Figure 3.12), Darton's map preceded large-scale pumping from this extensive aquifer. The current surface differs considerably (Case, 1984), with the equipotential lines now reflecting the nonuniform withdrawal of ground water throughout the entire region.

There is a theoretical point to be made in closing this section. To help this point along, let us briefly review a few statements from the preceding pages. A field was defined as a distribution of scalars, vectors, or tensors, described in terms of a space coordinate system and time. It is because some quantities can be specified at every point in space that the term "field" is used. A scalar field was further defined as a property that takes on a value (number) at any point of a specific region of space and time. Clearly, the potentiometric map satisfies this description, with the scalar property being hydraulic head. Further, the gradient was introduced as consisting of components in *x, y,* and *z,* each of which representing how fast the head changes in that respective direction. It was also stated that for the condition $K_x = K_y = K_z$, the gradient of the head is a vector perpendicular to lines of equal head. This is the same as saying the material is isotropic with respect to permeability. Thus, as the flow is presumed to be at right angles to the lines of equal potential as depicted on a potentiometric map, the gradient of the head or grad h is colinear with flow. The gradient of the head is thus seen to be the force that drives the flow, and the direction of this driving force and the flow is at right angles to the lines of equal head. The velocity of this flow, so directed, is given by Darcy's law. The connection between what Darcy observed and calculated at the experimental scale and what can be mapped and calculated at the field scale is thus complete.

Example 3.3

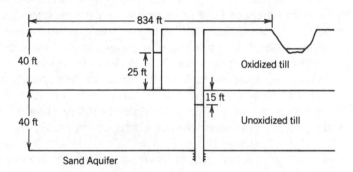

A waste disposal system is to be constructed in the materials given here. The facility will occupy a region 834 ft long (in the direction of ground water flow) by 600 ft wide. The trenches will extend about 40 ft into the oxidized till, which has a water table about 15 ft below land surface. The hydraulic conductivity of the oxidized zone, based on field tests, is about 10^{-7} m/sec (10 ft/yr). The underlying unoxidized till has a hydraulic conductivity based on laboratory measurements of 10^{-8} m/sec (1 ft/yr). A water table map has been prepared for the oxidized till layer and demonstrates a gradient of about 60/834 or 0.072 across the site. The water levels in the underlying sand layer are about 15 ft below the bottom of the contact between the oxidized and unoxidized layers. The materials underlying the small stream on the right-hand side of the diagram are to be excavated to the top of the unoxidized zone. This trench will be backfilled with gravel with some sort of collection system. The following are examples of the hydraulic calculations that are possible.

1. Velocity in oxidized till layer (assume effective porosity = 0.1)

$$v = \frac{K}{n_e}\left(\frac{\Delta h}{\Delta x}\right) = \left(\frac{10}{0.1}\right)(7.2 \times 10^{-2}) = 7.2 \text{ ft/yr}$$

2. Volumetric flow into trench

$$Q = KiA = (10 \text{ ft/yr})(7.2 \times 10^{-2})(15{,}000 \text{ ft}) = 10{,}800 \text{ ft}^3/\text{yr}$$

3. Velocity through unoxidized till layer (assume effective porosity of 0.1)

$$v = \frac{K_z}{n_e}\left(\frac{\Delta h}{\Delta z}\right) = \left(\frac{1 \text{ ft/yr}}{0.1}\right)(1) = 10 \text{ ft/yr}$$

4. Travel time for waste to reach sand aquifer

time $t = L/v = 40 \text{ ft}/10 \text{ ft/yr} = 4 \text{ yr}$

3.4 Flow in Fractured Rocks

In fractured rocks, the interconnected discontinuities are considered to be the main passages for fluid flow, with the solid rock blocks considered to be impermeable. Thus, on the scale of the field problem, one of two approaches might be followed when dealing with the flow of fluids in fractured rock: continuum or discontinuum (discrete). The continuum approach assumes that the fractured mass is hydraulically equivalent to a porous medium. The obvious advantage of treating fractured rocks as a continuum is that Darcy's law as developed can be applied so that no new theories are involved. If the conditions for a continuum do not exist, the flow must be described in relation to individual fractures or fracture sets.

Continuum Approach to Fluid Flow

Applying a continuum-based approach to fluid flow in fractures first requires an understanding of some of the basic assumptions implicit in a model for an intergranular porous medium. This understanding is essential in establishing to what extent these conditions can be achieved in some fractured rocks. Let us state at the outset that the conditions for a continuum approach to fluid flow in intergranular porous media are seldom, if ever, challenged. Such is not the case for fractured rocks.

Intergranular Porous Rocks

In Darcy's own words, he set out "to determine the laws of flow of water through sand." And thus he did, exclusively at the experimental scale. There are, of course, other scales from which it is possible to study certain phenomena. At the molecular scale, for example, one is interested in the behavior of molecules. This scale of behavior probably never crossed Darcy's mind because it is patently impossible to fully understand the behavior of fluids at this level. Besides, Darcy, like most of the rest of us, was more interested in the manifestations of molecular motion, or things that could be expressed in measurable terms, such as viscosity, density, temperature, concentration, and, of course, velocity.

If we step up a scale, we find ourselves at the pore scale, referred to as the microscopic level. The microscopic scale was clearly not for Darcy because nothing could be observed within the individual pore in 1856. Darcy was interested in a macroscopic law, and his only recourse was to develop it on a macroscopic scale. That is, he experimented with a volume of sand that was large with respect to a single pore but small with respect to the space within which significant variations of the macroscopic properties may be anticipated. That is to

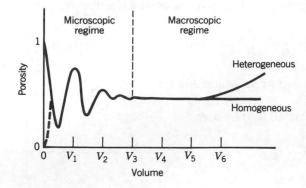

Figure 3.16
Diagram illustrating the representative elementary volume (from Hubbert, 1956). Reprinted with permission of the Amer. Inst. Mining, Met. and Petrol. Engrs.

say, he sought out the smallest possible sample that exhibited an acceptable level of homogeneity. We call this the macroscopic approach, more frequently referred to as a continuum approach.

In this regard, it is convenient to introduce a macroscopic control volume that is large with respect to an individual pore, but small with respect to the space within which significant variations of the macroscopic properties may be anticipated. Hubbert (1956) emphasized this point with a diagram such as Figure 3.16, which is a plot of porosity versus the volume of the sample in which it is measured at some point. At the microscopic level, the value of porosity varies widely. For example, the sample may be all solid or all pore. As the volume increases, a statistical average smooths out the microscopic variations, and we find there are no longer any variations with the size of the sample. Bear (1972) defines this limit as the representative elementary volume, defined as a volume of sufficient size such that there are no longer any significant statistical variations in the value of a particular property with the size of the element.

As noted in Figure 3.16 for a reasonably homogeneous medium, the property of interest (porosity) is adequately represented by some statistical average with little or no variance at any scale larger than the representative elementary volume. At larger scales in real rocks, a greater number of heterogeneities may be encountered, mostly as a result of the regional stratigraphic framework that includes several facies changes and resulting changes in the material properties. An average value can, of course, be obtained at any scale regardless of the degree of heterogeneity, but the variance about the mean will increase with the scale of the problem. This larger scale may be referred to as megascopic. It thus follows that in both the microscopic and megascopic regimes, there may be no single value that can be assigned to represent faithfully any one of our material properties. The continuum approach, then, is restricted exclusively to the macroscopic regime where properties are only a function of position, as defined by some appropriate coordinate system, and time, and do not vary with the size of the field.

Fractured Rocks

As with porous intergranular media, a small control volume of fractured rock may be filled entirely with openings or with solid rock. A larger volume will include cracks and solid rock, but the proportion of each will change as different points are sampled within the same small-scale feature. This variation presumably gets smaller as the volume increases so that the final representative volume is achieved. For many fractured rocks, it is likely that as the volume of rock increases beyond

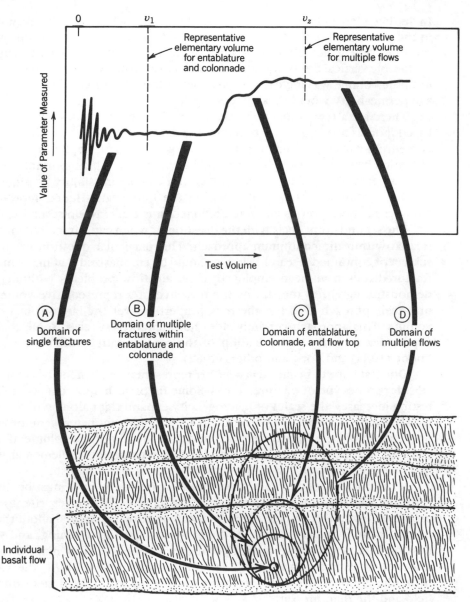

Figure 3.17
Schematic definition of a representative elementary volume as applied to basalt (from Department of Energy, 1986).

some representative volume, the parameters will start to vary again before becoming constant once more. Thus, the representative volume may exist on several scales. This hierarchy of scales is demonstrated in Figure 3.17 where a rock mass may be homogeneous on several scales. At small scales, say, the individual fracture scale of *A*, testing could encounter the intact rock, or the crack openings, with large variations in the measured parameter. As larger-scale sampling is conducted over the entire basalt interior, some representative volume may be achieved in that increasing the volume of the unit tested has no further effect on the value of the average property. This behavior is demonstrated at the *B* scale of Figure 3.17.

Including a few permeable flow tops in the testing scheme results in an increase in the average value of the permeability (C). Including many flow tops provides for homogeneity at yet a larger scale. If the rock mass was characterized by highly permeable vertical discontinuities, the rock again would exhibit nonhomogeneity until the volume was increased to include several such discontinuities. The average permeability would thus increase again.

There are a few points to be made here. If one wishes to apply continuum-based models to a large-scale problem, the data employed in that analysis should come from a testing program that is also large scale. This applies to all problems in ground water, not just in fractured rocks. Virtually all the important parameters such as permeability, dispersivity, and even rock compressibility are significantly smaller on the laboratory scale than on the field scale. Hence, representative volumes for both porous and fractured mediums can exist on several scales.

The second point deals with the question of when can a fractured medium be treated within the continuum approach. This particular question has been the subject of continued research. First, it should be emphasized that many models of fractured rock have been employed in certain flow problems without directly demonstrating either the size or the presence of a representative volume. One practical approach is to test the rock properties over the largest practical rock volume. This approach certainly does not guarantee success. Some theoretical papers dealing with determination of the representative volume are given by Endoe (1984) and Long and others (1982).

One last issue to be raised is whether representative elementary volumes even exist for some type of fractured rocks. Some theorists believe that as the scale of testing increases, the scales of heterogeneity expand right along with it. In other words, every block of rock will contain a few significant heterogeneities that are not representative of the fracturing found over the rest of the volume. If ongoing research indeed could support this idea that representative elemental volumes may not exist, support for continuum approaches will diminish.

If the condition for a continuum does not exist, the flow must be described through the individual fractures or fracture sets. This gives rise to several problems. First, to characterize the flow, information is required about the orientation, fracture density, degree of connectivity, aperture opening, and smoothness of fractures. Second, if the apertures are large, the flow may be turbulent rather than laminar, and Darcy's law no longer applies. Third, the values of hydraulic conductivity will vary with changes in the three-dimensional stress field and the fluid pressure.

An equivalent hydraulic conductivity or permeability may be calculated for a set of planar fractures with equations developed by Snow (1968)

$$K = \frac{\rho_w g N b^3}{12\,\mu} \qquad k = \frac{N b^3}{12} \tag{3.38}$$

where K is hydraulic conductivity, k is permeability, b is the aperture, and N is the number of joints per unit distance across the face of the rock (L^{-1}). For this case, the product Nb is the planar porosity. The expression for hydraulic conductivity given has been plotted for pure water at 20°C by Hoek and Bray (1981), where the equivalent conductivity is taken as parallel to the array (Figure 3.18). As an example, for a spacing of 1 joint per meter and an opening of 0.1 cm, the equivalent hydraulic conductivity is 8.1×10^{-2} cm/sec. This means that 1 m² of porous material with a hydraulic conductivity of 8.1×10^{-2} cm/sec will conduct just as

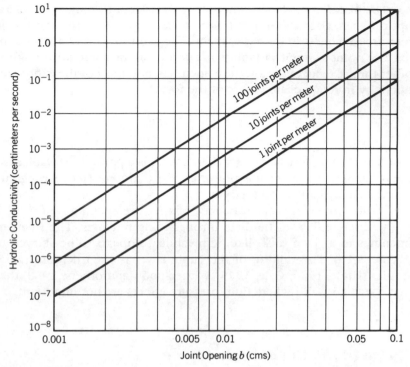

Figure 3.18
Influence of joint opening b and joint spacing on the hydraulic conductivity in the direction of a set of smooth, parallel joints in a rock mass (from Hoek and Bray, 1981). Reprinted with permission of the Institution of Mining and Metallurgy, London.

much water as will 1 joint per meter with an opening of 0.1 cm under identical hydraulic gradients.

The Cubic Law

The cubic law states that for a given gradient in head, flow through a fracture is proportional to the cube of the fracture aperture. For laminar flow between two smooth parallel plates, the volumetric flow rate can be expressed (Romm, 1966) as

$$Q = \frac{\rho_w g b^2}{12\mu} (bw) \frac{\partial h}{\partial L} \tag{3.39}$$

where Q is a volumetric flow rate, ρ_w is the density of water, g is the gravitational acceleration, μ is viscosity, b is the aperture opening, w is the fracture width perpendicular to the flow direction, and $\partial h/\partial L$ is the gradient in the flow direction. This equation is of the form $Q = KiA$, where i is the gradient and the area A is (bw). Hence, the flow rate is related to the cube of the fracture aperture. The hydraulic conductivity for this parallel plate model is

$$K = \frac{\rho_w g b^2}{12\mu} \tag{3.40}$$

The plates employed in experimental work confirming Eq. 3.39 were smooth optical glass. Gale and others (1985) review the several attempts to incorporate fracture roughness in the experiments. We can think of fracture roughness as the local or point-to-point variability in aperture along a fracture. Roughness, among other things, reduces the aperture openings, and most of the expressions for conductivity in rough channels are of the form

$$K = \frac{\rho_w g b^2}{12\mu[1 + C(x)^n]} \tag{3.41}$$

where C is some constant larger than one, X is a group of variables that describe the roughness, and n is some power greater than one. Hence, roughness causes a decrease in hydraulic conductivity.

It thus follows that the influence of aperture opening is of most importance in the discrete flow of fluids in throughgoing fractures. The frequency of such openings in a given rock, like permeability, appears to be skewed to the right (Figure 3.19). The validity of the cubic law has been demonstrated in several studies (Huitt, 1956; Gale, 1975; Witherspoon and others, 1980) that conclude that the law is valid where fluid pressure effects are not important.

3.5 Flow in the Unsaturated Zone

The driving force for ground water was demonstrated to be the potential gradient, where flow takes place from high to low potential. Darcy's law describes this flow. These same principles apply to the movement of water above the water table in the unsaturated zone. There are, of course, several differences and complications that have to be considered. One complication is that the concepts for unsaturated flow are not as fully developed as those for saturated flow, nor are they as easily applied.

Consider the simple situation shown in Figure 3.20 where the water table separates the unsaturated zone from the saturated zone. As noted in Chapter 1, the fluid pressure at the water table, which is a free surface, must be equal to atmospheric pressure. Thus, if a piezometer is placed exactly at the water table, the pressure head $P/\rho_w g$ is atmospheric, conveniently taken as zero, and the total head is equal to the elevation head z. If the pressure is atmospheric at the water table and greater than atmospheric below the water table, it is less than atmospheric in the unsaturated zone above the water table. This pressure distribution accounts for the

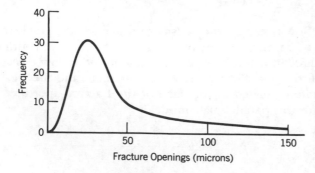

Figure 3.19
Statistical frequency curve of opening width (from Van Golf Racht, 1982). Reprinted with permission of Elsevier Scientific Publishing Company.

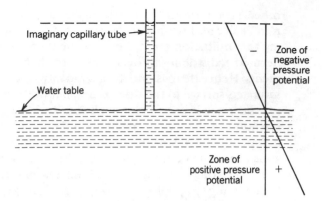

Figure 3.20
Pressure distribution above
and below a free water surface.

fact that water cannot move into boreholes in the unsaturated zone because the pressure in the borehole is atmospheric and is less than atmospheric in soil water surrounding the borehole. Pressures below atmospheric are commonly known as tension, or suction, and the prevailing potential is taken as negative. This negative potential results from the capillary forces that bind water to solids in the unsaturated zone. The lower the moisture content, the higher the suction and the lower the potential. As the moisture content increases, suction diminishes, and the potential increases. The total potential or total head H is thus given as

$$H = z + (-\psi) \tag{3.42}$$

where z is the elevation head and ψ is the matrix suction, taken as negative. The smaller the absolute value of the negative number, the smaller the suction (that is, the higher the moisture content). Assuming the same elevation head for two points, the gradient of the suction dictates direction of the water movement, where the water tends to move from low suction to high suction, that is, high moisture content to low moisture content.

The infiltration process may now be examined with these facts in mind. Infiltration capacity is illustrated on Figure 3.21. As noted in the figure, the rate of infiltration can equal the rainfall rate during the early phases of the rainfall event, but eventually tends to decrease and asymptotically approach some lower but constant rate. The wetter the initial condition of the soil, the smaller the initial rate of infiltration and the sooner the constant rate is achieved (Hillel, 1971). Rubin and

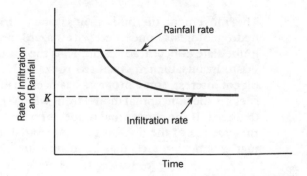

Figure 3.21
Infiltration rate versus time.

others (1963, 1964) demonstrated that the limiting rate of infiltration designated as *K* in Figure 3.21 is equal to the saturated hydraulic conductivity of the soil. Following the infiltration event, the movement of water from moist zones to drier zones (termed redistribution) decreases with time due to the diminishing suction gradients. Hence there is initially a rapid movement from the upper saturated or near-saturated surface to the drier zones, but this rapid movement does not persist due to a decrease in the suction gradient. With time, there appears to be a permanent retention of the water remaining in the upper zones. This behavior is commonly described in terms of the field capacity, that is, the maximum moisture that this zone can hold under the pull of gravity. Clearly, the transients we are discussing do not stop completely when the soil reaches field capacity. They do, however, become imperceptibly slow so that the concept of field capacity does serve some useful purpose.

Yet another complication with the unsaturated zone is that the hydraulic conductivity is strongly dependent in a nonlinear fashion upon the moisture content of the soil. When the soil is saturated, the hydraulic conductivity takes on its maximum value. As more pores become filled with air, that is, the degree of saturation decreases, the hydraulic conductivity decreases also. Thus, the form of Darcy's law generally applied to the unsaturated zone requires the hydraulic conductivity to be expressed as a function of moisture content. As the pressure (suction) head is also a function of moisture content, the hydraulic conductivity for unsaturated flow can also be depicted as a function of matrix suction (Hillel, 1971). Thus Darcy's law may be expressed as

$$\mathbf{q} = -K(\psi) \, \text{grad} \, H \tag{3.43}$$

where *H* may contain both suction and gravitational components or

$$\mathbf{q} = -K(\theta) \, \text{grad} \, H \tag{3.44}$$

where the symbol θ indicates moisture content, or the volume of water divided by the total volume, expressed as a percentage.

The foregoing relationships are demonstrated in Figure 3.22. Figure 3.22*a* shows the manner in which suction decreases at higher moisture content whereas Figure 3.22*b* shows the increase in hydraulic conductivity with increasing moisture content.

Detailed information on flow in the unsaturated zone can be found in Richards (1931) and Kirkham and Powers (1972).

Unsaturated Flow in Fractured Rocks

The movement of fluid in unsaturated fractured rock is of some interest to hydrogeologists concerned with hazardous and radioactive waste transport. Of immediate concern, for example, is the proposal for storage of high-level nuclear waste in unsaturated fractured rocks in Nevada. Although this is a problem of recent interest, some progress has been made. Of major concern for flow in such rocks is the conceptual model from which the physics of the flow process must be deduced. If the conceptual model is correct, mathematical solutions may capture the essence of the flow. If the conceptual model is incorrect or incomplete, the mathematical models may be misleading.

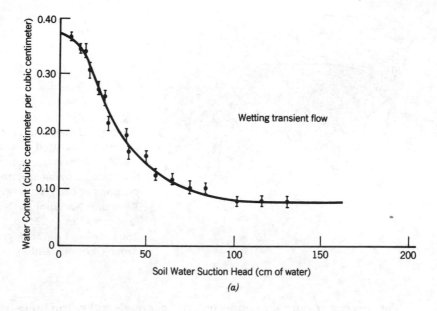

(a)

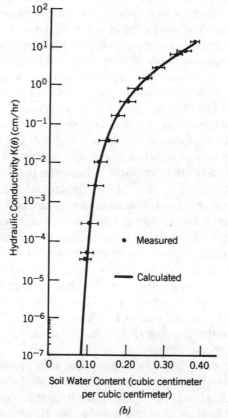

(b)

Figure 3.22

Soil water content–suction curve for Lakeland sand (*a*) and calculated and experimental values of hydraulic conductivity (*b*) for a range of soil water content in Lakeland sand (from Elzeftawy, Mansell, and Selim, 1976. Distribution of water and herbicide in Lakeland Pond during initial stages of infiltration. Soil Science, v.122, p. 297–307. © by Williams and Wilkins, 1976.)

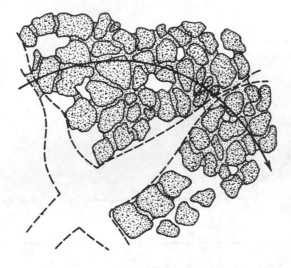

Figure 3.23
Conceptual model of partially saturated, fractured, porous medium showing schematically the flow lines moving around the dry portions of the fractures (from Wang and Narasimhan, Water Resources Res., v. 21, p. 1861–1874, 1985. Copyright by Amer. Geophys. Union)..

The presence of surface connected root channels and worm holes has historically served as an analog for fractures, prompting both theoretical and experimental work. Bevin and Germann (1982) show that these channels conduct water only during rainfall events where the infiltration rate through the soil matrix is less than the precipitation rate at the surface. The water enters the vertical cracks and can laterally infiltrate the matrix at depth. Ponding of water can occur in vertically discontinuous fractures. Davidson (1984, 1985) has modeled such a system, and Bouma and Dekker (1978) have performed experimental work using dyes. Bevin and Germann (1982) provide an excellent review.

When the infiltration rate at the surface exceeds the rainfall rate, the water at the surface is taken in by the pores in the rock matrix. The conceptual model of Wang and Narasimhan (1985) attempts to describe the flow process by a capillary theory that recognizes that large pores (fractures) desaturate first during the drainage process and small pores (matrix) desaturate last. Thus, fractures will tend to remain dry when the aperture is large, with some water films or continuous water contained in the micro aperture portions of the fracture. A dry fracture has a hydraulic conductivity smaller than the conductivity of the partially saturated matrix. The continuous air phase thus produces an infinite resistance to flow across the fracture except for those places that contain water in the micro apertures. The water in micro aperture parts of the fracture constitute pipelines for the transport of water from one matrix block to another. Thus, the flow lines avoid the dry portions of the fracture (Figure 3.23), which act as barriers to flow both along and normal to the fracture.

Peters and Klavetter (1988) recognize the relationships among vertical flux, hydraulic conductivity, and matrix-fracture flow. If the vertical flux in an unsaturated fractured rock is less than the saturated hydraulic conductivity of the matrix, the flow will tend to remain in the matrix. For this case, Wang and Narasimhan (1985) suggest that the steady-state flow field of a partially saturated fractured rock can be understood without detailed knowledge of the discrete fracture network. If the flux is greater than the saturated hydraulic conductivity of the matrix, the matrix will saturate and fractures or open fault zones will then accept the flow. Although this statement appears reasonable from the theory discussed thus far,

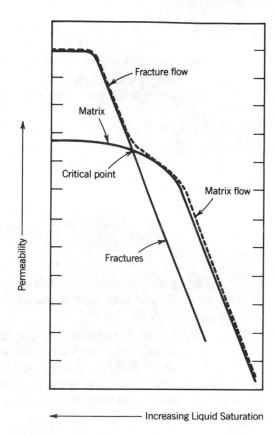

Figure 3.24
Hypothetical relationship between
permeability and the degree of
saturation as it influences matrix and
fracture flow (from Montazer and
Wilson, 1984).

field evidence suggests that fracture flow will commence at some critical matrix
saturation less than full saturation (Rasmussen and others, 1989). This is demonstrated in the conceptual model of Montazer and Wilson (1984) (Figure 3.24),
although the critical matrix saturation is not defined quantitatively.

Suggested Readings

Davis, S. N., 1969, Porosity and permeability of natural materials, in Flow
Through Porous Materials, ed., R. J. M. DeWiest: Academic Press, NY, p.
54–89.

Hubbert, M. K., 1940, The theory of groundwater motion: J. Geol., v. 48, p.
785–944.

Problems

1. Calculate the specific discharge in meters per second for a hydraulic
conductivity of 10^{-6} m/sec and a hydraulic gradient of 100 ft/mile.

2. Calculate the actual velocity in meters per second from the information
given above for an effective porosity of 0.1 and 0.0001.

3. What type of rock or lithology is likely being discussed in Problem 1?

4. What is one likely reason for the large differences in effective porosity cited in Problem 2?

5. Determine both the direction of water movement and the gradient of the flow from the following three points as measured in the field. The water levels are given in feet above mean sea level.

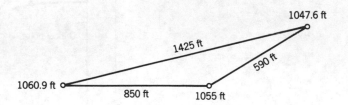

6. Prove that hydraulic head has the units of energy per unit fluid weight (L) and hydraulic potential has the units of energy per unit fluid mass (L²/T²).

7. In the accompanying diagram assume a hydraulic gradient of 100 ft/mile, with the water level at point *A* being at an elevation of 900 feet above sea level. Assume further that the water level in all the piezometers is at the top of the piezometers. The piezometers are located 1 mile apart. Calculate the following.

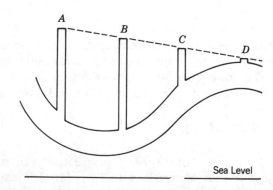

	A	B	C	D
Depth of piezometer	600 ft	575 ft	150 ft	25 ft
Total head	900 ft	_____	_____	_____
Pressure head, feet	_____	_____	_____	_____
Elevation head, feet	_____	_____	_____	_____

8. A geologic formation has a hydraulic conductivity of 10^{-6} m/sec.
   a. What is the conductivity in feet per second?
   b. What is the conductivity in gallons per day per ft²?

    c.   What is the permeability in square centimeters?

    d.   What is the permeability in darcys?

9. Draw or demonstrate the conditions under which Darcy's law can be expressed as $q_x = (k/\mu)(\partial P/\partial x)$, where k is the permeability, μ is viscosity, and P is the fluid pressure as measured at the bottom of a borehole.

10. Which of the following statements for Darcy's law are incorrect (if any)?

    a. $Q = KiA$

    b. $\mathbf{q} = K \operatorname{grad} h$

    c. $\mathbf{q} = (K/g) \operatorname{grad} \phi$

    d. $\mathbf{q} = \mathbf{v}/n_e$

    e. $\mathbf{v} = (K/n_e) \operatorname{grad} h$

    f. $\mathbf{v} = (k\rho g/\mu n_e) \operatorname{grad} h$

    g. $\mathbf{q} = (k\rho g/\mu) \operatorname{grad} h$

11. Three horizontal, homogeneous, and isotropic formations overlie one another, each of which is 20m thick, with horizontal conductivities of 10^{-6}, 10^{-7}, and 10^{-8} m/sec. Compute the horizontal and vertical components of hydraulic conductivity for an equivalent homogeneous, anisotropic formation.

12. A sample is subjected to a hydraulic gradient of $\partial h/\partial x = 0.1$ cm/cm in the x direction and a flux $\mathbf{q}_y$ of 0.001 cm/sec is measured in the y direction in response to this gradient (there is no gradient in the y direction). The same sample is then subjected to a hydraulic gradient $\partial h/\partial y$ of 0.01 cm/cm in the y direction and a flux $\mathbf{q}_x$ is measured in the x direction in response to this hydraulic gradient (there is no gradient in the x direction). What is the magnitude of the flux $\mathbf{q}_x$ in response to $\partial h/\partial y$?

13. If the hydraulic conductivity in area A is 10^{-6} m/sec, determine the hydraulic conductivity in the other areas. Assume the medium is isotropic and nonhomogeneous and that no flow is added to or lost from the system; that is, inflow equals outflow.

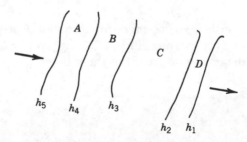

Potentiometric Surface

14. You are given the following piezometric surface:
For each of the conditions cited, give one reason that can account for the piezometric surface.

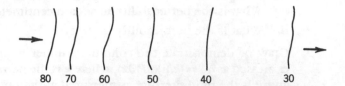

80 70   60   50   40   30

Condition 1. Inflow across $b = 80m = $ outflow across $b = 30m$.

Condition 2. The aquifer is homogeneous and isotropic.

Condition 3. The aquifer is homogeneous and isotropic.

15. The aquifer depicted in Problem 14 is underlain by a uniformly thick homogeneous clay layer. Below this clay layer is another aquifer. The head in this lower aquifer near the 70-m equipotential line shown in Problem 14 is on the order of 60m, or 10m lower than the measured 70-m equipotential line. In the vicinity of the 30-m equipotential line, the head in the lower aquifer is on the order of 25 m, or 5 m lower than the measured 30-m equipotential line. In which of these two regions is the velocity of vertical movement across the clay layer the greatest and why? Which way is the flow directed, upward or downward?

16. Calculate the specific discharge across the clay layer in cm/sec where the vertical hydraulic conductivity of the clay layer is 10^{-7} cm/sec. Which direction is the flow?

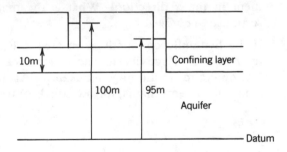

17. Consider an upper confined aquifer overlying a lower aquifer, separated by a low-permeability confining layer. Other than the information in the diagram, you are given the following:

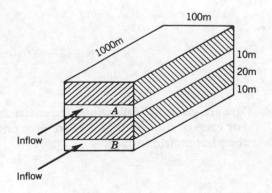

1. The hydraulic conductivity of both the upper and lower aquifer is on the order of 10^{-6} m/sec.

2. The vertical conductivity of the low-permeability confining layer is on the order of 10^{-9} m/sec.

3. The heads in the lower aquifer average about 10m higher than those in the upper aquifer.

4. The hydraulic gradient in the near vicinity of the inflow areas of both aquifers is on the order of 10^{-1} m/m.

   a. Calculate the inflow Q (m³/sec) into both aquifers.

   b. Calculate the outflow Q (m³/sec) out of both aquifers.

   c. Calculate the hydraulic gradient in the vicinity of the outflow areas of both aquifers.

   d. Sketch the general form for the piezometric surfaces of the upper and lower aquifers.

Chapter Four

Elastic Properties and
Main Equations of Flow

4.1 **Conservation of Fluid Mass**
4.2 **The Storage Properties of
     Porous Media**
4.3 **Boundary Conditions and
     Flow Nets**
4.4 **Deformable Porous Media**
4.5 **Dimensional Analysis**

From a field or laboratory perspective, we have pursued the concepts of fluid movement in porous rocks about as far as they can be carried. We have some idea about the range in magnitude of porosity and permeability in geologic materials, the meaning and measurement of hydraulic head, and the determination of flow rates and direction of ground water movement. Thus if we can map the hydraulic head, we can perform many useful calculations. To go further than this requires new information. This chapter is intended to provide this new information, and the next four chapters will put it into practice.

4.1 Conservation of Fluid Mass

In words, a conservation of fluid mass statement may be given as

mass inflow rate − mass outflow rate =
change in mass storage with time

in units of mass per time. In general, this statement may be applied to a domain of any size. Consider this statement as it applies to the small cube of porous material of unit volume where $\Delta x\, \Delta y\, \Delta z$ = unity (Figure 4.1). This box serves as a representative elementary volume. From this figure

$$\text{mass inflow rate through the face } ABCD = \rho_w q_x\, \Delta y\, \Delta z \tag{4.1}$$

The density ρ_w has the units M/L³ whereas the specific discharge q_x is a velocity L/T so that $\rho_w q_x \Delta y\, \Delta z$ has the units of mass per time.

In general, the mass outflow rate can be different than the input rate, and is given as

$$\begin{array}{l}\text{mass outflow rate through}\\ \text{the face } EFGH\end{array} = \left[\rho_w q_x + \frac{\partial(\rho_w q_x)\Delta x}{\partial x}\right]\Delta y \Delta z \tag{4.2}$$

The net outflow rate is the difference between the inflow and the outflow or, subtracting Eq. 4.2 from 4.1

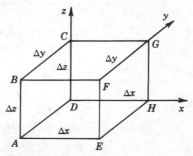

Figure 4.1
Representative volume.

$$\text{net outflow rate through } EFGH = -\frac{\partial(\rho_w q_x)\Delta x\,\Delta y\,\Delta z}{\partial x} \tag{4.3}$$

Making similar calculations for the remainder of the cube

$$\text{net outflow rate through the face } CDHG = -\frac{\partial(\rho_w q_y)\Delta x\,\Delta y\,\Delta z}{\partial y} \tag{4.4a}$$

$$\text{net outflow rate through the face } BCGF = -\frac{\partial(\rho_w q_z)\Delta x\,\Delta y\,\Delta z}{\partial z} \tag{4.4b}$$

Adding these results gives

$$\text{net outflow rate through all the faces} =$$
$$-\left[\frac{\partial(\rho_w q_x)}{\partial x} + \frac{\partial(\rho_w q_y)}{\partial y} + \frac{\partial(\rho_w q_z)}{\partial z}\right]\Delta x\,\Delta y\,\Delta z \tag{4.5}$$

Dividing by $\Delta x\,\Delta y\,\Delta z$

$$-\left[\frac{\partial(\rho_w q_x)}{\partial x} + \frac{\partial(\rho_w q_y)}{\partial y} + \frac{\partial(\rho_w q_z)}{\partial z}\right] = \tag{4.6}$$
$$\text{net outflow rate per unit volume}$$

The right-hand side of the conservation statement is merely a change in mass storage with respect to time. The mass M occupies the pores in the unit volume where density ρ_w is mass per unit volume of fluid (M/V_w) and porosity n for a fully saturated medium is the fluid volume per unit total volume (V_w/V_T) so that the product $\rho_w n$ is the mass per unit total volume. Thus, the word equation of mass conservation becomes

$$-\left[\frac{\partial(\rho_w q_x)}{\partial x} + \frac{\partial(\rho_w q_y)}{\partial y} + \frac{\partial(\rho_w q_z)}{\partial z}\right] = \frac{\partial(\rho_w n)}{\partial t} \tag{4.7}$$

which states that the net outflow rate per unit volume equals the time rate of change of fluid mass per unit volume. By making a further assumption that the density of the fluid does not vary spatially, the density term on the left-hand side can be taken out as a constant so that Eq. 4.7 becomes

$$-\left[\frac{\partial q_x}{\partial x} + \frac{\partial q_y}{\partial y} + \frac{\partial q_z}{\partial z}\right] = \frac{1}{\rho_w}\frac{\partial(\rho_w n)}{\partial t} \tag{4.8}$$

With the simple transformation from Eq. 4.7 to Eq. 4.8 our equation now deals with volumes of fluid per unit volume instead of mass per unit volume, where the two are related through $\rho_w = M/V_w$. For the left-hand side of Eq. 4.8, q_x is merely Q_x/A, which expresses a volumetric flow of fluid per unit time per unit area. Thus, the left-hand side of Eq. 4.8 expresses the net fluid outflow rate per unit volume. This being so, the right-hand side of Eq. 4.8 must describe the time rate of change of fluid volume within the unit volume, that is, $\partial(V_w/V_T)/\partial t$. Thus Eq. 4.8 states that the net fluid outflow rate for the unit volume equals the time rate of change of fluid volume within the unit volume.

Main Equations of Flow

As the q's on the left-hand side of Eq. 4.8 represent Darcy's specific discharge, we may directly substitute Darcy's law for anisotropic material, giving

$$\frac{\partial}{\partial x}\left(K_x \frac{\partial h}{\partial x}\right) + \frac{\partial}{\partial y}\left(K_y \frac{\partial h}{\partial y}\right) + \frac{\partial}{\partial z}\left(K_z \frac{\partial h}{\partial z}\right)$$

This expression is now of positive sign because the q's are negative. Assuming that the material is isotropic and homogeneous

$$K\left[\frac{\partial}{\partial x}\left(\frac{\partial h}{\partial x}\right) + \frac{\partial}{\partial y}\left(\frac{\partial h}{\partial y}\right) + \frac{\partial}{\partial z}\left(\frac{\partial h}{\partial z}\right)\right] = K\left[\frac{\partial^2 h}{\partial x^2} + \frac{\partial^2 h}{\partial y^2} + \frac{\partial^2 h}{\partial z^2}\right]$$

Some clarification is required here. The term $(\partial/\partial x)(\partial h/\partial x)$ represents a space rate of change in the gradient across the unit volume. Accordingly, there must follow velocity variations in the three component directions. If these velocity variations cancel each other, for example, increases in x are compensated by decreases in y, and so forth, the fluid mass per unit volume is not changing with time. For this condition, the right-hand side of Eq. 4.8 is zero, which gives,

$$\frac{\partial^2 h}{\partial x^2} + \frac{\partial^2 h}{\partial y^2} + \frac{\partial^2 h}{\partial z^2} = 0 \tag{4.9}$$

Equation 4.9 is Laplace's equation, one of the most useful field equations employed in hydrogeology. The solution to this equation describes the value of the hydraulic head at any point in a three-dimensional flow field. (Does the mapped potentiometric surface represent a "solution" to Laplace's equation for a two-dimensional flow field?)

Yet another possibility concerning the right-hand side of Eq. 4.8 is the case where the unit volume has some storage qualities. We can go no further here unless something can be said about how this storage takes place. Let us assume that the gains or losses in fluid volume within the unit volume are proportional to changes in hydraulic head, which turns out to be a pretty good assumption. That is to say, an increase in head suggests that water has gone into storage, and a decrease in head suggests just the opposite, that is, a removal of water from storage. To account for these gains and losses, the right-hand side of Eq. 4.8 must be of the form

$$\frac{1}{\rho_w}\frac{\partial(\rho_w n)}{\partial t} = S_s \frac{\partial h}{\partial t} \tag{4.10}$$

where S_s is some proportionality constant. That is, with fluid flow the substance of interest is the gains or losses in fluid volume, not hydraulic head, so that the constant S_s is needed to convert head changes to the amount of fluid added to or removed from storage in the unit volume. As Eq. 4.10 describes the time rate of change of fluid volume within the unit volume, S_s as a proportionality constant must be a measure of the volume of water withdrawn from or added to the unit volume when the head changes a unit amount. In this form, nothing can be ascertained about the physics of the proportionality constant S_s, but it obviously must have the units L^{-1} to make the equality of Eq. 4.10 dimensionally correct. We call S_s the specific storage, and we will have more to say about it later.

With this assumption, the conservation statement becomes

$$\frac{\partial^2 h}{\partial x^2} + \frac{\partial^2 h}{\partial y^2} + \frac{\partial^2 h}{\partial z^2} = \frac{S_s}{K} \frac{\partial h}{\partial t} \qquad (4.11)$$

which is called the diffusion equation. The quantity K/S_s is called the hydraulic diffusivity and has the units L^2/T. Equation 4.11 is another important field equation in hydrogeology and is used to describe unsteady, or transient, flow problems. The solution to this equation describes the value of the hydraulic head at any point in a three-dimensional flow field at any given time or, more precisely, how the head is changing with time.

There are now a few more, perhaps subtle, observations that can be made. For openers, let us compare the left-hand side of Eq. 4.7 with the expression for the gradient

$$\mathbf{i} \frac{\partial h}{\partial x} + \mathbf{j} \frac{\partial h}{\partial y} + \mathbf{k} \frac{\partial h}{\partial z} \qquad (4.12a)$$

$$\frac{\partial(\rho_w q_x)}{\partial x} + \frac{\partial(\rho_w q_y)}{\partial y} + \frac{\partial(\rho_w q_z)}{\partial z} \qquad (4.12b)$$

Both expressions describe the space rate of change of some quantity, in one case the hydraulic head (or a scalar) and in the other case a mass flux, M/L^2T, which is a vector. The abbreviation grad h or ∇h has been used for the gradient. To cover both cases, it is convenient to introduce the notation that is common to both

$$\frac{\partial(\ )}{\partial x} + \frac{\partial(\ )}{\partial y} + \frac{\partial(\ )}{\partial z}$$

which, of course, means nothing until something is put into the parenthesis signs. If we put in hydraulic head (or a scalar), this becomes the gradient, which has already been defined as a vector perpendicular to the equipotential lines, and is given the abbreviation grad h or ∇h. If we put in a mass flux (or a vector) this is called a divergence, and is given the abbreviation div $(\rho_w q)$ or $\nabla \cdot \rho_w q$ so that we have

$$- \nabla \cdot (\rho_w \mathbf{q}) = - \text{div } (\rho_w \mathbf{q}) =$$

$$- \left[\frac{\partial(\rho_w q_x)}{\partial x} + \frac{\partial(\rho_w q_y)}{\partial y} + \frac{\partial(\rho_w q_z)}{\partial z} \right] \qquad (4.13)$$

which is read "divergence of $\rho_w q$." Now, why the term divergence? Any useful dictionary will define divergence as a difference or deviation. And that is exactly what it is, a difference between the mass inflow rate and the mass outflow rate for the unit volume. Thus, divergence has the physical meaning of net outflow rate per unit volume.

The left-hand side of our conservation equation is, in mathematical terms, a divergence, that is, a net outflow rate per unit volume. With this foresight our main equation could have been immediately stated as

$$-\text{div } (\rho_w \mathbf{q}) = \frac{\partial \rho_w n}{\partial t} \qquad (4.14)$$

which is perfectly equivalent to Eq. 4.7 and which means the net outflow rate per unit volume equals the time rate of change of mass per unit volume. In addition, Eq. 4.8 could have been deduced immediately

$$-\text{div } \mathbf{q} = \frac{1}{\rho_w} \frac{\partial \rho_w n}{\partial t} \qquad (4.15)$$

which states that the net fluid outflow rate equals the time rate of change of fluid volume. In the absence of a time rate of change of fluid volume, Laplace's equation is recovered.

So we have a simple notation that not only is efficient in expressing long, sometimes cumbersome equations but contains physical meaning. Now, as long as we are on the topic of operators, there is yet another one that can be used for Laplace's equation (Eq. 4.9) and the left-hand side of the diffusion equation (Eq. 4.11). For these expressions we define

$$\nabla^2(\ ) = \frac{\partial^2(\ )}{\partial x^2} + \frac{\partial^2(\ )}{\partial y^2} + \frac{\partial^2(\ )}{\partial z^2}$$

where ∇^2 is a new operator, called a Laplacian. Again, this operator means nothing until something is placed in the parenthesis. Thus, Laplace's equation and the diffusion equation can be neatly and compactly expressed as

$$\nabla^2 h = 0 \qquad (4.16a)$$

$$\nabla^2 h = \frac{S_s}{K} \frac{\partial h}{\partial t} \qquad (4.16b)$$

respectively.

We have covered much ground in this section. We have learned not only about the conservation of mass, but have reformulated this conservation theorem in terms of the important field equations referred to as the diffusion equation and Laplace's equation. We have also introduced a shorthand set of operators that are useful to describe the three-dimensional problem. These mathematical ideas are among the most abstract that we will encounter in hydrogeology. Nevertheless, they can be understood conceptually, and the student is advised to study them well because they will appear time and again.

4.2 *The Storage Properties of Porous Rocks*

The specific storage was defined as a proportionality constant relating the volumetric changes in fluid per unit volume to the time rate of change in hydraulic head. From a field perspective, an immediate question arises as to how a fully saturated portion of a confined aquifer under pressure can release water from storage and still remain fully saturated, or, equally perplexing, how a fully saturated material can take in additional water? In what manner is space created in order to accommodate this added water?

A similar set of questions was pondered in 1925 by Meinzer (Meinzer and Hard, 1925) as a result of his study of the Dakota sandstone. At that time it was generally assumed that the fluid within confined aquifers was compressible, but that the solid matrix was incompressible, and changes in fluid pressure were not accompanied by changes in pore volume. Meinzer disagreed with this concept based on his observations of the pumping response of the Dakota sandstone. His calculations indicated that over a 38-year pumping period, more water was pumped and removed from the aquifer than the total recharge volume over that time span. In addition, this volumetric withdrawal could not be accounted for by the slight expansion of water when the pressure is lowered. Meinzer (1928) concluded that the interstitial pore space of the sandstone was reduced to the extent of the unaccountable volumetric withdrawals from storage. His problem, then, was to demonstrate that aquifers compressed when the fluid pressure was reduced, and expanded when the fluid pressure was increased. Several lines of evidence were explored, including the response of water levels in wells to ocean tides and passing trains, and subsidence of the land surface. These investigations established the groundwork for later work by Theis (1940) on the source of water derived from wells and by Jacob (1940) who developed most of the flow equations discussed in this section.

Compressibility of Water and Its Relation to Specific Storage

The isothermal compressibility of water may be defined as

$$\beta_w = \frac{1}{K_w} = -\frac{1}{V_w}\left(\frac{\partial V_w}{\partial P}\right)_{T,M} \tag{4.17}$$

where β_w is the fluid compressibility, with units of pressure^{-1}, K_w is the bulk modulus of compression for the fluid, V_w is the bulk fluid volume, and P is pressure. Here, β_w reflects the bulk fluid volume response to changes in fluid pressure at constant temperature T and mass M. The minus sign is used because the fluid volume decreases as the pressure increases. The isothermal compressibility β_w can be taken as 4.8×10^{-10} m^2/N (2.3×10^{-8} ft^2/lb) for ground water at 25°C.

As mass conservation requirements demand that the product of fluid density and fluid volume remain constant, that is, $M = \rho_w V_w$, a decrease in fluid density in response to a decrease in pressure is accompanied by an increase in fluid volume. That is, as $\rho_w V_w$ equals a constant

$$\rho_w dV_w + V_w d\rho_w = 0 \tag{4.18}$$

From this equation, and Eq. 4.17,

$$dV_w = -\frac{V_w d\rho_w}{\rho_w} \tag{4.19a}$$

$$dV_w = -V_w \beta_w dP \tag{4.19b}$$

Equating the right-hand side of these statements gives

$$\frac{\partial \rho_w}{\partial t} = \beta_w \rho_w \frac{\partial P}{\partial t} = \beta_w \rho_w^2 g \frac{\partial h}{\partial t} \tag{4.20}$$

The last term in this expression is arrived at by recognizing the pressure head is only one component of total head $h = z + P/\rho_w g$ so that if the elevation head z is invariant with respect to time, $\partial h/\partial t = (1/\rho_w g)\partial P/\partial t$.

Equation 4.20 states that fluid density decreases with decreasing pressure (the water expands) and increases with increasing pressure (the water contracts). Thus, a confined saturated volume will release water from storage when the pressure head is lowered. This idea is incorporated in the concept of specific storage where, from Eq. 4.10

$$\frac{1}{\rho_w} \frac{\partial(\rho_w n)}{\partial t} = \frac{1}{\rho_w}\left[n \frac{\partial \rho_w}{\partial t} + \rho_w \frac{\partial n}{\partial t} \right] = S_s \frac{\partial h}{\partial t} \tag{4.21}$$

Assuming that the matrix is incompressible, that is, the porosity does not change with time, substitution of Eq. 4.20 into this result gives

$$\frac{1}{\rho_w}\left[n \frac{\partial \rho_w}{\partial t} \right] = [\rho_w g \, n \, \beta_w] \frac{\partial h}{\partial t} \tag{4.22}$$

The bracketed quantity $[\rho_w g \, n \, \beta_w]$ is the specific storage of a rock with an incompressible matrix, defined as the volume of water released from or taken into a unit volume due exclusively to expansion or contraction of the water when the pressure head changes by a unit amount. For a porosity of 20% and the value of water compressibility cited, this amounts to about 9×10^{-7} m^3 of water released from each cubic meter of saturated sediment when the pressure head declines 1 m. This is not much, but this is the nature of confined aquifers.

Example 4.1

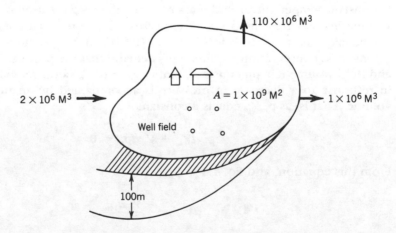

110×10^6 M^3

2×10^6 M^3

$A = 1 \times 10^9$ M^2

1×10^6 M^3

Well field

100m

The specific storage due exclusively to expansion or contraction of the water has been determined to be on the order of 9×10^{-7} m^{-1} for a sandstone with a porosity of 20%. How much of the total pumpage of 110×10^6 m^3 has been supplied by expansion of the water. Assume that the total head drop over the area was on the order of 100m.

According to the definition of specific storage, there are 9×10^{-7} m³ of water removed per cubic meter of rock when the head is lowered 1 m. The total volume of water removed from the part of the aquifer shown due to water expansion per unit decline in head is readily calculated

total volume/unit head decline $= S_s \times$ area $\times$ formation thickness

$\quad = 9 \times 10^{-7}$ m$^{-1} \times 1 \times 10^9$ m$^2 \times 1 \times 10^2$ m

$\quad = 9 \times 10^4$ m³/m

For a 100-m head drop, the total volume supplied by expansion of the water is

$$9 \times 10^4 \text{ m}^3/\text{m} \times 1 \times 10^2 \text{ m} = 9 \times 10^6 \text{ m}^3$$

This leaves 100×10^6 m³ unaccounted for, which was the essence of Meinzer's (1928) dilemma.

Compressibility of the Rock Matrix: Effective Stress Concept

For a porous rock to undergo compression, there must be an increase in the grain-to-grain pressures within the matrix; oppositely, for it to expand, there must be a decrease in grain-to-grain pressure. Without such pressure changes, no volumetric changes can occur. A simple analogy of this process has been given in several places (Terzaghi and Peck, 1948), one modification of which incorporates a spring, a watertight piston, and a cylinder.

In Figure 4.2a, a spring under a load σ has a characteristic length z. If the spring and piston are placed in a watertight cylinder filled with water below the base of the piston, the spring supports the load σ and the water is under the pressure of its own weight, as demonstrated by the imaginary manometer tube in Figure 4.2b. In Figure 4.2c, an additional load Δσ is placed on the system. Because water cannot escape from the cylinder, the spring cannot compress, and the additional load must be borne by the water. This is again demonstrated by the imaginary manometer, which shows the fluid pressure in excess of hydrostatic pressure. The term excess pressure is used in the sense that the pressure exceeds the original hydrostatic pressure.

As conservation laws of fluid mass state that no change in fluid pressure can occur except by loss (or gain) of water, a hermetically sealed cylinder will maintain an excess fluid pressure indefinitely. If some of the water is allowed to escape, the pressure of the water is lowered, and the spring compresses in response to the additional load it must support (Figure 4.2d). Hence, there has been a transfer of stress from the fluid to the spring. When the excess pressure is completely dissipated, hydrostatic conditions once more prevail, and the stress transfer to the spring is complete (Figure 4.2e).

The analogy to be made is that in any porous water-filled sediment, there are pressures in the solid phase by virtue of the points of contact, the resilient grain structure represented by the spring, and there are pressures due to the contained

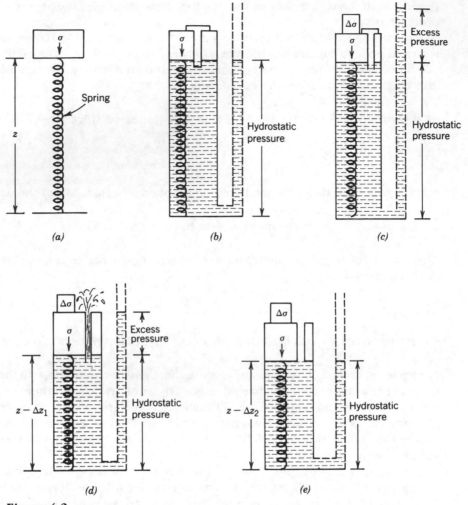

Figure 4.2
Piston and spring analogy showing the transfer of the support for the added load from water pressure to the spring.

water, represented by the watertight cylinder. The former are referred to as inter-granular pressures, or effective stresses, and the latter as pore-water pressures, or neutral stresses. The total vertical stress acting on a horizontal plane at any depth is resolved into these neutral and effective components

$$\sigma = \bar{\sigma} + P \tag{4.23}$$

where σ is the total vertical stress, $\bar{\sigma}$ is effective stress, and P is neutral stress. For the analogy cited, $\Delta\sigma$ equals ΔP at the initial instant of loading, and $\Delta\sigma$ equals $\Delta\bar{\sigma}$ at the terminal condition. Intermediate between these extremes, total vertical stress is always in balance with the sum of the effective and neutral stresses. Neutral stresses, whatever their magnitude, act on all sides of the granular particles, but do not cause the particles to press against each other. All measurable effects, such as compression, distortion, and a change in shearing resistance, are due exclusively to changes in effective stress.

A simple demonstration of the effective stress concept is shown in Figure 4.3. This figure shows an observation well in the vicinity of a railroad station. As a train approaches the station, the load of the train is added to the total stress as expressed on the left-hand side of Eq. 4.23. This added stress appears to be carried at least in part by the ground water, as witnessed by the rise in water levels. The water level in this confined aquifer then declines to a new position as the fluid pressure is dissipated by diffusion of the pore fluids to areas of lower pressure. The added weight of the train then appears to be carried entirely by the aquifer skeleton. Hence, there has been a rather rapid stress transfer from the water to the solids with a commensurate decrease in porosity. When the train leaves the area, the effective stress is decreased immediately, which causes now a vertical expansion of the aquifer; that is, the porosity has been recovered. Owing to this sudden increase in pore volume, the water level in the well declines sharply, but shortly returns to its original position.

To grasp the full significance of the effective stress concept, it is important to remember that any increases to the left-hand side of Eq. 4.23 are compensated by corresponding increases on the right-hand side. Witness the train example. On the other hand, a decrease in pressure caused by pumping a confined aquifer has no effect on the total stress. Instead, that part of the load carried by the grain structure must increase in proportion to the decrease in fluid pressure. Expressed mathematically, σ is constant and, from Eq. 4.23,

$$d\bar{\sigma} = -dP \qquad (4.24)$$

the negative sign indicating that a decrease in fluid pressure is accompanied by an increase in intergranular pressure.

Matrix Compressibility and Its Relation to Specific Storage

Earlier in this chapter, a relationship was developed between the compressibility of the fluid and the specific storage, or at least that part of the specific storage dealing

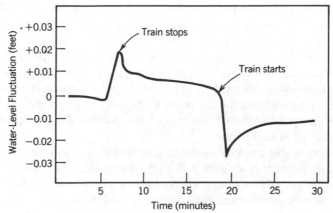

Figure 4.3
Water-level fluctuations observed in an observation well near a railroad station (from Jacob, Trans. Amer. Geophys. Union, v. 20, p. 666–674, 1939. Copyright by Amer. Geophys. Union).

with the fluid component. A similar approach may be taken for the component of storage due to pore volume changes. That is, we should like to state that the change in pore volume is proportional to the change in effective stress. If such is the case, effective stress can be expressed in terms of fluid pressure and, ultimately, in terms of total head. Accepting such an effective stress–pore volume relationship requires another proportionality constant, which in this case is an elastic property of the porous medium. This property is a material compressibility of some sort, and it is important to ascertain exactly what compressibility we require. For example, the compressibility of water as defined was expressed for the condition of constant mass, a volumetric dilation of water where the mass remained constant throughout. The definition we seek must be intimately related to the manner in which pore volume changes are related to changes in effective stress.

From a physical perspective, a change in effective stress as defined so far merely means that the individual grains (which are assumed to be incompressible) have been pushed closer together; that is, the pore volume has been reduced due to grain rearrangement. We note here for later purposes that the grains have also moved. For our purposes here, water must be permitted to drain freely while the sample is being compressed. In addition, as the stress giving rise to this reduction in pore volume is a vertical stress, the compression is assumed to take place only in the vertical direction. These conditions dictate the nature of the proportionality constant relating pore volume changes to changes in effective stress. The compressibility we seek must be described at constant pressure; that is, the water must be permitted to drain freely out of the deforming sample so that fluid mass is lost. Other constraints include incompressible grains and one-dimensional (vertical) compression. Within these constraints, the vertical reduction of the pore volume is exactly equal to the volume of pore fluid expelled. The required expression is

$$\beta_b = \frac{1}{K_b} = -\frac{1}{V_b}\left(\frac{\partial V_b}{\partial \bar{\sigma}}\right)_{P,T} = -\frac{1}{V_b}\left(\frac{\partial V_p}{\partial \bar{\sigma}}\right)_{P,T} = \beta_p = \frac{1}{H_p} \qquad (4.25)$$

where $\Delta V_b = \Delta V_p$ for the special case of incompressible grains. In this formulation, β_b is the bulk (total) compressibility in units of pressure^{-1}, K_b is a bulk modulus of compression, β_p is the vertical compressibility, H_p is a modulus of vertical compression referring to the pores only, V_b is the bulk volume, V_p is the pore volume, and $\bar{\sigma}$ is effective stress. Here, β_p reflects the pore volume decrease in response to changes in stress at constant temperature and pressure. The negative sign is a convention employed to account for a volume decrease with an increase in stress. Typical values for the vertical compressibility are given in Table 4.1. Note that the compressibility of water is of the same order of magnitude as the compressibility of sound rock.

As noted, the compressibility β_p is perfectly equivalent to bulk (total) compressibility β_b, that is, $\partial V_p/\partial \bar{\sigma} = \partial V_b/\partial \bar{\sigma}$. When the individual grains are compressible, the pore volume change is no longer equal to the bulk volume change.

Let us now turn to the time rate of porosity change in the expression

$$\frac{1}{\rho_w}\frac{\partial \rho_w n}{\partial t} = \frac{1}{\rho_w}\left[n\frac{\partial \rho_w}{\partial t} + \rho_w\frac{\partial n}{\partial t}\right] \qquad (4.26)$$

Table 4.1
Vertical compressibility (modified from Domenico and Mifflin, 1965)*

Material	Coefficient of vertical compressibility		
	ft²/lb	m²/N	bars⁻¹
Plastic clay	1×10^{-4}–1.25×10^{-5}	2×10^{-6}–2.6×10^{-7}	2.12×10^{-1}–2.65×10^{-2}
Stiff clay	1.25×10^{-5}–6.25×10^{-6}	2.6×10^{-7}–1.3×10^{-7}	2.65×10^{-2}–1.29×10^{-2}
Medium-hard clay	6.25×10^{-6}–3.3×10^{-6}	1.3×10^{-7}–6.9×10^{-8}	1.29×10^{-2}–7.05×10^{-3}
Loose sand	5×10^{-6}–2.5×10^{-6}	1×10^{-7}–5.2×10^{-8}	1.06×10^{-2}–5.3×10^{-3}
Dense sand	1×10^{-6}–6.25×10^{-7}	2×10^{-8}–1.3×10^{-8}	2.12×10^{-3}–1.32×10^{-3}
Dense, sandy gravel	5×10^{-7}–2.5×10^{-7}	1×10^{-8}–5.2×10^{-9}	1.06×10^{-3}–5.3×10^{-4}
Rock, fissured	3.3×10^{-7}–1.6×10^{-8}	6.9×10^{-10}–3.3×10^{-10}	7.05×10^{-4}–3.24×10^{-5}
Rock, sound	less than 1.6×10^{-8}	less than 3.3×10^{-10}	less than 3.24×10^{-5}
Water at 25°C	2.3×10^{-8}	4.8×10^{-10}	5×10^{-5}

*Water Resources Res. 4, p. 563–576. Copyright by Amer. Geophys. Union.

We have some problems here because effective stress has been related to changes in pore volume, which is not the same as changes in porosity. Notwithstanding these problems, the change in porosity may be expressed as

$$dn = d\left(\frac{V_p}{V_T}\right) = d\left(\frac{V_p V_T}{V_T V_T}\right) = \frac{V_T dV_p - V_p dV_T}{V_T^2} \tag{4.27}$$

giving

$$dn = \frac{dV_p}{V_T} - n\frac{dV_T}{V_T} \tag{4.28}$$

where V_p is the pore volume and V_T is the total volume. Because the total volume consists of the sum of the pore volume and the solid volume, the change in the total volume equals the change in the pore volume for the special case where the volume of solids is not changing with time, that is, $dV_T = dV_p$, so that

$$dn = (1 - n)\frac{dV_p}{V_T} \tag{4.29}$$

It is now clear that the change in pore volume per unit volume is simply $dn/(1 - n)$. Substituting for dV_p/V_T from the expression for compressibility (Eq. 4.25) gives

$$dn = -(1 - n)\,\beta_p d\bar{\sigma} = \beta_p(1 - n)(dP - d\sigma) \tag{4.30}$$

where the effective stress relationship of Eq. 4.23 has been used. Assuming that the total vertical stress remains constant

$$dn = \beta_p(1 - n)dP = \beta_p \rho_w g(1 - n)dh \tag{4.31}$$

where $dh = (1/\rho_w g)dP$ for an invariant elevation head z.

Assuming for the moment that the fluid is incompressible, substitution of this result into Eq. 4.26 gives

$$\frac{1}{\rho_w}\left[\rho_w \frac{\partial n}{\partial t}\right] = [\rho_w g \beta_p (1 - n)]\frac{\partial h}{\partial t} \tag{4.32}$$

In this development, the bracketed quantity on the right would appear to be the specific storage of a compressible rock body containing an incompressible fluid, and as such is defined as the volume of water released from or taken into storage due exclusively to compression or expansion of the rock matrix when the pressure head changes by a unit amount. Combining the result of Eq. 4.32 with the water expansion component given in Eq. 4.22 gives

$$\frac{1}{\rho_w}\left[n\frac{\partial \rho_w}{\partial t} + \rho_w \frac{\partial n}{\partial t}\right] = \rho_w g[(1 - n)\beta_p + n\beta_w]\frac{\partial h}{\partial t} = S_s^*\frac{\partial h}{\partial t} \tag{4.33}$$

where the specific storage S_s^* is taken as $\rho_w g[(1 - n)\beta_p + n\beta_w]$.

Now it will be noticed that we have included an asterisk in the description of the specific storage to differentiate it from the specific storage derived by Jacob (1940) in a somewhat different manner. Jacob's expression is given outright as

$$S_s = \rho_w g(\beta_p + n\beta_w) \tag{4.34}$$

So if we have two expressions for the same quantity, there are several questions that must now be addressed. First, and probably more important, which expression is the correct one for the problem we are addressing? The answer here is that Jacob's expression is unanimously taken as the correct one. Why then did we not proceed to this correct statement right away instead of taking what would appear to be an erroneous pathway? The answer here is that the pathway we took is the only one available for the information given thus far. It was the only pathway available to Jacob as well, but in his calculations, he erroneously assumed that the control volume itself was undergoing deformation. This error led, somehow, to an expression for the specific storage that today is accepted as correct. Had Jacob not made this incorrect assumption, he would have derived Eq. 4.33 just as we did. In fact, he very likely did derive this result because it is very straightforward, but then most likely rejected it on physical grounds. We suspect that Jacob intuitively knew the correct form for the specific storage but had to take some liberties with the mathematics to arrive at it.

Jacob's derivation stood for many years, and was first questioned by DeWiest in 1966. One main point of contention was that the control volume was first considered to be nondeforming to calculate the net inward flux, and then was allowed to deform to compute the rate of change of fluid mass in the volume. Cooper (1966), a close colleague of Jacob, demonstrated that Jacob's expression for what we call the specific storage is essentially correct when the vertical coordinate z is taken as a deforming coordinate. Today, there is no questioning the correctness of the statement, but the manner in which it was arrived at remains incorrect. Bear (1972), too, points out the incorrect assumptions for a deforming control volume.

How then may we correctly arrive at something that everyone agrees is correct but was arrived at by incorrect means? The answer here is that we need more information concerning the movement of the solids in a deforming medium. That is, to maintain a nondeforming control volume, the motion of both the fluids and the solids must be considered, and both must be conserved. To attack the problem correctly requires two conservation statements, one for the fluids and one for the solids

$$-\text{div}\,(\rho_w n\mathbf{v}_w) = \frac{\partial(\rho_w n)}{\partial t} \tag{4.35a}$$

$$-\text{div}\,[\rho_s(1 - n)\mathbf{v}_s] = \frac{\partial[\rho_s(1 - n)]}{\partial t} \tag{4.35b}$$

where $\mathbf{v}_w$ is the fluid velocity, $\mathbf{v}_s$ is the solid velocity, and ρ_s is the density of the solids. In this form $\rho_w n$ is the total mass of fluids per unit volume and $\rho_s(1 - n)$ is the total mass of solids per unit volume. In this development $\mathbf{v}_w$ and $\mathbf{v}_s$ are the average velocities of the fluids and the solids with respect to a stationary coordinate system. The Darcy flux q is described relative to the solid matrix

$$\mathbf{q} = n(\mathbf{v}_w - \mathbf{v}_s) \tag{4.36}$$

Now we must introduce the idea of a material derivative that follows the motion of the solid phase, that is, $Dn/Dt = \partial n/\partial t + \mathbf{v}_s \cdot \nabla n$, which we notice becomes equal to the partial derivative when the last term is small. Introducing Eq. 4.36 into Eq. 4.35a and assuming further that the solid grains are incompressible, $\rho_s =$ constant, Eq. 4.35 becomes

$$-\text{div}(\rho_w \mathbf{q}) - \rho_w n\, \text{div}(\mathbf{v}_s) = \frac{D(\rho_w n)}{Dt} \tag{4.37a}$$

$$(1 - n)\, \text{div}(\mathbf{v}_s) = \frac{Dn}{Dt} \tag{4.37b}$$

Eliminating the divergence of the solids between Eqs. 4.37 gives the final form for the conservation of fluid and solid mass (Palciauskas and Domenico, 1980)

$$-\left(\frac{1}{\rho_w n}\right) \text{div}(\rho_w \mathbf{q}) = \frac{1}{\rho_w} \frac{D\rho_w}{Dt} + \frac{1}{n(1-n)} \frac{Dn}{Dt} \tag{4.38}$$

If the fractional compression is small, the material derivatives are replaced by partial derivatives

$$-\left(\frac{1}{\rho_w}\right) \text{div}(\rho_w \mathbf{q}) = \frac{n}{\rho_w} \frac{\partial \rho_w}{\partial t} + \frac{1}{(1-n)} \frac{\partial n}{\partial t} \tag{4.39}$$

Now we may use the relationships discussed earlier to get the form of the diffusion equation promised in Eq. 4.11. First, it is assumed that the density does not vary spatially so that it can be taken out as a constant on the left-hand side of Eq. 4.39. Substituting Darcy's law in the first term of Eq. 4.39 gives us, as described previously

$$-\text{div } \mathbf{q} = K\nabla^2 h \tag{4.40}$$

The first term on the right-hand side of Eq. 4.39 becomes, by way of Eq. 4.20,

$$\frac{n}{\rho_w} \frac{\partial \rho_w}{\partial t} = n\beta_w \frac{\partial P}{\partial t} = n\rho_w g\beta_w \frac{\partial h}{\partial t} \tag{4.41}$$

In the last term of Eq. 4.39 we recognize that $dn/(1 - n)$ is merely the change in pore volume per unit volume dV_p/V_T (Eq. 4.29), where, in accordance with our definition of compressibility (Eq. 4.25) and effective stress, $dV_p/V_T = -\beta_p d\bar{\sigma} = \beta_p dP = \rho_w g\beta_p dh$. Thus, the last term of Eq. 4.39 becomes

$$\frac{1}{1-n} \frac{\partial n}{\partial t} = \rho_w g\beta_p \frac{\partial h}{\partial t} \tag{4.42}$$

Collecting Eqs. 4.40, 4.41, and 4.42 gives the diffusion equation

$$\nabla^2 h = \frac{\rho_w g(\beta_p + n\beta_w)}{K} \frac{\partial h}{\partial t} = \frac{S_s}{K} \frac{\partial h}{\partial t} \tag{4.43}$$

where the specific storage is defined as the volume of water that a unit volume releases from or takes into storage when the pressure head in the unit volume changes a unit amount. The water volume so derived is volumetrically equivalent to the volume produced from the volumetric expansion of the water as the pressure is lowered and the volumetric contraction of the pore space as the porosity is reduced. The specific storage has the units of length $^{-1}$. For a dense sandy gravel with a vertical compressibility of 10^{-9} m²/N (Table 4.1), the component of storage released due to compression of the matrix amounts to about 1×10^{-5} m³ of water for each cubic meter of sediment when the pressure head declines 1m. This volume is equivalent to the pore volume decrease. As with the volumes associated with fluid expansion, the amount of water produced is quite small.

Example 4.2

Returning to Example 4.1 determine how much of the pumpage is supplied by compression of the matrix for $S_s = 10^{-5}$ m^{-1} if one ignores expansion of the water

total volume/unit head decline = S_s × area × formation thickness

$\quad = 10^{-5}$ m^{-1} × 1×10^9 m² × 1×10^2 m

$\quad = 10^6$ m³/m

For a 100-m drop in head,

$$10^6 \text{ m}^3/\text{m} \times 1 \times 10^2 \text{ m} = 100 \times 10^6 \text{ m}^3$$

Thus the problem of Meinzer (1928) is resolved.

Equation for Confined Flow in an Aquifer

From developments given, the diffusion equation is expressed as

$$\nabla^2 h = \frac{\rho_w g(\beta_p + n\beta_w)}{K} \frac{\partial h}{\partial t} = \frac{S_s}{K} \frac{\partial h}{\partial t} \tag{4.44}$$

The conditions for steady flow are more readily seen from this expression. When both the fluid and the matrix are incompressible ($\beta_w = \beta_p = 0$), the right-hand side of Equation 4.44 becomes zero and the steady-state Laplace equation is obtained. In addition, both S_s and K may be multiplied by the formation thickness m, giving

$$\nabla^2 h = \frac{S_s m}{Km} \frac{\partial h}{\partial t} = \frac{S}{T} \frac{\partial h}{\partial t} \tag{4.45}$$

Here the product of S_s and m is called the storativity S and is dimensionless, whereas the product of the hydraulic conductivity and formation thickness is called transmissivity T, with units of L²/T. The storativity, or coefficient of storage, is defined as the volume of water an aquifer releases from or takes into storage per

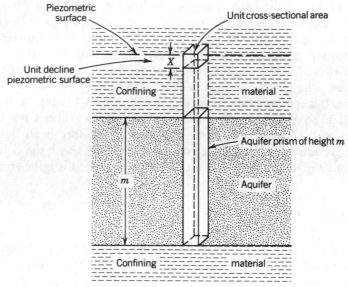

Figure 4.4
Diagram illustrating the storativity for confined conditions
(from Ferris and others, 1962).

unit surface area of aquifer per unit change in the component of pressure head normal to that surface. Stated in another way, the storativity is equal to the volume of water removed from each vertical column of aquifer of height m and unit basal area when the head declines by one unit. Figure 4.4 illustrates the field concept of storativity for confined aquifers.

Transmissivity, or the coefficient of transmissibility, is defined as the rate of flow of water at the prevailing temperature through a vertical strip of aquifer one unit wide, extending the full saturated thickness of the aquifer, under a unit hydraulic gradient. In Figure 4.5, it is the quantity of water flowing through opening B. It may be useful to note that the storativity is the volume of water removed from a prism of unit basal area extending the full saturated thickness of the aquifer under a unit head decline, whereas transmissivity is the volume of water flowing through one face of that same prism in a unit time under a unit hydraulic gradient. In this sense, both storativity and transmissivity are terms exclusively defined for field conditions in that they relate to the storage and transmissive properties of geologic formations of a specified thickness. From Figure 4.5, hydraulic conductivity is the quantity of water flowing in one unit time through a face of unit area (opening A) under a driving force of one unit of hydraulic head change per unit length. Hence, transmissivity is the product of the formation thickness and the hydraulic conductivity.

Specific Yield of Aquifers

The discussions thus far have been concerned exclusively with confined aquifers. Upon the lowering of head in such aquifers, they remain fully saturated so that no dewatering occurs, and the water released is volumetrically equivalent to the volumetric expansion of the water and contraction of the pore space. These processes

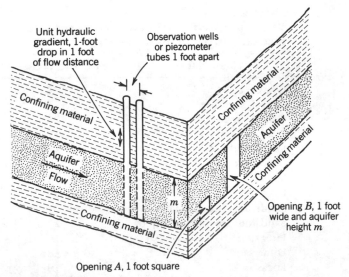

Figure 4.5
Diagram illustrating hydraulic conductivity and transmissivity
(from Ferris and others, 1962).

also occur in unconfined aquifers, but the water volumes associated with them are
negligibly small compared to the volumes obtained from drainage of the pores.
Figure 4.6 shows the concept of storativity associated with the unconfined condition, which is the volume of water drained from the x portion of the aquifer. The
storativity under unconfined conditions is referred to as the specific yield,
defined as the ratio of the volume of water that drains by gravity to the total
volume of rock. Specific retention, on the other hand, is the ratio of the volume

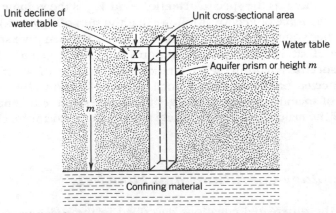

Figure 4.6
Diagram illustrating the storativity (specific yield) for unconfined conditions (from Ferris and others, 1962).

Table 4.2
Values of specific yield for various geologic materials (from Johnson, 1967)

Material	Specific yield (%)
Gravel, coarse	23
Gravel, medium	24
Gravel, fine	25
Sand, coarse	27
Sand, medium	28
Sand, fine	23
Silt	8
Clay	3
Sandstone, fine grained	21
Sandstone, medium grained	27
Limestone	14
Dune sand	38
Loess	18
Peat	44
Schist	26
Siltstone	12
Till, predominantly silt	6
Till, predominantly sand	16
Till, predominantly gravel	16
Tuff	21

of water the rock retains against the force of gravity to the total rock volume. These relationships may be expressed

$$S_y = \frac{V_{wd}}{V_T} \tag{4.46a}$$

$$S_r = \frac{V_{wr}}{V_T} \tag{4.46b}$$

where S_y is the specific yield, V_{wd} is the volume of water drained, V_T is the total rock volume, S_r is the specific retention, and V_{wr} is the volume of water retained against the force of gravity. It follows that the sum of V_{wd} and V_{wr} represents the total water volume contained within the interconnected pore space within a unit volume of saturated material. Total porosity is then equal to the sum of specific yield, specific retention, and the ratio of the volume of water contained in the unconnected pore space to the total volume. Johnson (1967) has summarized values of specific yield for a variety of rock types and sediments (Table 4.2). The coarser the material, the more closely the specific yield approaches total porosity.

4.3   *Boundary Conditions and Flow Nets*

Given the differential equations described in the previous section, there now remains one additional problem to be concerned with, namely, their solution or, more specifically, some additional information required in their solution. To appreciate this new information, let us first put down all the essentials of our flow theory:

1. A potential field is presumed to exist; that is, $h(x, y, z, t)$ is a well-defined scalar quantity in the field of interest.

2. The potential $h(x, y, z, t)$ changes over space and time. The mapping of this potential for one point in time constitutes the potentiometric map.

3. Because the potential changes over space, a gradient exists. This gradient has been demonstrated to be a vector perpendicular to the equipotential lines; that is, it is colinear with the flow for an isotropic porous medium.

4. Provided with this gradient, the water is in continual motion and is completely described by Darcy's law. With such motion, the fluid must be conserved. Thus we define the divergence, which is a net outflow rate per unit volume. If the net outflow rate is zero

$$\nabla^2 h = 0 \text{ (steady flow)}$$

If the net outflow rate is not zero

$$\nabla^2 h = \frac{S_s}{K} \frac{\partial h}{\partial t} \text{ (unsteady flow)}$$

These statements represent what this chapter has been mostly about. In their present form, the steady and unsteady flow equations constitute conservation principles. By themselves they mean little until they are solved. To this end we add a few more statements.

5. If the flow is steady, given the head or the gradient of head on the total boundary of the region of interest, it is possible to calculate $h(x, y, z)$ for the interior.

6. If the flow is unsteady, given the head or the gradient of the head along the total boundary of the region, along with additional information on the initial head before transients take place, it is possible to calculate $h(x, y, z, t)$ for the interior.

Statements 5 and 6 represent some important new information. To put this information in the proper context, it will be recalled that one way to obtain the head distribution for a region is to map it. We now learn that it can also be calculated if certain information is known along the boundaries of the flow region. Slichter first stated these ideas in a hydrogeologic context in 1899, but the impact of his effort was practically nil until the late 1930s. It was during that 25-year period from the late 1930s to the mid-1960s that hydrogeology was transformed from largely a descriptive practice to a highly quantitative and even predictive one.

Basically, there are two kinds of boundary conditions to deal with. The first, as discussed, is a specified head along a given boundary. An important special case for this boundary condition is where the specified head is constant, termed a constant head boundary. Any line along which the head is constant must be an equipotential line. This means that there can be no flow along the equipotential line, but instead, for isotropic material, the flow must be at right angles to the equipotential line. We conclude that for any region bounded in part by a constant head, the flow must be directed at right angles either away from or toward that boundary, and the boundary is an equipotential line.

A second type of boundary is where the normal component of the gradient is specified. Given the normal component of a gradient along a boundary, there must be a flux across the boundary. Thus, we use the term flux boundary or, more commonly, flow boundary, where flow either enters or leaves a region across a boundary. A special type of flow boundary is a no-flow boundary, that is, a region across which no flow is permitted to leave or enter a region. Expressed mathematically

$$\frac{\partial \phi}{\partial N} = 0 \tag{4.47}$$

where ϕ is the potential and N is the normal to boundary. This means there is no gradient normal to the boundary so that there is obviously no flow across the boundary. If there is no flow across a boundary in a two-dimensional flow region, then the flow must be tangential to the boundary. We conclude that a no-flow boundary must be aligned at right angles to the equipotential lines and must therefore be colinear with the lines of flow. No-flow boundaries are called flow lines, and they intersect the equipotential lines at right angles. Thus, the solution to our main differential equations is simply two intersecting families of curves. In isotropic media, the lines are mutually perpendicular, and one family of curves determines the other. Either may be taken as the solution. Such intersecting families of curves are called flow nets, with the boundaries as well as the interior consisting of flow lines and equipotential lines. For steady flow, the curves do not change over time; for unsteady flow, any given configuration represents an instantaneous condition. Unsteady flow is best viewed as a continuous succession of steady states, each of which is "steady" for short periods of time.

A few examples of flow nets are given in Figure 4.7. Figure 4.7a portrays flow in the x-z plane in a pervious rock unit beneath the base of a dam. This flow is established by the head of water behind the dam, the base of which can be taken as an equipotential line, that is, a constant head boundary across which the flow is directed downward. The contact between the impermeable and permeable rock unit is another boundary, in this case a no-flow boundary. Hence, this contact is a flow line. The base of the dam also functions as a flow line, whereas the stream below the dam that receives the ground water discharge functions as an equipotential line. As the flow pattern forms a system of squares or curvilinear squares, the material is assumed to be isotropic and homogeneous. Given this information along with the boundaries, the interior net has been sketched and represents a graphical solution to Laplace's equation.

Figure 4.7b is a set of flow lines and equipotential lines in plan view around a discharging well bordered by an infinite line source. The line source represents a constant head boundary so that the flow lines intersect it at right angles. This flow field is depicted in the x-y plane and represents a mathematical solution to the diffusion equation.

Figure 4.8 is from the classic study by Bennett and Meyer (1952). This flow net is for the Patuxent formation where the heads are below sea level (thus, when water is moving from the 20-ft equipotential to the 30-ft equipotential, it is moving in the direction of decreasing head). The numerous piezometric lows depicted on the map were caused by extensive pumping. Note that the surface forms a set

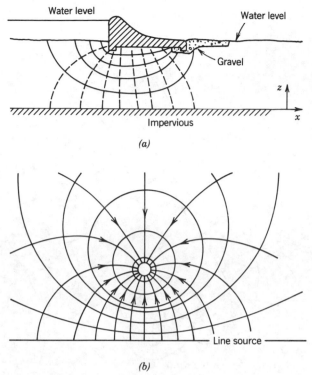

Water level

Water level

Gravel

z

x

Impervious

(a)

Line source

(b)

Figure 4.7

Examples of flow nets. Panel (*a*) represents flow in the *x*-
z plane of a pervious stratum underlying a dam. Panel (*b*)
shows flow toward a discharging well in the *x-y* plane as
influenced by a line source (constant head boundary).

of squares or curvilinear squares so it is represented as isotropic and homogeneous. However, the hydraulic conductivity for the various pumping regions differ
in accordance with the differences in the size of the squares. In this projection, the
head distribution was first mapped and the flow lines then superposed. Figure 4.8
thus represents a field or "mapped" solution to Laplace's equation for whatever
boundary conditions are prevalent in this area.

The volumetric flow rate within flow nets can be determined by summing the
individual Q's in each of the flow tubes of Figure 4.8. Another common approach
is derived from Darcy's law expressed for one section of a flow channel

$$\Delta Q = K \frac{\Delta b}{L} a \qquad (4.48)$$

where ΔQ is the flow in one flow channel, Δb is the head drop across a pair of
equipotential lines, L is the distance over which the head drop takes place, and a
is the distance between adjacent flow lines. For the homogeneous, isotropic system and its characteristic curvilinear squares, $L = a$. As the total Q in the system
is the ΔQ times the number of flow channels (n_f), and the total head drop across

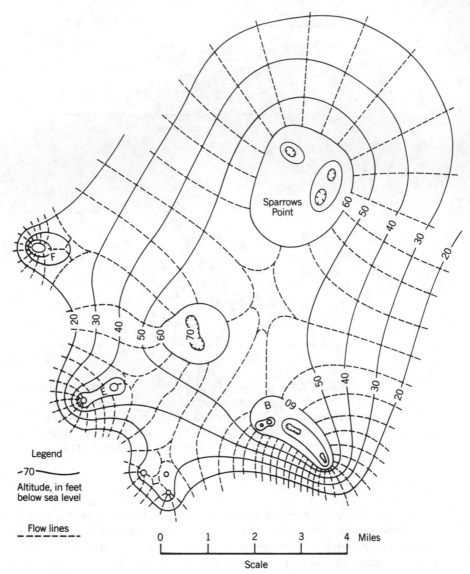

Figure 4.8
Flow net for the Patuxent formation (from Bennett and Meyer, 1952).

the flow system (ΔH) is the Δh times the number of head drops (n_d), the total flow is expressed

$$Q = \frac{n_f}{n_d} K \, \Delta H \tag{4.49}$$

The flow rate Q is expressed as the volumetric flow rate per unit time per unit thickness of the flow field. For Figure 4.7a this flow must be multiplied by the length of the dam (measured perpendicular to the page) to give the total flow in the previous statement. For Figure 4.8, the Q must be multiplied by the saturated aquifer thickness to give the total flow in the aquifer.

Example 4.3

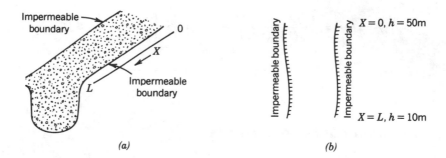

(a)   (b)

Consider the flow in a channel filled with gravel, cut into an impermeable rock. Panel (*b*) shows a plan view of the aquifer.

1. Given the head at $x = 0$ of 50m and the head at $x = L$ of 10m, the equipotential lines can be drawn for a homogeneous aquifer. Thus, at $x = L/2$, $b = 30$m, at $x = 3L/4$, the head is 20m, and so on. By superposing flow lines on the equipotential lines, it is noted that the impermeable boundaries are flow lines.
2. For the simple situation given, Laplace's equation is expressed

$$\frac{\partial^2 b}{\partial x^2} = 0$$

which is integrated to give

$$b = A_1 x + A_2$$

where A_1 and A_2 are constants that have to be evaluated for the boundary conditions. Assume $b = b_o$ at $x = 0$ and $b = b_L$ at $x = L$. Thus, from this equation, $A_2 = b_o$ and $A_1 = (b_L - b_o)/L$. The head distribution is thus expressed as

$$b(x) = (b_L - b_o)\frac{x}{L} + b_o$$

Thus, for Panel b, $b(x) = b_o$ at $x = 0$, and $b(x) = b_L$ at $x = L$. Given numerical values for the head at these boundaries, the interior may be calculated.

Example 4.4

In Figure 4.8, pumpage from the Sparrows Point District has been measured at 1.0×10^6 ft³ per day. The potentiometric surface is assumed to be steady; that is, the amount of water passing through the formation to supply the pumping wells is equal to the pumping rate. It is noted that there are 15 flow channels associated with the Sparrows Point pumping. For a total head drop ΔH of 30 ft, the number of drops is three so that $\Delta H/n_d = 10$. For a total head drop ΔH of 40 feet, the number of potential drops is 4 so that $\Delta H/n_d$ is again 10. Thus, $\Delta H/n_d$ will always be a constant for a region. The regional transmissivity is determined from Eq. 4.49

$$T = Km = \frac{1.0 \times 10^6 \text{ ft}^3/\text{day}}{(15)(10)\text{ ft}} = 6666 \frac{\text{ft}^2}{\text{day}} = 7.16 \times 10^{-3} \frac{\text{m}^2}{\text{s}}$$

4.4 Deformable Porous Media

Except for the treatment of specific storage, the discussions thus far and the mathematical statements resulting from them did not require detailed information on deformable media. With such media there is a coupling among stress, strain, and pore fluids, which is an important rock-water interaction that hydrogeologists tend to be concerned with. As fluids are ubiquitous within the crust, more and greater emphasis has been placed on the physics that describe fluid pressures in deformable porous rocks. Outstanding problems include the generation and dissipation of abnormally high fluid pressures in areas of active deposition, the generation of earthquakes due to fluid injections and in the vicinity of man-made reservoirs, landslides and slope failures involving high fluid pressure, and the role of fluid pressures in faulting and fracturing. This section focuses on the theoretical foundations for rock-water interactions in deformable media for they represent a primary ingredient in our understanding of ground water in the Earth's crust. Chapter 8 takes up some of the practical ideas concerning these concepts.

One-Dimensional Consolidation

The concept of disequilibrium compaction was discussed in Chapter 2 and was described in terms of a time lag between the loading of a sedimentary pile and the dissipation of the fluid pressure caused by the loading. This type of transient behavior is frequently associated with clays and other low-permeability sediments that have a relatively high specific storage and low hydraulic conductivity; that is, the hydraulic diffusivity is small. In addition to these fluid pressure transients, there are a host of others associated with high permeability materials subjected to natural loading, such as earth and oceanic tides, earthquakes, and even changes in atmospheric pressure. To examine the nature of these transients, it is necessary to reexamine Darcy's law and the concept of effective stress and to reformulate these ideas into a different form of a diffusion-type equation.

Development of the Flow Equation
An excess fluid pressure has been defined as a pressure in excess of some preexisting hydrostatic value. Within the concept of the effective stress statement, this may be stated as

$$\sigma + \Delta\sigma = \bar{\sigma} + (P_s + P_{ex}) \tag{4.50}$$

where the total pore water pressure consists of two parts, P_s and P_{ex}, where P_s is a hydrostatic pressure and P_{ex} is a transient pore water pressure in excess of the hydrostatic value, and is presumed to be caused by the increment in total stress $\Delta\sigma$. In a similar fashion, total head may be restated as

$$h = z + \frac{P_s}{\rho_w g} + \frac{P_{ex}}{\rho_w g} \tag{4.51}$$

For one-dimensional vertical flow, Darcy's law becomes

$$q_z = -K_z \frac{\partial h}{\partial z} = -K_z \frac{\partial}{\partial z}\left(z + \frac{P_s}{\rho_w g} + \frac{P_{ex}}{\rho_w g}\right) \tag{4.52}$$

For the problem under consideration, the flow occurs in response to a gradient in excess pressure, not total head, giving

$$q_z = -\frac{K_z}{\rho_w g}\frac{\partial P_{ex}}{\partial z} \tag{4.53}$$

as the appropriate form of Darcy's law.

The conservation equation required for further analysis has already been given as Eq. 4.39, reexpressed here for one-dimensional vertical flow of a homogeneous fluid

$$-\frac{\partial q_z}{\partial z} = \frac{n}{\rho_w}\frac{\partial \rho_w}{\partial t} + \frac{1}{(1-n)}\frac{\partial n}{\partial t} \tag{4.54}$$

where, as described earlier (Eqs. 4.20 and 4.30),

$$\frac{1}{\rho_w}\frac{\partial \rho_w}{\partial t} = \beta_w \frac{\partial P}{\partial t} \qquad \frac{1}{(1-n)}\frac{\partial n}{\partial t} = \beta_p\left(\frac{\partial P_{ex}}{\partial t} - \frac{\partial \sigma}{\partial t}\right) \tag{4.55}$$

In this development, $\partial\sigma/\partial t$ is taken as the vertical stress change that gives rise to the excess pressure development. It is noted that such stress changes were not required in our previous development of the flow equation for confined aquifers. Substituting Darcy's law along with Eq. 4.55 into the conservation equation gives the one-dimensional consolidation equation

$$K_z \frac{\partial^2 P_{ex}}{\partial z^2} = \rho_w g(n\beta_w + \beta_p)\frac{\partial P_{ex}}{\partial t} - (\rho_w g \beta_p)\frac{\partial \sigma}{\partial t} \tag{4.56}$$

Except for the one-dimensional flow condition and the inclusion of the stress term that gives rise to the fluid pressure production, the assumptions here are identical to those employed in the previous development of the diffusion equation. The quantity $\rho_w g(n\beta_w + \beta_p)$ is the specific storage, whereas the quantity $\rho_w g \beta_p$ is the component of specific storage due exclusively to pore compressibility.

Equation 4.56 incorporates the time rate of change of vertical stress as a pressure-producing mechanism, which is of some importance to many loading problems in hydrogeology. As there is a relationship between stress and strain, one would expect that there is a companion equation that incorporates the time rate of change of pore volume strain as a pressure-producing mechanism. For the simple case we are examining, where the fluids are compressible but the individual solid grains are not, this companion equation may be stated as

$$K_z \frac{\partial^2 P_{ex}}{\partial z^2} = (\rho_w g n \beta_w) \frac{\partial P_{ex}}{\partial t} + (\rho_w g) \frac{\partial \epsilon}{\partial t} \qquad (4.57)$$

where $\rho_w g n \beta_w$ is the specific storage due exclusively to expansion of the water and ϵ is the one-dimensional volume strain, or dilation, taken as the pressure-producing mechanism.

The Undrained Response of Water Levels to Natural Loading Events

Numerous cases of water-level fluctuations in wells in response to atmospheric pressure, ocean and earth tides, earthquakes, and passing trains have been reported (Jacob, 1939; Robinson, 1939; Parker and Stringfield, 1950; Bredehoeft, 1967; van der Kamp, 1972; van der Kamp and Gale, 1983). These fluctuations are generally accepted as evidence that confined aquifers are not rigid bodies, but are elastically compressible, an idea fostered by Meinzer (1928) and advanced by Jacob (1940). Currently, it is a routine exercise to compare the water-level response to the loads that cause them in order to calculate certain hydraulic properties of the medium.

In discussing and interpreting these fluctuations, two factors must be considered. First, there are loads that act not only on the rock matrix and its contained water, but also on the water level in an open observation well, and there are loads that act only on the rock matrix and its contained water. Examples of the former include changes in atmospheric pressure, and for the latter there are earth and oceanic tides. Second, it is important to consider the boundary conditions under which the loading takes place. Two types are possible. For a field deformation that is slower than the characteristic times for the diffusion of the pore fluid, the fluid pressure may rise above its ambient value and then dissipate rather quickly. Hence, for all practice purposes, the fluid pressure remains constant during the deformation, and such boundary conditions are called constant pressure, or drained. This behavior is an expected one in high-permeability materials. The effect of passing trains on water levels as demonstrated in Figure 4.3 is an example of a drained response, and the parameter β_p given in Table 4.1 provides an example of a drained compressibility. Conversely, when the load variations are rapid in comparison to the diffusion time for the pore fluid, the local fluid mass remains essentially constant during deformation. This response thus occurs at constant mass, or under undrained conditions. The undrained deformation is elastically stiffer than its drained counterpart so that the respective deformations are characterized by different coefficients. Hence, when a coefficient is sought to characterize a prescribed type of response, it is important to know the boundary conditions under which that response occurs.

All water-level fluctuations as observed at a point that are caused by rapid loading phenomena (tidal and atmospheric pressure variations) are treated as constant mass phenomena. As constant mass implies an absence of fluid flow, all the information we require about these deformations may be obtained by setting the fluid flow term in Eqs. 4.56 and 4.57 equal to zero, which gives, after minor rearrangement

$$\left(\frac{\partial P}{\partial \sigma}\right)_M = \frac{\beta_p}{\beta_p + n\beta_w} \tag{4.58a}$$

$$\left(\frac{\partial P}{\partial \epsilon}\right)_M = \frac{1}{n\beta_w} \tag{4.58b}$$

$$\left(\frac{\partial \epsilon}{\partial \sigma}\right)_M = \frac{\beta_p n\beta_w}{\beta_p + n\beta_w} \tag{4.58c}$$

In these developments, the minus sign has been ignored because we are interested only in absolute values.

Equation 4.58a is termed the pore pressure coefficient and is defined as the change in fluid pressure at constant fluid mass as the stress increases. Interpreted literally, this coefficient describes the percentage of an incremental load that is carried by the pore fluid provided the pore fluid is not permitted to drain. The pore pressure coefficient has been referred to as the tidal efficiency by Jacob (1940), an unfortunate choice of terms in that this equation describes the change in fluid pressure at constant mass in response to any blanket load, not just ocean tides. As used by Jacob (1940), tidal efficiency (T.E.) is defined as the ratio of a piezometric level amplitude as measured in a well to the oceanic tidal amplitude, or

$$\text{T.E.} = \frac{dP}{\gamma_w dH} = \frac{\beta_p}{\beta_p + n\beta_w} \tag{4.59}$$

where H is the height of the tide. Figure 4.9 shows an example of the correspondence between ocean tides and water levels in wells located in aquifers with an oceanic suboutcrop.

Equation 4.58b has been used by Bredehoeft (1967) to examine the effect of earth tides on water levels in wells. This dilation can cause water-level changes of about 1 to 2 cm. As noted in this expression, the volume change of a dilated rock-water system is represented by a volume change of water in the pores, where $n\beta_w = (S_s - \rho_w g\beta_p)/\rho_w g$. Figure 4.10 demonstrates the effect of earth tides on water levels.

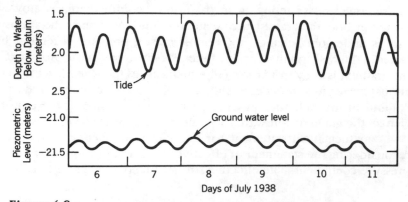

Figure 4.9
Tidal fluctuations and induced piezometric surface fluctuations observed in a well 30 m from shore at Mattawoman Creek, Maryland (from Meinzer, 1939).

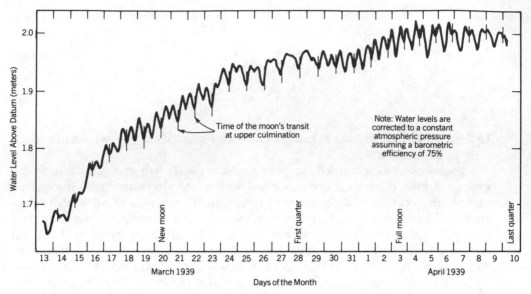

Figure 4.10
Water-level fluctuations in a well caused by earth tides (from Robinson, Trans. Amer. Geophys. Union, v. 20, p. 656–666, 1939. Copyright by Amer. Geophys. Union).

Barometric pressure has an inverse relationship with water levels in wells (Figure 4.11). The barometric efficiency (B.E.) is used to describe how faithfully a water level responds to changes in atmosphere pressure, and is given as

$$\text{B.E.} = \frac{\gamma_w db}{dP_a} \tag{4.60}$$

where P_a is atmospheric pressure. The response of water levels to changes in atmospheric pressure may be reasoned as follows. Incremental increases or decreases in atmospheric pressure acting on a column of water in a well are, respectively, added to or subtracted from the pressure of the water in the well. These same stresses acting on any part of the confined aquifer, however, are supported by both the grain structure and the fluid. As long as the atmospheric pressure is undergoing change, there must exist a pressure difference between the water in the well and that in the aquifer. When atmospheric pressure is increasing, the gradient is away from the well, which causes the water level to decline. When atmospheric pressure is decreasing, this decrease is fully removed from the water column in the well and only partially from the water in the confined aquifer. Hence, the gradient is toward the well, and the water-level rises.

A mathematical statement for barometric efficiency may be arrived at by reconsidering the statement of effective stress and our interpretation of the pore pressure coefficient. The effective stress concept may be restated as

$$1 = \frac{\partial P}{\partial \sigma} + \frac{\partial \bar{\sigma}}{\partial \sigma} \tag{4.61}$$

where the pore pressure coefficient $\partial P / \partial \sigma$ represents the percentage of the load that is carried by the fluid provided the fluid is not permitted to drain. Clearly, $\partial \bar{\sigma} / \partial \sigma$

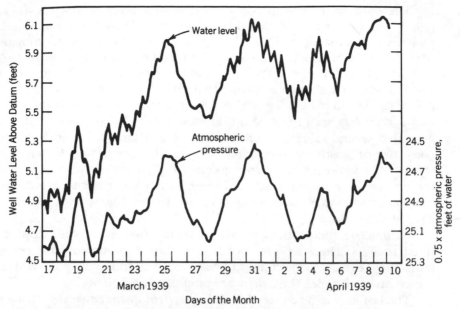

Figure 4.11
Water-level response to changes in atmospheric pressure (from Robinson, Trans Amer.
Geophys. Union, v. 20, p. 656–666, 1939. Copyright by Amer. Geophys. Union).

must represent the percentage of the stress that is carried by the solid matrix, again
under undrained conditions. This is the barometric efficiency, that is,

$$1 = \text{T.E.} + \text{B.E.} \tag{4.62}$$

so that, from Eqs. 4.59 and 4.60

$$\text{B.E.} = 1 - \frac{\beta_p}{\beta_p + n\beta_w} = \frac{n\beta_w}{\beta_p + n\beta_w} \tag{4.63}$$

Thus, as β_p approaches zero for a rigid aquifer, B.E. approaches one and T.E.
approaches zero. A barometric efficiency of one suggests that the well is a perfect
barometer, which is only possible for a perfectly incompressible aquifer ($\beta_p = 0$).
On the other hand, as β_p gets large, B.E. approaches zero and T.E. approaches one.
It follows that B.E. is closely related to the compressibility of the fluid and T.E. is
closely related to the compressibility of the matrix. The relationships among the
tidal efficiency, barometric efficiency, and specific storage may be stated as

$$\text{T.E.} = \frac{\rho_w g \beta_p}{S_s} \tag{4.64a}$$

$$\text{B.E.} = \frac{\rho_w g n \beta_w}{S_s} \tag{4.64b}$$

As the sum of the barometric and tidal efficiencies is unity, the percentage of
storage attributable to expansion of the water is B.E., and that attributed to compression of the matrix is T.E.

There are several examples of the application of the results just stated. Jacob (1941) was the first to employ these methods for a well in Long Island with a reported tidal efficiency of 0.42 and a porosity of 0.35. Given an estimated value of 0.58 for the barometric efficiency, the specific storage is readily calculated to be 8.6×10^{-7} ft^{-1}. Bredehoeft (1967) reported a barometric efficiency of 0.75 for a well in Iowa, along with an estimated porosity of 0.178. Specific storage in this case is calculated to be 2.8×10^{-7} ft^{-1}. An interesting application for a composite rock section has been reported by Carr and van der Kamp (1969). These authors measured several values of both barometric and tidal efficiencies in sediments consisting of mostly sandstone and siltstone. The barometric efficiency averaged about 0.37 and the tidal efficiency averaged 0.57. The difference of 0.06 between the sum of these values is reported to be within experimental error. Forty rock samples provided a mean porosity of 0.177. The average specific storage was calculated to be about 6.3×10^{-7} ft^{-1}.

Barometric fluctuations are not commonly observed in wells tapping unconfined aquifers. The reason for this is that changes in atmospheric pressure are transmitted equally to the column of water in a well and to the water table through the unsaturated zone. Thus, there are no pressure gradients.

The last item to be concerned with is to demonstrate that the tidal and barometric efficiencies as described thus far—although defined as an undrained response—consist of a combination of drained and undrained parameters. It will be recalled that the pore compressibility β_p could be expressed in terms of a modulus of compression $1/H_p$ where the subscript p means we are dealing with pore compressibility only. It is convenient now to introduce other notations for some of the parameters. These are

$$\frac{1}{R_p} = \beta_p + n\beta_w \tag{4.65a}$$

$$\frac{1}{Q_p} = n\beta_w \tag{4.65b}$$

where the subscript p indicates compressibilities for the case of incompressible solid grains. The relationship between these constants is given as

$$\frac{1}{Q_p} = \frac{1}{R_p} - \frac{1}{H_p} \tag{4.66}$$

With this new designation, it is clear that the tidal efficiency or pore pressure coefficient is equal to R_p/H_p and the barometric efficiency is equal to R_p/Q_p. In addition, Eq. 4.58c becomes

$$\left(\frac{\partial\epsilon}{\partial\sigma}\right)_M = \frac{1}{Q_p}\frac{R_p}{H_p} = \frac{1}{H_p}\left[1 - \frac{R_p}{H_p}\right] = \frac{1}{H_p}\text{ [B.E.]} \tag{4.67}$$

As $\partial\epsilon/\partial\sigma$ is a measurement at constant mass, it too is an undrained parameter and may be designated as $1/K_u$, where K_u is an undrained modulus of compression, still for the incompressible solid case. It thus follows that B.E. $= H_p/K_u$, which is a ratio of a drained to an undrained modulus, and T.E. $= 1 - H_p/K_u$. Further details on these relationships may be found in Domenico (1983).

The Drained Response of Water Levels to Natural Loading Events

There are numerous examples of the drained response of water levels to natural loading events, including the change in head in response to changes in river stage (Cooper and Rorabaugh, 1963), earthquakes (Cooper and others, 1965), and the inland propagation of sinusoidal fluctuations of ground water levels in response to tidal fluctuations of a simple harmonic motion (Ferris, 1951; Werner and Noren, 1951). These problems are generally treated within the framework of the conventional one-dimensional diffusion equation given as Eq. 4.11 where the pressure-producing mechanisms are normally formulated within the boundary conditions. In other instances, it is frequently assumed that the stress causing the excess pressure is applied rapidly and then held constant so that the stress term of the one-dimensional consolidation equation (Eq. 4.56) is taken as zero. A typical example of this is the so-called Terzaghi consolidation equation (Terzaghi and Peck, 1948). In the application of this equation to consolidation problems, it is frequently assumed that the pore compressibility is considerably larger than the compressibility of the water so that the latter may be ignored, giving from Eq. 4.56

$$\frac{\partial^2 P_{ex}}{\partial z^2} = \frac{\rho_w g \beta_p}{K_z} \frac{\partial P_{ex}}{\partial t} = \left(\frac{1}{C_v}\right) \frac{\partial P_{ex}}{\partial t} \tag{4.68}$$

where the hydraulic diffusivity C_v is termed the coefficient of consolidation and is given as

$$C_v = \frac{K_z(1 + e)}{a_v \rho_w g} = \frac{K_z}{S_s} \tag{4.69}$$

In Eq. 4.69 the specific storage for the case of incompressible fluids is $a_v \rho_w g / (1 + e)$, where e is the void ratio and a_v is a coefficient of compressibility, defined as the rate of change of void ratio to the rate of change of effective stress causing the deformation.

Problems in basin loading or crustal deformation are normally described in terms of stress or strain rates applied over geologic intervals of time. For these purposes, Eqs. 4.56 and 4.57 may be restated here as

$$\frac{\partial P_{ex}}{\partial t} = \frac{K_z}{\rho_w g(1/R_p)} \frac{\partial^2 P_{ex}}{\partial z^2} + \left(\frac{R_p}{H_p}\right) \frac{\partial \sigma}{\partial t} \tag{4.70}$$

and

$$\frac{\partial P_{ex}}{\partial t} = \frac{K_z}{\rho_w g(1/Q_p)} \frac{\partial^2 P_{ex}}{\partial z^2} - (Q_p) \frac{\partial \epsilon}{\partial t} \tag{4.71}$$

where R_p, H_p, and Q_p have been previously defined. It is noted that $1/R_p$ is incorporated within the specific storage for problems involving stress rates and $1/Q_p$ is incorporated within the specific storage for problems involving strain rates. Additionally, the pressure-producing stress term in Eq. 4.70 is multiplied by the pore pressure coefficient. Thus, the larger the pore pressure coefficient, the greater the percentage of the stress change that is initially carried by the fluid and, consequently, the greater the initial fluid pressure production. Thus, for a pore pressure coefficient close to one, virtually all the stress change is initially carried by the fluid. As mentioned in Chapter 2, the pore pressure coefficient decreases with increasing rock rigidity and is therefore rock dependent.

Three-Dimensional Consolidation

Elastic Properties in Deformational Problems

In our treatment of the consolidation equation, two assumptions stand out that may now be relaxed somewhat. First, it was assumed that the stress increment producing the abnormal fluid pressure was vertical only. Second, the individual solid components making up the rock matrix were incompressible. This last assumption means that the bulk volume changes are exactly equal to the pore volume changes. Or, stated another way, the volume of water squeezed out of a rock is equal exactly to the pore volume change of the rock. We have exploited this assumption, not only in our development of the consolidation equation, but in the development of the concept of specific storage.

The most complete model from which to obtain the important material properties is one that is fully compressible, that is, one in which all the components are compressible, including the solid grains β_s, the fluids β_w, the pores β_p, and the bulk volume β_b. When grain compressibilities are introduced into the material properties, the water volume squeezed out of a rock is invariably less than the bulk volume change. To incorporate such compressibilities in the material properties, it is necessary to introduce a proportionality constant between the bulk and pore volume change. This may be accomplished with the following expression

$$\beta_b - \beta_s = \beta_p = \xi \, \beta_b \qquad (4.72)$$

where β_b is the bulk compressibility, β_s is the grain compressibility, and the difference $\beta_b - \beta_s$ is taken as the pore compressibility. In this expression, ξ is a constant of proportionality and may be defined as the ratio of the water volume squeezed out of a rock to the total volume change of the rock if the deformation takes place at constant fluid pressure, that is the rock is free to drain. Thus, if ξ equals one, the bulk volume changes are equal to the pore volume changes, an assumption that has been inherent throughout our discussions of deformable media. From Eq. 4.72, the constant of proportionality may be described as

$$\xi = 1 - \left(\frac{K_b}{K_s}\right) = 1 - \left(\frac{\beta_s}{\beta_b}\right) \qquad (4.73)$$

where K_b is the bulk modulus of compression of the rock body and K_s is the bulk modulus of compression of the polycrystalline grains making up the porous rock. In this form, we note that if β_s is small with respect to β_b, ξ is approximately unity, and the volume of water removed from the rock is equal to the total volume change; that is, the incompressible grain case is recovered.

With the introduction of solid compressibilities, the elastic coefficients discussed in the previous section change considerably. These coefficients are given in Table 4.3. For the three-dimensional results, the stress term σ and the strain term ϵ refer to the mean confining stress and the volumetric dilation; that is,

Table 4.3
Definitions of the important elastic coefficients and their relation to the coefficients defined by Biot (1941) (from Palciauskas and Domenico, 1989)*

β	$=$	$\dfrac{1}{K}$	$=$	$-\left(\dfrac{1}{V}\right)\left(\dfrac{\partial V}{\partial \sigma}\right)_P$
β_s	$=$	$\dfrac{1}{K_s}$	$=$	$\left[\dfrac{(1-n)}{\rho_s}\right]\left(\dfrac{\partial \rho_s}{\partial \sigma}\right)_P$
$\beta_{s'}$	$=$		$=$	$-\left[\dfrac{(1-n)}{n\rho_s}\right]\left(\dfrac{\partial \rho_s}{\partial P}\right)_\sigma$
β_n	$=$	$\beta - \beta_s - n\beta_{s'}$	$=$	$\left(\dfrac{1}{V}\right)\left(\dfrac{\partial V_n}{\partial P}\right)_\sigma$
ξ	$=$	$\dfrac{K}{H}$	$=$	$\dfrac{(\beta - \beta_s)}{\beta}$
$\dfrac{1}{K_u}$	$=$	$\dfrac{1}{K} - \dfrac{R}{H^2}$	$=$	$-\left(\dfrac{1}{V}\right)\left(\dfrac{\partial V}{\partial \sigma}\right)_M$
$\dfrac{1}{H}$	$=$	$\beta - \beta_s$	$=$	$\left(\dfrac{1}{m_f}\right)\left(\dfrac{\partial m_f}{\partial \sigma}\right)_P$
$\dfrac{1}{R}$	$=$	$\beta_n + n\beta_f$	$=$	$\left(\dfrac{1}{m_f}\right)\left(\dfrac{\partial m_f}{\partial P}\right)_\sigma$
$\dfrac{1}{Q}$	$=$	$\dfrac{1}{R} - \dfrac{K}{H^2}$	$=$	$\left(\dfrac{1}{m_f}\right)\left(\dfrac{\partial m_f}{\partial P}\right)_\epsilon$
$\dfrac{R}{H}$	$=$		$=$	$\left(\dfrac{\partial P}{\partial \sigma}\right)_M$
ξQ	$=$		$=$	$\left(\dfrac{\partial P}{\partial \epsilon}\right)_M$
$\dfrac{\xi Q}{(R/H)}$	$=$		$=$	$\left(\dfrac{\partial \sigma}{\partial \epsilon}\right)_M$

*Water Resources Res., v. 25, p. 203–213. Copyright by Amer. Geophys. Union.

$$\sigma_m = \frac{\sigma_{xx} + \sigma_{yy} + \sigma_{zz}}{3} \tag{4.74a}$$

$$\frac{\Delta V}{V} = \epsilon_d = \epsilon_{xx} + \epsilon_{yy} + \epsilon_{zz} \tag{4.74b}$$

where σ_m is the mean stress, V is volume, and ϵ_d is the volume strain or dilation. In addition, following relationship now holds

$$\frac{1}{Q} = \frac{1}{R} - \frac{\xi}{H} \tag{4.75}$$

which reduces to Eq. 4.66 for ξ equal to one. Indeed, all expressions for the three-dimensional coefficients given in Table 4.3 reduce to the one-dimensional coefficients already cited for the case where $\xi = 1$. This implies, in addition, that the testing for the one-dimensional condition is conducted under constrained conditions where the concern is with vertical deformations only.

Biot and Willis (1957) discuss the measurement techniques for the various coefficients described in Table 4.3, whereas a limited set of values have been reported by Palciauskas and Domenico (1982), Domenico (1983), Zimmerman and others (1986), and Green and Wang (1986). Table 4.4 presents an overview of these values. Laboratory data for both the Kayenta sandstone and the Hanford basalts are reasonably good, whereas the clay and the limestone tabulations are "generic," that is, crude calculations based on tables of physical constants. The rocks are arranged from left to right in order of what is expected to be decreasing compressibility. Hence, ξ as a measure of the water volume squeezed out to the volume change of the material, decreases from unity to 0.23 as rock rigidity increases. For clays, there is virtually a one-to-one relationship between volume changes and associated water volumes released. The coefficients $1/H$, $1/R$, and $1/Q$ are compressibilities of sorts and, in general, decrease in the direction of rock rigidity. Thus, $1/H$ as a measure of the pore compressibility decreases consistently in the direction of rock rigidity, as does $1/R$ as a measure of the changes in fluid mass in response to changes in fluid pressure at constant stress. Except for a slight inconsistency with the generic limestone, this same statement holds for $1/Q$ as a measure in the change in fluid mass with respect to fluid pressure at constant strain. Also shown in the table are combinations of parameters, R/H and ξQ. The quantity R/H is already known as the change in fluid pressure with stress at constant fluid mass. As noted in the table, the more rigid the rock, the greater the proportion of the stress that is carried by the grain structure rather than the pore fluid. The combined parameter ξQ is defined as the change in fluid pressure with respect to strain at constant fluid mass. Here, this fluid pressure appears highest for rocks in the intermediate range, and decreases towards both the soft and rigid end members. The symbol K_u is used for an undrained modulus.

Table 4.4

Typical values for the elastic coefficients (from Palciauskas and Domenico, 1989)[*]

Parameter	Units	Clay	Mudstone	Sandstone	Limestone	Basalt
ξ		1	0.95	0.76	0.69	0.23
$\dfrac{1}{H}$	(Kbar)$^{-1}$	1.6	0.045	0.0079	0.0024	0.00052
$\dfrac{1}{R}$	(Kbar)$^{-1}$	1.62	0.054	0.012	0.0095	0.0044
$\dfrac{1}{Q}$	(Kbar)$^{-1}$	0.02	0.011	0.0059	0.0078	0.0043
$\dfrac{R}{H}$		0.99	0.83	0.67	0.25	0.12
ξQ	(Kbar)	50	83	128	88	54
$\dfrac{1}{K}$	(Kbar)$^{-1}$	1.6	0.047	0.011	0.003	0.0022
$\dfrac{1}{K_u}$	(Kbar)$^{-1}$	0.02	0.01	0.0052	0.0024	0.0022

[*]Water Resources Res., v. 25, p. 203–213. Copyright by Amer. Geophys. Union.

The coefficients described are for elastic, or reversible, deformations. A reversible compression is one that disappears completely upon release of the stress that caused it. With geologic materials, many of the substances we treat within an elastic or reversible framework are not elastic at all. The point is that for some problems, the materials need not be elastic in terms of reversibility, but the strains have to be proportional to the stresses. That is, Hooke's law must hold so that a constant ratio exists between the stress and the strains. The degree of variation in this ratio dictates how far the real behavior departs from the ideal one, with clays and other plastic materials generally falling into this nonideal category. Reversibility also implies that any compressions occurring under load with the pore fluid free to drain are completely recoverable by injecting fluid back into the pores. Indeed, the "coefficient" describing the volume changes associated with both compression and expansion should, in theory, be the same. Again, this reversibility does not coincide with our observations for many materials.

Flow Equations for Deformable Media

The flow equations in a completely deformable media are of an identical structure of those presented for the one-dimensional consolidation equation, with the major differences residing in the material properties, or coefficients. These equations may be expressed as (Palciauskas and Domenico, 1989)

$$\frac{\partial P_{ex}}{\partial t} = \nabla \cdot \left[\frac{K}{\rho_w g(1/R)} \nabla P_{ex} \right] + \frac{R}{H} \frac{\partial \sigma}{\partial t} \tag{4.76}$$

and

$$\frac{\partial P_{ex}}{\partial t} = \nabla \cdot \left[\frac{K}{\rho_w g(1/Q)} \nabla P_{ex} \right] - (\xi Q) \frac{\partial \epsilon}{\partial t} \tag{4.77}$$

where the coefficients R, Q, H, and ξ are described in Table 4.3. These equations may be compared with Eqs. 4.70 and 4.71.

As with the one-dimensional consolidation equation, the equations just given can be used to describe both the drained and undrained response of water levels to natural loads. Expressions for the undrained response to various loads were developed independently by van der Kamp and Gale (1983) and Domenico (1983). For the case of compressible solids, the tidal efficiency (pore pressure coefficient) and the barometric efficiency may be described as

$$\text{T.E.} = \left(\frac{R}{H} \right) = \frac{1}{\xi} \left(1 - \frac{R}{Q} \right) = \frac{1}{\xi} \left(1 - \frac{K_b}{K_u} \right) \tag{4.78}$$

and

$$\text{B.E.} = 1 - \frac{1}{\xi} \left(1 - \frac{R}{Q} \right) = 1 - \frac{1}{\xi} \left(1 - \frac{K_b}{K_u} \right) \tag{4.79}$$

respectively. Thus, these efficiencies are nothing more than partition coefficients that account for the distribution of incremental stress between the pore fluid and the solids under undrained conditions. For vertical strains only, the moduli K_b and K_u are described for the constrained condition. Similarly, Bredehoeft's

(1967) fluid-pressure response to earth tides stated as $\partial P/\partial \epsilon$ at constant fluid mass is given simply as ξQ, where Q is defined in Table 4.3.

4.5 Dimensional Analysis

Solutions to the diffusion equation will in general encompass a dimensionless group of terms that dictate either the nature of the diffusion process or demonstrate the competition between two rate processes. One of the more important uses of dimensional analysis is that it permits one to recognize the importance of parameters or, more precisely, groups of parameters in a fluid flow problem. To arrive at this dimensionless group for the diffusion equation, we first establish the following dimensionless variables that are noted with a superscript "+"

$$x^+ = x/L$$
$$y^+ = y/L$$
$$z^+ = z/L$$
$$t^+ = t/t_e$$
$$b^+ = b/b_e$$
$$\nabla^+ = L \nabla$$
$$\nabla^{2+} = L^2 \nabla^2$$

where L is some characteristic length, t_e is some characteristic time, and b_e is some characteristic head. Note that the operators are also written in dimensionless form. With these dimensionless quantities, the diffusion equation for fluid flow expressed as

$$\nabla^2 b = \frac{S_s}{K} \frac{\partial b}{\partial t} \tag{4.80}$$

becomes, in dimensionless form,

$$\nabla^{2+} b^+ = \left(\frac{S_s L^2}{K\, t_e} \right) \frac{\partial b^+}{\partial t^+} \tag{4.81}$$

where all terms are dimensionless, including the quantity in the brackets. This bracketed quantity is known as the Fourier number, N_{FO}, and occurs in all unsteady flow problems.

The Fourier number is more readily expressed as

$$N_{FO} = \frac{K/S_s}{L^2/t_e} \tag{4.82}$$

where the numerator has already been introduced as the hydraulic diffusivity. The quantity $1/u$ that determines the "well function" $W(u)$ used in well hydraulics is of this form (Chapter 5). Another useful form for the inverse of this dimensionless group is

$$N_{FO}^{-1} = \frac{S_s L^2/K}{t_e} = \frac{T^*}{t_e} \tag{4.83}$$

where T^* is the time constant for a basin. If the time t_e at which we wish to observe some transient is significantly larger than T^*, the transient will not be observable, and the basin would appear to be in some steady state. If the time of observation is less than the characteristic time T^*, the transient is readily observable. Note that the characteristic time T^* gets large for large values of S_s and L^2 and small values for K. This form for the dimensionless group is useful in studying problems in disequilibrium compaction (Chapter 8).

Suggested Readings

Biot, M. A., 1941, General theory of three-dimensional consolidation: Appl. Physics, v. 12, p. 155–164.

Jacob, C. E., 1940, On the flow of water in an elastic artesian aquifer: Trans. Amer. Geophys. Union, v. 21, p. 574–586.

Meinzer, O. E., 1928, Compressibility and elasticity of artesian aquifers: Econ. Geol., v. 23, p. 263–291.

Palciauskas, V. V., and Domenico, P. A., 1989, Fluid pressures in deforming porous rocks: Water Resources Research, v. 25, p. 203–213.

Parker, G. G., and Stringfield, V. T., 1950, Effects of earthquakes, rains, tides, winds, and atmospheric pressure changes on the water in the geologic formations of southern Florida: Econ. Geol., v. 45, p. 441–460.

Problems

1. Address the following problems as they pertain to Eq. 4.43:
   a. Express this equation for one-dimensional flow.
   b. Express this equation for a porous medium in which the fluid can be considered to be incompressible.
   c. Express this equation for a porous medium in which both the fluid and the matrix are considered to be incompressible.
   d. In what type of material (clay, siltstone, sandstone, gravel, limestone, granite) would you consider ignoring the term $\rho_w g n \beta_w$?

2. The storativity of an elastic artesian aquifer is 3×10^{-4}. The porosity of the aquifer is 0.3, the compressibility of water is 2.36×10^{-8} ft^2/lb, and the thickness of the aquifer is 200 ft.
   a. How many cubic feet of water are removed from storage under an area of 1×10^6 ft^2 when the head declines 1 ft?
   b. Ascertain the storativity if the aquifer matrix is incompressible.
   c. What is the value for the coefficient of vertical compressibility of the aquifer?

3. Consider a well pumping from storage at a constant rate from an extensive elastic aquifer.

   a. The rate of head decline times the storativity summed up over the area over which the head decline is effective equals _____ ?

   b. At any given time, the total head decline, summed up over the area over which the head change is effective, times the storativity equals _____ ?

4. A clay layer extending from ground surface to a depth of 50 ft is underlain by an artesian sand. A piezometer drilled to the top of the aquifer registers a head of 20 ft above land surface. The saturated unit weight of the clay is 125 lb/ft³. Assume a water table at land surface.

   a. What are the values of the total, neutral, and effective pressures at the bottom of the clay layer?

   b. If a trench is dug to a depth of 20 ft, what are the total, neutral, and effective pressures at the bottom of the clay layer beneath the trench? Assume the trench is filled with water to land surface.

   c. Is it possible to maintain the stability of the bottom of the trench in **b.** if the trench is dewatered?

   d. What is the approximate minimum depth of a trench excavated in the clay layer that will cause a quick condition or failure of its bottom? Assume the trench to be filled with water.

5. If the total withdrawal from pumping center B of Figure 4.8 is 1.14×10^6 ft³/day, compute the average transmissivity of the aquifer in the vicinity of pumping center B.

6. Designate whether the following are equipotential lines or flow lines and if they are constant head or no flow boundaries: AB, BC, CD, DE, FG.

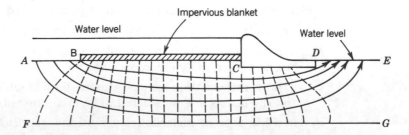

7. The barometric efficiency of a well in an elastic aquifer is 0.4. If the porosity is 0.2 and the thickness is 300 ft, estimate the storativity.

8. Reduce Eq. 4.76 to the more familiar form of the diffusion equation given as Eq. 4.43. State carefully all the relationships and assumptions required.

9. Derive Eq. 4.58 starting from Eqs. 4.56 and 4.57.

10. Considering sandstones, limestones, and basalts for the assumption of incompressible grains, prepare a chart that shows which of these rock

types have the largest response to atmospheric pressure and ocean tides and which have the smallest. Considering the compressible grain case, which of these rocks have the largest response to earth tides and which have the smallest?

Chapter Five

Hydraulic Testing: Models, Methods, and Applications

5.1 **Prototype Geologic Models in Hydraulic Testing**

5.2 **Conventional Hydraulic Test Procedures and Analysis**

5.3 **Single-Borehole Tests**

5.4 **Partial Penetration, Superposition, and Bounded Aquifers**

5.5 **Hydraulic Testing in Fractured or Low-Permeability Rocks**

5.6 **Some Applications to Hydraulic Problems**

Hydraulic testing as used in this book is a description of the field tests and testing procedures required to obtain certain hydraulic properties, pressure measurements, or indirect determinations of ground water velocity in rocks of various kinds. Such testing procedures were born out of the practical necessity of evaluating ground water as a water supply. Thus the early beginnings in this field dealt exclusively with pumping tests designed for permeable formations that were important from the perspective of water supply. In more recent times, due largely to concerns of hazardous waste migration and nuclear waste burial, interest has shifted to testing procedures suited to low-permeability rocks or rocks with a fracture-type permeability. For simplicity we classify the various tests as conventional or specialized. By conventional we mean those tests that virtually all hydrologists conduct in a routine fashion in relatively simple geologic environments. Specialized testing procedures are often required in low-permeability rocks and especially in fractured rocks.

5.1 Prototype Geologic Models in Hydraulic Testing

Every hydrologic model employed in hydraulic testing is patterned after some specific or prototype geologic environment. Three such environments are shown in Figure 5.1. Figure 5.1*a* shows a typical confined aquifer of large areal extent,

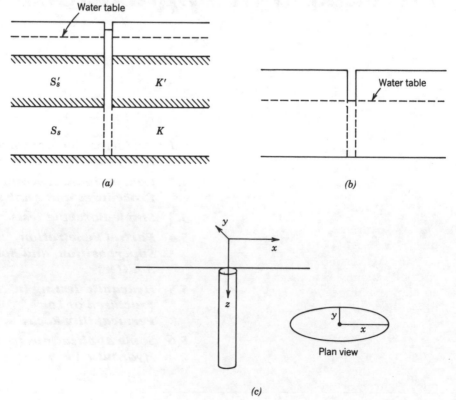

Figure 5.1
Prototype geologic models.

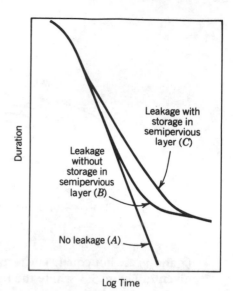

Figure 5.2
Comparison of time-drawdown response for
geologic prototypes of Figure 5.1a (from
Hantush, J. Geophys. Res., v. 65, p. 3713–
3725, 1960. Copyright by Amer. Geophys.
Union).

assumed to be isotropic and homogeneous, overlain by a confining layer over
which there is a water table. There are three variations of this prototype model.
First, if the confining layer is impermeable and contains an incompressible fluid
within an incompressible matrix ($S_s' = 0$), all the water removed from the aquifer
will come from storage within the aquifer, and the water-level change (drawdown)
versus time response at some observation point will be exponential, that is, will
plot as a straight line on semilogarithmic paper (curve *A*, Figure 5.2). For reasons
that will be clear later, this is referred to as the nonleaky response. If the confining
layer has some finite permeability but still contains an incompressible fluid within
an incompressible matrix, it may be possible to invoke the transfer of water across
the confining layer. The time-drawdown observation will reflect this additional
water source with an upward inflection at some point in time (curve *B*, Figure
5.2). This is referred to as the leaky response. Two things are worth noting here.
First, because of the time delay for water to enter the pumped aquifer, an aquifer
may appear nonleaky over several hours or days of pumping and will be so classified if the test is terminated before the leakage has been detected. This, however,
will not affect the analysis in that the leaky and nonleaky curves remain coincident or nearly so until the inflection point is realized (Figure 5.2). Second, leakage
across a confining layer may not be the only cause for the upward inflection. If
pumping is near a stream or a lake and surface water is brought into the aquifer
due to the established hydraulic gradients, a similar inflection will occur.

The last version of Figure 5.1*a* is for a permeable confining layer with a finite
specific storage. In this case, the confining layer itself can contribute water to the
aquifer, resulting once more in an upward inflection (Figure 5.2*c*). This is referred
to as leakage with storage in the confining layer. It is noted now that the time delay
is considerably shortened (Figure 5.2). For long times, the leakage with storage
case is coincident with the leaky case, but this coincidence evolves through a different time-drawdown pathway.

Yet another geologic prototype is the aquifer that is considered homogeneous
but anisotropic. In Figure 5.1*c* the aquifer is considered to be horizontal and one
of the principal directions of the hydraulic conductivity tensor is vertical, that is,
parallel to the well. In this case, the horizontal conductivities in a horizontal plane

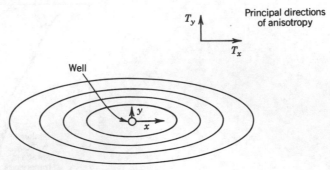

Figure 5.3
Plan view of drawdown response in an anisotropic aquifer.

(x and y) are not equal. The response of such a system is shown in the plan view given in Figure 5.3 where the transmissivity in the x direction is larger than the transmissivity in the y direction. Variations of this model include situations where the principal directions of the hydraulic conductivity tensor are neither vertical nor horizontal. Two- and three-dimensional anisotropic models may be useful in the analysis of fractured rock.

This brief discussion does not exhaust the prototype geologic model guiding the hydraulic testing of real rock. Others include a double-porosity concept where water can be removed from both the fractures and matrix of a fractured porous medium and discrete flow in individual fractures.

Operational hydraulic equations have been derived for the ideal prototype models discussed earlier. The solutions have been obtained from some form of the diffusion equation where time is a parameter. The essence of hydraulic testing calls for the matching of real response data with the theoretical response expected from the prototype condition. In most of the conventional methods, this matching is performed with a curve-fitting procedure or by some simple graphical technique. With some of the specialized testing procedures, the matching is frequently conducted with the aid of a computer.

5.2 Conventional Hydraulic Test Procedures and Analysis

The term "conventional" is used here in reference to hydraulic testing procedures that are most commonly used in formations of moderately high permeability with the stated purpose of determining the transmissive and storage properties of aquifers. In this section we will restrict our discussions to tests that require one pumping well and one or more observation wells or piezometers in which it is possible to measure the response to pumping. From an historic perspective, these procedures originated with the work of C. V. Theis in 1935.

The Theis Nonequilibrium Pumping Test Method

The mathematical expression for removal of heat at a constant rate from a homogeneous, infinite slab has provided a useful analogy for study of ground water flow to a pumping well. It is assumed, initially, that the slab is at some

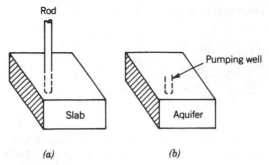

Figure 5.4
Schematic diagram of (*a*) an infinite slab and (*b*)
the well flow analogy.

uniform temperature. An infinitesimal rod of lower temperature parallel to the z axis is then allowed to draw off the heat (Figure 5.4*a*). In mathematical terminology, the rod represents a continuous line sink. The temperature change at some distance r from the rod is a function of the rate at which heat is withdrawn, the properties of the slab, and time. By analogy, the infinite, homogeneous slab is replaced by an extensive, homogeneous aquifer, and the rod by a well of infinitesimal diameter (Figure 5.4*b*). Similarly, the rate of pumping is analogous to the rate of heat withdrawal, and water-level change at any distance r from the pumping well is a function of the pumping rate, the properties of the aquifer, and time. A solution to the heat flow problem just stated is given by Carslaw and Jaeger (1959). The hydrologic analog was given by Theis (1935) and is of the form

$$b_o - b = s = \frac{Q}{4\pi T} \int_{\frac{r^2 S}{4Tt}}^{\infty} \frac{e^{-z}}{z} \, dz \tag{5.1}$$

where b_o is the original head at any distance r from a fully penetrating well at time t equals zero, b is the head at some later time t, s is the difference between b_o and b and is called the drawdown, Q is a steady pumping rate, T is the transmissivity, S is the storativity, and the exponential integral is a well-known tabulated function, the value of which is given by the infinite series

$$\int_{u}^{\infty} \frac{e^{-z}}{z} \, dz = -0.577216 - \ln u + u - \frac{u^2}{2 \cdot 2!} + \frac{u^3}{3 \cdot 3!} - \frac{u^4}{4 \cdot 4!} + \cdots \tag{5.2}$$

where

$$u = \frac{r^2 S}{4Tt} \tag{5.3}$$

Equation 5.1 is a solution to the polar coordinate form of the diffusion equation, or

$$\frac{\partial^2 b}{\partial r^2} + \frac{1}{r} \frac{\partial b}{\partial r} = \frac{S}{T} \frac{\partial b}{\partial t} \tag{5.4}$$

for the initial and boundary conditions

$$h(r, 0) = h_o$$

$$h(\infty, t) = h_o$$

$$\lim_{r \to 0} \left(r \frac{\partial h}{\partial r} \right) = \frac{Q}{2\pi T} \quad \text{for } t > 0$$

The first two of these are read: "The head at some radius r from the well at time zero equals the initial head h_o, and the head at an infinite distance at any time t equals the initial head h_o." The second condition stipulates that water withdrawal may affect distant parts of the aquifer but the exterior boundary is never encountered (as expected for an aquifer of infinite extent). The third condition provides for a constant withdrawal rate at r_w, which is the radius of the well, which, in turn, is taken as infinitesimal.

Equation 5.1 is referred to as the nonequilibrium equation. It is sometimes referred to as the nonleaky well flow equation in that all the water pumped is removed from a single aquifer with no contributions from other beds either above or below the pumped aquifer.

A verbal interpretation of the parameters incorporated in the nonequilibrium equation can provide insight into the shape and growth of a cone of depression in an ideal aquifer. It is first noted that the exponential integral of Eq. 5.2 is a function only of the lower limit of integration so that Eq. 5.1 can be written

$$s = \frac{Q}{4\pi T} W(u) \tag{5.5}$$

where u is defined in Eq. 5.3. The term $W(u)$ is termed the well function of u, which indicates its dependence on values of u expressed in Eq. 5.3. It follows that the value of the exponential integral may be ascertained and tabulated for each of several values of u. Such a tabulation is shown in Table 5.1 for values of u ranging from 10^{-15} to 9.0. If u equals 4.0×10^{-10}, then $W(u)$ equals 21.06, which is the value of the series of Eq. 5.2 for this particular value of u.

Of the variables comprising u, the storativity and transmissivity may be considered constant for a given set of conditions, and distance and time considered variables. Values of $W(u)$ may be plotted against values of u or $1/u$ in such a way that, for any given time, the plotted curve reveals the exact profile of a cone of depression as a function of distance r from a pumping well (Figure 5.5a), or such that for any fixed distance r, the plotted curve reveals the exact shape of the drawdown curve as a function of time t (Figure 5.5b). Actual values for drawdown depend on the pumping rate and the hydraulic properties (Eq. 5.5) and (1) the time of observation, in Figure 5.5a, and (2) the distance at which the observations are made, in Figure 5.5b. Curves showing the relation between $W(u)$ and u are termed "type" curves.

In that the value of u depends on time, distance, transmissivity, and storativity, u determines the radius of a cone of depression (Theis, 1940). The radius not only increases with increasing time but, for a given time, is larger for decreasing values of storativity and increasing values of transmissivity. By examining the complete statement for drawdown (Eq. 5.5), drawdown at any point for a given time is proportional to discharge and inversely proportional to transmissivity. The

Table 5.1
Values of $W(u)$ for values of u (from Wenzel, 1942)

u	1.0	2.0	3.0	4.0	5.0	6.0	7.0	8.0	9.0
$\times 1$	0.219	0.049	0.013	0.0038	0.0011	0.00036	0.00012	0.000038	0.000012
$\times 10^{-1}$	1.82	1.22	0.91	0.70	0.56	0.45	0.37	0.31	0.26
$\times 10^{-2}$	4.04	3.35	2.96	2.68	2.47	2.30	2.15	2.03	1.92
$\times 10^{-3}$	6.33	5.64	5.23	4.95	4.73	4.54	4.39	4.26	4.14
$\times 10^{-4}$	8.63	7.94	7.53	7.25	7.02	6.84	6.69	6.55	6.44
$\times 10^{-5}$	10.94	10.24	9.84	9.55	9.33	9.14	8.99	8.86	8.74
$\times 10^{-6}$	13.24	12.55	12.14	11.85	11.63	11.45	11.29	11.16	11.04
$\times 10^{-7}$	15.54	14.85	14.44	14.15	13.93	13.75	13.60	13.46	13.34
$\times 10^{-8}$	17.84	17.15	16.74	16.46	16.23	16.05	15.90	15.76	15.65
$\times 10^{-9}$	20.15	19.45	19.05	18.76	18.54	18.35	18.20	18.07	17.95
$\times 10^{-10}$	22.45	21.76	21.35	21.06	20.84	20.66	20.50	20.37	20.25
$\times 10^{-11}$	24.75	24.06	23.65	23.36	23.14	22.96	22.81	22.67	22.55
$\times 10^{-12}$	27.05	26.36	25.96	25.67	25.44	25.26	25.11	24.97	24.86
$\times 10^{-13}$	29.36	28.66	28.26	27.97	27.75	27.56	27.41	27.28	27.16
$\times 10^{-14}$	31.66	30.97	30.56	30.27	30.05	29.87	29.71	29.58	29.46
$\times 10^{-15}$	33.96	33.27	32.86	32.58	32.35	32.17	32.02	31.88	31.76

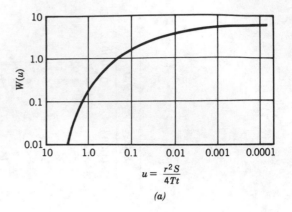

$$u = \frac{r^2 S}{4Tt}$$

(a)

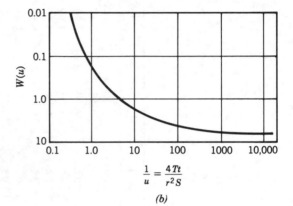

Figure 5.5
Values of $W(u)$ plotted against (*a*)
values of u and (*b*) values of $1/u$.

$$\frac{1}{u} = \frac{4Tt}{r^2 S}$$

(b)

lateral extent of a cone of depression at any given time and its rate of growth are independent of the pumping rate.

As a final point, the dimensionless variable u is of the form of an inverse Fourier number introduced in Chapter 4

$$u = \frac{r^2/t}{4(T/S)} = \frac{Sr^2/4T}{t} = \frac{T^*}{t} \tag{5.6}$$

where $Sr^2/4T$ has the units of time and may be referred to as a time constant T^*. Thus, if the time constant T^* is small compared to time t, u becomes small, $W(u)$ becomes large, and drawdowns become large (Figure 5.2*a*). Small values for the time constant are associated with small values of storativity, say, for confined conditions, small distances to the point of observation, or large values of transmissivity. If the time constant $Sr^2/4T$ becomes large with respect to a proposed time of observation t, the expected drawdown at the point of observation r becomes imperceptibly small. Large values for the time constant are frequently associated with large values for the storativity, as expected for unconfined conditions.

The Curve-Matching Procedure
The Theis (or nonequilibrium) equation is used extensively to determine the hydraulic properties of aquifers. For these purposes, it will be recalled that for a fixed distance r, drawdown versus time is of the form of the curve in Figure 5.5*b*.

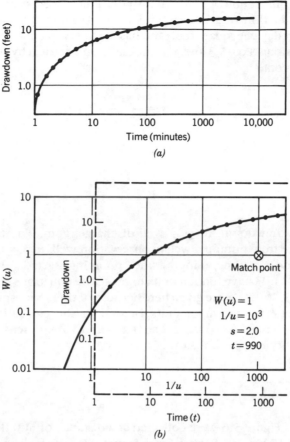

Figure 5.6
Graphs of (*a*) field data and (*b*) superposition of the
field data on the type curve to obtain the formation
parameters.

If drawdown s can be measured at one point r for several values of time t, and if
the discharge is steady and is known, the coefficients of transmissivity and storativity can be determined by a graphical method of superposition. Field data composed of drawdown versus time collected at a nonpumping observation well at a
known distance r from a pumped well are plotted on logarithmic paper of the
same scale as the type curve (Figure 5.6*a*). The field curve is superimposed on the
type curve, with the coordinate axes of the two curves kept parallel while matching field data to the type curve (Figure 5.6*b*). Any point on the overlapping sheets
is selected arbitrarily (the point need not be on the matched curves). The selected
point is defined by four coordinate values: $W(u)$ and s, and $1/u$ and t (Figure 5.6*b*).
Equation 5.5 can be solved for transmissivity by using the match-point coordinates s and $W(u)$ and the discharge Q. Equation 5.3 can be solved for storativity
by using the match-point coordinates $1/u$ and t, the distance r from the pumped
well to the observation well, and the value of transmissivity as determined earlier.

For the so-called American practical hydrology units, transmissivity is expressed
in gallons per day per foot and the storativity remains a dimensionless constant.
This requires a mixed system of volume measurement: gallons for transmissivity

and, inherently, cubic feet for storativity. This mixed system is the result of the early introduction of arbitrary units of measurement, such as the Meinzer unit (gallons per day per square foot) and has led to several dimensional versions of the nondimensional Eqs. 5.3 and 5.5. In American practical hydrologic units, Eqs. 5.5 and 5.3 become

$$s = \frac{114.6Q}{T} W(u) \tag{5.7}$$

and

$$u = \frac{2,693r^2S}{Tt} \tag{5.8}$$

where s = drawdown, ft; Q = well discharge, gpm; T = transmissivity, gpd/ft; r = distance from pumping well to observation well, ft; S = the storativity, dimensionless; and t = time, minutes since pumping started. The conversion factors 114.6 and 2,693 have hidden units to make the equations dimensionally correct. Equations 5.7 and 5.8 are given here because of their widespread use in the United States. Whatever system of units is used in the evaluation of the formation parameters, it must be consistent so that u and S are dimensionless quantities and transmissivity has the units of L²/T.

Example 5.1

The data of Figure 5.6 were collected at a distance of 500 ft from a well pumped at a rate of 192×10^3 ft³/day. From Figure 5.6

$$W(u) = 1 \qquad s = 2 \text{ ft}$$

$$\frac{1}{u} = 10^3 \qquad t = 990 \text{ min (about 0.7 day)}$$

Transmissivity is determined from Eq. 5.5,

$$T = \frac{Q}{4\pi s} W(u) = \frac{192 \times 10^3 \text{ ft}^3/\text{day} \times 1}{4 \times 3.14 \times 2 \text{ ft}} = 7.6 \times 10^3 \text{ ft}^2/\text{day}$$

$$= 8.17 \times 10^{-3} \text{ m}^2/\text{sec}$$

Storativity is determined from Eq. 5.3,

$$S = \frac{4uTt}{r^2} = \frac{4 \times 1 \times 10^{-3} \times 7.6 \times 10^3 \text{ ft}^2/\text{day} \times 7 \times 10^{-1} \text{ day}}{250 \times 10^3 \text{ ft}^2} = 8.5 \times 10^{-5}$$

Assumptions and Interpretations

The operational hydraulic equations were derived for an ideal model aquifer whereas time-drawdown data represent real aquifer response. The procedure requires a matching of real response data with the response expected under ideal conditions. The most conspicuous assumption is that the aquifer is infinite in areal extent. This means the following:

1. The cone of depression will never intersect a boundary to the system.
2. An infinite amount of water is stored in the aquifer.

Condition 1 is likely to be met for very short times, or for longer times in extensive aquifers of uniform material. In that the cone of depression is expanding with time, any geologic boundary within a short distance from the pumping well will immediately disrupt the postulated behavior. According to assumption 2, water levels will eventually return to their prepumping level once the wells are shut down.

In addition, it is assumed that the well is of infinitesimal diameter and fully penetrates the aquifer. This means that storage in the well can be ignored and that the well receives water from the entire thickness of the aquifer. Methods of treatment for deviations from full penetration have been presented in a number of papers (Muskat, 1937; Hantush and Jacob, 1955; Hantush, 1957a). A solution for drawdown in large-diameter wells that takes into consideration the storage within the well has been presented by Papadopulos and Cooper (1967). Yet another assumption is that water is released instantaneously with decline in head in the aquifer. This assumption may be met when water is released by compression of the aquifer and expansion of the water, but fails to describe adequately the gravity flow system of the unconfined case.

Homogeneity and isotropicity are major assumptions in the development of the descriptive differential equation of which the well flow equation is a solution. This means that the transmissivity and storativity are assumed to be constants, both in space and in time. Hence, the geologic medium is assumed to be the simplest type conceivable.

Modifications of the Nonequilibrium Equation

At least two important modifications of the nonequilibrium equation can be traced to a very simple observation made by Cooper and Jacob (1946), namely, that the sum of the series of Eq. 5.2 beyond ln u becomes negligible when u becomes small. This occurs for large values of time t or small value of distance r. By neglecting the series beyond ln u, Eq. (5.1) can be expressed

$$s = \frac{Q}{4\pi T}\left(-0.5772 - \ln \frac{r^2 S}{4Tt}\right) \tag{5.9}$$

or

$$s = \frac{Q}{4\pi T}\left(\ln \frac{4Tt}{r^2 S} - 0.5772\right) \tag{5.10}$$

As 0.5772 equals ln 1.78 and ln x equals 2.3 log x, Eq. 5.10 becomes

$$s = \frac{2.3Q}{4\pi T} \log \frac{2.25Tt}{r^2 S} \tag{5.11}$$

which is presented as the modified nonequilibrium equation. This equation may be applied to

1. Time-drawdown observations made in a single observation well, as in the original nonequilibrium method (modified nonequilibrium method).
2. Drawdown observations made in different wells at the same time (distance-drawdown method).

Time-Drawdown Method

If drawdown observations are made in a single well for various times, a plot of drawdown versus the logarithm of time will yield a straight line (Figure 5.7). At time t_1, the drawdown s_1 is expressed from Eq. 5.11

$$s_1 = \frac{2.3Q}{4\pi T} \log \frac{2.25Tt_1}{r^2S} \tag{5.12}$$

At time t_2, the drawdown s_2 will be

$$s_2 = \frac{2.3Q}{4\pi T} \log \frac{2.25Tt_2}{r^2S} \tag{5.13}$$

It follows that

$$s_2 - s_1 = \frac{2.3Q}{4\pi T} \log \frac{t_2}{t_1} \tag{5.14}$$

If t_1 and t_2 are selected one log cycle apart, $\log (t_2/t_1) = 1$. Equation 5.14 then becomes

$$\Delta s = \frac{2.3Q}{4\pi T} \tag{5.15}$$

where Δs is the drawdown per log cycle. The value of T can be obtained from Eq. 5.15.

The storativity can be obtained by selecting any drawdown on Figure 5.7 for a given time and substituting this value in Eq. 5.11. For convenience, s equals zero is selected, so that Eq. 5.11 becomes

$$s = 0 = \frac{2.3Q}{4\pi T} \log \frac{2.25Tt}{r^2S} \tag{5.16}$$

This requires that $2.25\, Tt_0/r^2S = 1$ so that

$$S = \frac{2.25Tt_0}{r^2} \tag{5.17}$$

where t_0 is the time intercept where the drawdown line intercepts the zero drawdown axis.

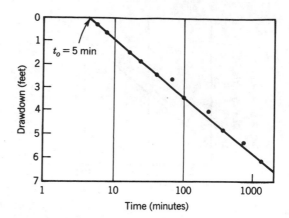

Figure 5.7
Semilogarithmic plot of drawdown
versus time in an observation well.

Example 5.2

The data of Figure 5.7 were collected at an observation well 1000 ft from a well pumping at a rate of 192×10^3 ft³/day. Drawdown per log cycle is determined from the graph to be 2.4 ft. Solving for T and S,

$$T = \frac{2.3Q}{4\pi\Delta s} = \frac{2.3(192 \times 10^3)\text{ft}^3/\text{day}}{4 \times 3.14 \times 2.4 \text{ ft}} = 14.6 \times 10^3 \text{ ft}^2/\text{day} = 1.57 \times 10^{-2} \text{ m}^2/\text{sec}$$

$$S = \frac{2.25Tt_o}{r^2} = \frac{2.25(14.6 \times 10^3 \text{ ft}^2/\text{day})(5 \text{ min})}{(1 \times 10^6 \text{ ft}^2)(1440 \text{ min/day})} = 1.1 \times 10^{-4}$$

Distance-Drawdown Method

The time-drawdown analysis just described requires one pumping well and one observation well, where a plot of drawdown versus time on semilogarithmic paper yields a straight line. This straight line demonstrates the exponential decline of water levels at a point in a single aquifer. It is also possible to obtain the hydraulic properties by examining drawdown at two or more points at one instant of time. At time t, the drawdown s_1 at a distance r_1 is, from Eq. 5.11,

$$s_1 = \frac{2.3Q}{4\pi T} \log \frac{2.25Tt}{r_1^2 S} \tag{5.18}$$

and the drawdown s_2 at a distance r_2 is

$$s_2 = \frac{2.3Q}{4\pi T} \log \frac{2.25Tt}{r_2^2 S} \tag{5.19}$$

It follows that

$$s_1 - s_2 = \frac{2.3Q}{4\pi T} \log \frac{r_2^2}{r_1^2} \tag{5.20}$$

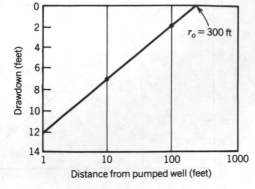

Figure 5.8
Semilogarithmic plot of drawdown
versus distance.

Recognizing that $\log(r_2^2/r_1^2) = \log(1/r_1^2) - \log(1/r_2^2)$ and that $\log(1/r^2) = 2\log(1/r)$, Eq. 5.20 becomes

$$s_1 - s_2 = \frac{2.3Q}{4\pi T} \log \frac{r_2}{r_1} \qquad (5.21)$$

The graphic procedure calls for the plotting of drawdowns observed at the end of a particular pumping period in two or more observation wells at different distances from the pumped well against the logarithms of the respective distances (Fig. 5.8). By considering drawdown per log cycle, Eq. 5.21 becomes

$$\Delta s = \frac{2.3Q}{2\pi T} \qquad (5.22)$$

By extrapolating the distance-drawdown curve to its intersection with the zero drawdown axis, the storativity can be determined in the same manner as described for the modified nonequilibrium method,

$$S = \frac{2.25Tt}{r_o^2} \qquad (5.23)$$

where r_o is the intersection of the straight-line slope with the zero drawdown axis.

Example 5.3

Consider the following data for Figure 5.8. Observation wells are placed 10 ft and 100 ft, respectively, from a pumping well. The well is pumped at a rate of 192×10^3 ft³/day. At the end of 200 min (1.39×10^{-1} days), the drawdowns in the observation wells are 7 ft and 2 ft, respectively. From Eqs. 5.22 and 5.23,

$$T = \frac{2.3Q}{2\pi\Delta s} = \frac{2.3(192 \times 10^3)\text{ft}^3/\text{day}}{2(3.14)(5 \text{ ft})} = 14 \times 10^3 \text{ ft}^2/\text{day} = 1.5 \times 10^{-2} \text{ m}^2/\text{sec}$$

$$S = \frac{2.25Tt}{r_o^2} = \frac{2.25(14 \times 10^3 \text{ ft}^2/\text{day})(1.39 \times 10^{-1} \text{ day})}{9 \times 10^4 \text{ ft}^2} = 5 \times 10^{-2}$$

Steady-State Behavior as a Terminal Case of the Transient Response

The straight line in Figure 5.8 represents two points on a profile of a cone of depression for a given instant of time. At some time later yet another curve can be constructed, and it will parallel the curve shown in the figure, with the same drawdown per log cycle. For very long periods of time, the drawdown in the observation wells will have approached their steady-state value and no longer decline with time. A plot of this drawdown on Figure 5.8 will again form a straight line, parallel to the other lines, with the same drawdown per log cycle. This problem involving steady-state drawdowns at two observation points was solved by Thiem (1906) completely independent of any transient flow considerations. The solution for Thiem's confined condition is identical to that given in Eqs. 5.21 and 5.22. However, as time is not a factor in Thiem's steady-state derivation, the storativity cannot be determined for the steady drawdown case. If the flow takes place under unconfined conditions, Thiem's solution changes somewhat and may be expressed as

$$s_1^2 - s_2^2 = \frac{2.3Q}{\pi K} \log \frac{r_1}{r_2} \tag{5.24}$$

As in the application of the confined case, any pair of distances may be employed. This equation, along with its steady-state counterpart for confined conditions, is frequently referred to as the equilibrium, or Thiem equation.

The Hantush-Jacob Leaky Aquifer Method

One major assumption of the Theis nonequilibrium solution is that all the water pumped is removed from storage within the aquifer. There are several ways in which this condition may be compromised in field conditions: direct recharge to the aquifer from streams, direct recharge across bounding low-permeability materials, and release of water from bounding low-permeability rocks that possess their own hydraulic characteristics (i.e., specific storage and hydraulic conductivity). The problem of leakage has been extensively investigated by Hantush and Jacob (1955) and Hantush (1956, 1960, 1964). In this section we will be concerned only with their earliest contribution, which provides methods most commonly used in field problems.

Figure 5.1*a* will help develop the pertinent ideas. If the artesian sand in the aquifer is pumped, and the bounding low-permeability rocks are in fact "impermeable," the anticipated response at an observation well can be described completely by the Theis equation. However, if water is entering the aquifer, say, by either leakage across the upper low-permeability unit or by release of water from storage in the low-permeability unit, we would anticipate some deviation from the Theis response. If we assume that the low-permeability unit has a negligible specific storage (that is, is incompressible) but a finite permeability, water can be transmitted across it from the overlying water table. Figure 5.9 shows a typical response at an observation well in such an aquifer. Note that the time-drawdown curve departs from the expected nonequilibrium behavior, resulting in less drawdown per log cycle of time.

There are two ways to analyze drawdown curves of the type shown on Figure 5.9, depending on how much information is desired. The simplest method

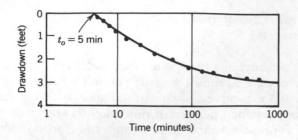

Figure 5.9
Time-drawdown response of leaky aquifer.

recognizes that it takes some length of time for water to permeate through the upper low-permeability material, and the upward inflection from the straight-line behavior of the Theis solution provides some estimate of this required time period. Hence, for all practical purposes, the time-drawdown behavior previous to the inflection represents a withdrawal of water from storage from the aquifer, no part of which was contributed from other sources. This part of the curve, along with its straight-line extension, may be analyzed with methods previously discussed for time-drawdown behavior. That is to say, over this period of the pumping test, the assumption that all the water pumped comes from storage in the aquifer is a reasonably valid one.

The second method requires a curve-matching procedure where the drawdown for all time is given as (Hantush and Jacob, 1955; Hantush, 1956)

$$s = \frac{Q}{4\pi T}\, W\left(u, \frac{r}{B}\right) \tag{5.25}$$

where $W(u, r/B)$ is the tabulated well function for leaky aquifers. It is thus recognized that short-term drawdowns previous to leakage are described by the Theis nonequilibrium equation and the well function $W(u)$ whereas long-term drawdowns are described by the well function $W(u, r/B)$. Values of $W(u, r/B)$ are given in Table 5.2, and a type curve is presented as Figure 5.10. Note that the family of curves converges on the nonleaky curve for small values of $1/u$ (which correspond to small values of time) and small values of r/B. For small values of time, $W(u, r/B) \approx W(u)$.

The graphic procedure for solving for the formation constants is similar to that described previously. Early drawdown data from the field curve are matched with the nonleaky part of the curve, but they soon deviate and follow one of the leaky r/B curves (Figure 5.11). The match point yields values of $W(u, r/B)$, $1/u$, t, and s. In addition, the r/B curve followed by the field data is noted. Transmissivity and storativity are readily determined from

$$T = \frac{Q}{4\pi s}\, W\left(u, \frac{r}{B}\right) \tag{5.26}$$

and

$$S = \frac{4uTt}{r^2} \tag{5.27}$$

In addition, the coefficient of vertical hydraulic conductivity can be determined from the relationship

Table 5.2
Values of $W(u, r/B)$ (after Hantush, 1956)*

u	\ r/B = 0.01	0.015	0.03	0.05	0.075	0.10	0.15	0.2	0.3	0.4
0.000001										
0.000005	9.4413									
0.00001	9.4176	8.6313								
0.00005	8.8827	8.4533	7.2450							
0.0001	8.3983	8.1414	7.2122	6.2282	5.4228					
0.0005	6.9750	6.9152	6.6219	6.0821	5.4062	4.8530				
0.001	6.3069	6.2765	6.1202	5.7965	5.3078	4.8292	4.0595	3.5054		
0.005	4.7212	4.7152	4.6829	4.6084	4.4713	4.2960	3.8821	3.4567	2.7428	2.2290
0.01	4.0356	4.0326	4.0167	3.9795	3.9091	3.8150	3.5725	3.2875	2.7104	2.2253
0.05	2.4675	2.4670	2.4642	2.4576	2.4448	2.4271	2.3776	2.3110	1.9283	1.7075
0.1	1.8227	1.8225	1.8213	1.8184	1.8128	1.8050	1.7829	1.7527	1.6704	1.5644
0.5	0.5598	0.5597	0.5596	0.5594	0.5588	0.5581	0.5561	0.5532	0.5453	0.5344
1.0	0.2194	0.2194	0.2193	0.2193	0.2191	0.2190	0.2186	0.2179	0.2161	0.2135
5.0	0.0011	0.0011	0.0011	0.0011	0.0011	0.0011	0.0011	0.0011	0.0011	0.0011

u	\ r/B = 0.5	0.6	0.7	0.8	0.9	1.0	1.5	2.0	2.5
0.000001									
0.000005									
0.00001									
0.00005									
0.0001									
0.0005									
0.001									
0.005									
0.01	1.8486	1.5550	1.3210	1.1307	0.9700	0.8409			
0.05	1.4927	1.3115	1.2955	1.1210	0.9440	0.8190			
0.1	1.4422	1.2955	1.1791	1.0505	0.9297		0.4271	0.2278	
0.5	0.5206	0.5044	0.4860	0.4658	0.4440	0.4210	0.3007	0.1944	0.1174
1.0	0.2103	0.2065	0.2020	0.1970	0.1914	0.1855	0.1509	0.1139	0.0803
5.0	0.0011	0.0011	0.0011	0.0011	0.0011	0.0011	0.0010	0.0010	0.0009

*Trans. Amer. Geophys. Union, 37, p. 702–714. Copyright by Amer. Geophys. Union.

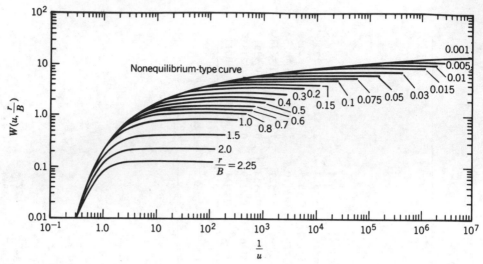

Figure 5.10
Type curve for leaky aquifers.

$$\frac{1}{B} = \left(\frac{k'/m'}{T} \right)^{1/2} \tag{5.28}$$

where $1/B$ is determined from the match, m' is the thickness of the bounding low-permeability formation, and k' is the vertical hydraulic conductivity.

Example 5.4

The data in Figure 5.11 are similar to those in Figure 5.6 except for the leakage. Assuming the data were obtained from an observation well located 500 ft from a well pumped at a rate of 192×10^3 ft³/day, the coefficient of transmissivity is calculated

$$T = \frac{Q}{4\pi s} W\left(u, \frac{r}{B}\right) = \frac{(192 \times 10^3 \text{ ft}^3/\text{day})(1)}{4(3.14)(2 \text{ ft})} = 7.6 \times 10^3 \text{ ft}^2/\text{day}$$
$$= 8.17 \times 10^{-3} \text{ m}^2/\text{sec}$$

The coefficient of storage is

$$S = \frac{4uTt}{r^2} = \frac{4(1 \times 10^{-3})(7.6 \times 10^3 \text{ ft}^2/\text{day})(7 \times 10^{-1} \text{ days})}{230 \times 10^3 \text{ ft}^2} = 8.5 \times 10^{-5}$$

Both results are identical with the numerical values obtained for Figure 5.6. Assuming the semipervious layer is 10 ft in thickness,

$$k' = \frac{Tm'(r/B)^2}{r^2} = \frac{(7.6 \times 10^3 \text{ ft}^2/\text{day})(5.6 \times 10^{-3})}{250 \times 10^3 \text{ ft}^2} = 1.7 \times 10^{-3} \text{ ft/day}$$
$$= 6 \times 10^{-9} \text{ m/sec}$$

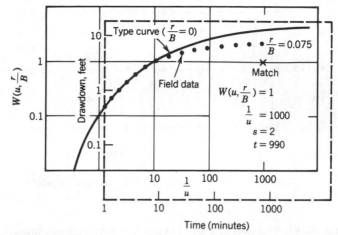

Figure 5.11
Graphic procedure for solving for formation constants for
leaky aquifers.

Water Table Aquifers

The tests just described are designed to test the confined flow condition. This means that during the testing, the aquifer undergoes no dewatering, provided the heads are not lowered below the uppermost confining layer. For this case water is released from storage due to elastic compression of the matrix and expansion of the water itself. For a water table, or unconfined aquifer, the lowering of water levels will actually cause some dewatering of the aquifer, and the value of the storativity should increase by a few orders of magnitude over its confined counterpart. In Chapter 4, the storativity for such cases was referred to as the specific yield, S_y.

Two major concerns are involved in the analysis of water table aquifers. First, dewatering of the aquifer can actually lead to a decrease in transmissivity during pumping in that $T = Km$, where m is the saturated thickness of the aquifer. The second problem relates to the assumption of the Theis equation, where "water is released from storage instantaneously with decline in head." If this condition is not satisfied, the time drawdown data will deviate from the expected response for the confined condition. The mathematical description of this response has been the subject of numerous papers starting with Boulton (1954, 1955, 1963), extended by Prickett (1965) based on the Boulton analysis, and advanced significantly by Dagan (1967), Streltsova (1972, 1973), and Neuman (1972, 1975).

In view of the attention given to this subject, it is informative now to obtain a semiquantitative perspective of a typical time-drawdown response in an unconfined aquifer. Actually, there are three distinct parts of the time-drawdown curve as obtained in the vicinity of a pumped well. The very early time-drawdown data should correspond to that predicted by the Theis equation where water is released from storage due to elastic compression and expansion of the water. Thus the time-drawdown data should follow the Theis curve where the storativity is on the order expected for the confined condition. With increased time, the effects of gravity drainage take over, with vertical flow components in the vicinity of the pumped well. During this period, the time-drawdown response will be a function

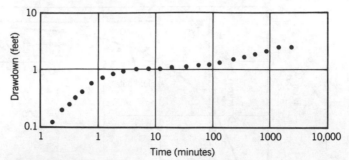

Figure 5.12
Time-drawdown response of an unconfined aquifer.

of the ratio of horizontal to vertical conductivity, the thickness of the aquifer, and the distance to the pumped well. If departures from the Theis-type curve do occur, they will be similar to those obtained when the slope decreases in the leaky aquifer test. Although this is occurring in response to vertical flow components, one may view this in a conceptual sense as an increase in the "storativity" over its confined value. For later time, the time-drawdown data will again follow some Theis-type curve, with the storativity now reflecting the specific yield S_y; that is, the "storativity" is no longer increasing with time.

On a logarithmic plot the curve has the shape of an elongated letter "s" (Figure 5.12). Given this information, there are two approaches to the interpretation of pumping tests performed in unconfined aquifers. First, let us recognize that previous to Boulton's contribution (1954), no distinction was made between the confined and unconfined response. Indeed, virtually hundreds of tests were conducted on unconfined systems using the Theis analysis with no apparent reference to the need for modifications. Jacob (1950) recognized the difficulties with unconfined flow to wells and recommended a procedure to account for an apparent increase in storativity with time. In this procedure, the transmissivity is first determined from the early time-drawdown data by the modified nonequilibrium method (Eq. 5.15). Field values of drawdown s for later time t are then introduced into Eq. 5.11 and the equation solved for the specific yield. If the time-drawdown curve actually contains three segments, as in Figure 5.12, the ultimate value of S_y can be readily determined. If only two segments are discernible, an intermediate value will be obtained.

A more exact analysis requires a curve-matching procedure. According to Neuman (1975) the solution to the flow problem is

$$s = \frac{Q}{4\pi T} \, W(u_A, \, u_B, \, \eta) \tag{5.29}$$

where $W(u_A, \, u_B, \, \eta)$ is the well function for the unconfined aquifer. The arguments of the well function are described as

$$u_A = \frac{r^2 S}{4Tt} \quad \text{(for early time-drawdown data)} \tag{5.30a}$$

$$u_B = \frac{r^2 S_y}{4Tt} \quad \text{(for late time-drawdown data)} \tag{5.30b}$$

$$\eta = \frac{r^2 K'}{m^2 K} \tag{5.30c}$$

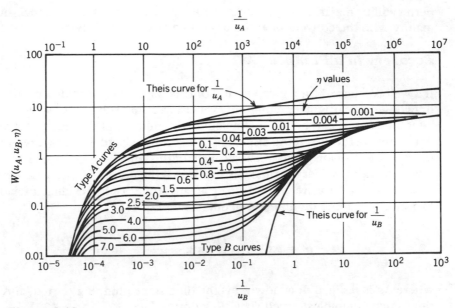

Figure 5.13
Theoretical curves of $W(u_A u_B, \eta)$ versus $1/u_A$ and $1/u_B$ for an unconfined aquifer (from Neuman, Water Resources Res., v. 11, p. 329–342, 1975. Copyright by Amer. Geophys. Union).

where m is the initial saturated thickness, K is the horizontal conductivity, and K' is the vertical conductivity. Figure 5.13 shows the type curve employed in the matching procedure where the type A curves merge out of the Theis curve shown on the left and the type B curves merge into the Theis-type curve shown on the right. The following method is recommended for the analysis. First, the early time-drawdown data should be matched to the type A curves, and any match point selected. This match point will have the coordinates $W(u_A, \eta)$, $1/u_A$, t, and s. The match will provide a value for η. The transmissivity is then determined with Eq. 5.29 and the storativity with Eq. 5.30a. The later time-drawdown data are then matched to the specific type B curve of the same η value determined for the A match. The match point yields values of $W(u_B, \eta)$, $1/u_B$, s, and t. Transmissivity is determined from Eq. 5.29, which should provide a value similar to that previously determined. The storativity can be determined with Eq. 5.30b and should yield a value somewhat larger than previously determined. As $K = T/m$ Eq. 5.30c can be used to determine the vertical hydraulic conductivity. Neuman (1975) provides tables for the function $W(u_A, u_B, \eta)$.

5.3 *Single-Borehole Tests*

The one feature shared by the methods described in the previous section is the necessity of one pumping well and one or more observation wells. Single-borehole tests offer some advantages in economy, frequently with a certain loss of information. In spite of this, the single-borehole test is becoming increasingly more common in hydraulic testing associated with waste disposal sites and in low-permeability material. As in the previous section, we will be concerned here mainly with the conventional or common methods of analysis.

Recovery in a Pumped Well

If a well is pumped for a given period of time t and then shut down, the residual drawdown (the original prepumping water level minus the water level at any time after shutdown) can be approximated as the numerical difference between the drawdown in the well if the discharge had continued and the recovery of the well in response to an imaginary recharge well, of the same flow rate, superimposed on the discharging well at the time it is shut down (Figure 5.14). Designating original head as h_o and the recovered head at any time as h', residual drawdown is expressed from the modified nonequilibrium Eq. 5.11

$$h_o - h' = \Delta s' = \frac{2.3Q}{4\pi T}\left(\log \frac{2.25Tt}{r^2S} - \log \frac{2.25Tt'}{r^2S}\right) \tag{5.31}$$

where $\Delta s'$ is residual drawdown, t is the time since pumping started, and t' is the time since pumping stopped. This equation reduces to

$$\Delta s' = \frac{2.3Q}{4\pi T}\log \frac{t}{t'} \tag{5.32}$$

Field procedure requires a drawdown measurement at the end of the pumping period (t) and recovery measurements during the recovery period (t'). The graphic procedure is to plot residual drawdown on the arithmetic scale and the value of t/t' on logarithmic scale. The time t includes the interval over the pumping plus recovery period whereas the time t' includes the recovery interval only. If calculations are made over one log cycle of t/t',

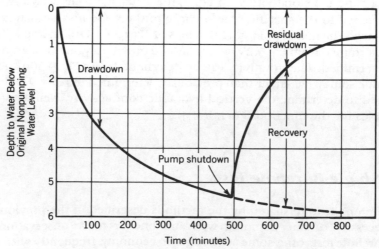

Figure 5.14
Arithmetic plot of drawdown and recovery curve versus time.

$$\Delta s' = \frac{2.3Q}{4\pi T} \tag{5.33}$$

where $\Delta s'$ is the residual drawdown per log cycle. The storativity is not determined directly with this method.

The Drill Stem Test

The drill stem test used in petroleum engineering is the equivalent of the recovery test. Figure 5.15 shows the main components of a typical drill stem test. The packer assembly is set at the isolated stratigraphic interval of interest. The bypass valve is initially opened and the formation flows under its head. This is equivalent to the pumping period of the recovery test. The valve is then closed, shutting in the formation pressure. The ensuing period is analogous to the recovery period of the single-borehole test previously discussed. The method of Horner (1951) is generally used to analyze the data

$$P_w - P_o = \frac{2.3Q\mu}{4\pi km} \log \frac{t}{t'} \tag{5.34}$$

where P_w is the pressure in the well bore, P_o is the undisturbed formation pressure, Q is the rate of production, μ is the viscosity of the fluid, k is the permeability,

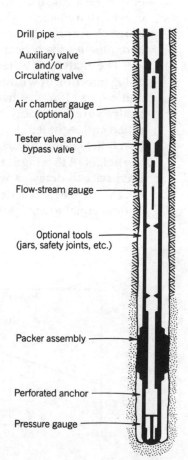

Drill pipe

Auxiliary valve
and/or
Circulating valve

Air chamber gauge
(optional)

Tester valve and
bypass valve

Flow-stream gauge

Optional tools
(jars, safety joints, etc.)

Packer assembly

Perforated anchor

Pressure gauge

Figure 5.15
Main components of a typical drill-stem test string
(from Bredehoeft, 1965). Reprinted by permission of
Ground Water. © 1965. All rights reserved.

m is the formation thickness, t is the time since the test began (the sum of the production and the recovery time), and t' is the time since the recovery portion of the test started.

A typical pressure buildup plot is shown in Figure 5.16. By considering the pressure change over one log cycle, Eq. 5.34 reduces to

$$\frac{km}{\mu} = \frac{2.3Q}{4\pi\Delta P} \tag{5.35}$$

where ΔP is the change in pressure over one log cycle.

By extrapolating the pressure curve to $t/t' = 1$, the undisturbed formation pressure P_o is determined; that is, in Eq. 5.34, log (1) = 0.

Pressure buildup tests in single boreholes have long been of interest to petroleum engineers, with the monograms of Matthews and Russel (1967) and Earlougher (1977) providing good reviews.

Slug Injection or Withdrawal Tests

The pressure recovery in a borehole after withdrawal of a known volume of water (slug), or the pressure decline after injection of a known volume of water (slug), is termed the slug test. The slug test is one of the more common methods employed in field practice in that it can be used in low- to moderately high-conductivity materials and requires no pumping apparatus. Thus, small-diameter boreholes normally used in geologic exploration may serve double duty in a hydraulic testing program.

The analysis of water level versus time data in a single borehole in response to a slug injection or withdrawal has been pioneered by Hvorslev (1951) and Cooper and others (1967). Whatever method is being employed, the procedures for the analysis are reasonably straightforward. The original head in the borehole is first noted and recorded. For the slug injection, water is instantaneously added to the borehole, raising the original head above its static level. This is termed h_o, that is, the height of the slug at time equal to zero, say, for example, 2 or 3 ft. The slug will then start to decay as water enters the formation. This change in head is noted over time, where h is designated the height of the slug for any given time. Thus, at time equal to zero, $h/h_o = 1$, and as time gets large, h/h_o approaches zero, that

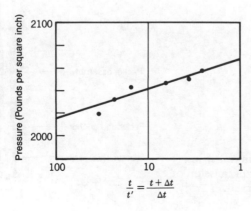

Figure 5.16
Drill-stem test result.

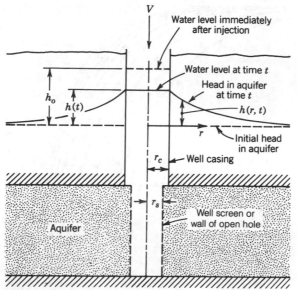

Figure 5.17
Well into which a volume *V*, of water is suddenly
injected for a slug test of a confined aquifer (from
Cooper and others, Water Resources Res, v. 3, p.
263–269, 1967. Copyright by Amer. Geophys, Union).

is, the slug has been completely dissipated (Figure 5.17). If the well or borehole is
screened above the water table, as is common for investigating oil spills or leaks,
a slug withdrawal procedure is necessary.

Cooper and others (1967) present a type curve solution for the interpretation
of slug tests. Papadopulos and others (1973) extended the range of the type curves
for greater use in practice (Figure 5.18). To determine the hydraulic parameters,
field data are plotted on semilogarithmic paper of the same scale as the type
curves shown in Figure 5.18. The field data consists of h/h_o versus time, or a time
plot of the height of the slug at any time *t* divided by the original height h_o (Figure
5.19). Keeping the arithmetic axes coincident, a best fit position between the field
data and the type curves is determined. The time value of the field curve that coincides with *W* equals one is determined in the matched position. The transmissivity is determined from

$$T = \frac{Wr_c^2}{t} = \frac{r_c^2}{t} \tag{5.36}$$

where r_c equals the radius of the well casing (Figure 5.17). The storativity may be
determined from

$$S = \frac{r_c^2\,\alpha}{r_s^2} \tag{5.37}$$

where r_s and α are defined in Figure 5.18. It has been noted that it is difficult to
achieve a definitive fit in the matching procedure described in that the field data

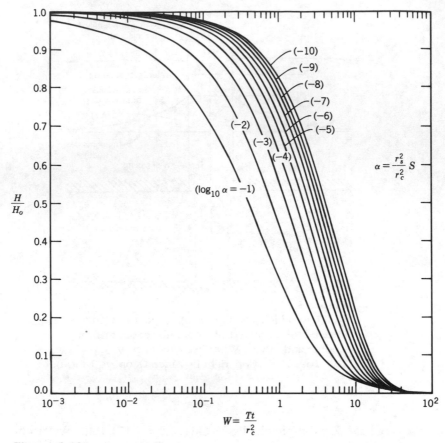

Figure 5.18
Type curves for instantaneous change in head in a well of finite diameter (from
Papadopulos and others, Water Resources Res., v. 9, p. 1087–1089, 1973.
Copyright by Amer. Geophys. Union).

frequently overlap or fits part of several of the type curves given in Figure 5.18.
This is one major drawback of the curve matching method.

The method of Hvorslev (1951) is probably the most widely used in field practice. The field data are plotted on semilogarithmic paper, with h/h_o on the log
scale and time t on the arithmetic scale (Figure 5.20). The hydraulic conductivity
may be determined from the relationship (Cedergren, 1967)

$$K = \frac{A}{F(t_1 - t_2)} \log_e \frac{h_1}{h_2} \tag{5.38}$$

where A is the area of the borehole and F is a shape factor that depends on the size
and shape of the intake area. Hvorslev (1951) has evaluated several values of the
shape factor F for a variety of conditions, as has the U.S. Navy Bureau of Yards and
Docks (1961) (see, for example, Cedergren, 1967). For a cased, uncased, or perforated extension into an aquifer of finite thickness for $L/r > 8$,

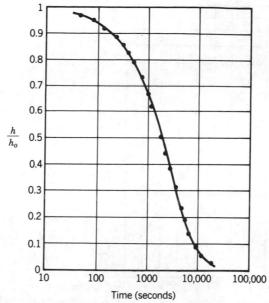

Figure 5.19
Field response of slug test.

$$F = \frac{2\pi L}{\ln(L/r)} \tag{5.39}$$

where L is the length of the intake area and r is the radius of the borehole. For a borehole area $= \pi r^2$, Eq. 5.38 becomes

$$K = \frac{r^2 \ln (L/r)}{2L} \frac{\log_e(h_1/h_2)}{(t_1 - t_2)} \tag{5.40}$$

It is convenient now to take $h_1 = h_o$ at $t = 0$ and $h_2 = 0.37h_o$ so that

$$\log_e \left(\frac{h_1}{h_2} \right) = \log_e \left(\frac{h_o}{0.37h_o} \right) = \log_e 2.7 = 1.0$$

Equation 5.40 then becomes

$$K = \frac{r^2 \ln(L/r)}{2LT_o} \tag{5.41}$$

where T_o is the time intercept on the field curve where $h/h_o = 0.37$ (Figure 5.20). T_o is termed the basic time lag and is determined where the head ratio $= 0.37$. The time lag refers to the fact that the transients in a well seldom correspond exactly to those occurring in the adjacent formation so that flows are associated with various time lags that depend on the shape factor F, the hydraulic conductivity, and the radius of the borehole.

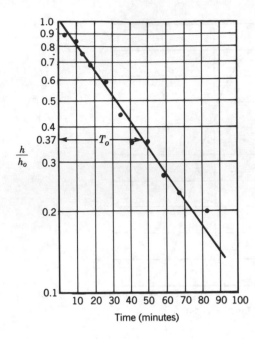

Figure 5.20
Field response of slug test.

Example 5.5

The field data curve of Figure 5.19 matches one of the type curves of Figure 5.18 for the position $W = 1$ and time $= 1000$ sec. For a casing radius $r_c = 2.54$ cm and a well screen length of 3.8m

$$T = \frac{Wr_c^2}{t} = (1)\frac{(2.54)^2}{1000} = 6.45 \times 10^{-3} \text{ cm}^2/\text{sec}$$

The field data of Figure 5.19 are plotted on Figure 5.20. For an observed value of $T_o = 44$ minutes, the hydraulic conductivity is determined from Eq. 5.41:

$$K = \frac{r^2 \ln(L/r)}{2LT_o} = \frac{(2.54)^2 \ln(381/2.54)}{(2)(381)(44)(60)} = 1.68 \times 10^{-5} \text{ cm/sec}$$

Response at the Pumped Well: Specific Capacity

Well capacity, or discharge per unit time, is often used as a measure of well yield. When comparing the strength of one well versus another, it is better to express well capacity in relation to some common standard. The accepted standard of comparison is the unit drawdown. Thus, dividing the pumping rate by the total drawdown in a well gives the specific capacity, that is, gallons or cubic feet per minute per foot of drawdown.

The drawdown observed in a pumping well is composed of two parts: (1) drawdown due to laminar flow of water in the aquifer toward the well, referred to as formation loss s and calculated with the well flow equations discussed in this chapter, and (2) drawdown resulting from the turbulent flow of water in the immediate vicinity of the well, through the well screen or casing openings and in the well casing, referred to as well loss s_w. Jacob (1950) stated that well loss is proportional to some power of the discharge exceeding the first power and approaching the second. As an approximation,

$$s_w = CQ^2 \tag{5.42}$$

where s_w is well loss (L), C is the well loss constant, and Q is the pumping rate. Total drawdown in a pumping well, s_t, is then expressed

$$s_t = s + s_w = \frac{Q}{4\pi T} W(u) + CQ^2 \tag{5.43}$$

The well loss constant C may be determined by comparing the actual drawdown in a pumping well with that predicted for a 100% efficient well ($s_w = 0$) as based on the Theis equation, where r is taken as the radius of the well r_w. The efficiency of the well is taken as the ratio of the predicted drawdown, where $s_w = 0$, to the actual measured drawdown s_t.

A theoretical specific capacity may be defined as the specific capacity that would be achieved for a 100% efficient well, that is, one in which the well loss $s_w = 0$. Theis and others (1963) demonstrated that the theoretical specific capacity of a well can be determined from the abbreviated nonequilibrium equation

$$T = \frac{Q}{4\pi s}\left(-0.5772 - \ln\frac{r^2 S}{4Tt}\right) \tag{5.44}$$

or, solving for Q/s in practical American hydrologic units,

$$\frac{Q}{s} = \frac{T}{264 \log\left(Tt/1.87 r_w^2 S\right) - 65.5} \tag{5.45}$$

where t is the pumping period in days, r_w is the effective radius of the well in ft, T is the transmissivity in gallons per day per foot, and Q/s is the specific capacity in gallons per minute per foot. The assumption here is that well loss is zero.

From this equation, the theoretical specific capacity of a well is directly proportional to T and inversely proportional to $\log t$, $\log 1/r_w^2$, and $\log 1/S$. Hence, large changes in T cause correspondingly large changes in specific capacity. Large changes in t, r_w, and S cause comparatively small changes in specific capacity. Given the actual specific capacity of a pumped well, the radius of the well, and the duration of pumping, and assuming some value for S, the transmissivity can be estimated with Eq. 5.45. Figure 5.21 has been prepared to aid in this determination. In this figure, the pumping period t has been taken as one day. To use the figure, the measured specific capacity (left-hand axis) is followed across to an intersection with an estimated storativity, and then the appropriate well diameter curve is followed to the right, yielding an approximate value of transmissivity in gallons per day per foot. If well losses are involved in the measured specific capacity, which appears inevitable, the transmissivity so determined represents an underestimated value. That is to say, with this method, the actual formation transmissivity is higher than the calculated one.

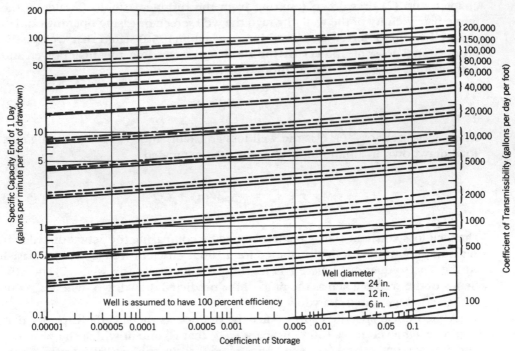

Figure 5.21
Graph showing interrelation of well diameter, specific capacity, transmissivity and storativity (from Theis and others, 1963).

Example 5.6

An aquifer has a transmissivity of 1340 ft²/day and a storativity of 1×10^{-4}. A 24-inch-diameter well in this aquifer has a drawdown of 300 ft after 24 hours of pumping at a rate of 192×10^3 ft³/day. The efficiency of the well is determined as follows:

For drawdown in the pumping well in the absence of well loss,

$$s = \frac{Q}{4\pi T} W(u) = \frac{192 \times 10^3 W(u)}{(4)(3.14)(1340)} = 11.4 \ W(u)$$

The value of u is determined for r equal to the well radius

$$u = \frac{r^2 S}{4Tt} = \frac{(1)(1 \times 10^{-4})}{(4)(1340)(1)} = 1.86 \times 10^{-8}$$

From Table 5.1, $W(u) = 17.2$ so that $s = (11.4)(17.2) = 196$ ft, and well efficiency is determined as $E = 196/300 = 65\%$.

The well loss constant C is determined from $s_w = CQ^2$

$$C = \frac{s_w}{Q^2}$$

$$= 104 \ \text{ft}/3.68 \times 10^{10} \ \text{ft}^6/\text{day}^2 = 2.82 \times 10^{-9} \ \text{day}^2/\text{ft}^5$$

$$= 8 \times 10^3 \ \text{sec}^2/\text{m}^5$$

Example 5.7

The well in Example 5.6 has an actual specific capacity (S.C.) of

$$\text{S.C.} = \frac{192 \times 10^3 \text{ ft}^3/\text{day}}{300 \text{ ft}}$$

$$= 640 \text{ ft}^3/\text{day per foot of drawdown}$$
$$\text{(or 3.32 gal/min per foot of drawdown)}$$

From the graph in Figure 5.21, this places the estimated transmissivity somewhat lower than 5000 (gal/day)/ft. The theoretical specific capacity of the previous example is

$$\text{S.C.} = \frac{192 \times 10^3 \text{ ft}^3/\text{day}}{196 \text{ ft}}$$

$$= 979 \text{ ft}^3/\text{day per foot of drawdown}$$
$$\text{(or 5 gpm/ft of drawdown)}$$

From the graph in Figure 5.21, this places the transmissivity at approximately 10,000 (gpd)/ft. The measured transmissivity is actually 10,023 (gpd)/ft. Hence the use of measured specific capacity data in conjunction with Figure 5.21 can result in large errors if actual well loss is substantial.

Direct Determination of Ground Water Velocity: Borehole-Dilution Tests

A direct determination of the seepage velocity v can be obtained with the borehole dilution test (Drost and others, 1968). In this procedure a nonreactive tracer is introduced into an isolated zone within a borehole at some concentration C_o. With natural flow through the formation, the rate of decrease in concentration in the isolated segment is proportional to the velocity $\bar{v}$

$$\frac{dC}{dt} = A \, \bar{v} \, \frac{C}{V} \tag{5.46}$$

where C is concentration, $\bar{v}$ is the ground water velocity in the center of the borehole, t is time, A is the vertical cross-sectional area through the center of the borehole over the length of the test segment, and V is the volume of the segment. Integrating for the condition that $C = C_o$ at time t equal to zero gives

$$\bar{v} = \frac{V}{A \, t} \ln \frac{C}{C_o} = \frac{\pi r}{2t} \ln \frac{C}{C_o} \tag{5.47}$$

where C is concentration at some later time t and r is the radius of the borehole. Normally the velocity $\bar{v}$ will be affected by the influence of the well and its gravel pack on the flow field. Drost and others (1968) recommend an adjustment factor F that ranges from 0.5 to 4 for sand or gravel aquifers, where the actual velocity $v = \bar{v}/F$. Clearly this range in some arbitrary adjustment factor is too large to provide dependable estimates of the actual seepage velocity. However, determinations of the velocity $\bar{v}$ at various isolated intervals in the borehole does provide information on the relative velocity of ground water through different permeable parts of the formation, with those zones of highest permeability resulting in the most rapid changes in concentration with time. Grisak and others (1977) describe a fluoride borehole dilution apparatus. Chloride is yet another readily available nonreactive tracer that can be used.

5.4 *Partial Penetration, Superposition, and Bounded Aquifers*

Partial Penetration

When the screened or open section of a well casing does not coincide with the full thickness of the aquifer it penetrates, the well is referred to as partially penetrating. With existing wells, this is the rule rather than the exception. Under such conditions the flow toward the pumping well (or observation point) will be three dimensional because of vertical flow components (Figure 5.22). Our experience with vertical flow components in the unconfined case previously discussed suggests upward inflection points in the time-drawdown response.

The topic of partial penetration has been the subject of numerous papers (Muskat, 1937; Hantush, 1961, 1964; Neuman, 1974). Two aspects of the problem are of interest here. First, solutions for partially penetrating wells have been derived and are discussed by Hantush (1961, 1964) and Walton (1970). These solutions may be applied directly if partial penetration is a factor affecting the response. However, with such solutions, it is necessary to know the actual degree of penetration of the aquifer, which may not be known in many cases. A second point is the effect of partial penetration on the water-level response. Hantush (1964) has conducted the most work in this area and provides some general guidelines for confined aquifers. The effects of partial penetration will not normally affect the pumping test result if the observation well is located at some distance $r > 1.5m\,(K/K')^{1/2}$ from the pumped well, where K is the horizontal conductivity, K' is the vertical conductivity, and m is the aquifer thickness. If the conductivities

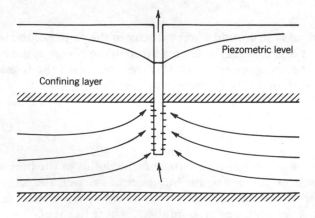

Figure 5.22
The effect of partial penetration on flow to a well.

K and K' are of the same order of magnitude, the condition $r > 1.5m$ is sufficient. If this condition is not satisfied, there will be an upward inflection in the response, similar to that obtained in the leaky method or for some sort of recharge boundary. In some cases, depending on the relative positions of the screens, more distant observation points can actually show more drawdown than closer ones. For long periods of time $t > Sm/2K'$, the time-drawdown slope of a partially penetrating response will be the same as expected if the well completely penetrated the aquifer. The reader is referred to Hantush (1961) or Walton (1970) for the methodology required to treat partial penetration.

Principle of Superposition

The Theis nonequilibrium equation represents a solution to the diffusion equation for a prescribed set of boundary and initial conditions. The pumping test procedure is relatively straightforward where a transient response to a steady pumping rate is observed over a limited period of time (say, 24 hours) at some known distance from the pumping well. Given the transmissive and storage properties as determined over this short pumping period, it is possible to predict the water level at later times and at other distances in response to any steady pumping rate Q. This assumes, of course, that the storativity and transmissivity do not change with time, whereas the postulated time of pumping t, the distance r, and the pumping rate Q may be taken as variables.

The Theis equation (for that matter the Hantush-Jacob equation) may also be used to obtain a theoretical drawdown for any time t at various distances r in response to a battery of wells pumping at various rates. This is because the diffusion equation is linear, that is, consists of a sum of linear terms where a linear term is first degree in the dependent variables and their derivatives. Linearity in differential equations is synonymous with the principle of superposition, which states that the derivative of a sum of terms is equal to the sum of the derivatives of the individual terms. Expressed in terms of the pumping response, the total effect resulting from several wells pumping simultaneously is equal to the sum of the individual effect caused by each of the wells acting separately.

Example 5.8

The principle of superposition is an important one in the application of hydraulic methods to real-world problems. The purpose of this example is to demonstrate an efficient manner in which well test data are organized to examine well interference problems. We shall assume that the data of Example 5.1 apply here; that is, a well is pumped for a relatively short period of time at a rate of 192×10^3 ft³/day and a transmissivity of 7.6×10^3 ft²/day and a storativity of 8.5×10^{-5} is determined. We assume these properties are constant over a large region. The table here demonstrates a procedure for tabulating values for the drawdown in this aquifer for times ranging from 1 to 100 days and over distances ranging from 1 to 1000 ft from a single pumping well. It is noted that these calculations are for the specific pumping rate of 192×10^3 ft³/day. The data in this table are conveniently placed in graphical form, as shown.

time = 1 day

$$u = \frac{r^2 S}{4Tt} = \frac{r^2 \times 8.6 \times 10^{-5}}{4 \times 7.7 \times 10^3 \times 1} = 2.74 \times 10^{-9} r^2$$

r, ft	u	W(u)	s, ft
1	2.74×10^{-9}	19.13	38.26
10	2.74×10^{-7}	14.53	29.06
100	2.74×10^{-5}	9.92	19.84
1000	2.74×10^{-3}	5.32	10.64

time = 10 days

$$u = \frac{r^2 S}{4Tt} = \frac{r^2 \times 8.6 \times 10^{-5}}{4 \times 7.7 \times 10^3 \times 1} = 2.74 \times 10^{-10} r^2$$

r, ft	u	W(u)	s, ft
1	2.74×10^{-10}	21.43	42.86
10	2.74×10^{-8}	16.83	33.66
100	2.74×10^{-6}	12.22	24.44
1000	2.74×10^{-4}	7.62	15.24

time = 100 days

$$u = \frac{r^2 S}{4Tt} = \frac{r^2 \times 8.6 \times 10^{-5}}{4 \times 7.7 \times 10^3 \times 1} = 2.74 \times 10^{-11} r^2$$

r, ft	u	W(u)	s, ft
1	2.74×10^{-11}	23.74	46.48
10	2.74×10^{-9}	19.13	38.26
100	2.74×10^{-7}	14.53	29.06
1000	2.74×10^{-5}	9.92	19.84

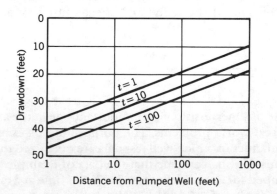

The graph may be referred to as a mathematical model for the determination of the response to pumping in this region of constant transmissivity and storativity. We note that the drawdown some 1000 ft from the pumping well at the end of 100 days of pumping will be on the order of 20 ft, whereas at 100 ft it will be 29 ft. On the other hand, if we are interested in the water-level response at one point 1000 ft distant from a pumping well and 100 ft distant from another pumping well, the principle of superposition states that we require the sum $20 + 29 = 49$ ft. These calculations are only valid for the "design" pumping rate of 192×10^3 ft³/day. However, we note that the pumping rate Q is not one of the variables making up the variable u and only occurs in the drawdown statement $s = (Q/4\pi T)W(u)$. Hence, the magnitude of drawdown is proportional to the pumping rate, so if we double the rate in the examples, we double the drawdown, and if we pump half as much from each well, we halve the drawdown.

With these points in mind, consider the following example. Two wells are pumping for 10 days at rates of 384×10^3 and 96×10^3 ft³/day. Five days after these wells start pumping, a third well turns on at a rate of 144 ft³/day. The pumping wells are at a distance of 1000, 100, and 10 ft, respectively, from an observation well. At the end of this 10-day period, the drawdown at the point of observation is determined from the preceding figure for the postulated pumping rates

$$s_1 = 30 \qquad s_2 = 12 \qquad s_3 = 23$$

giving a total drawdown of 65 ft.

Suppose now that the observation well is itself a pumping well pumping at a rate of 192×10^3 ft³/day for this same 10-day period. If the diameter of the well is about 2 ft, its self-induced drawdown, ignoring well loss, is about 43 ft. Hence, the total drawdown at this point is on the order of 108 ft.

Bounded Aquifers

An assumption employed throughout this chapter is that the aquifer is infinite in areal extent. Geologic boundaries limit the extent of real aquifers and serve to distort the calculated cones of depression forming around pumping wells. The method of images, which plays an important role in the mathematical theory of electricity and is employed in the solution of some geophysical problems, aids in the evaluation of the influence of aquifer boundaries on well flow. This theory, as described by Ferris (1959), permits treatment of the aquifer limited in one or more directions. However, the additional assumption of straight-line boundaries has been added. This gives aquifers of rather simple geometric form.

The image well theory can be explained as follows (Stallman, 1952): formation A is bounded by the relatively impermeable formation B, the boundary between the two located at a variable distance r from a pumping well (Figure 5.23a). As formation B is relatively impermeable, no flow can occur from it toward the pumping well, and the boundary is a no-flow (barrier) boundary. The effect of a barrier boundary is to increase the drawdown in a well. The problem now is to duplicate this physical situation by substituting a hydraulic entity that serves this purpose. As no flow occurs across a ground water divide, the barrier boundary is simulated by the supposition that formation A is infinite in areal extent and that an imaginary well is located across the real boundary, on a line at right angles thereto, and at the same distance from the boundary as a real pumping well. If the imaginary well is assumed to start pumping at the same time and at the same rate as the real well, the boundary will be transformed into a ground water divide.

Similarly, if the aquifer is bounded by a stream that provides recharge to the aquifer, the effect is to decrease the drawdown in a well. A zero-drawdown (constant-head) boundary can be simulated by an imaginary well, located as

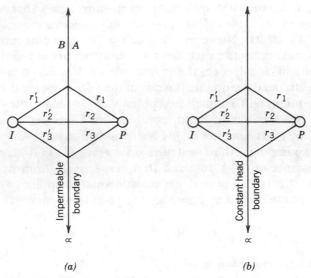

(a)   *(b)*

Figure 5.23
Simple two-well system for an aquifer bounded by (*a*) a
no-flow boundary and (*b*) a constant head boundary
(from Stallman, 1952).

earlier, with the exception that the imaginary well must recharge water at the
same rate as the pumping well (Figure 5.23*b*).

The drawdown at any point in the real aquifer or at the boundary for the simple two-well system is the sum of the effects of the real and imaginary wells operating simultaneously. From the principle of superposition

$$s = s_p \pm s_i = \frac{Q}{4\pi T}[W(u)_p \pm W(u)_i] = \frac{Q}{4\pi T}\Sigma\, W(u) \qquad (5.48)$$

where s is the observed drawdown at any point, consisting of the sum of the
effects of the real well s_p and the imaginary well s_i, the well function for the real
well is $W(u)_p$, and $W(u)_i$ is the well function for the imaginary well. If the imaginary well is a recharging well, the negative sign is used.

By similar reasoning,

$$u_p = \frac{r_p^2 S}{4Tt} \qquad u_i = \frac{r_i^2 S}{4Tt} \qquad (5.49)$$

where r_p is the distance from the pumping well to the observation point and r_i is
the distance from the imaginary well to the observation point. From the equality
expressed in Eq. 5.49

$$u_i = \left(\frac{r_i}{r_p}\right)^2 u_p \qquad (5.50)$$

or

$$u_i = \bar{K}^2 u_p \qquad (5.51)$$

where $\bar{K}$ equals the constant r_i/r_p. For any point on the boundary, r_p equals r_i ($\bar{K}$ equals 1), and u_i equals u_p. It follows that $W(u)_i$ equals $W(u)_p$ (Eq. 5.48), and the drawdown is either zero (constant-head boundary) or twice the effect of one well pumping in an infinite aquifer.

When a well in a bounded aquifer is pumped, water levels in observation wells will initially decline under the influence of the pumping well only. When the cone of depression reaches an exterior boundary, deviations from the ideal response will be noted in the observation well. Under these conditions, the early part of the time-drawdown curve unaffected by the boundary behaves as if the aquifer were infinite in areal extent, and this part of the curve can be used to determine the hydraulic properties.

Figure 5.24 shows the effects of pumping near a barrier and recharge boundary and the role of image wells in simulating this pumpage.

Aquifers are often bounded by two or more boundaries, and time-drawdown data will respond accordingly. As mentioned, the resulting geometry must be rather simple in order to apply the image theory. Hence, two converging boundaries delineate a wedge-shaped aquifer, two parallel boundaries intersected at right angles by a third boundary forms a semiinfinite strip, and four intersecting right-angle boundaries form a rectangular aquifer.

A number of image wells associated with each pumping well characterizes a multiple-boundary system. Clearly, primary image wells placed across a boundary will balance the effect of a pumping well at that boundary, but will cause an unbalanced effect at the opposite boundary in violation of the no-flow or constant-head requirement. It is then necessary to add secondary image wells at appropriate distances to satisfy the conditions of no flow or zero drawdown. Figure 5.25 shows some plan views of image well systems for some simple aquifers.

Once the boundaries of a finite aquifer have been simulated by means of image wells, analysis of drawdown effects can proceed as if the aquifer were infinite in areal extent. In other words, graphic models such as given in Example 5.8 can be used to find the drawdown in response to one or several wells pumping simultaneously. Once the distance from real and imaginary wells to the point of interest has been determined, drawdown (or recovery) may be obtained from the graphs. For example, the drawdown at any point in the real aquifer of Figure 5.25a is the algebraic sum of the effect of one real well, one discharging imaginary well, and two recharging imaginary wells. For the infinite array (Figure 5.25c), image wells are added until the most remote pair has negligible influence on water-level response.

Example 5.9

A 24-inch-diameter well is located in an aquifer with a transmissivity of 1340 ft^2/day and a storativity of 5×10^{-2}. A fault (barrier) is located 1000 ft from the well. The drawdown at the midpoint between the fault and the barrier at the end of one year's pumpage at a rate of 96×10^3 ft^3/day is determined

$$s = s_p + s_i = \frac{Q}{4\pi T} [W(u)_p + W(u)_i]$$

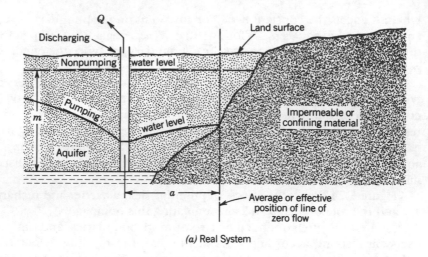

(a) Real System

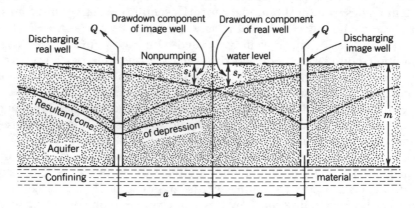

Note: Aquifer thickness m should be very large compared
to resultant drawdown near real well.

(b) Hydraulic counterpart of Real System

Figure 5.24
Diagrams of the effect of pumping near barrier and constant-head boundaries and appropriate hydraulic counterparts for image well theory (from
Ferris and others, 1962). Note: Aquifer thickness m should be very large
compared to resultant drawdown near real well.

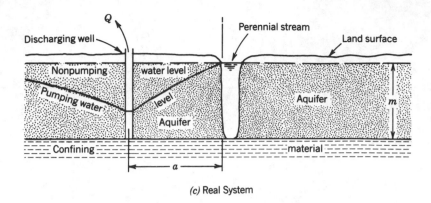

(c) Real System

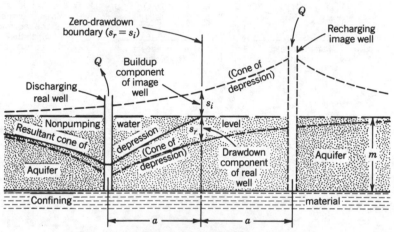

Note: Aquifer thickness m should be very large compared to resultant drawdown near real well.

(d) Hydraulic Counterpart of Real System

Figure 5.24
continued

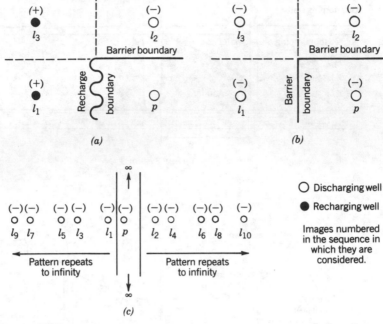

Figure 5.25
Image well systems for (*a*) wedge-shaped aquifer with both barrier and recharge boundaries, (*b*) wedge-shaped aquifer with barrier boundaries, and (*c*) infinite strip.

where

$$(u)_p = \frac{r_p^2 S}{4Tt} = \frac{(500)^2 \text{ft}^2 \, 5 \times 10^{-2}}{(4)(1340)\text{ft}^2/\text{day}(365)\text{day}} = 6.38 \times 10^{-3}$$

$$(u)_i = \frac{r_i^2 S}{4Tt} = \frac{(1500)^2 \text{ft}^2 \, 5 \times 10^{-2}}{(4)(1340)\text{ft}^2/\text{day}(365)\text{day}} = 5.75 \times 10^{-2}$$

From Table 5.1, $W(u)_p = 4.5$, $W(u)_i = 2.33$

$$s = \frac{96 \times 10^3 \text{ft}^3/\text{day}}{4 \times 3.14 \times 1.34 \times 10^3 \text{ft}^2/\text{day}} (6.83) = 39 \text{ ft} = 11.89 \text{ m}$$

Example 5.10

The accompanying figure depicts a time-drawdown response in an observation well affected by one pumping well and a barrier boundary. By construction, $s_1 = s_2$. It thus follows that $W(u)_1 = W(u)_2$ and

$$u_1 = \frac{r_1^2 S}{4Tt_1} \quad \text{and} \quad u_2 = \frac{r_2^2 S}{4Tt_2}$$

where $u_1 = u_2$. Thus

$$r_2 = r_1 \left(\frac{t_2}{t_1} \right)^{1/2}$$

where r_1 is the distance from the pumping well to the observation well, which is known, and r_2 is the distance from the observation well to the imaginary pumping well causing the inflection, which is readily calculated. However, because the exact location of the barrier is not known, r_2 merely defines a set of points that form a circle of radius r_2 with the observation well located at its center.

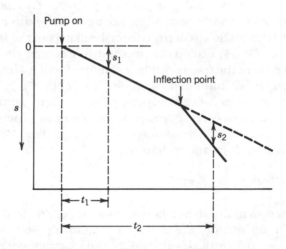

5.5 *Hydraulic Testing in Fractured or Low-Permeability Rocks*

Advances in hydraulic testing over the past decade have been born out of the practical necessity of evaluating rocks as hosts for chemical or nuclear wastes. In some cases, this requires testing in deep boreholes up to a few thousand feet in depth, often in rocks that are of a low permeability, and frequently in rock that is fractured. In many cases, the emphasis has been on the single-borehole test. These tests generally require special equipment such as packers to isolate testing zones, feed in apparatus whereby a slug of water may be removed from the isolated zone or the zone quickly pressurized, and pressure transducers to measure the response. With deep boreholes, temperature or salinity effects on fluid density must sometimes be accounted for. Hydraulic testing in this modern era is thus a field within itself, generally requiring a large degree of specialization. In this section, we will consider only a few of the methods in detail.

Single-Borehole Tests

The slug test discussed earlier is well suited for low-permeability testing. The pressure recovery or decline in a borehole can be measured after the removal or injection of a volume of water. The pulse test is a modification of the slug test whereby a testing interval within a single borehole is instantaneously under- or overpressured by removing or adding water. The time response of the pressure buildup or decay is then observed, with the analysis being similar to that used in the slug test. Bredehoeft and Papadopulos (1980) outline one analytical procedure using type curves of the form given in Figure 5.18. Neuzil (1982) presents a modified procedure for the testing arrangement involved with this type of test. "Skin" effects or problems due to well damage are addressed by Faust and Mercer (1984).

Wang and others (1978) extended the methods developed by Cooper and others (1967) to pulse testing in a single fracture. The method is based on the parallel plate model and assumes that the aperture width is independent of fluid pressure and that no flow is derived from the matrix of the fractured medium. Barker and Black (1983) extend this method to examine the effect of flow in the matrix as well as in the fractures. Their aquifer model is identical to that employed by de Swaan (1976) and Boulton and Streltsova (1977).

The determination of vertical permeability of low-permeability rocks in a single borehole has also been addressed in a number of papers (Burns, 1969; Pratts, 1970; Hirasaka, 1974). The monogram of Earlougher (1977) also discusses the available methodology and technology.

Multiple-Borehole Tests

Of the numerous multiple-borehole testing procedures available, only two will be discussed in this section: the method of Hantush (1960, 1964), which provides information on the vertical permeability and storativity of confining layers, and a modification of this method by Neuman and Witherspoon (1969a, 1969b, 1972).

The method of Hantush (1960, 1964) requires a pumping test procedure identical to the conventional pumping well–observation well setup discussed previously. The parameters obtained from the test include the storativity and transmissivity of the aquifer and the storativity and vertical permeability of the confining layer. The solution is an asymptotic one where the water-level response in the observation well is analyzed over a short time period and over a long time period. A curve-fitting procedure is involved for each of the designated time periods. For the case corresponding to Figure 5.1a, the short-term drawdown can be expressed as

$$s = \left(\frac{Q}{4\pi T} \right) H(u, \beta) \tag{5.52}$$

where u is defined as in all previous cases as $r^2 S / 4Tt$ and β is defined as

$$\beta = \left(\frac{r}{4B} \right) \left(\frac{S'}{S} \right)^{1/2} \tag{5.53}$$

where S' is the storativity of the confining layer and $1/B$ has already been defined as $(k'/m'T)^{1/2}$. By short times we mean times of observation less than $m'S'/10k'$. The quantity $H(u, \beta)$ is an infinite integral whose value is approximated as

$$H(u, \beta) = W(u) - \frac{4\beta}{(\pi u)^{1/2}} \left[0.2577 + 0.6931 \exp\left(\frac{-u}{2}\right) \right] \qquad (5.54)$$

where $W(u)$ is the familiar well function of u. Some typical curves for the well function $H(u, \beta)$ are shown in Figure 5.26. Hantush (1960) has tabulated relevant values for the function $H(u, \beta)$ that may be adequate in applications. Note from Eq. 5.54 that this function approaches $W(u)$ as β approaches zero.

The long-time unsteady drawdown is expressed as

$$s = \left(\frac{Q}{4\pi T}\right) W\left(u\delta, \frac{r}{B}\right) \qquad (5.55)$$

where

$$\delta = 1 + \left(\frac{S'}{3S}\right)$$

Note that as S' approaches zero, we recover the Hantush-Jacob leaky formula.
The procedure for analysis is as follows.

1. Early time-drawdown data are fitted to the family of curves shown on Figure 5.26, with a match point producing values of $H(u, \beta)$, u, s, and t. The value for β is also obtained as β is a parameter of the type curve in the same manner that r/B is a parameter of the leaky curves. The transmissivity is determined from Eq. 5.52 and the storativity S is determined from the relationship $u = r^2 S/4Tt$.

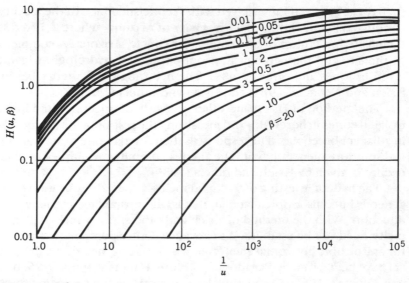

Figure 5.26
Type curves for a well in an aquifer confined by a leaky layer that releases water from storage (with permission from William Walton).

2. The usual procedures for leaky aquifers are then followed for the long-term data with the type curve of Figure 5.10. The match point will produce values of $W(u\delta, r/B)$, δu, s, and t. The value of r/B is again noted. This will produce values of transmissivity with Eq. 5.55 and the vertical hydraulic conductivity from the determined B value. As both β and B are now known, Eq. 5.53 may be solved for the storativity of the confining layer or, conversely, from known values of δS and S, the storativity of the confining layer may be determined. Details of analysis for other geometric configurations of leakage are given by Domenico (1972).

Interestingly, the solutions just described recover the leaky aquifer formula as the storativity of the confining layer approaches zero, as well as the Theis solution where both the storativity and permeability of the confining layer approach zero. Unfortunately, because of the similarity in the shape of the family of curves of Figure 5.26 used in the analysis, it is difficult if not impossible to achieve a definitive match. Figure 5.2 shows the expected time-drawdown response where leakage from storage is involved and compares this response with the conventional leaky and nonleaky drawdown curves.

The asymptotic short-term–long-term solution of Hantush (1960) has been solved over continuous time by Neuman and Witherspoon (1969a, 1969b). These results were organized into a field testing procedure whereby the hydraulic properties of the aquifer and bounding low-permeability material may be determined (Neuman and Witherspoon, 1972). The test, referred to as the "ratio test," requires an observation well in the confining layer. Testing the hydraulic response of a confining layer is not a new development and was part of the requirements of earlier work dealing with low-permeability response (Wolff, 1970; Wolff and Papadopulos, 1972). However, the so-called ratio method represents a sophisticated approach to testing low-permeability rock where low-permeability zones bound higher-permeability aquifers that can be pumped over an extended period of time.

Many fractured rocks qualify as low-permeability media, and the testing of such rocks has been long viewed with skepticism. Much of the recent research was conducted by the staff of the University of Arizona, where a field test facility was established in Precambrian granite near Oracle, Arizona. A complete description of the site and the geological, geophysical, and hydrological investigations are given in Jones and others (1985). The hydrologic test theory has been given by Hsieh and Neuman (1985) and Hsieh and others (1985).

The method of Hsieh and others consists of injecting fluid into a number of isolated zones in boreholes and measuring the response in the same isolated zones in adjacent boreholes. The response is then fitted to one of several possible analytical solutions, generally with the aid of a computer. A simple curve-matching procedure is given by Hsieh and others (1985).

The hydraulic testing of fractured rock is a special application of flow in homogeneous but anisotropic material; that is, the hydraulic conductivity changes with direction. With the method of Hsieh and others, the principal planes of anisotropicity need not be established in advance. Other pumping test methods designed for anisotropic horizontal aquifers in which one of the principal directions is vertical have been given by Papadopulos (1965), Hantush (1966a, 1966b), and Hantush and Thomas (1966). These models correspond to the geologic prototype given as Figure 5.1c. Much of the recent research refers to the contribution of Papadopulos (1965) as one of the first original contributions in this field, including Way and

McKee (1982), Loo and others (1984), and Hsieh and others (1985). A method of determining vertical conductivities has been given by Weeks (1969) and Way and McKee (1982).

5.6   *Some Applications to Hydraulic Problems*

There are various solutions or partial solutions to hydrologic problems that require either the results obtained from a hydrologic test or the utilization of the mathematical model upon which the formation response was predicated. For instance, any application of Darcy's law requires a priori knowledge of hydraulic conductivity. Hence, if we wish to make some calculations of flow rates or velocity of movement, the hydraulic conductivity must be known. If we wish to say something about travel time for a contaminant that moves at the speed of ground water, some determination of hydraulic conductivity is required. At yet another level, if we wish to apply some numerical model to ascertain the long-term response of a basin to prolonged pumping, information is required on the spatial distribution of hydraulic conductivity and storativity. Hence, the necessity of hydraulic testing.

On the other hand, there is also a need for the mathematical models employed in well testing. For example, a short-term pumping test (say, 24 hours) will yield the pertinent formation parameters T and S. As time and distance are variables in the mathematical model, it is possible to say something about what the drawdowns will be at later times and other distances. Due to the principle of superposition, it is possible to do the foregoing for a whole battery of wells pumping at different rates. Such predictions are invaluable in the design of dewatering operations, in water well interference considerations, in the design of well interceptors to contain contaminant migration, and for predicting pressure response and contaminant migration due to deep-well injection. As no new principles are involved in these applications, these points are best demonstrated with a few examples, most of which are based on actual case histories. That is to say, we already know the theory; now we need some rules for applying it.

Planning a Pumping Test

Several interdependent questions must be addressed when planning a pumping test with one or more observation wells:

1.   What pumping rate will be employed?
2.   What diameter well is required?
3.   What should be the duration of the test?
4.   At what distance should observation well(s) be placed?

Normally, if there are nearby wells in the formation of concern, some of these questions may be answered by merely observing the behavior of these wells, that is, specific capacity and well yield. In addition, for partially penetrating wells, the general rules regarding spacing and time of pumping already discussed should be taken into consideration. When very little is known about the formation to be tested, there will always be some uncertainty in the planning procedure. Much of the uncertainty can be removed through the following procedures.

Screen Diameter and Pumping Rates

The term "screen" is used to signify the diameter of the well that is open to the flow of water. It may be the perforated well casing or some manufactured steel cage apparatus. The larger the pumping rate we wish to employ, the larger the pumping apparatus required so that some minimal-diameter well should be associated with a given pumping rate. The following is a rule-of-thumb guideline. For a desired pumping rate of 125 gpm, the minimum well diameter should be 6 inches. The following pairs are recommended (300 gpm, 8 in.), (600 gpm, 10 in.), (1200 gpm, 12 in.), (2000 gpm, 14 in.), (3000 gpm, 16 in.), (5000 gpm, 18 in.).

It follows, of course, that utilizing an 8-inch-diameter well will not assure a well yield of 300 gpm; the actual yield will be controlled by the transmissivity. On the other hand, an 8-inch well will not produce 3000 gpm from a formation that is capable of delivering such amounts because of the limitations it places on the pumping equipment.

Well Yield: The Step-Drawdown Test

A step-drawdown test is one in which a well is operated during successive periods at constant fractions of its full capacity (Figure 5.27), and is generally recommended prior to running tests specifically designed to determine the formation parameters. Such tests are useful in determining the yield of a well and to establish the depth for the pump setting. The specific capacity may be ascertained for each of the steps, that is, from Figure 5.27, S.C. = 920/20 = 46 gpm/ft; S.C. = 1300/29 = 44.8 gpm/ft; S.C. = 1825/42.5 = 42.9 gpm/ft. The specific capacity thus decreases with increasing pumping rate due to increasing well loss at higher pumping rates.

Three useful types of information may be obtained from the test results given in Figure 5.27: pump setting requirements, the well loss constant C, and a preliminary estimate of transmissivity. Consider that the well in question penetrates 50 ft below the water table. The depth of penetration below the water table is called the maximum available drawdown. Thus, at a pumping rate of 920 gpm or 1300 gpm, only 20 ft and 29 ft of the maximum available drawdown will be utilized. With a pump setting at the bottom of the well, either pumping rate may be safely

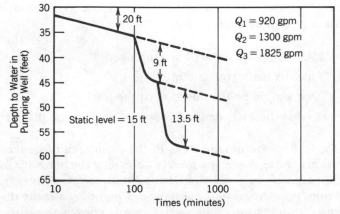

Figure 5.27
Semilogarithmic plot of time-drawdown curve obtained during a step-drawdown test.

employed in a long-term pumping test. At 1825 gpm, approximately 43 ft of the maximum available drawdown will be employed, that is, there remains only 7 ft of water in the well after a few hours of pumping. This is marginal so that this pumping rate is considered too high for long-term testing. We thus establish a reasonable upper bound for the pumping rate on the order of 1500 gpm.

The well loss constant C may be determined directly from equations presented by Jacob (1950), where

$$C = \frac{\Delta s_i / \Delta Q_i - \Delta s_{i-1} / \Delta Q_{i-1}}{\Delta Q_{i-1} + \Delta Q_i} \tag{5.57}$$

or, for steps 1 and 2 of Figure 4.27,

$$C = \frac{\Delta s_2 / \Delta Q_2 - \Delta s_1 / \Delta Q_1}{\Delta Q_1 + \Delta Q_2} = \frac{(9/380) - (20/920)}{920 + 380} = 1.5 \times 10^{-6} \text{ ft/(gpm)}^2$$

and for steps 2 and 3,

$$C = \frac{\Delta s_3 / \Delta Q_3 - \Delta s_2 / \Delta Q_2}{\Delta Q_2 + \Delta Q_3} = \frac{(13.5/525) - (9/380)}{380 + 525} = 2.2 \times 10^{-6} \text{ ft/(gpm)}^2$$

for an average of 1.85×10^{-6} (ft/gpm^2). Thus, the well loss associated with each of the pumping rates employed in the step-drawdown test is readily determined from the relationship $s_w = CQ^2$, where C has been determined to be 1.85×10^{-6} ft/(gpm)^2. For each of the pumping rates employed, the well loss is 1.56 ft, 3.3 ft, and 6.2 ft, respectively.

A preliminary estimate of the transmissivity may be obtained with the following reasoning. At a pumping rate of 1825 gpm, a total drawdown of 42.5 feet was registered, giving a measured specific capacity of 42.9 gpm/ft. It has been determined that of this total drawdown, 6.2 feet was well loss. Thus, the theoretical specific capacity for a 100% efficient well is actually 1825/36.3, or 50 gpm/ft. If the pumping well is 12 inches in diameter, the graph given as Figure 5.21 indicates that the transmissivity can vary from 150,000 gpd/ft for a storativity on the order of 1×10^{-5} to about 90,000 gpd/ft for a storativity of 0.1. This range is sufficient for further planning.

Distance to the Observation Well(s)

The last information required is the distance to the observation wells. At whatever distance the observation point is located, the drawdown can be described as $s = (Q/4\pi T)W(u)$, where $u = (r^2 s/4T)/t = T^*/t$ where r is the distance from the pumping well that concerns us, and T^* has been referred to as the time constant. For a pumping rate of 1500 gpm and the transmissivity range determined earlier, $s = 1.146\ W(u)$ for $T = 150,000$ gpd/ft and $S = 1 \times 10^{-5}$ and $s = 1.91\ W(u)$ for $T = 90,000$ gpd/ft and $S = 1 \times 10^{-1}$.

Let us now assume that we wish to run the pumping test for a period of at least 24 hours (1 day) and that we wish to register a drawdown of about 5 to 10 ft at the end of that pumping period. Thus, $W(u)$ must be on the order of 5 so that, from Table 5.1, u is on the order of 4×10^{-3}. Let us recognize further that u is dimensionless, so that if the distance r is desired in feet, the transmissivity must be expressed in square feet per day so that 150,000 gpd/ft/7.48 g/ft^3 = 20,053 ft^2/day and 90,000 gpd/ft/7.48 g/ft^3 = 12,032 ft^2/day. Thus, for $S = 1 \times 10^{-5}$,

$u = 4 \times 10^{-3} = r^2 \times 1 \times 10^{-5}/4 \times 20,053 \times 1$ so that the distance from the pumping well to the observation well is calculated to be 1.78×10^3 ft. On the other hand, for $S = 0.1$, $u = 4 \times 10^{-3} = r^2 \times 1 \times 10^{-1}/4 \times 1.2032 \times 10^4 \times 1$ so that the distance from the pumping well to the observation well is on the order of 44 ft. And herein lies the problem. In the absence of any information on the degree of confinement, a drawdown of 5 to 10 ft at the end of a 24-hour pumping period can occur at a distance of something on the order of 50 ft to something on the order of 1800 ft depending on the value of the storativity. Let us recognize that the storativities of 10^{-5} and 10^{-1} are extremes and can only occur in extremely rigid rocks or rocks that are completely unconfined. Thus, based on the geology, something has to be stated on the nature of confinement. For the situation under consideration, in the absence of geologic information, it is prudent to consider an observation well on the order of 100 ft from the pumping well so that some measurable drawdown is assured within a 24-hour pumping period. If the aquifer is known to be confined, a distance of several hundred feet may be used. If additional observation wells are desired, their distance from the pumping well may be ascertained after some information is obtained on the values of transmissivity and storativity.

Planning a Dewatering Operation

Figure 5.28 shows a situation where a tunnel must be constructed some 60 ft below a water table. To facilitate construction, the water table must be lowered below the tunnel during the construction period. Several questions must be addressed:

1. What pumping rates will be employed?
2. What is the depth and diameter of the pumping wells?
3. What is the duration of pumping?
4. How many wells will be required and how should they be spaced?

Many of these questions are interdependent, and some have to be answered on the basis of value judgment. The values for transmissivity, storativity, and specific capacity are noted in the diagram and were obtained by hydraulic testing.

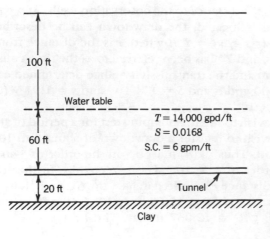

100 ft

Water table

$T = 14,000$ gpd/ft
$S = 0.0168$
S.C. $= 6$ gpm/ft

60 ft

20 ft   Tunnel

Clay

Figure 5.28
Schematic cross section of a proposed tunnel through an unconfined aquifer.

The first thing to note is that the maximum available drawdown for wells drilled to the top of the clay layer is on the order of 80 ft. Thus, with a measured specific capacity of 6 gpm/ft, the maximum possible pumping rate is calculated to be 480 gpm. This, however, is somewhat unrealistic in that such a pumping rate would readily remove all the water from the well. Restricting the drawdown in a single pumping well to 70 feet, the design pumping rate is calculated to be 420 gpm. From information provided in the previous example, this requires a design diameter of 8 inches for the dewatering wells. Thus, we have determined that the wells will be 180 ft deep, their diameter will be 8 inches, and their design pumping rate will be 420 gpm. All that remains is the determination of the pumping duration and well spacing.

Figure 5.29 shows some theoretical distance-drawdown curves for pumping durations of 10 and 100 days. Note that the difference in drawdown between the two curves is only about 8 ft. The longer-duration pumping period has the benefit of a greater spacing between the wells to obtain the desired drawdown. However, the shorter-duration period is the better one here for several reasons. First, in most dewatering operations, only a few hundred feet along the tunnel alignment are dewatered during construction. Once the construction activities are completed in a certain reach, the pumps are moved farther along the alignment. Thus, the whole alignment need not be dewatered in advance. In addition, a 100-day period would require the disposal of greater quantities of water. Thus, the 10-day design period is the better one.

Figure 5.30 shows a plan view of one feasible design. The 35-ft offset from the centerline of the tunnel was a contractual agreement. The water levels must be lowered some 65 ft all along the centerline. As a first guess, assume the wells in each row are 100 ft apart. The least drawdown will occur at points 1 and 3, and the most at point 2. Using superposition with the aid of Figure 5.29, the drawdown at points 1 and 3 is calculated to be on the order of 98 feet. Clearly, we do not have 98 ft of available drawdown, but the model calculation is not constrained and can produce this nonsensical result. For a 200-ft spacing, the drawdown is 79 ft at points 1 and 3 and 94 ft at point 2. For a 300-ft spacing the drawdown is

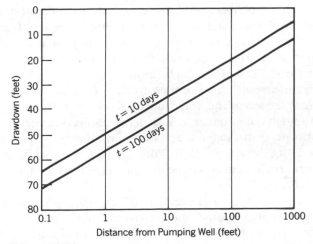

Figure 5.29
Time-drawdown data.

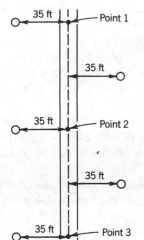

Figure 5.30
Plan view of well alignment for dewatering.

approaching 75 ft at points 1 and 3 and is slightly more at point 2. Thus it appears we are slowly moving toward an optimal spacing and this spacing is in excess of 300 ft. However, there is need for some caution here in that the calculated spacings have been based on a design pumping rate of 420 gpm, and drawdown is directly proportional to the pumping rate. With five interfering wells, this pumping rate cannot be maintained for the target 10-day period. For the 300-ft spacing, a decrease in the effective pumping rate from 420 gpm to 365 gpm will produce a 65-ft drawdown at points 1 and 3. For the 200-ft spacing, a decrease in the pumping rate to 345 gpm will likewise produce a drawdown of 65 ft at points 1 and 3. Thus the need for value judgments, with the 200-ft spacing being the more prudent one, at least for the preliminary design. This means that the design pumping rate of 420 gpm can fall off to as little as 345 gpm over the 10-day pumping period and the dewatering can still be accomplished.

A Problem in Water Supply

A power company plans to build a facility in the southwestern desert of the United States and requires a water supply for cooling purposes. The plant is to be constructed a few miles from a known ground water source, and the plan is to deliver the pumped water to the plant where it will be consumed in the cooling process (no return flows). The plant is to be built in three units, phased in every five years, with each unit requiring a continuous flow of water amounting to about 2100 gpm. A cross section of the water supply is shown in Figure 5.31, which is a rather narrow alluvial valley cut into low-permeability sediments. The company has access to property that extends across the entire valley (about 15,000 ft) and about 10,000 ft along the valley reach. Three wells are shown in the plan view of this property and their locations have been idealized to facilitate the required computations (Figure 5.32).

There are a few facts and a few questions of importance here. First, because of the geologic configuration of the alluvial valley, the problem should be treated as an infinite strip with image wells repeating to infinity (Figure 5.25c). Second, drawdowns in any given well should be limited to about 100 ft or the subsequent dewatering of the aquifer will destroy its transmissive properties. This means that

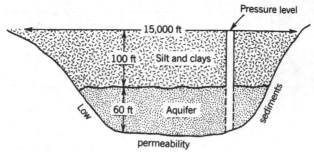

Figure 5.31
Cross section of alluvial aquifer.

the pumping should be distributed as equally as possible. With these constraints, can the power company obtain sufficient water for each of the units, each of which is phased in every 5 years? If not, can they supply one unit for the life of the plant (50 years)? If not, how long can they supply one unit without excessive dewatering of the aquifer? The transmissivity has been determined to be 40,000 gpd/ft, the storativity equals 10^{-3}, the well diameters are 14 inches, and the actual specific capacity of the wells is about 15 gpm/ft.

It is not intended to treat this entire complex problem, but merely to show some elements of the use of the pertinent hydraulic equations in addressing quantitative questions. Figure 5.33 shows the familiar distance drawdown plot for an infinite aquifer with a design pumping rate of 700 gpm. Let us focus on the centermost well. With a specific capacity of 15 gpm/ft and a pumping rate of 700 gpm, the drawdown is anticipated to be about 47 ft. This will certainly increase with

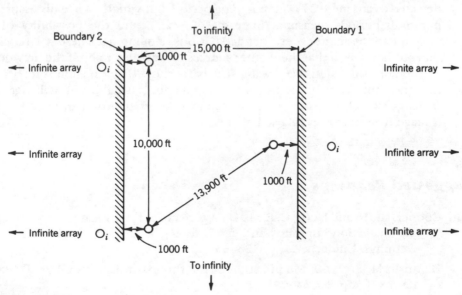

Figure 5.32
Plan view of well location relative to boundaries.

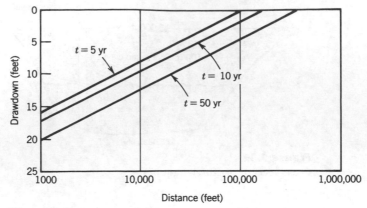

Figure 5.33
Time-drawdown response of an infinite aquifer.

time, but let us not dwell on this detail. The influence from the two neighboring real wells at the end of a five-year period will be on the order of 14 ft, giving a total drawdown of 61 ft. Consider now only four image wells associated with the boundaries affecting this well. These will be located 1000 ft and 29,000 feet from boundary 1 and 14,000 ft and 16,000 feet from boundary 2. These images are thus located the following distances from the pumped well: 2000 ft, 30,000 ft, 28,000 ft, and 30,000 ft. Their cumulative effect on the pumping well is about 52 ft, giving a total drawdown of 113 ft. Yet to be considered are the other images associated with this well, which have a measurable influence from as far away as 90,000 ft (Figure 5.33), plus the images associated with the other two pumping wells. Based on this preliminary analysis, an attempt to develop sufficient water for one unit over a five-year period will cause substantial dewatering of the aquifer. This conclusion must be tempered by other considerations, for example, the possibility of leakage from the overlying clays and silts (which was not detected over a mere 24-hour pumping period, but which is virtually assured over pumping periods measured in terms of years), and the possibility of some recharge to the aquifer over extended periods of pumping. These will require further studies. Given that these other sources of water act to limit the drawdown, it can still be concluded that a water supply for more than one unit is totally out of the question. The life of the water supply for a single unit will depend on recharge, which may not be substantial in a desert environment, and the rate of leakage from the overlying sediments.

Suggested Readings

Cooper, H. H., and Jacob, C. E., 1946, A generalized graphical method for evaluating formation constants and summarizing well field history: Trans. Amer. Geophys. Union, v. 27, p. 526–534.

Hantush, M. S., 1960, Modification of the theory of leaky aquifers: J. Geophys. Res., v. 65, p. 3713–3725.

Hantush, M. S., 1964, Hydraulics of wells. In Advances in Hydroscience, ed. V. T. Chow: Academic Press, NY, p. 281–432.

Hvorslev, M. J., 1951, Time lag and soil permeability in groundwater observations: U.S. Army Corps of Eng. Waterway Exp. Stat. Bull. 36, Vicksburg, MS.

Theis, C.V., 1935, The relation between the lowering of the piezometric surface and the rate and duration of discharge of a well using ground water storage: Trans. Amer. Geophys. Union, v. 2, p. 519–524.

Problems

1. After 24 hours of pumping a confined aquifer, the drawdown in an observation well at a distance of 320 ft is 1.8 ft, and the drawdown in an observation well at a distance of 110 ft is 3.5 ft. The pumping rate is 192×10^3 ft³/day. Find the transmissivity.

2. Time-drawdown data collected at a distance of 100 ft from a well pumping at a rate of 192×10^3 ft³/day are as follows.

t, min	s, ft
1	3.8
2	5.2
3	6.2
4	7.0
5	7.6
6	8.3
7	8.8
10	10.0
20	12.2
40	14.0
80	15.8
100	16.4
300	19.0
500	20.2
1000	21.6

Calculate the transmissivity and storativity.

3. A well 250 ft deep is planned in an aquifer with a transmissivity of 1340 ft²/day and a storativity of 1×10^{-2}. The well is expected to yield 960×10^2 ft³/day and will be 12 inches in diameter. If the nonpumping water level is 50 ft below land surface, estimate the depth to water after one year's operation and after three years' operation.

4. An 18-inch-diameter well within an aquifer with a transmissivity of 1070 ft²/day and a storativity of 7×10^{-2} is to be pumped continuously. What pumping rate should be used so that the maximum drawdown after two years will not exceed 20 ft?

5. You are asked to design a pumping test for a confined aquifer in which the transmissivity is estimated to be about 60,000 gpd/ft and the storativity about 1×10^{-4}. What pumping rate would you recommend for the test if it is desired that there be a drawdown of about 6 ft in the first five hours of the test at an observation well 100 ft from the pumping well?

6. Replace the fault in Example 5.9 with a fully penetrating stream, and calculate the drawdown at the stream and midpoint between the well and the stream for the same pumping rate and duration of pumping.

7. An aquifer has a transmissivity of 1000 ft²/day and a storativity of 1×10^{-4}. A 24-inch pumping well has a drawdown of 235 ft at the end of one day's pumping at a rate of 100×10^3 ft³/day. The efficiency of the well is determined as 58 percent. What is the efficiency when the well is pumped at 200×10^3 ft³/day for a one-day pumping period?

8. *a.* List three reasons that might explain an upward inflection of a semilogarithmic time-drawdown plot.

   b. List three reasons that might explain a downward inflection of a semilogarithmic time-drawdown plot.

9. Suppose the only type curve available to you was a plot of $W(u)$ versus u. How would you plot time-drawdown data obtained from the field in order to use this curve in the matching procedure?

10. A water well is located 1000 ft from a waste interceptor trench that acts to intercept a plume emitting from a uranium tailings pile. The well is to be pumped for 100 consecutive days at a rate of 192×10^3 ft³/day. The transmissivity of the formation is 5347 ft²/day and the storativity is 0.1. At the end of the 100-day pumping period, will the cone of depression reach the interceptor trench and what will be the drawdown?

11. Use the figure in Example 5.8 to answer the following:

   a. A 24-inch-diameter well is located about 10 ft from a fault (barrier) and is pumped at a rate of 192×10^3 ft³/day for a 10-day period. What is the approximate drawdown in the well at the end of the 10-day period (ignore well losses).

   b. What sort of shift would you expect in the curves given in Example 5.8 if the storativity was much greater, say, 1×10^{-1}?

12. A 24-inch-diameter well in an aquifer has a *measured* specific capacity Q/s of 5 gpm/ft. Assume a storativity of 1×10^{-4} and a 100% efficient well.

   a. Estimate the transmissivity of the aquifer.

   b. Later measurements show that the well is only 50% efficient. Is the original estimate of transmissivity in (a) too high or too low?

   c. What is a new reasonable estimate of the theoretical specific capacity (i.e., the specific capacity for a 100% efficient well)?

   d. What is a new reasonable estimate of the transmissivity?

13. A disposal well for liquid waste injection commences operation in a horizontal confined aquifer that has the following characteristics: thickness = 30 ft, porosity = 10%, hydraulic conductivity = 266 ft/day, specific storage = 10^{-6} ft^{-1}, injection rate 200×10^3 ft³/day. The disposal well has a radius of 1 ft.

   a. To what distance from the well will the front of the cone of impression have extended after 10 days' injection. (Assume, for all practical purposes, that 0.1 ft of head above the regional preinjection piezometric head marks the front of the cone of impression.)

b. Approximately how far will the injected contaminant move by the end of the 10-day injection period. Assume that the contaminant moves at the same speed as the ground water and that the induced hydraulic gradient due to the injection is imposed instantaneously and is linear.

14. Consider the enclosed figure to represent the pumping history of a single well over a six-year period. This pumping caused some draw-down in a nearby observation well as shown in the companion diagram. Explain very carefully how you would use superposition in this problem.

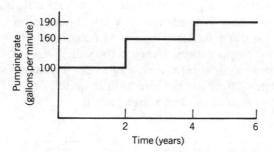 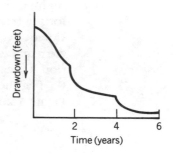

Start out with "I would pump the pumping well at a rate of _____ for _____ years to obtain a response, then I would pump it at a rate of _____ for _____ years, and (add) (subtract) the results to or from the previous response, then I would pump the pumping well at a rate of _____ for _____ years, and (add) (subtract) the results to or from the previous sum."

15. The accompanying diagram gives some pumping test results as measured in an observation well. You are given the following:

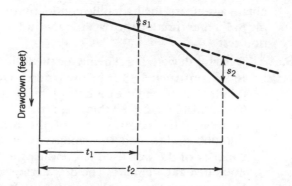

transmissivity $= 1 \times 10^4$ ft^2/day

storativity $= 1 \times 10^{-4}$

the distance from the observation well to the pumping well (r_1) is 1000 ft

the time (t_1) at which s_1 is measured is 0.1 day

the time (t_2) at which the drawdown s_2 is measured is 10 days

the drawdown s_2 is two times the drawdown s_1

Find the distance from this observation well to the image well responsible for this behavior.

16. An injection test in a single well produced the curve-fitting parameters for the transmissivity T, storativity S, vertical permeability k_v, and the storativity S' for the geometry shown. As no other facts are given, you are required to compute the recovery in the well in response to the injection. Calculate the short-term recovery in the injection well of radius 0.5 ft at the end of 0.22 days of injection at a rate of 1000 ft^3/day. (Hint: Figure 5.26 is required to obtain the result.)

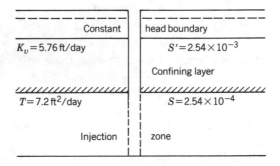

17. During a field investigation of a leaky underground storage tank for oil, a group of small-diameter observation wells is emplaced. In the interest of interception of some of the oil product floating on the water table, the perforations are placed both above and below the water table. Later, it is decided that a value of the hydraulic conductivity of the saturated zone is necessary for further calculations. Explain how the Hvorslev test for a withdrawal procedure would work. Use the data of Figure 5.20 to obtain a value for the hydraulic conductivity as has been done in Example 5.5, reinterpreting the data for a withdrawal as opposed to an injection test.

18. a. Design a dewatering scheme for the following situation. A tunnel is to be constructed 55 ft below land surface. The water table is at a depth of 40 ft, and the tunnel will reside 15 ft below the water table. The available saturated thickness from the water table to a clay layer is 40 ft. The transmissivity of the material to be dewatered is 300,000 gpd/ft, and the storativity is 0.01. The actual specific capacity of the well used to conduct the hydraulic testing has been measured at 64 gpm/ft. The dewatering wells will be finished 20 ft below the tunnel, and they must be offset 15 ft from the centerline of the tunnel. The wells will be placed alternately on each side of the

tunnel (see Figure 5.30). Consider four wells pumping simultaneously over a 10-day period at a pumping rate of 1200 gpm. Determine the well diameter and well spacing required to lower the water table by 20 ft along the tunnel alignment.

b. Repeat the exercise for paired wells, that is, four wells set in groups of two on the same line perpendicular to the tunnel alignment. Comment on the "best" design in terms of well spacing.

c. Your assignment is to type a report not to exceed one page on your selected design along with any contingencies as discussed in the example associated with Figures 5.28 through 5.30. Include two figures, one that shows your mathematical model such as depicted in Figure 5.29 and one that shows the optimal design (Figure 5.30).

19. Extend the boundaries of Figure 5.32 an additional 17,500 ft on both sides and analyze this problem for the centermost well and the pumping demand of 2100 gpm. Consider the effect of three real wells pumping and four image wells associated with each of the pumping wells. Include a sketch of your real and image well location and comment on the possibility of obtaining 2100 gpm for one unit for a 5-, 10-, and 50-year period. Type a report not to exceed one page on your analysis.

Chapter Six

Ground Water as a Resource

6.1 **Development of Ground Water Resources**

6.2 **Simulation of Aquifer Response to Pumping**

6.3 **Land Subsidence and Sea Water Intrusion**

6.4 **Simulation-Optimization Concepts**

If the history of the utilization of ground water in North America and Europe can be taken as typical, there are generally four stages in the ordered utilization of ground water for water supply: exploration, development, inventory, and management. Exploration requires geologic mapping on a regional or national scale and the delineation of formations or parts of formations that may serve as aquifers. The comprehensiveness of the data on such formations as presented by Meinzer (1923), Tolman (1937), Thomas (1951), and McGuinness (1963) suggests that the main exploration stage for ground water in North America is in the past.

A resource represents a supply of something that can be drawn upon for use. Like petroleum, water is a flow resource and thus can be readily transported to achieve a better balance between the location of the supply and the demand for its use. Unlike petroleum, ground water is not a minority fluid in the subsurface environment, and its value is not normally determined by the marketplace. However, ground water can also be nonrenewable—at least when viewed within a human time frame—and its exploitation is subject to supply and demand. In areas rather well endowed with abundant surface water, ground water is frequently an underexploited resource. Conversely, in areas void of surface water supplies, ground water is almost always overexploited. In areas intermediate between these extremes, the development of the total water resource will depend on the demand for water. The extent to which the total water resource system is developed and the manner in which it is operated depends not only on availability of supply but on legal, political, and socioeconomic precedents and constraints.

6.1 Development of Ground Water Resources

The Response of Aquifers to Pumping

In Chapter 5 we learned something about hydraulic testing, or measuring the response to pumping over short time frames in order to determine certain hydraulic parameters. A major assumption was that all the water pumped was removed from storage during the testing period. This assumption has its roots in the pioneering work of C. V. Theis, first in 1935 when he established his now-famous nonequilibrium equation, and later in 1938 and 1940 when he used identical hydraulic principles to determine the manner in which an extensive ground water basin responds to pumping when the area of pumping is far removed from the area of natural recharge and discharge.

Prior to the initiation of pumping, ground water recharge is just balanced by the natural discharge of water. Discharge by wells is an additional stress put on the ground water body that must be balanced by (1) an increase in recharge, (2) a decrease in natural discharge, (3) a loss of storage in the ground water body, or (4) a combination of these factors. Theis referred to this as a "state of dynamic equilibrium" whereby a new balance between inputs and outputs might be achieved at a new but lower storage level. As this may not happen within the short term, the basin can remain in a long-term transient state, with water levels falling in response to the pumping. Thus, in Theis's words, ground water may be classified as a renewable resource, but there are instances where it may not be so within a human time frame.

The essential features that determine the response of a basin to pumpage are

1. The distance to and character of the recharge.
2. The distance to the area of natural discharge.
3. The manner in which the ground water basin responds to pumping.

With reference to one, if recharge was rejected prior to pumping, there is a possibility of balancing well discharge by increased recharge. This is shown diagrammatically in Figure 6.1, where pumping in the recharge area will allow more water to infiltrate the underground portions of the basin. The well shown in the figure will also affect the nearby area of natural discharge after some water is taken from storage. To do so, the water table must be depressed in that area. Pumping in areas of rejected recharge, or natural discharge, permits the basin to act as a pirating agent to procure water from other components of the hydrologic cycle.

The manner in which a ground water basin responds to pumping was already discussed in hydraulic testing where

$$u = N_{FO}^{-1} = \frac{(r^2 S/4T)}{t} = \frac{T^*}{t} \tag{6.1}$$

where T^* is the time constant for a basin. If the time of pumping t is significantly larger than the time constant (that is, u is small, on the order of 0.001, Figure 5.5*a*), drawdowns are imperceptibly changing with time and the water levels appear to be approaching a steady state. The time constant T^* gets small with decreasing values of storativity and increasing values of transmissivity as is characteristic of confined

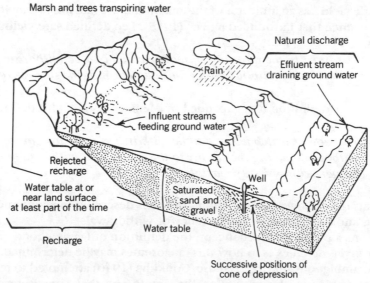

Figure 6.1
Factors controlling the response of a ground water basin to discharge by wells (from Theis, 1940). Reprinted with permission of The American Society of Civil Engineers.

aquifers. For the confined condition the cone of depression grows approximately 100 times faster than for the unconfined condition. Thus, confined aquifers excluding the very extensive ones, such as the Dakota sandstone, may readily approach a new steady state by either increasing recharge or decreasing discharge. Additionally we note that induced leakage across confining layers can also provide additional water to the aquifer, and this, too, tends to drive the system towards a new steady state. Thus, for $r/B = r(K'm'/T)^{1/2}$ on the order 2 (Eq. 5.28), a new steady state is approached for T^*/t as large as 0.5 (Figure 5.10).

Based on the reasoning cited, there is a need for adjusting the rate of exploitation and its spatial distribution in ground water basins (Theis, 1940):

1. The pumps should be placed as close as possible to areas of rejected recharge or natural discharge where water either cannot enter the ground or is being lost by evapotranspiration.

2. The pumps should be placed as uniformly as possible in areas remote from areas of rejected recharge or natural discharge.

3. The amount of pumping in any one locality should be limited.

Clearly, the objectives cited and the manner in which they may be achieved are not in any apparent conflict with sound hydrologic principles. However, conflicting factors arise when the reduction in natural discharge affects prior rights established on surface water resources, or on spring flow.

Yield Analysis

The objective of many ground water resource investigations is to determine how much water is available for pumping. Frequently, this is interpreted as the maximum possible pumping compatible with the stability of the supply. The term "safe yield" as an indicator of this maximum use rate has had an interesting evolution since first introduced by Lee (1915). Lee defined safe yield as

The limit to the quantity of water which can be withdrawn regularly and permanently without dangerous depletion of the storage reserve.

This definition was expanded by Meinzer (1923) who defined safe yield as

The rate at which water can be withdrawn from an aquifer for human use without depleting the supply to the extent that withdrawal at this rate is no longer economically feasible.

Thus the "dangerous depletion" of Lee is described in economic terms by Meinzer, and both speak of permanency of withdrawals.

As a philosophical concept, the definition of Lee is a good one. However, we are given no clues as to how this rate or rates may be determined, and it is still far too ambiguous for practical use. Conkling (1946) attempted to make the concept less ambiguous by specifying the conditions that constitute a safe yield. He described safe yield as an annual extraction of water that does not

1. Exceed average annual recharge.
2. Lower the water table so that the permissible cost of pumping is exceeded.

3. Lower the water table so as to permit intrusion of water of undesirable quality.

A fourth condition, the protection of existing rights, was added by Banks (1953).

The single-valued concept of safe yield as proposed by Conkling and modified by Banks encompasses hydrologic, economic, quality, and legal considerations. This overspecification of the term is not likely what Lee or Meinzer had in mind. The controversial nature of the concept as defined by Conkling is clearly demonstrated in 43 pages of discussion of his 28-page paper by no fewer than 10 authorities. In practice, safe yield has no unique or constant value, its value at any time depending of the spacing and location of the wells and their influence on the dynamics of interchange between ground water and other elements of the hydrologic cycle. In addition, it is not possible to develop a safe yield in the absence of developmental overdraft, a necessary first stage in ground water development where withdrawals cause a lowering of the water table in areas of natural discharge and recharge. Further, there are seasonal or cyclical overdrafts where water levels eventually return to their original levels during periods of limited withdrawals. Thomas (1951) and Kazmann (1956) have suggested abandonment of the term because of its indefiniteness. Freeze (1971) introduced a concept of maximum stable basin yield, determined from a three-dimensional saturated-unsaturated numerical model. Although the concept is a good one, determining its value is not a simple task.

Water Law

Ground water resources are appropriated in various manners according to laws and institutions in individual states. The appropriation doctrine is applied almost unequivocally in the most arid states where water is relatively scarce and most uses consumptive (Figure 6.2). Quotas, or rights to use, are based on priority in time of beneficial use. A quota in this sense implies that each user has a specific share. Centralized, or state, control comes into play in curtailing ground water development once all assigned quotas effectively exhaust safe yield, as in Nevada and Utah, in dictating minimum well-spacing restrictions in areas of overdraft or in the vicinity of surface water rights, in revoking rights in the event the criterion of beneficial use is not satisfied, in establishing priorities of use in the event water is scarce and additional water is required, or in authorizing withdrawals that limit water-level decline to a specified amount per year.

The riparian, or land ownership, doctrine is a common policy in eastern and midwestern United States, where water is more abundant and most uses nonconsumptive. A quota in the broader sense of the term is a reasonable use with respect to requirements of other riparians. Such a loosely defined quantity constitutes a poor definition of a property right and is only realistic if water is indeed abundant with respect to demands for its use. Under this policy, new rights may be established as long as new use is reasonable, priorities are not assigned on the basis of time, and centralized control is often reduced to a judicial matter in ascertaining the reasonableness of a particular use.

Several versions of this doctrine are applied in practice. The correlative rights doctrine of California recognizes all rights as equal, and appropriate shares or quotas are assigned on the basis of historic use in the event of shortages. The absolute ownership version, on the other hand, places no limitations on withdrawals or their effect on neighboring riparians. This doctrine provides a classic example

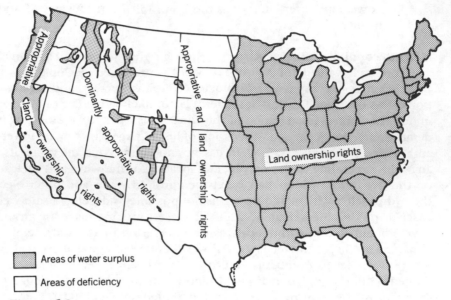

Figure 6.2
Water rights doctrines by states and areas of water surplus and deficiency (from
Thomas, 1955).

of common ownership of a natural resource and exploitation under competitive
pumping. Absolute ownership and reasonable-use doctrines do not generally differentiate between a renewable and nonrenewable supply.

A question now arises as to the effectiveness of water law in managing the
resource. We will examine this question from a historic perspective. In the late
1950s, ground water provided about 15% of the nation's water demand. At that
time, as is currently the case, there was no national water problem, only individual problems distributed throughout the states. One of the problems, ground
water mining, was examined by Bagley (1961), who investigated pumpage in
excess of replenishment in western United States. According to Bagley, nearly
60% of the total ground water withdrawals in 1955 were for irrigation in the 17
Western states. California, Arizona, New Mexico, and Texas pumped for irrigation
about one-half of the total ground water withdrawals for all purposes in the
United States. California, the largest user of ground water, had an estimated overdraft, or pumpage in excess of replenishment, equal to one-half of the total
ground water withdrawals in the state. On a nationwide basis, Bagley estimated
that over one-fourth of all ground water withdrawals were mined. A detailed
account of the aquifers involved is given by McGuinness (1963).

The amount of ground water used has increased steadily from 60 billion
gallons per day in 1965 to 88 billion gallons per day in 1980 (Solley and others,
1983). As noted on Figure 6.3, the largest increases have been for irrigation.
Unfortunately, there is no recent account of the proportion of these withdrawals
that is considered to be mined. As irrigation is normally associated with the more
arid western United States, it is reasonably certain that the mining episode of the
earlier years has not had any substantial effect on the water supplies. This is likely
because, by definition, overdraft is pumpage in excess of safe yield. As safe yield

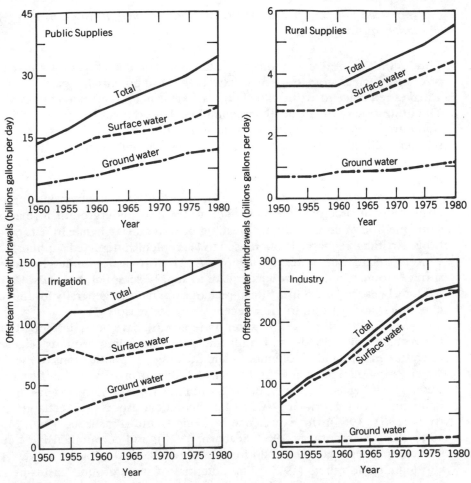

Figure 6.3
Trends in water withdrawals for public supplies, rural supplies, irrigation, and self-supplied industry. 1950–1980 (from Solley and others, 1983).

is frequently taken as annual recharge, somehow determined, it is more often than not grossly underestimated. Further, it would appear that annual replenishment expressed through a single-valued safe yield concept is unacceptable to private pumpers as a factor limiting economic growth.

Artificial Recharge and Conjunctive Use

The California State Department of Water Resources (1980) describes "ground water basin management" in the following terms:

> *Ground water basin management includes planned use of the ground water basin yield, storage space, transmission capability, and water in storage. It includes 1. protection of natural recharge and use of artificial recharge; 2. planned variation in amount and location of pumping over time; 3. use of ground water storage conjunctively with surface water from local and*

imported sources; and 4. protection and planned maintenance of ground-water quality.

This definition refers not only to managed aquifer use but to manageable resources such as storage space for alternative and potentially feasible uses. Item 1 makes reference to artificial recharge, whereas item 3 refers to conjunctive use. The utilization of artificial recharge implies that both surface water and ground water are being used conjunctively although all conjunctive use systems do not rely on artificial recharge as the means of augmenting a surface water supply.

Artificial Recharge

Artificial recharge has been defined by Todd (1980) as augmenting the natural infiltration of precipitation or surface water into underground formations by some method of construction, spreading of water, or a change in natural conditions. Artificial recharge is often used to (1) replenish depleted supplies, (2) prevent or retard salt water intrusion, or (3) store water underground where surface-storage facilities are inadequate to supply seasonal demands. California has long been the leader in artificial-recharge operations, generally for the reasons cited in (1) and (2) and, in more recent years, for reason (3).

Several methods of getting water underground have been developed, including recharge pits, ponds, and wells and water-spreading grounds (Figure 6.4). Literature in this field is voluminous and includes the publications of the Ground Water Recharge Center in California, which deal primarily with infiltration and water spreading (Schiff, 1955; Behnke and Bianchi, 1965); the extensive work of Baumann, dealing primarily with the theoretical aspects of recharge through wells (1963; 1965); the experience on Long Island (Brashears, 1946; Johnson, 1948; Cohen and others, 1968; Seaburn, 1970); and comprehensive annotated bibliographies prepared by Todd for the United States Geological Survey (1959a) and Signor and others (1970). The potential of underground basins as storage facilities is well demonstrated in the San Joaquin Valley, where the storage capacity has been estimated to be nine times the capacity of surface water reservoirs associated with the California Water Plan (Davis and others, 1959).

There are at least five interrelated questions associated with artificial recharge operations:

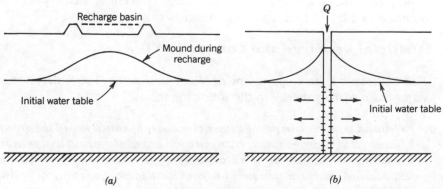

(a)   *(b)*

Figure 6.4
Schematic illustration of recharge through basins and wells.

1. What is the nature of the rechargeable water source?
2. What system of recharge will be utilized?
3. What are the expected injection rates?
4. How will the system respond hydraulically to injection?
5. How will the injected water be managed as part of the total water resource system?

There is considerable interdependency between the first two of these questions. For example, with water-spreading methods, water is generally diverted directly from surface water sources to topographically lower areas. On Long Island, New York, abandoned gravel pits are used to collect storm runoff that previously discharged to the ocean. On the other hand, in this same region, more than 1000 recharge wells are in operation in response to legislation that requires direct recycling of ground water pumped for air-conditioning and industrial cooling purposes. In the Santa Clara River Valley, flood waters are stored in reservoirs and later released at low rates so as to enhance their infiltration into the natural streambed. Alternately, municipal waste water subjected to secondary treatment is generally recharged by irrigation or spreading methods (Todd, 1980). Recharge wells for waste water must be accompanied by tertiary treatment. Bouwer (1985) and Sopper and Kardos (1974) discuss the waste water recharge program in Arizona.

Expected injection rates along with problems in rate decline have been reviewed by Todd (1959b). Table 6.1 gives some representative spreading basin

Table 6.1
Representative artificial recharge rates (from Todd, 1959b)*

Spreading basins		Recharge wells	
Location	Rate (m/day)	Location	Rate (m³/day)
California		California	
Los Angeles	0.7–1.9	Fresno	500–2200
Madera	0.3–1.2	Los Angeles	2900
San Gabriel River	0.6–1.6	Manhatten Bench	1000–2400
San Joaquin Valley	0.1–0.5	Orange Cove	1700–2200
Santa Ana River	0.5–2.9	San Fernando Valley	700
Santa Clara Valley	0.4–2.2	Tulare County	300
Tulare County	0.1		
Ventura County	0.4–0.5		
New Jersey		New Jersey	
East Orange	0.1	Newark	1500
Princeton	<0.1		
New York		Texas	
Long Island	0.2–0.9	El Paso	5600
		High Plains	700–2700
Iowa		New York	
Des Moines	0.5	Long Island	500–5400
Washington		Florida	
Richland	2.3	Orlando	500–51,000
Massachusetts		Idaho	
Newton	1.3	Mud Lake	500–2400

*Reprinted from *Groundwater Hydrology.* Copyright © 1959. John Wiley & Sons, Inc. Reprinted by permission of John Wiley & Sons, Inc.

recharge rates and some average well recharge rates. Recharge rates associated with spreading grounds vary from 15 m³/m²/day (m/day) for some gravels to as little as 0.5 m/day in sand and silt (Bear, 1979). Typically, the infiltration rates fall off with time due to swelling of the soil after wetting (Figure 6.5). There is some threshold value of infiltration where continued recharge is no longer economical so that the spreading ground is either abandoned or some scrapping and cleaning process must be put into operation. Decreases in well recharge rates can occur in response to silt introduced in the recharge water, dissolved air in the recharge water, which tends to decrease the permeability in the vicinity of the screen openings, and bacteria-induced chemical growth on the well screen.

The hydraulic response of recharge wells can be viewed as the inverse of the pumping response discussed in Chapter 5. The cone of depression associated with pumping becomes a cone of impression for confined aquifers, or simply a cone of recharge for unconfined aquifers. With spreading basins, the hydraulic response is in the form of a mound whose lateral extent exceeds the dimensions of the spreading grounds (Figure 6.4). The hydraulics of mound formation beneath recharge basins has received considerable attention in the literature (Baumann, 1965; Bittinger and Trelease, 1965; Hantush, 1967; Bianchi and Muckel, 1970). The solution of Hantush (1967) is for a circular spreading basin and provides a prediction of the height of the mound above the original water table elevation as a function of radial distance r and time t. Bianchi and Muckel (1970) consider a square recharge area and provide a dimensionless graph to determine the mound height as a function of time (Figure 6.6). In this figure, b is the height of the mound, S is the storativity, T is the transmissivity, W is the recharge rate, x is the distance from the center of the recharge area, L is the length of one side of the recharge area, and t is the time since recharge began. As can be noted, this figure is in the form of a profile of a ground water mound whose center is at $x/L = 0$. Note that the edge of a square recharge area coincides with $x/L = 0.5$. It will be noted further that the dimensionless parameter $L/(4Tt/S)^{1/2}$ of Figure 6.6 is of the form

$$\frac{(L^2 S/4T)}{t} = \frac{T^*}{t} \tag{6.2}$$

where T^* is again the familiar time constant for a basin. As T^* gets small with respect to recharge time t, as might be caused by a small S or large T, the mound is more readily dissipated.

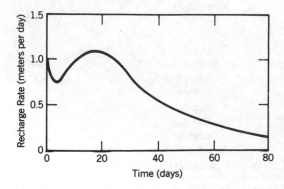

Figure 6.5
Time variation of recharge rate for water spreading on undisturbed soil (from Muckel, 1959).

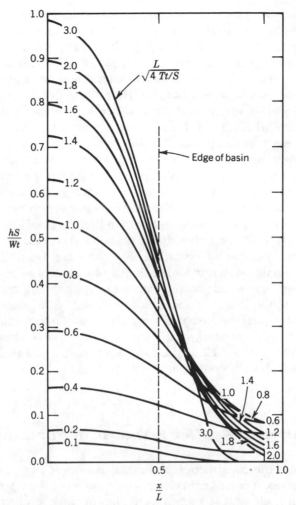

Figure 6.6
Graphical solution for the rise and horizontal spread
of a water table mound beneath a square recharge
area (from Bianchi and Muckel, 1970).

Conjunctive Use

Conjuctive use involves the coordinated use of surface and ground water to meet some specified water demand in a given area. Buras (1966) provides good insight into the general problem of conjunctive use:

1. What system has to be built in order to minimize the discrepancy (in time, space, and quality) between the natural supply of water and the demand for it?

2. To what extent should the water resource system be developed, and how extensive should be the region serviced?

3. How should the system be operated so as to achieve a given set of objectives in the best possible way?

Conjunctive operations may be of various kinds. The most common type that is of interest because of the hydraulics is the interconnected stream-aquifer system where the development of one affects the other (Figure 6.7). This is the type of system addressed by Bredehoeft and Young (1970) and Young and Bredehoeft (1972). An overall objective of such studies might be to utilize the total resource in such a way as to maximize benefits. Here, the hydraulic connection between the two sources plays a major role. Yet another type of operation has been described by Chun and others (1964) for the coastal plain of Los Angeles. This study proposes full use of underground supplies and storage capacity together with local and imported supplies where artificial recharge plays a sizable role in the project. The plan of basin operation is to provide water to the consumer at the lowest possible cost. Yet a third type of operation is where surface water is imported and is used directly to supplement the ground water supply with little or no connection between the two sources, other than their availability.

Cochran (1968) describes such a situation in Nevada where the objective was to minimize the cost of operation. For this case, Domenico (1972) derived a mathematical decision rule for the timing of surface water importation to supplement an overdeveloped ground water supply. As a marginal value rule, this decision rule states that the importation should take place when the cost of importation equals the cost of mining. The cost of importation includes the initial investment plus operating costs, whereas the cost of mining includes current pumping charges as well as the capitalized cost of all future pumping charges associated with a lowered ground water storage level.

6.2 Simulation of Aquifer Response to Pumping

Simulation is the construction and manipulation of an operating model of a system or process. It entails experiments on a model of the system rather than on the real system itself, and so permits scientists to study problems that would otherwise be impractical to study. Some of the first concepts of simulation for water resource systems were presented by Maass and others (1962) and Hufschmidt and Fiering (1966). The most recent addition to simulation techniques employed in ground water hydrology deal with contaminant transport.

Simulation techniques vary from discipline to discipline and are equally suited for both ground and surface water studies. The scientists applying the simulation designs an appropriate model, observes and evaluates the results, and comes to

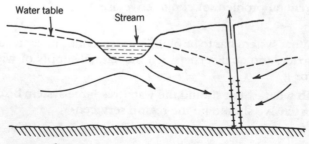

Figure 6.7
Induced recharge in the vicinity of a stream.

appropriate conclusions. This procedure is generally observed whether the simulation deals with natural runoff in a watershed, a release schedule for a surface water reservoir, a scheme for exploring the consequences of additional pumping in an overappropriated basin, pumping in aquifers subject to salt water intrusion or land subsidence, or some proposed remedial action for abatement of a contamination problem. The treatment of simulation in this section adheres to this general procedure but, in common with other resource simulation studies, is suited for one specific purpose, namely, ascertaining the water-level response of complex aquifers to pumping stresses.

Numerical Simulation

Assumptions employed throughout the chapter on hydraulic testing were that aquifers are homogeneous and isotropic, infinite in areal extent, and all the water pumped was removed from storage; that is, no provisions were allowed for recharge. Within the framework of these assumptions, we were able to make some simple predictions on the response of water levels to pumping, either for water supply or dewatering. Suppose now that the transmissive properties are not constant, the aquifer can be recharged through hydraulically connected streams or by other means, and there are several tens or even hundreds of operating wells at different pumping or recharging schedules within a complex bounded aquifer. How is this more difficult problem treated?

First, let us recognize that we do not have to search for an additional set of physical principles to deal with this problem, that is, the overall transient response of the system will still be described by a diffusion-type equation. What is required, however, is an additional set of mathematical tools, collectively referred to as numerical analysis. Numerical analysis means numerical methods for solving partial differential equations. Two methods are currently at the forefront; finite difference methods and finite element methods. In all cases, the computations are done with a computer.

Whichever method is utilized, the first step is to discretize the region of interest, that is, to replace the continuous region for which a solution is desired by an array of points (Figure 6.8). Three such grids are shown in Figure 6.8, two rectangular and one polygonal. The rectangular grids are associated with the finite difference method, which may be mesh centered (Figure 6.8*b*), or block centered (Figure 6.8*c*). The polygonal elements, whose shapes are determined by nodal points, are associated with the finite element method (Figure 6.8*d*). The replacement of a continuous field by an assemblage of discrete nodal points is accompanied by the generation of *n* equations, one for each node of an *n* node mesh, with a total of *n* unknown head values. Figure 6.9 gives a generalized approach to the methods of numerical analysis.

The Method of Finite Differences
In the method of finite differences, the governing differential equation is replaced by a difference equation in such a manner that the budgetary requirements of the original differential equation are approximately conserved. For a block-centered grid as given in Figure 6.10, the continuity equation states that the net inflow into any block equals the time rate of change of storage within the nodal block

$$Q_{1\text{-}0} + Q_{2\text{-}0} + Q_{3\text{-}0} + Q_{4\text{-}0} = S_{s_o} \, \Delta x \, \Delta y \, m \, \frac{\partial h_o}{\partial t} \tag{6.3}$$

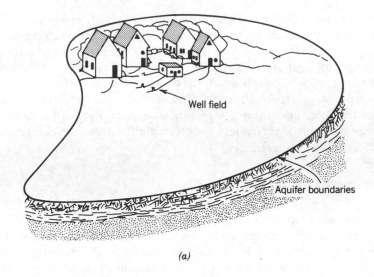

(a)

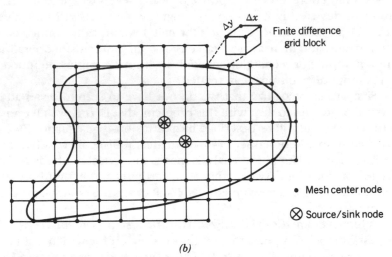

(b)

Figure 6.8
Finite difference and finite element grids (from Mercer and Faust,
1981). Reprinted by permission of Ground Water. © 1981. All rights
reserved.

where Q is the volumetric flow per unit time, S_{s_o} is the specific storage of nodal
block O, and m is aquifer thickness. For this discretized system Darcy's law may
be expressed as

$$Q_{1\text{-}0} = K_{1\text{-}0} \frac{b_1 - b_0}{\Delta y} \Delta x \, m \tag{6.4}$$

Here $K_{1\text{-}0}$ is the average hydraulic conductivity between nodes 1 and 0. Three
additional descriptions of Darcy's law may be given to describe the flow between
nodes 2 and 0, 3 and 0, and 4 and 0. For simplicity, it is assumed that the aquifer
is homogeneous and isotropic so that the hydraulic conductivity is constant, and

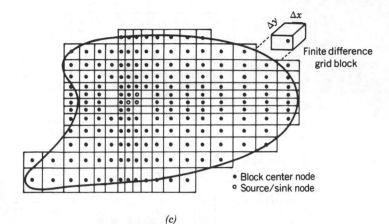

Finite difference grid block

• Block center node
○ Source/sink node

(c)

Triangular finite element

• Nodal point
○ Source/sink node

(d)

Figure 6.8
continued

the aquifer is of uniform thickness m. For a uniform grid spacing where $\Delta x = \Delta y$, substitution of Darcy's law into the continuity expression gives

$$h_1 + h_2 + h_3 + h_4 - 4h_0 = \frac{S}{T} \Delta x^2 \frac{\partial h_0}{\partial t} \qquad (6.5)$$

which is the finite difference form of the diffusion equation. The time derivative can be approximated by

$$\frac{\partial h_0}{\partial t} = \frac{h_0(t) - h_0(t - \Delta t)}{\Delta t} \qquad (6.6)$$

where Δt is a time step.

The steady-state version of Eq. 6.5 is readily expressed

$$h_1 + h_2 + h_3 + h_4 - 4h_0 = 0 \qquad (6.7)$$

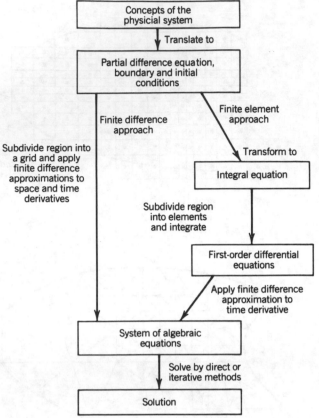

Figure 6.9
Generalized model development by finite difference and
finite element methods (from Mercer and Faust, 1981).
Reprinted by permission of Ground Water. Copyright ©
1981. All rights reserved.

Passing to the *ij* notation (Figure 6.11), Laplace's equation becomes

$$h_{ij} = \frac{1}{4}\left(h_{i,j\text{-}1} + h_{i+1,j} + h_{i,j+1} + h_{i\text{-}1,j}\right) \qquad (6.8)$$

whereas the diffusion equation becomes (Mercer and Faust, 1981)

$$h^{n}_{i,j\text{-}1} + h^{n}_{i+1,j} + h^{n}_{i-1,j} + h^{n}_{i,j+1} - 4h^{n}_{i,j} = \frac{S}{T}\frac{\Delta x^2}{\Delta t}\left(h^{n}_{i,j} - h^{n-1}_{i,j}\right) \qquad (6.9)$$

where n is the new time step, Δt is the time step, and Δx is the spacing.

Computational Techniques, Steady Flow   Equation 6.7 or 6.8 makes it possible to obtain values of the head at all nodal points such that the finite difference
equation is satisfied simultaneously at all nodes. In the case of Eq. 6.8, a satisfactory solution is obtained when the total head at all nodal points equals one-fourth
of the sum of the heads at the four adjacent nodes. As a first approximation, the

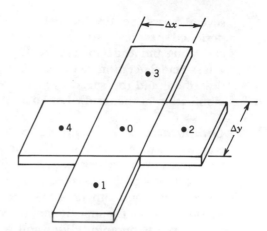

Figure 6.10
Block-centered grid.

heads along the boundaries and at each node are estimated. All assigned values will obviously be in error, with the exception of the boundary values. Beginning at nodes adjacent to the boundary, a new value for the head at each node is computed from the assigned values at the four adjacent nodes. This procedure is repeated until the computed heads converge to their limiting values.

Equation 6.7 for any node may be modified to account for the error of the assumed starting heads

$$h_1 + h_2 + h_3 + h_4 - 4h_0 = R_0 \qquad (6.10)$$

where R_0 is the residual amount representing the error at node zero for the assumed starting heads. The idea here is to remove the residual so that Eq. 6.10 converges to Eq. 6.7. The residual can be eliminated by adding to it an equal number of opposite sign. Clearly, the elimination of a residual at any one node changes the surrounding residuals at other nodes. Further, the change of a residual by -4 changes all adjacent residuals by $+1$. The procedure that follows entails the removal of all residuals node by node, starting with the highest.

It is thus seen that the methodology used to solve the steady-state finite difference expressions is one of iteration. The general literature in this area goes back

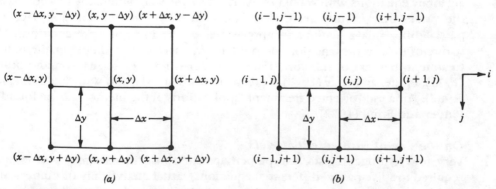

Figure 6.11
xy and *ij* notation for finite differences.

several years (Southwell, 1946; McCracken and Dorn, 1964). Modern texts designed specifically for ground water include those of Remson and others (1971) and Wang and Anderson (1982). The various techniques for solution include the Gauss-Seidel iteration method, which provides for the convergence requirements described, and the successive over relaxation method, which is employed for elimination of the residual of Eq. 6.10.

Computational Techniques, Unsteady Flow   The computational techniques employed to solve the finite difference approximations of the unsteady flow equation were developed originally by mathematicians for application to oil reservoirs and heat flow problems (Peaceman and Rachford, 1955; Douglas and Peaceman, 1955; Quon and others, 1965, 1966). The California Department of Water Resources made the first attempt to evaluate numerically the response of ground water levels to pumping (Tyson and Weber, 1964), followed by the work of Eshett and Longenbaugh (1965), Bittinger and others (1967), Pinder and Bredehoeft (1968), and Prickett and Lonnquist (1968). Five various computational schemes have been described by Quon and others (1965), the two chief methods being classified by whether they are explicit or implicit.

The replacement of a continuous field by an assemblage of discrete nodal points is accompanied by the generation of n equations, one for each node of an n-node mesh, with a total of n unknown head values. Initial procedure requires that all head values at all nodal points be estimated, or guessed. The explicit method then proceeds to solve for the correct heads at various time steps in terms of the known heads at adjacent nodes. This is an explicit iterative technique, in which at the kth iteration, a value of $h_{i,j}$ is determined without the simultaneous determination of a group of other values of head for other nodes (Forsythe and Wasow, 1960). The computer program presented by Prickett and Lonnquist (1968) is an example of an implicit method.

Textbooks by Remson and others (1971), Desai (1979), Wang and Anderson (1982), and Huyakorn and Pinder (1983) treat these methods in detail. In addition, the article by Mercer and Faust (1981) provides an excellent overview.

The Method of Finite Elements

The finite difference method replaces a continuous field with a descritized one and describes the governing partial differential equations by methods used in differential calculus. The finite element method likewise replaces a continuous field with a descritized one, and like the finite difference approach, results in N algebraic equations with N unknowns at each time step, where the unknowns are the values of hydraulic head. However, with the finite element approach the partial differential equations are approximated with integral representations. The nature of these representations do not lend themselves to a simple mathematical treatment in a text of this sort, and the reader is referred to Remson and others (1971), Pinder and Gray (1977), Mercer and Faust (1981), and Wang and Anderson (1982). For a user-oriented treatment, applications of the method can be found in Pinder and Frind (1972).

On the Use of Numerical Models

Very few hydrogeologists, except perhaps for those in research, will ever be required to solve partial differential equations, either analytically or numerically. With analytical solutions, most of the practical solutions are readily found in textbooks on quantitative hydrogeology. Similarly, there is a growing list of ready-made numerical solutions that can be adopted to most problems. Such a listing is provided by The International Ground Water Modeling Center of the Holcomb Research Institute of Butler University, Indianapolis, Indiana. In addition, the National Water Well Association, publishers of the journal "Ground Water," market some codes, as does a variety of commercial organizations. For the resource problem we have been discussing in this section, three codes by far outweigh the use of all others. These are the Prickett-Lonnquist model, a finite difference model developed at the Illinois State Water Survey (Prickett and Lonnquist, 1971), and two codes developed by the U.S. Geological Survey by Trescott and others (1976) and McDonald and Harbaugh (1984).

Last, let us recognize that the relatively small amount of space allocated to the topic of numerical analysis in no way reflects its importance in applied hydrogeology. The reason for this small allocation resides in the fact that numerical methods, like geochemistry, soil physics, soil mechanics, and mass transport, is a core course in any serious curriculum for hydrogeologists, and its full utilization can only be understood and appreciated in an extended coverage of at least one semester. It is the number one tool required by hydrogeologists to put the theory into practice.

6.3 Land Subsidence and Sea Water Intrusion

Land Subsidence

Subsidence of the land surface in many areas of the world has been ascribed to several causes: tectonic movement; solution; compaction of sedimentary materials due to static loads, vibrations, or increased density brought about by water table lowering; and changes in reservoir pressures with loss of fluids. Noteworthy cases of appreciable subsidence attributed to the last cause have been reported in Long Beach Harbor, California (Gilluly and Grant, 1949); the San Joaquin Valley, California (Poland and Davis, 1956; Poland, 1961); the upper Gulf coastal region, Texas (Winslow and Wood, 1959; Gabrysch and Bonnet, 1975); the Savannah area, Georgia (Davis and others, 1963); and Las Vegas Valley, Nevada (Domenico and others, 1966). Other localities include Mexico City (Cuevas, 1936) and London (Wilson and Grace, 1942). The geologic requisites and qualifying conditions for the occurrence of subsidence are so well adapted to alluvial basins that it is likely far more occurrences of this phenomenon are taking place than have been reported. The chief reason for this is the lack of close control of benchmarks necessary to detect small changes in land-surface altitude.

Serious problems can be caused by land-surface subsidence. The normal upward force of skin friction acting on piles or well casings may be reversed, which will subject these elements to "downdrag." This may result in failure or protrusion above land surface. Gradients of canals may be reduced, or even reversed, and so affect the normal flow of water. Cracking of concrete or brick structures is common in subsiding areas. Subsidence in coastal areas results in the conversion of pastures to tidelands, where areas previously elevated become subject to flooding from hurricanes (Winslow and Wood, 1959). Kreitler (1977) estimated that Hurricane Carla, which struck the Houston-Galveston region in 1961, would have flooded 146 square miles of land adjacent to Galveston Bay if it had struck in 1976. This is 25 square miles more than the area flooded in 1961.

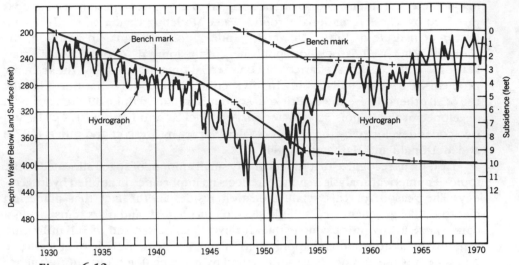

Figure 6.12
Subsidence and water-level decline in the San Joaquin Valley (from Poland and others, 1975).

Many of the observations of the rate of water-level decline and the rate of subsidence exhibit a fair degree of linearity. This linearity has been observed by Carrillo (1948) in Mexico City and by the Tokyo Institute of Civil Engineering (1975). Figure 6.12 demonstrates a reasonable linear trend between water-level decline and subsidence in the San Joaquin Valley over the period 1930 to mid-1940s and then again to the late 1950s. From this diagram it is noted that subsidence actually stopped in response to the rise in water levels. A similar linearity is noted for the volume of water pumped and the volume of resulting subsidence over the last 25-year period of record (Figure 6.13).

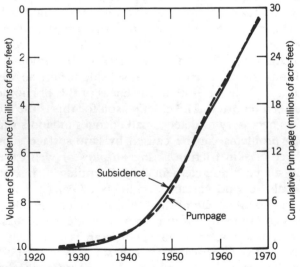

Figure 6.13
Volume of subsidence and cumulative pumpage, San Joaquin valley (from Poland and others, 1975).

Table 6.2
Summary of cases of subsidence

Name of place	Subsidence		Head decline (m)	Date of investigation
	Max. Settle. (m)	Area (sq. miles)		
United States				
San Joaquin Valley, CA	8.8	5212	90	1972
Santa Clara Valley, CA	4.0	251	25	1972
Houston–Galveston				
Area, TX	2.3	4710	100	1974
Eley–Picacho Area, AZ	1.1	—	30–60	1972
Las Vegas, NV	0.75	200	30	1972
Baton Rouge, LA	0.3	250	60	1970
New Orleans, LA	0.5	—	20	1968
Savannah, GA	0.1	19	27	1963
Japan				
Tokyo	4.6	77	30	1972
Nagoya	1.5	38	—	1976
Mexico				
Mexico City	8.5	58	20⁺	1964
Taiwan				
Taipei	1.35	Approx. 48	20	1969
United Kingdom				
London	0.1	—	90⁺	1942
Italy				
Venice	0.14	Approx. 3	5	1974

To investigate this linear trend further, consider Table 6.2 and Figure 6.14 constructed from it. It is important here to note the dates associated with the investigations reported in Table 6.2. As subsidence has continued beyond the period reported, it is not certain that the relationships demonstrated on Figure 6.14 still hold. At any rate, there appears to be two well-defined divisions, one where the subsidence–water-level decline ratio is larger than 0.09 (9m of subsidence to 100m of water-level decline) and the other where the ratio is equal to or less than 0.025 (2.5m of subsidence to 100m of water-level decline). One possible reason for this may be related to a component of subsidence caused by tectonic factors that have not been filtered out of the calculations. In Las Vegas Valley, Nevada, for example, there is a component of subsidence due to the elastic yield of the bedrock in response to the load of Lake Mead (Figure 6.15). This regional deformation has affected 8000 square miles in southern Nevada and adjoining states. Yet another possibility for the two divisions demonstrated on Figure 6.14 resides in the possibility of variations in the compressibility of the geologic units as well as in their thickness.

Physical Properties of Sediments
If compression is one dimensional in a vertical direction, the relative compression $\Delta H/H_o$ can be expressed as

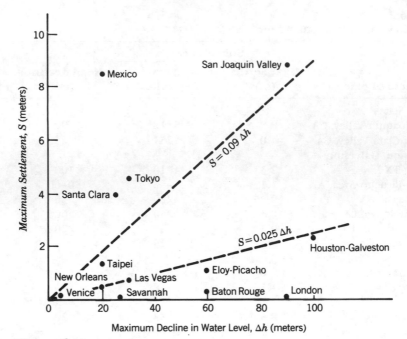

Figure 6.14
Relation between maximum settlement and maximum decline in water level.

$$\frac{\Delta V}{V_o} = \frac{\Delta H}{H_o} = \beta_p \Delta \bar{\sigma} \tag{6.11}$$

where ΔH is the change in height, H_o is the original height, ΔV is the change in volume, V_o is the original volume, β_p is the pore compressibility introduced in Chapter 4, and $\bar{\sigma}$ is the effective stress. Assuming that the solids are incompressible, changes in height are fully accounted for by changes in void ratio Δe or, from Figure 6.16,

$$\frac{\Delta H}{H_o} = \frac{\Delta e}{1 + e} \tag{6.12}$$

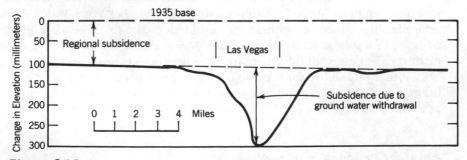

Figure 6.15
Subsidence profile, Las Vegas Valley, showing differential subsidence due to pumping superposed on regional subsidence (from Malmberg, 1960).

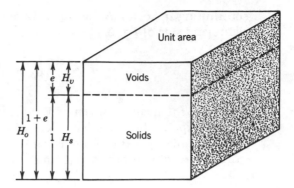

Figure 6.16
Schematic unit element of compressible material.

Equation 6.12 expresses the relative compression, or the relative amount of water expelled per unit height. Assuming the change in void ratio is proportional to the change in effective stress

$$\Delta e = a_v \Delta \bar{\sigma} \tag{6.13}$$

where a_v is termed the coefficient of compressibility in soil mechanics literature, defined as the rate of change in void ratio to the rate of change of effective stress causing the deformation. The coefficient a_v is determined as the slope of the line obtained by plotting void ratio versus pressure for test specimens (Figure 6.17). Thus, we have the uncomfortable situation where two different parameters are termed a coefficient of compressibility, one relating pore volume changes to changes in effective stress and one relating changes in the void ratio to changes in effective stress. The relationship between the two is found by

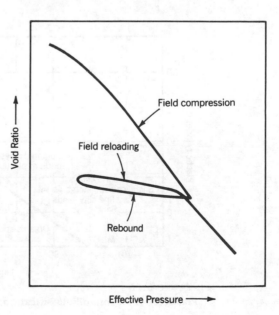

Figure 6.17
Void ratio versus effective pressure as obtained from laboratory consolidation.

substituting $a_v\Delta\bar{\sigma}$ for Δe in Eq. 6.12 and equating this new expression to the right-hand side of Eq. 6.11

$$\beta_p = \frac{a_v}{1 + e} \tag{6.14}$$

In words, the quantity given as Eq. 6.14 expresses the height of a pore water column expelled from a unit element when the effective stress is increased by one pressure unit. Ignoring fluid expansion, an alternate expression for specific storage is readily determined (Domenico and Mifflin, 1965)

$$S_s = \rho_w g \beta_p = \frac{a_v \rho_w g}{1 + e} \tag{6.15}$$

In soil mechanics literature, $a_v\rho_w g/(1 + e)$ is used almost exclusively in combination with the vertical hydraulic conductivity K_v, which gives (Chapter 4)

$$C_v = \frac{K_v(1 + e)}{a_v \rho_w g} = \frac{K_v}{S_s} \tag{6.16}$$

where C_v is termed the coefficient of consolidation, which is merely the hydraulic diffusivity.

Figure 6.18 in combination with Table 6.3 can be used to obtain approximate values of β_p, a_v, and the hydraulic diffusivity for various soil groups subject to consolidation. Figure 6.18 demonstrates the relationship between the Unified Soil Classification System and the plasticity chart. These same soil groups are represented on Table 6.3, which gives the results of over 1500 laboratory determinations by the Bureau of Reclamation (1960). The values for relative compression $\Delta H/H_o$ given in Table 6.3 are from drained tests on samples that were laterally confined; that is, the results indicate vertical compression.

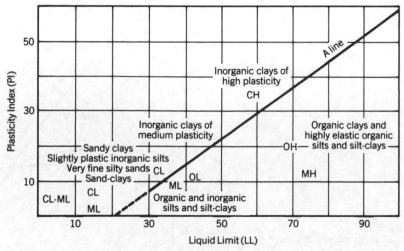

Figure 6.18
Plasticity chart for soil classification.

Table 6.3
Laboratory determination of physical properties of sediments (from Bureau of Reclamation, 1960)

Soil classification	Void ratio, e	Hydraulic conductivity (cm/sec)	$\frac{\Delta H}{H_o}$ at effective stress of 2880 lb/ft²	$\frac{\Delta H}{H_o}$ at effective stress of 7200 lb/ft²
GW	—	$2.6 \times 10^{-2} \pm 1.25 \times 10^{-2}$	<0.014	—
GP	—	$6.2 \times 10^{-2} \pm 3.3 \times 10^{-2}$	<0.008	—
GM	—	—	<0.012	<0.03
GC	—	—	<0.012	<0.024
SW	0.37	—	$0.014 \pm$	—
SP	0.5 ± 0.03	$>4 \times 10^{-6}$	0.008 ± 0.003	—
SM	0.48 ± 0.02	$7.2 \times 10^{-6} \pm 4.6 \times 10^{-6}$	0.012 ± 0.001	0.03 ± 0.004
SM-SC	0.41 ± 0.02	$7.7 \times 10^{-7} \pm 5.8 \times 10^{-7}$	0.014 ± 0.003	0.029 ± 0.01
SC	0.48 ± 0.01	$2.9 \times 10^{-7} \pm 1.9 \times 10^{-7}$	0.012 ± 0.002	0.024 ± 0.005
ML	0.63 ± 0.02	$5.7 \times 10^{-7} \pm 2.2 \times 10^{-7}$	0.015 ± 0.002	0.026 ± 0.003
ML-CL	0.54 ± 0.03	$1.26 \times 10^{-7} \pm 6.7 \times 10^{-8}$	0.01 ± 0.002	0.022 ± 0.0
CL	0.56 ± 0.01	$7.7 \times 10^{-8} \pm 2.9 \times 10^{-8}$	0.014 ± 0.002	0.026 ± 0.004
OL	—	—	—	—
MH	1.15 ± 0.12	$1.5 \times 10^{-7} \pm 9.6 \times 10^{-8}$	0.02 ± 0.012	0.038 ± 0.008
CH	0.8 ± 0.04	$4.8 \times 10^{-8} \pm 4.8 \times 10^{-8}$	0.026 ± 0.013	0.039 ± 0.015
OH	—	—	—	—

Example 6.1

A soil is classified as CL (organic silts and organic silt clays of low plasticity). At an effective stress increase of 2880 lb/ft², a first-order approximation for the values of various physical properties is determined as follows:

$$\beta_p = \frac{\Delta H/H_o}{\Delta \bar{\sigma}} = \frac{0.014}{2880} = 4.9 \times 10^{-6} \frac{ft^2}{lb}$$

$$\Delta e = \frac{\Delta H}{H_o}(1 + e) = (0.014)(1.56) = 0.0218$$

$$a_v = \frac{\Delta e}{\Delta \bar{\sigma}} = \frac{0.0218}{2880} = 7.6 \times 10^{-6} \frac{ft^2}{lb}$$

$$\frac{K_v}{S_s} = \frac{K_v}{\rho_w g \beta_p} = \frac{7.7 \times 10^{-8} \text{ cm/sec}}{1 \text{ gm/cm}^3 \times 10 \text{ cm/sec}^2 \times 10^{-6} \text{ cm}^2/\text{gm}}$$

$$= 7.7 \times 10^{-3} \frac{cm^2}{sec}$$

Mathematical Treatment of Land Subsidence

There are two related questions of general interest in the study of land subsidence. The first is concerned with the total amount of subsidence that might occur in a given hydrostratigraphic unit in response to a lowering of water levels. From what we have learned earlier, the amount can vary from point to point depending on the amount of water level decline, the thickness of the more compressible units, and their compressibility. Thus, differential subsidence will be the general rule. The second question deals with the time rate of subsidence in response to the time rate of head change within the aquifer. This problem is not unlike the problem of disequilibrium compaction described in Chapter 2. At one extreme, the subsidence can occur concurrently with the head decline; at the other extreme, the rate of subsidence can lag considerably behind the rate of water-level decline so that the basin can be in a state of progressive subsidence for decades.

Vertical Compression   We will focus on the first of these questions with an ideal depth-pressure diagram as given in Figure 6.19. Here an aquifer has an original head h that is lowered by the amount Δh, with water being removed from storage from the total system. Jacob (1940) was the first to attempt some quantification of this phenomenon when he postulated that the water so removed is derived from three sources: expansion of the confined water in the aquifer, compression of the aquifer, and compression of the adjacent and included clay beds. Jacob concluded that the third source is probably the chief one, a fact demonstrated by Poland in 1961. Jacob (1940) described the storativity of such systems as

$$S = \rho_w g m (\beta_p + n\beta_w + c\beta_p') \tag{6.17}$$

where all terms should by now be familiar to the reader except for β_p', which is the compressibility for the clayey beds, and c, which is a dimensionless constant that depends on the thickness, configuration, and distribution of the interbedded clay. For the assumption of Jacob (1940) that water is released from the clay beds

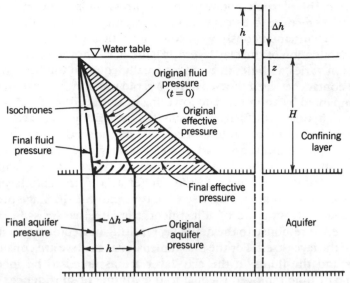

Figure 6.19
Depth-pressure diagram of the development of isochrones in a
confining layer in response to lowered aquifer pressure.

instantaneously with no time lag, c is indeed a constant and will be equal to H/m
where H is the thickness of the clayey beds and m is the thickness of the aquifers.
Where this condition is not satisfied, c is no longer a constant but will approach
a maximum value of H/m with time so that the storativity will increase with
increasing time. These points will be cleared up later.

The first two components of storage in Eq. 6.17 are already familiar to the
reader, at least in terms of what they mean with regard to the source of water in
confined aquifers. Lohman (1961) was the first to quantify their meaning in terms
of the elastic compression of aquifers. As the original head and the change in head
in the aquifer of Figure 6.19 is invariant with respect to depth, Eq. 6.11 may be
restated as

$$\frac{\Delta m}{m} = \beta_p \Delta \bar{\sigma} = \beta_p \Delta P \tag{6.18}$$

where P is the change in fluid pressure. The storativity for the elastic response
may be written as

$$\frac{S}{\rho_w g} = \beta_p m + n \beta_w m \tag{6.19}$$

Combining Eq. 6.18 and 6.19 and solving for Δm gives

$$\Delta m = \Delta P \left(\frac{S}{\rho_w g} - \beta_w n m \right) \tag{6.20}$$

This equation (or Eq. 6.18) gives the amount of vertical shortening of an aquifer
of thickness m in response to a pressure change ΔP. As aquifers are elastic, or

nearly so, this elastic component of compression is recoverable if the pressure is allowed to recover to its original (prepumping) value. Further, this compression takes place instantaneously with water-level decline.

Examination of the clay layer in Figure 6.19 shows that the original head distribution varies with depth z and is actually equal to the head within the aquifer at the contact between the two units. A rapid lowering of head in the aquifer is not accompanied by an immediate lowering of the head in the clay layer because of different hydraulic diffusivities. Thus, the head in the clay layer is converted to an "excess" head, at least with respect to the diminished head at its lower boundary. In response to this excess head, the clay layer will drain, with the head distribution at any time being shown by the so-called isochrones, which mean "same time." Eventually, the head in the aquifer and in the clay layer are once more matched at the boundary between the two, and the drainage process comes to a halt. During this period of adjustment, the effective stress in the clay layer has increased in response to the drainage, resulting in a commensurate volume reduction in the layer itself. For the one-dimensional case we are considering where the solids and the fluids in the clay layer are assumed to be incompressible, the volume of fluid removed is equal to the volume of subsidence.

As the head change in the clay layer ranges from zero at the top to Δh at the bottom (Figure 6.19), the volume of water produced from each element of the clay layer will be different, and can be expressed as

$$dq = S'_s h(z) \, dz \qquad (6.21)$$

where S'_s is the specific storage of the clay layer (ignoring fluid expansion) and $h(z)$ is the actual head change at some point z. The total volume of water extruded from a column of height H with a unit basal area is then

$$q = S'_s \int_O^H h(z) \, dz \qquad (6.22)$$

The solution of this equation is a general expression for the vertical shortening of a column of confining layer when steady flow conditions are reestablished within it. It is expressed as a volume of water passing through a unit surface area and has the units of length. As $h(z)$ varies linearly across the layer, $h(z) = \Delta h \, (z/H)$, where z varies from zero at the top to H at the bottom (Figure 6.19). Substituting this expression into Eq. 6.22 and integrating gives

$$q = S'_s \frac{\Delta h \, H}{2} \qquad (6.23)$$

Here we note that the quantity $\Delta h H/2$ represents the area of a triangle on the depth-pressure diagram of Figure 6.19 and is termed the "effective pressure area."

The result just given may be generalized for head changes in aquifers both above and below a compressible confining layer (Figure 6.20). The equation describing the head change in the confining layer is given as

$$h(z) = \Delta h_1 \left(\frac{z}{H} \right) + \Delta h_2 \left(\frac{H - z}{H} \right) \qquad (6.24)$$

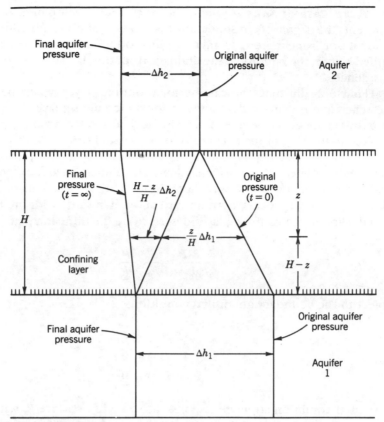

Figure 6.20
Depth-pressure diagram for a confining layer in response to
lowered pressure in adjacent aquifers (from Domenico and
Mifflin, Water Resources Res., v. 1, p. 563–576, 1965. Copyright
by Amer. Geophys. Union).

Substitution of Eq. 6.24 into Eq. 6.22 and performing the integration gives

$$q' = S'_s H \left(\frac{\Delta h_1 + \Delta h_2}{2} \right) \tag{6.25}$$

Examination of Eq. 6.25 and Figure 6.20 shows that when the transient flow
process is terminated, the vertical shortening of the confining layer equals the
product of the specific storage and the final area described by the increase in effective stress on a depth-pressure diagram. From Figure 6.20 and Eq. 6.25, this area
is again a triangle for $\Delta h_2 = 0$, or a rectangle when $\Delta h_1 = \Delta h_2$, or a trapezoid for
$\Delta h_1 \neq \Delta h_2$ (one-half the sum of the bases $\Delta h_1 + \Delta h_2$, times the height H).

The concepts described pertain to the permanent compression of low-
permeability layers in confined basins, the volume of water being a one-time
reserve associated with permanent compression. This process does not affect the
aquifer storage capacity in that the pore volume reduction takes place in the
confining units. According to Poland (1961), the process of subsidence makes
available a volume of water that would otherwise not be recoverable. Unlike elastic compressions, the lost pore volume is nonrecoverable.

The Time Rate of Subsidence   The isochrone development on Figure 6.19 represents the degree of consolidation at various depths as a function of time. The terminal isochrone represents full consolidation for the given head decline in the aquifer. Hence, the rapidity of the drainage from the clay layer controls the rate of subsidence.

Figure 6.21 illustrates these points for a confining layer separating two aquifers. The schematic isochrone development for the confining layer is produced for a rapid (instantaneous) lowering of head in the dual aquifer system, here for simplicity taken to be Δh. At time equal to zero, the head distribution through the three-layered system is assumed to be invariant with depth. When the pumping starts, the heads in the aquifers are immediately lowered an identical amount, converting the head in the confining layer to an "excess head" with respect to its boundaries. This starts the drainage process, both upward and downward, with the head (or pressure) decline in the confining layer described by the diffusion equation

$$\frac{\partial^2 P}{\partial z^2} = \frac{S'_s}{K_v} \frac{\partial P}{\partial t} \tag{6.26}$$

subject to the boundary and initial conditions

$$P(z,0) = P_i$$
$$P(H, t) = 0$$
$$P(-H, t) = 0$$

The initial condition stipulates a constant value of pressure P_i with depth. The two boundary conditions correspond to decreased pressure (assumed 0, with respect to P_i) at both the top H and bottom $-H$ of the confining layer.

The solution to Eq. 6.26 for the conditions stated is well known in consolidation theory (Terzaghi and Peck, 1948) and in heat flow theory (Carslaw and Jaeger, 1959). This solution provides for the pressure decline at different points in the confining layer, which for any time takes on maximum values near the two drainage boundaries and minimum values in the center. This is too much detail for this problem as we are here only interested in the average pressure decline (or consolidation) of the clay layer. This average is demonstrated in Figure 6.22. The

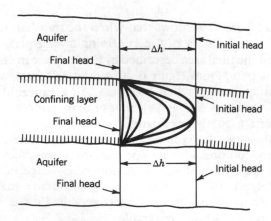

Figure 6.21
Schematic illustration of isochrome development in confining layer in response to instantaneous lowering of head at its upper and lower boundaries.

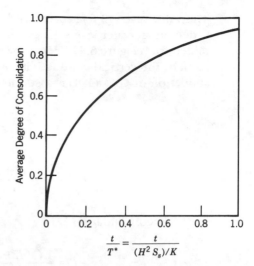

Figure 6.22
Average degree of consolidation versus the
ratio of time to the basin time constant.

$$\frac{t}{T^*} = \frac{t}{(H^2 S_s)/K}$$

ordinate represents the degree of consolidation, with values less than one indicating the presence of residual pressures. The abscissa is the familiar ratio of observation time and the time constant of the unit. As T^* gets small, corresponding to a small value for H and S_s in combination with a large value for K_v, the steady state is rapidly approached; that is, the unit approaches full consolidation so that the drainage process approaches termination shortly after the transient started. The quantity H in Figure 6.22 represents the distance to a drainage face and is taken as the thickness of the unit for single drainage (that is, the triangular effective pressure area of Figure 6.19) or $H/2$ for the double-drainage case given in Figure 6.21. Thus we notice that the time lag between the lowering of aquifer pressure and the consolidation of the interbedded clay layers contributing to subsidence is enhanced for thick clay layers of low permeability and high compressibility (storativity). Conversely, the time lag diminishes for thin units of moderate permeability and compressibility.

Last, we may now speculate that Jacob's c in Eq. 6.17 is of the form

$$c = \frac{t}{T^*}\,\frac{H}{m} \tag{6.27}$$

so that the storage component associated with the clay layer actually increases with time, reaching its maximum value when $t/T^* \cong 1$.

Example 6.2

Consider a confining layer between two aquifers. The specific storage of the confining layer is on the order of 1×10^{-3} ft^{-1}, and its thickness is 40 ft. The heads in the aquifers are lowered about 50 ft. Maximum vertical shortening of the confining layer is calculated

$$q = S_s' H \left(\frac{\Delta h_1 + \Delta h_2}{2} \right) = 1 \times 10^{-3} \times 40 \times \frac{100}{2} = 2 \text{ ft}$$

The value given (2 ft) is taken as the maximum subsidence in response to the 50 ft decline in water levels. The time to complete 50% of this subsidence is determined from Figure 6.22. Here, however, some additional information is required, namely, the vertical conductivity. Assume this value to be 1×10^{-3} ft/day. A consolidation of 50% (1 ft) is associated with a value of $t/T^* = 0.196$. Thus

$$\frac{t}{T^*} = \frac{t}{\dfrac{(H/2)^2\, S_s'}{K_v}} = 0.196$$

$$t = \frac{0.196\, (40/2)^2\, (1 \times 10^{-3})}{1 \times 10^{-3}} = 78 \text{ days}$$

For 95% consolidation

$$t = \frac{1\, (40/2)^2\, (1 \times 10^{-3})}{1 \times 10^{-3}} = 400 \text{ days}$$

For a water-level decline of 5 ft/yr in each of the aquifers

$$q = S_s' H \left(\frac{\Delta b_1 + \Delta b_2}{2} \right) = 1 \times 10^{-3} \times 40 \times \frac{10}{2} = 0.2 \text{ ft}$$

This represents an approximate measure of the subsidence to be expected with each years pumping. Note now that it still requires 78 days to achieve 50% of the annual subsidence (0.1 ft) and 400 days to achieve 95% of the subsidence. If benchmarks are releveled every year, the results would give the appearance that there is no appreciable time lag between the lowering of the head and the attainment of full consolidation.

Simulation of Subsidence   Frequently, subsidence in an area is sufficiently detrimental so as to warrant some detailed model study for prediction purposes. The first such study was conducted by Gambolati and Freeze (1973) who utilized a finite element model to obtain predicted drawdowns, and a finite difference model to determine the time rate of subsidence. The drawdowns obtained from the flow model were used as time varying boundary conditions to solve a set of one dimensional consolidation models. The model was applied to the subsidence problem in Venice, Italy, with the results reported by Gambolati and others (1974a, 1974b). Helm (1975) has also developed simulation models for application to land subsidence in California. Although the data demands for such models are large, the measured subsidence provides a means by which the models can be calibrated and improved over time. Such models are becoming more common where subsidence represents real as opposed to nuisance problems, for example, the Houston-Galveston area.

Sea Water Intrusion in Coastal Aquifers

Under natural conditions, the flow of fresh water toward the sea limits the landward encroachment of sea water. With the development of ground water supplies

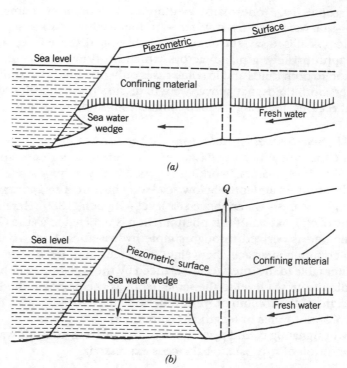

Figure 6.23
Hydraulic conditions near a coastline (*a*) not subject to sea
water intrusion and (*b*) subject to sea water intrusion with an
advancing sea water wedge.

and subsequent lowering of the water table or piezometric surface, the dynamic
balance between fresh and sea water is disturbed, permitting the sea water to
intrude usable parts of the aquifer (Figure 6.23). This phenomenon has been
reported in several parts of the world including the Netherlands (Ernest, 1969)
and Israel (Schmorak, 1967). The best documented cases in the United States
include Long Island (Lusczynski and Swarzenski, 1966), Miami (Kohout, 1961),
and many parts of California (California Department Water Resources, 1958). Salt
water intrusion is a special case of ground water contamination and is yet another
phenomenon subject to numerical simulation (Pinder and Cooper, 1970; Segol
and Pinder, 1976).

   Banks and Rictor (1953) suggest five methods of controlling salt water intrusion, four of which require control of the water table or piezometric surface at
or near the coast. The five methods include artificial recharge, a reduction or
rearrangement of the pumping wells, establishing a pumping trough along the
coast, thereby limiting the area of intrusion to the trough, formation of a pressure
ridge along the coast, and installation of a subsurface barrier. These methods have
been discussed at length by Todd (1980).

The Fresh Water–Salt Water Interface in Coastal Regions
There are many studies that attempt to ascertain the position of the fresh water–
salt water interface, both in coastal areas and underlying oceanic islands, and for
conditions of flow as might be induced in the vicinity of pumping wells. Many of

these studies have one major assumption in common, namely, that the two fluids—salt water and fresh water—are immiscible and are separated by a rigid interface. That is to say, the mixing zone discussed above is replaced by a line (or, more appropriately, a plane) across which no flow can occur. The interface is common to both fluids and in some cases only one fluid is permitted to move and in others both fluids may move. We will discuss these theories here as applied to coastal regions, focusing on their applications and limitations.

The Ghyben-Herzberg Relation   At the turn of the century, it was generally thought that salt water in coastal areas occurred at a depth approximating sea level. Two investigators, working independently, demonstrated that sea water actually occurred at depths below sea level equivalent to approximately 40 times the height of fresh water above sea level (Ghyben, 1889; Herzberg, 1901). The analytical explanation of this phenomenon is referred to as the Ghyben-Herzberg formula and is derived through simple hydrostatics. For two segregated fluids with a common interface, the weight of a column of fresh water extending from the water table to the interface is balanced by the weight of a column of sea water extending from sea level to the same depth as the point on the interface; that is, the weight of the column of fresh water of length $h_f + z$ equals the weight of the column of salt water of length z (Figure 6.24).

By designating ρ_f and ρ_s as the densities of fresh and salt water, respectively, the condition of hydrostatic balance is expressed

$$\rho_s g z = \rho_f g (h_f + z) \tag{6.28}$$

or

$$z = \frac{\rho_f}{\rho_s - \rho_f} h_f \tag{6.29}$$

where z is the depth below sea level to a point on the interface. If the density of fresh water is taken as 1.0 and sea water as 1.025

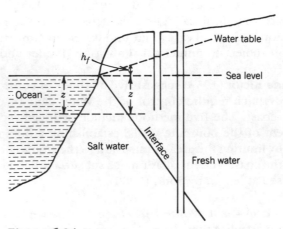

Figure 6.24
Hydrostatic conditions of the Ghyben-Herzberg relation.

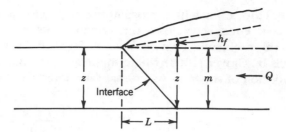

Figure 6.25
Idealized geometry to calculate the length of the salt
water wedge from the Ghyben-Herzberg relation.

$$z = 40h_f \tag{6.30}$$

as confirmed by Ghyben and Herzberg by observation. This means, where this condition is approximately correct, a fresh water level of 20 ft above sea level corresponds to 800 ft of fresh water below sea level. Stated in another way, a lowering of the water table by 5 ft will cause a 200-ft rise of salt water. It is further noted that the slope of the interface is 40 times greater than the slope of the water table.

The length of the prevailing salt water wedge under the hydrostatic conditions of the Ghyben-Herzberg may be established for the simple geometry of Figure 6.25. From Darcy's law of the form $Q = KiA = KimY$, where i is the hydraulic gradient, m is thickness, and Y is the length of the shoreline

$$Q' = \frac{Q}{Y} = K \frac{(h_f - 0)}{L} m \tag{6.31}$$

where Q' is the discharge per unit length of shoreline, and $(h_f - 0)/L$ is the hydraulic gradient where h_f is taken at the distance L from the shoreline. The Ghyben-Herzberg relation predicts that at a distance L from the shoreline

$$h_f = z \frac{(\rho_s - \rho_f)}{\rho_f} = m \frac{(\rho_s - \rho_f)}{\rho_f} \tag{6.32}$$

Solving Eq. 6.31 for h_f and equating this result to the right-hand side of Eq. 6.32 gives

$$L = \frac{(\rho_s - \rho_f)Km^2}{\rho_f Q'} \tag{6.33}$$

Thus, the length of protrusion of salt water under natural conditions is directly proportional to the hydraulic conductivity and thickness squared and inversely proportional to the flow of fresh water to the sea. As the flow of fresh water to the sea is reduced, say by pumping along coastal reaches, the length of the intruded salt water wedge increases.

The Shape of the Interface with a Submerged Seepage Surface The relationships just cited are only approximately correct because of the inherent hydrostatic assumptions. Further, the coincidence at the coastline of a zero fresh water and salt water head, along with the assumption of a rigid interface across which no flow can occur effectively closes the system at its discharge point. Another approach to this problem is demonstrated on Figure 6.26, which gives the results of an analysis by Glover (1964) where the total discharge takes place below sea level. The position of the fresh water–sea water interface is determined from

$$z^2 = \frac{2Q'\, x\, \rho_f}{K(\rho_s - \rho_f)} + \left[\frac{Q'\rho_f}{K(\rho_s - \rho_f)} \right]^2 \tag{6.34}$$

where z is the depth below sea level to the interface, x is the distance measured positive inland from the shoreline, and Q' is the fresh water flow per unit length of shoreline. The width of the fresh water discharge gap is determined where $z = 0$

$$x_o = - \frac{Q'\rho_f}{2K(\rho_s - \rho_f)} \tag{6.35}$$

whereas the depth to the interface below sea level at the coastline ($x = 0$) is given as

$$z_o = \frac{\rho_f Q'}{(\rho_s - \rho_f)\, K} \tag{6.36}$$

Last, the height of the water table at any distance x is given as

$$h_f = \frac{2Q'\, x\, (\rho_s - \rho_f)}{\rho_f K} \tag{6.37}$$

so that at $x = 0$, $h_f = 0$. Thus, the greater the flow to the sea, the deeper the interface, and the larger the required fresh water discharge gap.

For the geometry of Figure 6.27, the landward protrusion of the salt water wedge is determined directly from equation 6.34

$$L = \frac{1}{2} \left[\frac{m^2 K(\rho_s - \rho_f)}{Q'\rho_f} - \frac{\rho_f Q'}{(\rho_s - \rho_f)\, K} \right] \tag{6.38}$$

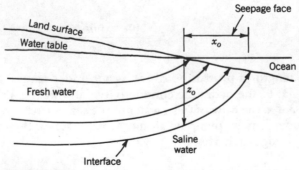

Figure 6.26
Flow pattern near the coast (modified from Glover, 1964).

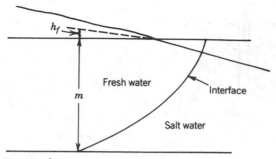

Figure 6.27
Idealized geometry to calculate the length of the
salt water wedge.

It is noted that the first term in the brackets is identical to Eq. 6.33 based on the
hydrostatic condition, whereas the second term is equal to the depth to the interface at $x = 0$ (Eq. 6.36). The role of fresh water flow to the sea as the major control
on retarding the advance of salt water is clearly shown in this development.

Upconing of the Interface Caused by Pumping Wells

The Ghyben-Herzberg relationship indicates that any lowering of head results in
a rise in the interface. When the lowering takes place by pumping wells that with-
draw water from above the interface, the interface can rise, a phenomenon
referred to as upconing (Figure 6.28). An approximate analytical solution for this
upconing has been given by Schmorak and Mercado (1969). Their solution gives
the new equilibrium elevation to an interface in direct response to pumping

$$z = \frac{Q\rho_f}{2\pi dK(\rho_s - \rho_f)} \tag{6.39}$$

where z is the new equilibrium elevation, Q is the pumping rate, and d is the distance
from the base of the well to the original (prepumping) interface (Figure 6.29). Both
laboratory and field observations suggest that this relationship holds only for very
small rises in the interface and there exists a critical elevation at which the interface
is no longer stable and salt water flows to the well (Schmorak and Mercado, 1969).

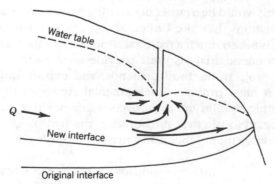

Figure 6.28
Fresh water flow to a well pumping above the
interface.

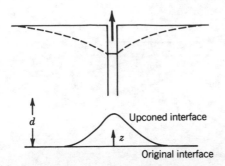

Figure 6.29
Upconing of interface in response to pumping.

Dagan and Bear (1968) suggest that the interface will be stable for upconed heights that do not exceed one-third of d given in Figure 6.29. Thus, if z is taken as $0.3d$, the maximum permitted pumping rate should not exceed

$$Q_{max} \leq 0.6\pi \ d^2K \left(\frac{\rho_s - \rho_f}{\rho_f} \right)$$

Dagan and Bear (1968) present other relationships dealing with upconing of the interface as well as estimates of maximum discharge to assure stability.

6.4 Simulation-Optimization Concepts

In general, management schemes in ground water basins seek to alter the resources for purposes of social welfare. The issues governing the management may be discussed under three categories: (1) analysis of the physical phenomena; (2) socioeconomic analysis; and (3) a system study that demonstrates the impact of water resource development on the social welfare of the inhabitants in the basin, on the region of which it is a part, or on the nation as a whole. In a developing agrarian economy where some agriculture products are imported, agriculture is important to the nation as a whole, and an evaluation must be made of the interaction between the existing agricultural activity and local, regional, and national food demands; population predictions; and even balance of payments. For example, if 25% of the agricultural activity in Mexico was threatened by salt water intrusion, this would be a rather devastating blow to the nation as a whole. In well-developed nations, like the United States, the ultimate loss of irrigation in the High Plains is not so much a national issue as it is a local and regional one. Hence, the socioeconomic analysis must provide information on resource availability, population projections, food demands, and export and balance of payments. Ultimately it must provide developmental strategies that either maintains the socioeconomic equilibrium or impacts it in a positive manner.

With regard to physical phenomena, the hydrologic system will respond to any water management scheme in a manner described by the hydrologic equation

$$\text{inflow} - \text{outflow} = \text{change in storage} \tag{6.40}$$

If one management objective is to prohibit further changes in storage in order to control salt water intrusion or land subsidence, a balance must be obtained between pumpage and recharge, both natural and artificial. If water is to be imported and used directly with a reduced ground water component, the problem is a relatively simple one and may be addressed with the simulation techniques discussed in this chapter. If the basin is void of surface water supplies and importation is not a likely possibility in the foreseeable future, the problem becomes more difficult because the alternatives are drastically reduced. Such a problem is compounded if the basin is subject to not only depletion of the storage reserve, but salt water intrusion as well.

It follows that the physical phenomena—water-level response to pumping, salt water intrusion, land subsidence—must be simulated in a management study of the type under discussion, and such simulation is well within the state of the art. Simulation, however, does not accomplish any optimization of decisions. Frequently, the stresses to be imposed in a simulation study are arbitrarily selected pumping decisions, usually developed through the judgment of experienced hydrologists. The simulation model produces the effects of these decisions and reveals which may be best from a physical perspective but not necessarily from a management perspective. Thus, it is frequently useful to design the simulation so as to provide a measure of the value of an objectives function for each of several alternative decisions. This idea was fostered for surface water studies by the Harvard water group (Maass and others, 1962; Hufschmidt and Fiering, 1966), and first applied to a ground water problem by Chun and others (1964). By incorporating an objectives function in simulation studies, one combines one of the stronger points of decision theory with a detailed ground water model. Alternative courses of action may be tested and quantitatively evaluated as to their influence on long-term costs and benefits. Hence, the required tasks include the following:

1. The listing of alternative courses of action

2. The determination of the consequences that follow from each of the alternatives

3. The comparative evaluation of these sets of consequences in terms of the value or values to be maximized

The mathematical model employed for investigating the various alternatives cited is a sequential linear program in conjunction with a ground water simulation model. The incorporation of linear programming provides a means to allocate the water resources and gives a measure of an objective function for each of several alternatives. It is not intended here to discuss linear programming or other optimization techniques as they are normally employed in water resource studies. A detailed account of these methods and optimization in general can be found in Domenico (1972).

The methodology involved in simulation-optimization studies has evolved over the past 20 years. Stults (1966) was the first to employ the idea, but not the methodology, by utilizing linear programming to allocate resources, and the simple hydrologic equation (Eq. 6.43) to transform the state variable (ground water storage). Martin and others (1969) extended these ideas by using linear programming as an allocation tool, and an electric analog simulation model to ascertain the effects of storage transformation. Bredehoeft and Young (1970) followed Martin's work, but employed a numerical model in conjunction with linear programming to ascertain the effects of a use tax and a quota system on the management of ground water. Their approach was intended to examine

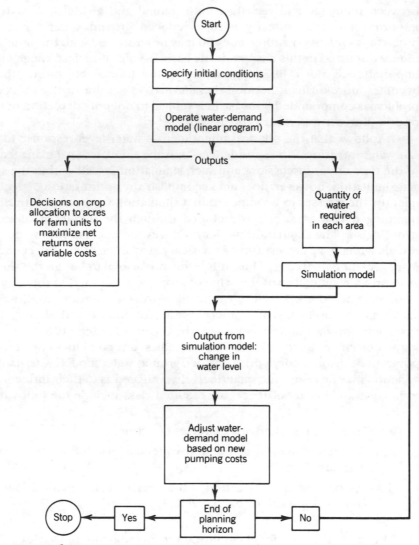

Figure 6.30
Flow diagram for a combined linear programming–simulation study (modified from Martin and others, 1969).

management of an interconnected surface water–ground water system both under certain and uncertain surface water supplies (Young and Bredehoeft, 1972; Daubert and Young, 1982; Bredehoeft and Young, 1983).

The flow diagram of Figure 6.30 illustrates the linking of the allocation and simulation models. Initial conditions include the net return over variable costs for each potential use of irrigation water, depth to water, and constraints on water use. These conditions are stipulated as parameters in the linear programming model, which yields two fundamental outputs: the activities required to maximize net returns and the amount of water required in the maximization process. The latter serves as an input to the simulation model. Simulation output is change in water levels in response to this pumping, which gives rise to changes in water demand. The feedback loop serves to operate the allocation model for the next stage. Hence, this procedure couples N single stages of ground water development for a problem similar to that devised by Stults (1966) for linear programming only.

Gorelick (1983) presents a review and classification of ground water management models, all of which involve some form of optimization. The linking of optimal resource allocation models and simulation models need not be restricted to problems of the type just described. Logical extensions of this approach can include problems dealing with land retirement due to an advancing salt water front. Domenico and others (1974) investigate such a problem in Hermosillio, Mexico, under the sponsorship of the World Bank. They utilized linear programming in conjunction with a simulation model for water-level response along with some simplified equations that described the salt water advance. The objective was to minimize water use and still obtain benefits that are more or less equal to those currently earned in agricultural production. This required investments in more efficient irrigation practices and some changes in crop activities. The reason for minimizing pumpage was to minimize the rate of salt water advance. The purpose here was to prolong agricultural production, a long-term holding action for a dwindling resource that is not likely to be supplemented with a surface water supply. Stated another way, the purpose was to maintain the socioeconomic equilibrium for as long as possible.

Suggested Readings

Glover, R. E., 1964, The pattern of freshwater flowing in a coastal aquifer: U.S. Geol. Survey Water Supply Paper 1613-C, p. 32–35.

Mercer, J. W., and Faust, C. R., 1981, Groundwater Modeling: National Water Well Assoc. Pub., Dublin, Ohio.

Poland, J. F., and Davis, G. H., 1969, Land subsidence due to withdrawal of fluids: Geol. Soc. Amer. Rev. Eng. Geol., 2, p. 817–829.

Theis, C. V., 1938, The significance and nature of a cone of depression in groundwater bodies: Econ. Geol., v. 33, p. 889–902.

Young, R., and Bredehoeft, J., 1972, Digital computer simulation for solving management problems of conjunctive groundwater and surface water systems: Water Resources Research, v. 8, p. 533–556.

Problems

1. A soil is classified as CH (inorganic clays). Estimate values for the following properties: β_p, Δe, a_v, K_v/S_s.

2. An elastic aquifer is overlain by 100 ft of clayey material. The aquifer has a storativity of 5×10^{-4} and a porosity of 0.3 and is 200 ft thick. The clayey material has a specific storage of 4×10^{-4} ft^{-1}.

    a. Calculate the elastic compression of the aquifer for a 100-ft decline in head.

    b.   Calculate the inelastic compression of the clayey material for a 100-ft decline in head.

    c.   Ascertain the bulk modulus of compression of the aquifer.

3.   Perform the calculations in example 6.2 for a clay layer 60 ft thick. If the benchmarks are releveled every year would the results give the appearance that there is no time lag between the lowering of the head and the attainment of full consolidation? Discuss your conclusions in terms of the time constant T^* and the observation time t.

4.   A salt water interface is stable at a position of about 40 ft beneath the base of a well in a formation with a hydraulic conductivity of 1×10^{-1} cm/sec. Calculate the maximum pumping rate so as not to cause upconing.

5.   The flow of fresh water toward the sea is 2×10^{-4} ft³/sec per foot of coastline in an aquifer that is 100 ft thick with a hydraulic conductivity of 3.3×10^{-3} ft/sec. Calculate the inland distance from the shoreline to the toe of the salt water wedge. What is the new position from the shoreline if the seaward flow is reduced by one-half.

6.   Select any aquifer in your state or region and discuss its exploitation, including pumpage and its distribution, recharge sources and their distribution, the general geology of the system, the effect of pumping on water levels, and any feedback effects such as subsidence or salt water intrusion.

Chapter Seven

Ground Water
in the Basin Hydrologic Cycle

7.1   Topographic Driving Forces

7.2   Surface Features of Ground
      Water Flow

7.3   Some Engineering and
      Geologic Implications of
      Topographic Drive Systems

The driving forces responsible for ground water movement were known before the turn of the century and were categorized by King (1899) as gravitational (topographic), thermal, and capillary. King did not mention the importance of tectonic strain as a driving force, although he clearly demonstrated an up-dip migration of water in response to the compaction of sediments in active depositional environments. Our interest in this chapter is not only on how driving forces can be manifested in the basin hydrologic cycle, but what physical consequences result from these manifestations. Such consequences are important in a variety of engineering projects and in understanding a host of processes operating near the Earth's surface.

7.1 Topographic Driving Forces

The Early Field Studies

"The basal principles of artesian wells are simple. The schoolboy reckons himself as their master." This is the opening statement in Chamberlin's (1885) classic treatment on the requisites and qualifying conditions for the occurrence of artesian wells in pervious rocks. The pervious stratum discussed by Chamberlin has been defined as an aquifer, and the watertight or impervious beds are known as confining beds, commonly referred to as aquitards. Chamberlin noted further that no stratum is entirely impervious so that the aquifer can progressively discharge its fluids down dip. Of special importance here is the fact that the elevation differences of the outcrops, or the outcrop and the subcrop, provide the topographic drive for ground water movement, hydraulically depicted by the potentiometric surface (Figure 7.1).

Other configurations of ground water flow were also being recognized at about this same time. This is documented in the United States Geological Survey's Nineteenth Annual Report (1899), part two of which is entitled "Papers Chiefly of a Theoretical Nature." Included in this volume is the classic work of Slichter (1899) on the principles and conditions for the movement of ground water, and a theoretical-field analysis by King (1899). Of importance here is King's observations regarding the influence of gravity on shallow ground water flow:

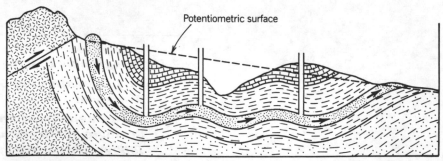

Figure 7.1
Schematic cross section showing trace of the potentiometric surface and areas of flowing and nonflowing wells (from Hubbert, 1953). Reprinted by permission.

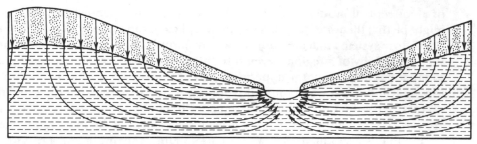

Figure 7.2
Diagrammatic section illustrating ground water flow in a watershed (from King, 1899).

> *The contours of the ground water level show that the (water) surface presents the features of the hills and valleys approximately conformable with the relief forms of the surface above, the water being low where the surface of the ground is low, and higher where the surface of the ground is high.*

In short, King stated that the water table was everywhere a subdued replica of the topography, and water moves from topographically high areas to topographically low areas. Figure 7.2, taken directly from King, diagrammatically illustrates these facts. As with confined systems, the topographic drive is provided (and limited) by differences in surface elevation.

The field evidence offered by Chamberlin (1885) and King (1899) still provides the basis for the categorization of ground water flow under confined and unconfined conditions. At one extreme is Chamberlin's concept of the confined condition, exemplified by dipping aquifers bounded by low but finite permeability units with continuous replenishment at the outcrop areas. If a well is placed in such an aquifer, the water will rise in the well to the elevation of the water table in the intake area, minus any head losses incurred from the point of intake to the point of measurement. The pressure-producing mechanism is the hydrostatic weight of the body of water extending down dip from the water table in the outcrop area. The dynamic mechanism required to maintain the high-pressure system is the continuous replenishment by precipitation. At the other extreme is the unconfined condition discussed by King (1899), where the water table is a subdued replica of topography. The dynamic mechanism required to maintain this flow is likewise a continuous replenishment by precipitation. Indeed, as water offers no resistance to deformation, its movement from high elevations to low elevations would, in the absence of recharge, result in drainage of the water contained in the topographic highs, thereby producing a flat surface of minimum potential energy (the hydrostatic or nonmoving condition). This tendency is opposed by continual replenishment. Elevation differences in the outcrops areas for the confined case, or in the land surface in the unconfined case, provide the topographic drive mechanism for movement of the contained fluids. Continual replenishment assures that this drive persists indefinitely.

Conceptual, Graphical, and Mathematical Models of Unconfined Flow

From what we have learned from the field studies discussed in the previous section, the question arises as to the possibility of establishing at least some features of a conceptual model that will always be valid. One such feature, for example, might be that the water table is a subdued replica of topography, at least for unconfined flow systems that are undisturbed by pumping. Thus, given information on the topography of a region—even if nothing else is known—it is at least possible to say something about the uppermost part of the saturated zone. In addition, following King's portrayal of flow, the flow appears to "branch" at the topographic divides, with the flow going in divergent directions. Thus, topographic divides appear to be ground water divides as well, this being the obvious result when the water table is a subdued replica of the topography. Again, in concert with King, streams appear to receive all the ground water discharge from the adjoining parts of the valley, with none of the flow passing underneath the stream. The streams, or at least the major ones occupying the major topographic lows, are ground water divides in that they prevent the intermingling of ground water from opposite flanks of the same basin. This, too, is inevitable for the lowest topographic points in a region where the water table reflects the topography.

If follows that a useful conceptual model of a flow field should incorporate those facts that have been learned from field investigations. Such a model was provided by Hubbert (1940) and is shown in Figure 7.3. Notice in this model that King's measurements on the relationship between the water table and the land surface are faithfully reproduced. In addition we find that King's observations at the topographic highs and lows are likewise incorporated in this model. This has the effect of setting up individual flow cells in each of the basins shown in Figure 7.3, with the flow in one basin being more or less independent of the flow in adjacent basins. It is noted also that the symmetrical portrayal of the land surface has the effect of producing a mirror image flow pattern for the flow in adjacent flanks of a given basin.

There are, of course, some differences between the presentations of King (1899) and those of Hubbert (1940). In Hubbert's conceptual diagram, the flow is presented by intersecting flow lines and equipotential lines. It is noted that the water table has a point of intersection with each of these equipotential lines, with the value of the hydraulic head anywhere on a given equipotential line taking on this water table elevation at its point of intersection with the equipotential line. This is an important detail and provides information that cannot be obtained from King's diagram. For example, consider a piezometer emplacement somewhere in the vicinity of the topographic high. After passing through the unsaturated zone, the piezometer will encounter the water table and the water will stand in the well

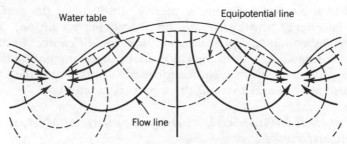

Figure 7.3
Topographically controlled flow pattern (from Hubbert,
1940). Reprinted by permission of the Journal of Geology,
University of Chicago Press. Copyright © 1940.

at the water table elevation. The piezometer is then set deeper, intersecting one of the equipotential lines. This means that the water will now stand in the well at an elevation equal to the elevation of the water table where that equipotential intersects the water table. Thus, the water level in the well will be at a lower elevation than the water table at the point of piezometer emplacement. With continued depth of emplacement, the water level in the well will continually occupy a lower elevation. The reason for this is obvious from Figure 7.3, which shows that in the vicinity of topographic highs, the potential (head) decreases with increasing depth as the flow is directed downward. Regions where the flow of water is directed downward with respect to the water table are called recharge areas (Tóth, 1963). The incorporation of recharge areas is thus part of our conceptual model; the delineation of such areas is a field problem, but we expect them to coincide with topographic highs.

Focus now on the topographic lows, where a piezometer emplacement encounters higher hydraulic heads with depth so that the water level in the well will actually be above the water table adjacent to the well. Hubbert (1940) noted that it is possible to obtain a flowing well for such conditions in that all that is required is the intersection of the piezometer with an equipotential line that has a larger value than the land surface at that point. Again, the reason for this behavior is shown in Figure 7.3, where the head increases with increasing depth so that flow is directed upward. Regions where the flow of water is directed upward with respect to the water table are called discharge areas, and the line that separates the recharge area from the discharge area is called a hinge line (Tóth, 1963). Clearly, the concept of a discharge area must be incorporated in our conceptual model; the field problem is to delineate such areas, and we expect then to coincide with topographically low areas.

Now let's stretch our basin out a bit, given it a flatter water table (topographic) expression (Figure 7.4). We note that the recharge and discharge areas are diminished somewhat, and we notice a large region where the equipotential lines are vertical or near vertical. If a piezometer is emplaced in this region, it will follow a single equipotential line. Thus, once the piezometer encounters the water table, the water level will not change with increasing depth. It will be noticed here

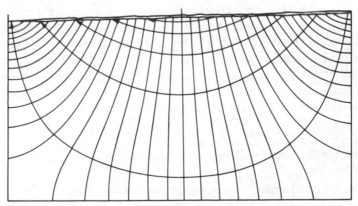

Figure 7.4
Topographically controlled flow pattern (from Tóth, J. Geophys. Res., v. 67, p. 4375–4387, 1962. Copyright by Amer. Geophys. Union).

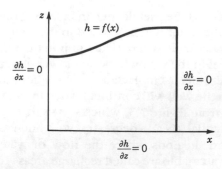

Figure 7.5
Two-dimensional region demonstrating typical boundary conditions for regional flow.

that this is the only set of conditions in the flow field where a piezometer emplaced below the water table actually provides a direct measurement of the water table elevation. Such a region can be called a region of lateral flow, which corresponds to Tóth's (1963) hinge line.

As long as the water table is not fluctuating too severely in response to recharge and discharge, the flow depicted in Figures 7.3 and 7.4 is steady, or nearly so. It follows that these figures represent graphical solutions to Laplace's equation in the two-dimensional region depicted. Thus, they are in fact mathematical models of the flow field. A question now arises as to the nature of the boundary conditions so that it is possible to reproduce other configurations of flow for other water table configurations and basin depth to basin length ratios. Because of the symmetry demonstrated on either side of a stream discharge area, one limb of a basin will serve these purposes. Because of the diverging flow at the topographic high and the confluence of the flow at the topographic low, these lateral boundaries are obviously of the no-flow type. That is to say, no flow crosses the topographic high into or out of the region, or crosses the topographic low into or out of the region. As no-flow boundaries, these may be represented by flow lines indicating that the flow is tangential to them. Provided we assume there is a real impermeable layer at some depth in the region, this impermeable layer is likewise a no-flow boundary. It, too, must be represented by a flow line. These boundaries can be placed in the general category; that is, they will always be present for the major topographic low and the major topographic high of a region, and for a specified depth. The upper surface or water table is neither a flow line nor an equipotential line, and is a specified head boundary where the head varies as a function of space in the same fashion that the elevation of the land surface varies as a function of space. Clearly, this boundary condition will be specific for a given topographic expression. If this head varies as a function of time as well, the problem is no longer one of steady flow, but of unsteady flow. The boundary conditions so described are given in Figure 7.5. An analytical solution to this problem for three fixed boundary conditions and a variable water table configuration is given in Example 7.1.

***Example 7.1** (From Domenico and Palciauskas, 1973)*

A solution for Laplace's equation in the region of Figure 7E.1 where the upper boundary is not specified but all other boundaries are of the no-flow variety is

$$\phi(x, z) = a_o + \sum_{n=1}^{\infty} a_n \cosh \frac{n\pi x}{L} \cos \frac{n\pi x}{L}$$

where ϕ is the hydraulic potential and the coefficients a_o and a_n are determined from the equation of the water table for various special cases. The coefficients are determined from

$$a_0 = \frac{1}{L} \int_O^L \phi(x, z_0) \, dx$$

and

$$a_n = \frac{2}{L \cosh (n\pi z_0/L)} \int_O^L \phi(x, z_0) \cos \frac{n\pi x}{L} \, dx$$

The equations given are general in that they apply to any water table configuration. For a specific case, assume that the equation of the water table $\phi(x, z_0) = A - B \cos \pi x/L$. The coefficients then become

$$a_o = A; \qquad a_1 = \frac{-B}{\cosh (\pi z_0/L)}; \qquad \text{other } a\text{'s} = 0$$

The solution for Laplace's equation then becomes

$$\phi(x, z) = A - \left[\frac{B \cosh (\pi z/L)}{\cosh (\pi z_0/L)} \right] \cos \frac{\pi x}{L}$$

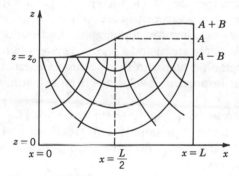

Figure 7.E1
Two-dimensional flow region.

This case permits a good understanding of the geometrical controls on the spatial distribution of the potential. It is noted that at the points $z = z_0$ for $x = 0$, $x = L$, and $x = L/2$, the value of the potential is $A - B$, $A + B$, A, respectively. Further, at $x = L/2$ for all z, the potential equals A; that is, a vertical equipotential line exists at $x = L/2$. This vertical equipotential corresponds to the midline of flow.

Effects of Basin Geometry on Groundwater Flow
The mathematical description of the boundary conditions given in Figure 7.5 was first stated by Tóth (1962, 1963) in two modern-day classic studies. This work was

the first significant extension of Hubbert's (1940) ideas in over 20 years. Tóth's work is based on the following assumptions:

1.   The medium is isotropic and homogeneous to a specified depth, below which there exists an impermeable basement.

2.   Flow is restricted to a two-dimensional vertical section. The topography can be approximated by simple curves, such as straight lines or sine waves, and the water table is a subdued replica of the topography.

3.   The upper boundary of the two-dimensional section is the water table, the lower boundary is the impervious basement, and the lateral boundaries are ground water divides.

Tóth was not content to investigate the individual flow cells demonstrated in Figure 7.3. Instead he considered a sinosoidal water table with a regional slope of the form (Figure 7.6)

$$b(x,\, z_o) = \left[z + \frac{B'x}{L} + b\, \sin \frac{2\pi x}{\lambda} \right] \tag{7.1}$$

where the second term on the right-hand side corresponds to the regional slope and the third term to the local relief superposed on the regional slope. In this third term, b is the amplitude of the sine wave and λ is the number of oscillations, so that L/λ is the number of flow cells. As noted in Figure 7.6, B' is the height of the major topographic high above the major topographic low and x is a variable ranging from zero to the basin length L. From this analysis, Tóth (1963) identified a local system, which has its recharge area at a topographic high and its discharge area at a topo-graphic low that are adjacent to each other; an intermediate system, which is characterized by one or more topographic highs and lows located between its recharge and discharge areas; and a regional system, which has its recharge area at the major topographic high and its discharge area at the major topographic low (Figure 7.7). The important conclusions from this study are as follows:

1.   If local relief is negligible ($b \sin 2\pi x/\lambda = 0$, Eq. 7.1) and there is a general slope of the topography, only regional systems will develop where ground water moves from the major topographic high to the major topo-graphic low.

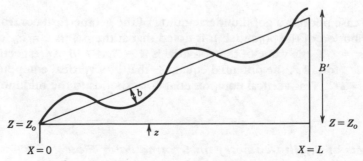

Figure 7.6
Sinusoidal water table with a regional slope.

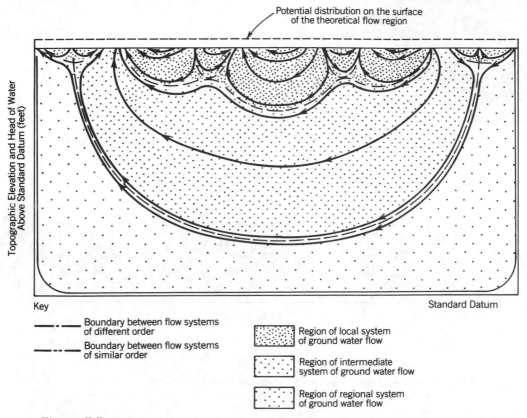

Figure 7.7

Two-dimensional isotropic flow model showing the distribution of local, intermediate, and regional ground water flow systems (from Tóth, J. Geophys. Res., v. 68, p. 4795–4812, 1963. Copyright by Amer. Geophys. Union).

2. If regional slope is negligible ($B'x/L = 0$, Eq. 7.1), only local systems will develop. The greater the relief the deeper the local systems that develop.

3. If both local relief and a regional slope are negligible, neither regional nor local systems will develop. Waterlogged areas will be common, and the ground water may discharge by evapotranspiration in response to a flat water table near the land surface.

4. Given both local relief and a regional slope, local, intermediate, and regional systems will develop.

Effects of Basin Geology on Ground Water Flow

The work of Tóth demonstrates the importance of basin geometry on ground water flow. By geometry, we mean the ratio of basin depth to length and the relief or topography making up the surface. As the medium was assumed isotropic and homogeneous, little can be inferred about the influence of basin geology on ground water flow, that is, the influence of the stratigraphic framework with its attendant permeability variations. This problem was addressed by Freeze and Witherspoon (1966, 1967) within the framework of the Tóth model. In their approach, *n* homogeneous and isotropic interlayered formations were treated within the two-dimensional vertical section (Figure 7.8). The water table

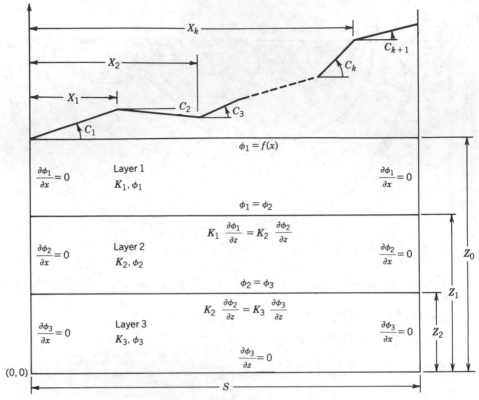

Figure 7.8
Two-dimensional, unconfined model with n homogeneous, isotropic layers (from Freeze and Witherspoon, Water Resources Res., v. 2, p. 641–656, 1966. Copyright by Amer. Geophys. Union).

is incorporated with a series of straight-line segments so that any configuration can be approximated as a function of the space variables.

The boundary value problem solved by Freeze and Witherspoon is actually n interrelated problems. When n equals 3, Figure 7.8 applies. The condition that $b_n = b_{n+1}$ ensures a continuous potential across the contact of two layers. A second condition is required so that fluid mass is conserved when flow crosses a boundary between adjacent strata of differing permeability. This is referred to as the tangent refraction law (Figure 7.9). Analytically, a refraction of the flow lines occur such that the permeability ratio of the two units equals the ratios of the tangents of the angles the flow lines make with the normal to the boundary.

$$\frac{k_1}{k_2} = \frac{\tan a_1}{\tan a_2} \tag{7.2}$$

In Figure 7.9b, the hydraulic gradient in the lower-permeability unit is steepened to accommodate the flow crossing the boundary from the unit of higher conductivity; in Figure 7.9c, both the hydraulic gradient and the cross-sectional flow area in the high-permeability unit are decreased to accommodate the flow crossing the boundary from the low-permeability unit.

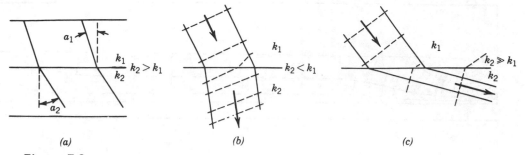

Figure 7.9
Diagrams of flowline refraction and conditions at the boundaries between materials of differing permeability.

Freeze and Witherspoon (1966, 1967) employed numerical methods to solve the flow problem discussed earlier. With analytical methods, both the problem and the solution are expressed in terms of parameters. This was the method used by Tóth (1962, 1963). With numerical solutions, both the problem and the solution are expressed in terms of numbers. Hence, with the latter method, each case is quite specific, and several cases must be solved to obtain some general information on the subject.

The main conclusions of Tóth (1963) regarding control of topography on the ground water flow were verified with the modeling effort of Freeze and Witherspoon. This is demonstrated in the sequence of figures given as Figure 7.10. Figure 7.10a shows that a lack of local relief results in development of a single regional system. Figure 7.10b demonstrates the effect of adding a slight amount of topographic relief, and Figure 7.10c demonstrates the local, intermediate, and regional flow system development when the relief is altered to that characteristic of hummocky terrain.

The primary effort of Freeze and Witherspoon (1967) dealt with determining the role of permeability contrasts in influencing various degrees of flow line refraction at the permeability boundaries between adjacent strata as well as the influence of this contrast in controlling the resulting equipotential and flow line distributions throughout the formations. These effects are shown in Figure 7.11. Figure 7.11a and 7.11b show the effects of a high-permeability layer underlying a layer of lower permeability. As the permeability of the lower layer increases, it acts like a pirating agent for the flow, forcing near-vertical flow through the uppermost unit. This is best demonstrated by the orientation of the equipotential lines in the lower-permeability unit for a 100-to-1 permeability contrast. Figures 7.11c and 7.11d show the effects of a high-permeability lens in the flow field. If the lens is in the recharge area, it too acts like a pirating agent, forcing vertical flow through the overlying unit and creating a discharge scenario at its terminus that would not be predicted on the basis of topography. Shifting the high-permeability lens to the discharge end of the system has similar effects on the overlying units.

The sloping stratigraphy case is demonstrated in Figures 7.12a and 7.12b. If the dip of the bed is in the direction of topographic drive (Figure 7.12a), the flow is down dip. With the hydraulic conductivity contrasts demonstrated, the behavior here is identical to what we would expect from the Chamberlin model. If the dip of the bed is opposite the topographic drive, the flow is up dip, with the large

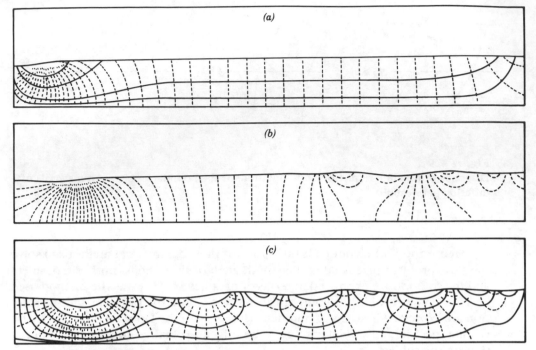

Figure 7.10
Effect of water table configuration on regional ground water flow through homogeneous isotropic mediums (from Freeze and Witherspoon, Water Resources Res. v. 3, p. 623–634, 1967. Copyright by Amer. Geophys. Union).

conductivity contrasts causing downward flow in the overlying-low permeability unit. Such a system has been postulated for part of the flow field in southern Maryland (Figure 7.13).

The points just discussed become important when ascertaining the influence of structure on ground water flow. For example, all diagrams illustrating flow in folded beds consisting of anticlines and synclines show the flow directed at right angles to the synclinal and anticlinal axes, that is, up and down the structural limbs (see, for example, Figure 7.1). This, of course, is appropriate for such folded rocks that obtain their topographic drive from the elevated outcrop areas. If, however, a series of anticlines and synclines have been uplifted into a meteoric recharge area and possess a distinct plunge to lower elevations, the flow will be parallel to the structural axes in the direction of the plunge. In this case, the elevated rocks in the recharge area provide the topographic drive, and the anticlinal axes serve as ground water divides separating the flow systems in adjoining synclines.

Based on the information presented thus far, it is possible to obtain a picture of regional flow for any basin for which the pertinent data are available. Data requirements include permeability distribution and geometry of the basin boundaries. Although the ideal homogeneous case is contrary to field observations, it is merely a condition in the mathematical development, not an assumption necessary to the general validity of the theory. Further, the steady-state solutions do not deny the existence of a fluctuating upper boundary of flow. The main argument is that the effect on flow patterns will be small if (1) the fluctuating zone is small

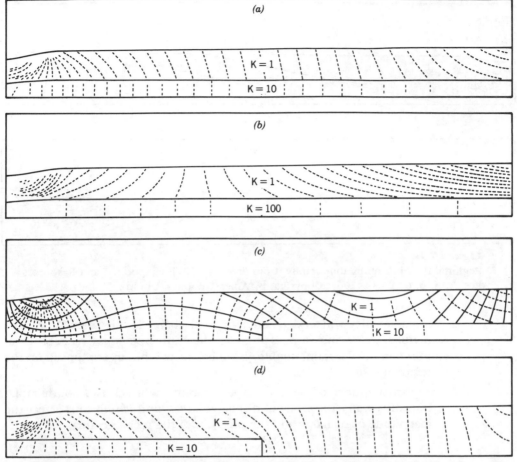

Figure 7.11
Regional flow showing the effect of permeability contrasts in adjacent layers (a) and (b) and the effect of a high permeability lens in the flow field *c* and *d* (from Freeze and Witherspoon, Water Resources Res., v. 3, p. 623–634, 1967. Copyright by Amer. Geophys. Union).

compared with total saturated thickness and (2) the relative configuration of the water table is unchanged throughout the cycles of fluctuation. Given that these conditions are reasonably satisfied, the value of the information provided by the flow net can best be appreciated if one is interested in the distribution of recharge and discharge areas along the water table, the depth and lateral extent of local systems in hummocky terrain, and the degree to which the zones of high permeability act as major conductors of water and their overall influence on the flow pattern. In summarizing the factors that influence these items, Freeze and Witherspoon (1967) cite the following as the most important:

1. The three most influential factors affecting the potential distribution are
    a. The ratio of basin depth to lateral extent.
    b. The configuration of the water table.
    c. Variations in permeability.

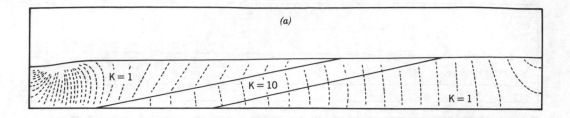

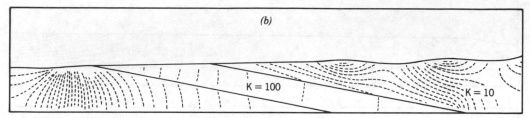

Figure 7.12
Regional flow in sloping topography (from Freeze and Witherspoon, Water Resources Res., v. 3, p. 623–634, 1967. Copyright by Amer. Geophys. Union).

2. A major valley will tend to concentrate discharge in the valley. Where the regional water table slope is uniform, the entire upland is a recharge area. In hummocky terrain, numerous subbasins will be superimposed on the regional system.

3. A buried aquifer of significant permeability will act as a conduit that transmits water to principal discharge areas, and it will thus affect the magnitude of recharge and the position of the recharge area.

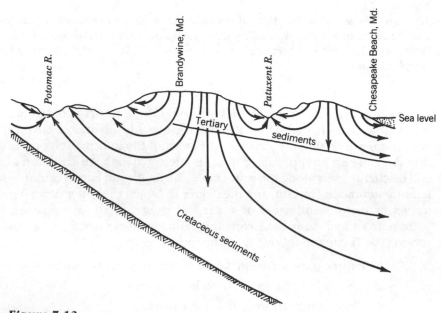

Figure 7.13
Diagrammatic cross-section through southern Maryland showing the lines of ground water flow (from Back, 1960).

4. Stratigraphic pinchouts at depth can create recharge or discharge areas where they would not be anticipated on the basis of water table configuration.

A field application of the ideas just presented is shown in Figure 7.14 for the Palo Duro Basin in Texas. The Palo Duro has been studied as a potential repository for the disposal of high-level radioactive waste by the U.S. Department of Energy. Figure 7.14*a* is a geologic cross section demonstrating the various geologic units in the section, where the salt beds have been considered to serve as the repository unit. Figure 7.14*b* shows the flow patterns as measured or inferred in the field, with the large hydraulic gradient across the salt beds reflecting their low permeability. Figure 7.14*c* is a numerical model of the flow field. Note that the downgradient lateral boundary is depicted as a constant head in that the flow presumably continues across this area to discharge in parts of the basin not depicted in this flow diagram.

Ground Water in Mountainous Terrain

Mountainous terrain occupies about 20% of the Earth's land surface and presents some special problems in hydrologic analysis, mostly because hydraulic head data are seldom available to confirm the suggested conformation between the land surface elevation and elevation of the water table. Further, mountainous regions promote deep circulation of ground water, denying access to ground water outcrops that aid in regional investigation. A further complication arises because mountainous regions are frequently fractured and may be in an active state of compression or extension, suggesting that fracture apertures may be functionally related to the state of stress in the Earth's crust. Thus, the elevation of the water table may be intimately related to a changing hydraulic conductivity of the region and the variable climatic factors that influence infiltration. For a given hydraulic conductivity distribution, the lower the infiltration rate, as controlled by climatic factors, the deeper the water table.

It follows that the simple relations given by King (1899) and Hubbert (1940) and substantiated in numerous studies in low-lying sediments may not apply to mountainous terrain characterized by a fracture permeability. In such cases, the water table may be considered a free surface whose depth and configuration depend on the interplay between infiltration and permeability distribution. Jamieson and Freeze (1983) were among the first to utilize a free surface technique as applied to mountainous terrain in British Columbia. They utilized a fixed infiltration rate to estimate the range of hydraulic conductivity that might be expected to produce a water table at a given depth. Each hydraulic conductivity pattern resulted in a different elevation of the free water surface. This study has been expanded by Forster and Smith (1988) to include the effects of varying infiltration rates, surface topography, topographic symmetry, and permeable fracture zones. They conclude that permeability has the greatest impact on the mountainous flow system; asymmetry can cause the displacement of the ground water divides from the topographic divides, and a relatively small increase in the vertical permeability of fractures relative to the horizontal permeability causes significant declines in the water table elevation (on the order of 400 m). The authors also state that high relief can promote deep ground water circulation to elevated temperature regions, requiring a modeling of both the fluid flow and the thermal regime. The thermal aspects of mountainous regions will be discussed in a later chapter.

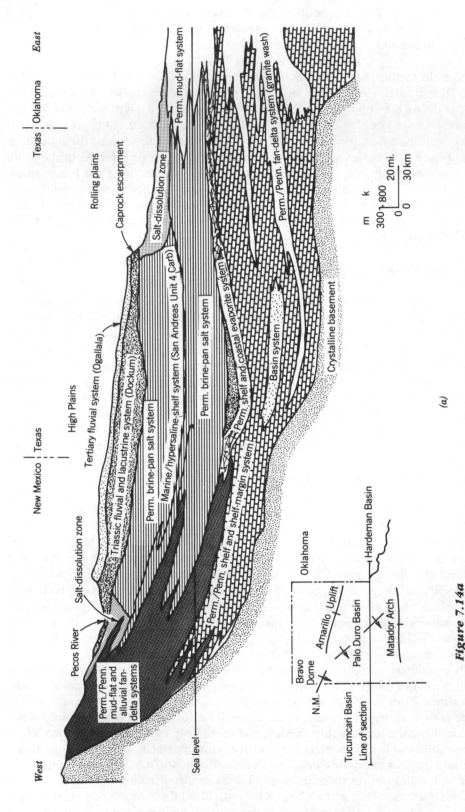

(a)

Figure 7.14a
Geology, regional flow, and simulation of flow in the Palo Duro Basin. Geology (*a*) and simulation of flow (*c*) from Senger and others, (1987); regional flow (*b*) from Bair (1987). Figure (*b*) reprinted by permission of Groundwater. Copyright © 1987. All rights reserved.

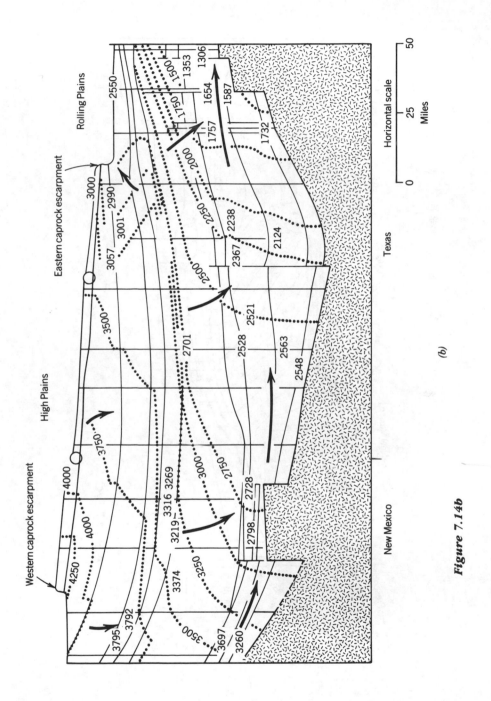

Figure 7.14b

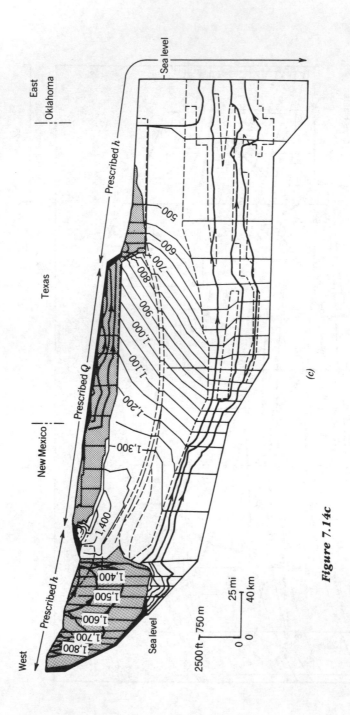

Figure 7.14c

One mountainous region that does not qualify as "data poor" is Yucca Mountain in southern Nevada, a volcanic tuff pile located north of Las Vegas. Yucca Mountain is currently under study as the first site in the United States for disposal of commercial nuclear waste. As part of the southern Great Basin, a distinct feature of the Yucca Mountain region is crustal extension, which suggests that the mountain has become "wider" during Late Miocene–Late Quaternary time, with the "widening" occurring in the form of vertical fracture development. The mountain is characterized by extremely low infiltration (less than 1 mm/yr), a deep (600 m) rather contorted water table configuration, and a layered permeability distribution that appears to decrease with increasing depth and which may reflect the tectonic history of the region (Szymanski, 1989).

There are several features worth noting about the hydraulic conductivity structure. In the lowermost zone, the field values of hydraulic conductivity are comparable with laboratory values of the saturated matrix hydraulic conductivity. The hydraulic conductivity zones are colinear, with a noted absence of stratigraphic control. Further, the boundaries between the layered structure are not at a uniform depth, but get closer to the surface from southeast to northwest across the Mountain. The uppermost zone of highest permeability constitutes the unsaturated domain.

Winograd and Thordarson (1975) have shown that the water table throughout the Death Valley flow system (which includes Yucca Mountain) is a series of "plateaus" separated by sharp steps and ranges in altitude from over 1900 m to below sea level in Death Valley. On the "plateaus," the gradients are small, measured in tens of meters per kilometer. Between the "plateaus," slopes as large as 10% and 30% are present. Szymanski (1989) relates this water table configuration to the hydraulic conductivity structure, with steep slopes occurring where the upper transmissive part is missing or thin, and small gradients in areas characterized by a thick upper transmissive unit (Figure 7.15). There are, of course, alternative causes such as faulted barriers to flow or highly transmissive zones conducting a limited flux. At any rate, it would appear that there is a propensity for deep circulation in mountainous terrain, and the position and configuration of the water table may be independent of surface topography.

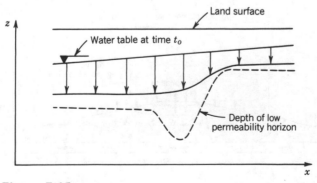

Figure 7.15
Schematic illustration of permeability control of the water table (from Szymanski, 1989).

7.2   Surface Features of Ground Water Flow

The saturated ground water system discussed thus far has been isolated from the basin hydrologic cycle by introducing the effect of its environment through boundary conditions. In the real world there is a recognizable interaction between ground water and the rest of the basin hydrologic cycle. Some of the manifestations of this interaction are readily seen at the surface, whereas other types are best examined with models of some sort.

Surface features of ground water flow include all observations that can be used to ascertain the occurrence of ground water, including springs, seeps, saline soils, permanent or ephemeral streams, ponds, or bogs in hydraulic connection with underground water. The areal occurrence of these features is invariably restricted to areas of ground water discharge, and their comprehension requires some knowledge of the nature of ground water outcrops (Meyboom, 1966b). The significance of these features may best be understood by examining a few studies that focus on their interpretation.

Recharge-Discharge Relations

The prairie profile (Figure 7.16), consisting of a central topographic high bounded on both sides by areas of major natural discharge, has been offered as a model of ground water flow to which all observable ground water phenomena in a prairie environment can be related (Meyboom, 1962, 1966b). Geologically, the profile consists of two layers, the uppermost being the least permeable, with a steady-state flow of ground water toward the discharge areas. The unconfined flow pattern has been substantiated by numerous borings for both small-scale systems, say, a typical knob and adjacent kettle common to rolling prairie topography, and for a scale of the magnitude of the prairie profile. Recharge and discharge areas have been delineated and are characterized by a decrease in head with increased depth and by an increase in head with increased depth, respectively. The occurrence of flowing wells is noted in parts of the discharge area.

As it is apparent that most of the natural discharge occurs by evapotranspiration, considerable attention has been given to this phenomenon and to the surface

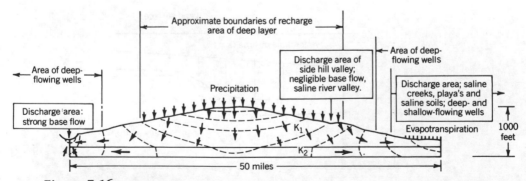

Figure 7.16
The prairie profile (from Meyboom, 1966b: Groundwater studies in the Assiniboine River drainage, Pt. I, The evaluation of a flow system in southern central Saskatchewan Geol. Survey Canada Bull. 139. Energy, Mines, and Resources Canada. Reproduced by permission of Supply and Services, Canada, 1990).

features observed where it occurs. Included among these observations are (1) the occurrence of willow rings and the chemical character of water bodies centered within them, (2) the distribution and types of vegetation of saline soils with respect to the occurrence of local and regional flow systems, and (3) the location of ponds and bogs with respect to ground water flow.

With regard to (1), Meyboom (1966a, 1966b) noted that most of the willows are located in the higher areas, which suggests that the recharge area is covered by numerous discharging points, each of which is characterized by a willow ring. Willows are phreatophytes, defined by Meinzer (1927) as plants whose roots extend to the water table so that they obtain water for transpiration directly from the saturated zone. As willows have a low tolerance for saline (alkali) conditions, their occurrence is associated with waters that have not moved very far in the system. Although their occurrence in the higher watershed areas is widespread, they do occur elsewhere in the basin. These other occurrences are regarded as possible manifestations of local flow systems superimposed on the regional ground water flow.

With regard to (2), extensive areas of saline soil occur in areas of regional ground water discharge, where a net upward movement of mineralized ground water takes place. A consistent transition noted is from willow vegetation on the watershed areas to halophytic plant communities within the discharge areas. A halophyte is a phreatophyte that thrives on saline waters. A related transition is noted for local flow systems that receive water from highly saline formations. Where the local system is not replenished by saline water formations, the ground water is relatively fresh, and saline zones fail to develop. In the latter case, fresh water phreatophytes occupy the discharge area and make possible its delineation as an end point of a local system. Tóth (1966) reported the results of a similar surface mapping method based on observations in the prairie environment of Alberta, Canada.

The mapping techniques discussed here are only of value in arid and semiarid regions, where surface water is not sufficiently abundant to mask or conceal the surface effects of ground water flow.

Investigations of the type discussed have not been restricted to the prairie environment. Extensive investigations in Nevada, starting perhaps with Meinzer (1922), have focused on surface features as they relate to the occurrence of ground water. In discussing basins in the Great Basin and the lakes that occupied them during the Pleistocene epoch, Meinzer (1922) recognized three types: (1) those in which lakes still exist, (2) those that no longer have lakes and are discharging ground water from the subterranean reservoirs into the atmosphere by evaporation and transpiration, and (3) those that do not have lakes and in which the water table is everywhere so deep that they do not discharge ground water except by subterranean leakage out of the basin.

The basin of type 2 is exemplified in early studies of Big Smokey Valley (Meinzer, 1917). Water enters the alluvial basin by influent seepage of streams of the alluvial fan areas, and at the contact between the surrounding mountain ranges and the alluvium. Ground water is discharged by evaporation and transpiration in the valley lowlands. Studies of the Big Smokey Valley and similar environments provided the first understanding of the surficial manifestations of ground water flow, including the distribution of soluble salts in discharge areas (Meinzer, 1927). The concentric arrangement of phreatophytes was first noted in these early studies, with salt grass occupying the inner belt where the water table is near land surface. Other species, such as greasewood in northern Nevada and mesquite in southern Nevada, occupy the outer belt, where the water table is further below

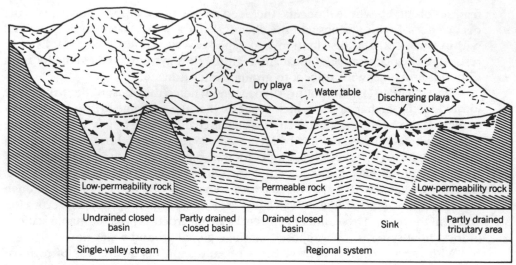

Figure 7.17
Flow systems in the Great Basin (from Eakin and others, 1976).

land surface (Meinzer, 1927). This type of basin is exemplified in Figure 7.17, where the basin alluvium is underlain by low-permeability rock.

The basins of type 3 have been investigated by Winograd (1962), Eakin (1966), and Mifflin (1968), which resulted in the delineation of two regional flow systems in carbonate terrain in southern and eastern Nevada. These systems are typified by large drainage areas encompassing several topographic basins, relatively long flow paths, and large spring areas of invariant discharge, where water temperature is several degrees higher than mean air temperature. In Figure 7.17 these basins are exemplified where the basin alluvium is underlain by highly permeable carbonate rock that syphons off the flow so as to eliminate ground water discharge at the surface.

One of these regional flow systems occurs in eastern Nevada in the White River area (Figure 7.18). The major discharge points occur as springs in White River Valley, Pahranagat Valley, and Moapa Valley, the last identified as the terminal point of the flow system. Boundaries to this flow system, at least to the south, appear to be controlled by thick clastic rocks whose permeability is considerably less than that of the carbonates. Eakin (1964) obtained closure of the water balance within a 13-valley area, with 78% of the recharge estimated as occurring in the four northern valleys, and 62% of the discharge estimated to be from spring areas in Pahranagat and Moapa Valley. A tentative analysis suggests a 15- to 20-year lag in spring discharge response to recharge from precipitation in the recharge areas (Eakin, 1964). This is demonstrated on Figure 7.19, where precipitation about 100 miles north of the springs in Moapa Valley is plotted against spring discharge. The two curves match for the sharp rise in precipitation during the period 1935–1941 and the rise in the discharge graph during the period 1956–1960. The above-average precipitation for that time period occurred regionally throughout eastern Nevada and western Utah, thus suggesting that above-average precipitation was widespread throughout the drainage area.

Most of Winograd's (1962) conclusions are based on the results of an extensive drilling program within the Nevada test site and adjacent areas. Hydraulic

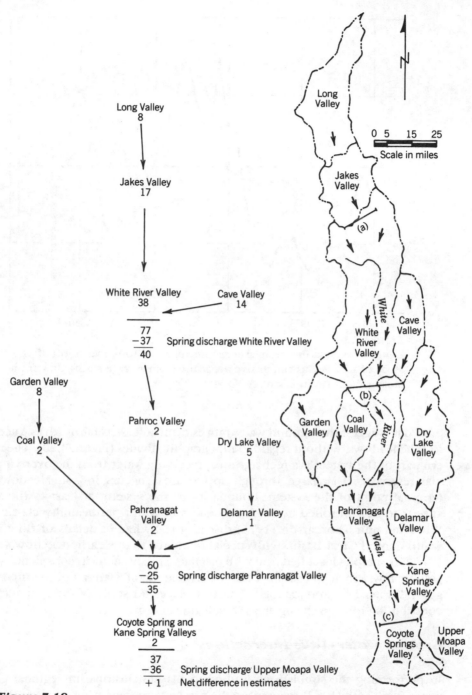

Long Valley
8

↓

Jakes Valley
17

↓

White River Valley   Cave Valley
38   ←   14

$\dfrac{77}{\underline{-37}}$ Spring discharge White River Valley
40

↓

Garden Valley
8

↓

Coal Valley   Pahroc Valley
2   2   Dry Lake Valley
5

Pahranagat   Delamar Valley
Valley   1
2

$\dfrac{60}{\underline{-25}}$ Spring discharge Pahranagat Valley
35

↓

Coyote Spring and
Kane Spring Valleys
$\dfrac{2}{}$

$\dfrac{37}{\underline{-36}}$ Spring discharge Upper Moapa Valley
+1   Net difference in estimates

Figure 7.18
Estimated average annual recharge to and discharge from a regional ground water system in eastern Nevada (from Eakin, Water Resources Res., v. 2, p. 251–271, 1966. Copyright by Amer. Geophys. Union).

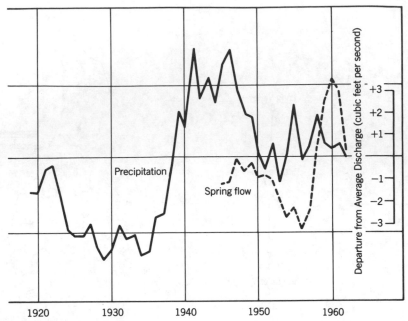

Figure 7.19
Cumulative departure from average precipitation 100 miles north of
Moapa Valley versus cumulative departure from average spring flow from
Moapa Valley (from Eakin, 1964).

head data suggest that ground water moves from at least as far north as Yucca Flat
to Ash Meadows, without regard to topographic divides (Figure 7.20). The northern part of the system is a recharge area, receiving water from the overlying alluvial reservoirs. Discharge through spring areas occurs in Ash Meadows, the
terminal point of the system. Boundaries to the system, at least to the south,
appear to be controlled by thick sequences of lower-permeability clastic rock.
Winograd and Thordarson (1975) discuss the hydraulic details of flow in the
South Central Great Basin, with special reference to this carbonate flow system.

The examples just cited deal with discharge phenomena in conspicuous topographic lows. Such areas reflect the observable manifestations of the theories of
ground water flow and are logical starting places for study of the system of flow
contributing to large spring areas in volcanic and carbonate terrain.

Ground Water–Lake Interactions

In performing the simulations leading to his conclusions on regional ground
water flow, Tóth (1963) recognized the presence of stagnation zones at the junc-
ture of flow systems. A stagnation point is a point of minimum head along a subsurface divide separating one flow system from another. This is shown
diagrammatically in Figure 7.21*a*, where the subsurface divide separates a local
flow system associated with a lake from a regional system. The minimum head at
the stagnation point is greater than the head in the lake. Thus, the lake of Figure
7.21*a* is termed a discharge lake in that it receives ground water discharge from
the aquifer. A recharge lake is one that recharges the ground water flow, and a
through-flow lake (Figure 7.21*b*) is one that both receives and releases water to the

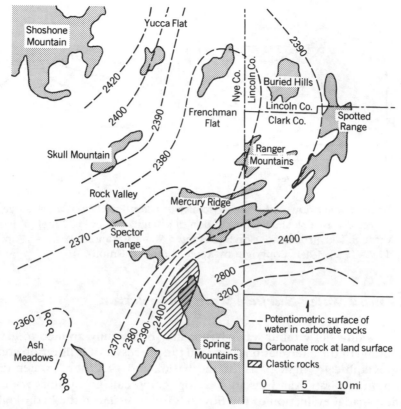

Figure 7.20
Regional flow systems in rocks underlying the Nevada test site and
adjoining area (from Winograd, 1962.)

ground water flow (Born and others, 1974, 1979). For a 63-lake sample studied by
Born and others (1974), most of which were in the Midwest region, 24 were discharge lakes, 23 were through-flow lakes, 6 were recharge lakes, 4 were placed in
some combined category, and 6 changed from one type to another in response to
seasonal changes (Anderson and Munter, 1981).

The search for stagnation points is aided by model studies of the type initiated
by Tóth (1963) and Freeze and Witherspoon (1967). One example is provided by
Winter (1976, 1978). Figure 7.22 gives results typical of these studies. In Figure
7.22*a*, a stagnation zone exists on the downgradient side of the single lake. The
lake is thus classified as a discharge lake and receives water from the flow local system. If a high-conductivity layer is introduced at depth (Figure 7.22*b*), the stagnation point is eliminated and the pirating effect of the high-conductivity layer
converts the lake into a recharge lake. Multiple-lake systems on a regional slope
have also been investigated by Winter (1976) and can be recharge, discharge, or
through-flow depending on basin geometry and geology. Steady-state models
have commonly been used to describe such behavior in the vicinity of lakes
(McBride and Pfannkuch, 1975; Larson and others, 1976). In addition to these
model studies, other investigators (Meyboom, 1966a, 1967; Anderson and Munter, 1981) have conducted field investigations in the vicinity of lakes for the purpose of determining seasonal changes in the flow pattern.

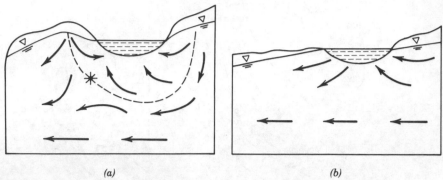

(a)   *(b)*

Figure 7.21
Ground water flow paths in the vicinity of lakes. Panel (*a*) is a discharge lake
where the asterisk shows the location of a stagnation point. Panel (*b*) is a flow-
through lake (from Anderson and Munter, Water Resources Res., v. 17, p.
1139–1150, 1981. Copyright by Amer. Geophys. Union).

Ground Water–Surface Water Interactions

Flow studies of ground water basins may be usefully applied to gain an understanding of the interrelations between the processes of infiltration and recharge at
topographically high parts of the basin, and of ground water discharge via
evapotranspiration and base flow at topographically low points. Such studies permit a spatial evaluation of the flow of ground water in the hydrologic cycle. For
example, at least some of the water derived from precipitation that enters the
ground in recharge areas will be transmitted to distant discharge points, and so
cause a relative moisture deficiency in soils overlying recharge areas. Water that
enters the ground in discharge areas cannot overcome the upward potential gradient, and therefore becomes subject to evapotranspiration in the vicinity of its
point of entry. The hinge line separating areas of upward and downward flow may
thus serve as a boundary common to areas of relative soil moisture surplus and
deficiency (Tóth, 1966). In nonirrigated agricultural areas, this may be reflected
by variations in crop yield. Further, the ramifications of human activities in discharge areas are immediately apparent. Some of these include waterlogging
problems associated with surface water irrigation of lowlands, or due to the destruction of phreatophytes.

The spatial distribution of flow systems will also influence the intensity of
natural ground water discharge. The main stream of a basin may receive ground
water from the area immediately within the nearest topographic high, and possibly from more distant areas if the region is characterized by both regional and local
systems. If base flow calculations are used as indicators of average recharge, significant error may be introduced because base flow may represent only a small
part of the total discharge occurring downgradient from the hinge line.

Other interesting aspects resulting from such studies involving the basin
hydrologic cycle deal with determining the actual amount of ground water that
effectively participates in the hydrologic cycle. Tóth (1963) calculated that about
90% of the recharge never penetrates deeper than 250 to 300 ft. Similar conclusions have been arrived at by tritium dating studies, which demonstrated a stratification of tritium and tritium-free waters (Carlston and others, 1960).

As might be expected, the majority of surface water–ground water studies
have been conducted with model calculations. Transient flows of water in the

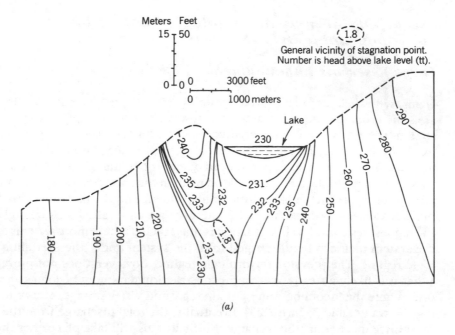

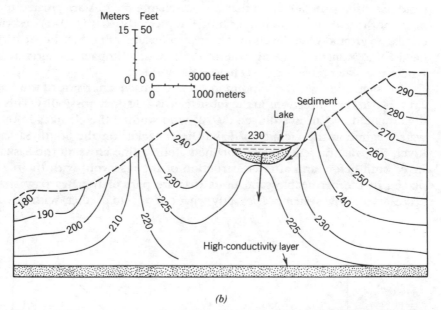

Figure 7.22
Flow conditions in the vicinity of a lake demonstrating the effect of a high permeability layer at depth (from Winter, 1976).

unsaturated zone have been examined by Rubin (1968), Hornberger and others (1969), and Freeze (1971). The models couple saturated-unsaturated conditions. The ideas of infiltration capacity have likewise been studied in a transient mode with models that incorporate coupled unsaturated and saturated behavior (Freeze, 1969). Models that couple surface and subsurface flow have been used to study the runoff components of the hydrologic cycle (Smith and Woolhiser, 1971) as well as aquifer-stream connections (Hornberger and others, 1970).

7.3   *Some Engineering and Geologic Implications of Topographic Drive Systems*

Large Reservoir Impoundments

Dams, with their attendant surface water reservoirs, are generally constructed in areas of ground water discharge. If a reservoir is to occupy a deeply incised stream, the water level in the reservoir can be higher than the regional ground water levels underlying the stream reach and in the adjacent flood plains that are inundated. With such an imposed head due to the filling of the reservoir, ground water discharge to the stream and flood plain in the entire region that is inundated is no longer possible. This has little or no influence on the rate of recharge taking place at adjacent topographic highs where the ground water levels are higher than the highest reservoir level. A situation is thus created where more water is entering the system than can be discharged from the system under the prevailing hydrologic regime. The prevailing hydrologic regime, however, does not prevail long. What was likely a steady state in preconstruction days now becomes a transient one, where the incoming water goes into ground water storage, everywhere raising the water table (Figure 7.23). Eventually, the total discharge from the system must once more equal the recharge. Normally this will take place after the water table has risen sufficiently so that new discharge regions are created to take the place of those that can no longer function properly. Cady (1941) describes such effects on ground water levels in response to surface storage in the vicinity of Flathead Lake, Montana, making note of the potential impact on agricultural lands where the water table is not far below the surface.

In association with an impoundment and subsequent rising of the water table, a reversal in flow directions in the subsurface is a distinct possibility. This has been documented by Van Everdingen (1972) who studied the changes in the ground water regime in the vicinity of Lake Diefenbacker on the South Saskatchewan River. Piezometric levels were obtained along a line crossing the Saskatchewan River both before and after construction of the reservoir, with the piezometers placed in the permeable sand units that are part of the Bearpaw shales. The prereservoir flow system was exactly what one would predict, with flow directed

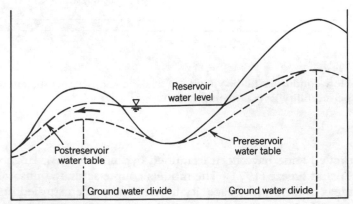

Figure 7.23
Changes in water levels due to construction of reservoirs (from Bryan, 1925).

toward the Saskatchewan River where it eventually ends up as ground water discharge. After construction, there were some notable changes. In the deeper unit, the transversal flow was maintained toward the river valley, but at much lower gradients. In the intermediate unit, the transversal flow was reversed away from the river valley. In the uppermost unit, all the transversal flow was unidirectionally away from the site.

Frequently, an increase in spring discharge has been observed in the vicinity of new reservoirs. This has been reported by Gupta and Suknija (1974), who measured increments in spring flow on the order of 0.145 liters/sec in springs some 15 miles from a dam in India. The dam is constructed on basalt. Gupta and Rastogi (1970) report fluctuations in 4 piezometers installed in limestone near a reservoir in Greece that were identical to those associated with fluctuations in the lake level. Snow (1972) has also reported increases in spring discharge in the vicinity of the same reservoir.

The impounding of reservoirs can also lead to some stability problems. Underneath the reservoir area itself, the increased fluid pressures as shown in Figure 7.24 are partially compensated by the increased weight of the water that must be added to the total stress. Downstream from the reservoir, however, no such compensation occurs. In Figure 7.24*b*, two principal effects are noted. These include increased deformation of the valley bottom and increased deformation and landslides in the valley walls.

Excavations and Tunnels: Inflows and Stability

The Sea-Level Canal
By their very nature, sea-level canals are built in coastal areas of low topographic relief, such as the Florida and Panama peninsulas. The natural equilibrium position of the fresh water–salt water interface is already at a fairly high elevation, although still below sea level. Excavation below the water table permits the canal to act as a natural drain, causing water levels to decline with an attendant rise of the interface (Figure 7.25). In a preliminary investigation of a sea-level canal across Flordia, Paige (1936) contended that a new ground water table would evolve and be steepest near the canal and approach the older one asymptotically at a distance not in excess of 15 miles. As the canal would be cut approximately 30 ft below sea level, it was taken as nothing more than an extension of the existing coast line of the Gulf of Mexico. The argument here was that freshwater is currently present in coastal wells along the Gulf of Mexico, and similar conditions will prevail along the canal cut. These conclusions were criticized by Brown (1937) and other notable experts in the profession at that time, including Thompson, Meinzer, and Stringfield (1938). Paige, however, was not a stranger among giants, being an eminent engineering geologist and later chairman and organizer of the prestigious Berkey volume honoring Charles Berkey.

In his investigation, Paige noted the general relationship of ground water to topography, focusing in particular on the various water table highs associated with topographic highs and the fact that the outcrop areas of the formation to be cut (Ocala limestone) actually served as discharge areas (Figure 7.26). He noted topographically high areas both south and north of the canal, between which there was generally lower ground along the canal route. Based on this information along with various borings and detailed investigations, Paige concluded that the canal route traversed the most important discharge area of the Floridian plateau. In his reply to his critics, Paige (1938) stated that this discharge area played an important part in

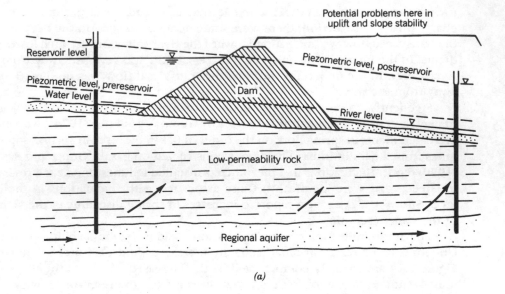

(a)

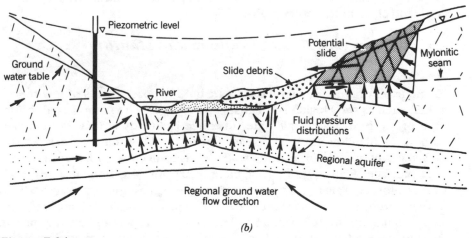

(b)

Figure 7.24
Effect of dams on water levels showing (*a*) possible stability problems created by
reservoir blocking regional ground water discharge and (*b*) effects of high fluid
pressures, on valley bottom and wells downstream from the reservoir (from Patton and
Hendron, 1974).

the selection of the canal route, as opposed to many others examined by the Army
Engineers. This was likely one of the first large-scale engineering projects where the
ground water flow system with its attendant recharge and discharge areas actually
figured in the decision process on site location.

One of the important issues was the potential effect on Silver Springs, the largest limestone spring in the United States and an important tourist attraction. Silver Springs is 3 miles north of the canal route and about 40 ft above sea level. Paige
argued that Silver Springs was an outlet of flow systems draining southward from
the north so that the canal should not affect the source of water supplied to the

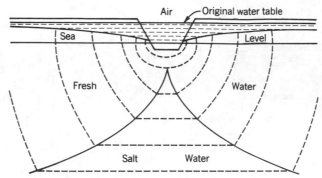

Figure 7.25
Salt water-fresh water relationship due to excavation along coastal areas (from Hubbert, 1940). Reprinted by permission of the Journal of Geology, University of Chicago Press. Copyright © 1940.

springs. In the final analysis, Paige (1936) recognized that there would be a substantial effect on water levels near the canal, but these effects would be of little consequence in terms of damage. As it turned out, the potential impact of the canal was probably not as severe as the impact of ground water development that took place in later years. Observations by Brown and Parker (1945) along a completed portion of the canal showed that salt water encroachment had been facilitated by hydraulic gradients created not by the canal but by pumping.

From this brief discussion, it is clear that most of the arguments for or against the canal were largely qualitative, based exclusively on the geologic controls of the hydrology. In view of the times it could be no other way. If such an event were to repeat itself today, the debate would not go on in a scientific journal, but in the courts, with a host of lawyers and their various consultants and more than likely more than one mathematical model that purportedly completely describes the projected response.

Ground Water Inflows into Excavations

Ibrahim and Brutsaert (1965) provide a simple analytical solution for predicting inflows and a projected water-level decline in the vicinity of excavations such as the sea-level canal just discussed. The case they considered is shown in Figure

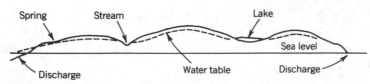

Figure 7.26
General relation of ground water to topography (from Paige, 1936). Reprinted from *Economic Geology*, 1936, v. 3, p. 537–570.

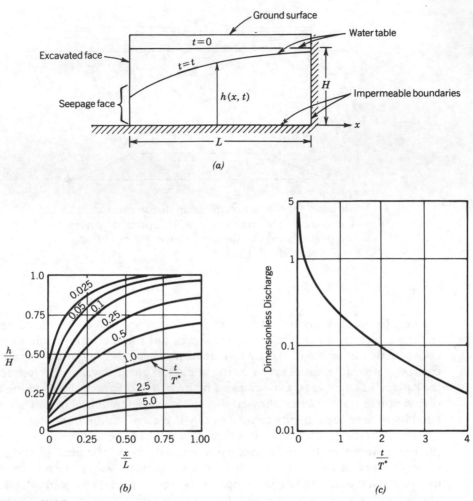

Figure 7.27
Prediction of groundwater inflows into an excavation (from Ibrahim and Brutsaert, 1965). Reprinted by permission of the American Society of Civil Engineers.

7.27 as a two-dimensional homogeneous isotropic region. The transient response of the water table is presented by the dimensionless graph of Figure 7.27*b* where

$$N_{FO} = \frac{t/S_y L^2}{KH} = \frac{t}{T^*} \tag{7.3}$$

where S_y is the specific yield and the product KH is the transmissivity, here obviously assumed not to be affected by the water-level decline, and T^* is the time constant for the basin. The dimensionless discharge χ is given in Figure 7.27*c* and is expressed

$$\chi = \frac{S_y L \, q'}{KH^2} = \frac{T^* \, q'}{HL} \tag{7.4}$$

where q' is the flow rate into the excavation per unit length of excavation and has the units L^2/T.

To apply this method, information is required for K, S_y, H, and L. Verma and Brutsaert (1970, 1971) provide numerical models for similar problems.

Example 7.2

Assume that the basin parameters S_y, L^2, K, and H are such that $T^* = 10^2$, in days. For a time period $t = 100$ days, $t/T^* = 1$. From Figure 7.27*b*, this means that the ground water head at the face of the excavation ($x = 0$) is about 0.8 of the original saturated thickness H, and at the ground water divide ($x = L$) it is about 0.5 of the original saturated thickness. The position of the water table may be readily determined at other points. For $t/T^* = 1$ and Figure 7.26*c*, the dimensionless discharge of Eq. 7.4 is determined to be about 0.25, from which q' can be determined. Note that for longer times, t/T^* increases and the discharge rate q' declines in response to the flatter hydraulic gradients.

The Stability of Excavations in Ground Water Discharge Areas

One of the most common problems in engineering construction is the bottom heaving of trenches or other excavations constructed in areas of ground water discharge. Several case histories are available, but we shall focus only on one to bring out the principles involved. This particular situation occurred during a project where cuts up to 65 ft deep were required in glacial materials. The major drainage in the region adjacent to the proposed excavations occupied an intermoranic sag, which are noted for their heterogeneity, variation in grain size, and general lenticular character of the sands and gravels. The sands and gravels that do occur in these geologic environments generally have a high permeability in that they have been well sorted by glacial melt water.

A lake occupying a gravel pit was located adjacent to the construction area. The lake was pumped prior to construction for a period of several months, lowering the lake level about 50 ft. Five observation wells indicated changes of 8 to 16 ft over this period. Four of the wells were bottomed in the underlying limestone, which is highly fractured with excellent hydraulic communication throughout. The one well in the glacial material showed a head drop of about 8 ft, and this well was closer to the gravel pit than some of the deeper wells that penetrated the limestone that showed a greater response to pumping. It is likely, then, that pumping from the pit did not materially affect the water levels in the glacial materials where the overall heterogeneity prevented a good hydraulic communication between the permeable lenses. The response in the underlying limestone was likely the result of pressure relief associated with pumping the lake.

Contract documents required the contractor to maintain the lake level about 35 ft below its natural level with the idea that this would sufficiently reduce pressures in the glacial material so that stability problems would not be encountered. However, during construction, the bottom of the excavations heaved, and the construction had to be stopped until dewatering wells were put in at an additional cost.

One interesting point in all this is that the position of the water table was entirely overlooked in these investigations. All the observation wells in the area indicated water levels that were at or a few feet below land surface. Indeed, one of the piezometers was flowing. This was termed the "water table" in the preconstruction investigations. However, backhoe data from several different locations indicated that the sediments were dry from land surface to in excess of 15 ft in depth; in none of these excavations was the water table encountered. Hence, we reemphasize the fact that water levels in piezometers that penetrate below the top of the zone of saturation do not indicate the position of the water table, but indicate a head or potential that may be greater than the water table, or less than the water table, depending upon where one is measuring in the flow system. In this case, as in all discharge areas, the former is the actual situation.

A simple calculation could have removed any doubt as to the need of dewatering or pressure relief prior to construction. Assume that the water table is at 15 ft below the surface, the dry unit weight of the material is on the order of 100 lb/ft^3, and the saturated unit weight is 110 lb/ft^3. The total vertical stress at the base of the glacial material (100 ft) is thus 10,850 lb/ft^2. If the water level in a piezometer finished at the base of the glacial material registered a head at land surface, the neutral stress is calculated to be about 6240 lb/ft^2. Hence, after removal of about 43 ft of material the effective stress at the base of the glacial material is reduced to zero. It would appear that cuts approaching 60 ft in depth are not likely to be stable without some pressure relief provided by pumping wells.

Ground Water Inflows into Tunnels

One of the potential hazards and important factors controlling the rate of advancement in driving a tunnel is the ground water inflow through its face and sides. This is generally dictated by local geologic and structural features along the tunnel alignment. Frequently, information on the lithology alone can give some indication of the potential magnitude of the problem. Limestone, for example, is more hazardous than sandstone because of the potential for solution channels that can provide a high concentration of inflow. On the other hand, weakly cemented sandstone may disintegrate completely under the high fluid pressures that may be encountered. Shale, along with low-permeability crystalline rock, may have little inflow. However, when these rocks have secondary openings such as fractures, the inflow potential increases.

Numerous studies are available on the factors found to control inflows for specific tunnels driven in various rock types. Hurr and Richards (1966) found that in igneous and metamorphic terrain, the inflows depend on the degree of fracturing, on the interconnection of the fractures, and on their continuity throughout the rock (Table 7.1). For reference purposes, Goodman and others (1965) consider flow rates of 100 to 500 gpm as low, 500 to 1500 gpm as moderate, and more than 1500 gpm as heavy.

According to Krynine and Judd (1957), water may enter a tunnel in three different ways:

1. Water may break through under large pressures at any place on the drift periphery.
2. Water may penetrate through the walls of the drift in the form of drops or continuous currents.
3. Water may drip from the top of the drift.

Table 7.1
Relation between joint spacing and groundwater discharge (from Hurr & Richards, 1966)*

Joint spacing (ft)	Active zone			Passive zone		
	Number of occurrence	Average discharge per occurrence (gpm)	Total discharge (gpm)[1]	Number of occurrences	Average discharge per occurrence (gpm)	Total discharge (gpm)[1]
Less than 0.1	None	—	—	8	6	45
0.1–0.5	7	78	549	62	55	3383
0.5–1.0	20	43	859	50	20	1001
More than 1.0	13	49	632	10	3	29

[1]Based on initial rate.
*Geol. Soc. America Publication, Eng. Geol., v. 3, p. 80–90.

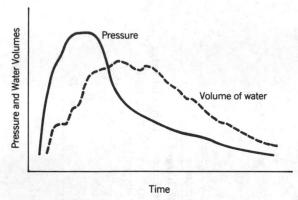

Figure 7.28
High pressure associated with excess discharge in
Kurobe Transportation Tunnel (from Haga, 1961).

The first case is the most dangerous one and can seldom be predicted in advance. Such inflows are frequently associated with gouge-containing fault zones. The low-permeability gouge can act as a dam for perched water so that the removal of the gouge during construction can cause the inflows. Such was the situation encountered during the construction of the Kurobe Transportation Tunnel (Haga, 1961), the San Jacinto Tunnel (Henderson, 1939), and the Awali Tunnel (Engineering News Record, 1960). Figure 7.28 shows schematically the rise and decline of the pressure and entering water volumes encountered in the Kurobe Tunnel.

Occasionally, inflows have very high temperatures, a result of active faults in the region or deeply circulating water. In the Tecolote Tunnel, water of 113°F was encountered (Rantz, 1961). The tunnel inflow was accompanied by a depletion in spring flow (Figure 7.29).

From a quantitative perspective there are three questions of most concern in tunneling, namely, what is the maximum anticipated inflow rate through the walls or through the face of the tunnel, how long will this maximum be sustained, and

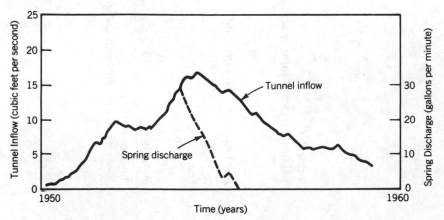

Figure 7.29
Depletion of spring flow due to driving the Tecolote Tunnel (from Rantz, 1961).

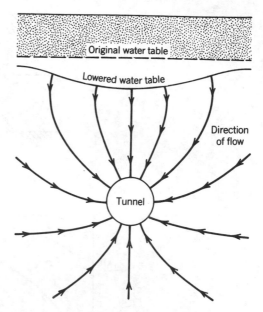

Figure 7.30
Seepage into a tunnel.

how will this inflow rate vary over time? Those are difficult questions that can never be answered in an absolute manner, but some progress has been made for a first-cut approximation.

Goodman and others (1965) consider three cases of interest. The first treats the tunnel as a steady-state drain, with interest focused on the steady rate of inflow through the sides. The second treats a cumulative inflow case for the very special circumstance where the overlying water table is drained to the extent that it intercepts the tunnel. The third case treats a transient inflow through the face of the tunnel when the tunnel breaks through into a highly permeable zone. All these solutions are extremely limited because of the usual assumptions of uniform thickness, isotropic and homogeneous materials, and infinite aquifers.

Figure 7.30 shows the effect of a tunnel on the flow paths in a simple geologic setting. For the steady-state drain, the water table remains at its original height, and the inflow per unit length of tunnel is (Goodman and others, 1965)

$$q' = \frac{2\pi K H_o}{2.3 \log(2H_o/r)} \tag{7.5}$$

where K is the hydraulic conductivity, r is the radius of the tunnel, and H_o is the depth of the centerline of the tunnel below the steady water table. The similarity between Eq. 7.5 and the steady-state response of pumping wells (Chapter 5) is apparent.

The geometry for application of the breakthrough inflow through the face is shown in Figure 7.31a. Goodman and others (1965) have solved this problem numerically but have generalized the solution in the form of a family of curves of dimensionless head versus distance behind the face for different values of the product of the hydraulic conductivity and time Kt (Figure 7.31b). The procedure for applying this graph is as follows. For a given time, Kt is determined and the head is noted as a function of distance x behind the face, with $x = 0$ corresponding to

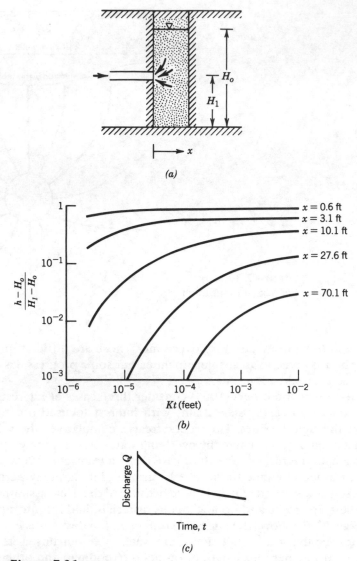

Figure 7.31
Predicting tunnel inflows through the face (from Goodman and
others, 1965. Geol. Soc. America Publication, Eng. Geol., v. 2,
p. 39–56).

the face. Given this spatially distribution of head for a single point in time, the gradient $\partial h/\partial x$ is noted and the inflow through the face is determined with Darcy's law $Q = KiA$, where A is the area of the tunnel face and i the gradient. A later time will give yet another head distribution and a new inflow rate. With increased time, the gradient becomes flatter and the inflow decreases (Figure 7.31c).

Landslides and Slope Stability

In his classic treatment of landslide development, Terzaghi (1950) noted the following causes for the relatively rapid movement of slopes: (1) external changes, (2) earthquake shocks, (3) lubrication by water, and (4) ground water level rises.

External changes refer to man-made or natural activities that disturb the stability of an existing slope, such as undercutting the toe, say, by meandering rivers or excavations. Earthquake shocks increase the shearing stresses relative to the shearing resistance as accelerations associated with earthquakes represent horizontal forces. The lubrication by water was dismissed by Terzaghi because water actually acts as an antilubricant in that there is more friction between "wet" mineral grains than between dry ones. The main emphasis of Terzaghi's paper is the effect of water-level changes in the interpretation of the Mohr-Coluomb equation, stated here as

$$\tau = \tau_o + (\sigma - P) \tan \phi \qquad (7.6)$$

where τ is the shearing resistance per unit area, τ_o is the cohesion per unit area, σ is the total normal stress at a point due to the weight of the solids and fluids, P is the fluid pressure at the point, and ϕ is the angle of sliding friction. The greater the fluid pressure P, the greater is the portion of the overburden that is carried by the water; when $\sigma - P = 0$, the overburden is in a state of floatation (Terzaghi, 1950), and the resistance to shear is reduced to zero for cohesionless material, and to the cohesive strength for materials that possess cohesive properties.

Terzaghi (1950) argued that seasonal variations in rainfall can give rise to seasonal variations in the fluid pressure, thereby reducing the shearing resistance independent of any effect on the angle of sliding friction. Thus, during periods of heavy or prolonged rainfall, slopes become more susceptible to failure because of the attendant increases in water levels and decreases in effective stress. This sometimes leads to a definite periodicity in slope failures. Seepage from sources of water such as newly built reservoirs or unlined canals can have similar results. Hence, if a slope starts to move, the means for stopping the movement must be adopted to the processes that started it (Terzaghi, 1950). As most slides are related to increases in fluid pressures, this means that both surface and subsurface drainage or dewatering is called for.

As with most original developments in science, once the preliminary physics of a problem are brought out in theory, others immediately rush in to embellish the original ideas. By embellish we mean to improve by adding detail. The detail added in this case is of a hydrologic and geologic nature. Deere and Patton (1967) put Eq. 7.6 in a hydrologic perspective, recognizing that slopes are part of a regional flow system that is characterized by elevation differences in water levels and in land surface. They recognize that small topographic deviations within a regional system produce local flow systems that are often the most critical with regard to fluid pressures and slope stability, especially in areas of ground water discharge. In addition, these authors demonstrated sequences of dipping beds susceptible to high pressures, the susceptibility of weathered shales to landslides when the slope is in a discharge area, and the pore water pressures that can develop at the base of low-permeability materials overlying more permeable rocks in a ground water discharge area. These same ideas are carried over by Patton and Hendron (1974).

Fractured rock masses are treated more or less along the same lines. There are, however, some differences, again, purely geological. In that the porosity of fractured rock may be two or three orders of magnitude smaller than that of sedimentary rock, a typical rainfall event should result in a relatively larger water-level response. Further, because of the prevailing shear stresses in a body of rock forming a slope at some inclination to the horizontal, the joints may be wider and more numerous than those located in otherwise similar flat lying rock. Hence, the quantity of water that can enter the rock mass is greater in the vicinity of the slope.

Numerous references are available on ground water and slope stability. These include Cedergren (1967), Morgenstern (1970), Kiersch (1973), and Royster (1979).

Suggested Readings

Eakin, T. A., 1966, A regional interbasin groundwater system in the White River area, southeastern Nevada: Water Resources Research, v. 2, p. 251–271.

Forster, C., and Smith, L. 1988, Groundwater flow systems in mountainous terrain: 2, Controlling factors. Water Resources Research, v. 24, p. 1011–1023.

Freeze, R. A., and Witherspoon, P. A., 1967, Theoretical analysis of regional groundwater flow: 2, Effect of water table configuration and subsurface permeability variation: Water Resources Research, v. 3, p. 623–634.

Tóth, J., 1963, A theoretical analysis of groundwater flow in small drainage basins: J. Geophys. Res., v. 68, p. 4795–4812.

Winograd, I. J., and Thordarson, W., 1975, Hydrogeologic and hydrochemical framework, South Central Great Basin, with special reference to the Nevada Test Site: U.S. Geol. Survey Prof. Paper 712-C.

Winter, T. C., 1976, Numerical simulation analysis of the interaction of lakes and groundwaters: U.S. Geol. Survey Prof. Papers 1001.

Problems

1. Briefly discuss the meaning of the boundary conditions given on Figure 7.5. What boundary conditions are included in Figure 7.8 that are not in Figure 7.5? Why?

2. Construct flow nets in the *x-z* plane for Figure 7.5 where the left-hand boundary is a flow (as opposed to a no-flow) boundary, for Figure 7.16 for the case where $K_1 = K_2$, for the alluvial undrained closed basin given in Figure 7.17, and for Figures 7.21*a* and 7.21*b*.

3. Briefly discuss the conditions under which local, intermediate, and regional flow systems may develop in homogeneous terrain.

4. Briefly discuss how the following geologic factors can contribute to interbasin flow:

   a. Rolling topography superimposed on a regional slope

   b. A lenticular high-permeability unit, extending in the subsurface over two or more basins

   c. A regionally extending high-permeability unit at considerable depth

5. What is meant by surficial features of ground water flow? Why are these features important in reconnaissance and mapping studies in regions where water wells are scarce?

6. Set up a problem similar to the one discussed for the sea-level canal. Assume a flat water table 40 ft above sea level, a canal that extends 30 ft below sea level, and a ground water divide (springs) approximately 3 miles from the canal. Assume a hydraulic conductivity of 1×10^2 ft/yr and a specific yield of 0.01.

    a. Calculate the head h at the canal and at the divide for times of 10, 20, 40, and 100 years.

    b. Calculate the flow rate q' into the canal for 10 years and 100 years.

7. Perform the calculations for the case history given in the section dealing with stability of excavations.

8. A tunnel of radius 5 ft penetrates a permeable rock unit with a saturated thickness of 50 ft. The tunnel is located 30 ft below the water table. The hydraulic conductivity of the rock is 10^{-3} ft/day.

    a. Calculate the head h in the penetrated rock at the end of one day.

    b. Calculate the approximate inflow into the face of the tunnel at the end of one day.

Chapter Eight

Ground Water
in the Earth's Crust

8.1 **Abnormal Fluid Pressures**
in Active Depositional
Environments

8.2 **Pore Fluids in Tectonic**
Processes

The role of ground water in performing geologic work has long been a topic of interest in the Earth sciences. We have seen this in discussions on disequilibrium compaction, pressure solution, and landform development in the form of karst. To pursue an understanding of this role further, we recognize that water does not reside in a passive porous solid, but instead there exists a complex coupling between the moving fluid, entities that might be carried by the fluid, and the solid matrix itself. In this chapter we focus on the coupling between stress, strain, and pore fluids. The moving "entities" (heat energy and mass) are discussed in later chapters.

8.1 Abnormal Fluid Pressures in Active Depositional Environments

Origin and Distribution

Sediment deposition and compaction occur concurrently in depositional environments, with a resultant porosity loss. Prior to porosity reduction, the added sediment load is carried at least in part by the water contained in the sediments, causing the fluid pressure to rise above its normal value. This has been schematically demonstrated with the spring analogy in Chapter 4 where, in the absence of fluid expulsion, the spring (rock matrix) cannot compress, and the added load is borne by the water. As demonstrated further with the example of water-level changes in response to passing trains, there is a diffusion of pore water to areas of lower pressure so that the sediment load is ultimately transferred to the matrix of skeletal grains. From the perspective of abnormally pressured environments, the main concern is with the length of time required for this stress transfer to take place, for this is the time period over which anomalously high fluid pressures may persist.

Bogomolov and others (1978) and Kissen (1978) have established the framework for categorizing abnormal pressure zones in depositionally active basins (Figure 8.1). The meteoric regime shown in the figure is the familiar topographic

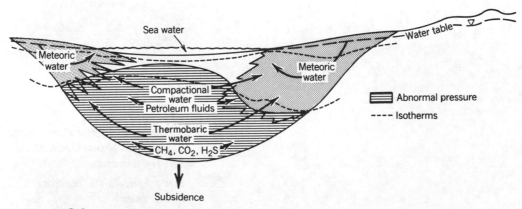

Figure 8.1
Conceptual diagram illustrating various hydrologic regimes in an active depositional environment (from Galloway and Hobday, 1983. Copyright © 1983 Springer-Verlag, New York. Reprinted with permission).

drive system where meteoric water enters the formations in the higher portions of the basins with discharge taking place at sea level. Hence, the shoreline is an active area of discharge, and this discharge area can recede or transgress depending on shifts in the shoreline. The compaction regime is characterized by upward migration of pore waters, as suggested in 1899 by King. The fluids so affected may have been buried with the sediment, and the abnormal fluid pressures are due in part to the added sediment load in parts of the basin undergoing active deposition.

To accumulate a thick sedimentary pile such as shown in Figure 8.1, the basement rock must be in some state of subsidence. Whatever the driving mechanism, such subsidence assures that the sediments and their contained fluids become subjected to a thermal field. This is referred to as the thermobaric regime (Figure 8.1), where the fluid pressures may be caused by the thermal expansion of water in a low-permeability environment or by trapped waters of mineral dehydration. Phase transformations such as gypsum to anhydrite or smectite to illite are generally temperature dependent and will produce free water that can take on the pressure of the overburden provided the permeability is sufficiently low so as to inhibit movement to areas of low pressure. In addition, the sedimentary pile may undergo horizontal compression, as might occur because of uplift adjacent to the deposits undergoing burial. Such compression or tectonic strain is yet another pressure-producing mechanism.

The environments just discussed will persist as long as the basin continues to take on added sediment. Once the depositional period ends and tectonic forces contribute to uplift, the flow system will evolve eventually to a topographic drive system. Thus, in the evolution of hydrologic basins, it is useful to think of two end members (Coustau and others, 1975): the subsidence-induced abnormal pressure environment, as might be expected in young basins undergoing rapid deposition, and the more mature topographically driven systems that eventually evolve from them after uplift. Figure 8.2 represents these two end members. As noted in the abnormal pressure environment (Figure 8.2*a*), the lower water zone contains original formation waters, whereas the upper zone is continually invaded by downward-moving meteoric water. Following uplift and erosion and the evolution of a topographic drive system, there is one distinct hydrologic system that is characterized by recharge at the outcrop areas, cross-formational discharge, and a definite transition from meteoric water to highly saline water or brines with distance from the outcrop areas.

The relationships between normal fluid pressure and geostatic pressure are demonstrated in Figure 8.3. At any given depth z, the difference between the geostatic pressure (vertical stress) and the normal fluid pressure (neutral stress) has been defined as the effective stress in accordance with the effective stress law. In this figure, the fluid pressure P at any given depth is 0.43 times the vertical (geostatic) stress. In some geologic environments, fluid pressures have been noted to be as high as 90% of the geostatic pressure. In such cases, the effective stress is reduced considerably, which is generally reflected by higher than normal porosities and lower than normal bulk densities of the water-saturated rock. Figure 8.4 shows the relationship between pressure and depth in some abnormally pressured environments reported by Watts (1948). At relatively shallow depths, many of these areas are characterized by normal or near-normal fluid pressures. With increasing depth, fluid pressures are seen to be in excess of normal, in some cases approaching the total overburden pressure.

According to Deju (1973) the occurrence of pressures in excess of normal is worldwide. In Europe, abnormal pressure has been noted in the Aquitaine Basin,

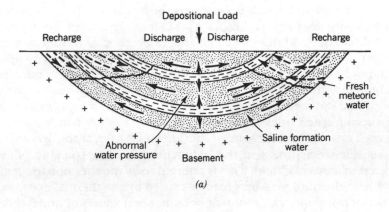

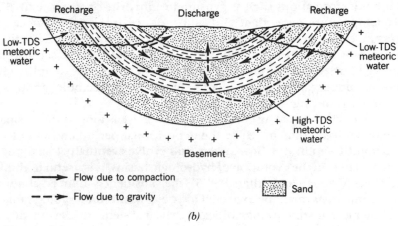

Figure 8.2
Hydrology of (*a*) compacting and (*b*) mature basins (modified from
Caustau and others, 1975).

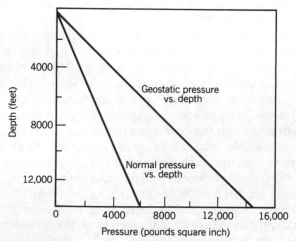

Figure 8.3
Relationship between normal fluid pressure and
geostatic pressure.

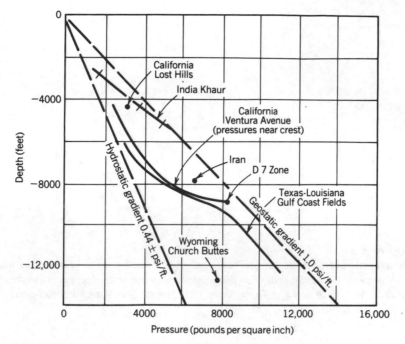

Figure 8.4
Relation of fluid pressure and depth in some excess-pressure oil pools
(from Watts, 1948). Reprinted by permission of the Amer. Inst. of Mining, Met., and Petrol. Engrs.

France, in the Alps, Italy, offshore Norway (0.8 psi/ft, 1.8×10^4 Pa/m), Holland,
northwest Germany (0.8 psi/ft, 1.8×10^4 Pa/m), the Carpathians (0.96 psi/ft,
2.17×10^4 Pa/m), and in the Soviet Union (0.8 psi/ft, 1.8×10^4 Pa/m). The sediment types include dolomite, evaporites, carbonates, marine shales, and sandstone. In Eastern Canada, an offshore zone was noted at about 12,000 ft with a
pressure gradient of 0.6 psi/ft (1.36×10^4 Pa/m). The occurrences in the United
States are well known and include California (both off and onshore), the Green
River Basin in Wyoming, the Uinta Basin in Utah (0.8 psi/ft, 1.8×10^4 Pa/m), and
several locations in Texas and Louisiana. Some detailed studies are provided by
Watts (1948) for the Ventura oil field in California and Dickinson (1953) for the
Louisiana Gulf Coast. Other occurrences have been reported in and around
Australia, Japan, West and East Africa, the Arabian Peninsula, South America,
South East Asia, and the Indian subcontinent.

On the basis of this worldwide survey, Deju (1973) presented the following
conclusions:

1. Two mechanisms appear to be responsible for most of the abnormal pressures observed: tectonic stress and rapid deposition of massive shales.

2. High fluid pressures of tectonic origin are usually associated with the
   major mountain systems of the world (Pyrenees, Alps, Himilayas, Andes).

3. High fluid pressures are often associated with mud volcanoes, diapirs,
   piercement folds, and dikes.

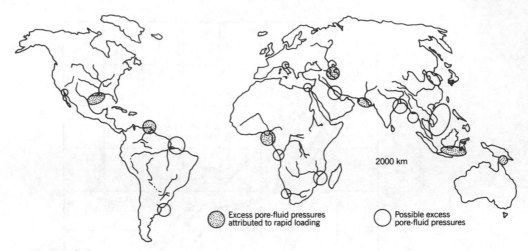

Figure 8.5
Geographic areas where geopressures have been attributed to continuous loading and compaction, or where sediment deposition rates are such that geopressures would be predicted (from Sharp and Domenico, 1976, Geol. Soc. Amer. Publication, from Geol. Soc. Amer. Bull., v. 87, p. 390–400).

4. Rapidly deposited bentonitic shales and thick evaporite beds seem to form the best seal for fluid in the subsurface.

5. Most high fluid pressure occurrences are restricted to Tertiary sediments, or younger. Older sediments have apparently had sufficient time to dissipate the pressure.

Modern depositional environments associated with abnormal pressures (or the potential for abnormal pressures) are shown in Figure 8.5. It is obvious that such environments are rather common and were likely equally widespread throughout geologic time.

Mathematical Formulation of the Problem

From a physiomathematical viewpoint, the study of abnormal pressure development is a study of competing rates of pressure production and dissipation. The complexity of the problem is well illustrated in Figure 8.6. Gravitational loading and tectonic forces act to compress the pore fluid by compressing or otherwise reducing the pore itself, thus leading to the generation of abnormal fluid pressures. With continued burial in a thermal field, phase transformations produce free pore water that can take on the pressure of the overburden, and this magnitude of pressure can be continually regenerated in the "window" that defines the critical temperature range over which the transformation takes place. The pressure produced by either of these mechanisms can be augmented by the thermal expansion of the fluid, which is most effective in low-permeability sediments under high confining pressures. With water flow to areas of lower pressure, some redistribution of the heat occurs by convective transport. Such convective transport can be of importance in the thermal evolution of the sedimentary basin. Given that the rate of pressure dissipation by fluid flow is slower than the rate of pressure generation, some sort of fracture is likely. This is demonstrated in Figure

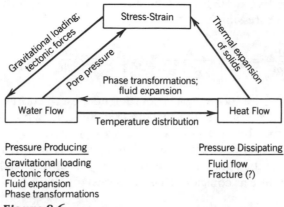

Figure 8.6
Coupling in abnormal pressure development.

8.6 with the coupling of pore pressure to the stress-strain characteristics of the rock. If the inelastic behavior is in the form of an extensive microfracture zone, the enhanced pore volume can accommodate the volumetric expansion of the heated water, providing pressure relief. If the microfracture zone coalesces to form macrofractures, the permeability may be enhanced and the pressures dissipate more readily. Thus, fracture generation may serve as a reserve mode of pressure dissipation that comes into play when the rate of pore water diffusion is inadequately slow compared to the rate of pressure generation.

The complexity and coupling of the processes described in Figure 8.6 are such that a complete physical description is not possible. First, the physics of many of the processes described are not sufficiently well known from a quantitative perspective. For example, if one wishes to describe the pressure production associated with phase transformations, the rate of transformation from layered or bound water at the low-temperature phase, to free water at a higher-temperature phase must be known. If one wishes to examine fracturing as a pressure dissipative mechanism, some threshold pressure must first be identified and incorporated in such a way that describes the rate at which the fracture dissipates pressure. Given that these mechanisms and rates can be described, they can in general be incorporated into a flow equation of the form of the diffusion equation. However, at this stage of our knowledge, we must focus on these pressure producing and dissipating processes that can be well described.

For stress-related problems, the appropriate equation is (Palciauskas and Domenico, 1989)

$$\frac{\partial P_{ex}}{\partial t} = \nabla \cdot \left[\frac{K}{\rho_w g(1/R)} \nabla P_{ex} \right] + \frac{R}{H} \frac{\partial \sigma}{\partial t} + (\alpha_m R) \frac{\partial T}{\partial t} \qquad (8.1)$$

where P_{ex} is the abnormal or excess fluid pressure, σ is the mean confining stress, and R and H are Biot's (1941) coefficients introduced in Chapter 4. A one- and three-dimensional isothermal version of this equation was developed in Chapter 4. This three-dimensional version includes the fluid pressure changes due to the thermal expansion of water, where T is temperature and α_m is the change in fluid mass content at constant stress and pressure when the temperature varies.

When bulk volume strain ϵ is taken as the pressure-producing mechanism, the appropriate expression becomes

$$\frac{\partial P_{ex}}{\partial t} = \nabla \cdot \left[\frac{K}{\rho_w g(1/Q)} \nabla P_{ex} \right] - (\xi Q) \frac{\partial \epsilon}{\partial t} + (\xi \alpha_b + \alpha_m) Q \frac{\partial T}{\partial t} \qquad (8.2)$$

where Q and ξ are Biot's (1941) coefficients described in Chapter 4. The coefficients associated with the thermal effect include α_b, which is the thermal expansion of the rock body at constant stress and pressure. The combined term $\alpha_m R$ represents the change in pressure with temperature at constant mass and constant stress, whereas the combined term $(\xi \alpha_b + \alpha_m)Q$ represents the change in pressure with temperature at constant mass and constant strain.

Isothermal Basin Loading and Tectonic Strain

In virtually all problems associated with basin loading, it is generally assumed that one of the principal stresses is oriented in a vertical direction, commonly expressed as

$$\sigma_v = \rho_b g z \qquad (8.3)$$

where the subscript v designates that the stress is vertical, ρ_b is the density of the water-saturated rock, where the saturation is often assumed to persist from land surface downward, g is the acceleration of gravity, and z is depth. The product $\rho_b g$ is taken as the unit weight of the overburden, where the density can vary with lithology changes. The stress σ_v is the pressure exerted by the total weight of the overburden (solids plus water) and is commonly referred to as the geostatic pressure. The overburden pressure may be a major σ_{11} or a minor σ_{33} principal stress, depending on whether the rocks are in a state of compression or extension. The geostatic pressure gradient σ_v/z is normally taken as 1 psi/ft (2.26×10^4 Pa/m).

There are a few other designations for the state of stress in the Earth's crust, many of which are useful in the formulation of theoretical models. One assumption is that the stresses at depth are hydrostatic, that is, the principal components σ_{11}, σ_{22}, σ_{33} are all equal. This implies that principal stress differences do not increase with depth, which is generally not realistic from the field perspective. Such a state of stress may exist for very soft rocks such as shales or salts where plastic or viscoelastic deformation permits a partial or complete equalization of the principal stresses.

Yet another assumption pertains to a laterally constrained situation where horizontal displacements are not permitted. In the laterally constrained case, the effective horizontal stress is determined from the relationship

$$\bar{\sigma}_b = \bar{\sigma}_v \frac{v}{1 - v} \qquad (8.4)$$

where v is Poisson's ratio. For most rocks, Poisson's ratio ranges from 0.25 to 0.33. In accordance with Eq. 8.4, the ratio of horizontal to vertical stress should lie between 0.33 to 0.5. This is contrary to most in situ measurements, which lie between 0.5 to nearly one, with the lower-range characteristic of hard rocks and the upper-range more appropriate for soft rocks such as shale and salt. Thus the horizontally constrained model may not always describe the field situation, possibly because of plastic or viscoelastic deformation of soft rocks, or because tectonic forces can affect the horizontal stress but not the vertical stress so that the condition $\bar{\sigma}_b/\bar{\sigma}_v > 1$ can be obtained.

As both horizontal and vertical stress generally increase with increasing depth (McGarr and Gay, 1978), it is sometimes possible to follow the suggestion of Jaeger and Cook (1969) and assume

$$\sigma_b = M\sigma_v \tag{8.5}$$

Here, M is some constant that ranges from greater than zero to in excess of one. Thus, if M ranges between 0.3 to 0.5, we recover the laterally constrained case of Eq. 8.4. If M is taken as one, the hydrostatic case is recovered. On the other hand, if M is greater than one, the horizontal stress becomes the maximum principal stress, as might be the case in areas of tectonic compression. By tectonic compression, we mean the presence of stresses that are induced by other than the simple weight of the overburden. The relationship given as Eq. 8.5 is an oversimplification, but does permit an increase in the deviatoric stress $(\sigma_v - \sigma_b)$ with depth, one of the few reliable facts known about principal stress differences (McGarr and Gay, 1978).

One-Dimensional Basin Loading

Consider the following assumptions applied to Eq. 8.1. Only fluid flow and gravitational loading will be considered in an isothermal environment. Both the flow and the compression take place in a vertical direction, and the fluids and the solids are incompressible so that $R/H = 1$. In addition, the stress rate $\partial\sigma/\partial t$ is assumed to be vertical and of constant value $\omega\gamma'$, where ω is a constant rate of sediment supply and γ' is the unit weight of the submerged sediment, that is, the difference between the bulk weight (sediments plus water) and the unit weight of water. Conceptually, the situation is depicted as a growing sedimentary pile in an oceanic environment (Figure 8.7). For these assumptions, Eq. 8.1 becomes

$$\frac{\partial P}{\partial t} = \left(\frac{K}{\rho_w g \beta_p}\right)\frac{\partial^2 P}{\partial z^2} + \gamma'\omega \tag{8.6}$$

where $\gamma'\omega$ is the stress rate or pressure-generating term.

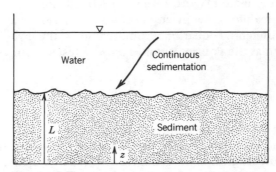

Figure 8.7
Conceptual diagram of continuous sedimentation in a fluid environment.

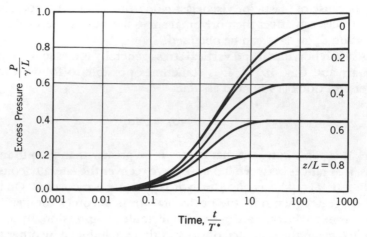

Figure 8.8
Graphic solution to the one-dimensional consolidation problem
where L is the thickness of the sedimentary pile, z/L is some
dimensionless depth ranging from zero at the bottom of the pile
to one at the top, t is time and T^* is the time constant for a basin,
and $P/\gamma'L$ is the dimensionless excess pressure (modified from
Bredehoeft and Hanshaw, 1968, Geol. Soc. Amer. Publication,
from Geol. Soc. Amer. Bull., v. 79, p. 1097–1106).

It is noted that the specific storage S_s equal to $\rho_w g \beta_p$ is now described in terms
of the vertical pore compressibility only in that the fluid and the solids were
assumed incompressible and the deformation vertical. In the absence of fluid
flow, the rate of pressure development $\partial P/\partial t$ would equal $\gamma'\omega$, suggesting that the
added load is transmitted totally and instantaneously to the water in the pores.
With fluid flow, the pressure that develops at any time will depend on the rate at
which fluid flow can dissipate this pressure.

Equation 8.6 was developed by Gibson (1958) and applied by Bredehoeft and
Hanshaw (1968) for the conditions shown in Figure 8.7. As noted, fluid is allowed
to escape from the top only, with the side and bottom boundaries being of the no-
flow type. Further, as mentioned previously, the basin is laterally constrained so
that no horizontal extensions are permitted. The solution to this problem is
shown graphically in Figure 8.8.

Prior to discussing Figure 8.8 and the conclusions that might be reached from
it, it is worthwhile to see if any useful information may be obtained from the
governing differential equation. From Chapter 4, the inverse of the dimensionless
group known as the Fourier number may be restated for this problem as

$$N_{FO}^{-1} = \frac{L^2/t}{K/S_s} = \frac{t}{K/S_s\omega^2} = \frac{t}{T^*} \tag{8.7}$$

where T^* is the time constant for a basin and is equal to $K/S_s\omega^2$, and t is the
depositional time to accumulate a given thickness of sedimentary pile. Note that
the dimensionless group t/T^* is the variable making up the abscissa of the graphical solution given as Figure 8.8. Thus, we see that for times $t <<< T^*$ (e.g., typically less than 10^{-1}, Figure 8.8) there is no significant abnormal pressure production and the basin is in the normal pressure regime. The smaller the specific

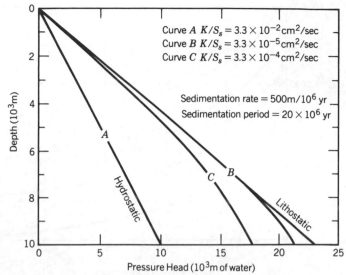

Figure 8.9
Pressure versus depth, Gulf Coast loading conditions (from Bredehoeft and Hanshaw, 1968, Geol. Soc. Amer. Publication, from Geol. Soc. Amer. Bull., v. 79, p. 1097–1106).

storage and the sedimentation rate, or the larger the hydraulic conductivity, the more likely the occurrence of normal pressure. For depositional times $t >>> T^*$ (e.g., typically greater than 100, Figure 8.8), the pressures will be approaching the overburden pressure.

The conditions described here are reflected in Bredehoeft and Hanshaw's (1968) analysis. These authors assumed a Gulf Coast rate of loading of 500 m/10^6 years for a period of 20×10^6 years, giving a sedimentary accumulation of 10,000 m. The results of this study are shown in Figure 8.9. For an upper limit of hydraulic diffusivity $K/S_s = 3.3 \times 10^{-2}$ cm²/sec, there is insignificant excess pressure production, and the fluid pressure remains near-normal throughout the sediment accumulation period (curve A). This corresponds to a time constant T^* equal to 4×10^8 years, whereas the depositional time t is on the order of 20×10^6, that is, $T^* >> t$. On the other hand, for a lower limit of hydraulic diffusivity of 3.3×10^{-5} cm²/sec, the fluid pressures approximate the overburden pressure (curve B). This corresponds to a time constant T^* equal to 4×10^5, that is, $t >> T^*$. Curve C lies intermediate between these extremes, with a hydraulic diffusivity of 3.3×10^{-4} and a time constant T^* equal to 4×10^6 years.

From the graphical solution of Figure 8.8, one can see that the limiting dimensionless numbers t/T^* of 10^{-2} and 10^2 apply to any case of loading irrespective of the hydraulic parameters used in the Bredehoeft-Hanshaw (1968) study. Values of t/T^* between 10 and 100 are associated with pressures approaching the overburden, whereas values on the order of 10^{-2} are associated with normal compaction.

To illustrate the relationships between the variables in the basin loading problem further, depositional time and the time constant have been plotted against various sedimentation rates for a wide range of hydraulic diffusivities (Figure 8.10). Also plotted is the time it takes to deposit 5, 10, and 20 km of sediments at the given sedimentation rate. Thus, we note that at a sedimentation rate of 500 m per million years (m/my) over a time period of 10 my, the sediment accumulation is 5 km. A

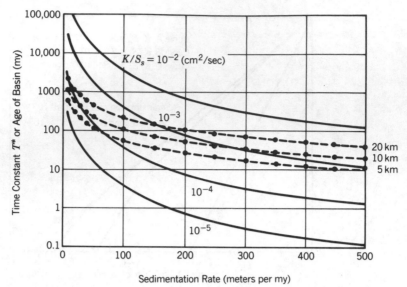

Figure 8.10
Time constant versus sedimentation rate. Solid lines represent the time
constant as a function of sedimentation rate for various values of K/S_s.
The dashed curves give the time required to accumulate 5, 10, and 20
km of sediment for the various sedimentation rates (from Palciauskas and
Domenico, Water Resources Res., v. 25, p. 203–213, 1989. Copyright by
Amer. Geophys. Union).

depositional rate of 100 m/my over a time span of 100 my results in a 10 km
deposit. It is noted that the time constant for sediments with a diffusivity of 10^{-2}
cm²/sec is significantly larger than the time it takes to deposit 20 km of sediments,
that is, the $K/S_s = 10^{-2}$ cm²/sec line lies above the 20 km line. As $T^* >> t$, these
sediments will be normally pressured. For a hydraulic diffusivity of 10^{-3} cm²/sec,
the time constant is greater than the depositional time only for sedimentation rates
less than 100 m/my, whereas for larger sedimentation rates, t and T^* are approximately equal, suggesting modest abnormal pressure development. Hydraulic diffusivity values on the order of 10^{-4} and 10^{-5} cm²/sec are characterized by time
constants that are considerably less than depositional times for most sedimentation
rates, and are expected to result in appreciably overpressured sediments.

Extensions of the One-Dimensional Loading Model
There are at least three extensions to the one-dimensional model presented that
may provide further insight into the physics pertinent to abnormal pressure development. The first of these is for fluids and solids considered to be compressible.
This seemingly insignificant addition has some important implications in areas of
active deposition, especially with regard to the establishment of vertical gradients
between units of differing lithology. The second of these extensions examines the
basin loading problem within the framework of a mean stress problem. This
implies the presence of deviatoric stresses and provides an opportunity to examine the possibility of rock fracture. The third extension is the one-dimensional
constant strain problem, which is the other extreme of the one-dimensional constant stress problem.

Vertical Compression with Compressible Components Given all the assumptions of the Bredehoeft-Hanshaw (1968) analysis with the exception that both the fluid and solid components are compressible ($R/H \neq 1$) gives the result developed in Chapter 4

$$\frac{\partial P}{\partial t} = \frac{K}{\rho_w g(1/R)} \frac{\partial^2 P}{\partial z^2} + \left(\frac{R}{H}\right) \gamma' \omega \tag{8.8}$$

For clays, the pore pressure coefficient R/H will be sufficiently close to unity (Table 4.4, Chapter 4) so that there is virtually no difference between this formulation and the previously described one. However, the coefficient R/H can range from 0.83 for mudstones to 0.25 for limestones. This suggests that any incremental load in a depositional environment containing various lithologies will not be totally borne by the fluid, but some part of it will be permanently carried by the solids, and the part carried by the solids will be different in different rocks. Depending on the magnitude of this parameter, it may be difficult to generate high fluid pressures in some rocks irrespective of the rate of loading and limitations on the rate of flow. For example, geologists have long noted that ancient limestones often show little sign of compaction during their diagenetic period. This may be the result of a pore pressure coefficient R/H considerably less than one if the lime mud crystallizes early in its depositional history so that most of the loading is carried by the grains. The increased stress on the grains can enhance pressure solution in such rocks.

The solution to Eq. 8.8 is likewise given by Figure 8.8, where the vertical axis is replaced by $P/\gamma'L$ (R/H). As R/H is less than one for rocks more rigid than clays, the initial fluid pressure will always be less than that predicted by the Bredehoeft-Hanshaw model. As the lithologies vary within the sedimentary pile, we anticipate large differences in vertical gradients as some units become more highly pressurized than others.

For compressible components, the inverse Fourier number is modified accordingly and can be expressed

$$N_{FO}^{-1} = \frac{t(R/H)}{T^*} \tag{8.9}$$

which indicates that the pressure production is not nearly as pronounced for situations where R/H is less than one. It is emphasized, however, that the limits established for the Fourier number from the results of Figure 8.8 still hold. That is to say, it still requires a number greater than 10^{-2} to generate abnormal pressure due to gravitational loading; however, with the numerator decreased by several percent for some rocks, it is just not as likely an event unless the rate of loading is very fast or the hydraulic diffusivity very small.

Horizontal Extension and Compression The analog field condition of the one-dimensional loading problem is the basin that is constrained laterally. This means that the basin cannot relax tectonically in response to the sediment load, nor can any form of tectonic compression occur during the sedimentation period. Normally in a tectonically relaxed state, there is little reason to expect significant horizontal resistance at the basin borders so that the sedimentary accumulation may react passively to movements of the subadjacent basement, with normal faults forming along the margins. At a later stage, crustal stress can become compressional, or give way to those that are epidermal in origin. The latter can be the result of large differences in elevation between the bordering lands and the bottom of the subsiding basin. This may promote compressional settling or large-scale slumping from each side of the subsiding basin, leading to compressional folding.

The problem of tectonic extension or compression during loading may be examined within the framework of a mean stress problem (Palciauskas and Domenico, 1980). For these purposes, the stress rate $\partial\sigma/\partial t$ is assumed to be the result of a mean principal stress rate, not solely the vertical one. For the case where the horizontal stresses are equal and are designated σ_3, the mean principal stress expressed in Eq. 4.74a becomes

$$\sigma_m = \frac{\sigma_1}{3} + \frac{2\sigma_3}{3} \tag{8.10}$$

The relationships between the horizontal and vertical stresses are given as Eq. 8.5

$$\sigma_3 = M\sigma_1 \tag{8.11}$$

Here, as discussed previously, M can range from slightly greater than zero for horizontal extension to greater than one for crustal or epidermal compression. For these conditions, Eq. 8.8 becomes

$$\frac{\partial P}{\partial t} = \frac{K}{\rho_w g(1/R)} \frac{\partial^2 P}{\partial z^2} + C\gamma'\omega \left(\frac{R}{H}\right) \tag{8.12}$$

where

$$C = \left(\frac{1}{3}\right) + \left(\frac{2M}{3}\right) \tag{8.13}$$

It is seen that C acts as a multiplier for the effective rate of loading $(R/H)\gamma'\omega$, which is the effective rate of application of effective vertical stress. The larger C, the greater the abnormal pressure development. It is noted that the graphical solution given in Figure 8.8 applies here also where the ordinate is expressed as $P/\gamma'L(2M/3 + 1/3)(R/H)$. Thus for $R/H = 1$ and $M = 0.6$ ($\sigma_3 = 0.6\sigma_1$), the mean stress problem yields 73% of the abnormal pressure that would result from the laterally constrained case. If $M = R/H = 1$, the laterally constrained case of Bredehoeft and Hanshaw (1968) is recovered.

An interesting feature of this development is that the smaller M, the larger the deviatoric stress, and the smaller the fluid pressure production. Later developments will show that failure can occur at low fluid pressures in association with large deviatoric stresses, or at higher fluid pressures in combination with smaller deviatoric stresses. Hence, some apparent conditions for failure would appear to be present for virtually all states of differential stress in compacting sediments.

Equation 8.12 is well suited for obtaining an approximate idea of the effects of horizontal compression and indicates that the pressure production can be quite large. If tectonic forces are involved, the multiplier C can exceed one. Thus, for $R/H = 1$ and $M = 1.5$ ($\sigma_3 = 1.5\sigma_1$), the pressure developed is 33% greater than one would expect from the laterally constrained case. Such pressures would never be achieved because of some pressure-induced inelastic rock failure during the compression.

For this mean stress problem, the Fourier number may be modified accordingly

$$N_{FO}^{-1} = \frac{tC(R/H)}{T^*} \tag{8.14}$$

where C is given in Eq. 8.13. It is emphasized again that the limits already established for the Fourier number hold here also. However, the numerator is again decreased as M gets small, that is, as the degree of horizontal extension is increased. Hence, the chances of exceeding the lower (10^{-2}) critical value may not be a likely event in some rocks.

Tectonic Strain as a Pressure-Producing Mechanism   The other extreme of the one-dimensional constant stress problem is the problem of constant one-dimensional strain, visualized as occurring in a layer whose lateral extent is very large compared to its thickness. The layer is overlain and underlain by high-permeability material that is likewise undergoing deformation, but remains at normal or hydrostatic pressure due to large values of the hydraulic diffusivity. The main fluid migration is taken as vertical due to the horizontal compression of the bed. Clearly, such a one-dimensional strain problem is only an approximation of reality, but should have some insightful features as the other extreme of the one-dimensional stress problem. The governing equation in this case may be stated as

$$\frac{\partial P_{ex}}{\partial t} = \left[\frac{K}{\rho_w g(1/Q)} \right] \frac{\partial^2 P_{ex}}{\partial z^2} - \xi Q \frac{\partial \epsilon}{\partial t} \tag{8.15}$$

Palciauskas and Domenico (1989) provide a solution to this problem for the boundary conditions of hydrostatic pressure at the upper and lower boundaries of the layer and an initial condition of $P_{ex} = 0$ at time $t = 0$. The details of this solution are not important if we are merely concerned with the magnitude of strain rates required to produce fluid pressures equivalent to those associated with basin loading as demonstrated in Figure 8.10. This may be done by equating the source terms of Eqs. 8.15 and 8.8

$$\gamma'\omega = \xi R \frac{\partial \epsilon}{\partial t} \tag{8.16}$$

Thus, for a sedimentation rate of 500 m/my and $1/R = 1.62$ Kbar^{-1} for clays (Table 4.4)

$$\frac{\partial \epsilon}{\partial t} = \frac{\gamma'\omega}{\xi R} = \left(0.130 \frac{\text{Kbars}}{\text{km}} \right) \left(0.5 \frac{\text{km}}{\text{my}} \right) \left(1.62 \frac{1}{\text{Kbars}} \right) \left(\frac{1 \text{ my}}{3 \times 10^{13} \text{ sec}} \right)$$

$$= \quad 3 \times 10^{-15} \text{ sec}^{-1}$$

Thus, given the information in Table 4.4 along with Figure 8.10, the relationship between the source terms given as Eq. 8.16 may be used to determine strain rates that develop excess pressures equivalent to those produced by basin loading. For the example above, strain rates on the order of 3×10^{-15} sec^{-1} will produce significant abnormal pressure, provided the hydraulic diffusivity is on the order of 10^{-4} cm^2/sec and smaller, and no abnormal pressure for diffusivities larger than 10^{-3} cm^2/sec (Figure 8.10). For this type of material (clay) with hydraulic diffusivities approaching 10^{-5} cm²/sec, strain rates as low as 7×10^{-16} sec⁻¹ can produce abnormal pressures. For mudstones with small hydraulic diffusivities and $1/R = 0.05$ (Table 4.4), strain rates of less than 1×10^{-16} sec⁻¹ can produce abnormal pressure.

Thermal Expansion of Fluids

When rocks undergo burial in a depositional environment, they are subjected to increases in temperature that promote a thermal alteration of the component parts of the sediment, that is, the fluid, the solids, and the pores. If the heating takes place without expulsion of the pore water, the pressure changes with temperature increases are classified as constant mass phenomenon. This assumption is at the heart of much of the research in this area. The concept was introduced originally by Barker (1972) and advanced by Bradley (1975) and Magara (1975). A review of this work is given by Gretener (1981).

In Barker's (1972) theory, a geologic formation must first become "sealed" at some depth and thereafter buried in a thermal field with no expulsion of its pore water. Additionally, the assumptions of Barker (1972) require that the thermal expansion of water acts against a rigid, nonyielding matrix that remains at constant pore volume throughout the burial. The pressure change with temperature is expressed as

$$\left(\frac{\partial P}{\partial T} \right) = \frac{\alpha_f}{\beta_w} \tag{8.17}$$

where α_f is the thermal expansion coefficient of the water and β_w is the isothermal compressibility of water, both of which are temperature dependent (Figure 8.11). On the basis of these arguments, a temperature increase of 25°C over a change in depth of approximately 1 km (25°C/Km) would cause the fluid pressure to increase by 300 bars (4400 psi). This corresponds to an incremental pressure gradient of 0.3 bars/m (1.34 psi/ft).

In examining Eq. 8.17, both Chapman (1980) and Daines (1982) suggested that the thermal expansion of water in depositional environments is probably not a very important pressure-producing mechanism. Their statements were not based on the physics of Eq. 8.17, but on heuristic arguments that discounted the possibility of maintaining constant fluid mass over long periods of geologic time. According to these authors, fluid flow would likely dissipate the pressures. Notwithstanding the importance of fluid flow, these arguments missed the main point of debate concerning the appropriateness of Eq. 8.17; that is, the equation depends solely on fluid properties and is medium independent. Thus, on the basis of Barker's (1972) result, one might conclude that the fluid pressure response to a geothermal gradient in a granite under a large confining stress is no different from that in a shale in a tectonically relaxed state. If constant mass considerations are to have any realism as upper-bound calculations for pressure production, they must include the expansive properties of the medium responsible for accommodating some of the pressure, as well as the boundary conditions under which the heating takes place. Three possible boundary conditions present themselves: heating at constant stress, at constant strain, or under conditions that are neither constant stress nor constant strain.

When addressing the pressure changes at constant stress or constant strain, two avenues are open to us. We may derive the appropriate expressions directly

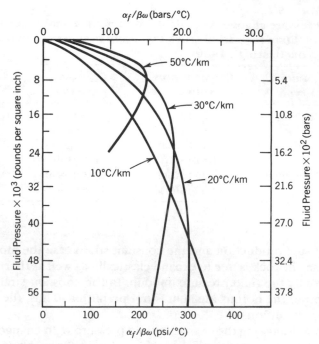

Figure 8.11
Values of α_f/β_ω as a function of hydrostatic pressure
for various geothermal gradients (modified from data
of Knapp and Knight, J. Geophys. Res., v. 82, p. 2515–
2522, 1977. Copyright by Amer. Geophys. Union).

independent of fluid flow considerations as has been done by Palciauskas and
Domenico (1982), or they may be obtained by simply ignoring fluid flow in Eqs.
8.1 and 8.2 and treating these expressions within the framework of constant stress
or constant strain. The end result is the same, giving the expressions

$$\left(\frac{\partial P}{\partial T}\right)_{M,\sigma} = R\alpha_m \tag{8.18}$$

and

$$\left(\frac{\partial P}{\partial T}\right)_{M,\epsilon} = Q(\xi\alpha_b + \alpha_m) \tag{8.19}$$

respectively. Equation 8.18 gives the change in pressure with temperature
increases at constant mass and constant stress, whereas Eq. 8.19 is the constant
mass formulation for the boundary conditions of constant strain. The coefficients
in these equations have been discussed earlier.

Table 8.1, taken from Hastings (1986), gives some representative values for
various rock types. For each of the cases shown, the change in pressure with temperature (bars/C°) was multiplied by a geothermal gradient of 25°C/km to give a
pressure change with depth. As expected, the constant pore volume case of Barker (1972) is rock independent, whereas the constant stress case yields the least

Table 8.1
Pressure changes with depth for a geothermal gradient of
25 °C/km for various rock types and boundary conditions
(from Hastings, 1986)

Rock type	Constant pore volume	Constant stress	Constant strain
Shale	0.423 bars/m (1.87 psi/ft)	0.0045 bars/m (0.02 psi/ft)	0.475 bars/m (2.1 psi/ft)
Sandstone	0.423 bars/m (1.87 psi/ft)	0.249 bars/m (1.1 psi/ft)	0.479 bars/m (2.12 psi/ft)
Limestone	0.423 bars/m (1.87 psi/ft)	0.362 bars/m (1.6 psi/ft)	0.471 bars/m (2.08 psi/ft)

fluid pressure production and the constant strain case the most. In the constant stress case, the rock is free to expand elastically as well as thermally to accommodate partially the fluid pressure buildup. In the constant strain case, volumetric expansion of the porous medium is not permitted. Thus, the importance of the boundary conditions.

Before dismissing the change in fluid pressure with changes in temperature at constant stress in soft rocks such as shale as insignificant, let us examine the nature of the coefficient ($1/R$). In elastic theory, $1/R$ is a compressibility that describes how the mass changes with pressure changes at constant stress (Table 4.3). In Eq. 8.18, $1/R$ represents an expansivity of the porous medium. Thus we are trapped in our elastic theory whereby reversibility is assumed a priori. Although this assumption of reversibility may be appropriate for most crystalline rocks and perhaps some sandstones and limestones, it hardly applies in the case of soft unconsolidated sediment. It is noted further that the constant stress case is part of the differential equation for basin loading (Eq. 8.1). Thus solutions to this flow equation or similar ones where temperature effects are included, such as those presented by Domenico and Palciauskas (1979), Sharp (1983), and Shi and Wang (1986), focus exclusively on the constant stress case and incorporate the assumption of reversibility.

The boundary conditions for constant stress or constant strain are not strictly valid in field situations where rocks at depth are being heated. Rather, the displacement field and the stress distribution should be derived from the stress equilibrium equation. This is generally a formidable task, but there is one situation that has been solved for the assumption that displacements far from a source of heat are zero. This permits the derivation of expressions that describe the change in fluid pressure P, the volume dilation ϵ, and the changes in hydrostatic stress σ purely as a function of temperature and the properties of the medium. Some typical results are presented in Table 8.2 for a variety of rock types. The fluid properties in these calculations were taken as averages over an approximate 80°C rise in temperature. The elastic coefficients for these rocks are the same as those employed in Table 4.4. In all lithologies, the fluid pressure increases with temperature at a faster rate than do increases in stress caused mainly by the thermal expansion of the solids. The greater the rock rigidity, the greater the net decrease in effective stress. For the sandstone or basalt, the net decrease in effective stress is about 4 bars °C⁻¹, compared to 1.54 bars °C⁻¹ for the mudstone and 0.09 bars °C⁻¹ for the clay. Further, the elastic strain with temperature is obviously greater for the

Table 8.2
Pressure, stress, and strain with temperature changes for various rock types (from Domenico and Palciauskas, 1988) *

	Clay	Mudstone	Sandstone	Limestone	Basalt
$\left(\dfrac{\partial P}{\partial T}\right)_{m,\sigma}$ (bars °C⁻¹)	0.0956	1.73	5.9	9.13	8.72
$\left(\dfrac{\partial P}{\partial T}\right)_{m,\epsilon}$ (bar °C⁻¹)	11	11.23	15.5	12.58	10.13
$\dfrac{dP}{dT}$ (bar °C⁻¹)	0.18	2.77	9.1	10.39	9.34
$\dfrac{d\sigma}{dT}$ (bar °C⁻¹)	0.09	1.23	4.7	5.51	5.4
$\dfrac{d\epsilon}{dT}$ (°C⁻¹)	1.88×10^{-4}	9.47×10^{-5}	5×10^{-5}	2.34×10^{-5}	1.5×10^{-5}

*Geol. Soc. Amer. Pub. Geology of N. America, 0–2, p. 435–445.

soft highly porous rocks, an indication of pressure accommodation by elastic volume increases. The small decrease in effective stress for soft rocks suggests that failure at some point may not be a likely event even under the maximum pressure-producing assumption of constant fluid mass. It is noted, however, that the concept of reversibility for the elastic coefficients is embedded in this analysis.

It is now useful to compare these results with other calculations for the Gulf Coast region. The calculations for the mudstone will suffice here. For a geothermal gradient of 30°C km⁻¹, the fluid pressure increase of 2.77 bars °C⁻¹ translates into a thermal pressure production of 0.0831 bars m⁻¹ (0.4 psi ft⁻¹). The fluid pressure production per meter of depth would obviously be much less for clays. Gretener (1981) notes that for the prevailing geothermal gradient, the thermal pressure development in the Gulf Coast can range between 0.255 bars m⁻¹ (1.25 psi ft⁻¹) to 0.66 bars m⁻¹ (3.2 psi ft⁻¹). Magara (1975) cites a value of 0.285 bars m⁻¹ (1.4 psi ft⁻¹). These calculations indicate predictions of thermal pressure development that are between three to eight times greater than determined from Table 8.2 for mudstones. As noted earlier, the calculations of Gretener (1981) and Magara (1975) are based on the constant pore volume assumption. We conclude that thermal pressure development in very soft rocks is not very significant (unless they are heated at constant strain), but does become more significant with increasing rock rigidity.

The constant mass case gives an upper bound for the development of thermal pressure in depositional environments. When fluid flow occurs, we once more have competition between rates of pressure production and pressure dissipation. The reader is referred to Domenico and Palciauskas (1979), Sharp (1983), and Shi and Wang (1986) for details on this topic.

Fluid Pressures and Rock Fracture

The Coulomb theory of failure given in Eq. 7.6 describes the shear failure mode exclusively in terms of total and fluid pressures. This is not the only failure mode, nor are other failure theories restricted to a simple effective stress relationship. Irrespective of the failure mode, abnormally high fluid pressures can lead to some sort of rock fracture. On one scale is the inelastic behavior referred to as hydraulic fracture, the accepted condition for which is fluid pressures in excess of the least compressive stress in a regional stress field. The complete theory states that incipient fracture propagation will occur when the fluid pressure exceeds the least principal stress by an amount equal to the tensile strength of the rock. The direction of fracture propagation has a preferred orientation, normal or near normal to the least compressive stress. In the tectonically relaxed (extensional) state, the horizontal stresses are smaller than the vertical stresses, and the fracture growth is vertical; in the tectonically compressed state, they are horizontal.

At yet another scale are the pore volume increases that materialize in the form of microfractures, commonly at fluid pressures well below the least compressive stress (Handin and others, 1963). As with hydraulic fracture, this dilatancy has a preferred orientation, normal or near normal to the least compressive stress, and is a likely precursor in the formation of large throughgoing fractures. In this section we will examine a few of these failure theories.

The first and best understood is hydraulic fracture as described by Hubbert and Willis (1957). According to these authors, the pressure required to initiate fracture is

$$P^* = \sigma_v M - P(M - 1) \tag{8.20}$$

where σ_v is the vertical stress, M is the ratio of horizontal to vertical stress and is some constant less than or equal to one, and P is the ambient (prefracture) fluid pressure. Dividing both sides by the depth z to obtain a critical pressure gradient gives

$$\frac{P^*}{z} = \frac{(\sigma_v M) - P(M - 1)}{z} \tag{8.21}$$

Thus, if the horizontal stress is one-third of the vertical, Eq. 8.20 predicts a fracture pressure

$$P^* = \frac{\sigma_v + 2P}{3} \tag{8.22}$$

This states that the fluid pressure required to initiate hydraulic fracture is one-third of the overburden pressure plus two-thirds of the ambient (prefracture) fluid pressure. As M approaches one, that is, as the deviatoric stress approaches zero, a greater fluid pressure is required to initiate fracture.

In addition to the tensional failure described, earth materials dilate in response to shearing strains, undergoing volume changes that may be positive or negative. This behavior has been observed in soils and both crystalline and water-saturated sedimentary rocks. For crystalline rocks, dilation begins at a deviatoric stress of one- to two-thirds of the stress difference ultimately required to fracture the rock (Brace and others, 1966). Fracture is accompanied by a porosity increase that ranges from 0.2 to 2 times the elastic volume change the material would have had were it perfectly elastic. This statement holds for both brittle and ductile rocks and for fracture at atmospheric pressure as well as under a large total confining pressure. For sedimentary rocks containing pore fluids, dilatancy has been observed when the ratio of the fluid to the confining pressure is equal to about 0.8 (Handin and others, 1963). This result is reported as being approximately correct for sandstone, limestone, siltstone, and probably shale.

Figure 8.12
Relation between effective confining
pressure, deviatoric stress, and failure.
Dilatancy reportedly initiated at or
previous to failure. Sandstone data from
Handin and others (1963); dolomite data
from Brace and others (1966) (from
Palciauskas and Domenico, 1980.
Reprinted by permission).

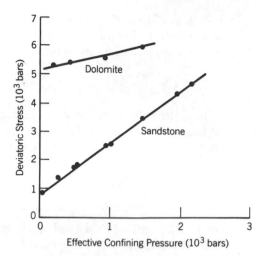

Some of the features of these tests are shown in Figure 8.12. In order to interpret these results, the effective confining pressure is defined as

$$\sigma_e = \sigma_c - P \tag{8.23}$$

where σ_e is the effective confining pressure, σ_c is the confining pressure in the
triaxial apparatus, and P is the fluid pressure in the pores of the sample. As noted
in the figure, when fluid pressure is low compared to the experimental confining
pressure (that is, a high effective confining pressure), the onset of dilatancy requires
a large deviatoric stress. As fluid pressure increases (that is, a lower effective confining pressure), the onset of dilatancy requires a smaller deviatoric stress. The important feature here is that failure is once more associated with high fluid pressures at
small deviatoric stresses or low fluid pressures at large deviatoric stresses.

There are several other factors that dictate the onset of dilatancy, including the
density of the sediments at the time of failure (Zoback and Byerlee, 1976) and the
degree of packing as reflected by the porosity (Scott, 1963). Ideally, one might
expect a physically based constitutive relation expressed in terms of a critical
porosity for a given deviatoric and mean stress. Unfortunately, such a relationship
is not available. There are, however, other approaches to the problem.

One approach is to consider the dilatant behavior in an empirical form suggested by Scott (1963). The basic assumption is that the inelastic volume change
is a function only of the deviatoric stress. Assuming σ_{22} and σ_{33} are equal and
designating the horizontal stress as σ_{33}, this may be expressed as

$$(dV/V)_{\text{inelastic}} = -D\,d\tau^* \tag{8.24}$$

where V is volume so that dV/V is dilatant strain, D is a dilation coefficient that can
be positive or negative depending on the material, and τ^* is the deviatoric stress.
The coefficient D is not likely to be a constant for a given material and will be
somewhat strain dependent. When D is negative, the inelastic volume change acts
to increase porosity with increasing deviatoric stress. However, to perform calculations, some threshold pressure has to first be identified and incorporated in such
a way that describes the rate at which the fracture dissipates pressure.

A useful limiting expression for the initiation of fracture is the experimental observation of Handin and others (1963), where the onset of dilatancy occurs when the fluid pressure approaches 0.8 of the confining pressure, that is,

$$P_{\text{critical}} = 0.8\sigma_c \qquad (8.25)$$

where the critical designation implies a fluid pressure necessary for the onset of dilatancy. If the confining pressure is taken as equivalent to the horizontal stress, the fluid pressure required for the initiation of microfractures become

$$P_{\text{critical}} = 0.8[\sigma_v M - P(M - 1)] \qquad (8.26)$$

where the bracketed quantity is the Hubbert-Willis (1957) fracture criteria. It will be noted here that as M gets large (approaches one), the deviatoric stress decreases and the fluid pressure required for the onset of dilatancy increases. This is in accord with the experimental observations given in Figure 8.12. This feature is also incorporated in the mean stress formulation for abnormal pressure development (Eq. 8.12). Palciauskas and Domenico (1980) utilize this fracture criterion to examine the initiation of microfractures in oil-bearing source rocks at different stages of maturation. Last, Eq. 8.26 suggests that an upper limit of fluid pressure is about 0.8 σ_v. That is, microfracture generation at fluid pressures on the order of 0.8 σ_v may represent an instability from which there may be no way to further increase fluid pressure.

Phase Transformations

Included among the main phase transformations that presumably lead to excess pressure development are the gypsum-anhydrite transformation (Heard and Ruby, 1966) and the montmorillonite-illite transformation (Powers, 1967; Burst, 1969; Prichett, 1980). Both transformations are thermally driven and result in the production of pore fluids. With montmorillonite, the expulsion of one interlayer of water decreases the layer spacing by about 2.5 Angstroms resulting in a possible 20% decrease in the mineral grain volume per layer. Hence, the potential water volume release is considerable. In the Gulf Coast, the alteration of montmorillonite to a mixed layer clay begins at temperatures on the order of 60° to 70°C. Although it has not yet been possible to quantify the rate at which pressure is produced in this transformation, it can be speculated that the rate of transformation is proportional to the decrease in effective stress per unit time interval.

Figure 8.13 illustrates this last argument where the contained water goes from the interlayered state in the lattice to a state of free water. Bound water in a solid structure supports a finite effective stress that goes to zero in the free water state. Assuming the pore water produced is not free to drain, it should immediately take on the weight of the overburden. It appears further that this magnitude of pressure can be continually regenerated in the "window" that defines the critical pressure range in a subsiding basin. If it can be assumed that the transformation is described by first-order kinetics of the Arrhenius form, the size and depth of this window is a function of certain kinetic factors along with temperature and the rate of burial. In Figure 8.14, curve I indicates the reaction occurs at near surface temperatures and the transformation is completed in the upper 1000 m of burial. In this development, N represents the number of expandable layers after time t, N_o is the original amount, A is the frequency factor, E is the activation energy, and ω

Figure 8.13
Schematic diagram of clay diagenesis showing (*a*) virtually all bound water with no porosity or permeability, (*b*) the generation of free water at the expense of the solids, (*c*) free water squeezed out upon completion of compaction (from Powers, 1967. Reprinted by permission).

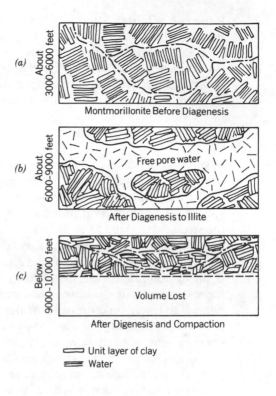

(*a*) About 3000–6000 feet

Montmorillonite Before Diagenesis

(*b*) About 6000–9000 feet

Free pore water

After Diagenesis to Illite

(*c*) Below 9000–10,000 feet

Volume Lost

After Digenesis and Compaction

⊂⊃ Unit layer of clay
≡ Water

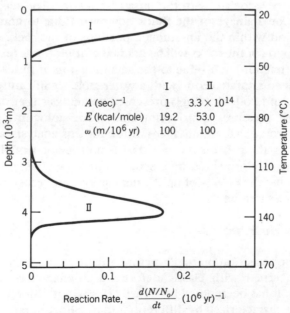

	I	II
A (sec)$^{-1}$	1	3.3×10^{14}
E (kcal/mole)	19.2	53.0
ω (m/10^6 yr)	100	100

Reaction Rate, $-\dfrac{d(N/N_o)}{dt}$ $(10^6 \text{ yr})^{-1}$

Figure 8.14
Transformation rate versus depth for smectite-illite (from Domenico and Palciauskas, 1988, Geol. Soc. Amer. Publication, from Geol. of N. Amer., 0–2, p. 435–445).

is the rate of burial. This rapid depletion of smectite is contrary to observations of smectite-illite occurrences in Gulf Coast sediments. Curve II, on the other hand, shows the reaction delayed to about 3000 m and then progressing to near completion over the next 1500 m of burial. Irrespective of the uncertainty regarding the exact location of the critical temperature range, Figure 8.14 suggests the presence of a rather narrow window associated with it.

Subnormal Pressure

If geologic loading can give rise to abnormally high fluid pressures, erosional unloading and eventual uplift may result in abnormally low pressures, referred to earlier as subnormal pressure (Russel, 1972). The mechanics can be explained by viewing the compaction process in reverse. With erosional unloading and uplift, the stress is removed from the porous matrix and from the solid grains. Hence, the pores may expand somewhat, as will the solid grains. As the horizontal stresses are generally increased relative to the vertical ones during erosional unloading, it is difficult to describe the details of this process. However, as the pores are generally more compressible than the solids, we anticipate a net increase in pore space. This will act to lower fluid pressure. At the same time, the fluid will expand due to the reduced stress, but will also contract somewhat due to a lowered-temperature environment. Both these processes will act to lower the fluid pressure even more. The major question, as yet not satisfactorily answered, is whether the increased pore space is sufficient not only to reduce the pressures to their normal value, but to create a state of suction wherein the pressures are less than atmospheric. If such a state can indeed be produced, the driving forces that take over are dominantly those that we associate with the unsaturated zone. For such cases, we anticipate movement or drainage of the fluids downward due to gravity and some sort of redistribution within the unsaturated portions of the rock. Ultimately, with time, the upper parts of the rocks will be drained of their fluids (except for those held by capillary attraction) and—due to the accumulation of fluids at depth—the formation of a deep saturated zone with a water table significantly below land surface.

Neuzil and Pollock (1983) presented an analysis that demonstrates that sub-atmospheric pressures can develop due to the unloading of hydraulically tight rocks. It is noted that an initial condition for this analysis was hydrostatic conditions; that is, the process did not start from the abnormally pressured state, but from a normally pressured one. Senger and Fogg (1987) and Senger and others (1987) discuss the effects of uplift, deposition, and erosion on the development of underpressured basins.

Irreversible Processes

Except for our section on rock fracture, the discussions thus far have been exclusively concerned with elastic deformations characterized by elastic material properties. It has been noted by Neuzil (1986) that different time scales of deformation are characterized by different moduli. Short-duration phenomena such as atmospheric pressure variations or tidal disturbances are described by elastic models that contain undrained moduli (Chapter 4). Intermediate-duration phenomena are likewise described with elastic models, but as there is sufficient time for fluid flow to take place in response to the deformation, the appropriate moduli are the drained coefficients. The deformations associated with the drained parameters are significantly greater than in the undrained case. It is noted that these same elastic coefficients have been used to describe phenomena that operate over geologic time scales. However, we are aware that over time periods measured in terms of tens of thousands of years, irreversible nonelastic processes will occur at the same applied stress that is driving the elastic response. These irreversible processes will result in considerably greater deformations, for the same applied stress condition, than would be expected from their elastic counterparts. For the most part, it appears that we have ignored these greater deformations and have focused on the lessor ones, and some explanation is in order.

The reason for this apparent shortcoming is that it is most difficult to determine the microscopical laws governing irreversible deformations. On the other hand, Neuzil (1986) suggests that such new laws may not be necessary, only different coefficients are required. These arguments suggest that both short- and long-term deformations can be described within elastic theory and the concept of effective stress, but that the constants applicable over long time frames will be different from those that apply to the short term. There still remains the problem of obtaining the constants for the long-term deformations. It is obvious that they cannot be measured in the laboratory but may be determined from field studies; for example, porosity versus depth determinations in thick sequences of rock. This has been the method of Shi and Wang (1986) with the utilization of Athy's (1930) relation for porosity versus depth in shales (Chapter 2). Yet another way is to devise a new microscopic law that "predicts" porosity variation with depth in response to some assumed process that is operating. This was the method of Palciauskas and Domenico (1989). Their calculations indicate that an irreversible "compressibility" coefficient for such deformation is almost two orders of magnitude greater than its elastic counterpart. This means that on a macroscopic scale, the irreversible deformational character of a rigid sandstone is similar to the elastic deformation of a rock whose stiffness lies intermediate between a soft clay and a mudstone.

8.2 Pore Fluids in Tectonic Processes

Walder and Nur (1984) posed three important questions regarding fluids in the earth's crust:

1. To what depths in the earth's crust is free water present?
2. What is the value of the fluid pressure at these depths?
3. What is the permeability of the rocks at these depths?

In that actual measurements of crustal properties exist only to depths of 2–3 km, any answers to these questions must be based on indirect evidence. Such evidence collected by these authors suggests that free water is present, at least episodically, at upper and midcrustal depths. The pressure of this fluid is thought by many to be abnormal, and by others to be at normal pressure. Brace (1980), for example, states that zones of permeability on the order of 10^{-15} m^2 must exist to depths of at least 10 km. He therefore concludes that this permeability is too high for the maintenance of abnormal fluid pressures. However, uniform values of permeability to midcrustal depths—high or low—are not to be expected, and not all rocks have direct permeable pathways to the earth's surface. Further, from what has been learned in the section on active depositional environments it is clear that if some process or processes can generate fluid pressure at a faster rate than fluid flow can dissipate it, the pressures will be maintained at abnormal proportions. We already discussed those processes that are likely responsible for such generation. Whether or not they or different ones operate at midcrustal depths is another question.

Fluid Pressures and Thrust Faulting

A thrust fault is a discontinuity in a rock mass characterized by a low dip and a net slip generally measured in miles, sometimes approaching 100 miles. The mechanism for overthrusting presents a problem in mechanics in that the force necessary for displacements of such magnitude far exceeds the strength of the rock itself. Thus, in the words of Smoluchowski (1909), "we may push the block with whatever force we like; we may eventually crush it, but we cannot succeed in moving it." With this constraint, two lines of reasoning were followed to explain such displacements; gravitational sliding down an inclined slope, and the lubrication of the thrust plane, usually by water. In other words, these propositions were concerned with reducing the sliding frictional resistance between two rock masses. As discussed by Terzaghi (1950) in our earlier treatment of landslides, water may not act as a lubricant but as an antilubricant.

Hubbert and Rubey (1959) invoked the Mohr-Columb theory and the same failure criteria as Terzaghi (1950) to present the case that high fluid pressures can sufficiently reduce the shearing resistance of a given rock mass so that large-scale displacements are possible. In their overthrust concept, fault blocks could be moved large distances by relatively small forces, provided the fluid pressures were sufficiently high to put the rocks in a state of floatation. The special geologic conditions required for the attainment and maintenance of such pressures were cited as (1) the presence of clay rocks; (2) interbedded sandstone; (3) large total thickness, and (4) rapid sedimentation. These conditions represent the essence of the depositional environments discussed in the previous section.

Since this original effort by Hubbert and Rubey, the embellishments added by others have taken two rather distinct lines. The first is a search for other causes of high fluid pressures needed in the overthrust concept within the same physical framework of the Mohr-Coulomb theory. Thus, Hanshaw and Zen (1965) propose that osmotically induced pressures across shale beds are an important cause of abnormal pressure where high- and low-pressure zones are induced on opposite sides of a shale bed. From a purely mechanistic viewpoint, it is not difficult to conceive of several other causes of abnormal pressure.

The second line of research is a search for mechanisms other than the fluid pressure hypothesis or, in some cases, a modification of the original concepts. In this regard, Chapman (1981) exhumes lubrication, and Guth and others (1982) argue that there are limits to the amount of fluid pressure development before some sort of fracture begins and releases the pressure. Gretener (1972) suggests that high fluid pressures could not be maintained under a thrust for long periods of time and argues that the thrust does not move all at once, but like a caterpillar, one segment at a time. The suggestion here is that abnormal pressure may be episodic.

It may be noted further that compressional forces associated with thrusting are sufficient to generate high fluid pressures that will inevitably fracture the rock, causing large-scale displacements. Thus, the origin of the fluid pressures may be tectonic compression as opposed to basin loading, which has the added attraction of containing the means by which the overthrust occurs. If the fracturing temporarily acts to relieve the pressure, continued compression may eventually build it up again, causing a repetition of the thrust and the caterpillar type of motion. In each event the pore volume that is generated at fracture may be just sufficient to equal the volume necessary to return the rock to its ambient pressure when the tensile strength of the rock was first exceeded. As long as the rocks remain in a state of compression, this process will continue, with the fractured state of the rock at any time representing an instability from which it is not possible to increase fluid pressures further. It is expected that with increased fracture, this instability will be realized at progressively lower fluid pressures over geologic time. Hence, the process may be a decaying one that ultimately dies out when the compression finally takes place at or near constant fluid pressure. The analogy to this terminal situation is the normal compaction concept put forth earlier.

Whatever the final consensus—if any—there is little doubt that the Hubbert-Rubey hypothesis as developed from the original work of Terzaghi (1950) generated a flurry of interest in the mechanics of the overthrusting and the strength of rocks in general.

Seismicity Induced by Fluid Injection

Among the many thousands of fluid injection wells in the world, there are a few that have drawn attention as a result of the earthquakes that occur in conjunction with them. Three such wells or well sites in particular have been studied rather extensively: a deep waste injection well at the Rocky Mountain Arsenal near Denver, Colorado; experimental injection wells in the Chevron Oil Field near Rangeley, Colorado; and a high-pressure injection well used for hydraulic mining of salt in the Attica-Dale region of western New York.

In 1961, a deep disposal well (3638 m) was constructed to penetrate highly fractured Precambrian schist and granite gneiss bedrock underlying the Denver Basin. Injection was initiated in March 1962. During the first month of operation over 15,000 m³ (4×10^6 gallons) were injected with well head pressure ranging from 0 to 72×10^5 Pa (72 bars, 10.4 psi). During the first month of operation, Denver experienced its first earthquake in 80 years (Simon, 1969). The injection period from 1962 to 1966 can be divided into four stages (Hsieh and Bredehoeft, 1981). From March 1962 to September 1963, the injection took place under pressure. Between October 1963 and September 1964, no injection took place. From October 1964 to March 1965, injection took place by gravity flow. Injection was stopped in April 1965, after D. Evans publicly suggested a relationship between fluid injection and over 700 earthquakes generated in the Denver area over this short three-year period (Figure 8.15).

Evans (1966) invoked the Mohr-Coulomb failure theory of the Terzaghi (1950) and Hubbert and Rubey (1959) hypothesis to explain the fault movements and induced seismicity along historically dormant faults. The effect of increased fluid pressure is to reduce the frictional resistance to fracture by decreasing the effective normal stress across the fracture (Byerlee, 1967). The continuous seismicity after injection stopped was explained by Healy and others (1968) who suggested that near vertical fractures, preearthquake in origin, have long been subjected to tectonic stress. This stored strain was released by the propagation of built-up fluid pressure throughout the reservoir after injections had ceased.

The association of the Denver earthquakes with fluid injection prompted a renewed interest in the response of earth materials to abnormal pressures. An

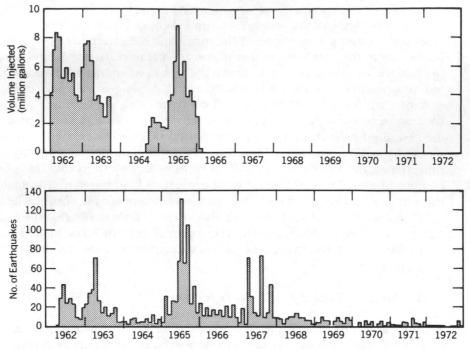

Figure 8.15

Comparison of fluid injected and the frequency of earthquakes at the Rocky Mountain Arsenal. Upper graph shows monthly volume of fluid waste injected in the disposal well. Lower graph shows number of earthquakes per month (from Hsieh and Bredehoeft, J. Geophys. Res., v. 86, p. 903–920, 1981. Copyright by Amer. Geophys. Union).

experiment was quickly established at the Rangeley Field where Chevron had been injecting fluids to enhance production. It was noted that a prevalent zone of earthquake activity persisted near the edge of the field. Four wells were chosen for the experiment, and testing started in 1969 and continued through 1973 where the fluid pressures were alternately raised by injection and lowered by pumping (Figure 8.16). As noted, the frequency of earthquake activity was dramatically increased when the fluid pressure within the reservoir exceeded 275×10^5 Pa (275 bars). The experiments are described by Raleigh and others (1972, 1976). Raleigh (1971) provides a good overview on this topic.

Fletcher and Sykes (1977) reported a rash of earthquakes of small magnitude in the vicinity of an injection well in western New York. This well is only 430 m deep. The relationship between the fluid pressures and the earthquakes are shown in Figure 8.17.

In examining the role of water in rock strength, Rojstaczer and Bredehoeft (1988) note that failure occurs at the three sites just cited when the fluid pressure is a mere one-third to three-fifths of the greatest principal stress. Such modest fracture pressures in rocks suggest that these rocks already possess a large deviatoric stress, perhaps to the extent that they were very close to failure at ambient fluid pressures. Hence, the perturbations necessary to induce failure were not large. On the other hand, with permeable materials, the pressure-producing

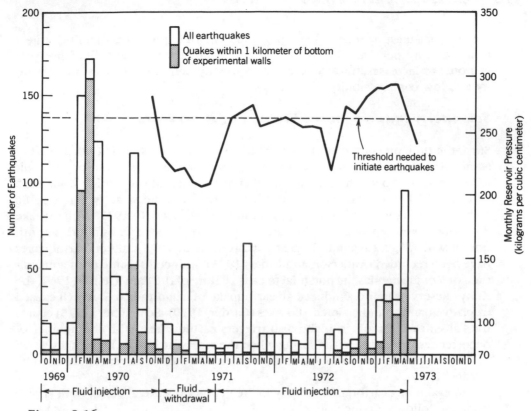

Figure 8.16
Earthquake frequency at Rangely Oil Field and its relation with induced reservoir pressures (from Wallace, 1974).

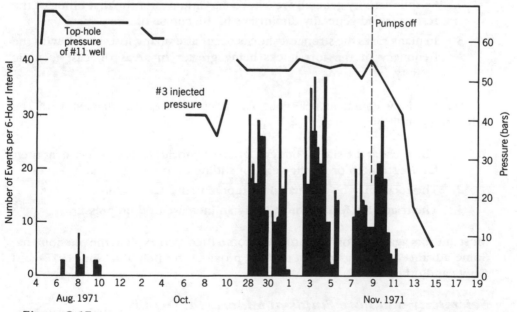

Figure 8.17
Number of earthquakes per 6-hour interval plotted with top hole pressure in injection well, western New York (from Fletcher and Sykes, J. Geophys. Res., v. 82, p. 3767–3780, 1977. Copyright by Amer. Geophys. Union).

mechanisms must operate rather quickly lest the fluid pressures drain off. Injection wells are perfectly suited for the establishment of a rapid fluid pressure response to increases in water volumes, especially in rocks of modest permeability and low compressibility.

Seismicity Induced in the Vicinity of Reservoirs

Since the fluid pressure mechanism of induced seismicity was first postulated in association with the Denver earthquakes, fluid pressure contributions to fault movements have been inferred at tens of sites throughout the world (Woodward and Clyde, 1979). The first such observations were at Lake Mead, where seismic activity started shortly after the filling period commenced. In 1937, with the lake 80% filled, 800 separate earthquakes were recorded over a two-month period, one of which had a magnitude greater than 5. As of 1975, 10,000 earthquakes have been recorded (Anderson and Laney, 1975). Induced seismicity of a magnitude of 5 or greater has the potential to cause failure of the dam itself. In 1967, the Konya Reservoir in India induced an earthquake of magnitude 6.5, which caused extensive damage and claimed 180 lives (Carder, 1970). Schleicher (1975) claims the Palisades Reservoir in Idaho also triggers earthquakes, with the number of occurrences increasing in the spring when the reservoir is at its highest level and in the fall when it is at its lowest level. Numerous other occurrences are cited by Carder (1970).

In examining earthquake activity in the vicinity of several dams, Roth (1969) drew the following conclusions:

1. Seismic activity seems to be more commonly associated with reservoirs deeper than 100 m, with the total water volume of less importance than the height.

2. Seismic activity generally reaches a maximum after the first filling of the reservoir and generally diminishes in the course of a few years.

3. In many cases the strongest shocks occur after many foreshocks, with the frequency of the foreshocks being greater in areas of least tectonic activity.

Seismic inducement at reservoirs has generally been associated with the following conditions:

1. The rocks at the site are under a large deviatoric stress and very near their failure strength or on the verge of sliding on preexisting fault planes.

2. The rocks may be associated with potentially active faults.

3. The triggering force is most likely an increase in fluid pressure.

Most authors again call on the Mohr-Coulomb theory to explain the phenomena. Some advanced reading on this topic is provided by Bell and Nur (1978) and Gupta and others (1972).

Seismicity and Pore Fluids at Midcrustal Depths

The information obtained from the "experiments" on induced seismicity suggests that failure along preexisting faults occurs at rather modest increases in fluid pressure. These "experiments" also suggest that in most if not all cases, the rocks are subjected to large deviatoric stresses, which accounts for the modest fluid pressure requirements. Given the same or similar states of deviatoric stress in the crust, fluid pressures that promote failure must be elevated at least to the levels as suggested by the induced seismicity studies. Given smaller deviatoric stresses, even greater fluid pressures are required. As little is known about the fluid pressure and stress conditions at crustal depths, little more can be stated about what is actually occurring there. Rojstaczer and Bredehoeft (1988) summarize those factors upon which failure at such depths will depend: (1) the rate and duration of the pressurization mechanisms, (2) the permeability and compressibility of the rock in which it operates, (3) the degree to which the process is isolated from the surface of the earth, (4) the orientation of the fault planes relative to the principal stresses, and (5) the degree of difference between the greatest and least principal stress. In this regard, items 1 and 2 dictate how close the pressure generation will take place to constant mass or constant pressure; item 3 refers to the ability of deep rocks to dissipate fluid pressures due to a hydraulic communication with the earth's surface, item 4 is purely a geometric argument, and item 5 puts constraints on the pressures required to promote failure.

The Phreatic Seismograph: Earthquakes and Dilatancy Models

Earthquakes occur when elastic strain in a deforming medium is released, resulting in intense seismic vibrations. When such earthquakes occur, seismic waves are released, some of which (so-called *P* and Rayleigh waves) have compressional and dilational components, which causes both compression and dilation of the materials through which the waves pass. Such waves can travel great distances and can cause water-level fluctuations in wells located far from the source of the earthquake.

Blanchard and Byerly (1935) were the first to suggest that a float connected to a recording instrument would record the dilational components of passing seismic waves. These authors recognized that different wells would produce different responses to the same wave, depending on the character of the well itself, and the hydraulic properties at the point of recording. Their solution was to correct these responses empirically by comparing records from various well responses and seismographs. In this regard, Eaton and Takasaki (1959) established empirical relations between the maximum amplitude of a water-level fluctuation, and the type, wavelength, and amplitude of the waves producing the response. They noted that the response of the water level decreases rapidly as the wavelength of the passing seismic waves decreases.

Cooper and others (1965) were the first to address this problem quantitatively with an analytical solution. This solution describes the manner in which a water level will respond to a passing seismic wave, and this response is a function of the properties of the well, the hydraulic properties of the aquifer, and the type, period, and amplitude of the wave.

The work of Cooper and others (1965) was extended by Bodvarsson (1970) who attempted to examine confined fluids as strain meters. Bodvarsson noted, like Cooper and others (1956), that a well's seismograph cannot be compared to true seismographs unless the well-aquifer system is well known. A few years later, Nur (1972) suggested that water-level fluctuations to such waves may be helpful in predicting the occurrence of earthquakes. Nur based his theory on the observation by Sadovsky and others (1969) as well as others that prior to earthquakes, the ratio of the seismic velocity of the primary and secondary waves decreased dramatically and, following the earthquake, resumed a normal level. The velocity of the primary wave is greatly affected by the degree of water saturation, but the velocity of the secondary wave is not. Thus, the intuitive conclusion that prior to an earthquake, some mechanism must be operative that causes a decrease in saturation of the rocks so affected. Preearthquake dilatancy is one mechanism that can account for this, where just prior to deformations that produce earthquakes, the rocks undergo inelastic volume increases produced by the formation of new fractures. As mentioned previously, dilation of crystalline rocks begins at a deviatoric stress of one- to two-thirds of the stress difference required to fracture the rock (Brace and others, 1966). This fracture is accompanied by a porosity increase that ranges from 0.2 to 2 times the elastic volume change the material would have had were it perfectly elastic. Hence, the flow of water from the preexisting pores or cracks into the newly formed pores or cracks can cause the rock to become temporarily undersaturated, which effects the velocity of the primary waves. Nur (1972) argued that as water moved into the dilatant region, the pore pressure would rise, lowering the effective confining pressure so that continued strain could cause the failure that triggers the earthquake. A further modification of this model is given by Scholz and others (1973).

It thus follows that if we had a water monitoring point in a rock mass during the sequence of events leading to an earthquake we might notice the following behavior. First, we assume the rocks are undergoing some tectonic strain and if the deformation is taking place at or near constant mass, or at least faster than the fluid can be expelled, the fluid pressure will rise accordingly. Thus we would notice a rise of the water level in the monitoring point. Now to go further we must speculate on the dilatancy model. We presume that the increased strain is manifested by an increase in the mean confining stress acting on a saturated unit volume. There is little question that the stress contributing to strain is increasing at a faster rate than the fluid pressure within the unit volume. This is demonstrated from Eqs. 8.1 and 8.2 for constant fluid mass

$$\frac{\partial P}{\partial \epsilon} = \xi Q \qquad \frac{\partial \sigma}{\partial \epsilon} = \frac{\xi Q}{R/H} \tag{8.30}$$

where R/H is generally less than one. This suggests that the effective stress is undergoing a net increase. However, if we speculate that the minimum principal stress is undergoing little change, the overall effect of this process is to increase the deviatoric stress progressively at a faster rate than the fluid pressure increase. When the critical deviatoric stress is achieved for the fluid pressure that is generated, the dilatant behavior is triggered. In this model, we have imposed the same conditions normally employed with the triaxial apparatus used in the laboratory. It is expected now that the water level in the well will fall as the fluid drains to fill the newly opened cracks. It is not possible to state the volume of new openings that are created during the dilatant period. One obvious estimate is that the pore volume that is generated is equal to the volume needed to return the fluid to its ambient pressure when the tensile strength of the rock was exceeded. With continued tectonic strain, the fluid pressure starts to rise again, but it can never rise to a level higher than it was when the rocks first became dilatant. That is, the fluid pressure that initiates dilatancy represents a theoretical upper limit to fluid pressure development lest the rocks become dilatant again. However, this is no longer the same rock with the same strength characteristics that it possessed previous to its dilatant behavior. For one thing, it appears likely that dilatancy represents a loss of cohesive strength, thereby reducing the resistance to shear. Hence, this second generation of fluid pressure rise may result in failure with consequent earthquake production.

Observations of the relationships between water well fluctuations and seismic events date back several years. Modern observations have been reported by Vorhis (1955), Gordan (1970), and Wesson (1981).

Suggested Readings

Bredehoeft, J. D., and Hanshaw, B., 1968, On the maintenance of anomalous fluid pressures: I, Thick sedimentary sequences: Geol. Soc. Amer. Bull., v. 79, p. 1097–1106.

Hubbert, M. K., and Willis, D. G., 1957, Mechanics of hydraulic fracture: Am. Inst. Min. Engr. Trans., v. 210, p. 153–168.

Hubbert, M. K., and Rubey, W. W., 1959, Role of fluid pressure in mechanics of overthrust faulting. Part I. Mechanics of fluid filled porous solids and its application to overthrust faulting: Bull. Geol. Soc. Amer., v. 70, p. 115–166.

Powers, M. C., 1967, Fluid release mechanisms in compacting marine mudrocks and their importance in oil exploration: Amer. Assoc. Petrol. Geol. Bull., v. 51, p. 1240–1254.

Raleigh, C. B., 1971, Earthquakes and fluid injection: Amer. Assoc. Petrol. Geol. Mem., No. 18, p. 273–279.

Problems

1. A sedimentary layer is 10,000 m thick and was deposited at a rate of 500 m/10^6 yr. The hydraulic diffusivity is 1 m^2/yr. Calculate t/T^*. Determine the excess pressure as a percentage of the overburden pressure at 2000 m and 10,000 m below the surface of the sediment.

2. Answer the following from Figure 8.10:
   a. For K/S_s = 10^{-3} cm^2/sec for a 20 km deposit, what is the largest sedimentation rate required to develop excess pressure?
   b. For K/S_s = 10^{-3} cm^2/sec for a 10 km deposit, what is the lowest sedimentation rate required to develop excess pressure?
   c. Comparing (a) and (b), which has the largest time constant T^*?
   d. What is the value of T^*/t for K/S_s = 10^{-2} cm^2/sec and a sedimentation rate of 500 m/my for a 5 km deposit?
   e. For problem (d), will this situation be characterized by excess pressure? Why?

3. What assumptions are required to reduce Eqs. 8.1 and 8.2 to Eqs. 4.56 and 4.57?

4. Reproduce curves *A*, *B*, and *C* on Figure 8.9. Take γ as 2.3 g/cm^3. Then reproduce curve *B* if R/H = 0.8.

5. A repository for nuclear waste is under construction at the 3000-ft level in low-permeability basalts. The total vertical stress at this depth is 248 bars. For a water table about 200 ft below land surface, the neutral stress is about 83 bars at the 3000-ft level. What temperature increase due to radioactive decay will reduce the effective vertical stress to zero at the 3000-ft level? Assume a constant fluid mass situation. Everything you are required to know about this basalt is given in Table 8.2

6. Prove the relations given in Eq. 8.30. Use Eqs. 8.1 and 8.2 as your starting point.

7. For basaltic rocks such as those in Table 4.4, calculate the change in total stress, fluid pressure, and effective stress with increased strain. Hint: See Eq. 8.30. Make similar calculations for the sandstone and mudstone of Table 4.4. Make some general statements regarding rigidity of rocks and the distribution of the total stress between the pore fluids and the matrix (effective stress).

Chapter Nine

Heat Transport in Ground Water Flow

9.1 Conduction, Convection, and Equations of Heat Transport

9.2 Forced Convection

9.3 Free Convection

9.4 Energy Resources

9.5 Heat Transport and Geologic Repositories for Nuclear Waste Storage

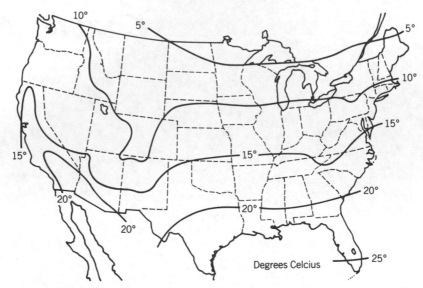

Figure 9.1
Temperature of ground water in the United States at depths of 10 to 20
meters (from Collins, 1925).

In the previous chapter, we discussed abnormal pressures that result from thermally expanded fluids and the role of temperature in phase transformations. For
the most part, however, our interest has been in isothermal flow, simply because
temperature effects were not important to the discussions. Our interest in this
chapter shifts to hydrogeologic investigations where the nonisothermal aspects
are the major ones that concern us.

Prior to getting to the main points of this chapter, a few well-known facts
should be established. Lovering and Goode (1963) show that the effective perturbation depth of temperature fluctuations at the earth's surface is on the order of
10 m. Thus, only very shallow ground water exhibits appreciable temperature
changes in response to seasonal variations in the amount of solar energy reaching
the earth's surface. This is reflected by the fact that the temperature of ground
water at depths ranging from 10 m to 20 m will generally be 1° to 2°C higher than
the local mean annual temperature (Figure 9.1). Below this depth, the temperature
increases more or less steadily. In general, the older and more compact the
rock, the lower the geothermal gradient. The geothermal gradient in the Canadian
Shield has been reported as 9.1°C/km, whereas gradients as high as 36°C/km
have been reported for the Mississippi Embayment in Louisiana. In performing
these measurements, it is normally assumed that an equilibrium exists between
the solid and liquid phases so that there is little or no temperature difference
between them.

As the conductance of heat is directly proportional to the gradient in temperature, the higher the geothermal gradient, the higher the conductive heat flow at the
surface. Heat flow on the shield areas has been determined to be on the order of 1
calorie per square centimeter per second (cal/cm²sec), whereas at the midoceanic
ridges, it is on the order of 8 cal/cm²sec or greater. Goguel (1976) reports that the
average heat flux by conduction at the earth's surface is on the order of 1.3
cal/cm²sec. Variations in surface heat flow and geothermal gradients have been

attributed to differences in the thermal conductance of various media, geologically recent intrusions of magma, and the influence of ground water flow on the thermal regime. Lachenbruch and Sass (1977) consider this last factor to be the most important in modifying the heat flow originating at great depths in the earth.

9.1 *Conduction, Convection, and Equations of Heat Transport*

Heat can be transported from point to point in a porous medium by way of three processes: conduction, convection, and radiation. Conductive transport may be described by a linear law relating the heat flux to the temperature gradient. Convective heat transport is the movement of heat by a moving ground water. Radiation, better known as thermal electromagnetic radiation, is the radiation emitted because of the temperature of a body.

Fourier's Law

Fourier's law describes the conduction of heat from regions where the temperature is high to where it is low. This process is called heat conduction, or simply conduction, and is described for a solid or a liquid as

$$\mathbf{H} = -\kappa \, \text{grad} \, T \tag{9.1}$$

where $\mathbf{H}$ is the heat flux, T is temperature, κ is the proportionality constant relating the two, referred to as the thermal conductivity of the substance, and grad T is the temperature gradient. In the metric system, the unit of heat is the calorie, with temperature expressed in centigrades and distance in centimeters. The thermal conductivity has the units cal $\text{sec}^{-1} \text{cm}^{-1} {}^\circ\text{C}^{-1}$. In the English system, the unit of heat is the BTU (British thermal unit), and temperature is measured in degrees Fahrenheit. The thermal conductivity is expressed in (BTU) $\text{hr}^{-1}\text{ft}^{-1}{}^\circ\text{F}^{-1}$. In SI units, the unit of heat is the joule with temperature expressed in degrees kelvin K so that thermal conductivity has the units joules $\text{m}^{-1}\text{sec}^{-1}{}^\circ\text{K}^{-1}$.

When dealing with fluids within solids, it is sometimes necessary to distinguish between energy transport in both the fluid and solid phase. We will assume that the solids and their contained fluids are at the same temperature so that we can define T in Eq. 9.1 as the average temperature of both phases. In addition, because both the fluids and the solids are conductors, it is necessary to introduce an effective thermal conductivity so that Eq. 9.1 is expressed

$$\mathbf{H}_e = -\kappa_e \, \text{grad} \, T \tag{9.2}$$

where $\mathbf{H}_e$ is an effective energy flux vector, κ_e is an effective thermal conductivity, and T is understood to be the average temperature of the solid and fluid mass.

In two phase mixtures consisting of water and solids, both of which are conductors, an effective thermal conductivity can be described by considering the volume fractions and conductivities of the individual phases. For a parallel arrangement of fluids and solids, the accepted relationship is

$$\kappa_e = n\kappa_f + (1 - n)\kappa_s \tag{9.3}$$

where κ_e is the effective thermal conductivity and the subscripts f and s refer to the fluids and the solids, respectively. In discussing this relationship, Slattery (1972) suggests that it must be amended to account for tortuosity (or a nonparallel arrangement of fluids and solids) in the porous medium. He distinguishes between two cases, one where the fluid is stationary and one where conduction is accompanied by fluid movement. For a nonmoving fluid, the effective thermal conductivity becomes

$$\kappa_e = n\kappa_f + (1 - n)\kappa_s - \kappa_s K^* \qquad (9.4)$$

where K^* accounts for the reduction in free transport because of the tortuosity of the porous medium. For a moving fluid, a convective effect referred to as thermal dispersion is noted. Thermal dispersion is a microscopic dispersal of heat due to the convective transport through the porous structure, and has the same effect, when viewed macroscopically, as an increase in the effective thermal conductivity. Dispersion will be taken up rather completely when we discuss mass transport. Here it is sufficient to say that this effect is normally incorporated in the transport parameter κ_e.

The special cases investigated by Slattery (1972) have been verified experimentally. Experimental data summarized by Green and Perry (1961) demonstrate that the value of the effective thermal conductivity is greater with a flowing fluid than with a stagnant one and, in general, increases with increasing velocity. Yagi and Kunii (1957) indicate that the effective thermal conductivity can be separated into two terms, one independent of fluid flow and the other dependent on fluid

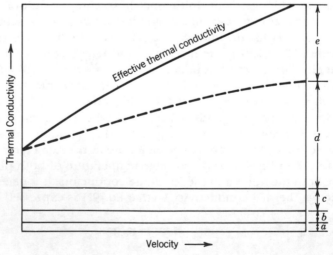

Figure 9.2
Relationship between effective thermal conductivity and the mechanisms that contribute to it. (*a*) Conduction by fluid, (*b*) conduction by solid particles, (*c*) particle to particle radiation, (*d*) conduction by a series mechanism from solid to fluid, and so on, (*e*) convection through fluid phase. (Modified from Schuler, Stallings, and Smith, Heat and mass transfer in fixed bed reactors, Chem. Eng. Prog. Symp. Ser., v. 48, p. 19–30, 1952). Reproduced by permission of the Amer. Inst. Chem. Engrs.

Table 9.1
Thermal conductivities of rocks

Material	Thermal conductivity (cal/m sec °C)
Quartz	2
Sandstone	0.9
Limestone	0.5
Dolomite	0.4–1
Clay	0.2–0.3
Water	0.11
Air	0.006

flow. In assessing the effect of several mechanisms on the thermal conductivity, Singer and Wilhelm (1950) include (1) molecular conduction through the fluid phase, (2) solid particle to particle molecular conduction, and (3) particle-to-particle radiation. None of these mechanisms is affected by fluid flow, and may be plotted as horizontal lines in a plot of thermal conductivity versus velocity (Figure 9.2). Two additional mechanisms control the dependence of the thermal conductivity on fluid velocity, (4) a series mechanism from fluid to solid to fluid, and so on, and (5) convection through the fluid phase. These become predominant at large velocities (Figure 9.2).

Some measured values for the effective conductivity in rocks are given in Table 9.1. In general the more porous the medium, the lower the effective thermal conductivity, mostly because water has a lower thermal conductivity than most solid minerals.

Because conductance can take place through both the solid and fluid phase, it is possible to propose upper and lower bounds of the effective conductivity. For macroscopically homogeneous and isotropic materials, such limits may be determined by the method of Hashin and Shtrikman (1962). For the upper and lower limits, respectively,

$$\kappa_{eu} = \kappa_s + n \left[\frac{1}{(\kappa_f - \kappa_s)} + \frac{(1-n)}{3\kappa_s} \right] \tag{9.5a}$$

and

$$\kappa_{e\ell} = \kappa_f + \frac{(1-n)}{1/(\kappa_s - \kappa_f) + n/3\kappa_f} \tag{9.5b}$$

If porosity approaches zero, the upper and lower limits approach the conductivity of the solids. As porosity approaches one, both limits go to the conductivity of the fluid. Equations 9.5 hold equally well for the electrical conductivity as well as other two-phase conductive properties of porous media.

Convective Transport

In systems where the fluid is moving there is a convective transport by the fluid motion. When the flow field is caused by external forces, the transport is said to occur by forced convection. Such is the case where ground water movement takes

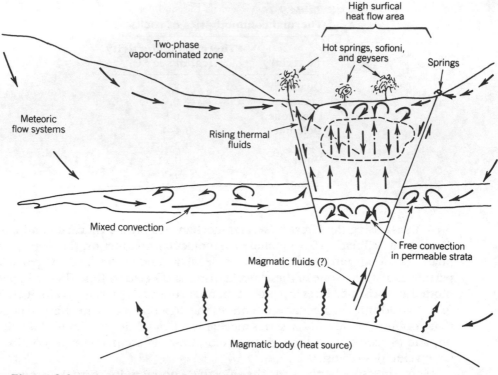

Figure 9.3
Forced, free, and mixed convection in a hydrothermal system (from Sharp and Kyle, 1988, Geol. Soc. Amer. Pub., from Geol. of N. America, 0–2, p. 461–483).

place in the absence of density gradients such that Darcy's law applies. A second type of transport, called free convection, occurs when the motion of the fluid is due exclusively to density variations caused by temperature gradients. Free convection is probably the dominant type of fluid motion in some hydrothermal systems where the bulk of liquid discharge is in the form of steam or hot water. A similar phenomenon can occur in mass transport, say, along fresh water–salt water interfaces in coastal areas where density differences reflect the various salinity differences. For these cases, there no longer exists a scalar potential so that Darcy's law does not describe the motion. It is emphasized that forced and free convection represent two limiting conditions. In the case of the former, buoyancy forces are assumed to be negligible. For the latter, fluid motion must be described entirely in terms of buoyancy. When buoyancy forces dominate, the velocity field and the energy field (temperature) are interdependent, and the equations must be solved iteratively.

The simplest system to demonstrate forced and free convection is shown in Figure 9.3, which is a hypothetical model for flow in hydrothermal areas. Note the transition from forced convection at shallow depths to a mixed convection pattern, and ultimately free convection. According to Bear (1972) the equation for ground water motion in such environments may be described in terms of two driving forces, fluid pressure changes and buoyancy, and is of the form

$$v = \frac{-k}{n\mu} \frac{\partial P}{\partial x} - \frac{kg\rho_o[1 - \alpha_f(T - T_o)]}{n\mu} \frac{\partial z}{\partial x} \qquad (9.6)$$

where ρ_o is some reference density, μ is the viscosity, and α_f is the coefficient of volume expansion for the water. If the pressure distribution is hydrostatic, $\partial P/\partial x = -\rho_o g \partial z/\partial x$, and Eq. 9.6 becomes

$$v = \frac{kg\rho_o\alpha_f(T - T_o)}{n\mu} \frac{\partial z}{\partial x} \qquad (9.7)$$

where, for vertical flow, $\partial z/\partial x = 1$. The quantity $\rho_o g \alpha_f (T - T_o)$ is the driving force per unit volume of fluid.

The rate equation combining conduction with a Darcy type of fluid motion (forced convection) may be expressed

$$H_x = -\kappa_{e_x} \frac{\partial T}{\partial x} + n\rho_w c_w T v_x \qquad (9.8a)$$

$$H_y = -\kappa_{e_y} \frac{\partial T}{\partial y} + n\rho_w c_w T v_y \qquad (9.8b)$$

$$H_z = -\kappa_{e_z} \frac{\partial T}{\partial z} + n\rho_w c_w T v_z \qquad (9.8c)$$

where v is the velocity and c_w is the specific heat of the fluid, defined as the heat necessary to raise the temperature of 1 g of fluid 1°C, with the units cal/g°C. The product $\rho_w c_w$ is the heat capacity per unit volume of fluid. For isotropic conditions

$$\mathbf{H} = -\kappa_e \text{ grad } T + n\rho_w c_w T\mathbf{v} \qquad (9.9)$$

where $\mathbf{v}$ is a vector with components v_x, v_y, v_z.

Equations of Energy Transport

The starting place for development of conservation statements pertaining to energy transport is a word equation of the form

$$\begin{array}{c} \text{energy inflow rate} - \text{energy outflow rate} = \\ \text{change in energy storage with time} \end{array} \qquad (9.10)$$

in units of energy per time. In general, this statement can be applied to a domain of any size. When applied to a representative unit volume with inflows through three sides, the left-hand side of this word equation is readily expressed as the divergence of the flux of interest (see chapter 4).

$$- \text{div } \mathbf{H} = \text{net energy outflow rate per unit volume}$$

It remains to describe the nature of the flux, which could be conduction (Eq. 9.2) or conduction combined with convection (Eq. 9.8). The right-hand side of the word equation represents the gains or losses in energy per unit time for the unit volume. It will be recalled that for the conservation equation for fluid flow, it was assumed that the time rate of change of fluid volume per unit volume was proportional to the time rate of change of hydraulic head, thereby introducing a proportionality constant termed the specific storage. For the equations of energy transport we assume the loss or gain of heat inside the unit volume is proportional to the temperature in the unit volume. Some proportionality constant is now

Table 9.2
Specific heat of substances

Material	$T(°C)$	Specific heat (kcal/kg°C)
Air	50	0.248
Water vapor	100	0.482
Methane	15	0.5284
Benzene	20	0.406
Ethyl alcohol	25	0.581
Basalt (dry)	20–100	0.20
Chalk (dry)	20–100	0.214
Clay (dry)	20–100	0.22
Granite (dry)	12–100	0.192
Quartz (dry)	12–100	0.188

needed to convert temperature changes to the gains or losses of heat. This proportionality constant is the heat capacity per unit volume, defined for water alone as the product of the fluid density ρ_w and the specific heat per unit volume of water c_w. The heat capacity per unit volume must express the quantity of heat gained or lost to or from the unit volume when the temperature changes by a unit amount (cal/L³°C). For the fluid component in the unit volume, this becomes $n\rho_w c_w$. For the solid components, the expression becomes $(1 - n)\rho_s c_s$, where the subscript s stands for solids. Thus we define an effective heat capacity for the unit volume

$$\rho'c' = n\rho_w c_w + (1 - n)\,\rho_s c_s \tag{9.11}$$

The conservative statement of Eq. 9.10 then becomes

$$- \text{div } \mathbf{H} = \rho'c'\frac{\partial T}{\partial t} \tag{9.12}$$

where $\rho'c'$ is the heat removed or gained from the unit volume when the temperature changes a unit amount. Equation 9.12 states that the net energy outflow rate per unit volume equals the time rate of change of energy within the unit volume. Table 9.2 gives some values for the specific heat of various substances.

The Heat Conduction Equation

For a conductive system defined by Fourier's law of the form of Eq. 9.2

$$\text{div } \mathbf{H} = \text{div}[\kappa_e \text{ grad } T] \tag{9.13}$$

Here, κ_e has been defined as an effective thermal conductivity. The conservation statement then becomes

$$\text{div}[\kappa_e \text{ grad } T] = \rho'c'\frac{\partial T}{\partial t} \tag{9.14}$$

By assuming that the effective thermal conductivity is constant

$$\frac{\partial^2 T}{\partial x^2} + \frac{\partial^2 T}{\partial y^2} + \frac{\partial^2 T}{\partial z^2} = \nabla^2 T = \frac{\rho' c'}{\kappa_e} \frac{\partial T}{\partial t} \tag{9.15}$$

This equation is called the heat conduction (diffusion) equation and is analogous to the unsteady flow equation introduced in Chapter 4. The solution to this equation describes the value of temperature at any point in a three-dimensional field, or more precisely, how the temperature is changing with time.

If the net outflow is zero

$$\frac{\partial^2 T}{\partial x^2} + \frac{\partial^2 T}{\partial y^2} + \frac{\partial^2 T}{\partial z^2} = \nabla^2 T = 0 \tag{9.16}$$

This equation is called Laplace's equation, as was the steady-state fluid flow equation introduced in Chapter 4. The solution of this equation describes the value of the temperature at any point in a three-dimensional field.

The Conductive-Convection Equation
For conduction and convection described by Eqs. 9.8

$$- \text{div } \mathbf{H} = - \text{div}[-\kappa_e \text{ grad } T + n\rho_w c_w T v] = \rho' c' \frac{\partial T}{\partial t} \tag{9.17}$$

or, with the usual assumptions of constant κ_e, n, ρ_w, c_w,

$$\kappa_e \nabla^2 T - n\rho_w c_w [\mathbf{v} \cdot \nabla T - T\nabla \cdot \mathbf{v}] = \rho' c' \frac{\partial T}{\partial t} \tag{9.18}$$

For steady ground water flow, $\nabla \cdot v = 0$

$$\kappa_e \nabla^2 T - n\rho_w c_w \mathbf{v} \cdot \nabla T = \rho' c' \frac{\partial T}{\partial t} \tag{9.19}$$

Note that when the velocity is zero we recover our transport equation for pure conduction (Eq. 9.15). When the temperature is steady

$$\kappa_e \nabla^2 T - n\rho_w c_w \mathbf{v} \cdot \nabla T = 0 \tag{9.20}$$

The equations developed here are called the conduction-convection equations. They are conservation expressions that describe the manner in which energy is moved from one point to another by means of bulk fluid motion and by conduction. The one-dimensional form of Eq. 9.19 is written

$$\frac{\kappa_e}{\rho' c'} \frac{\partial^2 T}{\partial x^2} - \frac{n\rho_w c_w}{\rho' c'} v_x \frac{\partial T}{\partial x} = \frac{\partial T}{\partial t} \tag{9.21}$$

where v_x is the mean ground water velocity in the x direction. The first term on the left describes energy transport by conduction, and the second describes energy transport by convection. In all cases, it is assumed that the temperature of the fluid and the solids are equal. Slattery (1972) describes the case where distinctions are required between the fluid and solid phases.

Table 9.3
Thermal diffusivities of some common rock and soils in cm²/sec (from Cartwright, 1973)

Soils and unconsolidated material	
Quartz sand, medium, dry	
Quartz sand, 8.3% moisture	0.0020
Sandy clay, 15% moisture	0.0033
Soil, very dry	0.0020–0.0030
Some wet soils	0.0040–0.0100
Wet mud	0.0022
Soil, Lexington, Kentucky	0.0021
Soil, Lexington, Kentucky (average 0–10 feet in place)	0.0072
Gravel	0.0057–0.0062
Rocks	
Shale	0.0040
Dolomite	0.0080
Limestone	0.0050–0.010
Sandstone	0.0113–0.0140
Granite	0.0060–0.0130

Last, the combined term $\kappa_e/\rho'c'$ has the units of L^2/T and is referred to as the thermal dispersivity λ. It is understood that the convective effect of thermodispersion is incorporated in the parameter κ_e. In the absence of fluid movement, this parameter (λ) becomes the thermal diffusivity. Table 9.3 gives some values for the thermal diffusivities of rocks.

Dimensionless Groups

The equations governing the transport of heat in ground water flow have been called the conduction-convection equations. The solution to these equations for certain boundary and initial conditions provides us with the temperature distribution of a region. By applying the nondimensionalizing procedures introduced in Chapter 4, it is easily seen that the behavior of a thermal system is controlled by a few combinations of dimensionless variables. We have already seen this in our discussions of the Fourier number.

To the cited dimensionless quantities in Chapter 4 we add now a dimensionless temperature $T^+ = T/T_e$, where T_e is some characteristic temperature and the superscript + indicates a dimensionless quantity. The conduction-convection Eq. 9.19 becomes

$$\nabla^{2+}T^+ - \left[\frac{n\rho_w c_w vL}{\kappa_e}\right]\nabla^+T^+ = \left[\frac{L^2(\rho'c'/\kappa_e)}{t_e}\right]\frac{\partial T^+}{\partial t^+} \tag{9.22}$$

where the Peclet number for energy transport is

$$N_{PE} = \frac{n\rho_w c_w vL}{\kappa_e} = \frac{\rho_w c_w qL}{\kappa_e} \tag{9.23}$$

where q is Darcy's specific discharge and L is some characteristic length. The Peclet number expresses the transport of energy by bulk fluid motion to the energy transport by conduction. In a practical sense this number reflects a competition between two rate processes, forced convection and conduction. Large numbers mean that convective transport dominates over conductive transport. The second dimensionless group on the right-hand side of Eq. 9.22 is recognized as the Fourier number for heat transport.

Suppose now that we assume a buoyancy-driven fluid system. A dimensional analysis of the steady-state transport Eq. 9.20 in combination with the buoyancy-driven velocity of Eq. 9.7 gives

$$\nabla^{2+}T^{+} - \left[\frac{g\rho_o c_w \rho_w Lka_f (T - T_o)}{\mu \, \kappa_e} \right] \nabla^{+}T^{+} = 0 \qquad (9.24)$$

where the Rayleigh number for energy transport is

$$N_{RA} = \frac{g\rho_o c_w \rho_w Lka_f \Delta T}{\mu \, \kappa_e} \qquad (9.25)$$

The Rayleigh number expresses the transport of energy by free convection to the transport by conduction and is generally used to establish the conditions for the onset of free convection. In most one-dimensional cases, the characteristic length is taken as equal to the thickness H of the formation over which the temperature difference ΔT is measured. In two-dimensional problems, it will be some length associated with fluid movement.

In problems of mixed convection, both the Peclet number and the Rayleigh number are important in the analysis. As we will see later, convective rolls at high Peclet numbers (substantial forced convection) take on a different geometry than those occurring at low Peclet numbers.

9.2 *Forced Convection*

Smith and Chapman (1983) recognize three main classes of forced convection problems of interest to hydrogeologists. The first of these is the vertical steady-state flow of ground water and its effect on a purely conductive vertical temperature distribution. If the resultant temperature profile can be measured, it can be matched against some curves representing the mathematical solution of the one-dimensional transport equation, and the Darcy velocity may be extracted. The second of the problems treats a two-dimensional vertical cross section of regional ground water flow where the flow field alters the conductive temperature profile. Two versions are possible here. First, for reasonably shallow systems where it can be assumed that the fluid density and viscosity are not affected by temperature, the velocity field can be determined independently of the transport problem and subsequently used in the solution to the transport problem. If the density or viscosity is affected by temperature, the velocity field and the temperature field are "coupled," and the equations must be solved iteratively. The third class of problems is for the two-dimensional geometry, but the temperature field or the velocity field is assumed outright and is not part of the problem. This was the method of Kilty and Chapman (1980) who were interested in conceptual models to account for heat flow variations in certain geologic settings. To this threefold classification, we may add yet a fourth, whereby the temperature field is affected by a three-dimensional flow field, and yet a fifth concerned with temperature distributions in evolving geologic basins.

Temperature Profiles and Ground Water Velocity

Consider the three ideal cases shown in Figure 9.4 where a one-dimensional velocity field is imposed on a one-dimensional purely conductive thermal field. In Figures 9.4*a* and *b*, the direction of ground water movement is taken as normal to the conductive isotherms. That is to say the hydraulic gradient and the conductive temperature gradient are collinear in the vertical direction so that the streamlines of fluid flow are normal to the isotherms of heat conduction. For these cases, a convection flux is established that will produce the greatest alteration of the conductive temperature distribution. In Figure 9.4*a*, the resultant temperature gradient will increase with increasing depth, whereas in Figure 9.4*b*, it will decrease with increasing depth. In addition, the heat flow at the surface will be greater for Figure 9.4*b*. In Figure 9.4*c*, the hydraulic gradient and the temperature gradient are normal so that the streamlines of fluid flow are collinear with the isotherms of heat conduction. Here convective transport is eliminated as it is impossible to transport heat along an isotherm. Mathematically, the convective term in our transport equation becomes

$$\mathbf{v} \cdot \text{grad } T = 0$$

For this case there is no alteration of the conductive gradient. In between the extremes of streamlines normal to isotherms and streamlines collinear with isotherms, some alteration of the conductive gradient will occur, with the maximum alterations associated with the former situation. In addition, either of these extremes can be viewed from the perspective of a one-dimensional transport problem. This is the basis of the Bredehoeft-Papadopulos (1965) model for determining ground water velocity from temperature profiles.

Figure 9.5 gives the conditions under which the model applies. In this diagram, T_o is an uppermost temperature measurement at $z = 0$, and T_L is a lower-

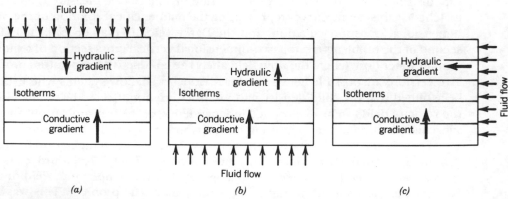

Figure 9.4
Three ideal cases where a fluid flow field is imposed on a conductive thermal gradient.

Figure 9.5
Diagrammatic sketch of typical leaky aquifer (from Bredehoeft and Papadopulos, Water Resources, Res., v. 1, p. 325–328, 1965. Copyright by Amer. Geophys. Union).

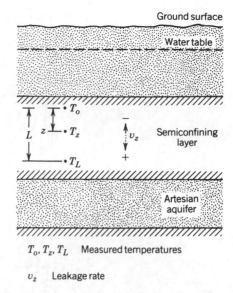

T_o, T_z, T_L Measured temperatures

v_z Leakage rate

most temperature measurement at $z = L$. These are the boundary conditions for the one-dimensional steady-state problem described by

$$\frac{\partial^2 T}{\partial z^2} - \frac{n\rho_w c_w v_z}{\kappa_e}\frac{\partial T}{\partial z} = 0 \tag{9.26}$$

The solution to this problem is (Bredehoeft and Papadopulos, 1965)

$$T_z = T_o + [T_L - T_o]\frac{\left[\exp\left(N_{PE}\frac{z}{L}\right) - 1\right]}{[\exp(N_{PE}) - 1]} \tag{9.27}$$

where T_z is the temperature at any depth over the thickness L (Figure 9.5) and

$$N_{PE} = \frac{\rho_w c_w q L}{\kappa_e} \tag{9.28}$$

which is the familiar Peclet number for heat transport.

A graphical solution to this problem is given by the type curves of Figure 9.6 with the Peclet number taken as the solution parameter. The curves are convex upward or downward. A suggested method of analyses by Bredehoeft and Papadopulos (1965) requires a field plot of $(T_z - T_o)/(T_L - T_o)$ against the depth factor z/L at the same scale as the type curves and superimposing this plot on the type curve set. The value of N_{PE} determined from this match is then used to calculate the velocity q from Eq. 9.28. This assumes information on κ_e.

Sorey (1971) reports some data by Kunii and Smith (1961) that have been employed to verify the mathematical model just described. The experiment involved the measurement of the temperature distribution in a column of glass beads and sand through which fluid was flowing counter to an imposed temperature gradient. Sorey (1971) constructed the dimensionless plot shown as Figure

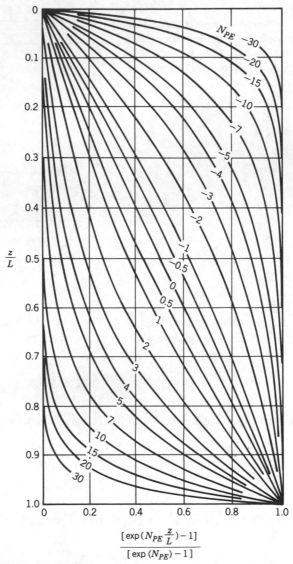

$$\frac{[\exp(N_{PE}\frac{z}{L})-1]}{[\exp(N_{PE})-1]}$$

Figure 9.6
Type curves for solution of one dimensional conductive advective equation (from Bredehoeft and Papadopulos, Water Resources Res., v. 1, p. 325–328, 1965. Copyright by Amer. Geophys. Union).

9.7, which matches the N_{PE} equals 3.3 curve of Figure 9.6. For the data reported by Kunii and Smith (1961) and the methods described, the mass flow rate was calculated to be $\rho_w q = 2.3 \times 10^{-4}$ g/cm²sec, which compares favorably with the measured mass flow rate of 2.6×10^{-4} g/cm²sec.

Field results have been reported by Cartwright (1970) and Sorey (1971). In Figure 9.8, Cartwright analyzes the temperature data from wells in the Illinois Basin and obtained a ground water flow rate of 4.9×10^{-8} cm/sec. The important limitations to this type of analysis reside in the one-dimensional assumptions for both the temperature field and the velocity field.

Figure 9.7
Type curve match for temperature profile in glass beads (from Sorey, Water Resources Res., v. 7, p. 963–970, 1971. Copyright by Amer. Geophys. Union).

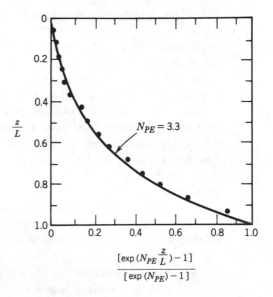

Figure 9.8
Type curve match for Illinois Basin (from Cartwright, Water Resources Res., v. 6, p. 912–918, 1970. Copyright by Amer. Geophys. Union).

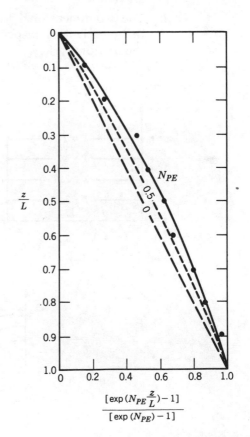

Heat Transport in Regional Ground Water Flow

Let us now continue the "experiment" started in the previous section concerning the thermal outcome of various orientations of streamlines of fluid flow and isotherms of pure conduction. A conductive field is shown in the two-dimensional region of Figure 9.9a. A two-dimensional flow field as shown in Figure 9.9b is superimposed on this conductive field, with Figure 9.9c demonstrating the alteration of the temperature field. The greatest alteration occurs where the streamlines of flow are normal or nearly so to the conductive isotherms, that is, the recharge and discharge areas, and the least alteration takes place in the region of lateral flow. Because of this convective alteration, the geothermal gradient will increase with increasing depth in the recharge area and decrease with depth in the discharge area, the latter associated with a greater amount of heat flow at the surface. This problem was solved by Domenico and Palciauskas (1973) for the following conditions:

1. The fluid properties are not affected by temperature and the medium is isotropic and homogeneous with respect to both fluid flow and heat conduction.

2. The two-dimensional flow problem is as described in Example 7.1 where the water table is unspecified but the lower and lateral boundaries are of the no-flow variety.

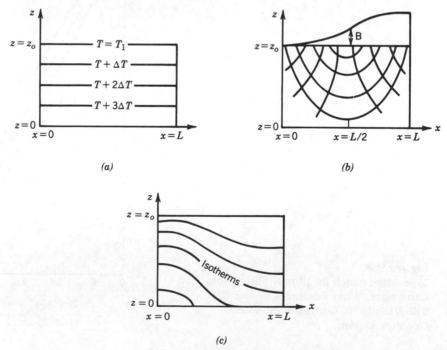

(a)

(b)

(c)

Figure 9.9
Diagrams illustrating (a) conductive heat transport in a two-dimensional region, (b) generalized flow system for the two-dimensional region and (c) alteration of the isotherms (modified from Domenico and Palciauskas, 1973, Geol. Soc. Amer. Pub., from Geol. Soc. Bull., v. 84, p. 3803–3814).

3. The temperature problem is described for the two-dimensional region of Figure 9.9 where the lateral boundaries are nonconductive ($\partial T/\partial x = 0$). The upper boundary is taken at some constant temperature T, and the lower boundary is described in terms of a constant temperature gradient T_o'.

Although the analytical solution to this problem was developed for any water table configuration, a special case was examined for the water table given in Figure 9.9b for which an analytical solution is presented in Example 7.1. For this special case, the solution to the two-dimensional transport equation is of the form

$$T(x, z) = [T_1 + T_o'(z - z_o)] - \frac{T_o'}{2} [N_{PE}] \left[f\left(\frac{z_o}{L}\right) \right] \qquad (9.29)$$

where T_o' is the temperature gradient across the bottom boundary, N_{PE} is the dimensionless Peclet number, and all other terms are described in Figure 9.9. In this form, the temperature distribution as described in Eq. 9.29 is controlled by three essential features. The first bracketed quantity describes the temperature distribution for pure conduction, that is, describes the conductive isotherms shown in Figure 9.9a. The other quantities obviously describe the perturbation of the purely conductive field and consist of a dimensionless Peclet number and a group of trigonometric and hyperbolic functions that deal exclusively with the ratio of basin depth z_o to basin length $[f(z_o/L)]$. For this case the Peclet number is described

$$N_{PE} = \frac{KB\rho_w c_w}{\kappa_e} \qquad (9.30)$$

where K is the hydraulic conductivity and B is the mean water table elevation as measured from the elevation of the discharge area (Figure 9.9). Thus if the hydraulic conductivity or the mean water table elevation B get large with respect to the thermal conductivity (large Peclet number), forced convection dominates over conduction and the conductive field is altered significantly. Note that B is the characteristic length in this problem, whereas for the one-dimensional problem of Bredehoeft and Papadopulos (1965), the characteristic length was determined to be the length of the vertical region over which perturbations were measured. The expression for $f(z_o/L)$ is cumbersome and will not be given here. However, as the basin length-to-depth ratio becomes large, the flow pattern is largely horizontal and $f(z_o/L)$ approaches zero so that there is a minimal interference with the conductive gradient, that is, $\mathbf{v} \cdot \text{grad } T \cong 0$. Small length-to-depth ratios produce flow systems that contain vertical components of flow throughout and are associated with large perturbations.

For an approximate measure of the potential for convective alterations of the geothermal gradient, a modified Peclet number may be given as

$$N_{PE} = \frac{BK\rho_w c_w(z_o/L)}{\kappa_e} \qquad (9.32)$$

Values less than one are associated with temperature distributions that are not far from the limiting case of pure conduction. Other geometric configurations of flow such as those describing local, intermediate, and regional flow (Figure 7.7)

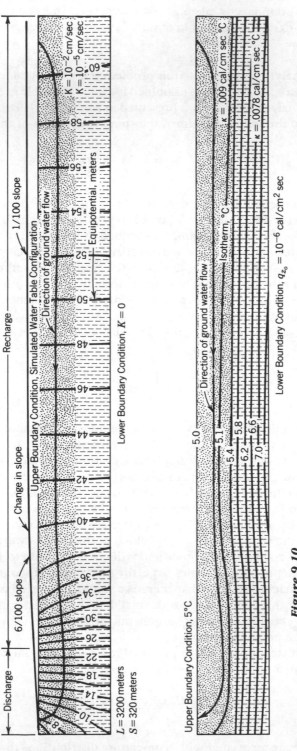

Figure 9.10
A simple hypothetical two-layered model of the ground water thermal regime. (*a*) Ground water basin. (*b*) Temperature distribution in the ground water basin (from Parsons, Water Resources Res., v. 6, p. 1701–1720, 1970. Copyright by Amer. Geophys. Union).

are also examined in terms of the interpretation of the characteristic length associated with Peclet numbers for two-dimensional flow systems (Domenico and Palciauskas, 1973).

The study described was a theoretical one where the main interest was in determining the effect of fluid flow on temperatures within the framework of the classic regional flow model introduced in Chapter 7. Morgan and others (1981) have used this model to investigate the threshold hydraulic conductivity where temperature perturbations become significant in basins of fixed geometry within the Rio Grande Rift. Betcher (1977) solved the problem numerically with a finite element model and conducted a sensitivity analysis to ascertain the role of the various parameters in controlling the temperature distribution. Parsons (1970) used a numerical model to investigate the effect of hydraulic conductivity contrasts and included a case where the upper boundary was taken as a variable temperature boundary. Some results of his study are shown in Figure 9.10. Note in particular that the three-order-of-magnitude hydraulic conductivity contrast associated with the lower-permeability region more or less preserves the pure conduction isotherms, as does the rather large area of lateral flow where $\mathbf{v} \cdot \mathrm{grad}\ T \cong 0$.

Smith and Chapman (1983) include the effects of temperature on fluid density and viscosity in heterogeneous anisotropic media. Figure 9.11a gives some of their results for homogeneous material, showing the convective influence on heat flow at the surface, which is most pronounced for higher values of permeability. For homogeneous but anisotropic material, the heat flow at the surface increases with decreasing anisotropic ratios K_x/K_z (Figure 9.11b).

Woodbury and Smith (1985) present a solution for a three-dimensional flow field. Their analysis was concerned with the variations and areal distribution in heat flow at the surface as affected by relief on the water table and permeability variations.

The studies described are concerned with temperature distributions in modern environments. Temperature distributions have also been ascertained for paleoflow systems that may have been important in the formation of ore deposits by topographically driven flow. Some results from Garven and Freeze (1984a, 1984b) are shown in Figure 9.12. Figures 9.12a and b depict temperature distributions at low and high Peclet numbers, respectively, due to differences in hydraulic conductivity. The advective perturbations are readily seen for the high conductivity condition. Figures 9.12c and d depict temperature distributions at high and low Peclet numbers, respectively, due to differences in the thermal conductivity. As noted, the higher the thermal conductivity the closer the temperature distribution approaches that of pure conduction. Figures 9.12e, f, and g depict temperature distributions for different water table configurations, where the effects of recharge and discharge along the water table are reflected in the various distributions.

The field observations that support the general interference between isotherms and streamlines discussed in the theoretical work are provided in studies by Van Orstrand (1934) and Schneider (1964).

Heat Transport in Active Depositional Environments

The pressure evolution in active depositional environments was seen in Chapter 8 to be a transient phenomenon. As expected, the transient velocity field is responsible for the convective transport of heat. Examining this process from the

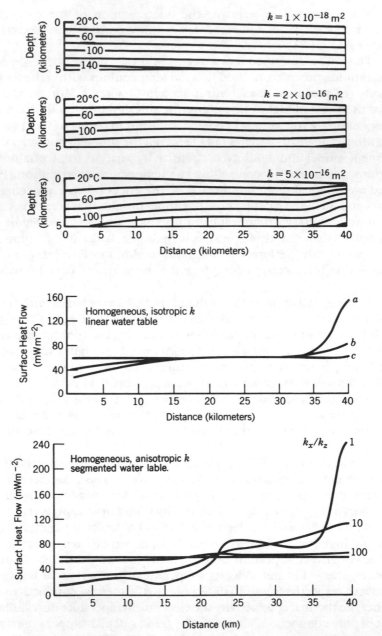

Figure 9.11
(*a*) Thermal effects of ground water flow in a basin of homogeneous, isotropic permeability with a linear water table. Surface heat flow expressed in mWm^{-2}. (*b*) Thermal effects of ground water flow in a basin of anisotropic homogeneous permeability. Horizontal permeability is 8×10^{-16} m^2, anisotropic ratio is 1, 10, and 100 for the three simulations shown. In all cases, ground water flow is from left to right on the diagrams (from Smith and Chapman, J. Geophys. Res., v. 88, p. 593–608, 1983. Copyright by Amer. Geophys. Union).

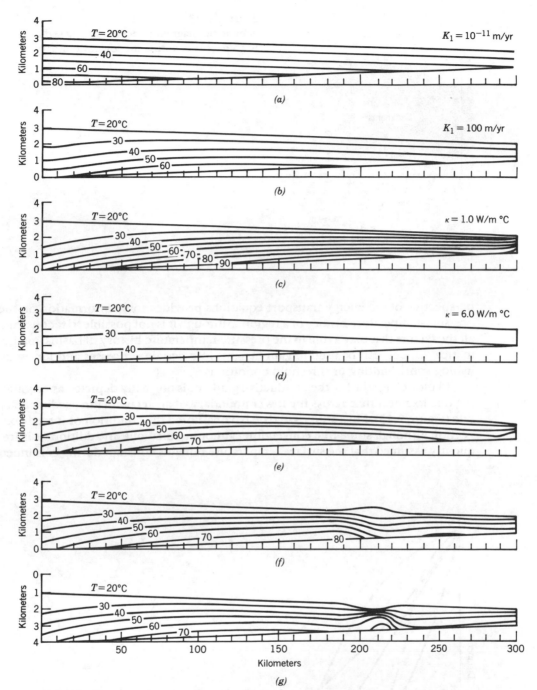

Figure 9.12
Temperature distributions at low (*a*) and high (*b*) Peclet numbers due to differences in
hydraulic conductivity; high (*c*) and low (*d*) Peclet numbers due to differences in thermal conductivity and for different water table configurations (*e*, *f*, and *g*). In all cases,
ground water flow is from left to right on the diagrams (from Garven and Freeze, 1984,
Amer. J. Sci., v. 284, p. 1085–1124. Reprinted by permission of the Amer. J. Sci.).

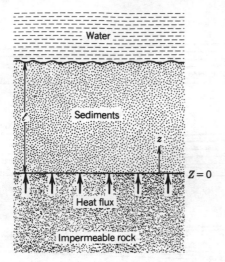

Figure 9.13
Schematic diagram of accumulating sediment with heat flux across the lower boundary.

perspective of the energy transport equations provides some information on the pressure-temperature history of a region. Although it is not possible to reconstruct all the factors that play a role in the pressure-temperature history of basins, mathematical models can be used to gain insights into certain processes and for establishing some limiting or threshold conditions.

In modeling studies, the conductive gradient is normally depicted as originating due to a heat flux across the lower boundary of an accumulating sedimentary pile (Figure 9.13). Most studies to date have assumed the condition of forced convection although some free convection cannot be ruled out. In addition, it is frequently assumed that the surface temperature remains constant during continued

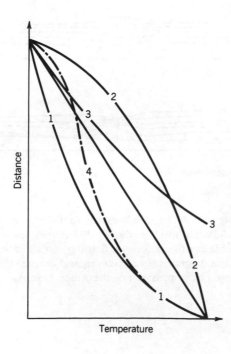

Figure 9.14
Temperature patterns with moving boundaries and convection (from Sharp and Domenico, 1976, Geol. Soc. Amer. Pub., from Geol. Soc. Amer. Bull., v. 87, p. 390–400).

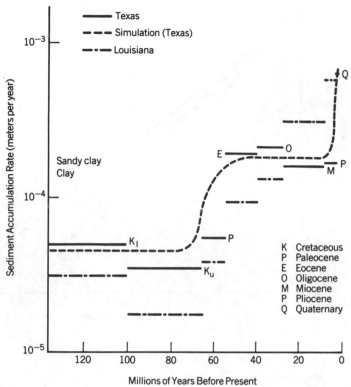

Figure 9.15
Estimated sediment accumulation rates for the offshore Texas
and Louisiana coasts (from Sharp and Domenico, 1976, Geol.
Soc. Amer. Pub., from Geol. Soc. Amer. Bull., v. 87, p.
390–400).

deposition of the sediment. Frequently, a moving coordinate system is used to
study this problem. The results of such studies demonstrate that with uniform
sedimentation—in the absence of fluid flow—the thermal gradient will constantly decrease, and for uniform erosion it will constantly increase. These facts are
shown in Figure 9.14 by profiles 1 and 2 when the distance axis represents the
depth from the sediment-water interface to the impermeable basement rock. Profiles 1 and 2 would also be expected from steady one-dimensional flow upward
and downward, respectively, where the sedimentary layer remains of uniform
thickness (Bredehoeft and Papadopulos, 1965). Profile 3 represents the combined
affect of one-dimensional forced convection upward and the separation of the
boundaries by the accumulation of sediment. In this case, a zero or near-zero
velocity at the lower boundary and a maximum velocity at the growing sediment-
water interface perturbs profile 1 into the reverse "S" pressure-depth profile 4.

Sharp and Domenico (1976) developed an energy transport model that incorporated one-dimensional fluid flow and one-dimensional heat transport. When
applied to Gulf Coast conditions for the loading history of Figure 9.15, they concluded that abnormal pressures were low for much of the geologic record and
were increased during the Eocene, the Oligocene, and most markedly during the
Late Pliocene-Quaternary period in response to rapid sedimentation (Figure
9.16a). The period of low abnormal pressure was accompanied by pressure-depth

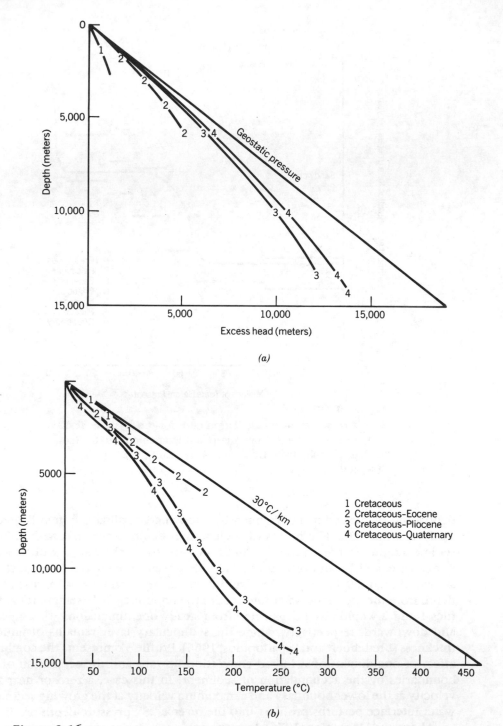

Figure 9.16
Calculated excess head and temperature distribution in Gulf of Mexico sediment at the
end of each of the four specified time periods (from Sharp and Domenico, 1976, from
Geol. Soc. Amer. Pub., from Geol. Soc. Amer. Bull., v. 87, p. 390–400).

profiles very close to the conductive geothermal gradient (assumed at 30°C/km). For the Late Pliocene-Quaternary period, the temperatures were uniformly lower than in the previous period below depths of 6000 m (Figure 9.16*b*). The reverse "S" profile dominates during this period of rapid sedimentation, mainly because of the moving boundary effect. They concluded that sediments presently in the near offshore are now at their maximum pore fluid pressure and minimum temperature at any depth. Chia (1979) extended this study by incorporating a three-dimensional loading or mean stress formulation within the confines of the vertical transport of fluid and heat. He also included the thermal expansion of fluids as a pressure-producing mechanism, and a fluid-pressure–fracture criterion to examine the conditions leading to fracture initiation.

Cathles and Smith (1983) developed a two-dimensional flow model in combination with energy transport to examine episodic basin dewatering and the genesis of mineral deposits. Bethke (1985) also has applied two-dimensional flow and energy transport models to accumulating sedimentary piles. His calculations indicate a tendency for fluids to migrate laterally toward the edge of the basins. Vertical flow occurs in the deeper parts of the basin where the ratio of vertical to lateral permeability exceeds the ratio of the lengths of vertical to lateral pathways to the surface. He demonstrated further that for small sedimentation rates the temperature gradient is not far removed from that expected from pure conduction, as would be expected for situations approximating normal compaction.

Heat Transport in Mountainous Terrain

The relief associated with mountainous terrain may promote an advective disturbance of the thermal regime that is somewhat more severe than discussed earlier in the chapter and may contribute to regional scale variations in surface heat flow. In the southern Great Basin, for example, Sass and others (1976) recognize a regional anomaly of low heat flux, termed the "Eureka low," which includes the Nevada Test Site and Yucca Mountain described earlier (Chapter 7). As the regional heat flow within the crust beneath the "Eureka low" is likely of the same intensity as occurs in the Great Basin in general, a hydrologic disturbance was postulated where "ground water is carrying much of the Earth's heat in the upper 3 km and is delivering it elsewhere" (Sass and Lachenbruch, 1982). Wherever that "elsewhere" is, there is likely to be yet another anomaly, this time a thermal high.

The nature of hydrologic disturbances in mountainous terrain has recently been investigated by Forster and Smith (1989). Their conceptual model is a two-dimensional region not unlike Figure 9.9, with the bedrock forming the upper surface and a conductive heat flux across the bottom boundary. As a continuation of their previous work described in Chapter 7 (Forster and Smith, 1988), the water table elevation is determined by using a free surface approach where an infiltration rate must be specified. Some of their results are shown in Figure 9.17*a*, *b*, and *c* for a fixed infiltration rate and a three-order-of-magnitude increase in permeability going from Figure 9.17*a* to 9.17*c*. As the permeability increases, the free water surface drops in response to higher rates of fluid flow, and the thermal disturbance is enhanced. By thermal disturbance we mean that the temperature represented by an isotherm at a given depth is reduced. Temperature within the shaded region of Figure 9.17*c* reflects the temperature of recharge because the upper region was not penetrated by the basal heat flux. Note also a possible thermal "high" forming at the restricted discharge area.

The flow system in Figure 9.17*d* is presumed to have the same permeability as the system in Figure 9.17*c*, but the infiltration rate has been reduced by a factor of

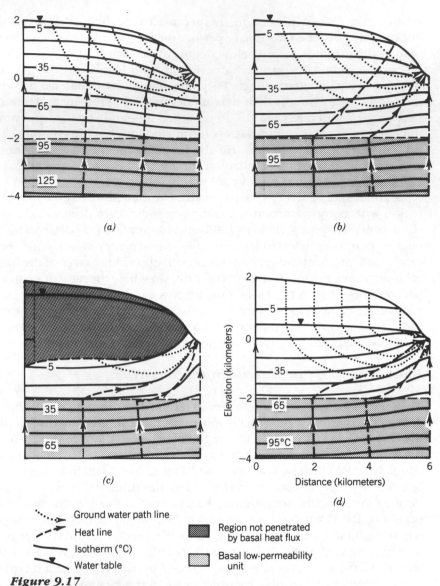

Figure 9.17
Patterns of ground water flow and heat transport with a three-order-of-magnitude increase in permeability when going from (*a*) to (*c*). Figure 9.17(*d*) has the infiltration rate decreased by a factor of five for the same permeability of Figure 9.17(*c*) (from Forster and Smith, J. Geophys. Res., v. 94, p. 9439–9451, 1989. Copyright by Amer. Geophys. Union).

5, giving a five fold decrease in the ground water flux. The depth of the water table is increased markedly, and the reduced flux produces a rather weak thermal disturbance and a warmer thermal regime than for the higher-infiltration case of Figure 9.17*c*.

Figure 9.18 shows the effect of a high-permeability fracture on the flow system and thermal regime. A large percentage of the flow and basal heat are captured by the fracture zone, resulting in an upwelling of isotherms in the fracture zone. Such features may produce thermal springs at the fault outcrop. If the hydraulic

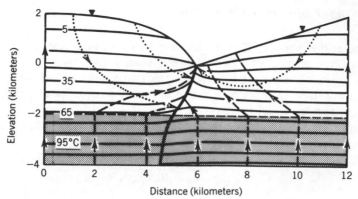

Figure 9.18
Patterns of ground water flow and heat transport as influenced
by a steeply dipping high permeability fracture zone (from Forster and Smith, J. Geophys. Res., v. 94, p. 9439–9451, 1989.
Copyright by Amer. Geophys. Union).

conductivity of the rock matrix is increased by one order of magnitude, a greater
flux is captured by the fracture, the 75 °C isotherm of Figure 9.18 is reduced to 35
°C, and the upwelling along the fault vanishes (Forster and Smith, 1989). In addition, the temperature in the upper regions approaches the temperature of the
recharge because the upper unit is not penetrated by the basal heat flux.

9.3 Free Convection

Free convection has been defined as a flow driven by density variations. As a
buoyancy-driven flow we have the visualization of rising warm water, a cooling
process, and the sinking of these same waters. Thus, with free convection we
visualize ascending and descending currents that form cells of various geometries.
The transport of heat is in the direction of the buoyant forces, and the isotherms
depart markedly from their expected conductive distribution. Figure 9.19 is a
typical example of a thermal profile established by free convection.

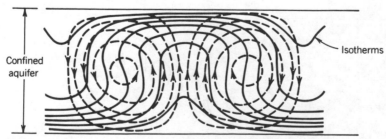

Figure 9.19
Free convection in a confined aquifer (from Donaldson, J. Geophys.
Res., v. 67, p. 3449–3459, 1962. Copyright by Amer. Geophys.
Union).

The Onset of Free Convection

Horton and Rogers (1945) were among the first to address the conditions for the onset of free convection in a porous medium. Their theoretical results were based on an imposed vertical temperature gradient across a horizontal layer, saturated with a nonmoving fluid, where the upper and lower boundaries were taken as perfect heat conductors. According to this work, convection currents may develop when the temperature gradient exceeds some critical value

$$\frac{\partial T}{\partial z} > \frac{4\pi^2\lambda^2\mu}{\kappa_e\rho_o g\alpha_f H^2} \tag{9.32}$$

where λ has been introduced as the thermal diffusivity. When compared with results based on the Rayleigh criteria for pure fluids, this minimum temperature gradient exceeds the required temperature gradient for the onset of convection in fluids by a factor of $(16/27)H^2/\pi^2\kappa_e\rho_o$. Thus, temperatures that establish free convection in fluids are not sufficient to establish free convection in fluids contained within porous solids, largely because of the viscous drag of the solids on the fluids.

Lapwood (1948) demonstrated that the onset of free convection in porous media occurs at Rayleigh numbers on the order of $4\pi^2$; that is,

$$N_{RA} = \frac{g\rho_o(c_w\rho_w)Hk\alpha_f\Delta T}{\mu\kappa_e} \cong 40 \tag{9.33}$$

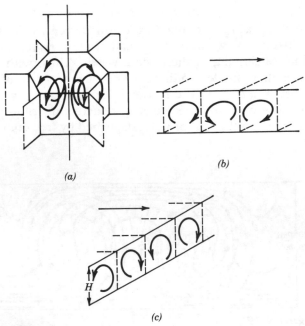

Figure 9.20
Forms of convective coils (from Combarnous and
Bories, 1975. Reprinted with permission of Academic
Press, from Advances in Hydroscience, v. 10, p.

This criterion is based on horizontal layers filled with a nonmoving fluid and upper and lower boundaries that are both impermeable to flow and isothermal. Other thermal and hydrologic boundary conditions for the horizontal layer case have been reported by Nield (1968).

Pratts (1966) investigated the effect of horizontal flow on the onset of free convection and found that the convection parameter $R_a \cong 40$ remains unchanged. The incorporation of horizontal flow, however, implies that both forced and free convection are occurring so it is expected that the nature of the convection rolls will be controlled in part by the magnitude of the horizontal flow. This is demonstrated in the review of Combarnous and Bories (1975), where for N_{PE} less than 0.75, the axes of convective cells appear to run perpendicular to the mean flow direction (Figure 9.20b). For higher values of N_{PE}, helicoidal rolls form, with their axes parallel to the mean flow direction (Figure 9.20c). These can be compared with the polyhedral cells associated with free convection operating in the absence of a bulk fluid motion (Figure 9.20a).

Example 9.1

The following is an example of the Rayleigh criteria as demonstrated in Eq. 9.33. The parameters are as follows. The coefficient of thermal expansion for the fluid is 2×10^{-4} °C^{-1}, mass density is 1 g/cm³, acceleration of gravity is 980 cm/sec², the permeability is 1 darcy or 10^{-8} cm², the sand thickness is 1000 ft or 3.048×10^4 cm, the temperature gradient is 0.023 °C/ft or 7.55×10^{-4} °C/cm so that the temperature change across 1000 ft of sand is 23°C, the heat capacity is 1 cal/cm³ °C, the fluid viscosity is 6×10^{-3} g/cm sec, and the thermal conductivity is 4×10^{-3} cal/cm sec °C.

$$N_{RA} = \frac{(980)(1)(1)(3.048 \times 10^4)(10^{-8})(2 \times 10^{-4})(23)}{(6 \times 10^{-3})(4 \times 10^{-3})} \cong 57$$

which exceeds the critical value of 40.

Sloping Layers

Combarnous and Bories (1975) discuss two important aspects of free convection in sloping layers. For such layers the required Rayleigh number is higher than the critical value of 40 so that for isothermal bounding planes

$$N_{RA} \cos \psi > 4\pi^2 \tag{9.34}$$

where ψ is the slope of the layer. For slopes of less than 15 degrees, the convective movement takes the form of polyhedral cells shown in Figure 9.20. For greater slopes, the cells are in the form of longitudinal coils.

A second point made by these authors is that convection will occur for all conditions where the isotherms are not parallel to the geopotentials, that is, where the temperature gradient is not collinear with the body forces. It is thus recognized that there is always a convective movement in a sloping layer bounded by isothermal planes of different temperature. The geometry of the roll is unicellular

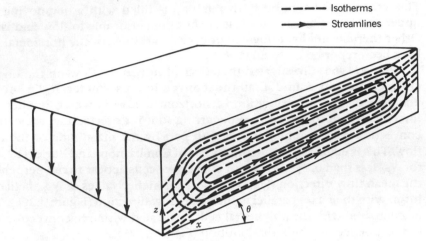

Figure 9.21
Geometry of unicellular flow in an inclined fluid layer (from Combarnous and Bories, 1975. Reprinted with permission of Academic Press, from Advances in Hydroscience, v. 10, p. 231–307).

and two dimensional (Figure 9.21). A hot current moves upslope along the bottom boundary and then turns and moves downslope along the upper cold boundary.

The three states of free convection just discussed may be viewed as transitional. For small Rayleigh numbers below the critical value, the unicellular convective flow occurs in sloping layers. At Rayleigh numbers above the critical value, the unicellular motion gives way to the polyhedral cells for slopes less than 15 degrees. For steep slopes and Rayleigh numbers above the critical value, the longitudinal coils form.

Geological Implications

In the study of sandstone diagenesis, it has become apparent that large amounts of mass must be transported to account for certain diagenetic changes. Free convection offers an attractive alternative as a mass transport mechanism as opposed to the "once-through" volumes of fluid associated with compaction or even the thousands of pore volumes of meteoric derived water through ancient uplifted rock. Horton and Rogers (1945) were probably the first to test this hypothesis for the Woodbine sand in Texas to account for the NaCl distribution. Their method was to calculate the minimum value of permeability for which convection currents might occur (Eq. 9.33). Blanchard and Sharp (1985) took the same approach with the Rayleigh criterion (Eq. 9.33) in an attempt to account for the silica distribution in the Frio sandstone in Texas. Wood and Hewett (1982) examined the stability analysis in sloping layers and made some general statements on the large-scale diagenetic effects that might be attributed to large-scale (kilometers) unicellular currents (Figure 9.22). They conclude that although the currents are slow (1 m per year), the mass flux associated with these currents operating over geologic time is sufficient to cause significant porosity changes. Cathles (1977) and Norton (1978) discuss convective fluid movements associated with intrusions and cooling plutons.

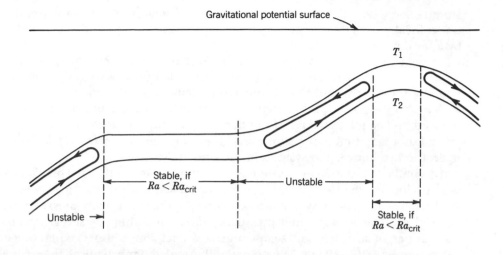

Figure 9.22
Schematic diagram illustrating regions of fluid stability in folded porous body.
Boundaries of body assumed isothermal and impermeable. Reprinted with permis-
sion of Geochim. Cosmochim. Acta, v. 46, Wood and Hewett, Fluid convection and
mass transfer in porous sandstone—a theoretical model. Copyright 1982. Pergamon
Press PLC.

9.4 Energy Resources

Geothermal Energy

Three types of geothermal systems are generally recognized: (1) hydrothermal convection, (2) hot igneous rock, and (3) conductive dominated. The hydrothermal
convective systems are either vapor dominated, which consist of dry or wet steam,
or hot water dominated. Dry steam fields are normally of low permeability so that
little recharge occurs. Temperatures on the order of 250°C are not uncommon. The
wet steam field is characterized by a greater amount of recharge and consists of
steam-water mixtures. The hot water systems may be classified as high ($>$ 150°C),
intermediate (90–150°C), and low ($<$ 90°C) temperature systems. Both the vapor-
and liquid-dominated systems require the following five characteristics: (1) a heat
source, (2) sufficient permeability, (3) a source of replenishment, (4) a low-
permeability bottom boundary, and (5) a low-permeability cap rock.

The temperature of the fluids in the foregoing classification dictates their use
as an energy source (Nathenson and Muffler, 1975). The high-temperature systems
are adequate for electrical generation. Nathenson and Muffler (1975) estimate the
heat stored in these sytems in the United States as 257×10^{18} calories. They suggest that this is five times the amount stored in undiscovered high-temperature
reservoirs. To provide a perspective of this energy source, 10^{18} calories is equivalent to 690 million barrels of petroleum (White and Williams, 1975). The intermediate temperature systems are adequate for space and process heating. They
have a stored heat content of about 345×10^{18} calories in the United States. For
both the high- and intermediate-temperature systems, only about 25% of the

stored energy is available at the surface. Low-temperature systems are useful only under locally favorable conditions, for example, the heating of homes or buildings by a local circulation system.

Hot igneous systems are of two types, molten magma and hot dry rock. The molten magma is currently not capable of exploitation due to the pressures and temperatures that would be encountered in drilling. Exploitation of the dry hot rock systems would require a circulation system for water to be injected and then recovered. The temperatures are under 650°C, and the stored energy is in the several thousand times 10^{18} calories range. Renner and others (1975) describe the hydrothermal convection systems in the United States.

A conductive-dominated system generally relies on the thermal blanket effect where a low-conductivity cap rock acts as an insulator for heat storage below. Yet another type of conductive-dominated system is the abnormally pressured reservoirs in the Gulf Coast. Fluid pressures can reach as high as several thousand pounds per square inch, and temperatures can get above 300°C when one considers depths of 15–20 km. The potential thermal and mechanical energy is estimated to be between 46,000 and 190,000 megawatts for 20 years (Papadopulos and others, 1975).

It may be noticed that most of the references cited here are mid-1970s in vintage. This is no coincidence and reflects the effort to develop alternative energy sources in response to the oil embargo of the early 1970s. With a return to pre-1971 prices for oil and gas, the interest in geothermal resources has diminished considerably. Geothermal systems are discussed rather extensively in proceedings of a United Nations conference (1976) and in a bibliography by ERDA (1976).

The development of geothermal resources often encounters problems not unlike those associated with the use of ground water as a water supply. Water injection in hot aquifers is often considered as a viable management scheme to maintain the supply (Goguel, 1976). In other cases, the exploitation results in both land subsidence and induced tectonic activity (Herrin and Goforth, 1975). In response to a daily production of 130 million gallons per day of hot water from a reservoir located in Texas, four earthquakes of magnitudes 3.3 to 4.4 occurred within one month. Hundreds of microearthquakes were recorded but they gradually diminished.

The modeling of geothermal reservoirs must address the coupled movement of heat and ground water. Mercer and others (1975) have solved such a problem for the Wairakei Field, New Zealand. Reviews on the modeling methods are given by Pinder (1979) and Gary and Kassoy (1981). A numerical model for a hot-water-dominated system is given by Sorey (1978).

Energy Storage in Aquifers

Hot water storage in aquifers has been suggested as an attractive means for storing energy produced during periods of low consumption for utilization during peaking hours, or even seasons. According to Sauty and others (1982a), the important parameters in an energy storage problem include the recovery factor, which is the ratio of the quantity of energy recovered to that injected, and the temperature of the recovered water. These factors depend on the properties of the aquifer, the water contained within the aquifer, and the operating conditions such as injection rates and duration. This problem has been investigated numerically by Tsang and others (1981) and Sauty and others (1982a and b). Moltz and others (1981) have described the results of a long-term field test that attempts to assess the feasibility of an energy storage system.

Some of the physics involved in energy storage have been discussed by Sauty and others (1982a). The warmer body of water that develops during injection is normally restricted within a cylinder of radius r where

$$r = \frac{c_w}{c_s} \left(\frac{Qt}{\pi H} \right)^{1/2} \tag{9.35}$$

where c_w is the specific heat of the fluid, c_s the specific heat of the solids, Q is the injection rate, and H is the aquifer thickness. The lower the thermal conductivity, the smaller the heat losses, and the higher the efficiency of extraction. These features are demonstrated through the use of dimensionless parameters.

9.5 Heat Transport and Geologic Repositories for Nuclear Waste Storage

The Nuclear Waste Program

For the past few decades, one of the largest applied earth science programs ever attempted throughout much of the world has been the investigations for safe disposal of nuclear waste. Many countries have such a program, including Sweden, Belgium, France, Italy, Switzerland, Canada, and the United States. The purpose of most of these programs is the disposal of nuclear waste associated with commercial nuclear power plants. The wastes can be classified into three major categories. The first is the fission products resulting from the splitting of uranium and other fissile nuclei used as fuel. These radioactive elements emit both beta and gamma radiation and produce most of the heat in high-level waste. They are both short lived and long lived, the latter including ^{129}I and ^{99}Tc. The second category includes the transuranic elements such as plutonium and neptunium, many of which are associated with extremely long half lives. Another category includes neutron activation products, for example, ^{63}Ni. These products are generated in the fuel claddings and structural materials in the core of the reactor. To provide some idea of the magnitude of the problem caused by these wastes, the Office of Technology Assessment (1982) has estimated that there will be something on the order of 72,000 metric tons of spent fuel by the year 2000. With reprocessing, these wastes would occupy a volume the size of a football field, one story high. Without reprocessing, the volume of an entire stadium could be filled.

The problem of nuclear waste disposal was first addressed in the United States in 1957 when the National Academy of Sciences released their report entitled "The disposal of radioactive wastes on land." Several options were considered, including disposal in space, in subseabeds, in ice sheets, in "very deep" holes, liquid injection in wells, rock melt disposal, transmutation, and a no-action alternative. The heat generation aspects associated with these alternatives did not go unnoticed. On the positive side the ice sheet disposal would utilize the heat for the waste to melt its way to the bottom. On the other hand, the seabed program ran into technical objections because of the uncertain thermal effects of the canisters on the sediments. With regard to the no-action alternative, a presidential

Table 9.4
Thermal conductivity of potential repository rocks

Material	Temperature (°C)	Thermal conductivity (W/m°C)
Quartz monzonite	—	2.1–3.4
Salt (Louisiana)	300	2.2
Salt (Louisiana)	40	4.0
Tuff	—	2.4
Tuff	—	1.55
Salt	110	2.08–6.11
Granite	—	1.99–2.85
Shale	—	1.47–1.68
Tuff (welded)	—	1.2–1.9
Tuff (unwelded)	—	0.4–0.8
Basalt	—	1.16–1.56

message on February 2, 1980 stated that "civilian waste management problems shall not be deferred to future generations" (Department of Energy, 1980). The mined geologic repository has emerged as the most viable alternative.

The Rock Types

The scientists in the early phases of the disposal program were far more concerned with "geologic media" than with specific hydrogeologic environments. Among the first two rock types proposed were salt and granite, largely because the thermal conductivity of these substances is quite high. In fact, salt has one of the highest thermal conductivities of all natural rock types (Table 9.4). Thus, from the early beginnings there was concern with the long-term production of heat due to decaying radionuclides. Salt formations, both domed and layered, have been selected as possible candidate sites in Denmark, West Germany, the Netherlands, and Spain. Crystalline rocks have been selected in Canada, France, Japan, Switzerland, and the Scandinavian countries. On the other hand, Belgium and Italy are considering a clay-type rock. Some early potential candidates for the first repository for commercial wastes in the United States included the Michigan Basin (salt), the Gulf Coast (salt), the Permian basin in Texas (salt), and Vermont–New Hampshire (granite). In addition, an active program for the disposal of defense nuclear wastes is being pursued in bedded salt deposits near Carlsbad, New Mexico. This program is referred to as the Waste Isolation Pilot Plant (WIPP). Some 9 miles or so of tunnels are already in place and testing has already begun.

In the United States, rock types and locations other than those cited earlier have been taken under consideration, including basalts at the Hanford Site in the state of Washington and volcanic tuff at the Nevada Test Site in southern Nevada. The Hanford Site has long been associated with subsurface storage and disposal of defense nuclear wastes and the Nevada Test Site has been the location of subsurface nuclear detonations for decades. Hence, both are polluted already and both have the advantage of possessing vast tracts of land that have been removed from the public domain. Further, some preliminary planning is underway for a second repository sometime after the turn of the century. A "granite" program was formed, but now dispanded, and a sedimentary rock (shale) program is still active.

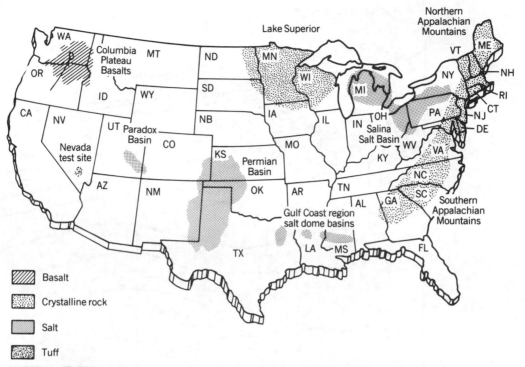

Figure 9.23
Regions considered for geologic disposal of radioactive waste.

Because of the political implications of nuclear waste disposal, whatever programs are active today may be abandoned in the future, or new ones added, but
it is reasonably certain that the disposal problem will be with us for many decades.

   Figure 9.23 shows the regions in the United States that were given consideration for the geologic disposal of nuclear waste. Nine sites were eventually selected
for an environmental assessment, seven in salt, one in basalt, and one in tuff. The
environmental assessments were published by the Department of Energy in 1986.
From these assessments, three were picked for extensive site characterizations,
and one of these three was to be selected in the 1990s as the nation's first repository for commercial waste. The three candidates selected included a salt site in
Texas and the two reservations just cited. In December 1987, Congress terminated
site characterization plans at the Hanford site and the Texas site. At this time, only
the Nevada test site remains as a candidate for the nation's first repository for commercial waste.

   Figure 9.24 shows a cross section of the Nevada test site where the design
repository is located in the unsaturated zone. Figure 9.25 shows the projected temperature within the repository rocks. In this plan, the nuclides are to be packed in
canisters of a design life between 300 and 1000 years, and the canisters are to be
put into individual boreholes. The heat pulse produced will propagate through
the host rock by conduction and convection. Other than establishing buoyancy-
driven flows, the heat generated is expected to alter the preemplacement condition of the host rock, and this is one of the main concerns. These concerns have
resulted in research in coupled phenomena, that is, the thermal-mechanical-
hydrochemical processes associated with a nuclear repository (for example,

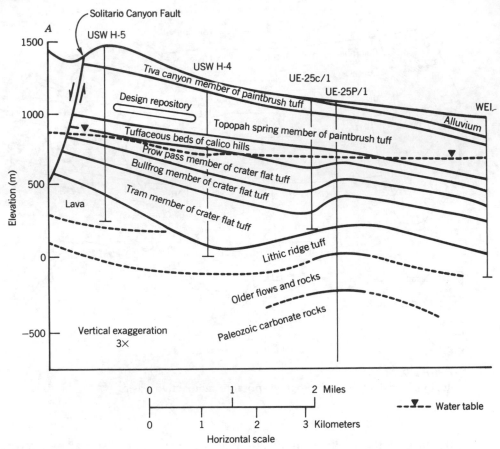

Figure 9.24
Cross section of the Nevada test site shows the location of proposed repository (from Department of Energy, 1988).

Lawrence Berkeley Laboratory, 1984). This research has not been limited to the unsaturated zone at the Nevada test site but includes work on all potential repository rocks. Thus, for salt, we are not surprised to learn that the main thermal-mechanical coupling is the deformation or creep (flow) of the salt in response to heating. In addition, brine pockets have been known to migrate up the thermal gradient, that is, toward the heat source. For saturated rocks, one of the main concerns is the potential for fracturing by thermally induced fluid pressures. For unsaturated zones in general, the problems are complex and include the potential for upward gas transport of radionuclides due to the buoyant rise of hot air along with temperature-induced vapor phase transport. For all rocks in general, there is concern for the change in mechanical strength of the rock body and stress corrosion of the canisters.

Thermohydrochemical Effects

From the study of hot springs and ore deposits, it has long been recognized that fluids in hydrothermal systems react and subsequently alter the rocks involved in hydrothermal circulation (Browne, 1978). As the repository and adjoining rock

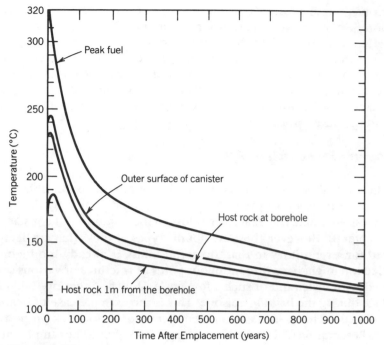

Figure 9.25
Temperature versus time after nuclear waste emplacement (from
Department of Energy, 1988).

will in effect be a "hydrothermal" system for a few to several hundred years after
emplacement, there is concern over the nature of these alterations. Of special concern is the effect on the sorbing minerals or those that undergo exchange, which
can provide a barrier to the migration of radionuclides. Many of these minerals are
unstable at elevated temperatures and may be altered by reversible dehydration or
mineral reactions. For example, certain zeolite minerals (hydrous aluminum silicates) are abundant in the volcanic tuffs at the Nevada test site. With few exceptions, natural zeolites are found in rocks that have formed at temperatures less
than 200°C. According to Smyth (1982) reactions that may occur in these rocks
below 200°C are of two types, simple reversible dehydration and mineralogical
phase inversions. The first of these involves a loss of bound water with contraction of the unit cell by as much as 10% with the development and propagation of
fractures and the loss of mechanical strength. Smyth (1982) concludes that this
constraint may limit to 50°C the maximum temperature rise in zeolite horizons
that occur 30 to 50 m from the emplacement horizons.

Another important mineral frequently associated with the retardation of
radionuclide migration is montmorillonite. As noted in Chapter 8, the alteration
of a montmorillonite to a mixed clay in the Gulf Coast begins at temperatures on
the order of 60–70°C. Based on first-order kinetics of an Arrhenius form,
Domenico and Palciauskas (1988) have examined the half life (the time required
for half of the expanded montmorillonite layer to transform to illite) as a function of temperature. At 40°C, the transformation requires 1.6×10^6 years; at
120°C, it requires 2400 years; and at 160°C, it requires 230 years. Thus, one might
conclude that the process is too slow in comparison to the time scales that are

involved in the thermal evolution of the host rock. However, the kinetics addressed the irreversible transformation only. Like zeolites, a reversible transformation can occur when the temperature reaches the boiling point of water. The energy required to break the bonds of the hydration envelope is significantly lower, so that the kinetics of the process involves time scales shorter than those of the heating process. The potentially large volume of pore fluid released and enhanced permeability could allow migration of the pore fluid in response to the fluid pressures produced during the heating episode.

Thermomechanical Effects

If heating takes place in fractured rock that has an appreciable permeability, the thermal expansion of the fluid is easily accommodated by fluid movement, and the response may take place very close to the condition of constant fluid pressure (Chapter 8). However, these are not the types of rocks that are likely to be considered for a repository in that the permeability is already too high. A more likely selection might have discontinuous pores or fractures of various orientation saturated with water and of such a low permeability so that no fluid expulsion takes place during the heating episode. This is the extreme case of heating at constant fluid mass. This case was examined theoretically for rock properties at the Hanford Reservation (Palciauskas and Domenico, 1982). The sample calculations indicate that fracturing is inevitable.

Yet, a third type of environment, and one that is presumably ideal from the perspective of nuclide migration, is the rock type that contains a few isolated pores or fractures and of such a low permeability that it contains essentially no water or at least is in a state of low fluid saturation. Heating in this case involves only the solid components and their thermal expansion. Significant porosity changes are anticipated only if the original porosity is very low, that is, on the order of 10^{-4} (Palciauskas and Domenico, 1982).

Other potential problems dealing with thermomechanical effects include the triggering of latent seismicity, spalling, and loss of mechanical strength. From the perspective of contamination, the safe disposal of nuclear waste is a very difficult problem; in the presence of heat generation from decaying nuclides, the problem is severely compounded.

Suggested Readings

Cathles, L. M., and Smith, A. T., 1983, Thermal constraints on the formation of Mississippi Valley type lead zinc deposits and their implication for eposodic basin dewatering and deposit genesis: Econ. Geol., v. 78, p. 983–1002.

Domenico, P. A., and Palciauskas, V. V., 1973, Theoretical analysis of forced convective heat transfer in regional groundwater flow: Geol. Soc. Amer. Bull., v. 84, p. 3803–3814.

Forster, C., and Smith, L., 1989, The influence of groundwater flow on the thermal regime in mountainous terrain: A model study. J. Geophys. Res., v. 94, p. 9439–9451.

Parsons, M. L., 1970, Groundwater thermal regime in a glacial complex: Water Resources Res., v. 6, p. 1701–1720.

Wood, J. R., and Hewett, T. A., 1982, Fluid convection and mass transfer in porous sandstone—a theoretical model: Geochim. Cosmochim. Acta, v. 46, p. 1707–1713.

Problems

1. Calculate upper and lower bounds for the thermal conductivity of saturated sandstone consisting largely of quartz with a porosity of 20% using Table 9.1. Repeat the calculations for a dry sandstone with the same porosity.

2. Figure 9.6 applies to both downward and upward moving ground water. Which half of the family of curves applies to upward moving ground water, and what is the significance of the curve where $N_{PE} = 0$?

3. From the type curve match of Figure 9.8, calculate the velocity q through a layer 10 m in thickness. Assume a thermal conductivity of 5×10^{-3} cal sec^{-1} cm^{-1} $°C^{-1}$.

4. From the data of Example 9.1

   a. Calculate the minimum value of permeability to initiate free convection.

   b. Repeat this calculation for a sand thickness of 500 ft.

   c. For a Rayleigh number of 57, would free convection occur in a rock body with a slope of 30°?

5. Compare the Fourier number of Eq. 9.22 with the Fourier number for fluid flow (Eq. 4.83). State the analogous quantities and provide an interpretation of T^* for heat flow. For a repository rock, what are the advantages of having a high-thermal conductivity and a low-heat capacity?

6. Consider one-dimensional steady state conduction across a layer of thickness L where the temperature at the bottom ($x = 0$) and at the top ($x = L$) is known. Solve this problem for the steady state temperature as a function of x, L, T_o, and T_L (Hint: see Example 4.3).

7. A soluble toxic chemical is moving downward through a thick unsaturated zone in sands and gravels threatening an underlying water supply. An experimental program is being devised whereby refrigerants will be circulated through bore holes penetrating the unsaturated zone in an attempt to freeze the water and the contaminant in the unsaturated zone. Based on what you have learned in this chapter and in previous chapters, is it possible to make some approximate calculation of the effectiveness of one bore hole in contributing to this feat by the process of conduction alone? What equation or physical concepts may be modified to describe this problem, and what is the role of the time constant in considerations of the rate at which freezing can be accomplished? (Hint: If you are confused, consult Carslaw and Yeager, 1959, the chapter on the use of sources and sinks in cases of variable temperature, the section on the continuous line source. Is this section (or equation) analogous to anything you have learned thus far?

Chapter Ten

Solute and Particle Transport

10.1   Advection

10.2   Particle Transport

10.3   Basic Concepts of
       Dispersion

10.4   Character of the Dispersion
       Coefficient

10.5   A Geostatistical Model of
       Dispersion

10.6   Mixing in Fractured Media

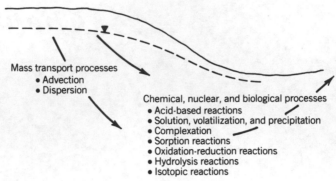

Figure 10.1
Conceptualization of mass transport in a ground water flow system.

Mass occurring in water as ions, molecules, or solid particles not only undergoes transport but also reactions. The reactions redistribute mass among ions or between liquid and solid phases. Figure 10.1 lists the important mass transport and mass transfer reactions in ground water. Over the next three chapters, we will study these processes in detail, establishing the background necessary to understand the classical problems involved with ground water geochemistry and contaminant hydrogeology. We begin here in Chapter 10 with the mass transport processes, advection and dispersion, that are responsible for moving mass in ground water systems. The term advection describes mass transport due simply to the flow of water in which the mass is dissolved. The direction and rate of transport coincides with that of the ground water. Dispersion is a process of fluid mixing that causes a zone of mixing to develop between a fluid of one composition that is adjacent to or being displaced by a fluid of another composition.

After several decades of work on this elusive process, particularly at the field scale, a reasonable consensus has developed about what dispersion is all about. The complexity of this topic, however, is reflected by the fact that more than half of this chapter is concerned with dispersion, although advection, which is better understood, is the more important transport process.

10.1   Advection

The intimate relationship between ground water flow and the process of advection has meant that knowledge about flow systems is directly transferable to understanding advection. For example, in the case of topographically driven flow systems (Chapter 7), the factors that influence ground water flow patterns—water table configuration, pattern of geologic layering, size of the ground water basin—control the direction and rate of mass flow. Thus, the background provided by the extensive earlier treatment on flow allows us to consider advection in just a few pages.

The two flow nets in Figure 10.2 show some basic features about advection. First, when only advection is operating, mass added to one or more stream tubes will remain in those stream tubes. Other processes (e.g., dispersion) operate to move mass between stream tubes. Second, the direction of mass spreading in steady-state

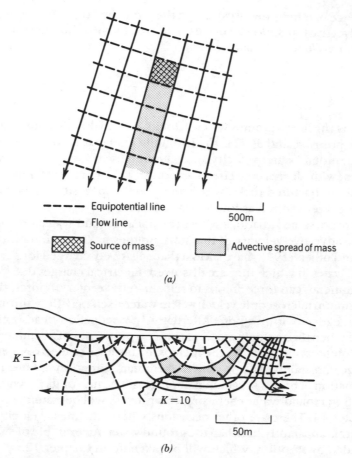

Figure 10.2
Mass spreading by advection alone in (*a*) a shallow, unconfined aquifer and (*b*) a local flow system.

systems is defined by path lines even in relatively complex flow systems. Because
advection is the dominant transport process, knowledge of ground water flow patterns provides the key to interpreting the pattern of contaminant migration.

Mathematically, the one-dimensional advective flux of the ith constituent
through a representative volume of porous medium is represented by the following equation

$$J_i = v_x C_i n \tag{10.1}$$

where J_i is the mass flux of i per unit area per unit time, v_x is the linear ground
water velocity in the x direction, n is porosity, and C_i is concentration in mass
per unit volume of solution. Note that J is a vector quantity with units of M/L^2T.

This simple model assumes that the transport of mass does not influence the
pattern of flow. When this assumption fails, for example, with contaminated
ground water having a density that is significantly different from the ambient
ground water, the flow of water and mass may diverge. However, for most practical problems, ground water and the dissolved mass will move at the same rate (in

the absence of other processes) and in the same direction. Accordingly, the velocity of advective transport can be characterized by the following form of the Darcy equation developed in Chapter 3

$$v = -\frac{K}{n_e}\frac{\partial b}{\partial l} \tag{10.2}$$

where v is the linear ground water velocity, K is hydraulic conductivity, n_e is the effective porosity, and $\partial b/\partial l$ is the hydraulic gradient. As discussed in Chapter 3, the linear ground water velocity and therefore the velocity of advective transport increases with decreasing effective porosity. This relationship is particularly important in fractured rocks, where the effective porosity can be much less than the total porosity, often as low as 1×10^{-4} or 1×10^{-5}.

There are some situations where the linear ground water velocity is different from the advective velocity of the mass. Studies by Corey and others (1963) and Krupp and others (1972) have shown that negatively charged ions can move faster than the water in which they are dissolved. Electrical charges due to the presence of clay minerals can force anions to remain in the center of pores, the location of the maximum microscopic velocity. The water itself may flow through the lower-velocity regimes within a pore. Alternatively, a reduction in advective velocity is possible when the geologic medium takes on properties of a semipermeable membrane. Solutes avoid entering the membrane because of electrokinetic or in some cases size constraints. The process of membrane filtration has been implicated in the formation of brines in sedimentary basins. Because of its potential importance in affecting ground water chemistry, this process will be examined in more detail in Chapter 14. There are other occasions where the linear velocity of advective transport is apparently less than the ground water. An example of such retardation is provided by sorption, which will be taken up in Chapter 12.

10.2 Particle Transport

Transport of particles in suspension is another way in which mass can move in a ground water system. As is the case with solute advection, transport occurs due to the motion of the ground water in which the particles are dispersed. The particles of greatest interest are in the clay and colloidal size ranges and include viruses, bacteria, clay minerals, and aggraded organic compounds (McDowell-Boyer and others, 1986).

These particles are small. For example, colloids have one dimension in a range from 1 to 1000 nanometers. The sizes of typical microorganisms are compared in Figure 10.3 with the grain size of gravels, sands and silts, and with molecules and atoms. Bacteria and viruses can be mobile in sand and gravel not containing a fine-grained fraction.

Particle transport is important because it can provide another way for mass to spread in the subsurface. This process may play a role in the genesis of soils or ore deposits (McDowell-Boyer and others, 1986), but it is of interest to hydrogeologists in relation to contaminant migration. The particles themselves may be contaminants as is the case with bacteria and viruses, or the contaminants can be trace metals, radionuclides, or organic compounds sorbed on to clay or organic parti-

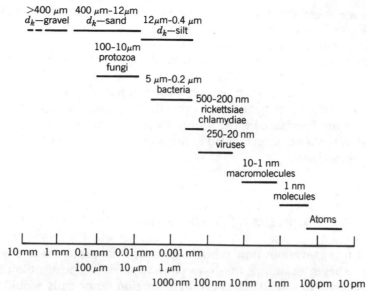

Figure 10.3

Comparison of the sizes of common microorganisms with the grain sizes of sediments, molecules, and atoms (modified from Matthess and Pekdeger, 1985, Groundwater Pollution Microbiology, eds. C. H. Ward, W. Giger, and P. L. McCarty). Copyright © 1985 by John Wiley & Sons, Inc. Reprinted by permission of John Wiley & Sons, Inc.

cles. Because of their small size, particles have relatively large surface areas, which makes them efficient contaminant collectors. In cases where these contaminants have a low mobility in solution, facilitated transport on particles can create unexpectedly high contaminant fluxes. Transport is controlled in these circumstances by the physical and chemical character of the particle rather than the chemical characteristics of the contaminant itself.

Particles that are unable to move or stop moving are also of interest. Shown in Figure 10.4 are two examples of physical filtration that limit particle migration. Filter cakes or surface mats form when particles of the same size or larger than the pore openings simply cannot penetrate into the medium. Straining occurs as particles small enough to enter the porous medium are caught in smaller pore spaces.

Figure 10.4

Two ways in which physical filtration limit particle migration (from McDowell-Boyer and others, Water Resources Res., v. 22, p. 1901–1921, 1986. Copyright by American Geophysical Union).

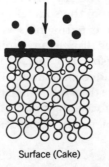

Surface (Cake)

Straining

However, there is not the same tendency for particle accumulation due to straining as compared to the formation of filter cakes. A third mechanism for filtration is the sorption of particles on to the porous medium.

The most important manifestation of these filtration mechanisms is a reduction in hydraulic conductivity as the solids clog the pores. These problems often develop when attempts are being made to put water back into the subsurface with artificial recharge through ponds or wells, the water flooding of petroleum reservoirs, or the disposal of waste waters. Particles plugging a formation may be contained in the water or generated in the porous medium due to the mobilization of pore-lining clays.

10.3   Basic Concepts of Dispersion

Thinking again about flow tubes in a steady-state ground water system is a good place to begin examining the concept of dispersion. The most important effect of dispersion is to spread mass beyond the region it normally would occupy due to advection alone. This point can be illustrated with a simple column apparatus (Figure 10.5a). The apparatus is similar to the Darcy column described earlier with additional plumbing to permit the controlled and continuous addition of a tracer to the water. Initially, a steady flow of water is set up through the column. The test begins with a tracer at a relative concentration $C/C_0 = 1$ added across the entire cross section of the column on a continuous basis. Monitoring the outflow from the column determines the relative concentration of the tracer as a function of time.

Figure 10.5a illustrates how the relative concentration of tracer varies in water going into and coming out of the column. These relative concentration versus time distributions are known as the source loading curve and breakthrough curve, respectively. The breakthrough curve does not have the same "step" shape as the source loading function. Dispersion creates a zone of mixing between the displacing fluid and the fluid being displaced. Some of the mass leaves the column in advance of the advective front, which is defined as the product of the linear velocity and time since displacement first started. The position of the advective front at breakthrough corresponds to a C/C_0 value of 0.5 (Figure 10.5a). In the absence of dispersion or other processes, the shapes of the loading function and the breakthrough curve would be identical.

A zone of mixing gradually develops around the position of the advective front (Figure 10.5b). Dispersion moves some tracer from behind to in front of the advective front. The size of the zone of mixing increases as the advective front moves farther from the source.

This column experiment is an example of one-dimensional transport involving advection and dispersion. Similar mixing will occur in two dimensions (Figure 10.6). Again, dispersion spreads some of the mass beyond the region it would occupy due to advection alone. There is spreading both ahead of the advective front in the same flow tube and laterally into the adjacent flow tubes. This dispersion is referred to as longitudinal and transverse dispersion, respectively. Dispersion in three dimensions involves spreading in two transverse directions as well as longitudinally.

There are a few other distinctions between Figures 10.5b and 10.6. At some distance behind the advective front for the case of one-dimensional dispersion,

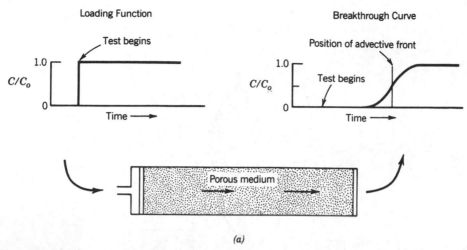

Figure 10.5(a)
Experimental apparatus to illustrate dispersion in a column. The test begins with a continuous input of tracer $C/C_0 = 1$ at the inflow end. The relative concentration versus time function at the outflow characterizes dispersion in the column.

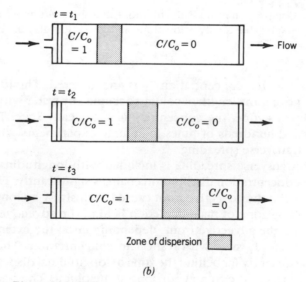

Figure 10.5(b)
Schematic representation of dispersion within the porous medium at three different times. A progressively larger zone of mixing forms between the two fluids ($C/C_0 = 1$ and $C/C_0 = 0$) displacing one another.

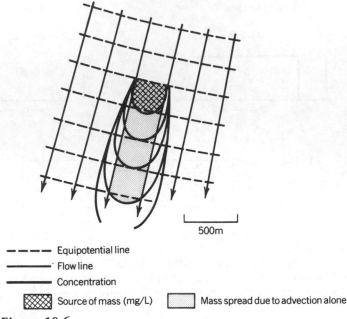

Figure 10.6
Comparison of mass distributions in a shallow unconfined
aquifer due to advection alone and advection-dispersion.

the original source concentration C_0 is encountered. The longer transport continues the greater the length of column occupied by water with a tracer concentration C_0. In this zone, the concentration is at steady state. This result has been demonstrated hundreds of times in column experiments where it is possible to eliminate transverse spreading.

When transverse spreading is included with longitudinal dispersion (Figure 10.6), the concentration distribution changes significantly. There is a zone of dispersion beyond the advective front caused exclusively by longitudinal dispersion and a resulting zone of mass depletion behind the advective front. At some distance behind the advective front, depending upon the extent of dispersion, the system is at steady state with respect to concentration. This part of the tracer plume is sufficiently far behind the zone of longitudinal dispersion that it does not contribute mass to the frontal portions of the plume. The steady-state concentrations are, however, less than the original concentration because mass is also spreading transversely and occupies an area (taken in the vertical plane) that is greater than the area where the mass was injected. Transverse dispersion will reduce concentrations everywhere behind the advective front while longitudinal dispersion will only do so at the frontal portions of a plume.

The relative concentration of a pulse of tracer added at a point changes as a function of time with transport in a constant velocity flow system. One important feature of the concentration distributions is that after a short time they become normal. The mean of the distribution defines the position due to transport at the linear ground water velocity, and the variance (σ_L^2), for example in a one-dimensional flow system (Figure 10.7a), characterizes the extent of longitudinal dispersion. The ratio of variance to time times two ($\sigma_L^2/2t$) is a constant termed the dispersion coefficient or

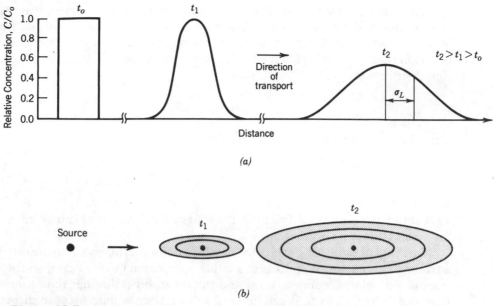

(a)

(b)

Figure 10.7
Variation in concentration of a tracer spreading in (*a*) one or (*b*) two dimensions in a constant velocity flow system.

$$D_L = \frac{\sigma_L^2}{2t} \qquad (10.3)$$

When the velocity is constant, Eq. 10.3 becomes

$$D_L = \frac{\sigma_L^2 v}{2x} \qquad (10.4)$$

where v is the linear ground water velocity and x is the distance from the source. Both theoretical arguments and column experiments support these relationships.

When mass spreads in two or three dimensions, the distributions of mass sampled normal to the direction of flow are also normally distributed with variances that increase in proportion to $2t$. Thus, coefficients of transverse dispersion are defined as

$$D_T = \frac{\sigma_T^2}{2t} \qquad (10.5)$$

The two-dimensional spread of a tracer in a unidirectional flow field, therefore, results in an elliptically shaped concentration distribution (Figure 10.7*b*) that is normally distributed in both the longitudinal and transverse directions. Typically, longitudinal dispersion is greater than transverse dispersion (Figure 10.7*b*). Concentration distributions in three dimensions form ellipsoids of revolution (football shapes) when the two components of transverse spreading are the same and longitudinal dispersion is larger. When the vertical transverse dispersivity is small, as is often the case, plumes take on a surfboard shape.

The variances we have discussed differ in a subtle but important way. Those in Eqs. 10.3 and 10.5 are spatial distributions, while the one calculated from a breakthrough curve is the variance of a temporal distribution. The following equation (Robbins, 1983) interchanges space/time statistics relating to concentration distributions

$$\sigma_L^2 = v^2 \sigma_t^2 \tag{10.6}$$

Thus, by recasting Eq. 10.3 in terms of σ_t

$$D_L = \frac{v^2 \sigma_t^2}{2t} \tag{10.7}$$

so that the dispersion coefficient in the longitudinal direction can be calculated for a breakthrough curve.

One way of estimating values for the dispersion coefficients is to observe how a tracer or contaminant spreads in a column experiment. For example, the variance in the relative concentration and the mean breakthrough time (time corresponding to $C/C_0 = 0.5$) can be used to estimate the microscopic dispersion coefficient. This same procedure with variations has been developed for use in the field.

Having looked at the manifestations and quantification of dispersion, we will now explore its origins. Dispersion or, more formally, hydrodynamic dispersion occurs as a consequence of two very different processes—diffusion and mechanical dispersion. These two contributions to hydrodynamic dispersion are represented mathematically as

$$D = D' + D_d^* \tag{10.8}$$

where D is the coefficient of hydrodynamic dispersion, D' is the coefficient of mechanical dispersion, and D_d^* is the bulk diffusion coefficient. The significance of the terms in Eq. 10.8 will become more apparent as we consider the process of dispersion in more detail.

Diffusion

In Chapter 3, Darcy's equation was introduced as a linear law relating fluid flow q to the gradient of hydraulic head. For mass transport, there is a similar phenomenological law that states that chemical mass flux is likewise proportional to the gradient in some scalar quantity, in this case concentration. This law is known as Fick's law and is expressed for a simple aqueous nonporous system as

$$\mathbf{J} = -D_d \, \text{grad}(C) \tag{10.9}$$

In this formulation, $\mathbf{J}$ is a chemical mass flux, with the negative sign indicating that the transport is in the direction of decreasing concentration. The proportionality constant, D_d, is termed the diffusion coefficient in a fluid environment. If the amount of diffusing material, $\mathbf{J}$, and the concentration, C, in Eq. 10.9 are expressed in terms of the same quantity, the diffusion coefficient D_d has the units of L^2/T. In other words, if $\mathbf{J}$ is expressed in units of moles/L^2T, and the concentra-

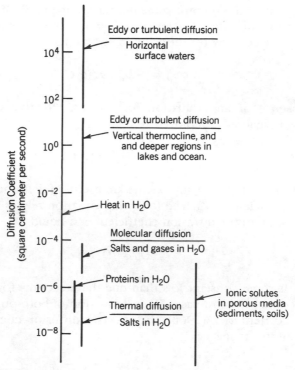

Figure 10.8

Diffusion coefficients characteristic of various
environments. Reprinted with permission from Lerman, A., in Non-equilibrium Systems in Water
Chemistry; Hem, J. D., ed.; ACS Advances in
Chemistry Series 106: American Chemical Society:
Washington DC, 1971, p. 32. Copyright 1971 American Chemical Society.

tion is in moles/L^3, the proportionality constant has the units L^2/T in that the gradient operator grad( ) has the units L^{-1}.

Molecular diffusion originates because of mixing caused by random molecular motions due to the thermal kinetic energy of the solute. Because of molecular
spacing, the coefficient describing this scattering is larger in gases than in liquids,
and larger in liquids than in solids. The diffusion coefficient in a porous medium
is smaller than in pure liquids primarily because collision with the solids of the
medium hinders diffusion. Figure 10.8 provides a range in values for diffusion
coefficients for various fluid environments.

In a porous medium, diffusion takes place in the liquid phase enclosed by a
porous solid. Averaging techniques provide the following rigorous statement for
Fick's law in the fluid phase of a porous sediment (Whitaker, 1967)

$$\mathbf{J} = -D_d\left[\text{grad}(Cn) + \frac{\tau}{V} \right] \tag{10.10}$$

where V is the averaging volume, n is porosity, and τ is a tortuosity vector to
account for the hindering of free diffusion by collision with the pore walls. The

form of Eq. 10.10 has prompted the definition of a bulk diffusion coefficient D_d^* to account for the effects of tortuosity, giving the statement most commonly referred to as Fick's law for diffusion in sediments

$$\mathbf{J} = -D_d^* n \, \text{grad}(C) \qquad (10.11)$$

In the absence of an evaluation for D_d^*, an effective diffusion coefficient $D_d' = D_d^* n$ is generally employed so that

$$\mathbf{J} = -D_d' \, \text{grad}(C) \qquad (10.12)$$

Several different empirical approaches are useful in defining a so-called effective diffusion coefficient (D_d'). The following equation relates D_d' for the fluid in a porous medium to the diffusion coefficient in a liquid D_d

$$D_d' = D^* D_d \qquad (10.13)$$

where D^* is some constant less than one that accounts for the structure of the porous medium. In most cases D^* is a function of both porosity and tortuosity. According to Helfferich (1966), the effective diffusion coefficients in exchange columns fall in a range

$$D_d' = \frac{n}{2} D_d \quad \text{to} \quad D_d \left[\frac{n}{(2-n)} \right]^2 \qquad (10.14)$$

where the factor D^* depends only upon porosity n. Other ways of expressing this relationship are (Greenkorn, 1983)

$$D_d' = u \, n D_d \qquad (10.15)$$

where u is a number less than one and (Greenkorn and Kessler, 1972)

$$D_d' = \frac{n}{\tau} D_d \qquad (10.16)$$

where τ is the tortuosity. In this latter formulation, the tortuosity τ is defined as the ratio of the length of a flow channel for a fluid particle (L_e) to the length of a porous medium sample (L). The value of (L_e/L) is always greater than one except for perfect capillary-type passages. Bear (1972) defines tortuosity as $(L/L_e)^2$ resulting in values less than one, and ranging between 0.56 and 0.8 in granular media. He states further that the effective diffusion coefficient is the product of tortuosity and the diffusion coefficient in the bulk fluid. Whichever of these equations is employed, effective diffusion coefficients in porous media increase with increasing porosity and decrease with increasing path length–sample length ratios.

Table 10.1 cites diffusion coefficients in water for common cations and anions. With the exception of H^+ and OH^- values of D_d range from 5×10^{-6} to 20×10^{-6} cm²/sec with the smallest values associated with ions having the greatest charge.

Table 10.1
Diffusion coefficients in water for some ions at 25°C
(from Li and Gregory, 1974)*

Cation	D_d (10^{-6} cm²/s)	Anion	D_d (10^{-6} cm²/s)
H^+	93.1	OH^-	52.7
Na^+	13.3	F^-	14.6
K^+	19.6	Cl^-	20.3
Rb^+	20.6	Br^-	20.1
Cs^+	20.7	HS^-	17.3
		HCO_3^-	11.8
Mg^{2+}	7.05		
Ca^{2+}	7.93	CO_3^{2-}	9.55
Sr^{2+}	7.94	SO_4^{2-}	10.7
Ba^{2+}	8.48		
Ra^{2+}	8.89		
Mn^{2+}	6.88		
Fe^{2+}	7.19		
Cr^{3+}	5.94		
Fe^{3+}	6.07		

*Reprinted with permission from Geochim. Cosmochim. Acta, 38, Y.-H. Li and S. Gregory, Diffusion of ions in sea water and in deep-sea sediments. Copyright 1974. Pergamon Press plc.

Mechanical Dispersion

Mechanical dispersion is mixing that occurs as a consequence of local variations in velocity around some mean velocity of flow. Thus, mechanical dispersion is an advective process and not a chemical one. With time, mass occupying some volume becomes gradually more dispersed as different fractions of the mass undergo transport in these varying velocity regimes. The main cause of the variability in the direction and rate of transport is the porous medium nonidealities. The most important variable in this respect is hydraulic conductivity. Table 10.2 lists some of the geological features that produce nonidealities. The impression that Table 10.2 should leave is that nonidealities can exist at a variety of scales ranging from microscopic to megascopic, to still an even larger scale (not included on the table) involving groups of formations.

Mechanical dispersion is generated as a consequence of nonidealities at all of these scales. Variability in velocity at the microscopic scale develops, for example, because of the differing flow regimes across individual pore throats or variability in the tortuosity of the flow channels (Figure 10.9a). The heterogeneities shown in Figures 10.9b and 10.9c produce dispersion by creating variability in flow at a macroscopic scale. The first case (Figure 10.9b) is an example of a nonideality created by layering. Contrasts in hydraulic conductivity among the various layers produce differing velocities. An example of macroscopic heterogeneity of a more complex character is shown on Figure 10.9c. This example, adapted from Sudicky (1986), represents the plotted results of a large number of hydraulic conductivity measurements for the shallow water table aquifer at Canadian Forces Base Borden. Dispersion could also be associated with heterogeneity among a group of formations. However, no detailed work has been conducted on these larger-scale systems.

Table 10.2

Geological features contributing to nonidealities in a porous medium (from Alpay, 1972)*

A. Microscopic heterogeneity: pore to pore
  1. Pore size distribution
  2. Pore geometry
  3. Dead-end pore space
B. Macroscopic heterogeneity: well to well or intraformational
  1. Stratification characteristics
     a. Nonuniform stratification
     b. Stratification contrasts
     c. Stratification continuity
     d. Insulation to cross-flow
  2. Permeability characteristics
     a. Nonuniform permeability
     b. Permeability trends
     c. Directional permeability
C. Megascopic heterogeneity: formational (either fieldwide or regional)
  1. Reservoir geometry
     a. Overall structural framework: faults, dipping strata, etc.
     b. Overall stratigraphic framework: bar, blanket, channel fill, etc.
  2. Hyperpermeability-oriented natural fracture systems

*Copyright 1972, Society of Petroleum Engineers Inc., JPT (July 1972).

(a) Microscale Nonidealities

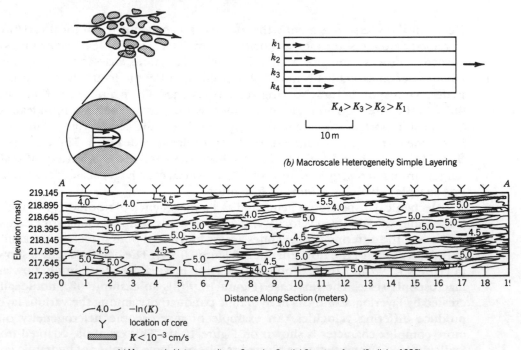

$K_4 > K_3 > K_2 > K_1$

10 m

(b) Macroscale Heterogeneity Simple Layering

$-4.0-$ $-\ln(K)$

Y location of core

$K < 10^{-3}$ cm/s

(c) Macroscale Heterogeneity — Complex Spatial Structure from (Sudicky, 1986)

Figure 10.9

Examples of nonidealities at different scales giving rise to mechanical dispersion. Panel *(c)* is from Sudicky, E. A., Water Resources Res., v. 22, p. 2069–2082, 1986. Copyright by American Geophysical Union.

10.4   Character of the Dispersion Coefficient

Studies at the Microscopic Scale

The coefficient of hydrodynamic dispersion (D) is the sum of the coefficients of bulk diffusion (D_d^*) and mechanical dispersion (D'). Values of the bulk diffusion coefficient can be estimated reasonably well within an order of magnitude for granular media. However, questions remain as to what is a realistic range of values for the mechanical component D' and what is its relative contribution to hydrodynamic dispersion. Fortunately, results from a variety of laboratory column experiments can address these issues.

The fluid flow velocity and grain size are the main controls on the longitudinal dispersion in a column. Pfannkuch (1962) explored these relationships by taking existing experimental data and casting them in the form of a series of dimensionless numbers D_L/D_d and vd_m/D_d, where v is the linear ground water velocity, d_m is the mean grain size, D_L is the longitudinal dispersion coefficient, and D_d is the diffusion coefficient. The first of these numbers normalizes the observed dispersion in the column by dividing it by the coefficient of diffusion of the tracer in water. The second number is the Peclet number (N_{PE}), a ratio expressing advective to diffusive transport.

Figure 10.10 is a simplified representation of the relationships observed by Pfannkuch (1962), which illustrates four classes of mixing related to the Peclet number. The first class (Class 1) is for small values of the Peclet number ($N_{PE} < 0.01$). For mixing in Class 1, D_L/D_d does not change as a function of N_{PE}, which

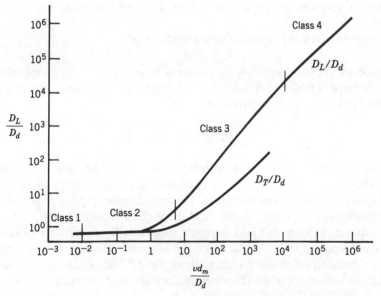

Figure 10.10
Behavior of D_L/D_d and D_T/D_d as a function of the Peclet number (N_{PE}). The classes identify regimes where mixing results from the same processes (after Pfannkuch, 1962 and Perkins and Johnston, 1963). Copyright 1963, Society of Petroleum Engineers Inc., SPEJ (March 1963).

indicates that diffusion is the main cause of mixing. The ratio D_L/D_d has a value of approximately 0.67 (Fried and Combarnous, 1971) because at this velocity D_L is equal to the bulk diffusion coefficient D_d^*, which is less than the diffusion coefficient (D_d) in water.

With increasing values of N_{PE} (Class 2, N_{PE} 0.1–4) mixing is influenced not only by diffusion but by mechanical mixing. As a result, D_L/D_d increases as a function of N_{PE} (Figure 10.10). With a further increase in velocity (Class 3, N_{PE} 4–10^4), the mechanical dispersion completely dominates the mixing process except at the low end of the range of N_{PE}. Values of D_L over this range are approximately proportional to $v^{1.2}$. Greenkorn (1983) explains this exponent of 1.2 as the outcome of several different mixing processes. With increasing N_{PE}, velocity profiles begin to develop within the pores. This effect produces mixing analogous to dispersion in a capillary with $D_L \sim v^2$. Other forms of mixing produce a velocity dependence with $D_L \sim v$. A case in point is mixing due to the tortuosity of the flow paths due to the presence of grains. For values of N_{PE} in the range 10^4–10^6 (Class 4), the effects of molecular diffusion are negligible. $D_L \sim v$ and the experimental data plot as a straight line with a slope of $45°$.

One of the important results from column experiments is the determination that longitudinal dispersion is proportional to velocity. This relationship has been generalized to describe dispersion at macroscopic and megascopic scales. Column experiments have also provided important information about transverse dispersion. Shown in Figure 10.10 is the ratio D_T/D_d plotted versus N_{PE} (from Perkins and Johnston, 1963). The two curves on Figure 10.10 are similar in shape, indicating that the process of transverse mixing is much the same as longitudinal mixing. However, longitudinal dispersion is greater than transverse dispersion at a given N_{PE}. For example when $N_{PE} > 100$, values of D_L are approximately ten times greater than D_T. This tendency for D_L to exceed D_T not only applies in general for microscopic dispersion but for larger scales of dispersion as well.

Dispersivity as a Medium Property

Laboratory experiments with columns have been useful in establishing relationships between the coefficients of mechanical dispersion and the linear groundwater velocity. The equations are

$$D_L' = \alpha_L v \quad \text{and} \quad D_T' = \alpha_T v \tag{10.17}$$

where v is the velocity in the direction of flow, and α_L and α_T are the longitudinal and transverse dispersivities of the medium. Dispersivities have units of length, and like hydraulic conductivity, are characteristic properties of a medium. In practice, they quantify mechanical dispersion in a medium.

In a homogeneous and isotropic medium, D_L and D_T are defined parallel and perpendicular to the direction of flow. Dispersivities can also be defined in relation to a Cartesian coordinate system (Bear, 1972). This situation develops in modeling of mass transport when dispersivities are defined in relation to a grid system.

Studies at Macroscopic and Larger Scales

Much of what is known about dispersion is based on early theoretical studies and column studies. However, important contributions have come from field studies at macroscopic and larger scales. The main emphasis in this research has been to

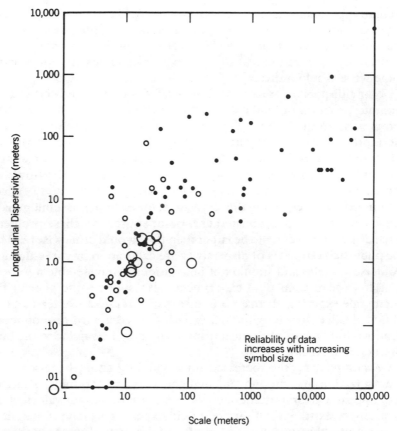

Figure 10.11

Scale of observation versus longitudinal dispersivity: reliability classification (from Gelhar and others, 1985). Copyright © 1985. Electric Power Research Institute. EPRI EA-4190. "A Review of Field-Scale Physical Solute Transport Processes in Saturated and Unsaturated Porous Media." Reprinted with permission.

estimate dispersivity values from field experiments. Gelhar and others (1985) undertook a critical review of field experiments at 55 sites around the world. Values of longitudinal dispersivity range from 0.01 m to 5500 m, apparently depending upon the scale of experiment. It appears that dispersivities increase indefinitely with scale. However, the detailed review of these experiments (Figure 10.11) suggested that only five studies (Lau and others, 1957; Mercado, 1966; Molinari and Peaudecerf, 1977; Valocchi and others, 1981; and Sudicky and others, 1983) yielded reliable estimates of dispersivity. A further 18 values could be considered of intermediate reliability. When the most reliable results are compared to the entire set of results, however, it is not clear that dispersivity will always increase linearly as a function of the scale of spreading. The most reliable dispersivity values are all at the low end of the range.

From these and other tests, a consistent view about macroscopic dispersion has emerged. Heterogeneity at the macroscopic scale contributes significantly to dispersion because it creates variability in velocity. Values of macroscopic dispersivity are in general two or more orders of magnitude larger than those from column experiments. A typical range of values in longitudinal dispersivity from column experiments is from 10^{-2} to 1 cm. In field experiments, values range from approximately 0.1 to 2m over relatively short transport distances. Although no reliable, large-scale studies have been carried out, it is probable that longitudinal dispersivity values in excess of 10m exist, but values are not nearly as large as some work would indicate.

Many dispersion studies have suffered from errors related to the collection and interpretation of field data. As will become apparent in later chapters, these errors arise from (1) poor definition of the plume due to a small number of monitoring wells and nonpoint sampling, (2) failure to account for temporal variations in the advective flow regime, which can lead to mass spreading, (3) incomplete information about the way the tracer is added to the flow system at the source, (4) inherent limitations in some test procedures, and (5) oversimplified techniques for interpreting the test results, for example, assuming that a source of finite size can be approximated as a point source. All these problems typically force dispersivity values to be larger then they would otherwise be. Thus, many of the published estimates of dispersivity are calibration or curve-fitting parameters (Anderson, 1984) and should not be considered transferable to other sites.

The results from field experiments also support the general finding from microscale experiments that transverse dispersivities are at least an order of magnitude smaller than longitudinal values. In media with pronounced horizontal stratification, values of vertical dispersivity may be similar to the bulk diffusion coefficient (Sudicky, 1986).

Accompanying the increased understanding about large-scale dispersion is the realization that the fundamental concepts may not be as straightforward as they appear. One important question is whether constant values of macroscopic dispersivity exist. Several different field experiments have apparently produced dispersivities that increase as a function of the travel distance of the plume. There are some instances where this variable dispersivity is only an artifact of the method of interpretation. For example, characterizing the two- or three-dimensional spread of a tracer with a one-dimensional model requires an apparent dispersivity that increases as a function of distance (Domenico and Robbins, 1984). However, the interpretation from a dimensionally correct model yields a constant dispersivity. There are nevertheless results to suggest that scale-dependent dispersivities do exist.

How can these nonconstant values be explained? One way is to consider them simply to be a result of incomplete spatial averaging. Consider a dispersion experiment run in a heterogeneous unit with relatively large lenses. The spread of tracer from a source to a nearby observation point in the same lens should yield a relatively small dispersivity that is comparable to a column value. The tracer has to move a substantial distance before it is able to interact fully with the heterogeneity to the extent necessary to produce a macroscale dispersivity. Thus, it is not surprising for dispersivity values to increase away from a source before becoming constant. Gelhar and others (1979) refer to this constant macroscale dispersivity as the asymptotic dispersivity. In practice, approaching an asymptotic dispersivity might require tracer spreading over tens or hundreds of meters.

Another way to explain spatially varying dispersivity values is in relation to some representative elemental volume (REV) for dispersivity. With the spreading of a tracer, eventually a transition will occur from the microscale to macroscale. The implicit assumption made is that for each scale a REV exists (Figure 10.12) (note the analogy with fracture flow, Chapter 3).

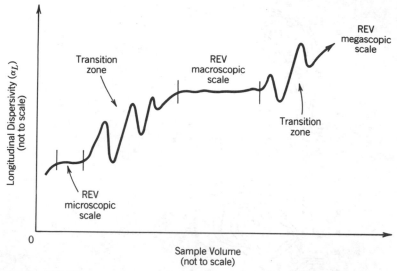

Figure 10.12
An example of the relationship between longitudinal dispersivity and sample volume with REVs defined at the microscopic and macroscopic scale.

In the zone of transition between the various scales, it may be difficult to define a continuum value for dispersivity. Spreading may not yet have encompassed a volume that is equivalent to a REV. As a consequence, small changes in volume could result in significant variability in the character of spreading. Also, mass may not be normally distributed as classical theory would predict, and as a result, dispersivity values may not be definable. An example of this behavior is the concentration distribution from a tracer experiment shown in Figure 10.13 (Sudicky and others, 1983). Because the tracer was being transported through different elements of the heterogeneous conductivity field in a sub-REV system, the tracer plume separated close to the source. Asymptotic values of dispersivity could conceivably exist for this system. However in the early stages of spreading, defining a dispersivity value may be difficult.

The tendency for dispersivity values to be scale dependent has some important implications about how macroscopic dispersivity values are estimated and used. For example, dispersivity values from tracer tests run over relatively short distances may not be representative of the unit. Further, it simply may not be practical to carry out dispersion experiments on a routine basis over distances large enough to yield an asymptotic value (Sudicky, 1986). This problem of scale transitions and how to account for them has implications for modeling of mass transport. Current practice still involves assuming that dispersivity values are constant. From what we now know about the relatively small spreading close to a source, this assumption overestimates the effects of dispersion. Using a dispersivity function that accounts for near-source behavior may be possible. However, there are practical problems in determining what this function might look like.

Problems associated with the definition and interpretation of macroscale dispersivities have not invalidated the use of these concepts. They remain, in spite of limitations, as a satisfactory model of complicated mixing processes. However, an

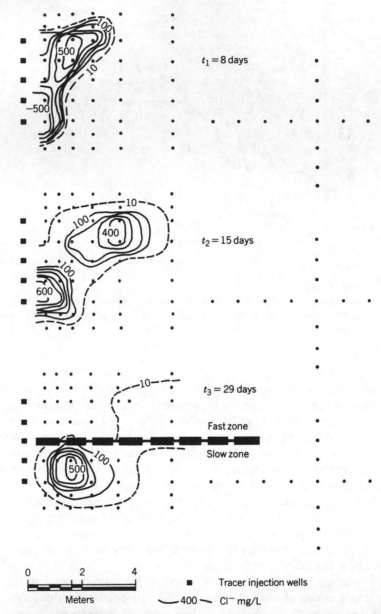

Figure 10.13

Measured Cl⁻ concentrations from a tracer experiment in the shallow aquifer at the Canadian Forces Base, Borden. After a short time, the Cl⁻ plume separated due to different rates of advection in a "fast" and "slow" zone (after Sudicky and others, 1983). Reprinted with permission of Elsevier Science Publishers from J. Hydrol., 63.

informed practitioner must be aware of some limitations in extending classical concepts of dispersion from microscale to macroscale and larger systems.

10.5   *A Geostatistical Model of Dispersion*

Exciting new contributions in the study of dispersion are quantitative approaches that relate asymptotic dispersivities to geostatistical models of hydraulic conductivity. Ultimately, these approaches should form the basis for a family of practical field techniques for characterizing dispersivity. To begin, a few basic ideas need to be developed concerning how a geostatistical model can represent heterogeneity in properties like hydraulic conductivity, porosity, or cation exchange capacity.

Medium properties in nature vary from place to place. The series of equally spaced hydraulic conductivity measurements (Figure 10.14), determined from core samples, can reflect this variability. Given such data, the goal of a geostatistical description is to represent the heterogeneity by a small number of statistical parameters. For the simplest case, three are sufficient: the mean, the variance, and the correlation length scales. This description implies the medium is stationary or, in other words, these parameters do not vary in space.

For a series of data, the mean is the level around which the parameter values fluctuate, while the variance describes the spread about this level (Box and Jenkins, 1976). When data vary over several orders of magnitude, they are often best described by a lognormal distribution (Freeze and others, 1989). In the case of hydraulic conductivity, the following equation describes the distribution

$$Y_i = \ln K_i \tag{10.18}$$

where Y_i is the log hydraulic conductivity. The mean and variance for the population are μ_Y and σ_Y^2, respectively.

The correlation length is a measure of the spatial persistence of zones of similar properties. For example, hydraulic conductivity often is correlated such that values immediately adjacent to one another are similar and those much further away are dissimilar. Because of the way in which sediments are deposited, the

Figure 10.14
A series of equally spaced ln K values (from Freeze and others, 1989).

Table 10.3
Summary of notation for the geostatistical description of a medium (from Freeze and others, 1989)

Parameter	Population statistic	Population estimate	Sample estimate
Mean	μ_Y	$\hat{\mu}_Y$	$\bar{Y}$
Variance	σ_Y^2	$\hat{\sigma}_Y^2$	S_Y^2
Autocovariance	τ_{Y_k} or $\tau_Y(h)$	$\hat{\tau}_{Y_k}$ or $\hat{\tau}_Y(h)$	C_{Y_k} or $C_Y(h)$
Autocorrelation	ρ_{Y_k} or $\rho_Y(h)$	$\hat{\rho}_{Y_k}$ or $\hat{\rho}_Y(h)$	r_{Y_k} or $r_Y(h)$
Correlation length	λ_Y	$\hat{\lambda}_Y$	

correlation is usually anisotropic. For example, spatial persistence in hydraulic conductivity is greater in the direction of bedding than across the bedding.

Mean and Variance

The population statistics for a heterogeneous field are usually estimated from a finite number of samples. In terms of notation, the statistics of the actual population are represented differently than the sample estimates (Table 10.3). Thus, for a series of log hydraulic conductivity values, the mean estimated from the samples is

$$\bar{Y} = \frac{1}{n} \Sigma \ Y_i, \qquad i = 1, \ldots, n \tag{10.19}$$

where $\bar{Y}$ is the mean hydraulic conductivity and Y_i are n individual log hydraulic conductivity values. The sample variance is

$$S_Y^2 = \frac{1}{n} \Sigma (Y_i - \bar{Y})_i^2 \qquad i = 1, \ldots, n \tag{10.20}$$

where S_Y^2 is the variance in hydraulic conductivity.

Autocovariance and Autocorrelation Functions

A population can be described in terms of autocovariance (τ) or autocorrelation (ρ) functions. These functions are related mathematically and describe how the correlation between any two hydraulic conductivity values decays with the separation or lag. A lag is simply some constant separation interval. For example, with lag 1, values being compared are a distance h apart and with lag 2, $2h$, and so on. Although there is flexibility in defining an autocorrelation function, an exponential model is used commonly (Freeze and others, 1989)

$$\rho(h) = \exp\left(\frac{-|h|}{\lambda}\right) \tag{10.21}$$

where $\rho(h)$ is the autocorrelation of the population at a separation h. The correlation length scale (λ), (or λ_x, λ_y, λ_z for statistically anisotropic media) is the separation in the given direction at which ρ takes on a value of e^{-1} or 0.37 (Figure 10.15), or τ declines to 0.37 of the covariance at lag zero.

Figure 10.15
Example of an autocorrelation function with the correlation length defined as separation at which $\rho = 0.37$ (modified from Freeze and others, 1989).

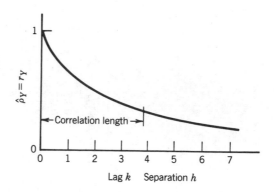

For example, with a finite number of vertical values of log hydraulic conductivity, sample autocovariance (C_Y) and autocorrelation (r_Y) functions can be calculated from pairs of equally spaced data. The autocovariance function at a separation h is

$$C_Y(h) = \frac{1}{m} \Sigma \left[Y(h) - \overline{Y} \right] \left[Y(y + h) - \overline{Y} \right], \qquad i = 1, \ldots, m \qquad (10.22)$$

where m is the number of pairs of sample points at a given separation. The autocorrelation function $r_Y(h)$ is formed by taking $C_Y(h)$ values and dividing by $C_Y(0)$ or the variance S_Y^2 of the sample. For unevenly spaced data, values are grouped into a series of class intervals with the separation distance given by the midpoint of each interval. The following example kindly provided by R. A. Freeze illustrates the calculation of the mean, variance, and correlation length for a stationary medium.

Example 10.1

Listed in Table 10.4 are values of hydraulic conductivity Y (where $Y = \ln K$) collected at 5 cm intervals along a core. Statistically, characterize the heterogeneous nature of the hydraulic conductivity.

The approach to solve this problem is outlined on Table 10.4. The mean and variance are 1.0 and 0.182, respectively. The autocovariance coefficients at separations of 5 and 10 cm are 0.128 and 0.07, respectively, which when divided by the variance, give autocorrelation coefficients 0.70 and 0.38, respectively. Because the autocorrelation coefficient at 10 cm is nearly 0.37, λ is about 10 cm.

Estimation of Dispersivity

Work by Gelhar and Axness (1983) provides the following theoretical equation for predicting longitudinal dispersivity from a geostatistical description of the hydraulic conductivity field

$$A_L = \frac{\sigma_Y^2 \lambda}{\gamma^2} \qquad (10.23)$$

Table 10.4
Work sheet for Example 10.1

t	Y_t	$Y_t - \bar{Y}$	$(Y_t - \bar{Y})^2$	$(Y_{t+1} - \bar{Y})$	$(Y_{t+2} - \bar{Y})$	$(Y_t - \bar{Y})(Y_{t+1} - \bar{Y})$	$(Y_t - \bar{Y})(Y_{t+2} - \bar{Y})$
1	1.5	0.50	0.25	0.70	0.50	0.35	0.25
2	1.7	0.70	0.49	0.50	0.10	0.35	0.07
3	1.5	0.50	0.25	0.10	-0.2	0.05	-0.10
4	1.1	0.10	0.01	-0.20	-0.60	-0.20	-0.06
5	0.8	-0.20	0.04	-0.60	-0.40	0.12	0.08
6	0.4	-0.60	0.36	-0.40	-0.50	0.24	0.30
7	0.6	-0.40	0.16	-0.5	-0.10	0.20	0.04
8	0.5	-0.50	0.25	-0.1	0.0	0.05	0.0
9	0.9	-0.10	0.01	0.0		0.0	
10	1.0	0.0	0.0				
Σ	10	0.0	1.82			1.16	0.58
$\frac{1}{n}\Sigma$	1		0.182			0.128	0.07

where A_L is the asymptotic longitudinal dispersivity, σ_Y^2 is the variance of the log-transformed hydraulic conductivity (i.e., $Y = \ln K$), λ is the correlation length in the mean direction of flow, and γ is a flow factor that Dagan (1982) considers equal to one. This equation assumes (1) unidirectional mean flow and (2) an exponential covariance. The longitudinal asymptotic macrodispersivity (A_L^*) includes contributions from (1) the heterogeneous structure of the medium (A_L), (2) a local or pore-scale component (α_L), and (3) diffusion

$$A_L^* = A_L + \alpha_L + \frac{D_d^*}{v} \tag{10.24}$$

where D_d^* is the bulk diffusion coefficient and v is the linear ground water velocity. The transverse asymptotic macrodispersivity is

$$A_T^* = \alpha_T + \frac{D_d^*}{v} \tag{10.25}$$

The transverse asymptotic dispersivity (A_T) is zero implying that the heterogeneous structure of the medium does not create transverse dispersion. Section 18.4 presents a practical example in the use of the geostatistical approach to parameter estimation.

Given the variance in $\ln(K)$ and the correlation length, one can estimate the overall contribution to dispersion that is related to the heterogeneity of the system. The other two components, α_L and D_d^*/v are relatively small and can be estimated or in many cases neglected. The sum gives us A_L^*. Unfortunately, the characterization of the statistical properties of a medium requires considerable hydraulic conductivity data, which practically are not available in most studies. Work is ongoing to use borehole flow meters or geophysical approaches to provide ways of characterizing the statistical structure of porous media.

10.6 Mixing in Fractured Media

Many of the concepts of mixing from the previous sections apply to fractured media. There is, nevertheless, sufficient added complexity to warrant treating these media separately. The discussion here will be focused on the only type of fractured medium that has been studied in detail. This is a case where fractures form the only permeable pathways through the rock. The hydraulic conductivity of the unfractured part of the medium is so low that fluid flow and hence advective transport are negligible. However, mass can diffuse into the unfractured rock matrix in response to concentration gradients.

The macroscopic outcome of the transport in a fracture system is the result of both mechanical mixing and diffusion. Concentrations in the fractures will be affected by diffusion into the matrix, a chemical process (Figure 10.16a); variability within individual fractures caused by asperities, an advective dispersive process (Figure 10.16b); fluid mixing at the fracture intersections, a diluting or possibly diffusive process (Figure 10.16c); and variability in velocity caused by differing scales of fracturing (Figure 10.16d) or by variations in fracture density (Figure 10.16e). In addition, some of these processes are coupled in a complex fashion and compete with each other.

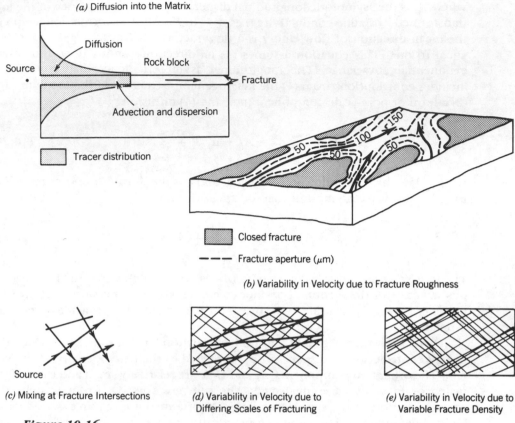

Figure 10.16
A series of sketches illustrating the different causes of dispersion in fractured rocks.

Diffusion into the matrix controls concentration distributions to an important extent. This result contradicts experience with porous media, where the contribution of diffusion to dispersion is generally swamped by mechanical dispersion. Diffusion is more important in fractured media because localizing mass in fractures provides the opportunity for large concentration gradients to develop. Theoretical studies by Grisak and Pickens (1980) and Tang and others (1981) explored this process using numerical and analytical models of a single fracture bounded by an infinite porous matrix. Dispersion in a single fracture is caused by the variability in fracture aperture. This variability develops due to the roughness of the fracture walls and the precipitation of secondary minerals. At many locations, the fracture may be closed to flow and transport. This kind of aperture topology (see Figure 10.16b) gives rise to channeling (Neretnieks, 1985), where mass moves predominantly along networks of irregularly shaped pathways in the plane of a fracture.

This conceptual model of a fracture has developed from tracer tests carried out in fractured rocks. For example, studies at the Stripa mine (Neretnieks, 1985) showed how the inflow of water to tunnels was extremely localized. Approximately one-third of the flow entered from approximately 2 percent of the fractured rocks. These results are strongly indicative of localized channel flow within individual fractures.

Table 10.5

Tabulation of dispersivity values based on tracer tests in fractured rocks (from Neretnieks, 1985)

Experiments	Migration distance (m)	Dispersivity (m)	Fissure width (mm)
Laboratory experiment with natural fissure (Neretnieks and others, 1982)	0.3	0.025	0.18
Laboratory experiment with natural fissure (Moreno and others, 1984)	0.19	0.005	0.15
	0.27	0.011	0.14
Stripa natural fissure, two different channels (Abelin and others, 1982)	4.5	2.0	0.11
	4.5	0.62	0.14
Studsvik site 2 (Landström and others, 1982)	14.6	—	—
Finnsjön site (Moreno and others, 1983)	30	0.35/5	0.47–0.98
Studsvik site 1 (Landström and others, 1978)	22	6.1	—
	51	7.7	—
French site (Lallemand-Barrès, 1978)	11.8	0.8	—
U.S. site (Webster and others, 1970)	538	134	—

Channels can be so poorly interconnected that they may not interact with one another for appreciable distances (Neretnieks, 1985). This kind of fracture development is termed pure channeling. From this discussion, it is easy to understand why the smooth, parallel plate model used by some investigators is really a simplified representation of a fracture.

Neretnieks (1985) has tabulated values of dispersivity from tracer experiments in fractured rocks (Table 10.5). Because of the relatively short transport distances, these values reflect mixing caused by diffusion into the matrix and mechanical dispersion within individual fractures. Thus, they are not true advective dispersivities. The values in general increase as a function of travel distance as in porous media. The larger dispersion lengths for the last four tests, however, may be artifacts of the interpretive model, which did not account for matrix diffusion (Neretnieks, 1985). Here is yet another example that points to the need to examine published dispersivity numbers in relation to the methods used for interpretation.

As is the case with porous media, there are conceptual problems that confound the application of classical theory. If pure channeling does in fact develop, the dispersivity will increase continuously as a function of travel distance and never reach an asymptotic value (Neretnieks, 1985). The lack of spatial averaging in channeled fracture systems in addition may also inhibit the development of normal concentration distributions.

Dispersion at the next larger scale occurs when the geometry of the three-dimensional network begins to influence mass transport. Mass moving along a fracture to an intersection (Figure 10.16c) can be partitioned into two or more fractures. As the carrier fluid is likewise partitioned and as concentration is the mass per unit volume of solution, partitioning by itself does not affect the concentrations. However, concentrations will be indeed affected if there is mixing with water that does not contain the tracer. This dilution depends directly on the type of mixing process in the intersection and quantity of water able to flow along individual fractures.

The dispersion illustrated on Figures 10.16d and e is analogous to that caused by heterogeneities in porous media. Differing scales of fracturing or spatial variability in fracture density create variability in local velocity. Even larger-scale dispersion could develop if individual fractures or discontinuities exist on a regional scale or the variability in fracture density includes several different units. Many of the concerns about the applicability of classical concepts of dispersion apply to fractured rocks. For example, is it meaningful or even possible to describe asymptotic dispersivities? How does dispersivity change as a function of distance? Is it even possible in real networks to define a REV? We cannot answer these questions now because the necessary field and theoretical studies are just starting. Looking into the crystal ball for a moment, it is quite possible that many of the existing ideas about dispersion may have to be extended or reworked for fractured rocks.

Suggested Readings

Anderson, M. P., 1979, Using models to simulate the movements of contaminants through groundwater flow systems: Crit. Rev. Environ. Controls, v. 9, no. 12, p. 97–156.

Anderson, M. P., 1984, Movement of contaminants in groundwater: groundwater transport-advection and dispersion: Groundwater Contamination, NRC Studies in Geophysics, National Academy Press, Washington, D.C., p. 37–45.

Gelhar, L. W., Mantoglou, A., Welty, C., and Rehfeldt, K. R., 1985, A review of field-scale physical solute transport processes in saturated and unsaturated porous media: Electric Power Research Institute EPRI EA-4190 Project 2485-5, 116 pp.

Sudicky, E. A., 1986, A natural gradient experiment on solute transport in a sand aquifer: Spatial variability of hydraulic conductivity and its role in the dispersion process: Water Resources Res., v. 22, no. 13, p. 2069–2082.

Problems

1.   Consider two different media, one fractured and the other unfractured. The hydraulic conductivity is the same for each (1×10^{-8} m/s) but the effective porosities are different (3×10^{-4} and 0.30 respectively). Under an imposed hydraulic gradient of 0.01, determine how much time is required for a tracer to be advected 30m, both for fractured and porous media.

2.   The hydrogeologic cross-section on the figure illustrates the pattern of ground water flow along a local flow system. Assume that a source of contamination develops with advection as the only operative transport process. Describe the pathway for contaminant migration and estimate at what time in the future the plume will reach the stream.

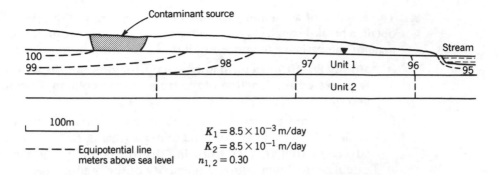

$K_1 = 8.5 \times 10^{-3}$ m/day
$K_2 = 8.5 \times 10^{-1}$ m/day
$n_{1,2} = 0.30$

3. The approximate coefficient of diffusion for Cl⁻ in water is 2×10^{-5} cm²/s. Estimate its value in a saturated porous medium having a porosity of 0.25 and a tortuosity of 1.25.

4. Label the following statements true or false.

_____ The higher the flow velocity the higher the dispersivity.

_____ Mechanical dispersion does not occur in a stationary (nonmoving) ground water.

_____ Molecular diffusion does not occur in a stationary ground water.

_____ The coefficient of hydrodynamic dispersion increases with increasing values of the dispersivity or diffusion coefficient.

_____ With pure channeling, dispersivity never reaches an asymptotic value.

5. A useful rule of thumb is that mass has to spread several ten's of correlation lengths before the asymptotic dispersivity is reached. Explain why.

6. With longitudinal dispersion in the absence of transverse dispersion, the maximum concentration one will observe will always be (a) equal to, (b) less than, (c) greater than, or (d) cannot be determined, the initial concentration of the source.

7. Ground water flows through the left face of a cube of sandstone (1m on a side) and out of the right face with a linear ground water velocity of 10^{-5} m/s. The porosity of the sandstone is 0.10, and the effective diffusion coefficient is 10^{-10} m²/s. Assume that a tracer has concentrations of 120, 100, and 80 mg/l (1.2, 1.0, and 0.8×10^5 mg/m³) at the inflow face, the middle of the block, and the outflow face respectively. Calculate the mass flux through the central plane due to advection and diffusion.

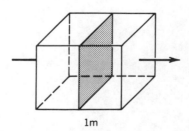

1m

8. The essence of a column experiment for measuring dispersivity is to obtain a breakthrough curve, which can be evaluated to provide an estimate of the variance in concentration as a function of time (σ_t^2).

    a. Modify Eq. 10.7 to a form in which the longitudinal dispersivity α_L is expressed as a function of the length of the column (L), time (t), and σ_t^2.

    b. A breakthrough curve for a column experiment provides a σ_t^2 value of 5.5 min^2 at a time of 82 minutes (that is, time at which the advective front arrives). Knowing that the column is 56 cm long, calculate the longitudinal dispersivity of the medium (α_L).

9. A contaminant is added as a point source to ground water flowing with a constant velocity of 4×10^{-6} m/s. Assuming longitudinal and two transverse dispersivities (y and z directions) of 1.0, 0.1, and 0.01m, determine the spatial standard deviations ($\sigma_{x, y, z}$) in the plume size after 400m of transport.

10. Following is a series of 15 hydraulic conductivity measurements (K) measured on cores collected at a spacing of 0.2 m. Estimate the sample mean, variance, and vertical correlation length (using two lags). Assume an exponential autocorrelation function.

    Hydraulic Conductivity $\times 10^{-4}$ cm/s

1 4		8 36
2 13		9 12
3 48		10 13
4 64		11 18
5 62		12 36
6 58		13 48
7 38		14 52
		15 64

Chapter Eleven

Principles of
Aqueous Geochemistry

11.1 **Introduction to Aqueous Systems**

11.2 **Structure of Water and the Occurrence of Mass in Water**

11.3 **Equilibrium Versus Kinetic Descriptions of Reactions**

11.4 **Equilibrium Models of Reaction**

11.5 **Deviations from Equilibrium**

11.6 **Kinetic Reactions**

11.7 **Organic Compounds**

11.8 **Ground Water Composition**

11.9 **Describing Chemical Data**

Chapters 11 and 12 discuss the theory and processes of mass transfer. These are the important reactions operating in the fluid phase or between the mass in water and the rock matrix that contains it. A prerequisite to the detailed discussions of chemical hydrogeology that follow in the book is an examination of fundamental ideas about the properties of water, the way mass occurs in water, and equilibrium and kinetic concepts in aqueous systems. Another topic considered in this chapter is the chemistry of organic compounds. There has been in recent years an increasing awareness of the importance of these compounds in both contaminated and uncontaminated systems. Finally, we examine the constituents commonly found in natural ground water and some of the unique graphical and statistical techniques developed to describe their occurrence.

11.1 Introduction to Aqueous Systems

Mass can exist in the subsurface as (1) separate gas or solid phases, for example, CO_2 gas in the soil zone or minerals that form the porous medium; (2) a separate liquid phase, for example, crude oil and liquid organic contaminants; or (3) mass dissolved in the water itself (that is, solutes), for example, Na^+ or Cl^-. While all these occurrences are important in a geologic context, hydrogeologists have typically focused on the third type. It is for this reason that our discussion will concentrate on mass dissolved in water. We cannot, however, avoid treating the nonaqueous phase liquids, gases, and solids because these are often the sources of solutes in ground water or potential contaminants.

Mass occurring in water as ions, molecules, or solid particles not only undergoes transport (Figure 10.1) but also reactions, which redistribute mass among various ion species, or between the liquid and solid phases. Our list of reactions is not exhaustive but as will become apparent in Chapter 12, it includes the most important ones affecting the chemistry of ground water. These chemical processes operate in a system consisting of one or more solid phases, an aqueous solution phase, and a gas phase (Figure 11.1) and redistribute mass within a phase or between phases.

Most inorganic substances are electrolytes, dissolving in water to form ions. Positively charged species, such as Ca^{2+} or K^+, are called cations. Negatively charged species, such as HCO_3^- or Cl^- are called anions. Complex ions form by combining simpler cations and anions. Organic substances can also dissolve to form organic cations or anions. However, organic liquids are usually nonelectrolytes and dissolve in water as nonionic molecules. In the aqueous phase, the distribution of species is described in terms of a molar concentration or an activity, which is an effective concentration accounting for chemical nonidealities in reactions. The notation adopted throughout the book is to represent molar concentrations with round brackets for example (Ca^{2+}) and activities with square brackets $[Ca^{2+}]$. Gases are represented in terms of partial pressures of constituent gases (for example, P_{CO_2}) or in nonideal systems by fugacities f_{CO_2}. Solid phase compositions (Figure 11.1) are represented as mole fractions and activities.

Concentration scales quantitatively describe mass distributions. The next two sections will examine the commonly used scales, and provide scale conversions for concentrations.

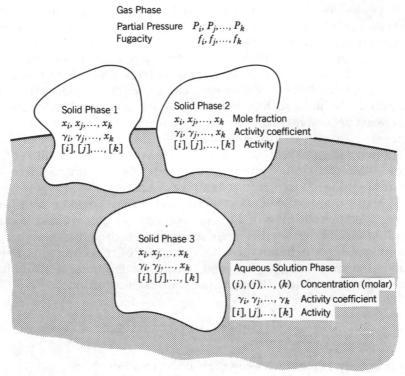

Figure 11.1
Generalized model of a chemical system for ground water at some temperature and pressure (modified from Stumm and Morgan, 1981, Aquatic Chemistry). Copyright © 1981 by John Wiley & Sons, Inc. Reprinted by permission of John Wiley & Sons, Inc.

Aqueous Solution Phase

Molar concentration (M) defines the number of moles of a species per liter of solution (mol/L). A mole is the formula weight of a substance expressed in grams. For example, a 1 liter solution containing 1.42 g of Na_2SO_4 has a (Na_2SO_4) molarity of $1.42/(2 \times 22.99 + 32.06 + 4 \times 16.00)$ or 0.010 M. Because Na_2SO_4 dissociates completely in water according to the following reaction

$$Na_2SO_4 = 2Na^+ + SO_4^{2-}$$

the molar concentrations of Na^+ and SO_4^{2-} are 0.02 and 0.01 M, respectively. This reaction states that 1 mole of Na_2SO_4 dissolves to produce 2 moles of Na^+ and 1 mole of SO_4^{2-}. Because substances react in molar proportions, many of the thermodynamic calculations use concentrations expressed in moles.

Molal concentration (m) defines the number of moles of a species per kilogram of solvent (mol/kg). This scale for concentrations in dilute solutions is almost the same as molar concentrations because a 1 liter solution has a mass of approximately 1 kg. For more concentrated solutions, the two scales become increasingly different. Pytkowicz (1983a, p. 9) provides equations for converting between these two scales.

Equivalent charge concentration is the number of equivalent charges of an ion per liter of solution with units such as eq/L or meq/L. The equivalent charge for an ion is equal to the number of moles of an ion multiplied by the absolute value of the charge. For example with a singly charged species such as Na^+, 1 M Na^+ = 1 eq/L, and with a doubly charged species such as Ca^{2+}, 1 M Ca^{2+} = 2 eq/L. Equivalent concentrations can also be represented as equivalent charges per unit mass of solution with units such as eq/kg or equivalents per million (epm).

Mass per unit mass concentrations define a scale in terms of the mass of a species or element per total mass of the system. Many older analyses have been reported using this scale with concentrations in parts per million (ppm) or parts per billion (ppb). More recently, these units of concentration have given way to corresponding concentrations in mg/kg or μg/kg.

Mass per unit volume concentration is the most common scale for concentration. It defines the mass of a solute dissolved in a unit volume of solution. Concentrations are reported in units such as mg/L or μg/L. There is again a close correspondence between these last two scales of concentrations. For dilute solutions 1 ppm = 1 mg/kg = 1 mg/L. As the total salinity of a solution begins to exceed approximately 10^4 mg/L, the conversion equation

$$mg/kg = \frac{mg/L}{\text{solution density}} \tag{11.1}$$

is necessary to move between these concentration scales.

Concentration conversions involve a few simple equations. The most common scale change takes the reported values of chemical analyses in mg/L (also mg/kg or ppm) and converts them to molar concentrations

$$\text{molarity} = \frac{mg/L \times 10^{-3}}{\text{formula weight}} \tag{11.2}$$

Conversion from mg/L to meq/L is sometimes necessary to present chemical data graphically or to check chemical analyses. This conversion equation is

$$meq/L = \frac{mg/L}{\text{formula wt/charge}} \tag{11.3}$$

As an example in the conversion of scales, consider the following.

Example 11.1

Given a water analysis with an SO_4^{2-} concentration of 85.0 mg/L, express this concentration first in terms of molarity and second in meq/L.

$$mol/L = \frac{85 \times 10^{-3}}{32.06 + 4 \times 16.0} = 0.89 \times 10^{-3}$$

$$meq/L = \frac{85}{(32.06 + 4 \times 16.0)/2} = 1.77$$

Gas and Solid Phases

Although interest in most chemical investigations is with the water, the gas and solid phases are important because of the possible interactions with the solution phase. Further, the volatiles associated with some organic liquids may be contaminants. In thermodynamic calculations, concentrations of gases are given as partial pressures. In a mixture of gases at a fixed temperature and in equilibrium with the aqueous phase, the partial pressure of a gas is the pressure the gas would exert if it occupied the whole volume of the mixture by itself. For example, CO_2 gas in the atmosphere has a P_{CO_2} of about $10^{-3.5}$ atmospheres (atm). It is also quite common to report gas concentrations using a scale such as ppm or mg/L, and so on, when reporting the results of laboratory determinations of gas compositions.

Concentrations for solids are represented as a mole fraction, which is the ratio of the number of moles of the phase of interest to the total number of moles in the system. However, thermodynamic calculations often do not require the solid phase concentrations because by convention the solid phase concentration or activity for a pure solid is one.

11.2 Structure of Water and the Occurrence of Mass in Water

The structure of water is important in controlling the chemistry of ground water because of its influence on the solubility of solids and organic liquids. The water molecule is comprised of two hydrogen atoms and one oxygen atom. Shared electrons form covalent bonds between atoms. Hydrogen atoms are located asymmetrically with respect to the oxygen atom (Figure 11.2a). Because of this structure and the covalent bonding, the electrical center of the negative charges

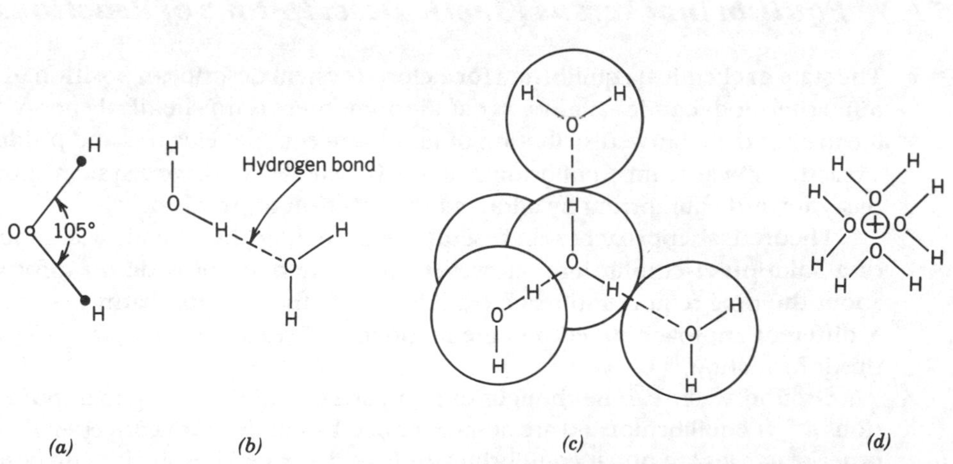

(a)   *(b)*   *(c)*   *(d)*

Figure 11.2
Essential features of the structure of water and adjustments due to the presence of an ion (adapted from Pytkowicz, 1983): (*a*) the position of the hydrogen and oxygen atoms in the polar water molecule; (*b*) hydrogen bonding between water molecules, giving rise to (*c*) the tetrahedral structure of water, (*d*) water molecules rearrange themselves around a cation due to the polar nature of the water.

(electrons) has a different location than does that of the positive charges. This separation makes the molecule polar in terms of electrical charge. When a molecule is perfectly symmetrical (e.g., carbon tetrachloride, benzene, and other hydrocarbons), there is no charge separation and the molecule is nonpolar.

The water molecules are joined in a series of hydrogen bonds (Figure 11.2*b*) formed by electrostatic interaction between a hydrogen and an oxygen atom. As a result of these interactions, the water molecules form a tetrahedral network (Figure 11.2*c*) creating a series of larger molecules. This molecular structure explains some of the unique characteristics of water such as its freezing and boiling points.

This structure also contributes to the relative solubility of electrolytes and the insolubility of nonelectrolytes in aqueous solution. From an energetic viewpoint, it is advantageous for charged species (e.g., ions or molecules) to be accommodated in the water structure, overcoming the energy that tends to keep atoms in a solid or another liquid. In water, a charged species forces a local rearrangement in the structure (Figure 11.2*d*). Water molecules immediately adjacent bind to the charged species. Those nearby reorient themselves because of the charge. The zone of influence immediately adjacent to the ion is known as the hydration sheath. A so-called random region separates the hydration sheath from the bulk solution. Within the random region, the water molecules are less oriented than they are in either the bulk solution or the hydration sheath.

In the case of an uncharged species, such as many nonpolar organic molecules, it is energetically more favorable for mass to remain in an organic liquid rather than the aqueous solution. The nonpolar molecule in effect has to jam itself into the spaces in the water structure. As we will examine in Section 12.2, some large molecules do not fit well. Thus, the structure of water, on the one hand, favors the solubility of electrolytes yet, on the other hand, contributes to the low solubility of many nonpolar, organic molecules.

11.3   *Equilibrium Versus Kinetic Descriptions of Reactions*

The state of chemical equilibrium for a closed system describes a position of maximum thermodynamic stability. At equilibrium, there is no chemical energy available to alter the relative distribution of mass between the reactants and products in a reaction. Away from equilibrium, energy is available to drive a system spontaneously toward equilibrium by allowing the reaction to progress.

Theoretical approaches are useful for modeling the chemical composition of a solution at equilibrium. However, these methods provide no information about the time required to reach equilibrium or the reaction pathways involved. A different approach based on kinetic models of reaction is required to provide this information.

Ground water can be thought of as a partial equilibrium system. Some reactions are at equilibrium and are best described by equilibrium concepts. However, other reactions are not at equilibrium and are described best by kinetic concepts. How a reaction is finally modeled depends upon its rate. An equilibrium reaction is "fast" in relation to the mass transport processes that are working to change concentrations. A kinetic reaction is "slow" in relation to the transport processes. Thus, in applying an equilibrium model to a reaction, we assume that mass is transferred instantaneously between reactants and products to move the system to equilibrium. If the system transfers mass in a reaction at a rate that is slower than physical transport, it requires a kinetic description.

The degree of competition between the rates of reaction and the mass transport processes influences how one might model a system. For example, it may be appropriate to model ion exchange in a sluggish regional flow system as an equilibrium process. However, the same exchange process operating in an aquifer with high-flow velocities may require a kinetic description.

Not all the reactions found in a ground water system are of the equilibrium type. Irreversible reactions proceed in the forward direction until the reactants are used up. Examples include radioactive decay, some redox reactions, and some organic reactions. Reactions of this type require a kinetic description.

The kinetic approach to modeling reactions seems to be much more general than an equilibrium one. Nevertheless, most geochemical work involves models based on equilibrium approaches. The greatest limitation in the more general application of kinetic models is the lack of basic data concerning the order and rate of the reactions. The strength of the equilibrium approaches is the ability to apply concepts of chemical thermodynamics to estimate equilibrium and other constants as a function of temperature (Nordstrom and Munoz, 1986).

Reaction Rates

There are sufficient data on reaction rates to provide guidance in deciding whether to use an equilibrium or a kinetic approach. Figure 11.3 compares the half-time or half-life of many of the reactions on our list in Figure 10.1. As a generalization, the fastest reactions are the solute-solute or solute-water reactions with half-times of fractions of seconds to at most a few minutes. These reactions are examples of homogeneous reactions, reactions that occur in a single phase. Of the list in Figure 10.1, the acid-base and complexation reactions are homogeneous. The dissolution-precipitation reactions are examples of heterogeneous reactions or reactions involving more than one phase. These reactions have half-times varying from days to 10^6 years (Figure 11.3). Heterogeneous solution or exsolution reactions involving a gas phase have relatively larger half-times than do liquid-

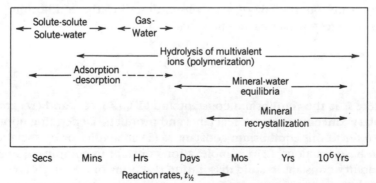

Figure 11.3
A comparison of reaction half-times ($t_{1/2}$) for many of the common reactions in aqueous systems (from Langmuir and Mahoney, 1984). Reprinted by permission First Canadian/American Conference on Hydrogeology. Copyright © 1984. All rights reserved.

solid reactions. Reactions on surfaces, which Langmuir and Mahoney (1984) term adsorption-desorption reactions, have half-times ranging from seconds to days. The larger half-times describe surface reactions in the porous matrix of rock fragments. Radioactive decay and isotopic fractionation reactions are not included in the figure because the range in half-times is extremely variable.

Of our list (Figure 10.1) only redox and organic reactions remain to be discussed. Redox reactions are in general relatively slow because they are mediated by microorganisms (Morel, 1983). Typical half-times for these reactions range from hours to a few years based on data for marine systems (Morel, 1983; Berner, 1980). Transformation reactions involving organic contaminants in ground water depend only on the physical and chemical properties of the particular compounds (for example, hydrolysis), or in addition on the abundance of microbes (for example, biodegradation). In the case of halogenated organic compounds, the hydrolysis reactions are relatively slow with extremely variable half-times ranging from months to several thousand years. Because of the effects of microbial activities, the biodegradation reactions tend to be faster. However, the range in rates is relatively large with half-times for a variety of organic compounds ranging from days to months (Bouwer and McCarty, 1984).

This summary of reaction rates illustrates why the equilibrium approaches have proven so successful in modeling the chemistry of natural ground water systems. Almost all the important reactions have rates that are faster than the typical rates of ground water flow.

11.4   *Equilibrium Models of Reaction*

Equilibrium concepts are useful in modeling many homogeneous and heterogeneous reactions. Consider the following reaction where components C and D react to produce ions Y and Z

$$cC + dD = yY + zZ \qquad (11.4)$$

with c, d, y, and z representing the number of moles of these constituents. For a dilute solution, the law of mass action describes the equilibrium distribution of mass between reactants and products as

$$K = \frac{(Y)^y (Z)^z}{(C)^c (D)^d} \qquad (11.5)$$

where K is the equilibrium constant and (Y), (Z), (C), and (D) are the molal (or molar) concentrations for reactants and products. Depending upon the reaction, we refer to this equilibrium constant as (1) an acidity or dissociation constant in acid-base reactions, (2) a complexation constant in complexation reactions, (3) a solubility constant in solid dissolution reactions or a Henry's law constant in gas dissolution, and (4) an adsorption constant for surface reactions.

Standard geochemical texts including Garrels and Christ (1965), Stumm and Morgan (1981), Drever (1982), and Morel (1983) tabulate values for these constants. Equilibrium constants expressed as a function of temperature are also included in the data bases of computer codes for modeling aqueous systems such as SOLMNEQ (Kharaka and Barnes, 1973), EQ3/6 (Wolery, 1979), and PHREEQE (Parkhurst and others, 1980). Values for equilibrium constants typically come from laboratory experiments, thermodynamic calculations, and calculations based on field measurements.

Activity Models

Activity models for dissolved species represent the nonideal behavior of components in nondilute solutions. Nonidealities result from electrostatic interactions among ions in solution. Mass action equations account for nonidealities by replacing molar concentrations by activities or thermodynamically effective concentrations. For the case of species Y, the following equation relates activity to molar concentration

$$[Y] = \gamma_Y(Y) \tag{11.6}$$

where γ_Y is the activity coefficient. Typically, γ is close to one in dilute solutions and decreases as salinity increases. Activities of species with more than one charge usually are smaller than those with a single charge. At relatively high salinities, the activity coefficients in many cases begin to increase and even exceed one.

We can calculate γ for ions from one of several different activity models. The simplest is based on the Debye-Hückel equation

$$\log \gamma_i = -Az_i^2(I)^{0.5} \tag{11.7}$$

where A (Table 11.1) is a constant that is a function of temperature, z_i is the ion charge. and I is the ionic strength of the solution. Mathematically, the ionic strength in mols/L is

Table 11.1
Parameters used in the extended Debye-Hückel equation at 1 atmosphere pressure

Temperature (C)	A	B ($\times 10^8$)	$\mathring{a}_i \times 10^{-8}$ cm	Ion
0	0.4883	0.3241	2.5	Rb^+, Cs^+, NH_4^+, Ag^+,
5	0.4921	0.3249	3.0	K^+, Cl^-, Br^-, I^-, NO_3
10	0.4960	0.3258	3.5	OH^-, F^-, HS^-, BrO_3^-
15	0.5000	0.3262	4.0–4.5	Na^+, HCO_3^-, $H_2PO_4^-$,
20	0.5042	0.3273		HSO_3^-, Hg_2^{2+}, SO_4^2,
25	0.5085	0.3281		HPO_4^{2-}, PO_4^{3-}
30	0.5130	0.3290	4.5	Pb^{2+}, CO_3^{2-}, MoO_4^{2-},
35	0.5175	0.3297	5.0	Sr^{2+}, Ba^{2+}, Ra^{2+}, Cd^{2+},
40	0.5221	0.3305		Hg^{2+}, S^{2-}, WO_4^{2-}
50	0.5319	0.3321	6	Li^+, Ca^{2+}, Cu^{2+}, Zn^{2+},
60	0.5425	0.3338		Sn^{2+}, Mn^{2+}, Fe^{2+}, Ni^{2+}, Co^{2+}
			8	Mg^{2+}, Be^{2+}
			9	H^+, Al^{3+}, Cr^{3+}, trivalent rare earths
			11	Th^{4+}, Zr^{4+}, Ce^{4+}, Sn^{4+}

Reprinted with permission from Manov, G.G. and others, J. Am. Chem. Soc., v. 65, 1943, p. 1765–1767. Copyright © 1943 American Chemical Society.

From Klotz, 1950 with permission.

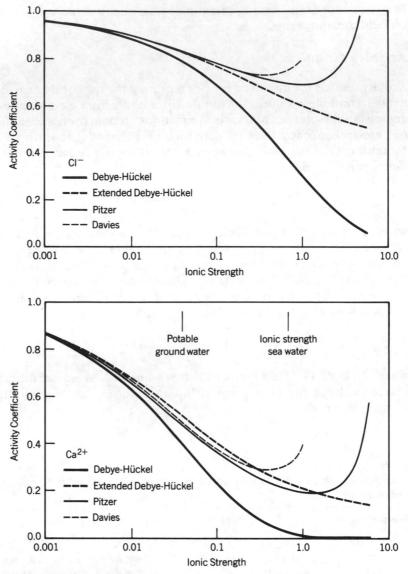

Figure 11.4
Comparison of activities of Cl^- and Ca^{2+} as a function of ionic strength, calculated using the Debye-Hückel extended Debye-Hückel, Davies, and Pitzer equations (unpublished data from A. S. Crowe).

$$I = 0.5 \ \Sigma \ M_i z_i^2 \tag{11.8}$$

with (M_i) as the molar concentration of species i having a charge z. Ionic strength is a measure of the total concentration of ions that emphasizes the increased contribution of species with charges greater than one to solution nonideality.

Use of the Debye-Hückel equation is limited to solutions with ionic strengths less than 0.005 M, which is fresh, potable ground water. In general, this equation is inappropriate for use with "typical" ground waters.

The extended Debye-Hückel equation can be used to estimate values of γ_i to a maximum ionic strength of about 0.1 M or a total dissolved solids content of approximately 5000 mg/L (Langmuir and Mahoney, 1984). This equation is

$$\log \gamma_i = \frac{-Az_i^2(I)^{0.5}}{1 + Ba_i(I)^{0.5}}$$ (11.9)

where A and B are temperature-dependent constants and a_i is the radius of the hydrated ion in centimeters (Table 11.1).

Geochemical studies of basinal brines require other activity models to deal with the higher-salinity water. The simplest model takes an equation like Eq. 11.9 and adds additional correction or curve-fitting parameters. The Davies equation is an example of this approach. Establishing activities in even more saline water requires the use of ion interaction models of a type proposed by Pitzer and Kim (1974). This sophisticated activity model is more difficult to use than the Debye-Hückel or Davies-type equations, but provides accurate estimates of activity up to 20 M (Langmuir and Mahoney, 1984).

On an activity versus ionic strength plot for Cl^- and Ca^{2+}, note the typical curve shapes for monovalent and divalent ions (Figure 11.4). We have shown activities calculated using the Pitzer model, but they are not strictly comparable to the other three. At low ionic strengths, all four methods give approximately the same activity coefficients. However, as the ionic strength increases (Figure 11.4), estimates of activity from the Debye-Hückel, extended Debye-Hückel, and Davies equations differ noticeably.

Because the extended Debye-Hückel equation or the Davies equation is applicable to charged ionic species, they cannot determine γ's for neutral species existing as ion pairs or other complexes. For solutions with ionic strengths less than about 0.1 M, we assume activity coefficients have a value of one (Morel, 1983). In solutions with higher ionic strengths, it is difficult to estimate activity coefficients. Possible approaches include making activities a proportion of the ionic strength or equivalent to the activity of CO_2.

For most thermodynamic calculations, activity coefficients are calculated using the extended Debye-Hückel equation (Eq. 11.9) or the Davies equation. The following example illustrates the use of the extended Debye-Hückel equation.

Example 11.2

Water contains ions in the following molar concentrations Ca^{2+} 3.25×10^{-3}, Na^+ 0.96×10^{-3}, HCO_3^- 5.75×10^{-3}, and SO_4^{2-} 0.89×10^{-3}. Calculate activity coefficients for Na^+, and Ca^{2+} at 25°C using Eq. 11.9.

$$I = 0.5 \times [(3.25 \times 2^2) + (0.96 \times 1^2) + (5.75 \times 1^2) + (0.89 \times 2^2)] \times 10^{-3}$$
$$= 0.012 \text{ M}$$

This value of I is at the upper end of the range of applicability of the extended Debye-Hückel equation.

$$\log \gamma_{Na^+} = \frac{-0.508 \times 1^2(0.012)^{0.5}}{1 + 0.328 \times 10^8 \times 4.0 \times 10^{-8}(.012)^{0.5}}$$
$$= -0.049$$
$$\gamma_{Na^+} = 10^{-0.049}$$
$$= 0.89$$

$$\log \gamma_{Ca^{2+}} = \frac{-0.508 \times 2^2 (0.012)^{0.5}}{1 + 0.328 \times 10^8 \times 6.0 \times 10^{-8} (.012)^{0.5}}$$

$$= -0.183$$

$$\gamma_{Ca^{2+}} = 10^{-0.183}$$

$$= 0.66$$

Nonidealities relate to both gas and solids. However, for most problems, gas phase nonidealities can be neglected. Nonideal behavior in minerals is due to elemental substitution in a crystal lattice. Some applications may require treatment of nonidealities. However, all the problems in this book assume that pure solids have an activity of one. Readers interested in a discussion of solid activities should refer to Garrels and Christ (1965) and Stumm and Morgan (1981).

11.5   Deviations from Equilibrium

Viewing ground water as a partial equilibrium system implies that some reactions may not be at equilibrium. Examples are the heterogeneous reactions involving the dissolution or precipitation of minerals. The departure of a reaction from equilibrium is determined as the ratio of the ion activity product to the equilibrium constant (*IAP/K*). We calculate the ion activity product by substituting sample activity values in the mass law expression for a reactions. For example, given a ground water with known activities of [*C*], [*D*], [*Y*], and [*Z*], the ion activity product (*IAP*) for the reaction in Eq. 11.4 is

$$IAP = \frac{[Y]^y [Z]^z}{[C]^c [D]^d} \tag{11.10}$$

If the *IAP* > *K*, the reaction is progressing from right to left reducing [*Y*] and [*Z*] and increasing [*C*] and [*D*]. If *IAP* < *K*, the reaction is proceeding from left to right. The *IAP* at equilibrium is equal to the equilibrium constant.

This approach determines the saturation state of a ground water with respect to one or more mineral phases. When *IAP/K* < 1, the ground water is undersaturated with respect to the given mineral. When *IAP/K* = 1, the ground water is in equilibrium with the mineral, and when *IAP/K* > 1 the ground water is supersaturated. Undersaturation with respect to a mineral results in the net dissolution of the mineral provided it is present. Supersaturation results in the net precipitation of the mineral should suitable nuclei be present.

A different way of expressing the saturation state of a ground water is in terms of a saturation index (SI), defined as log(*IAP/K*). When a mineral is in equilibrium with respect to a solution, the SI is zero. Undersaturation is indicated by a negative SI and supersaturation by a positive SI.

Example 11.3 illustrates how to calculate the state of saturation.

Example 11.3

Given a ground water with the molar composition shown here, calculate the saturation state with respect to calcite and dolomite. The activity coefficients for Ca^{2+},

Mg^{2+}, and CO_3^{2-} are 0.57, 0.59, and 0.56, respectively. The equilibrium constants defining the solubility of calcite and dolomite are 4.9×10^{-9} and 2.7×10^{-17}, respectively.

$(Ca^{2+}) = 3.74 \times 10^{-4}$: $(Mg^{2+}) = 4.11 \times 10^{-6}$: $(Na^+) = 2.02 \times 10^{-2}$:
$(K^+) = 6.14 \times 10^{-5}$: $(H^+) = 10^{-7.9}$
$(HCO_3^-) = 1.83 \times 10^{-2}$: $(CO_3^{2-}) = 5.50 \times 10^{-5}$:
$(SO_4^{2-}) = 1.24 \times 10^{-3}$: $(Cl^-) = 1.19 \times 10^{-3}$

$$IAP_{cal} = [Ca^{2+}][CO_3^{2-}]$$
$$= 0.57 \times 3.74 \times 10^{-4} \times 0.56 \times 5.50 \times 10^{-5}$$
$$= 6.56 \times 10^{-9}$$

$$\{IAP/K\}_{cal} = \frac{6.56 \times 10^{-9}}{4.90 \times 10^{-9}} = 1.34$$

The sample is slightly oversaturated with respect to calcite.

$$IAP_{dol} = [Ca^{2+}][Mg^{2+}][CO_3^{2-}]^2 = 4.89 \times 10^{-19}$$

$$\{IAP/K\}_{dol} = \frac{4.89 \times 10^{-19}}{2.7 \times 10^{-17}} = 0.018$$

The sample is strongly undersaturated with respect to dolomite.

Calculations are easily carried out using one of several different computer codes such as EQ3 (Wolery, 1979) or WATEQF (Plummer and others, 1976). The codes correct equilibrium constants for temperature and account for the reduced activities of free ions due to the formation of complexes. We will discuss the use of the computer codes in more detail in Section 15.3 and the basic concepts of complex formation in Section 12.3.

11.6 Kinetic Reactions

Chemical kinetics provide a useful framework for studying reactions in relation to time and pathways. A kinetic description is applicable to any reaction, but is necessary for irreversible reactions or reversible reactions that are slow in relation to physical transport. Consider this reaction

$$aA + bB + \cdots \text{(other reactants)} \underset{k_1}{\overset{k_2}{=}} rR + sS + \cdots \text{(other products)} \tag{11.11}$$

where k_1 and k_2 are the rate constants for the forward and reverse reactions, respectively. Each constituent in Eq. 11.11 has a reaction rate (for example, r_A) that describes the rate of change of concentration as a function of time $\{d(A)/dt\}$. Because of the stoichiometry, these rate expressions are related

$$-\frac{r_A}{a} = -\frac{r_B}{b} = \cdots = \frac{r_R}{r} = \frac{r_S}{s} = \cdots \tag{11.12}$$

where a, b, r, and s are the stoichiometric coefficients for the reaction.

Reaction rate laws describe the time rate of change of concentration as a function of rate constants and concentration. The rate law for a component (say, A) in Eq. 11.11 has this form

$$
\begin{aligned}
r_A = &-k_1(A)^{n1}(B)^{n2} \cdots \text{(other reactants)} \\
&+ k_2(R)^{m1}(S)^{m2} \cdots \text{(other products)}
\end{aligned}
\tag{11.13}
$$

where $n1$, $n2$, ... and $m1$, $m2$, ... are empirical or stoichiometric coefficients. This equation expresses the rate of change in (A) as the difference between the rate at which the component is being used in the forward reaction and generated in the reverse reaction.

Although reactions are written as a single, overall expression, they usually occur as a series of elementary reactions. In other words, a reaction like Eq. 11.11 represents what is really a set of reactions that involves intermediate species not included in the overall reaction. Elementary reactions are defined as reactions that take place in a single step. Examples of these reactions from Langmuir and Mahoney (1984) include

$$H^+ + OH^- = H_2O$$
$$CO_2(aq) + OH^- = HCO_3^-$$
$$H_4SiO_4 = SiO_2(quartz) + 2H_2O$$

For an elementary reaction, the coefficients in Eq. 11.13 are the stoichiometric coefficients of the reaction. Thus, rate laws for elementary reactions are easy to establish. Unfortunately, most reactions are not of this type and the rate laws contain empirical coefficients. Their values are established through laboratory experiments.

Many of the rate laws involve simple forms of (Eq. 11.13). For example, (1) the irreversible decay of ^{14}C

$$^{14}C \rightarrow {}^{14}N + e \tag{11.14}$$

has this rate law

$$\frac{d(^{14}C)}{dt} = -k_1(^{14}C) \tag{11.15}$$

and (2) the oxidation of ferrous iron between pH 2.2 and 3.5

$$Fe^{2+} + \tfrac{1}{4}O_2 + \tfrac{1}{2}H_2O \rightarrow FeOH^{2+} \tag{11.16}$$

has this rate law in Fe^{2+} (Langmuir and Mahoney, 1984)

$$\frac{d(Fe^{2+})}{dt} = -k_1(Fe^{2+})(P_{O_2}) \tag{11.17}$$

Both Eqs. 11.14 and 11.16 are examples of irreversible reactions, which progress until the reactants are consumed by the reaction. For this reason, the rate law expressions for ^{14}C and Fe^{2+} in Eqs. 11.15 and 11.17 describe the behavior of these components as a consequence of the total reaction.

Kinetic reactions are described according to the order of the reaction. For the reaction,

$$2A + B \rightarrow C \tag{11.18}$$

the rate law in terms of (A) is

$$\frac{d(A)}{dt} = -k_1(A)^{n1}(B)^{n2} \tag{11.19}$$

The coefficients $n1$ and $n2$ define the order of the reaction in A and B. The order of the overall reaction is the sum of all the n's. Thus, Eq. 11.19 is of order $n1 + n2$. Applying this scheme to the previous examples, Eq. 11.14 is a first-order reaction, and Eq. 11.16 is second order.

As Eq. 11.13 illustrates, the rate law for an equilibrium reaction depends not only on how fast it is being consumed in the forward reaction but how fast it is being created in the reverse reaction. For this reaction (Langmuir and Mahoney, 1984)

$$Fe^{3+} + SO_4^{2-} \overset{k_2}{\underset{k_1}{=}} FeSO_4^+ \tag{11.20}$$

rate laws for Fe^{3+} and SO_4^{2-} in the forward direction are of second order and written

$$\frac{d(Fe^{3+})}{dt} = \frac{d(SO_4^{2-})}{dt} = -k_1(Fe^{3+})(SO_4^{2-}) \tag{11.21}$$

The reverse reaction has the following rate law

$$\frac{d(FeSO_4^+)}{dt} = k_2(FeSO_4^+) \tag{11.22}$$

The change in (Fe^{3+}) with time for the overall reaction is

$$\frac{d(Fe^{3+})}{dt} = -k_1(Fe^{3+})(SO_4^{2-}) + k_2(FeSO_4^+) \tag{11.23}$$

Rate laws are even more complex for reactions in parallel or series that cannot be written as a single overall reaction.

Solutions to rate equations like Eq. 11.23 describe how the concentrations of various components change from some initial condition. This exercise is useful mainly for characterizing reactions in laboratory experiments, but not in modeling chemical changes in ground water systems. In practice, kinetic equations like Eq. 11.23 must be integrated into a complete equation for mass transport, which includes all of the other processes.

11.7   *Organic Compounds*

The classical approach to classifying organic compounds involves defining what are known as functional classes. These classes are distinguished by particular functional groups and the unique structural relationships between the functional group and other atoms in the compound. A functional group (e.g., hydroxyl group, OH; carbonyl group, C=O) is a simple combination of two or more of the following atoms: C, H, O, S, N, and P that gives the chemical compound unique chemical and physical properties.

The large number of different organic compounds, the complexity of their structure, and the specific requirements of particular studies, have prompted the use of a variety of classification schemes. For example, Thurman (1985) proposes a scheme based on 12 functional classes, while a more comprehensive scheme by Garrison and others (1977) involves 24 major classes and more than 100 subclasses. Our scheme (Figure 11.5) is a condensed and slightly reordered version of the Garrison and others (1977) classification. It consists of 16 major classes with all subdivisions except the first defined on the basis of functional classes.

While this scheme is not all encompassing, it accommodates most important natural or man-made organic compounds found in ground water. The actual steps involved in classifying a compound are complicated because most can be included in more than one class. Garrison and others (1977) for this reason have proposed a simple set of rules. A hierarchical search of classes begins with Class 1 and proceeds through Class 16. For example any hydrocarbon, defined as a compound containing only carbon and hydrogen, containing a halogen atom (chlorine, fluorine, bromine, or iodine) is placed in Class 2 except the few that fall in Class 1. Any compound containing phosphorous is placed in Class 4 and so on. Readers can refer to Garrison and others (1977) for a more complete discussion of rules for classification and a summary of subclasses.

The brief description of the various major classes includes a sketch of the various functional classes where appropriate, a brief discussion of typical compounds in each class and their origin, and a sketch illustrating how the functional class is represented in the structure of various organic compounds (Figure 11.5).

1. Miscellaneous nonvolatile compounds are a diverse class of compounds that are difficult to classify in terms of functional classes. Naturally occurring compounds include fulvic acid, humic acid, chlorophyll, xanthophylls, and enzymes. Industrially produced compounds include tannic acid, most dyes, and optical brighteners. Fulvic and humic acids are major components of a complex set of compounds called aquatic humic substances. According to Thurman (1985) 40% to 60% of the dissolved organic carbon occurring naturally in ground water is comprised of aquatic humic substances. The natural plant pigments—chlorophylls and xanthophylls—generally have a low solubility and are minor constituents in ground water as are enzymes. Tannic acids, dyes, and optical brighteners occur in ground water mainly as the result of contamination.

2. Halogenated hydrocarbons are one of the largest and most important groups of contaminants found in ground water. This class is divided into the aliphatic and aromatic subclasses and is characterized by the presence of one or more halogen atoms (Figure 11.5). The term aliphatic describes a structure where carbon atoms are joined in open chains. An

1. Miscellaneous Nonvolatile Compounds

2. Halogenated Hydrocarbons

Aliphatic

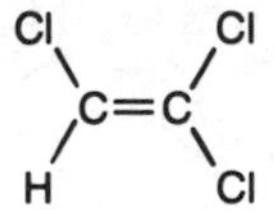

Trichloroethylene

Aromatic

Chlorobenzene

3. Amino Acids

Basic structure

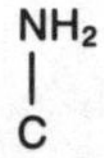

Aspartic acid

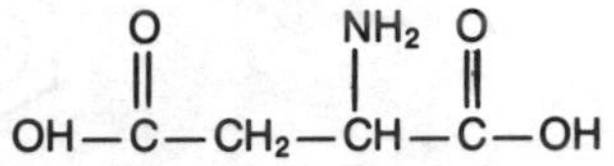

4. Phosphorous Compounds

Basic structure

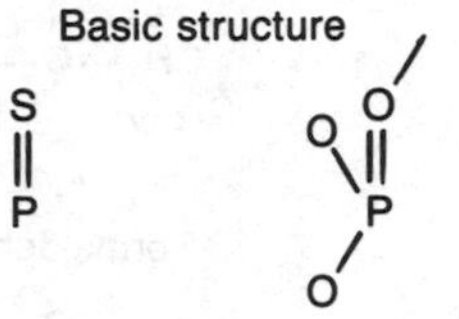

Malathion

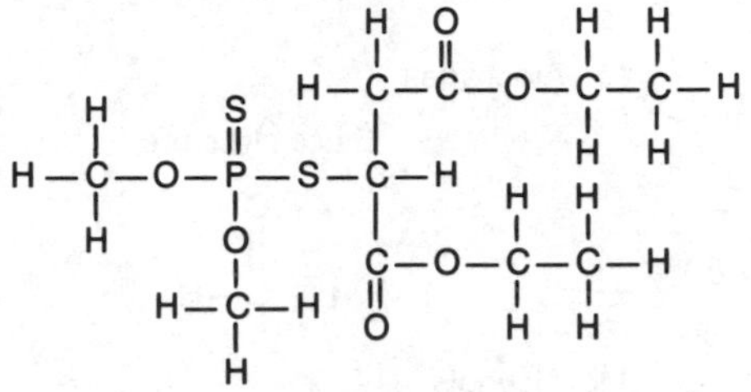

5. Organometallic Compounds

Tetraethyllead

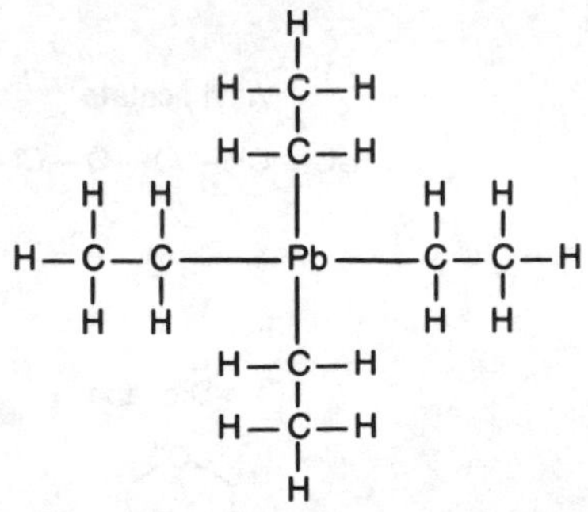

6. Carboxylic Acid

Basic structure

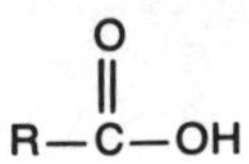

Acetic acid

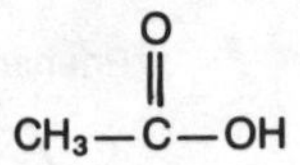

Figure 11.5
Classification of organic compounds (*continued on next page*).

7. Phenols

           Basic structure   Cresol

8. Amines

           Basic structure   Dimethylamine

    Aliphatic   Aromatic

    N— NH_2 CH_3

    |

   CH_3 $CH_3—N—H$

9. Ketones

           Basic structure   Acetone

              O   O

              ‖   ‖

          R—C—R′ $CH_3—C—CH_3$

10. Aldehydes

           Basic structure   Formaldehyde

              O   O

              ‖   ‖

           R—C—H   H—C—H

11. Alcohols

           Basic structure   Methanol

             R—OH $CH_3—OH$

12. Esters

           Basic structure   Vinyl acetate

              O $H_2C=CH—O—C—CH_3$

              ‖

          R—C—OR′   O

13. Ethers

           Basic structure   1,4-Dioxane

             C—O—C

14. Polynuclear Aromatic Hydrocarbons

           Phenanthrene

Figure 11.5
(*Continued.*)

15. Aromatic Hydrocarbons

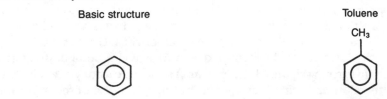

Basic structure

Toluene

16. Alkane, Alkene, and Alkyne Hydrocarbons

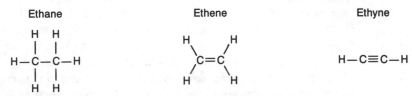

Ethane

Ethene

Ethyne

R = aliphatic backbone

Figure 11.5
(Continued.)

aromatic hydrocarbon has a molecular structure based on that of benzene, C_6H_6, characterized by six carbon atoms arranged in a ring and radially bonded hydrogen atoms. Three of the carbons are doubly bonded and three are singly bonded. Because the bonds change continually, all the carbon-carbon bonds are assumed to be intermediate between a single and double bond.

Included in the aliphatic subclass are solvents such as methylene chloride, chloroform, carbon tetrachloride, 1,1,1-trichlorethane, trichloroethylene; pesticides such as aldrin and dieldrin; and industrial chemicals such as vinyl chloride, methyl chloride, and methyl iodide. The aromatic subclass contains common pesticides such as DDD and DDE and industrial chemicals such as 1,2,4-dichlorobenzene and the various PCBs.

3. Amino acids are characterized structurally by an amino group, NH_2 on the carbon atom adjacent to the carbonyl carbon, COOH (Figure 11.5). These compounds, in general, represent a relatively small proportion of the naturally occurring organic matter in water. However, this group is important because of the relatively large numbers of studies involving these compounds. According to Thurman (1985), the dominant amino acids in ground water (for example, glycine, serine, alanine, and aspartic acid) are present at concentrations of a few hundred micrograms per liter.

4. Phosphorous compounds are distinguished by the presence of phosphorous in phosphorous-sulfur, phosphate, or phosphite bonds (Figure 11.5). The most important group of compounds within this class is a large group of pesticides. The unhalogenated compounds include diazinon, malathion, parathion, and tributyl phosphate and halogenated compounds include chlorothion, dicapthon, and dursban.

5. Organometallic compounds are a highly toxic group of organic contaminants. They have a broadly diverse structure, distinguished by the presence of a metal atom. Examples of compounds in this class are tetraethyllead, diethylmercury, and copper phthalocyanine.

6. Carboxylic acids are an important class of naturally occurring compounds and contaminants. These organic acids are part of a family of functional classes that includes ketones, aldehydes, and esters, which are characterized by the presence of a carbonyl COO functional group. Carboxylic acids are distinguished by a hydroxyl group bonded to the carbonyl group (Figure 11.5). In natural systems, carboxylic acids ionize and therefore are quite soluble, contributing 5% to 8% of the organic carbon in surface water (Thurman, 1985). Common compounds in this class include acetic acid, benzoic acid, butyric acid, formic acid, palmitic acid, propionic acid, stearic acid, and valeric acid. Carothers and Kharaka (1978) report concentrations of carboxylic acid in excess of 5000 mg/L in formation waters. Carboxylic acids are also important components of landfill leachates due to anaerobic fermentation reactions.

   Many of these acids are produced for industrial uses. The most important products are phenoxy acetic pesticides such as 2,4-D, silvex, and 2,4,5-T. These compounds are also widely used in the food industry.

7. Phenols are a small but important class of naturally occurring and manufactured compounds. Structurally, the phenols are characterized by an aromatic ring with an attached hydroxyl group. In uncontaminated ground water, Thurman (1985) cites natural abundances of 1 μg/L or less with compounds such as phenol, cresol, and syringic, vanillic, and p-hydroxybenzoic acids. While phenols are commonly associated with petroliferous rocks, they originate in ground water mostly as contaminants from industrial wastes or biocides. Some of the wastes are cresol, phenol, pyrocatechol, and napthol, while pesticides are dinitrocresol, 2,4-dinitrophenol. Chlorinated phenolic pesticides, included in Class 2, are important contaminants near wood-preserving facilities.

8. Amines considered here are mainly a group of industrial contaminants. In this system of classification, some of the naturally occurring amino acids contain the amine functional class. The aliphatic functional class can be considered as derivatives of ammonia (NH_3) with one, two, or three of the hydrogens replaced by alkyl groups (CH_3) (Figure 11.5). In the aromatic amines, a hydrogen is replaced by an aromatic ring (Figure 11.5). Sources of potential contamination in ground water are wastes from herbicide and synthetic rubber production. Some common contaminants are diethylamine, dimethylamine, ethylamine, methylamine, aniline, benzidine, napthylamine, and pyridine.

9. Ketones are also distinguished by the presence of the carbonyl group (Figure 11.5). In the ketones, the carbonyl group is bonded to two alkyl groups that may be alike or different. These compounds occur naturally but are mainly contaminants in industrial effluent and municipal sewage. Acetone, fluorenone, fenchone, 2-butane, butyl propyl ketone, and methyl propenyl ketone are examples of these compounds.

10. Aldehydes have a structure similar to the ketones except that the carbonyl group is bonded to one alkyl group and one or two hydrogens (Figure 11.5). They occur naturally at relatively low concentrations and are occasionally contaminants in ground water. Formaldehyde and isobutyraldehyde are examples of these compounds.

11. Alcohols are some of the most soluble organic compounds in water. They can be considered as derivatives of a hydrocarbon in which a hydroxyl group replaces a hydrogen (Figure 11.5). In natural systems, alcohols are present only in trace quantities because they are nontoxic and are broken down easily in the food chain. They occasionally are found as contaminants in ground water. A few of the alcohols are methanol, glycerol, terpinol, ethyleneglycol, and butanol.

12. Esters like the carboxylic acids, ketones, and aldehydes contain the carbonyl functional group and have the basic structure shown in Figure 11.5. When esters are found in ground water, it is almost always the result of contamination. Some commonly occurring compounds include pesticides such as butyl mesityl oxide, dimethrin, dimethy carbate, dinobuton, omite, tabutrex, vinyl acetate, and warfarine plus a variety of industrial chemicals.

13. Ethers occur in water both naturally and as contaminants. Structurally, the ether functional group consists of an oxygen bonded between two carbons. Shown in Figure 11.5 are two examples, diethyl ether and diphenyl ether. Examples of ethers found as contaminants are tetrahydrofuran and 1,4-dioxane.

14. Polynuclear aromatic hydrocarbons (PAH) form from a series of benzene rings (Figure 11.5). These compounds are important components of ancient sediments and crude oils. They occur in potable water in trace concentrations and probably originate from coal tar or creosote contamination when detected in ground water. Thurman (1985) lists phenanthrene, fluoranthene, anthracene, and benzopyrene as examples of PAHs found in rivers and rain water.

15. Aromatic hydrocarbons are compounds with a molecular structure based on that of benzene, C_6H_6 (Figure 11.5). These hydrocarbons are a major constituent of petroleum and in ground water are usually indicative of contamination from a spill of crude oil or a petroleum distillate. Some typical benzoid hydrocarbons are benzene, ethylbenzene, toluene, and o-xylene.

16. Alkane, alkene, and alkyne hydrocarbons are important components of oil and natural gas. Alkanes are characterized by the CH_3 functional class and conform to the general formula C_nH_{2n+2} (where n is the number of carbon atoms). Common alkanes are methane, ethane, propane, and butane and the various structural isomers (same chemical formula but different structure). The cycloalkanes (C_nH_{2n}) are characterized by a ring structure that contains single carbon-carbon bonds. Cyclopropane and cyclohexane are examples.

   The alkenes are aliphatic hydrocarbons characterized by a carbon-carbon double bond (Figure 11.5). Alkenes have a general formula C_nH_{2n} with members of the series including ethene and propene, named similarly to the alkane series.

   The alkynes are characterized by a carbon triple bond and the general formula C_nH_{2n-2} (Figure 11.5). Again the alkyne series is named in a manner similar to the alkanes giving, for example, ethyne, propyne, and butyne, and so on. When this class of hydrocarbons is found in potable ground water, it is usually the result of ground water contamination from crude oil or refined petroleum products.

11.8   Ground Water Composition

Eventually some of the inorganic and organic solids, organic liquids, and gases found in the subsurface dissolve in ground water. Thus, the variety of solutes in ground water (Table 11.2) is not surprising. The dissolved inorganic constituents are classified as major constituents with concentrations greater than 5 mg/L (Table 11.2), minor constituents with concentrations in a range from 0.01 to 10 mg/L, and trace elements with concentrations less than 0.01 mg/L (Davis and DeWiest, 1966). Naturally occurring, dissolved organic molecules in ground water could number in the hundreds of individual compounds. They are typically present in minor or trace quantities. By far the most abundant organic compounds in shallow ground water are the humic and fulvic acids. However, other organic compounds are also present (Table 11.2). Deep ground water or formation water can contain concentrations of organic matter (for example, 2000 mg/L, Hull and others, 1984). Examples of organic acids are acetate and propionate, anions that contribute to titratable alkalinity. The important ground water gases include oxygen, carbon dioxide, hydrogen sulfide, and methane. What determines the maximum concentration for all constituents is their solubility in water.

The Routine Water Analysis

It is not feasible or really necessary to measure the concentration of all constituents that conceivably might occur in water. Measuring the concentration of a standard set of constituents produces what is known as the routine analysis for water quality. This standard test forms the basis for assessing the suitability of water for human consumption or various industrial and agricultural uses. The routine analysis typically includes the major constituents with the exception of silicon and carbonic acid, and the minor constituents with the exception of boron and strontium. Laboratory results are reported as ion concentrations in mg/kg or mg/L. The reported concentrations for metals (e.g., Ca, Mg, Na) are the total concentration of metals irrespective of whether they are complexed or not.

A routine analysis often includes a few other items in addition to concentrations (Table 11.3). Of note are pH, total dissolved solids (TDS) reported in mg/L, and specific conductance reported in microsiemens per centimeter. The TDS content is the total quantity of solids when a water sample is evaporated to dryness. Specific conductance is a measure of the ability of the sample to conduct electricity and provides another approximate measure of the total quantity of ions in solution. The measurement is approximate because the specific conductance of a fluid with a given TDS content varies depending upon which ions are present.

The routine analysis identifies nearly all the mass dissolved in a sample. Unanalyzed ions and organic compounds usually represent a negligible proportion. One simple check on the quality of a routine analysis is to compare the total concentrations of cations and anions in milliequivalents per liter. Given that water is electrically neutral, the ratio of the sums should be one. Table 11.4 illustrates this calculation for the previous example. While not exactly one, the value is

Table 11.2

The dissolved constituents in groundwater classified according to relative abundance*

Major constituents (greater than 5 mg/L)

Bicarbonate	Silicon
Calcium	Sodium
Chloride	Sulfate
Magnesium	Carbonic acid

Minor constituents (0.01–10.0 mg/L)

Boron	Nitrate
Carbonate	Potassium
Fluoride	Strontium
Iron	

Trace constituents (less than 0.1 mg/L)

Aluminum	Molybdenum
Antimony	Nickel
Arsenic	Niobium
Barium	Phosphate
Beryllium	Platinum
Bismuth	Radium
Bromide	Rubidium
Cadmium	Ruthenium
Cerium	Scandium
Cesium	Selenium
Chromium	Silver
Cobalt	Thallium
Copper	Thorium
Gallium	Tin
Germanium	Titanium
Gold	Tungsten
Indium	Uranium
Iodide	Vanadium
Lanthanum	Ytterbium
Lead	Yttrium
Lithium	Zinc
Manganese	Zirconium

Organic compounds (shallow)

Humic acid	Tannins
Fulvic acid	Lignins
Carbohydrates	Hydrocarbons
Amino acids	

Organic compounds (deep)

Acetate
Propionate

*Modified from Davis, S.N., and R.J.M. DeWiest, 1966, Hydrogeology. Copyright © 1966 by John Wiley and Sons, Inc. Reprinted by permission of John Wiley & Sons, Inc.

Table 11.3
Example of a routine water analysis

Parameter	mg/L	Parameter	mg/L
pH	7.7	Conductivity	2300
Calcium	1[1]	Magnesium	1
Sodium	550	Potassium	3.5
Iron	8.7	NO_2, NO_3[2]	0.1[1]
Nitrite	0.1[1]	Chloride	45
Sulfate	59	Fluoride	0.25
Bicarbonate	1315	Hardness, T	8
Alkalinity, T	1078		
TDS[3]	1321		

Balance 1.01

[1]Indicates concentration "less than." Conductivity reported in
μS, pH in pH units. All metal parameters expressed as totals.
Alkalinity and hardness expressed as calcium carbonate. Nitrate,
nitrite, and ammonia expressed as N.
[2]NO_2 = nitrite, NO_3 = nitrate.
[3]Total dissolved solids.

within the ± 0.05 range of acceptability used by most laboratories. The cation/
anion ratio is often reported by the laboratory (Table 11.4). This number provides
one check that concentration determinations are not grossly in error.

Large numbers of routine analyses collected over time provide basic data for
research studies in many areas. These data must be used with care for sometimes
they may be in error because of a failure to measure rapidly changing parameters
in the field, to preserve the samples against deterioration due to long storage, and
to assure the quality of the laboratory determinations. These issues are discussed
in more detail in Section 16.4.

Specialized Analyses

Laboratories can carry out a routine analysis in a rapid and cost-effective manner.
Occasionally, there is a need for more specialized analyses for example, trace
metals, radioisotopes, organic compounds, various nitrogen-containing species,

Table 11.4
**Evaluating the electroneutrality of the example routine
analysis**

	Cation concentration			Anion concentration		
	mg/L	meq/L			mg/L	meq/L
Ca^{2+}	1.0	0.05	HCO_3^-		1315	21.6
Mg^{2+}	1.0	0.08	SO_4^{2-}		59	1.22
Na^+	550	23.9	Cl^-		45	1.27
K^+	3.5	0.09	F^-		0.25	0.01
Fe	8.7	0.31		Total		24.1
	Total	24.4	cation/anion ratio = 1.01			

environmental isotopes, or gases. Such work is often related to problems of ground water contamination or research and regulatory needs. Commercial laboratories should be able to analyze for all but the most exotic constituents. However, given the cost of this more specialized work, serious attention has to be paid to quality assurance issues.

11.9 Describing Chemical Data

Describing the concentration or relative abundance of major and minor constituents and the pattern of variability is part of almost every ground water investigation. Over time, different graphical and statistical techniques have been developed to assist with this task. Each technique has particular advantages and disadvantages in representing features of the data. Thus in working with a set of analytical results, one might examine alternative ways of presenting results to select the most appropriate.

The methods are divided into two major groups. First is a group of graphical approaches for describing abundance or relative abundance. Second is a group that concentrates on presenting patterns of variability in addition to abundance. This second group subdivides further into graphical/illustrative and statistical approaches.

Abundance or Relative Abundance

Several different graphical approaches can depict the abundance or relative abundance of ions in individual water samples. The most common are: (1) the Collins (1923) bar diagram, (2) the Stiff (1951) pattern diagram, (3) the pie diagram, and (4) the Piper (1944) diagram. As the examples (Table 11.5) will illustrate, all these approaches require concentrations in units of meq/L or %meq/L. The Collins and Stiff diagram require absolute concentrations (meq/L), while the pie and Piper diagrams require relative concentrations (%meq/L). Figures 11.6a, b, c, and d illustrate how the sample data look plotted in these different ways. The Collins, Stiff, and pie diagrams are relatively simple to construct. They require only that concentrations be plotted as a bar segment, a point on a line, or a percentage of the pie. The appropriate fields are shaded and possibly labeled in the case of the Collins and Piper diagram (Figure 11.6). The Stiff diagrams may or may not include the labeled system of axes.

Plotting data for individual samples on the Piper diagram is more complicated because there are three individual diagrams (Figure 11.6d). The relative abundance of cations with the %meq/L of $Na^+ + K^+$, Ca^{2+}, and Mg^{2+} assumed to equal 100% is first plotted on the cation triangle. Similarly the anion triangle displays the relative abundance of Cl^-, SO_4^{2-}, and $HCO_3^- + CO_3^{2-}$. Straight lines projected from the two triangles into the quadrilateral field define the position of the point on the third field (Figure 11.6d). To provide some indication of the absolute quantity of dissolved mass in the sample, the size of the data point is sometimes related to the salinity (TDS). One advantage of all four techniques is that they present much of the major and minor ion data for a single sample on one figure. However, with exception of the Piper diagram, these approaches are most useful in displaying the results for a few analyses, which are often "type" waters from an area.

Table 11.5
Sample chemical data used to demonstrate various techniques for plotting chemical data (from Zapovozec, 1972)*

Chemical analyses	Sample 1: Tertiary, Czechoslovakia			Sample 2: Upper Cretaceous, Czechoslovakia			Sample 3: Upper Cambrian, Wisconsin		
	mg/L	meq/L	meq (%)	mg/L	meq/L	meq (%)	mg/L	meq/L	meq (%)
Cations									
$Na^+ + K^+$	266.2	10.68	68.24	1913.7	81.54	70.24	7.9	0.34	4.4
Mg^{2+}	21.9	1.80	11.4	132.8	10.95	9.41	43.0	3.54	45.6
Ca^{2+}	61.7	3.08	18.5	468.5	23.38	20.50	78.0	3.89	50.1
Mn^{2+}	Traces	–	–	0.67	0.02	0.01	0.14	0.04	0.5
Fe	2.3	0.08	–	0.15	0.005	0.004	0.11	–	–
Sum	–	15.64	100.0	–	115.89	100.0	–	7.77	100.0
Anions									
Cl^-	11.3	0.32	2.05	850.0	23.98	20.63	17.0	0.48	6.4
NO_3^-	0.0	–	–	0.0	–	–	0.7	0.0	0.1
HCO_3^-	906.1	14.85	95.1	2568.5	42.10	36.23	364.0	5.96	79.5
SO_4^{2-}	21.2	0.44	2.82	2406.5	50.10	43.11	50.0	1.04	13.9
Sum	–	15.61	100.0	–	116.18	100.0	–	7.50	100.0

*Reprinted by permission of Ground Water. Copyright © 1972. All rights reserved.

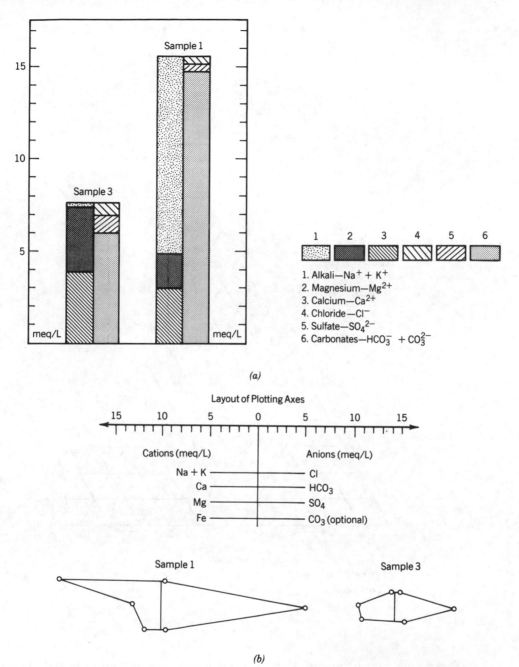

Figure 11.6
Four different ways of plotting major ion data (modified from Zaporozec, 1972). Reprinted by permission of Ground Water. Copyright © 1972. All rights reserved. (*Continued on next page.*)

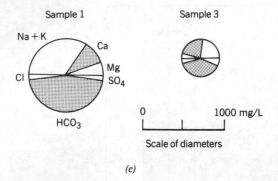

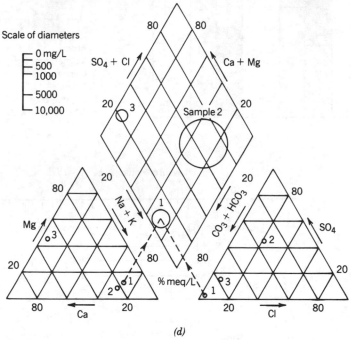

Figure 11.6
(*Continued.*)

Presenting a large number of these diagrams together is confusing and not much more helpful than a table of concentrations.

Abundance and Patterns of Change

Graphical/illustrative-type diagrams or statistics can define the pattern of spatial change among different geologic units, along a line of section, or along a pathline. One simple way of representing spatial change in a single geological unit is to take the single-sample diagrams (for example, pie or Stiff) and place them on a map. Such maps can convey a sense of how the pattern of ion abundances change within a unit (for example, Swenson, 1968). Including all the constituents on a

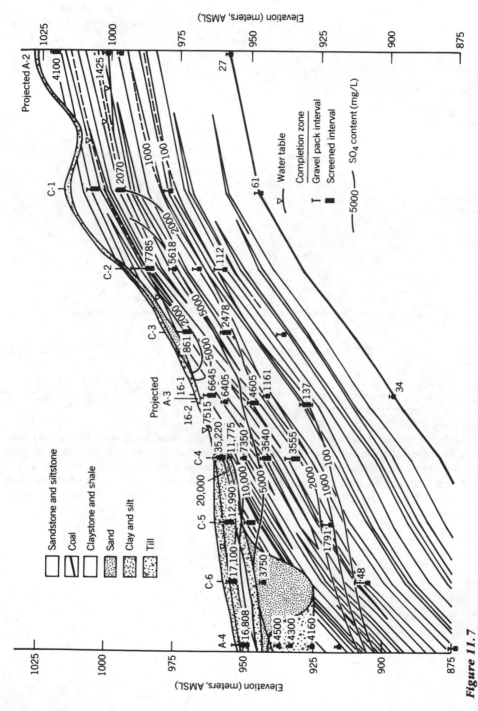

Figure 11.7

Example of how concentration data for an ion can be represented on a cross-section. Shown here are SO_4^{2-} data from Black-spring Ridge, Alberta, Canada (from Stein and Schwartz, 1989).

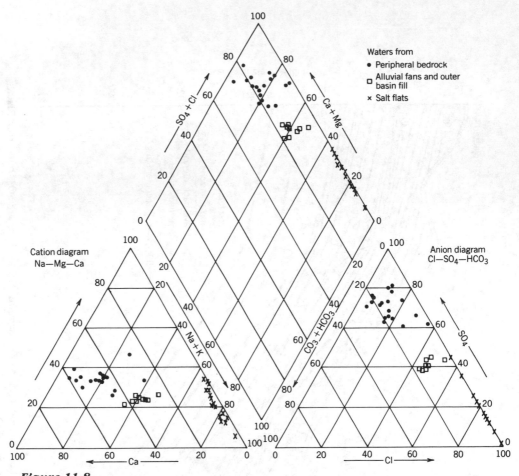

Figure 11.8
Example of the use of a Piper diagram for classifying ground water and defining a pathway of chemical evolution. The diagram shows how samples from the Salt Basin in West Texas can be classified according to setting for example, limestone (dots), alluvial fans, and basin fills (squares), and the salt flats (crosses). The evolutionary pathway is defined by ground water moving from the limestone, through the alluvium to the salt flats (from Boyd and Kreitler, 1986).

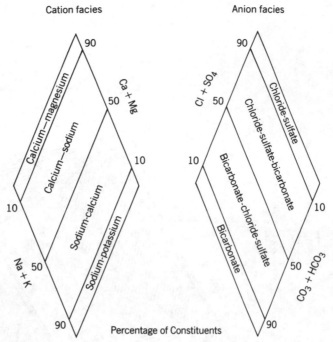

Figure 11.9
Templates for classifying waters into facies for cations and
anions (from Back, 1961).

single map can be advantageous. However, to interpret the geometric shapes quantitatively requires a considerable effort.

When chemical data vary systematically in space, it is often best to plot and contour concentrations (or other data) on maps or cross sections (Figure 11.7). This presentation makes it obvious how individual parameters vary. Any measured or calculated chemical parameter can be represented in this way, which is an advantage over other approaches that involve specific combinations of ions. One problem is the large number of figures that could be required to describe fully the chemistry of an area. However, given the usefulness of these diagrams, this limitation is not a serious one.

Another much more serious problem arises with "noisy" data. Simply contouring a set of such data could produce a complex and cluttered figure that is not really useful. A Piper diagram is best when the data are noisy. By classifying samples on the Piper diagram, one can identify geologic units with chemically similar water, and define the evolution in water chemistry along a flow system (Figure 11.8).

Noisy data can also be smoothed before plotting on a map or cross section. The facies mapping approach (Back, 1961) is one way of smoothing chemical data. Samples are classified according to facies with two templates for the Piper diagram (Figure 11.9), one for the cations and the other for the anions. The limited number of possibilities for classifying the chemical data effectively eliminates local variability yet preserves broad trends. The example (Figure 11.10) shows the progressive change in cation chemistry from calcium-magnesium water in the upland part of the Atlantic coastal plain to sodium water in deeper units located at the downstream end of the flow system.

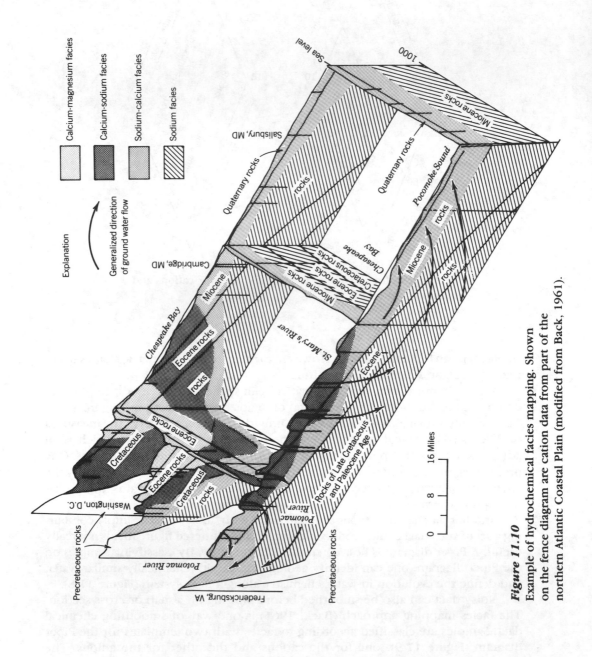

Figure 11.10

Example of hydrochemical facies mapping. Shown on the fence diagram are cation data from part of the northern Atlantic Coastal Plain (modified from Back, 1961).

In closing this discussion, we would like to consider briefly a few possible statistical approaches. Summary statistics (for example, mean and standard deviation) for various subsets of a chemical data set provide a simple way of describing patterns of variability. Another technique that has proven to be extremely useful in classifying chemical data is cluster analysis (Davis, 1986). The term cluster analysis refers to a group of analytical procedures to sort a set of individual samples into smaller groups. Presenting the results on a two-dimensional diagram called a dendogram enables one to represent the pattern of clustering graphically down to individual pairs of samples and the natural breaks between major groups. One important advantage in using this approach is that water can be classified on the basis of any desired combination of chemical data. Unlike the Piper diagram, one is not limited to the same group of major ions. For a complete discussion of the method, readers can refer to Davis (1986).

The discussion of statistical techniques has not addressed the spectrum of possible approaches. For example, regression or trend surface analyses could quantify observed chemical patterns, while still other approaches might serve to quantify the correlation between various constituents.

Suggested Readings

Langmuir, D., and Mahoney, J., 1985, Chemical equilibrium and kinetics of geochemical processes in ground water studies: Proc. First Canadian/American Conference on Hydrogeology, eds. B. Hitchon and E. I. Wallick, National Water Well Assoc., Dublin, Ohio, p. 69–95.

Morel, F. M. M., 1983, Principles of aquatic chemistry: New York, John Wiley, 446 p.

Problems

1. The routine analysis of a water sample provides the following concentrations (as mg/L): Ca^{2+} − 93.9; Mg^{2+} − 22.9; Na^+ − 19.1; HCO_3^- − 334; SO_4^{2-} − 85.0; Cl^- − 9.0 and a pH of 7.20.

    a. Calculate the concentrations of Ca^{2+}, Na^+, HCO_3^-, and Cl^- in terms of molarity and milliequivalents per liter (meq/L).

    b. What is the ionic strength of the sample?

    c. Calculate the activity coefficients for Ca^{2+} and HCO_3^- using the extended Debye-Hückel equation.

2. Write mass law expressions for the following equilibrium reactions.

    a. $CaCO_3 = Ca^{2+} + CO_3^{2-}$

    b. $CO_2(g) = CO_2(aq)$

    c. $Mn^{2+} + Cl^- = MnCl^+$

3. For the water sample in problem 1, determine the saturation index for calcite at 25°C given that $[CO_3^{2-}] = 0.34 \times 10^{-5}$ and that the equilibrium

constant for calcite dissolution is 4.9×10^{-9}. What does the saturation index indicate about the state of saturation with respect to calcite?

4. Assess in a preliminary way the quality of the analytical results in Problem 1 by determining the cation/anion balance.

5. Determine the order of the following kinetic expressions.

   a. Degradation of organic matter (org):

   $$\frac{d(org)}{dt} = - k_1(org)$$

   b. Oxidation of pyrite:

   $$\frac{d(Fe^{3+})}{dt} = -k_1 S(Fe^{3+})$$

   where S is surface area of reacting mineral

   c. General reaction:

   $$\frac{d(A)}{dt} = -k_1$$

   d. General reaction:

   $$\frac{d(C)}{dt} = -k_1 (A) (B)^2$$

6. Represent the analytical results in Problem 1 graphically, using (a) a Collins diagram and (b) a Piper diagram. Classify the water using the concept of hydrochemical facies.

Chapter Twelve

Chemical Reactions

12.1 **Acid-Base Reactions**

12.2 **Solution, Exsolution, Volatilization, and Precipitation**

12.3 **Complexation Reactions**

12.4 **Reactions on Surfaces**

12.5 **Oxidation-Reduction Reactions**

12.6 **Hydrolysis**

12.7 **Isotopic Processes**

This chapter is concerned with reactions—processes that operate in the sub-surface to transfer mass among the fluids, gases, and solids found there. What we want to do here is to classify and describe the reactions in detail, breaking down this relatively large and complex topic into manageable chunks. Further, this chapter will develop a sense of why reactions are important in problems of con-tamination and the geochemistry of natural waters.

12.1   Acid-Base Reactions

Acid-base reactions are important in ground water because of their influence on pH and the ion chemistry. pH is a master variable controlling chemical systems. It is defined as the negative logarithm of the hydrogen ion activity and describes whether a solution is acidic (pH < 7), neutral (pH = 7), or basic (pH > 7). Hydro-gen ion activity is represented here as $[H^+]$ so that $pH = -\log[H^+]$. In solution, hydrogen ion exists as a proton associated with a molecule of water (H_3O^+). Thus, we can express hydrogen ion activity as $[H_3O^+]$. Writing the hydrogen activity as H_3O^+ differentiates between a proton (a hydrogen atom that has lost an electron, H^+) and a hydrogen ion in solution.

An acid is a substance with a tendency to lose a proton. A base is a substance with a tendency to gain a proton. Acids react with bases in what are called acid-base reactions. In general, because no free protons result from an acid-base reac-tion, there must be two acid-base systems involved:

$$\text{Acid}_1 + \text{Base}_2 = \text{Acid}_2 + \text{Base}_1 \tag{12.1}$$

In the forward reaction, the proton lost by Acid_1 is gained by Base_2, and in the reverse reaction the proton lost by Acid_2 is gained by Base_1. The following equilib-rium reaction illustrates this point

$$HCO_3^- + H_2O = H_3O^+ + CO_3^{2-} \tag{12.2}$$

In Eq. 12.2, protons are transferred from HCO_3^- to H_2O, which in this reaction functions as a base, and from the acid H_3O^+ to the base CO_3^{2-}.

Applying the law of mass action to Eq. 12.2 yields

$$K = \frac{[H_3O^+][CO_3^{2-}]}{[HCO_3^-][H_2O]} \tag{12.3}$$

This equation simplifies by assuming the $[H_2O] = 1$ and $[H_3O^+] = [H^+]$, which gives

$$K = \frac{[H^+][CO_3^{2-}]}{[HCO_3^-]} \tag{12.4}$$

By making this substitution, the acid-base reaction involving H_2O-H_3O^+ is neglected. This simplification is usually made in writing acid-base reactions in water. Nevertheless, thinking of acid-base reactions in terms of reaction pairs helps to understand these reactions.

In the previous example, the solvent water functions as a base. However, bases such as ammonia (NH_3) also ionize in water

$$NH_3 + H_2O = NH_4^+ + OH^-$$
$$\text{Base}_1 \quad \text{Acid}_2 \quad \text{Acid}_1 \quad \text{Base}_2 \tag{12.5}$$

which implies that water also functions as an acid. The concept of an acid or base, thus, can be quite abstract.

The strength of an acid or base refers to the extent to which protons are lost or gained, respectively. For our purposes, it is sufficient to characterize strength using the qualitative terms strong or weak. Consider the following generalized ionization reaction for an acid HA in water (Glasstone and Lewis, 1960)

$$HA + H_2O = H_3O^+ + A^- \tag{12.6}$$

where A^- is an anion like Cl^-. The strength of the acids and bases depends on whether equilibrium in the reaction is established to the right or left side of Eq. 12.6. For example, a strong acid will ionize and establish an equilibrium to the right because of a significant proton transfer to H_2O in spite of the fact that H_2O is a weak base. When the acid (HA) is strong, its conjugate base A^- will be weak. If HA is a weak acid, then its conjugate base will be strong. This relationship with just a few exceptions applies to acids and bases of all types (Glasstone and Lewis, 1960).

Natural Weak Acid-Base Systems

The few weak acid-base reactions that are important in natural ground waters are listed in Table 12.1. The first reaction describes the dissociation of water into hydrogen and hydroxyl ions, where the constant K_w is the ion product for water. This reaction is a simplified form of the acid-base reaction

$$H_2O + H_2O = H_3O^+ + OH^- \tag{12.13}$$

Writing the reaction this way implies that protons will be transferred between water molecules and between H_3O^+ and OH^- even in pure water. The dissociation reaction for water is fundamental in establishing the relationship between (H^+) and (OH^-) according to the law of mass action.

Next in Table 12.1 are weak acid-base reactions that originate mainly from adding CO_2 gas to water. The first reaction describes how CO_2 gas dissolves to produce carbonic acid $H_2CO_3^*$. By convention, the concentrations of $CO_2(aq)$ and H_2CO_3 are represented by the single species $H_2CO_3^*$ or

$$(H_2CO_3^*) = (CO_2 \cdot aq) + (H_2CO_3) \tag{12.14}$$

The next two reactions (Table 12.1) describe the two-stage ionization of carbonic acid, which is controlled by the first and second dissociation constants K_1 and K_2, respectively. The last group of reactions describes the dissociation of silicic acid in water. This acid originates mainly from the dissolution of silicate minerals.

Table 12.1
Important weak acid-base reactions in natural water systems*

Reaction	Mass law equation	Eqn	$-\log$ $K(25°C)$
$H_2O = H^+ + OH^-$	$K_w = (H^+)(OH^-)$	12.7	14.0
$CO_2(g) + H_2O = H_2CO_3^*$	$K_{CO_2} = \dfrac{(H_2CO_3^*)}{P_{CO_2}(H_2O)}$	12.8	1.46
$H_2CO_3^* = HCO_3^- + H^+$	$K_1 = \dfrac{(HCO_3^-)(H^+)}{(H_2CO_3^*)}$	12.9	6.35
$HCO_3^- = CO_3^{2-} + H^+$	$K_2 = \dfrac{(CO_3^{2-})(H^+)}{(HCO_3^-)}$	12.10	10.33
$H_2SiO_3 = HSiO_3^- + H^+$	$K = \dfrac{(HSiO_3^-)(H^+)}{(H_2SiO_3)}$	12.11	9.86
$HSiO_3^- = SiO_3^{2-} + H^+$	$K = \dfrac{(SiO_3^{2-})(H^+)}{(HSiO_3^-)}$	12.12	13.1

*From Morel, F.M.M., 1983. Principles of Aquatic Chemistry. Copyright © 1983 by John Wiley & Sons, Inc. Reprinted by permission of John Wiley and Sons, Inc.

CO_2-Water System

The reactions involving CO_2 in Table 12.1 show that CO_2 dissolved in water partitions among $H_2CO_3^*$, HCO_3^-, and CO_3^{2-}. When the pH of a solution is fixed, the mass law equations on Table 12.1 let us determine the concentrations of the species. Following is an example illustrating this calculation.

Example 12.1

Assume that CO_2 is dissolved in water so that $(CO_2)_T = 10^{-3}$ M and

$$10^{-3} = (H_2CO_3^*) + (HCO_3^-) + (CO_3^{2-}) \tag{12.15}$$

What is the concentration of the carbonate species at a pH of 6.35? We chose this particular pH to simplify the calculation.

Substitution of the known pH into Eq. 12.9 gives

$$10^{-6.35} = \frac{(HCO_3^-)(10^{-6.35})}{(H_2CO_3^*)} \tag{12.16}$$

Simplification of this equation gives $(HCO_3^-) = (H_2CO_3^*)$. Substitution into Eq. 12.10 provides

$$10^{-10.33} = \frac{(CO_3^{2-})(10^{-6.35})}{(HCO_3^-)} \tag{12.17}$$

and shows that $(HCO_3^-) >> (CO_3^{2-})$. Assuming for the moment that (CO_3^{2-}) is negligible, a solution to Eq. 12.15 gives $(HCO_3^-) = (H_2CO_3^*) = 10^{-3.30}$ M. Substitution of this result into Eq. 12.10 gives $(CO_3^{2-}) = 10^{-7.28}$ M.

Similar calculations over a broad range of pH values would show how the concentrations of the carbonate species change. A useful way of illustrating this relationship is with a Bjerrum plot. It is a plot of the logarithm of activities (or concentrations) of various species versus pH. A log C–pH diagram for the carbonate system is shown in Figure 12.1 for $(CO_2)_T = 10^{-4}$. The arrows on the figure depict the cross-over points, where at pH 6.35, $(H_2CO_3^*) = (HCO_3^-)$ and at 10.33, $(HCO_3^-) = (CO_3^{2-})$. These are known as the pK_a values for aqueous CO_2 (that is, $pK_1 = 6.35$ and $pK_2 = 10.33$).

A Bjerrum diagram illustrates the dominance of particular species over some pH range. For example, below pH 5, $H_2CO_3^*$ is the dominant carbonate species (Figure 12.1). In the range from approximately pH 7 to 9, HCO_3^- is the most abundant species. The carbonate ion (CO_3^{2-}) is dominant when pH is above 10.

Alkalinity

So far, we have only discussed cases where the pH determines the distribution of carbonate species. In actual ground water systems, the situation is more complex. The pH and carbonate speciation are interdependent, a function of not only the ionization equilibria for the carbonate species and water but also strong bases added through the dissolution of carbonate and silicate minerals.

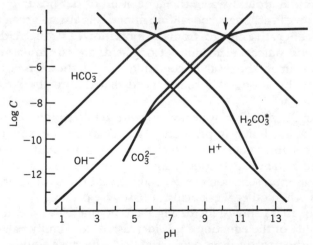

Figure 12.1
Log C–pH diagram for the carbonate system with $(CO_2)_T = 10^{-4}$ M (from Morel, F.M.M., 1983, Principles of Aquatic Chemistry). Copyright © 1983 by John Wiley & Sons, Inc. Reprinted by permission of John Wiley & Sons, Inc.

Formulating these processes mathematically requires an understanding of the concept of alkalinity. For our purposes, alkalinity is defined as the net concentration of strong base in excess of strong acid with a pure CO_2-water system as the point of reference (Morel, 1983). When CO_2 is dissolved in water at a fixed P_{CO_2}, the charge balance is

$$(H^+) = (OH^-) + (HCO_3^-) + 2(CO_3^{2-}) \tag{12.18}$$

This charge balance will change, for example, if we add a strong acid such as HCl and a strong base such as NaOH to the solution (Morel, 1983). Assuming that NaOH and HCl ionize fully, the charge balance is

$$(Na^+) + (H^+) = (OH^-) + (HCO_3^-) + 2(CO_3^{2-}) + (Cl^-) \tag{12.19}$$

Rearranging Eq. 12.19 to group together ions contributed from the strong acids and bases yields

$$(Na^+) + (Cl^-) = -(H^+) + (OH^-) + (HCO_3^-) + 2(CO_3^{2-}) \tag{12.20}$$

The left side of Eq. 12.20 is the net concentration of strong base in excess of strong acid and defines alkalinity when $(Na^+) > (Cl^-)$. A more general form of this equation is

$$Alk = \Sigma(i^+)_{sb} - \Sigma(i^-)_{sa} = -(H^+) + (OH^-) + (HCO_3^-) + 2(CO_3^{2-}) \tag{12.21}$$

where (i^+) and (i^-) are the positively and negatively charged species from the strong acids and bases. When the alkalinity is zero, Eq. 12.21 becomes the charge balance equation for a pure CO_2-water system.

In natural ground water, the generation of net positive charges through the dissolution of carbonate and silicate minerals is always greater than the contribution of net negative charges from the ionization of strong acids. Thus, as a general rule, ground waters are alkaline. Strong acids are rare in natural waters and when they do occur are the result of contamination. In these cases, the charge contribution from the strong acid may exceed that due to the strong base and create mineral acidity.

So far, the pure CO_2-water system with zero alkalinity is the point of reference in defining alkalinity. In practice, it may be more convenient to use other reference points and to define carbonate or caustic alkalinity (Morel, 1983). Further, establishing the relative contribution of noncarbonate alkalinity may be useful. Morel (1983) presents a more general form of the basic charge balance equation for cases when other weak acids or bases are present.

As Eq. 12.21 shows, increasing alkalinity increases the net positive charge on the left side of the equation. This increase is not simply balanced by an increase in one of the negative ions on the right side, because equilibrium relationships in the solution must be maintained. The increase in the alkalinity is matched by an increase in the concentration of negatively charged species that comes from the ionization of HCO_3^- to CO_3^{2-} and an increase in pH. Thus, increasing the alkalinity with a strong base ultimately leads to an increase in pH. This behavior is commonly observed as ground waters evolve by dissolving minerals along a flow path.

12.2   *Solution, Exsolution, Volatilization, and Precipitation*

Because water is an excellent solvent, it dissolves gases, liquids, and solids in the subsurface. Dissolution more than any other process is responsible for the large solute loading to ground water. Other processes like gas exsolution, volatilization, and mineral precipitation remove mass from water.

Gas Solution and Exsolution

Gas solution and exsolution can transfer significant quantities of mass between soil gases and ground water. Commonly, we model these processes using equilibrium concepts based on Henry's law. This mass law equation relates the concentration of dissolved gas in solution to the partial pressure of the same gas in an atmosphere in contact with the solution. Henry's law strictly speaking does not apply to gases such as CO_2 or NH_3 that react in solution. However, in the case of CO_2, so little of the $CO_2(aq)$ reacts that Henry's law approximates the distribution of gas between the two phases. Thus, the concentration of $H_2CO_3^*$ is almost all $CO_2(aq)$ with a negligibly small concentration of H_2CO_3.

The Henry's law equation for CO_2 is

$$K_H = \frac{(CO_2 \cdot aq)}{P_{CO_2}} \tag{12.22}$$

where K_H is the Henry's law constant (M atm^{-1}), P_{CO_2} is the partial pressure of CO_2 (atm) in the gas phase and $(CO_2 \cdot aq)$ is the molar concentration of CO_2 gas in the solution (M). In applications, the concentration of $CO_2(aq)$ in Eq. 12.22 is the same as $(H_2CO_3^*)$.

Example 12.2 illustrates how Henry's law determines the concentration of CO_2 in a solution in equilibrium with an atmosphere of a given composition.

Example 12.2

What is the concentration of $H_2CO_3^*$ dissolved in a small drop of rain at 25 °C falling through the atmosphere. Assume that the CO_2 in the droplet is in equilibrium with the atmosphere with a P_{CO_2} of $10^{-3.5}$ atm. From Eq. 12.22,

$$(H_2CO_3^*) = (CO_2 \cdot aq) = K_H P_{CO_2}$$
$$= 10^{-1.47} 10^{-3.5}$$
$$= 10^{-4.97} M$$

This example shows that CO_2 gas is related to the carbonate system through $H_2CO_3^*$. Thus, a change in the partial pressure of CO_2 gas not only changes $(H_2CO_3^*)$ according to Henry's law but also (HCO_3^-), (CO_3^{2-}), (H^+), and (OH^-), which are related through the carbonate equilibria. The concentrations of some cations may also change because of equilibria with solid phases.

The addition or removal of gases from solution, thus, can play a major role in controlling the chemistry of ground water. For example, a significant step in the chemical evolution of natural ground water involves the solution of CO_2 gas in the soil zone. This process not only increases the concentration of HCO_3^- but also enhances the aggressiveness of the water in dissolving solids.

Solution of Organic Solutes in Water

Liquids other than water in the subsurface can include oil or organic contaminants. These nonaqueous components can migrate as a separate liquid phase (see Section 16.3) or dissolved in ground water. This latter process is of interest here.

Organic compounds differ widely in their overall solubility (Mackay et al., 1985). Some solutes like methanol are extremely soluble, while others such as PCBs or DDT are sparingly soluble or hydrophobic. As a general rule, the most soluble organic compounds are charged species or those containing oxygen or nitrogen. Examples of this latter group are alcohols or carboxylic acids, which hydrogen bond with the water and fit easily into the structure of water. Thus, one important factor influencing the solubility of organic compounds is the functional groups that are part of the molecule. Further, even minor changes in the structure of an organic compound such as the position of the Cl atom on a benzene ring can have a noticeable influence on solubility.

Without hydrogen bonding, solubility is diminished. Forcing a nonpolar molecule into the tetrahedral structure of water requires considerable energy. The extent to which the hydrogen bonding of the water is disrupted ultimately determines the solubility of the organic compound. Thus, the larger the organic molecule is, the larger the space that is required in the water structure, and the less soluble the compound is. We notice this effect, for example, with aromatic compounds (Table 12.2) where the solubility is inversely proportional to the molecular mass or size of the molecules. Temperature and the presence of other organic liquids are factors that also influence solubility.

Verschueren (1977) tabulated solubility data for many organic compounds. When data are unavailable or unreliable, measured octanol/water partition coefficients provide a basis for estimating solubility. The octanol/water partition coeffi-

Table 12.2
A comparison of molecular mass versus solubility for various aromatics[1]*

Compound	Molecular mass (g/mol)	Solubility (g/m³)
Benzene (C_6H_6)	78.0	1780
Toluene (C_7H_8)	92.0	515
o-Xylene (C_8H_{10})	106.0	175
Cumene (C_9H_{12})	120.0	50
Naphthalene ($C_{10}H_8$)	128.0	33
Biphenyl ($C_{12}H_{10}$)	154.0	7.48

*Reprinted with permission from Mackay, D., and Leinonen, P.J., Environ. Sci. Technol., v. 9, 1975, p. 1179. Copyright © 1975 American Chemical Society.

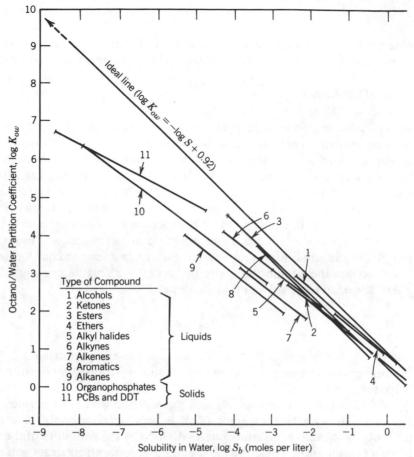

Figure 12.2

Relationship between solubility in water and octanol-water partition coefficients for various organic compounds. Reprinted with permission from Chiou, C.T., Schmedding, D.W., and Manes, M., Environ. Sci. Technol., 16, 1982, p. 6. Copyright © 1982 American Chemical Society.

cient (K_{ow}) is a dimensionless equilibrium constant that characterizes partitioning of an organic solute (org) between octanol (an organic liquid) and water. The mass law expression for this reaction is

$$K_{ow} = \frac{(\text{org} \cdot \text{oct})}{(\text{org} \cdot \text{aq})} \tag{12.23}$$

where (org·oct) is the concentration of the solute in octanol and (org·aq) is the concentration in water. Values of K_{ow} for many organic compounds are tabulated by Hansch and Leo (1979). Because the partitioning of an organic solute between water and octanol is conceptually not much different than the partitioning of an organic compound between itself and water, a correlation exists between K_{ow} and solubility. This relationship is illustrated for various classes of liquid and solid organic compounds (Figure 12.2 from Chiou and others, 1982). The regression equation for all compounds is

$$\log K_{ow} = 7.30 - 0.747 \log Sb \qquad (12.24)$$

where Sb is the solubility (M) of the compound. Chiou and others (1982) develop equations of this form for all the classes of compounds shown on Figure 12.2.

Volatilization

Volatilization is a process of liquid or solid phase evaporation that occurs when contaminants present either as nonaqueous phase liquids or dissolved in water contact a gas phase. This situation commonly arises with organic contaminants in the saturated and unsaturated zones (Figure 12.3), or during the sampling and analysis of volatile compounds. The process itself is controlled by the vapor pressure of the organic solute or solvent. The vapor pressure of a liquid or solid is the pressure of the gas in equilibrium with respect to the liquid or solid at a given temperature. Vapor pressure represents a compound's tendency to evaporate and is essentially the solubility of an organic solvent in a gas. At equilibrium, Raoult's law describes the equilibrium partial pressure of a volatile organic in the atmosphere above an ideal solvent (like benzene)

$$P_{org} = x_{org}\, P_{org}^{o} \qquad (12.25)$$

where P_{org} is the partial pressure of the vapor in the gas phase, x_{org} is the mole fraction of the organic solvent, and P_{org}^{o} is the vapor pressure of the pure organic solvent.

Volatilization of dissolved organic solutes from water is described by Henry's law. The Henry's law constant (K_H), expressed commonly as atmospheres-m³/mole, is equal to P^o in atmospheres divided by the solubility of the compound in water (moles/m³). Thus, the most common convention treats volatilization of pure solvents (e.g., benzene) using Raoult's law, and the volatilization of solutes

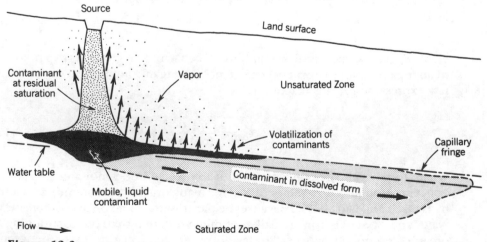

Figure 12.3
Migration of volatiles in the soil atmosphere away from an organic contaminant spill (modified from Schwille, 1985). Reprinted by permission Second Canadian/American Conference on Hydrogeology. Copyright © 1985. All rights reserved.

Table 12.3
Vapor pressure data for a selected group of organic contaminants

Compound	Formula	V.P.[1]	Henry's law[2] constant
HALOGENATED HYDROCARBONS			
Dichloromethane	CH_2Cl_2	349	3×10^{-3}
Trichloromethane	$CHCl_3$	160	4.8×10^{-3}
Tetrachloromethane	CCl_4	90	2.3×10^{-2}
Bromoform	$CHBr_3$	5.6 (25°C)	5.8×10^{-4} c
1,1-Dichloroethane	$CHCl_2CH_3$	180	4.3×10^{-3} c
1,2-Dichloroethane	CH_2ClCH_2Cl	61	9.1×10^{-4} c
1,1-Dichloroethane	$H_2C{=}CCl_2$	500	—
Trans-1,2-dichloroethene	$CHCl{=}CHCl$	200 (14°C)	4.2×10^{-2} c
1,1,1-Trichloroethane	CCl_3CH_3	100	1.8×10^{-2}
Trichloroethene	$Cl_2C{=}CHCl$	60	1×10^{-2}
1,1,2-Trichloroethane	$CH_2ClCHCl_2$	19	7.4×10^{-4} c
Tetrachloroethene	$Cl_2C{=}CCl_2$	14	8.3×10^{-3}
AROMATIC HYDROCARBONS			
Benzene	C_6H_6	76	5.5×10^{-3}
Phenol	C_6H_5OH	0.2	3.0×10^{-7} c
Chlorobenzene	C_6H_5Cl	8.8	2.6×10^{-3} c
Ethylenebenzene	$C_6H_5C_2H_5$	7	8.7×10^{-3}
Toluene	$C_6H_5CH_3$	22	5.7×10^{-3}
o-Xylene	$C_6H_4(CH_3)_2$	5	5.3×10^{-3}
OTHER ORGANIC SOLVENTS			
Acetone	$CH_3{-}CO{-}CH_3$	89 (5°C)	—
Diethyl ether	$C_2H_5OC_2H_5$	442	5.1×10^{-4} c
Tetrahydrofuran	C_4H_8O		—
1,4-Dioxane	$O(CH_2{-}CH_2)_2O$	30	—
BIOCIDES			
Pentachlorophenol	C_6Cl_5OH	1.1×10^{-4}	3.4×10^{-6}
DDT	$(ClC_6H_4)_2CHCCl_3$	1×10^{-7}	3.8×10^{-5}
Lindane	$C_6H_6Cl_6$	9.4×10^{-6}	4.8×10^{-7}

[1]Vapor pressure in mm Hg at 20°C; 1 atm = 760 mm Hg.

[2]atmospheres-m³/mole; c indicates values calculated from vapor pressure and solubility data.

Modified from Jackson and others, 1985, Contaminant Hydrogeology of Toxic Organic Chemicals at a Disposal Site, Gloucester, Ontario. 1. Chemical Concepts and Site Assessment, IWD Scientific Series 141, Environment Canada, 114p. Reproduced with the permission of The Minister of Supply and Services Canada, 1990.

(e.g., benzene dissolved in water) using Henry's law. Complex mixtures such as gasoline are treated with modified forms of Raoult's law to account for the nonideal effects of the mixture.

Table 12.3, from Jackson and others, 1985, lists vapor pressures, and Henry's law constants for a selected group of organic contaminants. Even a limited set of data shows vapor pressures ranging over 9 or 10 orders of magnitude from the most volatile compounds such as dichloromethane or 1,1-dichloroethane to non-volatile compounds such as pentachlorophenol or DDT. Verscheuren (1977) provides data on solubility and vapor pressure.

Volatilization is important for problems of organic contamination within or close to the unsaturated zone, and for issues related to sample collection and analysis. Soil gases containing volatiles, when allowed to accumulate in the basements of structures, can cause fires, explosions, or health problems. Contamination can be so serious that venting of the soil may be required. Alternatively, the tendency for some organic contaminants to volatilize forms the basis of a useful technique for detecting and monitoring contamination in the subsurface (Section 15.4), and for in situ approaches for recovering contaminants like gasoline from the subsurface (Section 19.1).

Volatilization also causes problems in sampling and analysis. When samples have access to the atmosphere, volatilization reduces the concentration of the contaminant exponentially. If exposed to the atmosphere long enough, the concentration of the contaminant can fall below detection limits.

Dissolution and Precipitation of Solids

Of all the processes that influence solute transport, the dissolution and precipitation of solids are two of the most important in terms of their control on ground water chemistry. Extremely large quantities of mass can be transferred under some conditions between the ground water and solid mineral phases. For example, recharge derives almost its entire solute load through the dissolution of minerals along flow paths. Mineral precipitation removes much of the metal present in a low-pH contaminant plume as dispersion and other processes increase the pH of the ground water. These examples show how ground water proceeds toward chemical equilibrium with respect to various minerals from undersaturation in the case of evolving natural ground water and from supersaturation in the case of metal transport.

Solid Solubility

The solubility of a solid reflects the extent to which the reactant (solid) or products (ions and/or secondary minerals) are favored in a dissolution-precipitation reaction. In many reactions where the activity of the reacting solid is equal to one, a comparison of the relative size of the equilibrium constant provides an indication of the solid solubility in pure water. For example, Table 12.4 shows the chloride and sulfate salts to be the most soluble phase and the sulfide and hydroxide groups to be the least soluble. Minerals in the carbonate and the silicate and aluminum silicate groups have a small but significant solubility.

When other ions are present in a solution, the solubility of a solid can be different from its value in pure water. Solubility increases due to solution nonidealities and decreases due to the common ion effect. The previous chapter explains this issue of nonidealities in relation to activity models. Generally, the solubility of a solid increases with increasing ionic strength, because other ions in solution reduce the activity of the ions involved in the dissolution-precipitation reaction.

The common ion effect occurs when a solution already contains the same ions that will be released when the solid dissolves. The presence of the common

Table 12.4

Some common mineral dissolution reactions and associated equilibrium constants

Mineral or solid	Reaction	$\log K(25°C)$	Source
CHLORIDES AND SULFATES			
Halite	$NaCl = Na^+ + Cl^-$	1.54	1
Sylvite	$KCl = K^+ + Cl^-$	0.98	1
Gypsum	$CaSO_4 \cdot 2H_2O = Ca^{2+} + SO_4^{2-} + 2H_2O$	-4.62	1
CARBONATES			
Magnesite	$MgCO_3 = Mg^{2+} + CO_3^{2-}$	-7.46	1
Aragonite	$CaCO_3 = Ca^{2+} + CO_3^{2-}$	-8.22	1
Calcite	$CaCO_3 + Ca^{2+} + CO_3^{2-}$	-8.35	1
Siderite	$FeCO_3 = Fe^{2+} + CO_3^{2-}$	-10.7	1
Dolomite	$CaMg(CO_3)_2 = Ca^{2+} + Mg^{2+} + 2CO_3^{2-}$	-16.7	2
HYDROXIDES			
Brucite	$Mg(OH)_2 = Mg^{2+} + 2OH^-$	-11.1	1
Ferrous hydroxide	$Fe(OH)_2 = Fe^{2+} + 2OH^-$	-15.1	1
Gibbsite	$Al(OH)_3 = Al^{3+} + 3OH^-$	-33.5	1
SULFIDES			
Pyrrhotite	$FeS = Fe^{2+} + S^{2-}$	-18.1	1
Sphalerite	$ZnS = Zn^{2+} + S^{2-}$	-23.9	3
Galena	$PbS = Pb^{2+} + S^{2-}$	-27.5	1
SILICATES AND ALUMINUM SILICATES			
Quartz	$SiO_2 + H_2O = H_2SiO_3$	-4.00	1
Na-Montmorillonite	$\text{Na-Mont} + \frac{11}{2} H_2O =$		
	$\frac{3}{2} \text{Kaol} + HH_4SiO_4 + Na$	-9.1	2
Kaolinite	$\text{Kaol} + 5H_2O = 2Al(OH)_3 + 2H_4SiO_4$	-9.4	2

[1]Morel (1983).

[2]Stumm and Morgan (1981).

[3]Matthess (1982).

ion means that less solid is able to dissolve before the solution reaches saturation with respect to that mineral. Thus, the solubility of a solid is less in a solution containing a common ion than it would be in water alone.

12.3   Complexation Reactions

As interest developed in modeling the chemistry of aqueous systems, hydrogeologists became aware of the importance of complexation reactions in determining the saturation state of ground water. However, because complexation reactions are most significant in reasonably saline ground water, the hydrogeological literature does not treat this topic extensively. With the explosion of knowledge in contaminant hydrogeology, the prevailing view about the importance of complexation reactions has changed. For example, complexation facilitates the transport of potentially toxic metals such as cadmium, chromium, copper, lead, uranium, or plutonium. Such reactions also influence some types of surface reactions. Thus, a process that was mainly of academic importance now has taken on practical significance in contamination problems.

A complex is an ion that forms by combining simpler cations, anions, and sometimes molecules. The cation or central atom is typically one of the large number of metals making up the periodic table. The anions, often called ligands, include many of the common inorganic species found in ground water such as Cl^-, F^-, Br^-, SO_4^{2-}, PO_4^{3-}, and CO_3^{2-}. The ligand might also be an organic molecule such as an amino acid.

The simplest complexation reaction involves the combination of a metal and ligand such as Mn^{2+} and Cl^- as follows:

$$Mn^{2+} + Cl^- = MnCl^- \tag{12.26}$$

A more complicated manifestation of complexation is the reaction series that forms when complexes themselves combine with ligands. An example is the hydrolysis reaction of Cr^{3+}

$$Cr^{3+} + OH^- = Cr(OH)^{2+}$$
$$Cr(OH)^{2+} + OH^- = Cr(OH)_2^+ \tag{12.27}$$
$$Cr(OH)_2^+ + OH^- = Cr(OH)_3^0$$

and so on. The metal is distributed among at least three complexes. Such series not only involve reactions with $(OH)^-$ but other ligands such as Cl^-, F^-, and Br^-.

Stability of Complexes and Speciation Modeling

Most inorganic reactions involving complexes are kinetically fast. Thus, they can be examined quantitatively using equilibrium concepts. For example, the mass law equations for Eqs. 12.26 and 12.27 are

$$K_{MnCl^-} = \frac{[MnCl^-]}{[Mn^{2+}][Cl^-]}$$

and

$$K_{Cr(OH)^{2+}} = \frac{[Cr(OH)^{2+}]}{[Cr^{3+}][OH^-]}$$

$$K_{Cr(OH)_2^+} = \frac{[Cr(OH)_2^+]}{[Cr(OH)^{2+}][OH^-]}$$

and so on...

Calculation of the distribution of metals among various complexes involves the solution of a series of mass law equations. We write the reactions in terms of the formation of the complex from some combination of the metal and an appropriate number of ligands. For example, writing Eq. 12.27 as a series of association reactions gives the series of reactions and mass law equations on Table 12.5. These so-called mononuclear complexation reactions have this general form

$$M + lL + hH = ML_lH_h: \quad \beta_i = \frac{[ML_lH_h]}{[M][L]^l[H]^b} \tag{12.28}$$

where M is a metal, L is a ligand, and H is a hydrogen ion.

The values of β_i are the stability constants for the reaction and determine the strength of the complex. Large values of β_i are associated with the stronger or more stable complexes. Stability constants are related to the equilibrium constants for the series of reactions. Morel (1983, p. 242) tabulates stability constants for a host of metal-ligand reactions.

In some cases, polynuclear complexes form. This complex is characterized by the presence of more than one metal, for example, $Cr_3(OH)_4^{5+}$ or $Cu_2(OH)_2^{2+}$. However, because these complexes are usually rare and difficult to include in speciation models, they are not often treated.

The stability constants for a series of metal-ligand reactions provide the basic information necessary to determine how the total concentration of a metal in

Table 12.5
Association reactions and mass law expressions describing the hydrolysis of Cr

Equation	Mass law expressions
$Cr^{3+} + OH^- = Cr(OH)^{2+}$	$\beta_1 = \dfrac{Cr(OH)^+}{(Cr^{3+})(OH^-)}$
$Cr^{3+} + 2OH^- = Cr(OH)_2^+$	$\beta_2 = \dfrac{(Cr(OH)_2^+)}{(Cr^{3+})(OH^-)^2}$
$Cr^{3+} + 3OH^- = Cr(OH)_3^0$	$\beta_3 = \dfrac{(Cr(OH)_3^0)}{(Cr^{3+})(OH^-)^3}$

.
.
.
.
.

solution $(M)_T$ is distributed as a metal ion and various complexes. The following example illustrates the speciation calculation when Cr hydrolyses.

Example 12.3

A solution with a pH of 5.0 contains a trace quantity of Cr, $(Cr)_T = 10^{-5}$ M. Determine the speciation for Cr among the various hydroxy complexes and Cr^{3+}. Stability constants (as base 10 logarithms) are 10.0, 18.3, and 24.0 for $Cr(OH)^{2+}$, $Cr(OH)_2^+$, and $Cr(OH)_3^0$, respectively. Assume that the pH does not change with the addition of the metal, no solids form, and the solution behaves ideally.

The mole balance equation for $(Cr)_T$ is

$$(Cr)_T = (Cr^{3+}) + (Cr(OH)^{2+}) + (Cr(OH)_2^+) + (Cr(OH)_3^0)$$

Substitution of the appropriate mass law equations for the association reactions into this equation gives

$$(Cr)_T = (Cr^{3+}) + \beta_1(Cr^{3+})(OH^-) + \beta_2(Cr^{3+})(OH^-)^2 + \beta_3(Cr^{3+})(OH^-)^3$$
$$(Cr)_T = (Cr)^{3+}\{1 + \beta_1(OH^-) + \beta_2(OH^-)^2 + \beta_3(OH^-)^3\}$$

Because pH is fixed, $(OH^-) = 10^{-9}$ M and all the terms in the brackets are known. Substitution of the known values solving for (Cr^{3+}) gives

$$
\begin{aligned}
(Cr^{3+}) &= \frac{10^{-5}}{\{1 + 10^{10}(10^{-9}) + 10^{18.3}(10^{-18}) + 10^{24}(10^{-27})\}} \\
&= \frac{10^{-5}}{\{1 + 10 + 10^{0.3} + 10^{-3}\}} \\
&= 10^{-6.12} \text{ M}
\end{aligned}
$$

The concentration of the complexes is calculated by substituting known values of (Cr^{3+}) and (OH^-) in the mass law equations

$$
\begin{aligned}
(Cr(OH)^{2+}) &= \beta_1(Cr^{3+})(OH^-) \\
&= 10^{10.0}(10^{-6.12})(10^{-9}) \\
&= 10^{-5.12} \text{ M} \\
(Cr(OH)_2^+) &= 10^{-5.82} \text{ M} \\
(Cr(OH)_3^0) &= 10^{-9.12} \text{ M}
\end{aligned}
$$

Thus, most of the Cr occurs as $Cr(OH)^{2+}$ and $Cr(OH)_2^+$. Only about 7.5% of $(Cr)_T$ is the free ion Cr^{3+}.

In Example 12.3, the speciation of Cr is controlled by solution pH. The log C–pH plot for $(Cr)_T = 10^{-5}$ (Figure 12.4) shows that at a pH below 4, Cr^{3+} is the dominant Cr species. As the pH increases, the various hydroxy complexes dominate.

This use of stability constants to determine the speciation of metals can be extended to a mixed group of mononuclear complexes. For example, if trace

Figure 12.4

Log *C*–pH diagram for chromium hydroxide complexes.

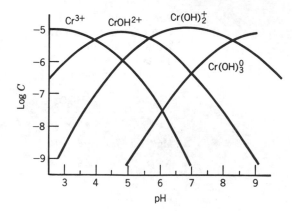

quantities of Pb are present in a ground water, a mixed group of chloride, hydroxy, and carbonate complexes can form. Beginning again with the appropriate mole balance equation

$$(Pb)_T = (Pb^{2+}) + (PbCl_2^0) + (PbCl_3^-) + (PbOH^+) + (PbCO_3^0) \qquad (12.29)$$

and substituting the appropriate mass law equations lets us determine the speciation. This approach requires that the quantity of metal dissolved in the ground water be sufficiently small so as not to influence the major ion chemistry or pH of the solution. At some point, however, adding large quantities of metal affects the pH and carbonate chemistry. The speciation becomes much more difficult to determine in this case because the system of mole balance and mass law equations is larger and more complex. An example of this more complex situation is the case where the chemistry of a ground water changes when the most abundant ions, Ca^{2+}, Mg^{2+}, Na^+, K^+, and H^+ complex with the common ligands SO_4^{2-}, Cl^-, HCO_3^-, and OH^-.

Major Ion Complexation and Equilibrium Calculations

Modeling the speciation involved with ion complexation is possible with computer codes such as SOLMNEQ, PHREEQE, or others of this type (Section 15.5). Without a computer, these calculations are tedious and relatively difficult. Most of the computer-based approaches involve an ion association model for sea water developed by Garrels and Thompson (1962) that accounts for complexation of the major species. Given the total concentration of individual metals $(M_i)_T$ and ligands $(L_j)_T$ in solution from a chemical analysis, the model predicts how this mass is distributed among the free metals, the ligands, and the complexes. A simple overview of this procedure is given by Garrels and Christ (1965).

So far, in calculating mineral saturations, we have not considered that the free ion or ligand concentration is less than the total concentration once complexes form. The calculated *IAP/K* ratios for some minerals can decrease depending upon the extent of complex formation. Let us reexamine Example 11.2 to illustrate this point. Assume now that the molar concentrations for the ions in Example 11.2 are total concentrations. Table 12.6 presents the concentrations of the major complexes. Even though the concentrations of the complexes are relatively small, the reduction in $[Ca^{2+}]$ and $[CO_3^{2-}]$ due to complexation reduces the *IAP/K*

Table 12.6
Calculated molar concentrations of complexes formed from major ion species (based on data from Example 11.2)

Metal	Ligand		
	HCO_3^-	CO_3^{2-}	SO_4^{2-}
Ca^{2+}	5.46×10^{-5}	8.03×10^{-6}	2.16×10^{-5}
Mg^{2+}	4.98×10^{-7}	1.46×10^{-7}	2.78×10^{-7}
Na^+	1.54×10^{-4}	1.13×10^{-5}	6.66×10^{-5}

ratio for calcite from 1.34 to 1.03. For dolomite, the ratio decreases from 0.018 to 0.011. This example shows that even in relatively fresh ground water (ionic strength = 0.023M), the error in determining mineral saturation can be substantial when complexation is not considered. This error can become even larger as the water becomes more saline. In sea water, for example, only approximately 40% of $(SO_4)_T$ exists as SO_4^{2-}, while the remainder exists as $NaSO_4^-$ (37%) and $MgSO_4^0$ (19%) (Morel, 1983).

Enhancing the Mobility of Metals

In general, metals in ground water are most mobile in waters with a low pH, where most of the mass occurs as a charged metal ion. Ignoring for a moment the effects of sorption, the mobility begins to decline at the pH where a solid, usually a metal-hydroxide, metal-carbonate, or metal-sulfide, becomes the dominant phase. At this point, equilibrium with the solid determines that most of the metal will be associated with the solid phase.

Over the pH range common to most natural ground waters, the concentration of most metals is small reflecting this control. Complexes in these circumstances can enhance the solubility of metals. The transport of uranium in ground water is a good example. Figure 12.5 shows the relative abundance of various uranyl complexes in ground water with a composition typical of the Wind River Formation in Wyoming (Langmuir, 1978). Across the entire range of pH represented on the figure, a significant proportion of the total uranium is complexed. Below a pH of about 4, a uranyl-fluoride complex is the most dominant species (Figure 12.5). Over the range of typical ground water for this formation (pH 6.6–8.3), both phosphate and carbonate complexes are important. This extensive complexing especially over the neutral to alkaline pH range increases the solubility of some uranium minerals by several orders of magnitude. Such increases in solubility apparently are necessary to form uranium ore deposits.

Organic Complexation

Complexation can also involve organic ligands present naturally in ground water or added as contaminants. For example, at a site near Ottawa Canada, Killey and others (1984) showed how up to 80% of the ^{60}Co present as a contaminant in ground water occurred as weakly anionic complexes. These complexes formed with naturally occurring organic compounds. This and similar work emphasizes

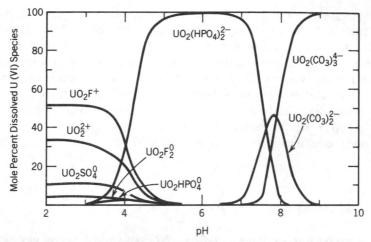

Figure 12.5
Distribution of uranyl complexes vs pH for some typical ligand
concentrations in ground waters of the Wind River Formation at
25°C. $P_{CO_2} = 10^{-2.5}$ atm, $\Sigma F = 0.3$ ppm, $\Sigma Cl = 10$ ppm, $\Sigma SO_4 =$
100 ppm, $\Sigma PO_4 = 0.1$ ppm, $\Sigma SiO_2 = 30$ ppm (from Langmuir,
1978). Reprinted with permission from Geochim. Cosmochim.
Acta, 42, D. Langmuir, Uranium solution-mineral equilibria at low
temperatures with applications to sedimentary ore deposits.
Copyright 1978, Pergamon Press plc.

the need to describe organic systems quantitatively. The task is difficult due to the
diversity of organic materials existing in water and problems in identifying them
in laboratory analyses. However, it may be essential. Killey and others (1984) indicate that concentrations of organic matter as low as 2 mg/L can increase contaminant mobility.

We will restrict the overview of organic complexation of metals to specific
ligand groups, including humic substances, artificial complexing agents such as
nitrilotriacetic acid (NTA) or ethylenediaminetetraacetic acid (EDTA), and specific
compounds such as amino or carboxylic acids. The term humic substance (Section 11.7) refers to a group of organic acids that comprises the most abundant
fraction of naturally occurring organic matter dissolved in water. Complexation of
humic substances is usually modeled as a simple mixture of one or two "type"
ligands. These could include well-characterized components such as fulvic acid
and humic acid or ideal classes of ligands within the humic acid group. Perdue and
Lytle (1983) describe a more sophisticated approach that involves a continuous
ligand distribution model to deal with the inherent variability and changeability
of this ligand mixture.

NTA and EDTA are used in detergents and cleaning agents. By virtue of their
strong complexing capability, these two organic compounds are useful in studying metal-organic complexation. It is unlikely that these ligands would be encountered in ground water except perhaps in problems related to waste water disposal
into the subsurface. However as Morel (1983) points out, these ligands are important end members approximating the behavior of the most powerful ligands one
would likely encounter in nature.

Varieties of other naturally occurring organic compounds such as amino acids and carboxylic acid have significant complexing capabilities. However, the most common amino acids, serine, glycine, and aspartic acid are not present in sufficient concentrations to influence metal transport in ground waters. In deep ground water systems, the concentration of organic compounds can be relatively large. For example, carboxylic acids, forming as organic matter undergoes diagenesis in a geologic basin, can reach concentrations as high as 10,000 ppm (Surdam et al., 1984). The presence of these acids could be important in enhancing the solubility of aluminum in formation waters.

12.4   Reactions on Surfaces

Reactions between solutes and the surfaces of solids play an extremely important role in controlling the chemistry of ground water. In natural systems, these reactions can completely change the cation chemistry particularly in the case of the water-softening reactions. When contaminants are being transported, they can retard the spread of some constituents or even immobilize them.

Sorption Isotherms

When water containing a trace constituent with a concentration C_i is mixed with a solid medium and allowed to equilibrate, mass often partitions between the solution and the solid. The following equation represents this process

$$S = \frac{(C_i - C)(\text{solution volume})}{sm} \tag{12.30}$$

where C_i and C, the equilibrium concentration, have units such as milligrams per liter (mg/L) or micrograms per liter (µg/L), sm is the sediment mass (grams), and S is the quantity of mass sorbed on the surface (e.g., mg/g or µg/g). The results of this experiment provide the single point a on the S versus C plot (Figure 12.6). By repeating the procedure at the same temperature (hence, isotherm) with different values of C_i, the family of points forms a sorption isotherm (Figure 12.6). This experiment is known as a batch test.

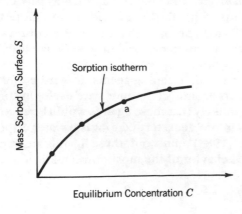

Figure 12.6
Example of a simple sorption isotherm constructed using data points from batch tests.

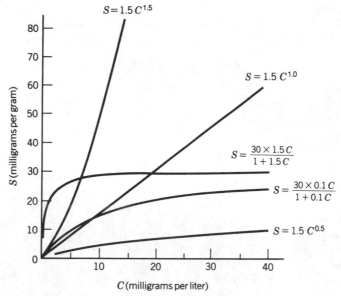

Figure 12.7
Example of Langmuir and Freundlich isotherms.

Real isotherms have no prescribed shape. They can be linear, concave, convex or a complex combination of all these shapes. Sorption is modeled by fitting an experimentally derived isotherm to one of several theoretical equations. Two of the most common relationships are the

$$\text{Freundlich isotherm:} \quad S = KC^n \tag{12.31}$$

and

$$\text{Langmuir isotherm:} \quad S = \frac{Q^0 KC}{1 + KC} \tag{12.32}$$

where K is a partition coefficient reflecting the extent of sorption, n is another constant usually ranging between 0.7 and 1.2 and Q^0 is the maximum sorptive capacity for the surface.

Figure 12.7 depicts Freundlich isotherms calculated with a K value of 1.5 and values of n of 0.5, 1.0, and 1.5 and Langmuir isotherms with values of K of 0.5 and 1.5 and a Q^0 value of 30 mg/g. These examples illustrate the range in curve shapes that one might fit with these two equations. If the fit is not satisfactory, there are other equations.

A Freundlich isotherm with $n = 1$ is a special case because this linear isotherm is easy to incorporate into mass transport models. The following equation relates S to C

$$S = K_d C \tag{12.33}$$

where K_d with units like milliliters per gram (ml/g) is the distribution coefficient (i.e., the slope of the linear sorption isotherm). Increasing values of K_d are indicative of a greater tendency for sorption.

Table 12.7

**Empirically derived K_d(ml/g) values from batch experiments at
25°C, 0.1 MPa for selected radionuclides (from Moody, 1982)**

Element	Salt	Basalt	Tuff	Granite
Tc	2^1	20^2, 0	10^2, 0	4^1
Pu	500^2, 50	200^2, 100	500^2, 40	500^2, 100
Np	30^2, 7	50^2, 3	50^2, 3	50^2, 1
I	0	0	0	0
U	1^1	6^1	4^1	4^1
Cs	1^3, 800	300	100	300
Ra	5	50	200	50
Sr	5	100	100	12
C	0	0	0	0
Am	300	50	50	200
Sn	1^2, 50	10^2, 100	50^2, 500	10^2, 500
Ni	6	50	50	10
Se	20^2; 100^2, 20	20^2, 5	2	2
Cm	300	50	50	200
Zr	500	500	500	500
Sm	50	50	50	100
Pd	3	50	50	10
Th	50^3, 100	500	500	500
Nb	50	100	100	100
Eu	50	50	50	100
Pa	50	100	100	100
Pb	2	25	25	5
Mo	0, 5^2, 1	10^2, 4	10^2, 4	5^2, 1

[1]No significant difference between value measured in oxidizing and reducing E_H.

[2]Reducing conditions, second value oxidizing conditions.

[3]First value for dome salt, second for bedded salt.

Modeling sorption with a single parameter K_d is the most common approach.
One of the most concerted efforts to apply this model was for the evaluation of
geological units as potential host rocks for nuclear wastes. The extent to which
sorption retards the spread of radionuclides has an extremely important bearing
on the suitability of a given site. Listed in Table 12.7 are estimates of K_d values for
selected radionuclides for a variety of different rock types. In many cases, the
values are large reflecting the significant capability for sorption.

The difficulty in obtaining consistent results from experiments has frustrated
the effort to characterize K_d values. In reality, K_d is not a constant but changes as
a function of the grain size and surface area, the agitation rate in batch experiments, the ground water composition, the pH and E_H, the method of adding
tracer in the experiment, the nuclide concentration, temperature and pressure,
the mineralogy of the porous medium, and the solid:solution ratio in the experiment (Moody, 1982). Undetected processes (e.g., mineral precipitation) can also
reduce concentrations and produce uncertain results. Thus, in an experiment, it
is difficult to control all the pertinent variables, or fully characterize the chemistry
of the ground water. Not surprisingly, the same sorption experiment conducted in
different laboratories can yield different K_d values.

In spite of its appealing and apparent simplicity, one should be cautious in using models based on distribution coefficients. The problem is that the processes are far too complex to be represented accurately by a simple one-parameter model. Modeling these systems using isotherms is an empirical approach that is not physically realistic. Thus, the limitations in their applicability are not too surprising. Nevertheless, as the next section shows, there are cases of sorption that can be modeled using a K_d approach.

Hydrophobic Sorption of Organic Compounds

The hydrophobic character of nonpolar organic molecules explains the sorption of many organic compounds. This same characteristic results in the diminished solubility of organic compounds in water. Nonpolar molecules tend to partition preferentially into nonpolar environments provided by small quantities of solid organic matter (e.g., humic substances and kerogen). This organic matter can occur as discrete solids, as films on individual grains, or as stringers of organic material in grains. Overall, the more hydrophobic an organic compound is the greater its tendency to partition into the solid phase.

This sorption process often can be modeled with a linear isotherm. One of the reasons why this simplified theory works is because the process is not surface controlled but driven by the energetics related to the changes in the structure of water that occurs when a nonpolar molecule dissolves. Several studies (e.g., Karickhoff and others, 1979; Schwarzenbach and Westall, 1981) have shown that K_d is proportional to the weight fraction of organic carbon (f_{oc}) such that

$$K_d = K_{oc}f_{oc} \qquad (12.34)$$

where K_{oc} is the partition coefficient of a compound between organic carbon and water with units such as (g solute sorbed/g soil organic carbon)/(g solute/m³ solution) (Karickhoff, 1984).

When values of f_{oc} and K_{oc} are known, Eq. 12.34 predicts the extent of partitioning. Organic carbon contents can be measured in the laboratory on porous-medium samples. However, this parameter is not well characterized. The range of values reported for geologic materials is extremely variable (Table 12.8). In estimating K_{oc}, it is fortunate that a good correlation exists between log K_{oc} and log K_{ow}, the octanol/water partition coefficient. This correlation is expected because the partitioning of an organic compound between water and organic carbon is not much different than between water and octanol.

Regression equations in practice describe the relationship between K_{oc} and K_{ow} (Griffin and Roy, 1985)

Karickhoff et al., 1979: $\log K_{oc} = -0.21 + \log K_{ow}$ (12.35)

Schwarzenbach and Westall, 1981: $\log K_{oc} = 0.49 + 0.72 \log K_{ow}$ (12.36)

Hassett et al., 1983: $\log K_{oc} = 0.088 + 0.909 \log K_{ow}$ (12.37)

The similarity in these equations is evident in Figure 12.8. Overall, they provide a preliminary estimate of K_{oc} when other information is lacking. In field studies, these initial estimates of K_{oc} need to be refined with field or laboratory experiments.

Table 12.8
A synthesis of some data on the organic carbon content of sediments

Site name	Type of deposit	Texture	Organic carbon content
Borden, Ontario[1]	Glaciofluvial	Fine-medium sand	0.0002
Gloucester, Ontario[2]	Glaciofluvial	Sands and gravels	0.0006
North Bay, Ontario[3]	Glaciofluvial	Medium sand	0.00017
Woolwich, Ontario[3]	Glaciofluvial	Fine-medium sand	0.00023
Chalk River, Ontario[3]	Glaciofluvial	Fine sand	0.00026
Cambridge, Ontario[3]	Glaciofluvial	Medium sand	0.00065
Rodney, Ontario[3]	Glaciofluvial	Fine sand	0.00102
Wildwood, Ontario[3]	Lacustrine	Silt	0.00108
Palo Alto, Baylands[4]	?	Silty sand	0.01
River Glatt, Switzerland[5]	Glaciofluvial	Sand, gravel	<0.000–0.01
Oconee River, Georgia[6]	River sediment	Sand,	0.0057
		coarse silt,	0.029
		medium silt,	0.02
		fine silt	0.0226

[1]Mackay and others (1986).

[2]Jackson (personal communication, 1989).

[3]J. Barker, University of Waterloo (personal communication, 1987).

[4]Mackay and Vogel (1985).

[5]Schwarzenbach and Giger (1985).

[6]Karickhoff (1981).

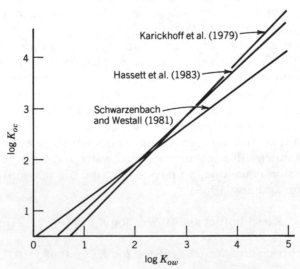

Figure 12.8
Correlation between the log octanol-water partition
coefficient (K_{ow}) and the log organic carbon/water
partition coefficient as determined by three different
studies (modified from Griffin and Roy, 1985).
Reprinted by permission of the Environmental Instit.
for Waste Management Studies, The Univ. of Alabama.

Because K_{ow} also correlates with solubility (S), the following relationships exist between K_{oc} and S (Griffin and Roy, 1985)

$$\text{Karickhoff et al., 1979:} \quad \log K_{oc} = 0.44 - 0.54 \log S_1 \quad (12.38)$$

$$\text{Kenaga, 1980:} \quad \log K_{oc} = 3.64 - 0.55 \log S_2 \quad (12.39)$$

$$\text{Hassett et al., 1983:} \quad \log K_{oc} = 3.95 - 0.62 \log S_2 \quad (12.40)$$

where S_1 is solubility in water expressed as a mole fraction and S_2 as mg/L.

The following example illustrates the application of the empirical equations in estimating a distribution coefficient.

Example 12.4

An aquifer has an f_{oc} of 0.01. Estimate a log K_d describing the sorption of 1,2-dichloroethane having a log $K_{ow} = 1.48$. Starting with the basic equation $K_d = f_{oc}K_{oc}$ and taking logs of both sides gives

$$\log K_d = \log f_{oc} + \log K_{oc}$$

Substitution for log K_{oc} with the Schwarzenbach and Westall (1981) equation yields

$$\log K_d = \log 0.01 + 0.49 + 0.72(1.48) = -0.445$$

It is tempting to extend the concept of hydrophobic sorption to more polar and less hydrophobic (hydrophilic) compounds (for example, methanol). Such compounds have large aqueous solubilities and relatively low values of log K_{ow} (for example, -0.66 for methanol). In contrast to the nonpolar compounds, these are preferentially partitioned to the aqueous phase with much less affinity for organic carbon. Karickhoff (1985) recommends that the model of hydrophobic sorption should only be used for compounds with solubilities less than 10^{-3} M.

When organic molecules carry a charge, hydrophobic sorption represents only part of the total sorption. Electrostatic forces caused by interaction with charged surfaces are also important. Organic acids and bases, for example, ionize to some extent depending upon the pH of the ground water. Above a specified pH, a significant proportion of the compound will occur in ionic form. The pH range over which ionization is significant depends upon the equilibrium constant for the reaction (pK_a or pK_b). For example, with the following acid and base dissociation reactions

$$HA = H^+ + A^- \qquad BOH = B^+ + OH^-$$

the organic acid dissociates to a negatively charged compound (A^-), and the base to a positively charged compound (B^+). The extent of acid or base dissociation is defined of in terms of the ionization fractions (α and β) where

$$\alpha = \frac{[A^-]}{([A^-] + [HA])} \quad \text{and} \quad \beta = \frac{[B^+]}{([B^+] + [BOH])} \qquad (12.41)$$

By substitution of the mass action equations

$$K_a = \frac{[H^+][A^-]}{[HA]} \quad \text{and} \quad K_b = \frac{[B^+][OH^-]}{[BOH]}$$

Eq. 12.41 becomes

$$\alpha = \frac{K_a}{(K_a + [H^+])} \quad \text{and} \quad \beta = \frac{K_b}{(K_b + [OH^-])}$$

These expressions provide a convenient way to determine the ionization fractions for the organic acid or base. For example, in the case of organic acids, HA is sorbed but A^- moves through the system without retardation. Thus a large ionization fraction (say, >0.5) implies a deterioration in the ability for hydrophobic sorption to retard contaminant spread. In the case of organic bases, BOH is sorbed and B^+ can undergo cation exchange. Thus, the effect of ionization on transport is more complicated for a base than an acid.

Another instance where the hydrophobic model of sorption begins to break down is with small f_{oc} values. Organic compounds can sorb to a small but significant extent on inorganic surfaces. This sorption is negligible in media with a large f_{oc}. At the critical level of organic matter (McCarty and others, 1981), the sorption due to organic and inorganic solids is equal. The critical level of organic matter is not constant but a function of K_{ow} and the surface area of inorganic solids. Thus, for organic compounds with a strong affinity for organic carbon (high K_{ow}), the critical level is smaller than it would be for less readily sorbed compounds. Similarly, the critical level will be greater for a medium with a larger surface area (i.e., a large clay content). McCarty and others (1981) have developed an empirical equation to estimate the critical organic fraction with K_{ow} and surface area as controlling parameters.

There are other limitations in the model of hydrophobic sorption. A linear isotherm does not describe all sorption especially at large concentrations. Commonly, nonlinear forms of the Freundlich or other isotherms best describe experimental data, especially when the concentration of the organic compound is relatively high. Another problem relates to the question of desorption, the partitioning of organic compounds back into the aqueous phase. Often, the isotherms for desorption may be quite different than the comparable ones for sorption. This failure for reversible processes to behave in the same way is termed hysteresis. Commonly, hysteresis results in an isotherm that during desorption exhibits a greater affinity for partitioning to the solid than was the case during sorption. Other limitations of the model include problems of competition between compounds and kinetic effects in sorption desorption reactions.

Multiparameter Equilibrium Models

A more rigorous framework for surface reactions exists that can account for properties of the solution and the solid surfaces. To illustrate this approach, we will examine classical cation exchange reactions on clay minerals and trace metal sorption on variably charged surfaces.

Table 12.9
Properties of the common clay minerals (from Yong, 1985)*

Clay mineral	Lattice description	C.E.C. (meq/100 g)	Surface area (m²/g)	Source of charge	Charge characteristics
Kaolinites	1:1, strong H bonds	5–15	15	Edges, broken bonds (hydroxyl- ated edges)	Variable charge
Illites	1:2, strong K bonds	25	80	Isomorphous substitution, some broken bonds at edges	Mostly fixed charge
Chlorites	2:2, strong bonds	10–40	80	Isomorphous substitution	Fixed charge
Vermiculites	1:2, weak Mg bonds	100–150	n.d.	Isomorphous substitution	Fixed charge
Montmorillonites	1:2, very weak bonds	80–100	800	Isomorphous substitution, some broken bonds at edges	Mostly fixed charge

n.d. – not determined.

Cation exchange is the most well known of the surface reactions. Driving the
reaction is the electrostatic attraction between charged cations in solution and the
surface charge on clay minerals or other oxide surfaces. Clay minerals in particular have a significant fixed negative charge on basal surfaces available to ions
because of substitutions of cations of a lower valence in the lattice, and to a lesser
extent because of broken bonds at the edges of the mineral (Table 12.9). Kaolinite
differs from the rest of the clay minerals. It has a variable charge as hydroxylated
edges change their charge as a function of pH.

Cations binding to the exchange sites balance the negative surface charges of
the mineral. These cations can exchange with other cations present in solution.
The term cation exchange capacity describes the quantity of exchangeable cations
sorbed on the surface or indirectly the negative charge of the soil (Yong, 1985).
Cation exchange capacity (CEC) has units of milliequivalents per 100 g of sample,
and for the fixed-charge clay minerals is an important descriptive parameter.
Values of CEC vary considerably depending upon the type of clay mineral (Table
12.9). Illite and chlorite fall at the lower end of the range of observed values while
the vermiculite and smectite groups are at the upper end of the range. Much of the
variability in CEC is a function of the surface area of the exchanger (Table 12.9).

Clay minerals exhibit a preference as to which ion occupies an exchange site.
Attempts have been made over the years to establish a selectivity sequence, where
equivalent amounts of cations are arranged according to their relative affinity for
an exchange site. Many variations of this sequence have been published, such as
the following one proposed by Yong (1985)

$$Li^+ < Na^+ < H^+ < K^+ < NH_4^+ < Mg^{2+} < Ca^{2+} < Al^{3+}$$

In general, the greater the charge on a cation, the greater the affinity for an exchange site. Although the idea of an affinity sequence provides a useful concept for addressing exchange preference, it is too simplified to be applied rigorously to evaluate exchange processes (Spositio, 1984).

The general form of a cation exchange reaction is

$$nMX + mN^{n+} = nM^{m+} + mNX \tag{12.42}$$

where M and N are metal cations with charges m^+ and n^+, respectively, and MX and NX represent the metals sorbed on the solid phase. The mass law expression for this reaction is

$$K = \frac{[M^{m+}]^n[NX]^m}{[N^{n+}]^m[MX]^n} \tag{12.43}$$

where the bracketed quantities are the activities of species in solution and on the exchanger. As with ions in solution, activity models are required for ions on the exchanger. The most common model assumes that the activities of adsorbed species are equal to their mole fraction, for example, $[NX]^m = x_{NX}^m$, and $[MX]^n = x_{MX}^n$, where the mole fraction of NX is

$$x_{NX} = \frac{(N^*)}{(N^*) + (M^*)}$$

with (N^*) and (M^*) representing the molar concentration of species sorbed on the solid. With an appropriate activity model, the mass law expression is

$$K_s = \frac{[M^{m+}]^n x_{NX}^m}{[N^{n+}]^m x_{MX}^n} \tag{12.44}$$

where K_s is now known as a selectivity coefficient. In general, once an activity model is assumed for the adsorbed species, the equilibrium constant and the selectivity coefficient are no longer equal.

As an example of how this theory is applied, consider the following exchange reaction between an ion in solution P^+ and an ion Q^+ sorbed on the surface as Q-clay

$$P^+ + Q\text{-clay} = Q^+ + P\text{-clay}$$

The mass law expression assuming that concentration is equal to activity is

$$K_s = \frac{(Q^+)\, x_{P\text{-clay}}}{(P^+)\, x_{Q\text{-clay}}} \tag{12.45}$$

where P^+ and Q^+ are molar concentrations and, $x_{P\text{-clay}}$ and $x_{Q\text{-clay}}$ are mole fractions.

We can manipulate Eq. 12.45 further to learn more about the distribution coefficient, K_d. Assume that $(P^+) << (Q^+)$ and that no other ions are present in solution. Because both ions are singly charged, the molar concentrations and equivalent concentrations of the sorbed species are the same so that $CEC = (P^*) + (Q^*)$ and because (P^+) is small $CEC \simeq (Q^*)$ and $\tau = (P^+) + (Q^+) \simeq (Q^+)$. If P^+ and

Q^+ are the only ions in solution, $x_{P\text{-clay}} = (P^*)/(P^* + Q^*)$, and $x_{Q\text{-clay}} = (Q^*)/(P^* + Q^*)$. Substitution of the approximations into Eq. 12.45 gives

$$K_s = \frac{\tau(P^*)}{(P^+)CEC} \tag{12.46}$$

Given that $K_d = (P^*)/(P^+)$, Eq. 12.46 becomes

$$K_d = \frac{K_s CEC}{\tau} \tag{12.47}$$

Thus, K_d is a function of the properties of the exchanger and the solution. Values are not unique constants and must be determined for particular ions on the exchanger as a function of differing solution concentrations. This is one of the main weaknesses in the K_d approach for modeling sorption.

One important example of cation exchange is the natural water softening reactions. Ground water with relative high concentrations of Ca^{2+} and Mg^{2+} (hard water) and other minor species such as Fe^{2+} changes its chemical composition as it moves into marine clay or shale with Na^+ on the exchange sites. The following reaction describes this change

$$
\begin{matrix}
Ca^{2+} & & Ca \\
Mg^{2+} + 2Na\text{-clay} = 2Na^+ + & Mg\text{-clay} \\
Fe^{2+} & & Fe
\end{matrix}
$$

Selectivity coefficients for these reactions are such that in clay-rich sediments, the ground water is softened with nearly all the Ca^{2+} and Mg^{2+} replaced by Na^+. Iron, which is often associated with problems of taste and staining, is also removed. Chapter 15 will present more specific examples of how this process affects the composition of natural ground waters.

Another important type of sorption reaction involves solids whose surface charges change as a function of ground water composition. Examples of such solids include kaolinite, metal oxides (for example, SiO_2, Al_2O_3), and metal oxyhydroxides (e.g., $Al(OH)_3$, $Si(OH)_4$).

Typically, the surface on one of these solids is positively charged at low pHs. However, at higher pHs, the surface becomes negatively charged and functions as a cation exchanger. There is a pH in between, known as the zero point of charge (ZPC), where the surface has zero net charge. These surface-charge relationships are unique for given solids and solutions, and usually must be determined experimentally.

The variable surface charge is explained by the presence on the surface of species such as XOH, XO^-, and XOH_2^+, where XOH represents the hydroxylated surface and XO^- the ionized surface site (Morel, 1983). The following set of equilibrium reactions describes the concentrations of these surface species

$$XOH = H^+ + XO^- \qquad XOH + H^+ = XOH_2^+$$

These equations illustrate how the species comprising the surface depend on the pH of the ground water. At low pH, XOH_2^+ is the dominant surface species, while at high pHs XO^- is dominant. The surface species are free to react with metals or metal complexes. For example, the sorption of a metal ion (M^{2+}) is described by an equilibrium reaction similar to the preceding one or

$$M^{2+} + XO^- = XOM^+$$

By solving the resulting system of equilibrium reactions, one can calculate the concentration of the metal on the surface (XOM^+). As Morel (1983) shows, this or other surface species bind the metal or metal complexes. Choosing the appropriate species to include usually involves matching experimental data to theoretical models. This model for exchange is sufficiently general to account for the sorption of metal complexes.

The state of the art in sorption modeling is a much more complex version of this model. The surface ionization and complexation-type models not only account for variations in surface charge but also the energetics involved with transporting constituents back and forth from the bulk solution to the surface. To apply these models rigorously requires a great deal of information about the chemistry of the ground water; the charge, potential, and capacitance in the double layer; and equilibrium constants for the surface reactions. It is feasible to model surface reactions in a realistic way. However, these approaches are complex and require considerable data. Such models are just now being incorporated into state-of-the-art contaminant transport models and being evaluated in relation to actual field problems.

12.5 Oxidation-Reduction Reactions

The oxidation-reduction or redox reactions are unlike any of the other reactions so far because they are mediated by microorganisms. The microorganisms act as catalysts speeding up what otherwise are extremely sluggish reactions. Microorganisms, mainly bacteria, occur ubiquitously in the subsurface. While they can be found in water, they usually form colonies attached as films on the porous medium or fracture surfaces. Microorganisms use redox reactions as a source of energy. In spite of the help from microorganisms, redox reactions are slow in relation to other reactions and typically are treated from a kinetic viewpoint.

Oxidation Numbers, Half-Reactions, Electron Activity, and Redox Potential

Our overview of basic concepts begins with an explanation of the oxidation number. To take into account bond polarity, charges defining the oxidation number are assigned to a compound. For example, in the case of CO_3^{2-}, the oxidation number of carbon is (+IV) and that for oxygen (−II) giving the molecule a net negative charge of −2. By convention, oxidation numbers are written as roman numerals.

The term oxidation refers to the removal of electrons from an atom forcing a change in the oxidation number of an element. For example, the following reaction

$$Fe^{2+} = Fe^{3+} + e^-$$

describes the oxidation of Fe^{2+}, where e^- is an electron. In this reaction, the oxidation number for iron changes from (+II) to (+III).

Table 12.10

Some elements found with more than one oxidation state and examples of ions or solids formed from those elements*

Element and oxidate state in brackets	Example
C(+IV)	HCO_3^-, CO_3^{2-}
C(O)	CH_2O, C
C(−IV)	CH_4
Cr(+IV)	CrO_4^{2-}, $Cr_2O_7^{2-}$
Cr(+III)	Cr^{3+}, $Cr(OH)_3$
Fe(+III)	Fe^{3+}, $Fe(OH)_3$
Fe(+II)	Fe^{2+}
N(+V)	NO_3^-
N(+III)	NO_2^-
N(O)	N
N(−III)	NH_4^+, NH_3
S(+VI)	SO_4^{2-}
S(+V)	$S_2O_6^{2-}$
S(+II)	$S_2O_3^{2-}$
S(−II)	H_2S, HS^-

*From Morel, F.M.M., 1983. Principles of Aquatic Chemistry. Copyright © 1983 by John Wiley & Sons, Inc. Reprinted by permission of John Wiley & Sons, Inc.

Reduction refers to the addition of an electron to lower the oxidation number

$$Fe^{3+} + e^- = Fe^{2+}$$

All redox reactions transfer electrons and involve elements with more than one oxidation number. Listed in Table 12.10 are some of the more important of these elements, their typical oxidation states, and some of the ions and solids that form. A host of trace metals not included in the table also have variable oxidation numbers.

There is a direct analogy between oxidation-reduction reactions and the acid-base reactions studied in Section 12.1. Instead of a transfer of protons from an acid to a base, there is a transfer of electrons from a reductant (electron donor) to an oxidant (electron acceptor). No free electrons result from a redox reaction because each complete reaction involves a pair of redox reactions (Morel, 1983)

$$Ox_1 + Red_2 = Red_1 + Ox_2 \tag{12.48}$$

where Ox and Red refer to oxidants and reductants, respectively. This equation is analogous to Eq. 12.1, which applies to acid-base reactions.

Following is an example of a complete redox reaction

$$O_2 \quad + \quad 4Fe^{2+} \quad + 4H^+ = 2H_2O \quad + \quad 4Fe^{3+}$$
$$O(0) \quad Fe(+II) \qquad\qquad O(-II) \quad Fe(+III) \tag{12.49}$$
$$Ox_1 \qquad Red_2 \qquad\qquad\quad Red_1 \qquad Ox_2$$

consisting of two separate half-redox reactions Ox_1-Red_1 and Red_2-Ox_2:

$$O_2 + 4H^+ + 4e^- = 2H_2O$$
$$4Fe^{2+} = 4Fe^{3+} + 4e^- \tag{12.50}$$

Adding these two reactions together gives the original reaction (12.49). Half-reactions are unique in that electrons are reactants or products. Otherwise, these reactions are similar to other equilibrium reactions.

Mass law equations can be written in terms of the concentration or activities of reactants and products (including the electrons) and an appropriate equilibrium constant. For example, the mass law expression for the half-reaction

$$Ox + ne^- = Red$$

is

$$K = \frac{[Red]}{[Ox][e^-]^n} \tag{12.51}$$

Rearrangement of Eq. 12.51 gives the electron activity $[e^-]$ for a half-reaction

$$[e^-] = \left\{ \frac{[Red]}{[Ox]K} \right\}^{1/n} \tag{12.52}$$

In the same way that pH defines $[H^+]$, pe defines electron activity, $[e^-]$. Rewriting Eq. 12.52 by taking the negative logarithm of both sides gives

$$pe = -\log[e^-] = \left(\frac{1}{n} \right) \left\{ \log K - \log \frac{[Red]}{[Ox]} \right\}$$

When a half-reaction is written in terms of a single electron or $n = 1$, the log K term is written as pe^o so that

$$pe = pe^o - \log \frac{[Red]}{[Ox]} \tag{12.53}$$

Morel (1983, p. 319) and Pytkowicz (1983b, p. 249) tabulate values of pe^o or log K. Some of the important half-reactions together with values of pe^o are presented in Table 12.11.

Another way of characterizing redox conditions is in terms of E_H, the redox potential. Values of E_H have units of volts, acknowledging that a redox reaction involves the transfer of electrons. E_H is related to pe by the following equation

$$E_H = \frac{2.3RT}{F} pe \tag{12.54}$$

where F is the Faraday constant defined as the electrical charge of 1 mole of electrons (96,500 Coulombs) with $2.3RT/F$ equal to 0.059 V at 25 °C. In practice, pe is taken to be the calculated value of electron activity and E_H the measured electrode potential for an electrochemical cell. In other words, pe is a calculated quantity and E_H is a measured one (Stumm and Morgan, 1981). The following example will illustrate how we calculate the pe of a ground water assuming a redox couple is in equilibrium.

Table 12.11
Some important half redox reactions*

	$pe^o = \log K$
HYDROGEN	
$H^+ + e^- = \frac{1}{2} H_2(g)$	0
OXYGEN	
$\frac{1}{2} O_3(g) + H^+ + e^- = \frac{1}{2} O_2(g) + \frac{1}{2} H_2O$	$+35.1$
$\frac{1}{4} O_2(g) + H^+ + e^- = \frac{1}{2} H_2O$	$+20.75$
$\frac{1}{2} H_2O_2 + H^+ + e^- = H_2O$	$+30.0$
(Note also $HO_2^- + H^+ = H_2O$; $\log K = 11.6$)	
NITROGEN	
$NO_3^- + 2H^+ + e^- = \frac{1}{2} N_2O_4(g) + H_2O$	$+13.6$
(Note: $N_2O_4(g) = 2NO_2(g)$; $\log K = -0.47$)	
$\frac{1}{2} NO_3^- + H^+ + e^- = \frac{1}{2} NO_2^- + \frac{1}{2} H_2O$	$+14.15$
(Note $NO_2^- + H^+ = HNO_2$; $\log K = 3.35$)	
$\frac{1}{3} NO_3^- + \frac{4}{3} H^+ + e^- = \frac{1}{3} NO(g) + \frac{2}{3} H_2O$	$+16.15$
$\frac{1}{4} NO_3^- + \frac{5}{4} H^+ + e^- = \frac{1}{8} N_2O(g) + \frac{5}{8} H_2O$	$+18.9$
$\frac{1}{5} NO_3^- + \frac{6}{5} H^+ + e^- = \frac{1}{10} N_2(g) + \frac{3}{5} H_2O$	$+21.05$
$\frac{1}{8} NO_3^- + \frac{5}{4} H^+ + e^- = \frac{1}{8} NH_4^+ + \frac{3}{8} H_2O$	$+14.9$
SULFUR	
$\frac{1}{2} SO_4^{2-} + H^+ + e^- = \frac{1}{2} SO_3^{2-} + \frac{1}{2} H_2O$	-1.65
[Note also $(SO_3^{2-} + H^+ = HSO_3^-$; $\log K \cong 7)$]	
$\frac{1}{4} SO_4^{2-} + \frac{5}{4} H^+ + e^- = \frac{1}{8} S_2O_3^{2-} + \frac{5}{8} H_2O$	$+4.85$
$\frac{1}{6} SO_4^{2-} + \frac{4}{3} H^+ + e^- = \frac{1}{48} S_8^0(s. ort.) + \frac{2}{3} H_2O$	$+6.03$
[Note also $\frac{1}{8} S_8^0(s. ort.) = \frac{1}{8} S_8^0(s. col.)$; $\log K = -0.6$]	
$\frac{3}{19} SO_4^{2-} + \frac{24}{19} H^+ + e^- = \frac{1}{38} S_6^{2-} + \frac{12}{19} H_2O$	$+5.41$
$\frac{5}{32} SO_4^{2-} + \frac{5}{4} H^+ + e^- = \frac{1}{32} S_5^{2-} + \frac{5}{8} H_2O$	$+5.29$
(Note also $S_5^{2-} + H^+ = HS_5^-$; $\log K = 6.1$)	
$\frac{2}{13} SO_4^{2-} + \frac{16}{13} H^+ + e^- = \frac{1}{26} S_4^{2-} + \frac{8}{13} H_2O$	$+5.12$
(Note also $S_4^{2-} + H^+ = HS_4^-$; $\log K = 7.0$)	
$\frac{1}{8} SO_4^{2-} + \frac{5}{4} H^+ + e^- = \frac{1}{8} H_2S(aq) + \frac{1}{2} H_2O$	$+5.13$
(Note also $H_2S(g) = H_2S(aq)$; $\log K_h = 1.0$, and other acid-base, coordination, and precipitation reactions)	
TRACE METALS	
Cr $\quad \frac{1}{3} HCrO_4^- + \frac{7}{3} H^+ + e^- = \frac{1}{3} Cr^{3+} + \frac{4}{3} H_2O$	$+20.2$
(Note $HCrO_4^- = H^+ + \frac{1}{2} Cr_2O_7^{2-} + \frac{1}{2} H_2O$, $\log K = -15$; $HCrO_4^- = H^+ + CrO_4^{2-}$, $\log K = -6.5$; and various Cr(III) precipitation and coordination reactions)	
Mn $\quad \frac{1}{5} MnO_4^- + \frac{8}{5} H^+ + e^- = \frac{1}{5} Mn^{2+} + \frac{4}{5} H_2O$	$+25.5$
$\frac{1}{2} MnO_2(s) + 2H^+ + e^- = \frac{1}{2} Mn^{2+} + H_2O$	$+20.8$

Table 12.11
(Continued)

		$pe^\circ = \log K$
TRACE METALS *(cont.)*		
Fe	$Fe^{3+} + e^- = Fe^{2+}$	$+13.0$
	$\frac{1}{2}Fe^{2+} + e^- = \frac{1}{2}Fe(s)$	-7.5
	$\frac{1}{2}Fe_3O_4(s) + 4H^+ + e^- = \frac{3}{2}Fe^{2+} + 2H_2O$	$+16.6$
Co	$Co(OH)_3(s) + 3H^+ + e^- = Co^{2+} + 3H_2O$	$+29.5$
	$\frac{1}{2}Co_3O_4(s) + 4H^+ + e^- = \frac{3}{2}Co^{2+} + 2H_2O$	$+31.4$
Cu	$Cu^{2+} + e^- = Cu^+$	$+2.6$
	$\frac{1}{2}Cu^{2+} + e^- = Cu(s)$	-5.7
Se	$\frac{1}{2}SeO_4^{2-} + 2H^+ + e^- = \frac{1}{2}H_2SeO_3 + \frac{1}{2}H_2O$	$+19.4$
	$\frac{1}{4}H_2SeO_3 + H^+ + e^- = \frac{1}{4}Se(s) + \frac{3}{4}H_2O$	$+12.5$
	$\frac{1}{2}Se(s) + H^+ + e^- = \frac{1}{2}H_2Se$	-6.7
	(Note also $H_2Se = H^+ + HSe^-$, $\log K = -3.9$; $H_2SeO_3 = H^+$ $+ HSeO_3^-$, $\log K = -2.4$; $HSeO_3^- = H^+ + SeO_3^{2-}$, $\log K =$ -7.9; $SeO_4^{2-} + H^+ = HSeO_4^-$, $\log K = +1.7$)	
Ag	$AgCl(s) + e^- = Ag(s) + Cl$	$+3.76$
	$Ag^+ + e^- = Ag(s)$	-13.5
Hg	$\frac{1}{2}Hg^{2+} + e^- = Hg(l)$	-14.4
	$Hg^{2+} + e- = \frac{1}{2}Hg_2^{2+}$	$+15.4$
Pb	$\frac{1}{2}PbO_2 + 2H^+ + e^- = \frac{1}{2}Pb^{2+} + H_2O$	$+24.6$
	(Note many other reactions for Mn, Fe, Co, Cu, Se, Ag, Hg, Pb)	

*From Morel, F.M.M., 1983. Principles of Aquatic Chemistry. Copyright © 1983 by John Wiley & Sons, Inc. Reprinted by permission of John Wiley & Sons, Inc.

Example 12.5

A ground water has $(Fe^{2+}) = 10^{-3.3}$ M and $(Fe^{3+}) = 10^{-5.9}$ M. Calculate the pe at 25 °C assuming that the activities of Fe species are equal to their concentrations. What should be the measured E_H of this solution?

From Table 12.11, the half-reaction for the reduction of Fe^{3+} to Fe^{2+} is

$$Fe^{3+} + e^- = Fe^{2+} \quad \text{with } pe^o = 13.0$$

Substitution into Eq. 12.53 gives

$$pe = pe^o - \log\left\{\frac{(Fe^{2+})}{(Fe^{3+})}\right\}$$

$$pe = 13.0 - \log\left\{\frac{10^{-3.3}}{10^{-5.9}}\right\}$$

$$pe = 13.0 - 2.6 = 10.4$$

$$E_H = \frac{2.3RT}{F}\, pe = 0.059 \times 10.4 = 0.61 \text{ V}$$

Kinetics and Dominant Couples

So far, we have considered redox phenomena as equilibrium reactions. The implied assumption is that at least some of the half-redox reactions are in equilibrium. However, redox reactions often are slow relative to physical processes because the number of active microorganisms is small or some of the reactants are not metabolized easily by microorganisms. Another feature of many redox reactions is that with extremely large equilibrium constants they are essentially irreversible. Take for example a reaction where an excess of dissolved oxygen oxidizes organic matter (as CH_2O)

$$CH_2O + O_2 \rightarrow CO_2 + H_2O$$

This reaction progresses to the right until all the organic matter disappears. This example raises the obvious question of how redox phenomena apply to nonequilibrium systems. A constant value of pe or E_H exists when the concentration of one of the couples, O_2/H_2O in this case, is much greater than the other. Because utilization of all the organic matter will require a negligibly small proportion of the oxygen, the redox potential defined by the O_2/H_2O couple does not change as a function of the reaction progress. Thus, the measured E_H may be dominated by one particular couple, although mixed potentials are the rule (Stumm and Morgan, 1981).

The concept of a dominant couple is useful in understanding the redox chemistry of natural and contaminated systems. However, research has shown that redox equilibrium among all the couples does not generally occur (Lindberg and Runnells, 1984). Effectively, pe values computed from each couple might span a broad range of values. Thus while it is tempting to use the concept of master pe to predict ion speciation, this practice is not recommended (Lindberg and Runnells, 1984). As Stumm and Morgan (1981) point out, calculations of pe provide indications of the direction in which a system is evolving. This general absence of internal equilibrium implies that care must be taken in assessing pe values. Further, the difficulty in interpreting E_H values measured with electrodes means that a quantitative description of redox reactions remains elusive for ground water.

Control on the Mobility of Metals

Redox reactions also influence the mobility of metal ions in solution. One example is the behavior of metals in sulfur systems involving the sulfate/sulfide couple. Assume that the total sulfur in a system (S_T) exists as four possible species, SO_4^{2-}, H_2S, HS^-, and S^{2-}, where in the first sulfur occurs as $S(+VI)$ and in the last three as $S(-II)$. The relationship among H_2S, HS^-, and S^{2-} is fixed by equilibrium relationships, much like the carbonate system. In oxic environments, SO_4^{2-} is the dominant species of the couple with essentially none of the $S(-II)$ species present. Sulfate ion is reduced in an anoxic environment, and the other three species dominate.

As the sulfur species change with changing redox conditions, so too do the solids that precipitate. When SO_4^{2-} dominates, metal concentrations can be relatively high because there are no solubility constraints to remove solids from solution. Once the system becomes reducing, almost all the metal and sulfur is removed from the system as metal sulfides, which are often relatively insoluble. Thus, in one case, the total concentration of the metal is found as a metal ion and in the other as a solid.

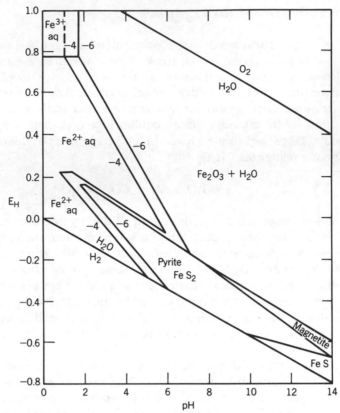

Figure 12.9
Example of an E_H-pH diagram showing the stability relationships of iron oxides and sulfides (from Macumber, 1984).
Reprinted by permission First Canadian/American Conference
on Hydrogeology. Copyright © 1984. All rights reserved.

In a more complex system where pH can change, other ions are present, and
one of several solids could form, it would be useful to have a way of presenting the
essential features of the redox and solid phase chemistry. E_H-pH or pe-pH diagrams serve this purpose nicely. Shown in Figure 12.9 is an E_H-pH diagram for
iron. The upper and lower lines on the field represent the oxidation of water to O_2
and the reduction of water to H_2. These two half-reactions define the upper and
lower limits for oxidation or reduction. The other lines are boundaries for what
are known as stability fields. These fields are labeled with the one dominant ion
or solid for the specified E_H and pH. When an ion dominates (e.g., Fe^{2+}), the element is mobile because a relatively small proportion of the metal occurs as a solid.
In the stability fields for solid phases (e.g., FeS_2 and Fe_2O_3), nearly all the iron will
exist as that phase, making the element essentially immobile.

For many problems, the preparation of an E_H-pH or pe-pH diagram is a first
step to understanding how oxidation-reduction reactions can influence transport.
Morel (1983) and Garrels and Christ (1965) provide a step-by-step description of
how to construct a diagram. Otherwise, one can use published figures.

Biotransformation of Organic Compounds

A large number of half-reactions involve the oxidation of organic compounds into simpler inorganic forms such as CO_2 and H_2O. These reactions are referred to as biodegradation or biotransformation reactions because they are microbially catalyzed. In some cases, oxygen acts as the electron acceptor

$$\tfrac{1}{4} CH_2O + \tfrac{1}{4} O_2(g) = \tfrac{1}{4} CO_2(g) + \tfrac{1}{4} H_2O$$

However, when O_2 is unavailable, other species, for example, NO_3^-, Fe($+$III), Mn($+$IV), SO_4^{2-}, and CO_2, accept electrons. A few of the most important of these biotransformation reactions are

Fe(III) reduction: $\quad \tfrac{1}{4} CH_2O + Fe(OH)_3 + 2H^+ = \tfrac{1}{4} CO_2(g)$

$$+ Fe^{2+} + \tfrac{11}{4} H_2O$$

denitrification: $\quad CH_2O + \tfrac{4}{5} NO_3^- + \tfrac{4}{5} H^+ = CO_2(g)$

$$+ \tfrac{2}{5} N_2 + \tfrac{7}{5} H_2O$$

sulfate reduction: $\quad CH_2O + \tfrac{1}{2} SO_4^{2-} + \tfrac{1}{2} H^+ = \tfrac{1}{2} HS^-$

$$+ H_2O + CO_2(g)$$

methane formation: $\quad CH_2O + \tfrac{1}{2} CO_2(g) = \tfrac{1}{2} CH_4 + CO_2(g)$

where CH_2O is a "type" organic compound.

Biotransformation reactions are important when the organic compound is a ground water contaminant. If biotransformation occurs rapidly, the reaction can attenuate contaminant concentrations and perhaps alter the major ion chemistry. Unfortunately, the broad diversity in the organic compounds involved and the complexity of microbiological reactions in general make it difficult to deal with these processes.

We can conceptualize the kinetics of biotransformation reactions in relation to a model biofilm (Figure 12.10). The biofilm consists of bacteria held together

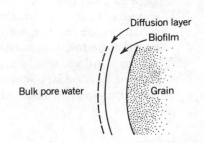

Figure 12.10
Conceptualization of a biofilm at a pore scale.

and to the porous medium by extracellular polymers (McCarty and others, 1984). The biotransformation reactions take place within the biofilm. In terms of this biofilm system, there are two coupled processes of interest, the transport of substrates or organic compounds from the bulk ground water into the biofilm (Figure 12.10) and the growth (or decline) in the mass of the biofilm. The processes are coupled because transport of the substrate depends in part on how much biofilm there is and the mass of the biofilm depends in part upon the availability of the substrate as a source of energy (McCarty and others, 1984).

Rates Limited by Substrate Availability

Rittmann and McCarty (1980a) developed a model to simulate biofilm processes where the concentration of a single substrate is sufficiently small to be rate limiting. Later, we will examine a case where the availability of the electron acceptor (e.g., O_2) limits the reaction rate.

The organic compound moves by diffusion through the diffusion layer and into the biofilm (Figure 12.10) and ultimately is lost by bacterial utilization. The following empirical, kinetic law describes the rate of loss (Bouwer and McCarty, 1984)

$$\frac{dS_f}{dt} = \frac{k\,X_f S_f}{K_s + S_f} \tag{12.55}$$

where S_f is the concentration of the rate limiting substrate, X_f is the active cell density in the biofilm (mass of bacteria L^{-3}), k is the maximum specific rate of substrate utilization (mass of substrate/mass of bacteria per unit time), and K_s is the Monod half-maximum-rate concentration (mass of substrate L^{-3}). K_s is the concentration at which the specific rate of utilization is one-half of k.

Substrate utilization is a function of two constants (k and K_s), the substrate concentration, and the active cell density X_f. McCarty and others (1984) present expressions for bacterial growth, in essence describing how X_f changes with time as the difference between growth rates (that is, substrate conversion) and loss rates. By solving the coupled mathematical problem of substrate transport and bacterial growth, it is possible to model a steady-state biofilm system.

Based on theoretical studies and experimental work, Rittmann and McCarty (1980a, 1980b) developed the concept of a minimum substrate concentration. Once the concentration of the organic substrate falls below S_{min}, the oxidation reaction does not yield sufficient energy to support the growth of the biofilm, which shrinks to zero thickness. Thus, below some minimum or irreducible concentration, the biofilm loses its capability to degrade an organic compound. Interestingly, if a second compound is present, which supports growth of the biofilm, the concentration of a trace substrate can be reduced below the minimum concentration. The term primary substrate describes one compound or the aggregate of many compounds present at concentration above the minimum that supports long-term bacterial growth (Bouwer and McCarty, 1984). Secondary substrates are trace organic compounds that by themselves are not capable of supporting the microorganisms. The processing of these substrates is referred to as primary and secondary utilization.

For a secondary substrate with S_f much less than K_s, Eq. 12.55 simplifies to the following first-order rate law (Bouwer and McCarty, 1984)

$$\frac{dS_f}{dt} = \frac{k\,X_f S_f}{K_f} \tag{12.56}$$

This equation states that substrate utilization within the biofilm is proportional to the concentration of bacteria and the ratio k/K_s. Under some conditions, bacterial utilization, as described by Eq. 12.56, can be the rate limiting process controlling transport of the secondary substrate within the biofilm. In other words, diffusion is relatively fast through the biofilm and boundary layer relative to this loss rate. In this situation, we can model the biofilm at a macroscopic scale by a simple kinetic expression (that is, Eq. 12.56). The reaction is characterized by a half life of substrate decomposition (Bouwer and McCarty, 1984)

$$t_{1/2} = \frac{ln\ 2}{[k\ X_f/K_s]} \tag{12.57}$$

where Eq. 12.57 comes from the integration of Eq. 12.56. This model is appropriate when the concentration of active organisms is small and ground water movement is slow (Bouwer and McCarty, 1984).

The discussion so far has implied that organic compounds in half-reactions simply biodegrade to simple inorganic species like CO_2 and H_2O. In some cases, however, one compound transforms to other intermediates following combined biotic and abiotic pathways. In the following section, we use the various transformations of 1,1,1-trichloroethane to illustrate the potential difficulties in defining reaction pathways.

There has been a good start toward describing the impact of these biotransformation reactions on problems of contamination. Bouwer and McCarty (1984) found that secondary utilization of chlorinated benzenes such as chlorobenzene or 1,2-dichlorobenzene and nonchlorinated aromatic compounds such as ethylbenzene, styrene, or naphthalene can occur relatively rapidly under aerobic conditions. Halogenated aliphatic compounds such as chloroform or carbon tetrachloride, however, persist in aerobic environments. When a system becomes more reducing and anaerobic, the nonchlorinated aliphatic and chlorinated benzene compounds persist and the halogenated aliphatic compounds may be biodegraded. This sensitivity to redox conditions implies that the extent of biotransformation varies within a flow system where E_H is zoned.

Rates Limited by the Availability of Electron Acceptors
Conditions exist when the biodegradation of a substrate is limited by the availability of an electron acceptor. For example, the aerobic biodegradation of an organic liquid in water can utilize the small quantities of oxygen naturally dissolved in the water. In spite of an abundance of organic substrate, the lack of sufficient oxygen, as dictated by the stoichiometry of the redox reaction, effectively limits biodegradation. The reaction rate increases when additional oxygen is provided.

The mathematical description of material transport to the biofilm in this case would involve two transport equations, one for the substrate and one for the electron acceptor. Again, both reactants move by diffusion in the diffusion layer and film and disappear through utilization by bacteria. The two governing transport equations are coupled through the reaction term because the rate of substrate utilization (following Bailey and Ollis, 1977) is a function of both the substrate and electron acceptor. This rate expression is in form a modified version of Eq. 12.55. The kinetic equation for the utilization of the electron acceptor also depends upon both concentrations as well.

Work is ongoing to develop more sophisticated biofilm models that can account for contributions of both the substrate and electron acceptors to cell growth (Borden and Bedient, 1986 and Molz and others, 1986).

12.6   *Hydrolysis*

Hydrolysis is another kind of transformation reaction that operates on organic compounds. Unlike biodegradation, hydrolysis is not catalyzed by microorganisms. In the reaction, an organic molecule reacts with water or a component ion of water, for example

$$R-X \xrightarrow{H_2O} R-OH + X^- + H^+ \tag{12.58}$$

where *R-X* is an organic molecule with *X* representing an attached halogen, carbon, phosphorous, and nitrogen. According to Neely (1985), the introduction of a hydroxyl group into the parent molecule makes the product more susceptible to biodegradation, and more soluble.

Table 12.12

Summary of the susceptibility of various functional groups to hydrolysis

Resistant to hydrolysis[1]	
Alkanes	Aromatic nitro compounds
Alkenes	Aromatic amines
Alkynes	Alcohols
Benzene	Phenols
Biphenyl	Glycols
Polycyclic aromatic hydrocarbons	Ethers
	Aldehydes
Halogenated aromatics	Ketones
Halogenated hydrocarbon pesticides (DDT, etc.)	Nitriles
	Carboxylic acid
	Sulfonic acid

[1]Reprinted from Harris, J. C., *Handbook of Chemical Property Estimation Methods*, Lyman, W. J., Reehl, W. F., and Rosenblatt, D. H., eds. Originally McGraw-Hill, 1982, subsequently published by American Chemical Society, 1990. Copyright © 1990 American Chemical Society.

Susceptible to hydrolysis[2]	
Alkyl halides	Epoxides and lactones
Amides[1]	Phosphoric acid esters
Carbamates[2]	Sulfonic acid esters
Carboxylic acid esters	

[1]Most amides have a half life greater than a year at pH 7 and 25°C. Structures such as $Cl_2CHCONH_2$ have a half life less than a year.

[2]All carbamates having only alkyl or aromatic on N and O are persistent (i.e., half life > 1 year).

[2]Reprinted with permission from Neely, W.B., 1985, Hydrolysis: in Environmental Exposure from Chemicals, eds. W.B. Neely and G.E. Blau, CRC Press, p. 157–173. Copyright CRC Press, Inc. Boca Raton, FL.

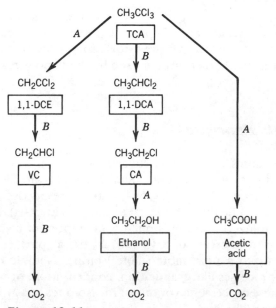

Figure 12.11
Probable fate of TCA under methanogenic conditions: biotic transformation pathways are denoted by lines marked *B* and abiotic transformation pathways by lines marked *A*. Reprinted with permission from Vogel, T.M., and McCarty, P.L., Environ. Sci. Technol., v. 21, 1987, p. 1209. Copyright 1987 American Chemical Society.

Not all organic compounds will undergo hydrolysis. Table 12.12 lists the tendency of various functional groups to hydrolyze. First-order rate laws generally describe hydrolysis with the rate of disappearance of *RX* proportional to the concentration of *RX*. As is the case with many kinetic descriptions of reactions, this kind of simple rate law oversimplifies what is often a much more complicated relationship (Neely, 1985).

Both biotic and abiotic transformation reactions can operate on organic compounds in a complicated way. Such is the case for the common industrial solvent 1,1,1-trichloroethane (TCA), which is often a contaminant in ground water (Vogel and McCarty, 1987). Under abiotic conditions, TCA transforms to 1,1-dichloroethylene (1,1-DCE) along one reaction pathway and to acetic acid along another (Figure 12.11). The pseudo-first-order rate constants for the formation of 1,1-DCE and acetic acid from TCA are about 0.04 yr^{-1} and 0.20 yr^{-1}, respectively (Vogel and McCarty, 1987). These intermediates will transform biologically under methanogenic conditions to vinyl chloride (VC) and CO_2, respectively (Figure 12.11). In the presence of a highly active population of methanogenic bacteria, most of the TCA would biotransform along the middle pathway (Figure 12.11) by reductive dehalogenation to 1,1-dichloroethane (1,1-DCA) and chloroethane (CA). The next reaction along this pathway, the hydrolysis of CA to ethanol, has a pseudo-first-order rate constant of about 0.37 yr^{-1}, giving CA a half life of about 1.9 yr (Vogel and McCarty, 1987). In biologically active samples, ethanol would be transformed to CO_2.

Exactly which of these pathways will be dominant and which intermediates of TCA are found at a site depend upon the potential for methanogenic biotransformation. Irrespective of the pathway, however, the intermediates should be present because of the rate limitations imposed by the various reactions.

12.7   *Isotopic Processes*

Isotopes are atoms of the same element that differ in terms of their mass. For example, hydrogen with an atomic number of 1 has three isotopes, $_1^1H$, $_1^2H$, and $_1^3H$, with mass numbers (superscripts) of 1, 2, and 3, respectively. The first of these isotopes is stable, while the last one, $_1^3H$ (usually written ^3H), decays radioactively to $_2^3He$. Thus, radioactive decay is one of the important reactions that involve isotopes. In radioactive decay, atoms of a particular isotope change spontaneously to a new, more stable isotope. Isotopic concentrations change also due to processes like evaporation, condensation, or water-rock interactions. Typically, these processes favor one of the isotopes of a given element over others producing fractionation.

Radioactive Decay

Radioactive decay occurs mainly by the emission of an α particle ($_2^4He$) or a β particle (electron $_{-1}^0e$). Often accompanying the emission of these particles is γ radiation, which is electromagnetic energy of short wavelength. This radiation forms when nuclides produced in an excited state (noted by *) revert back to a so-called ground state. As the following examples illustrate, α-decay changes both the mass number and the atomic number, β-decay changes the atomic number only, and γ-emission changes neither

$$\text{α-decay:} \quad _{90}^{232}\text{Th} \rightarrow _{88}^{228}\text{Ra} + _2^4\text{He}$$

$$\text{β-decay:} \quad _{88}^{228}\text{Ra} \rightarrow _{89}^{228}\text{Ac} + _{-1}^0e$$

$$\text{γ-emission:} \quad _{92}^{236}\text{U}^* \rightarrow _{92}^{236}\text{U} + \gamma$$

The arrows in these equations indicate that radioactive decay is irreversible. The amount of the reacting or parent isotope continually decreases, while the product or daughter isotope increases. Reactions become more complex when the daughter itself decays through a series of other products until a stable form is finally created. We refer to these reactions as decay chains or disintegration series (for example, $_{90}^{232}\text{Th}$, $_{92}^{235}\text{U}$, and $_{92}^{238}\text{U}$ series).

The decay of radioactive isotopes is independent of temperature and follows the first-order rate law

$$\frac{dN}{dt} = -kN \tag{12.59}$$

where N is the number of atoms of radioactive material, t is time, and k is the rate constant for radioactive decay. Integration of Eq. 12.59 yields an expression for the quantity of parent remaining at any time following its formation

Table 12.13
Radioactive half lives for elements of interest in age dating and contaminant studies (modified from Moody, 1982)

Nuclide	$t_{1/2}$ (yr)	Nuclide	$t_{1/2}$ (yr)
^{3}H	12.3	^{231}Pa	3×10^4
^{14}C	5×10^3	^{241}Am	432
^{36}Cl	3.1×10^5	^{243}Am	7×10^3
^{63}Ni	100	^{79}Se	6.5×10^4
^{90}Sr	29	^{93}Mo	3.5×10^3
^{93}Zr	1.5×10^6	^{99}Tc	2×10^5
^{94}Nb	2×10^4	^{99}Tc	2×10^5
^{107}Pd	7×10^6	^{126}Sn	1×10^5
^{129}I	2×10^7	^{151}Sm	90
^{135}Cs	3×10^6	^{147}Sm	1.3×10^{11}
^{137}Cs	30	^{106}Ru	1.0
^{154}Eu	8.2	^{235}U	7×10^8
^{210}Pb	22	^{238}U	4.5×10^9
^{226}Ra	1.6×10^3	^{237}Np	2×10^6
^{227}Ac	22		
^{230}Th	8×10^4		
^{232}Th	1.4×10^{10}		

$$N(t) = N_0 e^{-kt} \tag{12.60}$$

where N_0 is the initial number of atoms of parent isotope.

The decrease in the activity of a radioactive substance is commonly expressed in terms of radioactive half life ($t_{1/2}$), which is the time required to reduce the number of parent atoms by one-half. The half life relates to the rate constant for decay as follows:

$$t_{1/2} = \frac{0.693}{k} \tag{12.61}$$

Listed in Table 12.13 are radioactive half lives for many radionuclides of interest in ground water investigations.

Radioactive decay is important for two reasons. When radioactive isotopes occur as contaminants, decay through several half lives can reduce the hazard when the residence time in a flow system is much greater than the half-life for decay. This capacity to attenuate radioactivity provides the rationale to support the subsurface disposal of some radioactive contaminants. Radioactive decay also forms the basis for various techniques used to age-date ground water. The most important radioactive isotopes used for this purpose are ^{3}H (tritium), ^{14}C, ^{32}Si, and ^{36}Cl. Because the interpretation of ground water ages in general requires the evaluation of radioactive decay in relation to other chemical and physical processes, we will postpone detailed discussion of these techniques until Chapter 15.

Isotopic Reactions

The isotopes of hydrogen, oxygen, carbon, and sulfur are useful in studying chemical processes. Related to these elements are seven isotopes of environmental significance (Table 12.14). Commonly ground water studies involve ^2H or D, ^{18}O, ^{13}C, and ^{34}S.

   Because one isotope dominates the rest in terms of relative abundances (Table 12.14) and the changes due to fractionation are too small to measure accurately, isotopic abundances are reported as positive or negative deviations of isotope ratios away from a standard. This convention is represented in the following general equation (Fritz and Fontes, 1980)

$$\delta = \frac{R_{sample} - R_{standard}}{R_{standard}} \times 1000 \tag{12.62}$$

where δ, reported as permil (‰), represents the deviation from the standard and R is the particular isotopic ratio (for example, $^{18}O/^{16}O$) for the sample and the standard. For example, we would express sulfur:isotope ratios as

$$\delta^{34}S_{sample} = \frac{(^{34}S/^{32}S)_{sample} - (^{34}S/^{32}S)_{standard}}{(^{34}S/^{32}S)_{standard}} \times 1000$$

A $\delta^{34}S$ value of $-20‰$ means that the sample is depleted in ^{34}S by 20‰ or 2% relative to the standard.

Table 12.14
Isotopes of environmental significance and their relative abundances (modified from Fritz and Fontes, 1980[3])

Element	Isotopes		Average abundance % of stable isotopes
Hydrogen	^1_1H		99.984
	^2_1H[1]		0.015
	^3_1H[2]	radioactive	10^{-14} to 10^{-16}
Oxygen	$^{16}_8O$		99.76
	$^{17}_8O$		0.037
	$^{18}_8O$		0.10
Carbon	$^{12}_6C$		98.89
	$^{13}_6C$		1.11
	$^{14}_6C$	radioactive	$\sim 10^{-10}$
Sulfur	$^{32}_{16}S$		95.02
	$^{33}_{16}S$		0.75
	$^{34}_{16}S$		4.21
	$^{36}_{16}S$		0.02

[1]Deuterium, often referred to by D.

[2]Tritium, often referred to by T.

[3]Reprinted by permission of Elsevier Science Publishers from Handbook of Environmental Isotope Geochemistry, v. 1.

This way of expressing isotopic compositions takes advantage of the ability of mass spectrometers to measure isotopic ratios accurately. The errors involved in determining δ values are typically a small proportion of the possible range of values.

Deuterium and Oxygen-18

This section discusses the occurrence of D and ^{18}O in water, gases (e.g., CH_4), and ions (e.g., HCO_3^- and SO_4^{2-}). Deuterium and oxygen-18 compositions of water are usually measured with respect to the SMOW (Standard Mean Ocean Water) standard (Fritz and Fontes, 1980). This choice of standard is particularly appropriate because precipitation that recharges ground water originates from the evaporation of ocean water. However, the isotopic composition of rain or snow in most areas is not the same as ocean water. Evaporation and subsequent cycles of condensation significantly change the isotopic composition of water vapor in the atmosphere.

Water vapor in equilibrium with water is depleted in δD and $\delta^{18}O$ by $80\%_0$ and $10\%_0$, respectively (Wallick et al., 1984). Thus, for ocean water (δD and $\delta^{18}O \sim 0\%_0$), a vapor in equilibrium should have $\delta D = -80\%_0$ and $\delta^{18}O = -10\%_0$. The first few drops of rain falling from this vapor theoretically should have an isotopic composition that is the same as the starting water. This change in isotopic ratios (for example, $^{18}O/^{16}O$) because of a chemical reaction (for example, water = water vapor) is termed fractionation. This process is described mathematically by a fractionation factor (α) (Fritz and Fontes, 1980), for example,

$$\alpha = \frac{(^{18}O/^{16}O)_{\text{water}}}{(^{18}O/^{16}O)_{\text{vapor}}}$$

or in terms of del notation

$$\alpha = \frac{1000 + \delta^{18}O_{\text{water}}}{1000 + \delta^{18}O_{\text{vapor}}} \tag{12.63}$$

The isotopic ratios of water vapor in air masses associated with oceans are, however, not in equilibrium with water. The vapor is more depleted than expected. The extent of depletion is correlated mainly with temperature (Gat, 1980). Rainfall sampled in coastal areas near the equator has an isotopic composition most like sea water, but it is still slightly depleted. Moving away from the equator to higher latitudes, the rainfall becomes more and more depleted relative to sea water. For example, along the western coast of the United States and Canada, values of δD range from approximately -22 to $-101\%_0$ and $\delta^{18}O$ from -4 to $-14\%_0$ (Gat, 1980). This so-called "latitude effect" is related to progressive temperature-controlled depletion of heavier isotopes in the vapor as they are preferentially removed during precipitation events. This continuous removal of a condensate enriched in heavier isotopes is known as a Rayleigh process.

Because fractionation effects at lower temperatures are increasingly more pronounced, the depletion becomes more marked as air masses begin to move across cooler, continental land masses, or are forced to higher and colder elevations over mountain ranges. The temperature control on fractionation also explains why snow is more depleted in D and ^{18}O than rain.

If we assume that differences in the isotopic composition of rain water reflect the increasing effects of "rainout" from the vapor, then a relationship could exist

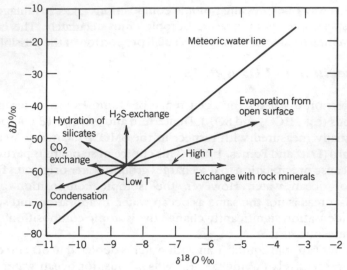

Figure 12.12
Deviations in isotopic compositions away from the meteoric
water line as a consequence of various processes (from IAEA
Report No. 288, 1983).

among samples. When the δD and $\delta^{18}O$ content of rain water from sampling sites
around the world are plotted together, they in fact lie along a straight line (Figure
12.12), known as the meteoric water line (Craig, 1961). The equation for this line
is approximately $\delta D = 8\delta^{18}O + 10‰$.

The meteoric water line provides an important key to the interpretation of deuterium and oxygen-18 data. Water with an isotopic composition falling on the
meteoric water line is assumed to have originated from the atmosphere and to be
unaffected by other isotopic processes. Deviations from the meteoric water line
result from other isotopic processes. In most cases, these processes affect the relationship between δD and $\delta^{18}O$ in such a unique way that the position of the data
points can help to identify a process. Figure 12.12 illustrates the direction away
from the meteoric water line that various processes push the composition of water.

Two of the more commonly observed processes, evaporation from open
water and exchange with rock minerals (Figure 12.12), exhibit this deviation
(Fontes, 1980). When water in a pond or lake evaporates, there is an enrichment
in the heavier isotopes. However because of the dynamics of the process, the isotopic composition of the pond follows an "evaporation line" with a slope ranging
from one to five depending upon the local rates of evaporation. The point of intersection of the evaporation line with the meteoric water line usually is taken as the
isotopic composition of unaltered precipitation. In general, the farther along the
evaporation line the data points lie, the greater the evaporation.

Isotopic exchange between minerals and ground water is important in deep,
basinal flow systems or in geothermal systems. The relatively high temperatures
enable oxygen and hydrogen to exchange between phases and to achieve an equilibrium distribution. This equilibrium is described by the fractionation factor for
the mineral. Savin (1980) has summarized fractionation factors for a variety of different minerals and described how they change as a function of temperature.

Exchange between meteoric water and minerals containing oxygen results in the kind of deviation depicted on Figure 12.12. Because only ^{18}O is involved in exchange, the samples typically fall along a horizontal line, reflecting an "oxygen shift" away from the meteoric water line. Truesdell and Hulston (1980) indicate that the size of the oxygen shift is proportional to the difference in original $\delta^{18}O$ between the water and rock, the temperature, and the time of contact, and inversely proportional to the water-rock ratio. Yellowstone Park, Steamboat Springs, Wairaki, and Salton Sea are well-known examples. Oxygen and hydrogen isotopes are extremely valuable in hydrogeologic studies. The most important applications include flow system tracing and indirect age determinations, which are discussed later.

Carbon-13 and Sulfur-34

In terms of carbon and sulfur, ^{13}C and ^{34}S are the main stable isotopes of interest in ground water applications. Typically, their behavior is studied in the carbonate (for example, HCO_3^-) and sulfate-sulfide systems. Because these systems are complex in the subsurface, their most important application is in process identification.

Dissolution of solids containing sulfur in the soil zone and the ground water system can change the $\delta^{34}S$ of SO_4^{2-} $\{\delta^{34}S(SO_4^{2-})\}$ of recharge and ground water depending upon whether the source of sulfur includes, for example, evaporite gypsum, organic matter (coal or peat), or sulfide minerals such as pyrite, or combinations of all these. Sulfate reduction is the most important fractionation process involving sulfur that operates in a ground water system. As a consequence of this process, the SO_4^{2-} can become enriched in $\delta^{34}S$ and the mineral sulfides that precipitate become depleted. By characterizing the $\delta^{34}S$ contents of these potential reactants and products, one can identify the processes contributing to the evolution of the sulfur chemistry.

The $\delta^{13}C$ of ground water is also controlled by dissolution, precipitation, and fractionation processes. Initially, the ^{13}C chemistry is controlled by the $\delta^{13}C(CO_2)_{gas}$ in the atmosphere, which falls in a range from -7 to $-10‰$ with respect to the carbonate PDB standard. Additional carbon enters the water in the soil zone from the dissolution of carbonate minerals, organic carbon, and the solution of CO_2. Carbon dioxide occurs in the soil atmosphere due to plant root respiration and bacterial decay. The $\delta^{13}C$ of this soil gas varies geographically depending upon the types of plants growing (Wallick and others, 1984). There is some indication that isotopic exchange between CO_3^{2-} and calcite, or repeated dissolution and precipitation of calcite, which has the same effect, can be important as well (Mook, 1980).

The $\delta^{13}C(HCO_3^-)$ can change deeper in the zone of saturation due to (1) sulfate-reduction carbon generated from a source of organic matter such as methane, (2) the dissolution of additional carbonate minerals when undersaturation results from the removal of Ca^{2+} in cation exchange reactions with Na^+-clay minerals, or (3) still other sources of carbon being added, for example, from the thermal maturation of deeply buried organic matter.

Interpreting ^{34}S and ^{13}C data is much more difficult than was the case for D and ^{18}O. To work with ^{34}S and ^{13}C requires a systematic, site-specific characterization of the important reactions in terms of the sources and sinks for sulfur and carbon. In many instances, this exercise in tracing the pathways by which carbon and sulfur enter or leave ground water is the most practical use for these isotopes.

Suggested Readings

Bitton, G., and Gerba, C. P., eds, 1984, Groundwater pollution microbiology: New York, John Wiley, 377 p.

Bouwer, E. J., and McCarty, P. L., 1984, Modeling of trace organics biotransformation in the subsurface: Ground Water, v. 22, no. 4, p. 433–440.

Drever, J. I., 1988, The geochemistry of natural waters: Englewood Cliffs, Prentice-Hall, 437 p.

Fritz, P., and Fontes, J. C., eds, 1980, Handbook of environmental isotope geochemistry, v. 1, The terrestrial environment A: New York, Elsevier, 545 p.

Hitchon, B., and Wallick, E. I., eds, 1984, First Canadian American Conference on Hydrogeology: National Water Well Assoc., Ohio, Dublin, 323 p.

Matthess, G., 1982, The properties of groundwater: New York, John Wiley, 406 p.

Morel, F. M. M., 1983, Principles of aquatic chemistry: New York, John Wiley, 446 p.

Problems

1. Often (CO_3^{2-}) is not reported in water quality analyses because the concentration is too low to measure directly. Develop an expression from the appropriate mass laws to calculate the concentration of (CO_3^{2-}) given pH and (HCO_3^-).

2. A water sample has the following ionic composition at 25 °C.

			Ionic Concentration (M) $\times 10^{-3}$			
Ca	Mg	Na	HCO$_3$	SO$_4$	Cl	pH
4.17	1.32	11.7	12.0	5.49	0.08	7.60

Given that the activity coefficients for Ca^{2+}, Mg^{2+}, and HCO_3^- are 0.52, 0.55, and 0.85 respectively, answer the following questions.

   a. What is the partial pressure of CO_2 in equilibrium with the sample?

   b. Assuming no complexation, what is the saturation state of the sample with respect to calcite?

   c. Assume that the concentration reported for calcite represents $(Ca)_T$ and that three significant ion pairs form $CaCO_3^0$, $CaHCO_3^+$, and $CaSO_4^0$. Develop an equation to calculate the concentration of free Ca^{2+} in the sample, assuming the solution is ideal.

3. a. Given that the concentration of benzene (molecular weight 78.1) dissolved in groundwater at the water table is 280 µg/L, calculate the equilibrium partial pressure in the adjacent soil gas. The Henry's law constant for the partitioning reaction is 5.5×10^{-3} atm-m³/M.

   b. If benzene was present as the pure liquid having a vapor pressure of 0.1 atm, what would the partial pressure be?

4. Following are the results of a batch sorption experiment in which two

grams of porous media were mixed with 10 ml of a solution containing various concentrations of Cd.

Initial (Cd) Concentration (mg/ml)	Equilibrium (Cd) Concentration (mg/ml) $\times 10^{-4}$
0.0005	0.040
0.001	0.093
0.010	1.77
0.050	39.3
0.096	205.2

a. Find the equation for the Freundlich isotherm that best fits these data. *Hint*: Transform the basic equation and data in terms of the logarithm of concentration.

b. Assuming that only the lowest concentrations are of interest, fit the same set of data with a linear Freundlich isotherm (that is, $n = 1$) to obtain a distribution coefficient (K_d). Be sure to indicate the units for K_d.

5. At a site, the following contaminants are found (i) parathion (log K_{ow} 3.80), (ii) chlorobenzene (log K_{ow} 2.71), and (iii) DDT (log K_{ow} 6.19).

a. Using the equation of Hassett and others (1983) in Eq. 12.37, estimate the partition coefficient between solid organic carbon and water (K_{oc}).

b. With an f_{oc} of 0.01, estimate the distribution coefficients (K_d) for these compounds.

c. On the basis of sorption alone, which of these compounds is the most mobile and which is the least mobile?

6. Compare the range in variability of the distribution coefficients for carbon tetrachloride (log K_{ow} 2.83) estimated using the three different regression equations (12.35, 36, and 37) and an f_{oc} of 0.01.

7. The properties of six organic compounds are listed here.

Compound	log K_{ow}	Henry's Law Constant (atm-m^3/M)
#1	2.04	5×10^{-3}
#2	0.89	5×10^{-4}
#3	0.46	7×10^{-4}
#4	-.27	3×10^{-4}
#5	2.83	5×10^{-3}
#6	1.48	4×10^{-3}

a. List the organics in order of decreasing solubility.

b. List the organics in order of decreasing mobility.

8. Consider the following half redox reaction involving NO_3^- with a concentration of 10^{-5} M and ammonium with a concentration of 10^{-3} in groundwater with a pH of 8.

$$\tfrac{1}{8} NO_3^- + \tfrac{5}{4} H^+ + e^- = \tfrac{1}{8} NH_4^+ + \tfrac{3}{8} H_2O$$

What is the pe and E_H of the system at 25°C assuming redox equilibrium.

9. The couple O_2/H_2O is found to be the dominant couple producing a pe of 12.0. By assuming redox equilibrium, determine how $(Fe)_T = 10^{-5}$ M would be distributed as Fe^{2+} and Fe^{3+}.

10. For an aerobic reaction, k/K_s is estimated to be 5.0 L/mg cells-day. Given a concentration of microorganisms of 0.01 mg cell/L or 10^4 bacteria/ml, estimate the half life of substrate utilization.

11. An organic compound has a tendency to hydrolyze with $t_{1/2} = 3.5$ yr. What would an initial concentration of 100µg/L be reduced to after 10 years?

12. Explain what is meant by the SMOW standard and the convention for reporting isotopic results.

13. These isotopic measurements were made on several prairie lakes and potholes.

Sample	$\delta^{18}O$ $(^0/_{00})$	δD $(^0/_{00})$ (SMOW)
#1	−6.5	−98.0
#2	−8.5	−102.0
#3	−2.6	−83.0
#4	−4.0	−86.0

What is the best explanation for the deviation in these results from the meteoric water line? What is the estimated concentration of unaltered meteoric water in this area?

Chapter Thirteen

The Mathematics of Mass Transport

13.1 **Mass Transport Equations**

13.2 **Mass Transport with Reactions**

13.3 **Boundary and Initial Conditions**

As emphasized in earlier chapters, hydrogeologists traditionally have worked to extend their understanding of ground water flow through the precise language of mathematics. Not surprisingly, this same approach has been applied to problems of mass transport. The mass transport equations developed in this chapter provide a comprehensive framework for quantitatively describing mass transport with and without accompanying chemical reactions. These equations are at the heart of approaches used to model the chemical evolution of natural ground water and contaminant migration, topics to be taken up later in the book.

13.1 Mass Transport Equations

As with flow, the starting point for developing the mass transport equations is a conservation equation. This equation in words is

mass inflow rate − mass outflow rate = change in mass storage with time

in units of mass per unit time. In general, this statement applies to a domain of any size. For a representative volume with inflow through three sides, the left-hand side of this word equation is the divergence of the flux of interest (see Chapter 4); that is,

− div **J** = net mass outflow rate per unit volume

It remains to describe the nature of the flux **J**, which could be diffusive transport or advective transport combined with diffusion.

The right-hand side of the conservation statement assumes that the gains or losses in mass inside the volume are proportional to changes in concentration in mass inside the unit volume. Thus, a proportionality constant is required that expresses the mass lost or gained when the concentration changes a unit amount. In this case, the proportionality constant is unity because the mass occupies the pores, where concentration is mass per unit volume of fluid (m/V_w) and porosity n for a fully saturated medium is the fluid volume per unit total volume (V_w/V_T). Thus the product Cn is mass per unit total volume, and the conservation statement becomes

$$- \text{div } \mathbf{J} = \frac{\partial(Cn)}{\partial t} \tag{13.1}$$

This equation states that the net mass outflow rate per unit volume equals the time rate of change of mass within the unit volume.

The Diffusion Equation

For a diffusive system defined by Fick's law of the form in Eq. 10.11, the conservation Eq. 13.1 becomes

$$\text{div}[D_d^* n \text{ grad } C] = \frac{\partial(Cn)}{\partial t} \tag{13.2}$$

With the further assumption that porosity does not vary spatially or temporally, and that the diffusion coefficient is likewise constant spatially, we can write

$$D_d^* \, \text{div} \, [\text{grad} \, C] = \frac{\partial C}{\partial t}$$

or

$$\frac{\partial^2 C}{\partial x^2} + \frac{\partial^2 C}{\partial y^2} + \frac{\partial^2 C}{\partial z^2} = \nabla^2 C = \left(\frac{1}{D_d^*}\right)\frac{\partial C}{\partial t} \tag{13.3}$$

This diffusion equation is analogous to the unsteady flow equation introduced in Chapter 4 and the unsteady conductive heat transport equation in Chapter 9. The solution to this equation determines how the concentration at any point in a three-dimensional field is changing with time. If the net outflow is zero so that the concentration is not changing in time

$$\frac{\partial^2 C}{\partial x^2} + \frac{\partial^2 C}{\partial y^2} + \frac{\partial^2 C}{\partial z^2} = \nabla^2 C = 0 \tag{13.4}$$

This equation is Laplace's equation as was the steady-state fluid flow equation and the steady-state heat conduction equation. The solution of this equation describes the value of the concentration at any point in a three-dimensional field.

The Advection-Diffusion Equation

For systems with fluid motion, mass transport can be due to both advection and diffusion. This combined flux can be described mathematically by combining the advective flux of Eq. 10.1 and the diffusive flux of Eq. 10.11, or

$$J_x = -nD_d^*\left(\frac{\partial C}{\partial x}\right) + v_x C n \tag{13.5a}$$

$$J_y = -nD_d^*\left(\frac{\partial C}{\partial y}\right) + v_y C n \tag{13.5b}$$

$$J_z = -nD_d^*\left(\frac{\partial C}{\partial z}\right) + v_z C n \tag{13.5c}$$

For isotopic conditions,

$$J = -nD_d^* \, \text{grad} \, C + \mathbf{v} C n \tag{13.6}$$

where v, the linear ground water velocity, is a vector with components, v_x, v_y, v_z. Substitution of the rate expression Eq. 13.6 into Eq. 13.2 and carrying out the necessary operations gives

$$\nabla \cdot (nD_d^*\nabla C) - \nabla \cdot \mathbf{v}Cn = \frac{\partial(Cn)}{\partial t} \tag{13.7}$$

Expanding the advection terms and assuming constant D_d^* and n gives

$$D_d^* \nabla^2 C - \mathbf{v} \cdot \nabla C + C \nabla \cdot \mathbf{v} = \frac{\partial C}{\partial t} \qquad (13.8)$$

For steady ground water flow, $\nabla \cdot \mathbf{v} = 0$ and

$$D_d^* \nabla^2 C - \mathbf{v} \cdot \nabla C = \frac{\partial C}{\partial t} \qquad (13.9)$$

Note that when the velocity goes to zero, that is, no advection, we recover the diffusion equation for a stationary fluid (Eq. 13.3). When the concentration profile is steady, Eq. 13.9 becomes

$$D_d^* \nabla^2 C - \mathbf{v} \cdot \nabla C = 0 \qquad (13.10)$$

Equations 13.9–13.10 are called the advection-diffusion equations. They describe the manner in which mass is moved from one point to another by advection as modified by diffusion. The one-dimensional form of Eq. 13.10 is

$$D_d^* \frac{\partial^2 C}{\partial x^2} - v_x \frac{\partial C}{\partial x} = \frac{\partial C}{\partial t} \qquad (13.11)$$

The first term on the left describes mass transport by diffusion and the second describes mass transport by advection.

Before leaving the advection-diffusion equation, we would like to consider the concept of dimensional analysis as applied to mass transport. In Chapter 4, the method of dimensional analysis was introduced and applied to the unsteady flow equation. The result was a dimensionless group known as the Fourier number. Applying these same ideas to the advection-diffusion equation, let us define a dimensionless concentration $C^+ = C/C_e$, where C_e is some characteristic concentration and the superscript $+$ indicates a dimensionless quantity. The diffusion-advection equation in dimensionless form is

$$\nabla^{+2} C^+ - \left(\frac{vL}{D_d^*}\right) \nabla^+ C^+ = \left(\frac{L^2}{t_e D_d^*}\right) \frac{\partial C^+}{\partial t^+} \qquad (13.12)$$

where the dimensionless quantity vL/D_d^* is the Peclet number for mass transport (N_{PE}) with L as some characteristic length. The Peclet number expresses the transport by bulk fluid motion or advection to the mass transport by diffusion.

The dimensionless quantity $L^2/t_e D_d^*$ is the Fourier number for mass transport. This number can be used to approximate the diffusion length or the distance of mass diffusion over a specified time interval

$$L \propto (D_d^* t_e)^{1/2}$$

Thus, if $D_d^* = 10^{-5}$ cm²/sec and $t_e =$ one year (3.15×10^7 sec), the diffusion length is approximately 18 cm.

The Advection-Dispersion Equation

Diffusion alone does not fully account for mixing during mass transport. Scale-dependent, mechanical dispersion in many systems is usually more important. Logically, if dispersion is to be incorporated in the advection-diffusion equation, it should be reflected in the velocity term. Unfortunately, it has not been possible to do this in a simple way. Instead the coefficient of hydrodynamic dispersion is introduced to incorporate the combined effect of diffusion and mechanical dispersion. If the dispersion coefficient is constant, this equation is

$$D\nabla^2 C - \mathbf{v} \cdot \nabla C + C \nabla \cdot \mathbf{v} = \frac{\partial C}{\partial t} \tag{13.13}$$

where $D = D' + D_d^*$. The justification for treating dispersion in this manner is purely a practical one and stems from the fact that the macroscopic outcome is the same for both diffusion and mechanical dispersion. The actual physical processes, however, are entirely different. By replacing the diffusion coefficient by a coefficient of hydrodynamic dispersion, the mixing process is assumed to be Fickian. Thus, in cases where mechanical dispersion is small compared to diffusion, Eq. 13.13 reverts to the advection-diffusion equation. In the absence of ground water flow, Eq. 13.13 reduces to the diffusion equation.

   The advection-dispersion equation is the workhorse of modeling studies in ground water contamination. We will have much to say about its solution and application later in the book.

13.2   Mass Transport with Reaction

The equations presented so far describe the processes of advection, mechanical dispersion, and diffusion. With reactions, the equation is modified by source or sink terms depending upon whether a constituent is being added or removed by chemical processes. The appropriate statement of mass conservation when reactions are considered is

mass inflow rate − mass outflow rate ± mass production rate

= change in mass storage with time

The plus or minus term designates either a source or a sink and takes on different forms for different reactions. At this point, it is convenient to represent such sources or sinks in a symbolic form. For example, in the case of one-dimensional transport, the word equation becomes

$$D\frac{\partial^2 C}{\partial x^2} - v_x \frac{\partial C}{\partial x} \pm \frac{r}{n} = \frac{\partial C}{\partial t} \tag{13.14}$$

where r is taken symbolically as the mass produced or consumed per unit volume per unit time, moles/L^{3}T. This equation applies only to a single constituent. Involving several dissolved species in the transport requires a system of such equations, one for each constituent.

Source terms are usually specified in terms of a rate law. This law, for example, can be expressed in terms of the rate of decrease of a reactant, or the rate of increase of a product depending on which constituent is being described by the transport equation. It may be necessary to track both a reactant and a product using two coupled transport equations. One example is with biodegradation where both the reactant and the products are contaminants.

Prior to examining some rate laws, let us consider the concept of dimensional analysis as applied to transport with reaction. In order to do so we replace the quantity r/n in Eq. 13.14 by the product of a reaction rate coefficient k and concentration C, that is, kC where k is the volume reacted per unit volume per unit time. Applying the dimensionalizing procedures to the steady state form of Eq. 13.14, we arrive at two dimensionless groups

$$\frac{kL}{v} \text{ (Damköhler number I)}$$

$$\frac{kL^2}{D} \text{ (Damköhler number II)}$$

where the Damköhler numbers may be taken as a measure for the tendency for reaction to the tendency for transport. Damköhler number I is important at high Peclet numbers and Damköhler number II is important at low Peclet numbers. Note the ratio of number II to number I reduces to the Peclet number.

Let us now examine some rate laws and show how they are included in mass transport equations. Before proceeding, however, we will review the terminology used to organize the discussion. Reactions are classified as homogeneous, operating within a single phase, or heterogeneous, operating between two phases. In addition, reactions can be described from an equilibrium or kinetic viewpoint depending upon the rate of the reaction relative to the mass transport process.

First-Order Kinetic Reactions

One example of a simple kinetic reaction is the first-order decay of a constituent due to radioactive decay, biodegradation, or hydrolysis

$$r = \frac{d(nC)}{dt} = -\lambda nC \tag{13.15}$$

where λ is the decay constant for radioactive decay, or some reaction rate coefficient for biodegradation or hydrolysis. In all cases, λ has units of time^{-1} and is the volume reacted (or disintegrated) per unit volume per unit time. With this formulation, the one-dimensional transport equation becomes

$$D_x \frac{\partial^2 C}{\partial x^2} - v_x \frac{\partial C}{\partial x} - \lambda C = \frac{\partial C}{\partial t} \tag{13.16}$$

Equilibrium Sorption Reactions

An example of a heterogeneous equilibrium reaction is the sorption of mass from solution. The rate law is generally expressed as

$$r = \frac{\partial C^*}{\partial t} \tag{13.17}$$

where C^* is the concentration of the solute on the solid phase. The one-dimensional transport equation that incorporates this reaction is

$$D_x \frac{\partial^2 C}{\partial x^2} - v_x \frac{\partial C}{\partial x} - \frac{1}{n} \frac{\partial C^*}{\partial t} = \frac{\partial C}{\partial t} \tag{13.18}$$

The rate law incorporated in Eq. 13.18 is a general one appropriate for both equilibrium or kinetic sorption reactions. A solution to this equation requires that a specific rate law be incorporated. For kinetic nonequilibrium sorption reactions,

$$\frac{\partial C^*}{\partial t} = f(C, C^*) \tag{13.19}$$

or, in words, the rate of sorption is a function of both the concentration of mass in solution and the mass sorbed on the solid. For equilibrium sorption reactions,

$$C^* = f(C) \tag{13.20}$$

Because the concentration of sorbed mass is a function of mass in solution, the $\partial C^*/\partial t$ term in Eq. 13.18 can be expressed in a more tractable $\partial C/\partial t$ term. This substitution produces a single differential equation containing one dependent variable, which is solvable by analytical methods.

The following derivation explains how the transport equation for the simple case of linear sorption is obtained. The rate of mass sorption per unit volume of porous medium is

$$\frac{\partial C^*}{\partial t} = \rho_b \frac{\partial S}{\partial t} \tag{13.21}$$

where S is the quantity of mass sorbed on the surface and ρ_b is the bulk density. The bulk density can also be defined as

$$\rho_b = \rho_s(1 - n) \tag{13.22}$$

where ρ_s is the mass density of the minerals making up the rock or soil, normally 2.65g/cm^3 for most sandy soils. The quantity $\rho_s (1 - n)$ is the total mass of solids per unit volume of porous medium.

What now is required is an expression for an equilibrium sorption isotherm. Although several are available, we will use the simplest (and likely most useful) one, the linear Freundlich isotherm. By taking this isotherm and differentiating with respect to time, we obtain

$$\frac{\partial S}{\partial t} = K_d \frac{\partial C}{\partial t} \tag{13.23}$$

where K_d is the distribution coefficient. Combining Eqs. 13.21, 13.22, and 13.23 gives

$$\frac{\partial C^*}{\partial t} = (1 - n)\rho_s K_d \frac{\partial C}{\partial t}$$ (13.24)

which is a function of the fluid concentration alone. Substitution of Eq. 13.24 into the general transport equation and rearranging terms gives

$$D_x \frac{\partial^2 C}{\partial x^2} - v_x \frac{\partial C}{\partial x} = \frac{\partial C}{\partial t} \left[1 + \frac{(1 - n)}{n} \rho_s K_d \right]$$ (13.25)

The bracketed quantity on the right-hand side is a constant termed the retardation factor, R_f or

$$R_f = \left[1 + \frac{(1 - n)}{n} \rho_s K_d \right]$$ (13.26)

Equation 13.25 becomes

$$\frac{D_x}{R_f} \frac{\partial^2 C}{\partial x^2} - \frac{v_x}{R_f} \frac{\partial C}{\partial x} = \frac{\partial C}{\partial t}$$ (13.27)

Thus, the retardation factor merely serves to decrease the magnitude of the transport parameters D and v. As local equilibrium implies that the net rate of reaction is zero, we anticipate similar effects for other equilibrium controlled isotherms of the form $f(C)$, although the mathematics do not work out so simply. Aris and Amundson (1973) cite several other forms of sorption isotherms, which all have this property. Equilibrium-controlled reactions form a basis for much of the modeling conducted in contaminant hydrogeology. Similar assumptions have also been employed by geochemists in models describing the diagenesis of marine sediments (Berner, 1974), by chemical engineers interested in the effects of diffusion and longitudinal dispersion in ion exchange and chromatographic columns (Lapidus and Amundson, 1952; Kasten et al., 1952), and by soil scientists interested in the effects of dispersion and ion exchange in porous media (Lai and Jurinak, 1971; Rubin and James, 1973).

Heterogeneous Kinetic Reactions

The general case of a reaction between a solid and a solution involves five steps in series, with the slowest being rate controlling. The rate-determining step can be transport controlled or surface controlled, depending upon the relative magnitude between the rate at which the mass moves to or reacts on the surface.

The starting point for discussing transport-controlled reactions is a model of the reaction process. Assume that a thin stationary layer occurs at the interface between a solid and the bulk fluid and that diffusion is slow relative to actual reactions at the surface so that the coming and going of atoms is rate controlling. If the process of transport across the thin layer is diffusion, there must be a concentration gradient. This gradient can be assumed linear and represented as $(C_{eq}^* - C)/\sigma$, where C_{eq}^* is the saturation concentration at the solid surface, C is the concentration in the bulk fluid, and σ is the thickness of the stationary layer separating the two. The flux J across the surface in moles vol^{-1} time^{-1} is

$$J = \frac{D_d A(C_{eq}^* - C)}{\sigma} \qquad (13.28)$$

where A is the surface area of the material per unit volume of water and D_d is the diffusion coefficient. The ratio D_d/σ is the mass transfer coefficient k_m and has the units of velocity. In this form, the rate of dissolution of the solid is the product of a driving force, which is the departure from saturation ($C_{eq}^* - C$), a mass transfer coefficient, and a surface area of the material per unit volume. Diffusion-controlled reactions of the form represented in Eq. 13.28 can be expressed as

$$\left(\frac{1}{n}\right)\frac{\partial C^*}{\partial t} = \left(\frac{D_d S^*}{\sigma}\right)(C_{eq}^* - C) \qquad (13.29)$$

where S^* is the specific surface of the porous medium (Domenico, 1977). It is possible to include Eq. 13.29 in a mass transport model.

Now let us consider the surface-controlled reactions. The best examples of these reactions are the sorption and desorption reactions. According to Klotz (1946) one of the earliest formulations assumes that the irreversible local rate of sorption is

$$\left(\frac{1}{n}\right)\frac{\partial C^*}{\partial t} = k_1 C(C_{eq}^* - C^*) \qquad (13.30)$$

where k_1 is a constant and C_{eq}^* is the saturation concentration on the solid. Thomas (1944) considered the form

$$\left(\frac{1}{n}\right)\frac{\partial C^*}{\partial t} = k_1 C(C_{eq}^* - C^*) - k_2 C^* \qquad (13.31)$$

where k_2 is a constant. For a porous medium, the following equation describes a reversible sorption process

$$\left(\frac{1}{n}\right)\frac{\partial C^*}{\partial t} = S^*(k_1 C - k_2 C^*) \qquad (13.32)$$

where k_1 is a forward (sorption) rate constant and k_2 is a backward (desorption) rate constant with units of velocity. This equation states that sorption and desorption occur simultaneously and at different rates. This equation can be rewritten as

$$\left(\frac{1}{n}\right)\frac{\partial C^*}{\partial t} = S^* k_1 \left[C - \left(\frac{k_2}{k_1}\right) C^* \right] = S^* k_1 (C - C_{eq}^*) \qquad (13.33)$$

for the case where $(k_1/k_2)C^* = C_{eq}^*$. This equation is of the same form as the diffusion-controlled reaction discussed earlier. Here the rate constant k_1 has the units LT^{-1} and the specific surface has the units L^{-1} so that the combined term has the units of T^{-1}. With diffusion-controlled reactions, the combined parameter D_d/σ is referred to as the mass transfer coefficient LT^{-1} and is equivalent to k_1 in Eq. 13.33. The combined parameter DS^*/σ has the units T^{-1}.

The utilization of the reversible equation for sorption is widespread in transport problems of an academic nature. Transport equations incorporating this term have been solved first for the condition of advection only (Amundson, 1950) and later for advection and dispersion (Lapidus and Amundson, 1952). More recent theoretical studies dealing with sorption in relation to contamination problems includes the work of Relyea (1982), Melnyk and others (1983), Valocchi (1985), and Bahr and Rubin (1987). However, kinetic models for sorption are rare in field applications because of the difficulty in determining the rate expressions.

13.3 Boundary and Initial Conditions

Solving any time-dependent differential equation requires boundary and initial conditions. The job of the boundary conditions is to account for the effects of the system outside of the region of interest on the system being modeled. Effectively, once the boundary conditions are specified, the larger system can be ignored. The boundary conditions discussed earlier for flow have their counterparts in relation to mass transport. The three are (1) a fixed concentration boundary, (2) a fixed gradient boundary, or (3) a variable flux boundary. Sometimes, these boundary conditions are referred to as first, second, and third types, respectively. The one-dimensional problem in Figure 13.1a, involving contaminant migration in an aquifer away from a long narrow ditch, provides an example of boundary conditions. At the upper boundary at $x = 0$, contaminants enter the system. A lower boundary not shown on the figure is located far down the system at $x = \infty$. The

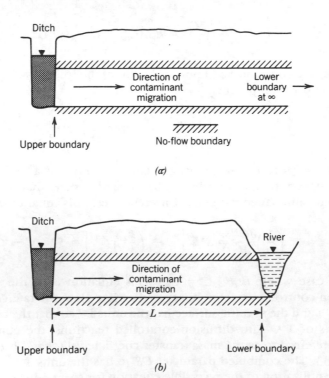

Figure 13.1
Contaminant migration in a one-dimensional system
that is (a) semi-infinite or (b) finite in length.

location of this boundary makes the system semi-infinite. A finite system is one in which the lower boundary is a specified distance from the origin (Figure 13.1*b*). When a region is extended to two or three dimensions, additional boundaries must be specified.

In many mass transport problems (e.g., Figures 13.1*a*, *b*), contaminants enter the system through one or several boundaries. This loading can be specified using either a fixed concentration or else a variable flux boundary condition. An example of a concentration condition for a one-dimensional case is

$$C(0, t) = g(t) \tag{13.34}$$

This equation indicates that at the boundary ($x = 0$), the concentration is given for all time as the function $g(t)$. An example of the variable flux boundary condition is

$$-D \frac{\partial C}{\partial x} + vC = v\,g(t) \qquad (x = 0) \tag{13.35}$$

where $g(t)$ again is some known function. The variable flux boundary condition is a statement of continuity equating the mass flux across the boundary to the total the system is able to move to the boundary by dispersion and advection. These boundary conditions represent the different kinds of loading discussed in Chapter 10. Examples of some commonly used boundary conditions for one-dimensional systems are listed on Table 13.1. The second condition shown on the table is an example of pulse loading. The boundary condition changes after some time (t_o) has elapsed. Thus, it is necessary to provide two boundary conditions and the time for which each applies.

Table 13.1
Examples of some boundary conditions that are used to add contaminants to a one-dimensional system

Name	Type	Form
Constant concentration	Fixed concentration	$C(0, t) = C_o$
Pulse-type loading with constant concentrations	Fixed concentration	$C(0, t) = \begin{cases} C_o, & 0 < t \leqslant t_o \\ 0, & t > t_o \end{cases}$
Exponential decay with source concentration $\to 0$	Fixed concentration	$C(0, t) = C_o e^{-\alpha t}$
Exponential decay with source concentration $\to C_a$	Fixed concentration	$C(0, t) = C_a + C_b e^{-\alpha t}$
Constant flux with constant input concentration	Variable flux	$\left(-D\dfrac{dc}{dx} + vC \right)\bigg|_{x=0} = vC_o$
Pulse-type loading with constant input fluxes	Variable flux	$\left(-D\dfrac{dc}{dx} + vC \right)\bigg|_{x=0} = vC_o,\ 0 < t \leqslant t_o$ $\left(-D\dfrac{dc}{dx} + vC \right)\bigg|_{x=0} = 0,\ t > t_o$

C_o, C_a, C_b = various constant concentrations.
α = decay constant.
t_o = time at which concentration changes due to pulse loading.

The lower boundary in most one-dimensional problems is specified by a constant gradient (constant flux) condition as

$$\frac{\partial C}{\partial x} = g(t) \qquad (x = \infty, \text{ or } x = L) \tag{13.36}$$

The no flux condition $\partial C/\partial x = 0$ is the commonest boundary of this type for one-dimensional problems. In two or three dimensions, we need to specify the boundary conditions for all boundaries.

An initial condition describes the distribution of a contaminant within the domain when the simulation begins (time $= 0$). For a one-dimensional problem, a general form of initial condition is written

$$C(x, 0) = f(x) \tag{13.37}$$

where $f(x)$ is a function describing the one-dimensional variation in concentration in the x-direction. Some common initial conditions are $C(x, 0) = 0$, and $C(x, 0) = C_i$, which provide for a constant concentration along the system.

Formulating the transport problem in terms of equations and boundary conditions is the first step in modeling mass transport. The step of solving the resulting equation using various modeling approaches will be taken up in Chapter 17.

Suggested Readings

Bear, J., 1972, Dynamics of fluids in porous media: New York, Elsevier, p. 617–626 p.

Cherry, J. A., Gillham, R. W., and Barker, J. F., 1984, Contaminants in groundwater: chemical processes: Groundwater Contamination, National Academy Press, Washington, D.C., 179 p.

Problems

1. Consider one-dimensional steady-state diffusion across a layer of thickness L where the concentration C at the bottom ($x = 0$) and at the top ($x = L$) is known. Solve this problem for the steady-state concentration as a function of x, L, C_0, and C_L. (Hint: see Example 4.3)

2. Provide a dimensional analysis of the steady-state version of Eq. 13.14 (where r/n is replaced by kC) and obtain the two Damkohler numbers.

3. The diffusion coefficient has the units of L^2/T. Can you name two other parameters or groups of parameters that have these same units, one treating fluid flow and the other for the transport of heat by conduction.

4. Compare the Fourier number of Eq. 13.12 with the Fourier number for fluid flow (Eq. 4.83) and the Fourier number for heat conduction (Eq. 9.22). State the analogous quantities and provide an interpretation of T^* for mass transport by diffusion.

Chapter Fourteen

Mass Transport in Ground Water Flow: Aqueous Systems

14.1 **Mixing as an Agent for Chemical Change**

14.2 **Inorganic Reactions in the Unsaturated Zone**

14.3 **Organic Reactions in the Unsaturated Zone**

14.4 **Inorganic Reactions in the Saturated Zone**

14.5 **A Case Study of the Milk River Aquifer**

14.6 **Quantitative Approaches for Evaluating Chemical Patterns**

14.7 **Age Dating of Ground Water**

The mass transport processes and chemical concepts presented in the last four chapters provide a framework for studying the chemically related problems in hydrogeology. This framework is general and applies equally well to a variety of core hydrogeological problems like the chemical evolution of ground water along flow systems or ground water contamination. Moreover, this processes approach also lets us address other important problems such as the diagenesis of carbonate rocks, karst formation, the origin of some kinds of ore deposits, and the secondary migration of petroleum, which historically have resided in other areas of geology.

The next three chapters, thus, will be concerned in large measure with explaining how all these problems are conceptualized in mass transport terms. Chapter 14 examines the processes that control the chemical evolution of natural ground water and techniques for interpreting these processes. In Chapter 15, the focus shifts from the water to the rocks, linking the classical geological problems to the mass transport processes. Chapter 16 culminates the discussion on the applications of mass transport concepts with an examination of processes that affect the movement of contaminants in ground water.

This process-oriented way of conceptualizing chemical problems has had a particularly significant impact on studies related to the geochemistry of natural ground water. Early thinking was that geochemical processes alone controlled the composition of ground water. This orientation is not surprising given the important role that chemical reactions do play. The legacy of this work is detailed knowledge about the reactions important in the chemical evolution of water and a variety of quantitative tools to interpret chemical patterns. However, as the following section will show, mass transport, particularly mixing, often also plays a major role in controlling the chemistry of ground water.

The remainder of this chapter identifies the chemical reactions important in the evolution of ground water. We also consider the quantitative approaches designed to interpret reactions and explain their impact on water chemistry, namely, (1) computer codes capable of modeling complex equilibria and the thermodynamic state of homogeneous and heterogeneous systems; (2) mass balance models that provide a way of estimating the sources and sinks of mass in a ground water flow system; (3) reaction path models capable of describing pathways of evolution; and (4) models used in the dating of ground water with tritium, carbon-14, and chlorine-36. The development and refinement of these quantitative techniques marked the transition of ground water geochemistry from purely a descriptive science to a quantitative one.

14.1 Mixing as an Agent for Chemical Change

All the solutes of interest to us and the stable isotopes in the water itself are affected by advection and dispersion. These processes become apparent along a flow system as a water of one chemical composition displaces or simply ends up next to water with a different composition. Mechanical dispersion and diffusion create a zone of mixing between the two waters. Accordingly, phenomena thought of most often in relation to laboratory columns operate on a much larger scale and strongly influence the chemistry of natural ground waters.

The Mixing of Meteoric and Original Formation Waters

There are several reasons why ground water mixing can occur. A common situation is related to large-scale structural changes in a sedimentary basin; when following uplift, meteoric water has the opportunity to flush units that previously contained connate water or formation water. Meinzer (1923) defined connate water as water that is trapped in the pores of a sediment during deposition. For marine deposits, this suggests formation water with the composition of sea water. Whether or not this composition is retained up to the beginning of burial depends strongly on the depositional environment as well as other numerous geologic factors. For example, Land and Prezbindowski (1981) point out that the Edwards Formation was massively altered by meteoric diagenesis prior to burial, where the original connate water was first mixed with several pore volumes of meteoric water.

Whatever its chemical composition at the time of burial, the displacement of formation water will generally begin when the following two conditions are satisfied: (1) the permeable unit is uplifted so that its outcrop occupies a position that permits recharge by meteoric water, and (2) the down-dip portions of the unit have outlets through which the formation water can be displaced. When permeable beds pinch out with depth, definite outlets are not readily available and the water must escape through overlying less permeable formations.

The typical chemical pattern is for fresh, unaltered meteoric water to be found in and near regions of outcrop with progressive increases in salinity occurring down the flow system. Fresh water or "noses" of relatively fresh water grading into saline water have been noted quite frequently. Some examples are the Madison Formation west and south of the Williston Basin (Downey, 1984b), the Inyan Kara Formation in southwestern North Dakota (Butler, 1984), and the Dakota-Newcastle Formation of the Dakota Sandstone in western South Dakota (Peter, 1984). All these formations have limited recharge areas that coincide with the uplifted portions of the aquifers.

Two field examples of the development of fresh water noses are shown on Figures 14.1 and 14.2. Figure 14.1 from Downey (1984b) demonstrates how freshwater is beginning to invade regions around areas of limited outcrop in Western United States. Figure 14.2 from Mitsdarffer (1985) shows a well-developed fresh water nose in the Mission Canyon Formation in the Williston Basin with the source of recharge probably in the Black Hills of South Dakota. An interesting question is whether the entire fresh water portions of the Dakota Sandstone represent an immense fresh water nose surrounded by more saline waters in the neighboring basins.

Domenico and Robbins (1985a) present some theoretical arguments regarding displacement patterns in uplifted rocks (Figure 14.3). A steady-state concentration pattern will evolve early in the mixing history, that is, after a few ten's of pore volumes have passed through the rock. The steady state is characterized by water approaching the concentration of meteoric recharge when the area of intake is extensive (Figure 14.3a). When the inlet for meteoric recharge is small, the steady state is characterized by a spatial distribution in concentrations that can range from meteoric water composition at the recharge end to highly concentrated water at the discharge end (Figure 14.3b). The role of transverse dispersivity is noted on Figure 14.3. Small values produce tight concentration distributions that are characterized by minimal lateral spreading but more extensive protrusions of relatively fresh water (compare Figures 14.3b and 14.3c).

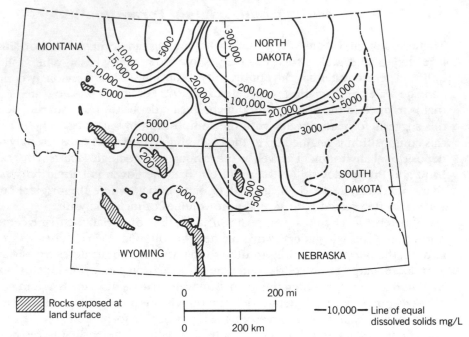

Figure 14.1
Dissolved solids concentration in water from the Cambrian–Ordovician aquifer (after Downey, 1984b).

Another type of mixing occurs whenever brines are forced upward in a compaction-driven flow system. An example of this mixing occurs in the Gulf of Mexico Basin (Bethke and others, 1988), where abnormal fluid pressure is driving pore fluids out of the basin. In the last 2 million years, when the effects of over-pressuring have been most pronounced, brine migration has pushed back an apparently much more extensive fresh water incursion into the basin (Bethke and others, 1988). A schematic illustration of this mixing zone is presented in Figure 14.4. As will be discussed in the next chapter, the mixing zone of converging flow gives rise to hydraulic traps for hydrocarbon entrapment.

The implication of this process in terms of ground water chemistry is that brine can move into parts of a basin where it otherwise may not have formed. An example is the Na-Ca-Cl brine found in Lower Cretaceous shelf carbonates of the Edwards Group in south-central Texas (Land and Prezbindowski, 1981). This brine may have formed in the overpressured portion of the Gulf of Mexico Basin from evaporites known to occur there and subsequently migrated into shallower, up-dip parts of the Edwards Group.

Not all mixing in ground water is related to the large-scale tectonic deformation of sedimentary basins. In units having a low hydraulic conductivity, the pore water can often be relatively old and chemically different than modern-day recharge. Examples of such materials are the clay-rich tills found in the northern part of the United States and in Canada (Desaulniers and others, 1981; Bradbury, 1984). These deposits often have hydraulic conductivities of 10^{-8} cm/s and lower and because of the relatively short time since deposition may contain water incor-

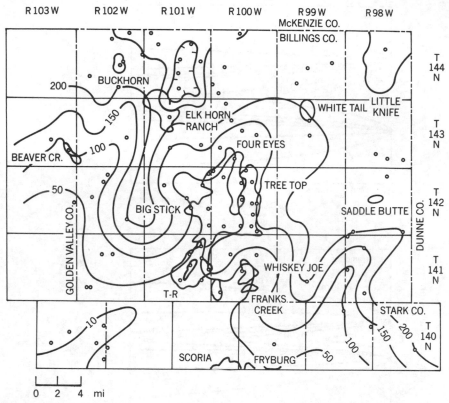

Figure 14.2
Dissolved solids concentration for the Mission Canyon Formation, in parts per thousand (after Mitsdarffer, 1985).

porated when the till was deposited. Even where this is not the case, the low permeability of till means that any long-term changes in the chemistry of recharge will be reflected in the chemical patterns along the flow system because of the age differences of the water. For example, at a site in Wisconsin, modern-day recharge is enriched by a maximum of 7‰ in $\delta^{18}O$, has a higher HCO_3^- concentration, and has a lower Cl^- concentration (Bradbury, 1984). Both Desaulnier and others (1981) and Bradbury (1984) feel that in low-permeability materials like these that diffusion may be the most important transport process.

Diffusion in Deep Sedimentary Environments

Diffusion has been described as the transport of mass in response to concentration gradients. As the diffusive flux is proportional to the concentration gradient, this process can cause significant mass redistribution whenever large concentration gradients can occur and can be maintained over geologic time. Thus, this process should be most effective where zones of hypersalinity underlie brackish water. Diffusion acts to eliminate the concentration gradient and, thus, becomes less significant as this equalization occurs.

The occurrence of brines in sedimentary basins has been attributed to three causes: dissolution of original evaporites, original formation waters buried with

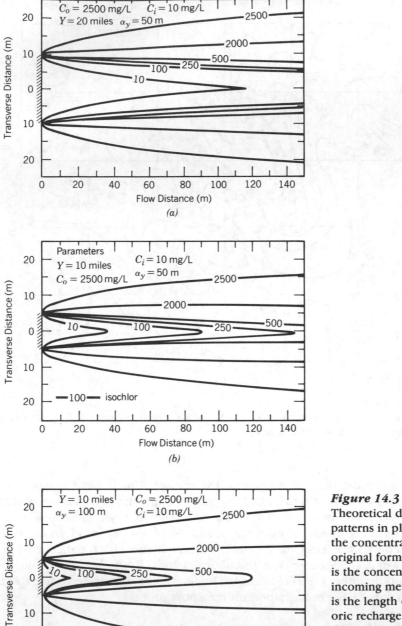

Figure 14.3
Theoretical displacement patterns in plan view. C_o is the concentration of the original formation water, C_i is the concentration of incoming meteoric water, Y is the length of the meteoric recharge area, and α_y is the transverse dispersivity (from Domenico and Robbins, 1985a. Geol. Soc. Amer. Pub., from Geol. Soc. Amer. Bull., v. 96, p. 328–335).

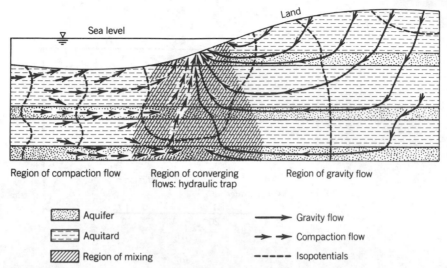

Figure 14.4
Hydraulic trap created by convergence of continental gravity flow and marine compaction flow of formation waters (from Tóth, 1988, Geol. Soc. Amer. Pub., from the Geology of North America, v. 0–2, p. 485–501).

the sediment, and membrane filtration. Whatever the mechanism, the resulting concentration gradients can give rise to significant mass redistribution over geologic time. The common pattern that emerges is a consistent increase in pore water salinity (mainly Na^+ and Cl^-) with depth (Figure 14.5). The data from north Louisiana, in particular, show a linear increase in NaCl over 2700 m to a depth at which salt is known to occur.

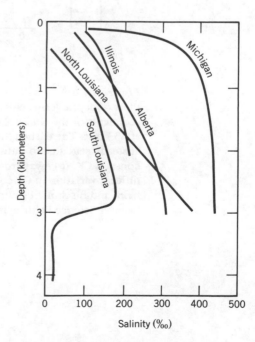

Figure 14.5
Variation in maximum pore water salinity as a function of depth in the Alberta, Illinois, and Michigan basins as well as the Louisiana Gulf Coast (from Ranganathan and Hanor, 1987). Reprinted with permission of Elsevier Science Publishers from J. Hydrol., v. 92.

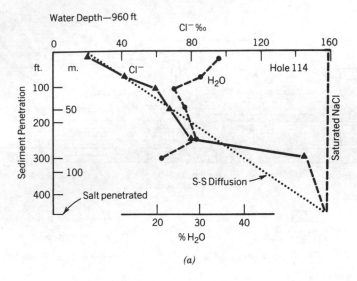

(a)

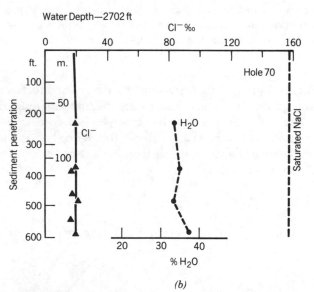

(b)

Figure 14.6

Plots of depth versus chlorinity and water content in shallow
sediments of the Gulf of Mexico. (*a*) Rock salt is encountered
at 433 feet. The dashed line represents the distribution of Cl⁻
at steady state due to diffusion between two boundaries of
constant Cl⁻. (*b*) Away from diapiric structures there is no significant variation in Cl⁻ concentration with depth (from Manheim and Bischoff, 1969). Reprinted with permission of
Elsevier Science Publishers from Chem. Geol., v. 4.

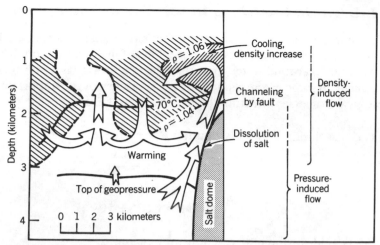

Figure 14.7
Postulated flow regime near a salt dome in the Gulf Coast (after Bennett and Hanor, 1987). Reprinted with permission of Academic Press from Dynamical Geology of Salt and Related Structures.

Speculation as to how these chemical patterns have evolved is based on theoretical studies by Ranganathan and Hanor (1987) and field studies by Manheim and Bischoff (1969) and Manheim and Paull (1981). Ranganathan and Hanor (1987) solved one-dimensional forms of the diffusion and diffusion-advection equation with realistic sets of transport parameters and concluded that salinity-depth relationships develop over several tens of millions of years that are similar to those observed in a variety of sedimentary basins. Manheim and Bischoff (1969) examined the pore water composition from six holes on the continental slope of the Gulf of Mexico and showed how the presence or absence of salt diapirs affected pore water salinity. Samples from holes drilled near salt plugs showed systematic increases in salinity with depth that was interpreted as being caused by salt diffusion (Figure 14.6a). The development of the diffusion profile on Figure 14.6a required about 400,000 years. Samples from holes drilled away from diapiric structures showed little change in chemistry as compared to sea water except for minor diagenetic changes and a loss of SO_4^{2-} (Figure 14.6b). Similar studies along the southeastern Atlantic coast of the United States showed the same result over a greater range in depth (Manheim and Paull, 1981).

Detailed work by Bennett and Hanor (1987) on the relationship among hydraulic head, pore water density, temperature, and salinity has led to postulated pathways for solute transport in the vicinity of salt domes. These authors noted a zone of high salt concentration above the salt dome and attributed this to active salt dissolution at depth and transport by upward moving geopressured fluids along a fault (Figure 14.7). The density stratification near the top of the dome is unstable and gives rise to large-scale convective flow. This pattern of fluid migration should result in an elevation of the isotherms in the upwelling section, which correlates reasonably well with a temperature anomaly noted at a depth of 2 km. It has been suggested that the salinity patterns induced by this mechanism have permitted diffusion to operate in the uppermost kilometer, where salinities decrease markedly toward the overlying fresh water zone.

14.2   *Inorganic Reactions in the Unsaturated Zone*

In the unsaturated zone, it is mainly the mass transfer processes that ultimately exert the greatest control the concentrations of major and minor ions and organic compounds in ground water. Almost all the reactions that we discussed in Chapter 12 operate to some extent. Following is a summary of the important chemical and biological processes affecting the inorganic constituents.

(1) Gas dissolution and redistribution, for example,

$$CO_2(g) + H_2O = H_2CO_3^*$$
$$H_2CO_3^* = HCO_3^- + H^+$$
$$HCO_3^- = CO_3^{2-} + H^+$$

(2) Weak acid–strong base reactions, for example,

calcite: $CaCO_3(s) + H^+ = Ca^{2+} + HCO_3^-$

anorthite: $CaAl_2Si_3O_8(s) + 2H^+ + H_2O = $ kaolinite $+ Ca^{2+}$

albite: $2NaAlSi_3O_8(s) + 2H^+ + 5H_2O = $ kaolinite $+ 4H_2SiO_3 + 2Na^+$

enstatite: $MgSiO_3(s) + 2H^+ = Mg^{2+} + H_2SiO_3$

(3) Sulfide mineral oxidation, for example,

$$4FeS_2(s) + 15O_2 + 14H_2O = 4Fe(OH)_3 + 16H^+ + 8SO_4^{2-}$$

(4) Precipitation-dissolution of gypsum,

$$CaSO_4 \cdot 2H_2O(s) = Ca^{2+} + SO_4^{2-} + 2H_2O$$

(5) Cation exchange, for example,

$$Ca^{2+} + 2Na\text{-}X = 2Na^+ + Ca\text{-}X$$
$$Ca^{2+} + Mg\text{-}X = Mg^{2+} + Ca\text{-}X$$
$$Mg^{2+} + 2Na\text{-}X = 2Na^+ + Mg\text{-}X$$

where Na-X is Na adsorbed onto a clay mineral.

This list is generally the conceptual model for unsaturated zone processes proposed by Moran and others (1978*a*). While their model was not intended to be general, combinations of these processes do explain the chemistry of water in different settings.

(1)   Gas Dissolution and Redistribution

The dissolution and redistribution of $CO_2(g)$ are important soil-zone processes. Rain water or melted snow contains relatively small quantities of mass, is somewhat acidic, and has a P_{CO_2} of about $10^{-3.5}$ atm. As this water moves downward, it rapidly dissolves CO_2, which occurs in soil at partial pressures larger than the atmospheric value. Elevated CO_2 pressures are due primarily to root and microbial respiration and, to a lesser extent, the oxidation of organic matter (Palmer and

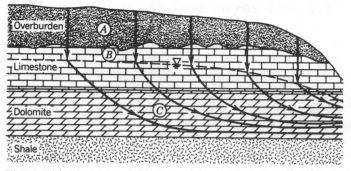

Figure 14.8
Idealization of flow through a hypothetical carbonate
sequence. The circled letters are reference points for calculations discussed in the text (modified from Palmer and Cherry,
1984). Reprinted with permission of Elsevier Science Publishers from J. Hydrol., v. 75.

Cherry, 1984). Values for soils range generally from $10^{-3.5}$ atm to more than 500
times larger (Palmer and Cherry, 1984). CO_2 dissolved in water is further redistributed among the weak acids of the carbonate system. For a pH range of 4.5 to
5.5, $H_2CO_3^*$ is the dominant carbonate species with HCO_3^- and H^+ the most
dominant anion and cation, respectively.

One direct result of dissolving $CO_2(g)$ in water is a rapid increase in the total
carbonate content of the water and a decrease in pH. Calculations show the extent
to which the P_{CO_2} in the unsaturated zone influences the evolving pore water
chemistry (Palmer and Cherry, 1984). Consider the hypothetical system on Figure
14.8. In moving to position A (Figure 14.8), the only process is the open system
dissolution of CO_2 (P_{CO_2} remains fixed). As expected, increasing the P_{CO_2} from
$10^{-3.5}$ to $10^{-2.5}$ to $10^{-1.5}$ atm increases the total amount of carbonate in solution by
about two orders of magnitude and lowers the pH from 5.62 to 4.62 across the
range of partial pressures (Table 14.1).

Another important soil zone–atmospheric process is the dissolution of $O_2(g)$.
The resulting levels of dissolved oxygen are sufficiently large at least initially to
control the redox chemistry in shallow ground water.

Table 14.1
**Calculated pH and total carbonate concentration
for water in equilibrium with specified P_{CO_2}
values (modified from Palmer and Cherry, 1984)**[*]

Partial pressure CO_2 (atm)	pH	Total carbonate[1] (mM)
$10^{-3.5}$	5.62	0.019
$10^{-2.5}$	5.12	0.178
$10^{-1.5}$	4.62	1.72

[1]$(H_2CO_3^*) + (HCO_3^-) + (CO_3^{2-})$ in millimoles per liter; temperature = 10°C.

[*]Reprinted with permission of Elsevier Science Publishers
from J. Hydrol., v. 75.

(2) Weak Acid–Strong Base Reactions

CO_2-charged ground water is effective in dissolving minerals. The most common reactions involve the weak acids of the carbonate and silicate systems and strong bases from the dissolution of carbonate, silicate, and aluminosilicate minerals. As we showed in Section 12.2, this process causes the weak acids to dissociate. In the carbonate system, the relative abundance of HCO_3^- and CO_3^{2-} increases at the expense of $H_2CO_3^*$. Overall, both the alkalinity and cation concentrations increase.

Calculations again illustrate these effects (Palmer and Cherry, 1984). Referring to Figure 14.8, the point of interest is now *B*, where water encounters limestone bedrock. The system is assumed open (that is, the P_{CO_2} remains fixed) with sufficient residence time to be at saturation with respect to calcite. P_{CO_2} again is a key variable in determining the dissolved carbonate content, the total Ca content (mainly Ca^{2+}), and the total quantity of calcite dissolved to reach equilibrium. As the P_{CO_2} is increased from $10^{-3.5}$ to $10^{-2.5}$ to $10^{-1.5}$, pH decreases from 8.30 to 7.64 to 6.99 (Table 14.2). The dissolved carbonate content (mainly HCO_3^-) increases, as does total Ca. The tendency for weak acids to ionize is reflected by the dominance of HCO_3^-. This simple system only involves the dissolution of one mineral. Langmuir (1971) calculated how HCO_3^- and pH change as the water approaches equilibrium with respect to a second mineral (dolomite). For a system that is open with respect to CO_2, water evolves to equilibrium with calcite along lines of constant P_{CO_2} (Figure 14.9). For example, moving along the P_{CO_2} line of $10^{-2.5}$ atm, the pH of a solution in equilibrium with calcite is 7.65. The second solubility curve labeled "dolomite" illustrates the pH and HCO_3^- concentration for soil water in equilibrium with dolomite alone or calcite and dolomite together. At the temperature of calculation, the solubility curves for dolomite and dolomite plus calcite are the same (Langmuir, 1971).

Another feature of this diagram is the series of paths labeled "no CO_2 added." These lines describe how the concentrations of HCO_3^- and pH change under closed system conditions. In calculating these pathways, the P_{CO_2} decreases as CO_2 is depleted by the dissolution of carbonates. Whether the system is open or closed with respect to CO_2 can have an important bearing on the pore water chemistry. For example, an open system ($P_{CO_2} = 10^{-2}$ atm) has a HCO_3^- concentration of about 260

Table 14.2
Calculated pH, total calcium concentration, total carbonate concentration, and amount of calcite reacted in water at equilibrium with respect to calcite and with specified P_{CO_2} values (modified from Palmer and Cherry, 1984)[*]

Partial pressure CO_2 (atm)	pH	Total Ca (mM)	Total[1] CO_3 (mM)	Calcite[2] reacted (mM)
$10^{-3.5}$	8.30	0.628	1.26	0.628
$10^{-2.5}$	7.64	1.40	2.96	1.40
$10^{-1.5}$	6.99	3.22	8.13	3.22

[1]$(H_2CO_3^*) + (HCO_3^-) + (CO_3^{2-})$ in millimoles per liter.

[2]Amount of calcite dissolved in the unsaturated zone; temperature = 10°C.

[*]Reprinted by permission of Elsevier Science Publishers from J. Hydrol., v. 75.

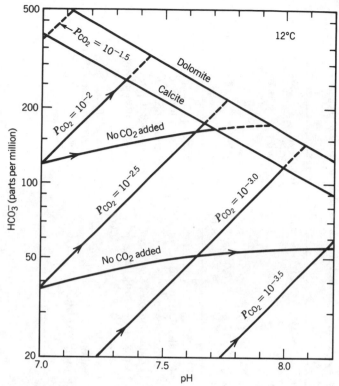

Figure 14.9
Possible approaches to equilibrium of ground water in contact
with calcite and (or) dolomite at 12°C. The solubility curves of
dolomite and calcite plus dolomite are the same at this temperature. Reprinted with permission from Geochim. Cosmochim.
Acta, v. 35, D. Langmuir, The geochemistry of some carbonate
ground waters in central Pennsylvania, Copyright 1971. Pergamon Press plc.

ppm and pH of about 7.34 at calcite saturation (Figure 14.9). Under closed-system
assumptions, these values are 170 ppm and 7.70, respectively.

A model involving just the two processes considered so far, the dissolution of
CO_2 gas accompanied by calcite dissolution, describes the chemistry of water in
the unsaturated zone in carbonate terranes. Shown in Table 14.3 is a chemical
analysis for water from a near surface spring in a limestone in central Pennsylvania
(Langmuir, 1971). This analysis probably approximates the composition of water
from the unsaturated zone. The most abundant ions are Ca^{2+} and HCO_3^-, with relatively small concentrations of Na^+ and Cl^-. The calculated P_{CO_2} for this sample is
$10^{-2.08}$ atm, and like nearly all of the spring water, the sample is undersaturated
with respect to calcite and dolomite. Presumably, transport through fractures has
been so rapid that calcite equilibrium was not achieved. Farther along the flow
system, calcite equilibrium finally constrains increases in Ca^{2+} and HCO_3^- concentrations.

As our sample list of weak acid–strong base reactions shows, silicate and
aluminosilicate minerals will also react to some extent when they are present. The
relatively low solubility of these minerals, however, means that their contribution

Table 14.3
Comparison of the chemistry of waters from four different recharge environments and the unsaturated zone processes giving rise to these waters

Study	Ca^{2+}	Mg^{2+}	Na^+	K^+	HCO_3^-	SO_4^{2-}	Cl^-	pH	$\log P_{CO_2}$	SI_c	SI_d
1. Langmuir (1971)	41.	4.4	6.5	2.1	134	—	12	7.13	−2.08	−0.78	−1.16
2. Garrels and MacKenzie (1967)[1]	3.1	0.7	3.0	1.1	20	1.0	0.5	6.2	−1.8	—	—
3. Kimball (1984)	57	21	26	0.4	210	57	2.6	7.9	−2.57	0.26	0.14
4. Moran et al. (1978b)[1]	62	25	469	7.9	748	577	4.7	8.0	—	—	—

1. Processes
- CO_2 dissolution and redistribution
- weak acid–strong base
  - calcite

2. Processes
- CO_2 dissolution and redistribution
- weak acid–strong base
  - plagioclase
  - biotite
  - K-feldspar

3. Processes
- CO_2 dissolution and redistribution
- O_2 dissolution
- weak acid–strong base
  - calcite
  - dolomite
  - albite
- pyrite oxidation

4. Processes
- CO_2 dissolution and redistribution
- O_2 dissolution
- weak acid–strong base
  - calcite
- pyrite oxidation
- precipitation/dissolution of gypsum
- cation exchange

Notes: Concentration in milligrams per liter.

Partial pressures in atmospheres.

SI, saturation indices for calcite and dolomite.

[1]Mean values for several samples.

to the mass dissolved in pore water will be relatively small when soluble minerals like calcite are present. In system where there are no carbonates present, these reactions control the ion chemistry.

Garrels and MacKenzie (1967) looked at a case where waters rich in CO_2 reacted with a suite of minerals found in igneous rock (Table 14.3). Again by using water from ephemeral springs as a surrogate for soil water, we can observe the behavior of a system with no carbonate minerals (Table 14.3). The concentration of all ions is low because only plagioclase feldspar, biotite, and K-feldspar dissolve. In fact, the concentration of most ions is so low that the initial composition of precipitation had to be considered when Garrels and MacKenzie worked with the data. Dissolution apparently takes place under closed-system conditions with about half of the CO_2 consumed.

(3)   Sulfide Oxidation

Sulfide oxidation is one of the important redox reactions within the unsaturated zone. Minerals like pyrite or marcasite are oxidized to produce $Fe(OH)_3(s)$, SO_4^{2-}, and H^+. In fact, pyrite oxidation is one of the most important acid-producing reactions in geological systems (Moran and others, 1978a). In coal mining areas like the Appalachians, this reaction can be the cause of serious acid–mine drainage problems (Moran and others, 1978a).

An example of how various chemical reactions, including pyrite oxidation, control recharge chemistry is a study by Kimball (1984) in the Piceance Creek Basin of northwestern Colorado. The initial acquisition of solutes in the recharge to the Uinta Formation is controlled by three processes considered so far, namely, open-system dissolution of CO_2 and O_2, weak acid–strong base reactions, and sulfide mineral oxidation (Table 14.3). The main difference as compared to the essentially monomineralic system described by Langmuir (1971) is that the dissolution of dolomite and albite and the oxidation of pyrite add Mg^{2+}, Na^+, and SO_4^{2-} to the water (Table 14.3).

(4)   Gypsum Precipitation and Dissolution

Conditions exist when the solute load in soil waters can be much higher than we have considered so far. The cyclical precipitation and dissolution of gypsum is probably the most important process in this respect. Over much of the Great Plains region of the United States and Canada, annual potential evaporation exceeds annual precipitation by a considerable amount. Thus, water that infiltrates in normal precipitation years evaporates and deposits a small quantity of gypsum. With repeated rain or snowmelt, gypsum accumulates in the upper part of the soil horizon (Moran and others, 1978a). Exceptional recharge can dissolve some of this soluble material and move it down and into the ground water system (Figure 14.10). In some arid areas, recharge water could have SO_4^{2-} concentrations in excess of 5000 mg/L (Hendry and others, 1986).

(5)   Cation Exchange

The precipitation and dissolution of gypsum in the Great Plains region is also accompanied by cation exchange (Moran and others, 1978a). The most important exchange reactions are the water-softening reactions where Ca^{2+} and Mg^{2+} in the

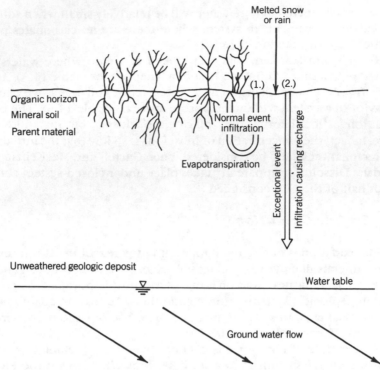

Figure 14.10
A conceptual model of subsurface flow in the plains region of the United States and Canada. (1) Annual potential evapotranspiration greatly exceeds annual precipitation and most infiltration is lost by evapotranspiration (2) Exceptional precipitation events produce recharge (from Moran and others, 1978a).

water exchange with sorbed Na^+ as ground water moves through clayey material. Shallow ground water from a study area in North Dakota (Moran and others, 1978a) provides an example of how all the processes we have considered so far work together. The chemical data on Table 14.3 represent mean values for 39 water samples collected from various near surface drift units. The abundance of Na^+ and SO_4^{2-} in the shallow ground water is attributed to gypsum dissolution, sulfide oxidation, and cation exchange. The cation exchange process is important here because it has effectively increased Na^+ concentrations at the expense of Ca^{2+} and Mg^{2+}.

14.3   *Organic Reactions in the Unsaturated Zone*

Work by Wallis and others (1981), Thurman (1985), and Hendry and others (1986) show the important organic reactions to be

1.   dissolution of organic litter at the ground surface
2.   complexation of Fe and Al

3. sorption of organic compounds
4. oxidation of organic compounds

$$CH_2O + O_2 = CO_2 + H_2O + \text{energy}$$
$$2(R\text{-}SH) + 3O_2 + 2H_2O = 2R + 4H^+ + 2SO_4^{2-}$$

(1) Dissolution of Organic Litter

The dissolution of organic litter at or close to the ground surface is the major source of dissolved organic carbon (DOC) in soil water and shallow ground water. DOC concentrations typically fall in a range from 10 to 50 mg/L in the upper soil horizons and less than 5 mg/L deeper in the unsaturated zone (Thurman, 1985). Concentrations are highest at their source and decline with depth through sorption and oxidation. The most important fraction of the DOC is a group of humic substances, consisting mainly of humic and fulvic acids. Tannins and lignins, amino acids, and phenolic compounds are often present in smaller concentrations (Wallis and others, 1981).

(2) Complexation of Fe and Al

The complexation of Fe and Al with organic matter is an important process facilitating the transport of these poorly soluble metals from the *A* horizon of the soil to the *B* horizon (Thurman, 1985). This is one key feature of the soil-forming process called podzolization (Thurman, 1985). In terms of ground water systems, this reaction is not particularly important because most of these complexed metals sorb in the *B* horizon of the soil.

(3) Sorption of Organic-Metal Complexes

Dissolved organic compounds originating in the upper part of the soil horizon are not particularly mobile due to sorption. Many of the sorption models we discussed in Chapter 12 operate within the soil zone. Hydrophobic sorption occurs because of the relatively large quantities of solid organic matter present in the upper part of many soil horizons (Thurman, 1985). Similarly, the abundance of metal oxides, hydroxides, and clay minerals leads to surface complexation reactions and electrostatic interactions. Again in terms of the chemistry of shallow ground water, this is one of the processes that keeps the quantity of DOC in recharge at relatively low concentrations.

(4) Oxidation of Organic Compounds

Oxidation reactions involving organic matter can influence the chemistry of shallow ground water. For example, the oxidation of dissolved organic matter (represented as CH_2O) provides a source of CO_2 gas within the unsaturated zone, which is readily dissolved in soil water. A second reaction, involving the oxidation of a sulfur-containing compound (represented by the amino acid cysteine), is thought to play a major role in the accumulation of gypsum in shallow soils (Hendry and others, 1986). This reaction is the organic counterpart to the pyrite oxidation reaction. In arid areas, this process can also contribute to recharge with large SO_4^{2-} concentrations (Hendry and others, 1986).

14.4 *Inorganic Reactions in the Saturated Zone*

The chemistry of ground water not only depends upon the chemistry of the recharge but also the reactions operating within the flow system itself. Processes in the saturated zone are more complex than in the unsaturated zone because of the possibilities of dispersive mixing and because geologic, hydrogeologic, and geochemical settings are much more diverse. However, most of the same processes affecting ion concentrations in the unsaturated zone are also operative in the saturated zone, including

(1) weak acid–strong base reactions, for example,

$$\text{carbonate minerals} + H^+ = \text{cations} + HCO_3^-$$
$$\text{silicate minerals} + H^+ = \text{cations} + H_2SiO_3$$
$$\text{alumino-silicate minerals} + H^+ = \text{cations} +$$
$$H_2SiO_3 + \text{secondary minerals (for example, clay minerals)}$$

(2) dissolution of soluble salts, for example,

halite: $NaCl(s) = Na^+ + Cl^-$
anhydrite: $CaSO_4(s) = Ca^{2+} + SO_4^{2-}$
gypsum: $CaSO_4 \cdot 2H_2O(s) = Ca^{2+} + SO_4^{2-} + 2H_2O$
carnalite: $KCl \cdot MgCl_2 \cdot 6H_2O(s) = K^+ + Mg^{2+} + 3Cl^- + 6H_2O$
kieserite: $MgSO_4 \cdot H_2O(s) = Mg^{2+} + SO_4^{2-} + H_2O$
sylvite: $KCl(s) = K^+ + Cl^-$

(3) redox reactions, for example,

$$\tfrac{1}{4}O_2(g) + H^+ + e^- = \tfrac{1}{2}H_2O$$
$$\tfrac{1}{2}Fe_2O_3(s) + 3H^+ + e^- = Fe^{2+} + \tfrac{3}{2}H_2O$$
$$\tfrac{1}{2}MnO_2(s) + 2H^+ + e^- = \tfrac{1}{2}Mn^{2+} + H_2O$$
$$\tfrac{1}{8}SO_4^{2-} + \tfrac{9}{8}H^+ + e^- = \tfrac{1}{8}HS^- + \tfrac{1}{2}H_2O$$
$$\tfrac{1}{8}CO_2(g) + H^+ + e^- = \tfrac{1}{8}CH_4(g) + \tfrac{1}{4}H_2O$$
$$\tfrac{1}{4}CO_2(g) + H^+ + e^- = \tfrac{1}{4}CH_2O + \tfrac{1}{4}H_2O$$

(4) cation exchange, for example,

$$\begin{array}{lcl} Ca^{2+} & & Ca \\ Mg^{2+} + 2\text{Na-clay} = 2Na^+ + & & \text{Mg-clay} \\ Fe^{2+} & & Fe \end{array}$$

(1) *Weak Acid–Strong Base Reactions*

If the ground water is not yet in equilibrium with carbonate, silicate, and alumino-silicate minerals, they will continue to dissolve in the saturated zone. Because of their relative abundance, reasonably fast reaction rate with H^+, and reasonable solubilities, these reactions increase the cation concentrations, alkalinity, and pH.

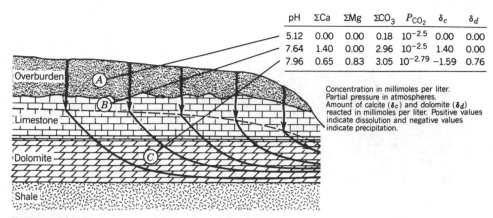

pH	ΣCa	ΣMg	ΣCO_3	P_{CO_2}	δ_c	δ_d
5.12	0.00	0.00	0.18	$10^{-2.5}$	0.00	0.00
7.64	1.40	0.00	2.96	$10^{-2.5}$	1.40	0.00
7.96	0.65	0.83	3.05	$10^{-2.79}$	-1.59	0.76

Concentration in millimoles per liter.
Partial pressure in atmospheres.
Amount of calcite (δ_c) and dolomite (δ_d) reacted in millimoles per liter. Positive values indicate dissolution and negative values indicate precipitation.

Figure 14.11
Pattern of mass transfer as water proceeds through an idealized overburden-carbonate rock sequence (modified from Palmer and Cherry, 1984). Reprinted with permission of Elsevier Science Publishers from J. Hydrol., v. 75.

The examples presented in the previous section are again useful in illustrating these effects. Consider the flow system on Figure 14.11, where water flows through a sequence of overburden, limestone, and then dolomite (Palmer and Cherry, 1984). Recall that as the water moved to reference point A it dissolved CO_2 at a fixed partial pressure, and at point B it had become saturated with respect to calcite. In the saturated zone at point C, ground water reaches equilibrium with respect to both calcite and dolomite. In moving to C, the pH, the total Mg concentration, and the total CO_3 concentrations increase (Figure 14.11). The P_{CO_2} declines from $10^{-2.5}$ atm to $10^{-2.79}$, because once the system is closed below the water table, CO_2 is depleted by the dissolution of dolomite. The total Ca concentration decreases because in dissolving dolomite the ground water became supersaturated with respect to calcite. In moving from B to C, 1.59 mM of calcite precipitates and 0.76 mM of dolomite dissolves (Figure 14.11).

The chemistry of the ground water at point C not only depends upon the processes involved but also the order of encounter (Freeze and Cherry, 1979). A slight change in geology, placing the dolomite above the limestone, has a significant impact (compare the data on Figure 14.11 with those on Figure 14.12). The pH in the second case is lower, and ΣCa, ΣMg, ΣCO_3, and P_{CO_2} are all higher. Holding the system open while dolomite dissolves to equilibrium enables more mass to dissolve. Furthermore, in moving from point B to point C, the chemical composition of the ground water does not change (Figure 14.12).

The example of evolving composition in a predominantly carbonate terrane (Langmuir, 1971) illustrates how the composition simply proceeds toward equilibrium with respect to those minerals available for dissolution (for example, calcite and dolomite). Once in the zone of saturation, the ground water approaches saturation with respect to both calcite and dolomite because of the longer residence time (Figure 14.13, Table 14.4). Also, water from a predominantly limestone source rock more closely approaches saturation with respect to calcite than dolomite. This fact is illustrated on Figure 14.13 by the abundance of data points for limestone source rocks (triangles) lying above the dashed line ($SI_c = SI_d$). Even in this simple carbonate system, there are differences in the

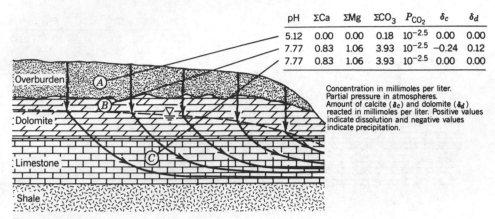

pH	ΣCa	ΣMg	ΣCO_3	P_{CO_2}	δ_c	δ_d
5.12	0.00	0.00	0.18	$10^{-2.5}$	0.00	0.00
7.77	0.83	1.06	3.93	$10^{-2.5}$	−0.24	0.12
7.77	0.83	1.06	3.93	$10^{-2.5}$	0.00	0.00

Concentration in millimoles per liter.
Partial pressure in atmospheres.
Amount of calcite (δ_c) and dolomite (δ_d)
reacted in millimoles per liter. Positive values
indicate dissolution and negative values
indicate precipitation.

Figure 14.12
Pattern of mass transfer as water proceeds through an idealized overburden-carbonate rock sequence (modified from Palmer and Cherry, 1984). Reprinted with permission of Elsevier Science Publishers from J. Hydrol., v. 95.

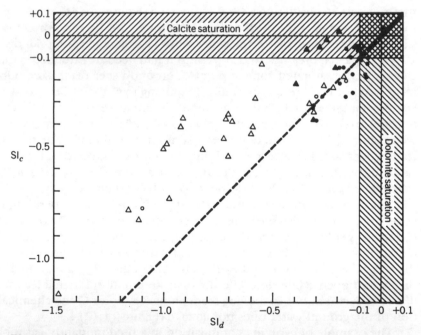

Figure 14.13
Saturation index for calcite versus the saturation index for dolomite. Spring waters are designated by open symbols; well waters by solid symbols. Triangles denote a limestone source rock; circles a dolomitic source rock. Cross-hatched area shows the limits of uncertainty in SI_c and SI_d.
Reprinted with permission from Geochim. Cosmochim. Acta, v. 35, D. Langmuir, The geochemistry of some carbonate ground waters in central Pennsylvania, Copyright 1971. Pergamon Press plc.

Table 14.4

Averages of various chemical parameters for spring and well waters from a carbonate terrane (modified from Langmuir, 1971) [*]

Parameter	Average value	
	Spring water	Well water
Specific conductance[1]	347	499
Ca^{2+} (ppm)	47.5	55.0
Mg^{2+} (ppm)	13.9	29.6
HCO_3^- (ppm)	183	265
pH	7.37	7.46
SI_c[2]	−0.41	−0.15
SI_d	−0.63	−0.18

[1]millimhos per centimeter.

[2]Saturation index.

[*]Reprinted with permission from Geochim. Cosmochim. Acta, v. 35, D. Langmuir, The geochemistry of some carbonate ground water in central Pennsylvania. Copyright 1971. Pergamon Press plc.

pathways of evolution, apparently depending upon what carbonate minerals were encountered and in what order.

Looking further at the example presented by Garrels and MacKenzie (1967), evolution through acid-base reactions increases the quantity of mass dissolved in the ground water. The effects of increased residence time can be evaluated by comparing the chemistry of ephemeral and perennial springs (Table 14.5). The system behaves as expected. The concentrations of cations increase due to the continued hydrolysis of biotite and plagioclase as does the alkalinity (reflected in the HCO_3^- concentration and pH). An important source of Ca^{2+} is the dissolution of small quantities of carbonate minerals. In the deeper parts of the system, montmorillonite occurs in addition to kaolinite as a weathering product of plagioclase.

Table 14.5

Chemical composition of ephemeral and perennial springs of the Sierra Nevada[*]

Sample source	Ca^{2+}	Mg^{2+}	Na^+	K^+	HCO_3^-	SO_4^{2-}	Cl^-	SiO_2	pH
Ephemeral springs	3.11	0.70	3.03	1.09	20.0	1.00	0.50	16.4	6.2
Perennial springs	10.4	1.70	5.95	1.57	54.6	2.38	1.06	24.6	6.8

[1]Concentrations in milligrams per liter.

[*]Reprinted with permission from Garrels, R.M. and MacKenzie, F.T., in Equilibrium Concepts in Natural Water Systems; Gould, R.F. ed., American Chemical Society: Washington D.C., 1967, p. 224–225. Copyright 1967 American Chemical Society.

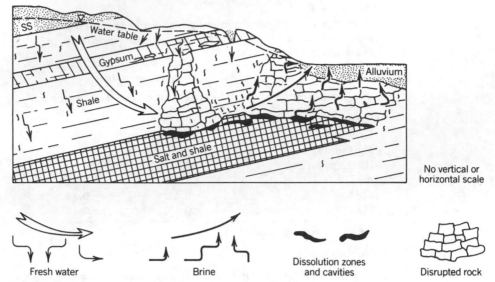

Figure 14.14
Schematic representation of the circulation of fresh water and brine in areas of salt dissolution in western Oklahoma. There is no scale, but the length of the section could range from 1–15 km and the thickness from 30–300 m (modified from Johnson, 1981). Reprinted with permission of Elsevier Science Publishers from J. Hydrol., v. 54.

(2)   Dissolution of Soluble Salts

We saw in Section 14.1 how the presence of evaporites in a sequence could affect the water chemistry through diffusion. There are other situations (for example, Figure 14.14) where active ground water flow directly encounters evaporites. Mineral salts are extremely soluble and dissolve to produce a brine whose composition depends upon the particular minerals present (for example, halite, anhydrite, gypsum, carnalite, kieserite, and sylvite).

This process (Figure 14.14) leads to the formation of saline brines in the shallow ground water of western Oklahoma and the southeastern part of the Texas Panhandle (Johnson, 1981). Fresh water recharged through permeable units moves downward until it encounters salts at depths ranging from 10–250 m. The dissolving salt produces cavities at the up-dip limit or the top of the salt (Johnson, 1981). Periodically, the rocks overlying the cavities collapse. The process is apparently self-perpetuating because the collapse and fracturing of overlying units provides improved access to the salt for fresh water.

The evaporites, mainly halite and gypsum/anhydrite, are interbedded with a thick sequence of red beds. Given the particular salts involved, high concentrations of Na^+ and Cl^- are not surprising. Clearly, when ground water encounters large quantities of soluble salts in the subsurface, the impact on the chemistry is considerable.

(3)   Redox Reactions

The redox conditions encountered along a flow system are important in controlling the chemistry of metal ions and solids (for example, Fe^{2+}, Mn^{2+}, and Fe_2O_3), species or solids containing sulfur (for example, SO_4^{2-}, H_2S, and FeS_2), and dis-

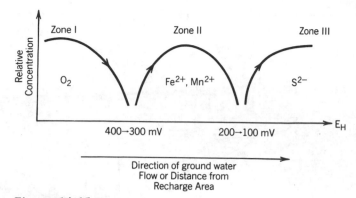

Figure 14.15
Different redox zones existing along a confined aquifer. Reprinted from Jackson and Inch, 1980. Hydrogeochemical processes affecting the migration of radionuclides in a fluvial sand aquifer at Chalk River Nuclear Laboratories, NHRI Scientific Series, Paper 7, Environment Canada, 58p. Reproduced with the permission of the Minister of Supply and Services Canada, 1990.

solved gases containing carbon (for example, CO_2, CH_4). It is possible in some flow systems to define redox zones. These zones are parts of an aquifer in which E_H is controlled by a dominant redox couple. Field studies (for example, Champ and others, 1979; Jackson and Patterson, 1982) have shown that oxygen, iron-manganese, and sulfides zones will often be present. The probable half-reactions controlling E_H in these zones are the reduction of oxygen to water, the reduction of iron or manganese oxides, and sulfate reduction to HS^- or H_2S. In a few cases a methane zone, formed from the reduction of CO_2, could even be present.

Reduced sulfide and methane zones tend to develop in confined flow systems containing excess oxidizable DOC. The favorable hydrodynamic condition is when water does not mix continually with other recharge containing oxygen. Thus, redox zones will generally be observed in an extensive artesian aquifer that receives recharge from a limited area of outcrop or in an aquifer with confining units.

Figure 14.15 summarizes the common changes in water chemistry from zone to zone (Jackson and Inch, 1980). In zone I, oxygen initially present in recharge will decline through reduction by organic carbon

$$CH_2O + O_2 = CO_2 + H_2O \tag{14.1}$$

The concentrations of Fe^{2+} and Mn^{2+} in zone II increase because the Fe(III) and Mn(IV) minerals, which are oxidized solids, are not stable in the more reducing environment. These reactions are

$$CH_2O + 8H^+ + 4Fe(OH)_3(s) = 4Fe^{2+} + 11H_2O + CO_2 \tag{14.2}$$

$$CH_2O + 4H^+ + 2MnO_2(s) = 2Mn^{2+} + 3H_2O + CO_2 \tag{14.3}$$

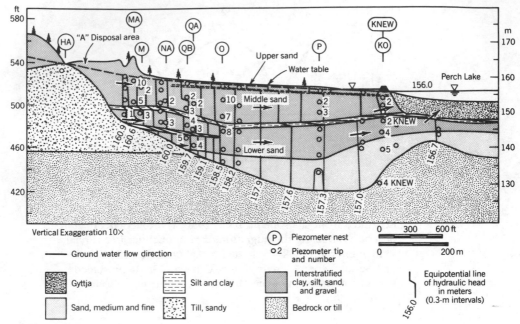

Figure 14.16
Cross section through the lower Perch Lake basin aquifer at Chalk River Nuclear Laboratories, Canada, showing equipotential lines, piezometer tips, and the position of the water table. Hydraulic head measurements averaged over the period of 1973–1975 (from Jackson and Patterson). Water Resources Res., v. 18, p. 1255–1268, 1982. Copyright by American Geophysical Union.

Once the E_H is sufficiently reduced, sulfide species appear from the reduction of SO_4^{2-} or

$$2CH_2O + SO_4^{2-} + H^+ = HS^- + 2H_2O + 2CO_2 \qquad (14.4)$$

When the rate of sulfide production exceeds the dissolution rate of iron and manganese oxides, Fe^{2+} and Mn^{2+} concentrations fall (Figure 14.15) as metal sulfides precipitate as the stable phase.

A study at Chalk River Nuclear Laboratories near Ottawa, Canada, demonstrates how redox conditions can change along a flow system. By careful measurements of pH, E_H, dissolved oxygen (DO), and the total concentration of sulfides (S_T^{2-}) as well as other key chemical parameters, Jackson and Patterson (1982) defined three redox zones. The main geologic units of interest were two fluvial sand aquifers separated by a thin layer of interbedded clay about 1 m thick (Figure 14.16). Ground water flows southward from the upland area toward Perch Lake. Inflow to the aquifers is near the area labeled "disposal area" (Figure 14.16), with none further downgradient.

The lower aquifer in the recharge area contains dissolved oxygen and has an E_H of about 0.55 V (Figure 14.17), which is quite close to the theoretical range of 0.71 to 0.83 V. Both Fe^{2+} and Mn^{2+} concentrations are low and sulfide is undetectable. The declining E_H along the deep-flow system coincides with the reduction of

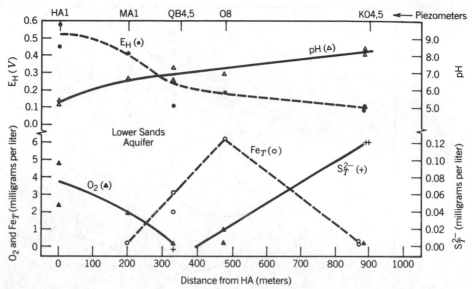

Figure 14.17
The pH, E_H, O_2, and total dissolved iron and sulfide values for the deep, confined Lower Sands aquifer. The piezometer numbers are shown on the top abscissa (from Jackson and Patterson). Water Resources Res., v. 18, p. 1255–1268, 1982. Copyright by American Geophysical Union.

oxygen. High concentrations of iron and manganese are evident once the oxygen is depleted (Figure 14.17). Zone II is probably not extensive because sulfide is becoming relatively abundant at E_H values below 0.2 V. The downstream end of the flow system is a sulfide zone with E_H values approaching the theoretical range for the sulfate/sulfide couple (that is, -0.21 to -0.32 V). Concentrations of Fe^{2+}, Mn^{2+}, and SO_4^{2-} decline in zone III from the precipitation of ferrous sulfides (Jackson and Patterson, 1982).

The reactions that oxidize organic matter (Section 12.5) all generate CO_2 as a product. This CO_2 is redistributed among $H_2CO_3^*$, HCO_3^- and CO_3^{2-}. In aquifers where Ca^{2+} and Mg^{2+} are exchanged onto clay minerals for Na^+, the possibility exists for carbonate dissolution and even higher HCO_3^- concentrations. With CO_2 generated by redox reactions, ion exchange, and carbonate dissolution, the water will evolve chemically to a sodium bicarbonate type like that from the Atlantic coastal plain (Foster, 1950), the Eocene aquifers of the east Texas basin (Fogg and Kreitler, 1982), and the Milk River aquifer (Hendry and Schwartz, 1990).

Chapelle and others (1987) investigated the source of CO_2 in ground water from the Atlantic coastal plain in Maryland. The units of interest included the Magothy-Upper Patapsco aquifer and the Lower Patapsco aquifer. These aquifers crop out between Washington and Baltimore (Figure 14.18) and dip toward the east. In general, recharge occurs along the outcrop areas with flow toward the Potomac River and under Chesapeake Bay (Figure 14.18). Although these units contain no calcareous minerals, HCO_3^- concentrations increase significantly in the direction of flow. Where the Patapsco aquifer crops out, HCO_3^- concentrations range from 0–50 mg/L. Approximately 30 km downgradient, they have increased to

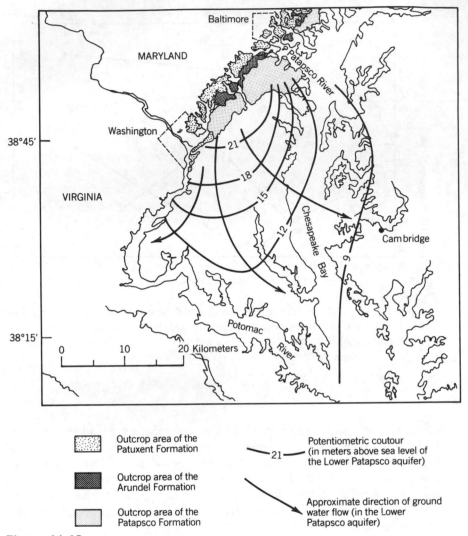

Figure 14.18
Map showing the regional outcrop area for the Patuxent, Arundel, and Patapsco Formation on the Atlantic coastal plain in Maryland. Ground water flow in the Lower Patapsco Formation is away from the outcrop area (from Chapelle and others). Water Resources Res., v. 23, p. 1625–1632, 1987. Copyright by American Geophysical Union.

about 150–200 mg/L, and finally near Cambridge, Maryland (Figure 14.18), HCO_3^- concentrations range from 400 to 500 mg/L (Chapelle and others, 1987). This pattern is similar to that observed by Foster (1950) farther south in Virginia.

The potential sources of CO_2 are the bacterially mediated oxidation of solid organic matter in the aquifers or the abiotic decarboxylation of these materials. The presence of both sulfate reducing and methanogenic bacteria in cores, however, points to the generation of CO_2 in redox reactions. Although sulfate-reducing bacteria are present, the relatively small quantities of SO_4^{2-} present makes the contribution of CO_2 from sulfate reduction small relative to that from methanogenesis. The bacteria facilitating these reactions occur together because the theoretical pe of a system is nearly the same with either reaction.

(4) Cation Exchange

The most important exchange reactions are the natural water-softening reactions, which take Ca^{2+} and Mg^{2+} out of water and replace them with Na^+. The main requirement for this process is a large reservoir of exchangeable Na^+, which is most often provided by clay minerals deposited in a marine environment. One does not have to go far in the United States or Canada to find clays or shales capable of ion exchange. A case in point is bedrock in the Wabamun Lake area of central Alberta, Canada (Schwartz and Gallup, 1978). Shale, sandstones, and coal of the Edmonton Group are overlain by two younger bedrock units that include the Pembina coals and sandstone of the Paskapoo Formation. Bedrock in turn is overlain by a thin veneer of till, sand, and gravel. The shales of the Edmonton group and those interbedded with the Pembina coals were deposited in a marine environment. Figure 14.19 is a plot of Ca^{2+} versus Na^+ concentrations for samples from the drift, Pembina coal, and Edmonton Group. It shows how the pattern of cation dominance shifts from Ca^{2+} (and Mg^{2+}, not shown) to Na^+ as water moves deeper into the sequence.

When ion exchange takes place, its effects on the cation chemistry of water should be unmistakable. However, as the following case study shows, an equivalent increase in Na^+ concentration that is matched by an increasing Cl^- concentration may mean that other processes are at work.

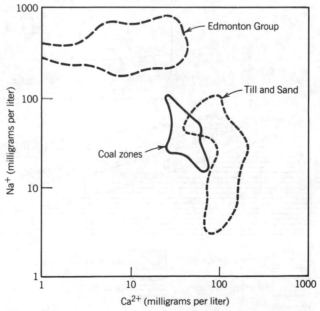

Figure 14.19
Fields defined by plotting Ca^{2+} versus Na^+ data for three different geologic units in the Wabamun Lake area of central Alberta, Canada (from Schwartz and Gallup, 1978).

14.5   Case Study of the Milk River Aquifer

We have singled out the Milk River aquifer for detailed discussion because the interpretation of chemical patterns requires the integration of concepts of mass transport and mass transfer. Several studies have documented the geologic and hydrogeologic setting including Meyboom (1960), Schwartz and Muehlenbachs (1979), and Hendry and Schwartz (1988, 1990). The aquifer is part of a thick sequence of Cretaceous rocks (Figure 14.20). Units of interest, from oldest to youngest, are the shale of the Colorado Group, the Milk River Formation, whose lower member is the aquifer, and shale of the Pakowki Formation (Figure 14.20). The aquifer pinches out to the north and east as the result of a facies change.

Ground water recharge occurs where the aquifer crops out in northern Montana and along an area of outcrop-subcrop in southern Alberta. Flow is generally northward or down-dip in the aquifer. Leakage from the aquifer is through the Pakowki Formation and overlying units. Before extensive development of the ground water resource, flowing wells were common.

An exciting feature of this aquifer system is the well-defined patterns of variability in the major and minor ions and the stable or radiogenic isotopes. Cl⁻, $\delta^{18}O$, and δD change markedly along the flow system (Figure 14.21) and appear to act

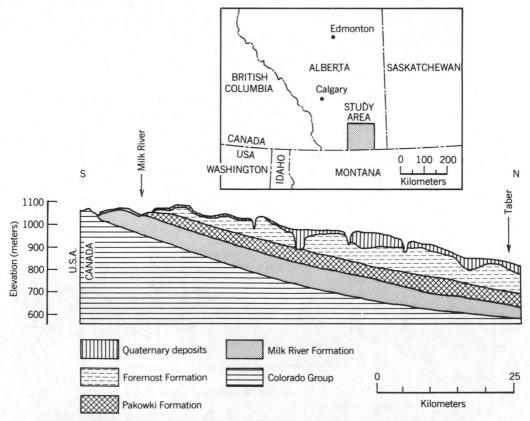

Figure 14.20
Study area for the Milk River aquifer investigation. The geological cross section illustrates the most important drift and bedrock units in southern Alberta, Canada (from Hendry and Schwartz, 1990). Reprinted by permission of Ground Water. Copyright © 1990. All rights reserved.

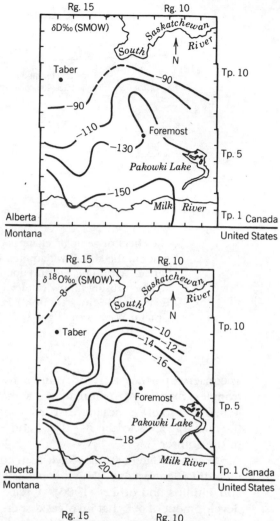

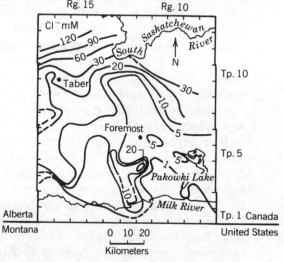

Figure 14.21
Areal variation in Cl⁻, δ¹⁸O, and
δD in water from the Milk River
aquifer (from Hendry and
Schwartz). Water Resources Res.,
v. 24, p. 1747–1764, 1988.
Copyright by American
Geophysical Union.

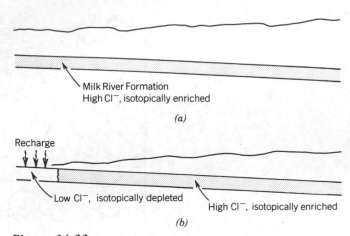

Figure 14.22

The effect of geologic changes on the chemistry of ground water in the Milk River aquifer. In (*a*) the aquifer is deeply buried. With time, erosion uncovers upstream end of the aquifer, causing inflow of meteoric water (modified from Hendry and Schwartz, 1990). Reprinted by permission of Ground Water. Copyright © 1990. All rights reserved.

as conservative tracers (Hendry and Schwartz, 1988). Cl^- concentrations increase from less than 1 mM to more than 100 mM at the downstream end of the flow system. $\delta^{18}O$ and δD values increase from about $-20\permil$ and $-150\permil$, respectively, in the recharge areas to about $-8.5\permil$ and $-85\permil$, respectively, at the northern end of the aquifer.

Several different studies have been concerned with explaining these patterns, including Schwartz and Muehlenbachs (1979), Domenico and Robbins (1985a), and Phillips and others (1986). A reexamination of these ideas prompted the development of one that involves dispersion and advection in the aquifer and the diffusion of mass from the aquitard (Hendry and Schwartz, 1988). This model takes into account the effects of both mass transport processes and geologic change in controlling ground water chemistry.

Tóth and Corbet (1986) in this area showed how the configuration of the land surface changed with time. As recently as 5 million years ago, the land surface was probably 700 m higher than now. Even in the early Pleistocene, the land surface was about 200 m higher than present (Tóth and Corbet, 1986). When the aquifer was deeply buried (Figure 14.22), ground water probably was enriched isotopically relative to present-day meteoric water ($\delta^{18}O \simeq -20\permil$ and $\delta D \simeq -150\permil$) and had a mean Cl^- concentration of about 85 mM (Hendry and Schwartz, 1988). Continuing erosion eventually exposed a relatively large recharge area and provided a new source of water to the aquifer. Proximity to the surface would result in recharge with a low Cl^- concentration (Figure 14.22*b*). This change in the flow dynamics of the aquifer probably occurred about 1×10^6 years ago. Somewhat later, the stable isotope composition of the precipitation became similar to present-day meteoric water due to a changing climate.

This chemically different water with time has displaced the original formation water. However, in so doing, it has created diffusion gradients between the isotopically depleted, low-chlorinity water in the aquifer and the original water remaining

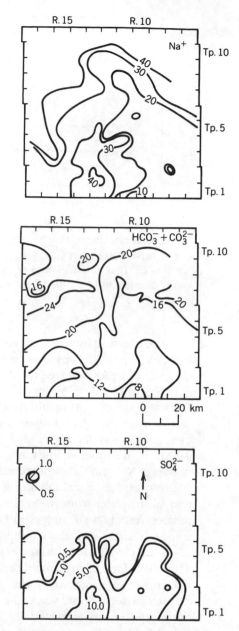

Figure 14.23
Concentration distribution (mM) of Na^+, $HCO_3^- + CO_3^{2-}$, and SO_4^{2-} in waters of the Milk River aquifer (modified from Hendry and Schwartz, 1990). Reprinted by permission of Ground Water. Copyright © 1990. All rights reserved.

in the shale of the Colorado Group. Thus, water moving down the aquifer experienced an increase in Cl^- concentration and enrichment in $\delta^{18}O$ and δD due to acquitard diffusion. This simple model of advection and dispersion in the aquifer and diffusion in the aquitard was tested mathematically by Hendry and Schwartz (1988).

In general, Na^+ and HCO_3^- concentrations increase in the direction of flow (Figure 14.23). SO_4^{2-} behaves oppositely with the highest concentrations in the recharge area and a systematic decrease downgradient in the confined part of the aquifer (Figure 14.23). Not shown are maps for Ca^{2+} and Mg^{2+}. Except for a few areas where the Milk River crops out, concentrations of these ions are generally less than a few milligrams per liter.

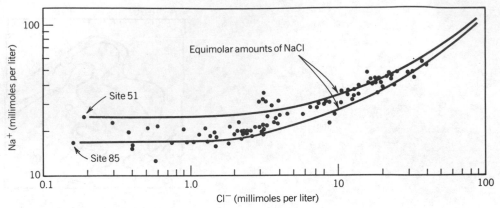

Figure 14.24
Scatter diagram illustrating how Na$^+$ concentration varies in relation to Cl$^-$. The lines depict how water from sites 51 and 85 would evolve by adding equivalent amounts of Na$^+$ and Cl$^-$ (from Hendry and Schwartz, 1990). Reprinted by permission of Ground Water. Copyright © 1990. All rights reserved.

The classical interpretation of these data would point to sulfate reduction and ion exchange as the key processes responsible for generating these patterns. However, other processes are probably more important. The aquitard diffusion model explaining the distribution of Cl$^-$ also constrains the behavior of the cations. Diffusion of only Cl$^-$ would produce a charge imbalance. Thus, counter ions must diffuse from the shale to maintain electroneutrality. Na$^+$ would be the likely species, given its abundance in formation waters in the basin (Hendry and Schwartz, 1990). If the Na$^+$ distribution is in fact related to Cl$^-$, a relationship should exist between Na$^+$ and Cl$^-$ concentrations. The lines on Figure 14.24 are hypothetical pathways of chemical evolution generated by adding equivalent amounts of Na$^+$ and Cl$^-$ to water from the recharge area. The lines fit the observed trends in the data extremely well, suggesting that the process controlling the Na$^+$ ion concentration is also the one affecting Cl$^-$.

Sulfate reduction cannot explain the increase in HCO$_3^-$ and decline in SO$_4^{2-}$. There is a lack of measurable sulfide in either ionic or gaseous forms. The most reasonable explanation of the pattern of variability in SO$_4^{2-}$ is a geologic one. Approximately 35,000 years ago, glaciation continued to open up the recharge area of the aquifer and deposited till. The sulfate chemistry of the aquifer in areas of subcrop simply reflects unique processes that operate within the till to produce high SO$_4^{2-}$ (Hendry and others, 1986). Thus, high-SO$_4^{2-}$ water is limited to the vicinity of the subcrop because time has not been sufficient for this chemically distinctive water to move further. On average, the transport of a tracer through the confined part of the aquifer requires about 3×10^5 yr (Hendry and Schwartz, 1988).

Mass transfer calculations for the confined part of the aquifer (that is, downgradient of the SO$_4^{2-}$ bulge) suggest that minor cation exchange and sulfate reduction are occurring, but not nearly to the extent suggested initially by the major ion distributions. The increasing HCO$_3^-$ concentrations and large concentrations of methane gas in the down-dip end of the aquifer come from the reduction of organic matter. It is not certain whether the reaction is occurring in the aquifer or whether both CH$_4$ and HCO$_3^-$ are diffusing from the Colorado shale. Given that a major gas field occurs at the down-dip end of the aquifer, it is most likely that the shale unit is the source rock for the gas.

14.6 Quantitative Approaches for Evaluating Chemical Patterns

Quantitative approaches for interpreting concentration distributions in ground water have developed along two different pathways. On the one hand, there are what can be termed the chemically oriented approaches. These methods were developed mainly by geochemists with a view to modeling extremely complex aqueous systems and involve modeling distributions of ions and isotopes in terms of chemical processes. On the other hand, there are the transport-oriented approaches practiced by hydrogeologists. These approaches emphasize mass transport, usually at the expense of the chemical mass transfer processes. The single-species models of contaminant transport are typical examples. In recent years, the gap between these approaches has narrowed with the integration of chemical models in a complete mass transport framework (Miller and Benson, 1983; Cederberg, 1985). However, such models are not now used routinely.

The state of practice is that studies involved with the interpretation of natural chemical patterns rely on the chemically oriented approaches and those involved with ground water contamination use the transport-oriented approaches. The size and complexity of natural systems have made it computationally infeasible to treat most natural chemical problems in a rigorous transport framework. Besides, the tendency for large-scale systems to be at approximate equilibrium with respect to many chemical processes makes mass transfer models extremely useful. Here, we will emphasize the chemically oriented approaches. Chapter 17 discusses the transport-oriented approaches.

Homogeneous and Heterogeneous Equilibrium Models

Equilibrium models provide a means to (1) estimate the equilibrium distribution of mass in the weak acid systems and among complexes and redox couples (that is, the speciation problem for homogeneous systems), (2) determine the saturation indices for various minerals from a water analysis, and (3) study more complex heterogeneous systems that involve equilibria among aqueous, solid, and gas phases. The models in (3) help to answer questions such as, given minerals a, b, and c and surfaces e and f, what is the chemical composition of pore water, assuming that equilibrium exists among the weak acids, ions, complexes, redox couples, minerals, and surfaces?

The development of chemical models has been truly evolutionary. Models used today have roots that can be traced back to Barnes and Clark (1969) and Morel and Morgan (1972), the WATEQ and REDEQL families of models, respectively (Figure 14.25). The WATEQ family historically has addressed the problem of aqueous speciation and the calculation of mineral saturation. A particular strength of these codes is their extensive and well-documented thermodynamic data bases. Models of the REDEQL family let us calculate the composition of complex heterogeneous systems. For example, MINTEQ (Felmy and others, 1983) can account for equilibrium mass transfer from a variety of reactions, including

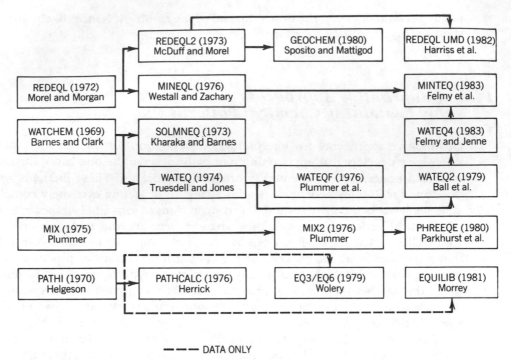

Figure 14.25
Genealogical summary of geochemical models (from Kincaid and others, 1984).
Copyright © 1984. Electric Power Research Institute. EPRI EA-3417, Vol. 1. "Geo-
hydrochemical Models for Solute Migration." Reprinted with permission.

complexation, oxidation-reduction, mineral precipitation-dissolution, gas
solution-exsolution, and sorption.

These are not the only codes suitable for modeling equilibrium systems.
For example, some of the reaction path models we discuss shortly do determine
speciation-mineral saturations (for example, PHREEQE; Parkhurst and others,
1980) and heterogeneous equilibria (for example, PHREEQE; EQ3/EQ6, Wolery,
1979).

Both the WATEQ and REDEQL families are formulated in terms of mole
balance equations for the metals and ligands. In the case of complexation reactions, these equations have this form

$$(M_i)_T = (M_i{}^{n+}) + \Sigma(M_i \cdot \text{ligand complexes})$$
$$(L_j)_T = (L_j{}^{m-}) + \Sigma(\text{metal} \cdot L_j \text{ complexes}) \tag{14.5}$$

For example, the total Ca concentration $(Ca)_T$ might include free calcium ion and
several complexes

$$(Ca)_T = (Ca^{2+}) + (CaHCO_3^+) + (CaCO_3^0) + (CaSO_4^0) \tag{14.6}$$

When solids or gas phase components enter into a calculation with REDEQL, the
mole balance equations have extra terms to account for mass associated with the
additional phases (Morel and Morgan, 1972). When H^+ is present as a component,

it is not included as a mole balance equation. Usually a charge balance or proton balance equation serves instead.

With $(M_i)_T$ and $(L_j)_T$ known from a chemical analysis, the solution of the system of mole balance equations provides the concentration of free ions and complexes. Normally, the number of unknown concentrations in the equation set is reduced by substituting appropriate mass law expressions. Effectively, this step removes the complexes from the calculation at least initially. For example, Eq. 14.6 would appear as follows when the mole balances are written only in terms of the metals and ligands in the system (the basis or principal set of variables).

$$(Ca)_T = (Ca^{2+}) + K_1(Ca^{2+})(HCO_3^-) + K_2(Ca^{2+})(CO_3^{2-}) + K_3(Ca^{2+})(SO_4^{2-})$$

where K_1, K_2, and K_3 are the equilibrium constants for the ion association reactions:

$$Ca^{2+} + HCO_3^- = CaHCO_3^+$$
$$Ca^{2+} + CO_3^{2-} = CaCO_3^0$$
$$Ca^{2+} + SO_4^{2-} = CaSO_4^0$$

Solving the system of equations and associated mass law expressions for concentration is difficult. Morel (1983) provides a useful repertoire of approximate techniques that work for surprisingly complex problems. However, a complete speciation calculation for water containing organic compounds might involve literally hundreds of separate aqueous species and be all but impossible to solve without a computer. The computer techniques are all iterative, where an initial guess at the equilibrium composition is refined through a succession of repeated calculations. Adding to the complexity is the need to recalculate activity coefficients for each species with each iteration to account for small changes in ionic strength.

The following applications will illustrate how these models can be used in practice. The speciation problem is a good place to start. Table 14.6 is a chemical analysis of river water provided as a test case by Nordstrom and others (1979). With the help of SOLMNEQ or WATEQF, we can determine the speciation in terms of concentration ($-\log M$) of metals, ligands, and complexes (Table 14.7). The concentration of the minor species was calculated but not shown. Differences in the results that the two models provide stem from differences in the two thermodynamic data bases (Nordstrom and others, 1979). While the differences are not large for the results on the table, there are major differences among the calculated concentrations for the minor species. One always has to be aware of the problems that can develop as a consequence of inconsistencies or approximations that find their way into the data bases of these types of models.

Having determined the speciation, the calculations of ion activity products (*IAP*) and saturation indices (SI) as $\log(IAP/K)$ are straightforward. Shown on Table 14.8 is a selected set of SI values obtained from SOLMNEQ and WATEQF for the test data. Differences in some values again reflect differences in the data bases of solubility constants (Nordstrom and others, 1979).

The speciation/mineral saturation calculation is a standard method for processing chemical data. Many studies in fact include data on saturation in either tabular form (for example, Kimball, 1984) or as maps or cross sections (for example, Back

Table 14.6

Set of test data used as input to speciation calculations *

Parameter	Value[1]	Parameter	Value[1]
Na	12.	Zn	0.00049
K	1.4	Cd	0.0001
Ca	12.2	Hg	0.00001
Mg	7.5	Pb	0.00003
Si	8.52	Cu	0.0005
HCO_3^-	75.2[2]	Co	0.0005
Cl	9.9	Ni	0.0018
SO_4	7.7	Cr	0.0005
B	0.050	Ag	0.00004
Br	0.006	Mo	0.0005
I	0.0018	As	0.002
F	0.10	H_2S	0.002
PO_4	0.210	DO	10.94
NO_3	0.898	E_H (V)	0.440
NO_2	0.019	DOC	2.5
NH_4	0.144	T°C	9.5
Fe(II)	0.015	pH	8.01
Fe(III)	0.0007	Density	1.00
Mn	0.0044		
Al	0.005		

[1]Concentrations are reported are milligrams per liter.

[2]Titration alkalinity as HCO_3^-.

*Reprinted from Nordstrom, D.K. and others, in Chemical Modeling in Aqueous Systems; Jenne, E.A., ed.; ACS Symposium Series 93: Washington DC, 1979, p. 870. Copyright 1979 by American Chemical Society.

Table 14.7

Results of the speciation calculation for selected major species in the river water test case *

Species	Concentration − log M		Species	Concentration − log M	
	SOLMNEQ	WATEQF		SOLMNEQ	WATEQF
Ca^{2+}	3.527	3.525	KSO_4^-	7.942	7.935
$CaSO_4^0$	5.545	5.578	SO_4^{2-}	4.122	4.121
$CaHCO_3^+$	5.518	5.722	Cl^-	3.558	3.554
$CaCO_3^0$	5.959	6.000	HCO_3^-	2.918	2.917
Mg^{2+}	3.519	3.519	CO_3^{2-}	5.328	5.334
$MgSO_4^0$	5.726	5.767	$B(OH)_3^0$	5.379	5.354
$MgHCO_3^+$	6.027	5.495	$B(OH)_4^-$	6.726	6.690
$MgCO_3^0$	5.622	6.156	Br^-	—	7.124
Na^+	3.283	3.283	F^-	5.282	5.284
$NaSO_4^-$	6.617	6.819	$H_4SiO_4^0$	3.523	3.851
$NaHCO_3^0$	6.496	6.495	H^+	7.990	7.987
K^+	4.446	4.446	OH^-	6.525	6.502
			Ionic strength	0.00240	0.00240

*Reprinted from Nordstrom, D.K. and others, in Chemical Modeling in Aqueous Systems; Jenne, E.A., ed., ACS Symposium Series 93: Washington DC, 1979, p. 871. Copyright 1979 by American Chemical Society.

Table 14.8
Saturation indicies for a selected group of minerals calculated using SOLMNEQ and WATEQF*

Mineral	Formula	Saturation index	
		SOLMNEQ	WATEQF
Calcite	$CaCO_3$	−0.765	−0.634
Dolomite	$CaMg(CO_3)_2$	−1.329	−1.384
Siderite	$FeCO_3$	−3.377	−7.347
Rhodochrosite	$MnCO_3$	−2.136	−2.180
Gypsum	$CaSO_4 \cdot 2H_2O$	−2.942	−3.057
Celestite	$SrSO_4$	—	—
Hydroxyapatite	$Ca_5(PO_4)_3OH$	+5.046	−1.784
Fluorite	CaF_2	−3.338	−3.079
Ferric hydrox. (Am)	$Fe(OH)_3$	−7.584	+1.304
Goethite	$FeO(OH)$	−1.484	+7.810
Hematite	Fe_2O_3	−3.252	+15.144
Gibbsite (crypt.)	$Al(OH)_3$	−0.058	−0.336
Birnessite	MnO_2	—	−4.114
Chalcedony	SiO_2	+0.217	−0.142
Quartz	SiO_2	+0.697	+0.405
Kaolinite	$Al_2Si_2O_5(OH)_4$	—	+1.638
Sepiolite	$Mg_2SiO_{7.5}(OH) \cdot 3H_2O$	−5.734	−3.699
FeS amorphous	FeS	—	−7.644
Mackinawite	FeS	—	−6.928

*Reprinted from Nordstrom, D.K. and others, in Chemical Modeling in Aqueous Systems; Jenne, E.A. ed.; ACS Symposium Series 93: Washington DC, 1979, p. 879. Copyright 1979 by American Chemical Society.

and others, 1966; Plummer and others, 1976). The choice of how to present the saturation calculations often depends upon whether systematic patterns of variability exist. One important key to success in using this modeling approach is the quality of the chemical data. In the next chapter, we identify problems that can influence the quality of chemical data.

One final application is a modeling exercise involving a heterogeneous system (Kincaid and Morrey, 1984) which tested how well different codes modeled single mineral solubility as a function of ionic strength. Minerals were selected so that experimental data were available for comparison. Here, we will examine the calcite solubility test at 25°C. All the models (that is, MINTEQ, PHREEQE, EQ3/EQ6, and EQUILIB—Morrey and Shannon, 1981) do a reasonably good job in matching the experimental data (Figure 14.26). However, as before, different models give different results. For example, MINTEQ and PHREEQE predict the experimental results more accurately than do EQUILIB and EQ3/EQ6 because the data bases of the first two codes contained an ion pair $NaHCO_3^0$ not present in the other two. Still other discrepancies are related to differences in the calculation of ionic strength (Kincaid and Morrey, 1984). Thus, what might seem like a straightforward calculation is actually relatively difficult because of the complex speciation and the uncertainty in calculating activity coefficients at high ionic strengths.

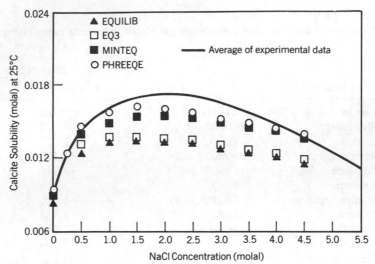

Figure 14.26
Computed solubility of calcite ($CaCO_3$) as a function of NaCl concentration (modified from Kincaid and Morrey, 1984). Copyright © 1984. Electric Power Research Institute. EPRI EA-3417, Vol. 2. "Geohydrochemical Models for Solute Migration." Reprinted with permission.

Mass Balance Models

Plummer and others (1983) and Plummer (1984) describe another modeling approach that uses chemical data along a flow system to establish (1) which chemical reactions have occurred (for example, mineral dissolution, precipitation, or ion exchange), (2) the quantity of mass transfer through each reaction, (3) the conditions under which the reaction took place (for example, open versus closed system, and constant versus variable temperature), and (4) how the water quality and mineralogy will change in response to natural processes and perturbations of the system.

A simple example illustrates how the method works. Assume we have chemical data from two wells along a flow path (Figure 14.27). Between the wells, the Na^+ concentration increases by 2.5 mM or $(\Delta Na^+) = 2.5$ mM, $(\Delta SO_4^{2-}) = 1.0$ mM, and $(\Delta Cl^-) = 0.5$ mM. A possible explanation for the change in concentration is the dissolution of thenardite (Na_2SO_4) and halite (NaCl). Mass balance equations for Na and Cl define the mass transfer from dissolution of these minerals ($\alpha_{mineral}$)

$$\alpha_{halite} = (\Delta Cl) = 0.5 \text{ mM}$$
$$\alpha_{halite} + 2\alpha_{thenardite} = (\Delta Na) = 2.5 \text{ mM}$$

which are trivial to solve, giving $\alpha_{halite} = 0.5$ mM and $\alpha_{thenardite} = 1.0$ mM. The following mass transfer reaction represents the quantity of minerals dissolved

$$\text{well water } A + 0.5 \text{ halite} + 1.0 \text{ thenardite} = \text{well water } B$$

where the reaction coefficients are in millimoles per liter (mM).

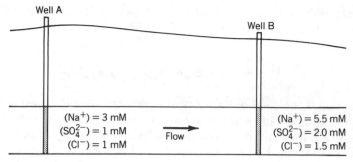

Figure 14.27
Schematic representation of the change in ground water composition between two wells along a flow system.

The equations describing the conservation of mass have the following general form (Plummer, 1984)

$$\sum_{p=1}^{P} \alpha_p b_{p,k} = \Delta m_{T,k} \qquad k = 1, j \tag{14.7}$$

where p represents the mineral or process acting as a source or sink for mass, α_p is the unknown quantity of mass transferred from the mineral or process (mM), $b_{p,k}$ is the stoichiometric coefficient of the kth element in the pth mineral or process, $\Delta m_{T,k}$ is the change in net total concentration of the kth element between the two wells. Equations for J-2 elements, excluding hydrogen, can be written for the P phases with Eq. 14.7. For redox problems, another equation comes from a conservation of electrons expression or

$$\sum_{p=1}^{P} u_p \alpha_p = \Delta RS \tag{14.8}$$

where u_p is the operational valence of the pth phase (refer to Plummer and others, 1983, for a discussion of the concept). The change in redox state of the system (ΔRS) is determined from field data (Plummer, 1984) with the following equation

$$\Delta RS = \sum_{i=1}^{1} v_i m_i (\text{well } B) - \sum_{i=1}^{1} v_i m_i (\text{well } A) \tag{14.9}$$

where v_i is the operational valence and m_i is the molality of the ith species in solution. The electron balance equation is simply a statement that when redox reactions occur, electrons are conserved among the dissolved species.

Occasionally, it may be possible to utilize an isotope balance equation with the general form

$$\sum_{p=1}^{P} \alpha_p b_p \delta I_p = \Delta m_T I \tag{14.10}$$

where b_p is the stoichiometric coefficient of the element (for example, C or S) in the pth phase, δI_p is the isotopic composition (per mil) of the pth phase, and $\Delta m_T I$ is

$$\Delta m_T I = (m_{T,k} \delta I_T)_{\text{well } B} - (m_{T,k} \delta I_T)_{\text{well } A}$$

where $m_{T,k}$ is the total molarity of the element in solution (for example, in the case of S, the sum of sulfate and sulfide species) and δI_T is the average isotopic composition (per mil) of the total dissolved element in solution.

In developing the governing system of equations, the number of equations must be the same as the number of phases or processes. If there are more plausible phases than equations, it is necessary to construct several mass balance models, each having the correct number of phases. One of these models can be selected on the basis of the following criterion (Plummer, 1984). The reaction model should preserve consistency with observed saturation indices. For example, if the model indicates that a given mineral should be dissolving then in the field there should be zones that are undersaturated with respect to that mineral. Another way to select a unique set of processes is with petrographic data. Minerals that the model predicts to be dissolving or precipitating should appear that way under the microscope. One final possibility is the use of isotopic data to provide an independent check (Plummer, 1984). In essence, the change in isotopic composition that results from the predicted mass transfer is compared to that observed at the downstream well. The test is an independent one when the mass balance equations exclude the particular isotope. The system of algebraic equations can be solved by hand even for moderately complicated problems. A computer code called BALANCE (Parkhurst and others, 1982) implements the method.

Plummer and others (1983), Plummer (1984), Powell (1984), and Hendry and Schwartz (1990) describe applications of this approach. Here, we will briefly examine Plummer's (1984) study of the Madison Limestone aquifer of Montana, Wyoming, and South Dakota. The chemical data are for water from two wells about 130 miles apart (Table 14.9). The upstream sample is a composite of recharge entering the aquifer in the Bighorn Mountains of north central Wyoming. The downstream sample is water from the Mysse well in east central Montana from a depth of about 5000 feet.

On the basis of mineralogical analyses of rocks in the Madison Group, observed trends in water composition, and computed saturation indices, Plummer (1984) identified the following 10 plausible phases or processes: calcite ($CaCO_3$), dolomite ($CaMg(CO_3)_2$), gypsum/anhydrite ($CaSO_4$), organic matter (CH_2O), carbon dioxide (CO_2), ferric hydroxide ($FeOOH$), pyrite (FeS_2), Ca-Na ion exchange, halite ($NaCl$), and sylvite (KCl). Of the 10 resulting equations (Table 14.10), 8 are elemental mass balance equations, 1 is a conservation of electron equation, and 1 is a sulfur isotope balance equation. This system can be solved algebraically (Table 14.10) or by using BALANCE. The net transfer of mass as ground water moves between the two wells is

$$\text{recharge water} + 20.15 \text{ gypsum} + 3.54 \text{ dolomite} + 0.87 \text{ CH}_2\text{O}$$
$$+ 15.31 \text{ NaCl} + 2.52 \text{ KCl} + 0.09 \text{ FeOOH} + 8.28 \text{Na}_2 X$$
$$= 5.33 \text{ calcite} + 0.09 \text{ pyrite} + 8.28 \text{ Ca}X + \text{Mysse water}$$

where the reaction coefficients are in millimoles liter and X denotes an ion exchange site.

Table 14.9
The chemistry of ground water used by Plummer (1984) to demonstrate the use of the geochemical code BALANCE*

Constituent	Recharge	Mysse
Calcium[1]	48.0	450.
Magnesium	24.5	110.
Sodium	0.5	730.
Potassium	0.85	99.
Iron	0.06	0.02
Chloride	0.6	630.
Sulfate	15.0	1900.
Alkalinity	245.	320.
Total inorganic carbon	—	—
Redox state	—	—
H_2S	—	8.8
Tritium (TU)	27.6	—
$\delta^{34}S_{SO_4^{2-}}$ (‰)	9.73	16.30
$\delta^{34}S_{H_2S}$ (‰)	—	−22.09
$\delta^{13}C$ (‰)	−6.99	−2.34
^{14}C (‰)	33.1	0.8
T°C	9.9	63.
pH	7.55	6.61
SI calcite	−0.05	
SI dolomite	−0.27	
SI gypsum	−2.47	
log P_{CO_2}	−2.26	

[1]Major ion concentrations in mg/L.

Reaction Path Models

Of the models discussed so far, reaction path models are the most powerful. Unlike the heterogeneous reaction models, a system need not be at equilibrium with respect to all the solid phases. Reaction path models treat partial equilibrium systems, which are systems at equilibrium with respect to some minerals yet undersaturated with respect to others (Helgeson, 1968). In this respect, they realistically portray ground water systems. Unlike the mass balance models, reaction path models describe in detail the chemical pathways—a "stop action" description of a complex reaction.

To explain how this approach works, refer again to Figure 14.19 illustrating the change in Na^+, SO_4^{2-}, and Cl^- between two wells. A reaction path model requires the following information: (1) the composition of the starting water, (2) the reacting phases, and (3) a mass transfer coefficient for each reactant. In practice, these coefficients are normalized, with the largest one being set equal to one. For the example problem, the reactants and the mass transfer coefficients in equation form are

Table 14.10
System of mass balance equations describing the plausible reaction model and the algebraic solution (from Plummer, 1984)[*]

1. MASS BALANCE

$$\alpha_{calcite} + 2\alpha_{dolomite} + \alpha_{CH_2O} + \alpha_{CO_2} = \Delta m_{T,C}$$
$$\alpha_{gypsum} + 2\alpha_{pyrite} = \Delta m_{T,S}$$
$$\alpha_{calcite} + \alpha_{dolomite} + \alpha_{gypsum} - \alpha_{exchange} = \Delta m_{T,Ca}$$
$$\alpha_{dolomite} = \Delta m_{T,Mg}$$
$$\alpha_{halite} + 2\alpha_{exchange} = \Delta m_{T,Na}$$
$$\alpha_{sylvite} = \Delta m_{T,K}$$
$$\alpha_{halite} + \alpha_{sylvite} = \Delta m_{T,Cl}$$
$$\alpha_{FeOOH} + \alpha_{pyrite} = \Delta m_{T,Fe}$$

2. CONSERVATION OF ELECTRONS

$$4\alpha_{calcite} + 8\alpha_{dolomite} + 6\alpha_{gypsum} + 4\alpha_{CO_2} + 3\alpha_{FeOOH} = \Delta RS$$

3. SULFUR ISOTOPE BALANCE

$$\alpha_{gypsum}\, \delta^{34}S_{gypsum} + 2\alpha_{pyrite}\, \delta^{34}S_{pyrite} = \Delta m_T{}^{34}S$$

4. ALGEBRAIC SOLUTION

$$\alpha_{gypsum} = \frac{\Delta m_T{}^{34}S - \delta^{34}S_{pyrite}\, \Delta m_{T,S}}{(\delta^{34}S_{gypsum} - \delta^{34}S_{pyrite})}$$

$$\alpha_{pyrite} = \frac{(\Delta m_{T,S} - \alpha_{gypsum})}{2}$$

$$\alpha_{KCl} = \Delta m_{T,K}$$

$$\alpha_{NaCl} = \Delta m_{T,Cl} - \alpha_{KCl}$$

$$\alpha_{exchange} = \frac{(\Delta m_{T,Na} - \alpha_{NaCl})}{2}$$

$$\alpha_{dolomite} = \Delta m_{T,Mg}$$

$$\alpha_{FeOOH} = \Delta m_{T,Fe} - \alpha_{pyrite}$$

$$\alpha_{calcite} = \Delta m_{T,Ca} + \alpha_{exchange} - \alpha_{gypsum} + \alpha_{dolomite}$$

$$\alpha_{CO_2gas} = \frac{(\Delta RS - 3\alpha_{FeOOH} - 4\alpha_{calcite} - 8\alpha_{dolomite} - 6\alpha_{gypsum})}{4}$$

$$\alpha_{CH_2O} = \Delta m_{T,C} - \alpha_{CO_2gas} - 2\alpha_{dolomite} - \alpha_{calcite}$$

[*]Reprinted by permission First Canadian/American Conference on Hydrogeology. Copyright © 1984. All rights reserved.

$$\tfrac{1}{2}NaCl + 1Na_2SO_4$$

The modeling transfers a small quantity of mass ($\Delta\xi$) from the reactants to the products, which in this case are Na^+, SO_4^{2-}, and Cl^-. Thus, after one progress step, $\Delta\xi/2$ moles/liter of NaCl and $\Delta\xi$ moles/liter of Na_2SO_4 moles dissolve with $\Delta\xi$ being a small number (for example, 10^{-7} M). The computer code calculates how this mass is distributed among the products and the resulting change in the composition of the ground water. In our simple example, the concentration of Na^+ increases by $2.5\Delta\xi$ M, Cl^- by $0.5\Delta\xi$ M, and SO_4^{2-} by $\Delta\xi$ M after each reaction step. These results can easily be verified by writing and balancing the equation.

The new ion concentrations after one reaction step are calculated by taking their values in the starting water plus the increment in concentration. After a large number of progress steps, water should evolve chemically to what was observed at well *B*. The changing composition with each reaction step defines the reaction path.

A reaction path model is not really necessary for our simple example because the path is linear. Complexities inherent in real problems make the calculation intractable without a computer. In most true reaction path models (for example, PATHI and EQ3/EQ6), the code checks the aqueous solution for saturation with respect to a large number of minerals in the data base at the end of each reaction step. If saturation occurs, the mineral automatically becomes one of the products. Similarly, if the solution becomes undersaturated with respect to a solid, the mineral becomes a reactant. One of the reasons why the quantity of mass transferred in a reaction step (that is, $\Delta \xi$) is kept so small is to avoid leaping past the solid phase boundaries as the solution evolves. The reaction paths often change significantly as a consequence of mineral saturation.

A code also has to maintain equilibria in aqueous phase reactions. For example, if $CaCO_3$ is a reactant, the CO_3 transferred in the reaction has to be redistributed properly among $H_2CO_3^*$, HCO_3^-, and CO_3^{2-}. Similarly, the code has to redistribute metals among several complexes. In some cases, it may be necessary to adjust the mass transfer coefficients during the simulation to account for changing temperatures along the reaction paths.

It is beyond the scope of our brief overview to discuss the mathematical formulation of reaction path models. The approaches have been implemented in three codes: PATHI (Helgeson and others, 1970), EQ3/EQ6 (Wolery, 1979), and PHREEQE (Parkhurst and others, 1980). The first two codes are true reaction path models in the sense that they include the precipitation of minerals automatically once saturation is reached. In running PHREEQE, a user preselects the saturation constraints on the reaction path (Plummer, 1984). This approach avoids the much larger execution times that the other codes require. Each of the computer packages includes a speciation algorithm to fully characterize the starting water before the simulation begins. PHREEQE also contains other simulation options for calculating mineral saturation or the composition of fluid mixtures.

Reaction path modeling has not been used extensively in the analysis of chemical problems. Schwartz and Domenico (1973) applied the reaction path concept to model advective mass transport accompanied by reaction in regional flow systems. Plummer (1984) discusses in detail how PHREEQE can be used to examine the pathways of chemical evolution in the Madison Limestone aquifer and how these results compare to those obtained with BALANCE.

14.7 *Age Dating of Ground Water*

All the direct dating techniques interpret the distribution of a radioactive species in terms of a first-order kinetic rate law for decay. The residence time of mass in the system or the ground water age (t) is described mathematically as

$$t = \frac{t_{1/2}}{\ln 2} \ln \left(\frac{A_0}{A_{obs}} \right) \tag{14.11}$$

where $t_{1/2}$ is the half life for decay, A_0 is the activity assuming no decay occurs, and A_{obs} is the observed or measured activity of the sample.

In terms of ground water studies, tritium (^3H, $t_{1/2} = 12.26$ yr) and carbon-14 (^{14}C, $t_{1/2} = 5730$ yr) are commonly used for age dating. However, both suffer from limitations. The relatively short half life for ^3H makes it only useful for dating water less than about 40 years old. In addition, the cessation of nuclear testing in the atmosphere has eliminated the global source of new tritium. Within a few more decades the tritium levels in precipitation will decline to an extent that tritium will be less useful as a tracer. ^{14}C, with a much longer half life, has the potential to date water up to about 40,000 years old. However, as will become clear shortly, interpreting ^{14}C data is difficult because of the need to account for other processes besides radioactive decay that influence the measured ^{14}C activity.

Phillips and others (1986) have demonstrated the potential of using chlorine-36 (^{36}Cl) in dating water. Particularly attractive with this radionuclide are the long half life ($t_{1/2} = 3.01 \times 10^5$ years) and the smaller number of potential reactions (as compared to ^{14}C) that need to be accounted for in calculating an age date (Bentley and others, 1986). One limitation with this radioisotope is the need for tandem accelerator mass spectrometry for measurements, which at present limits the availability of analyses.

Tritium

Tritium concentrations are reported in terms of tritium units (TU), with 1 TU corresponding to one atom of ^3H in 10^{18} atoms of ^1H (Fontes, 1980). Tritium occurs naturally in the atmosphere with concentrations in precipitation usually less than 20 TU. However, tritium generated by thermonuclear testing in the atmosphere between about 1952 and 1963 has swamped the natural production of tritium. The long-term record of ^3H in precipitation at Ottawa, Canada, shows that, during the period of nuclear testing, concentrations were often greater than 1000 TU (Figure 14.28). Tritium levels declined once weapons testing stopped in 1963, but present-day levels remain far above natural background. Details on the seasonal

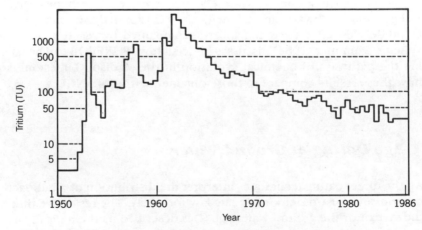

Figure 14.28
Tritium levels in precipitation at Ottawa, Ontario (modified from Robertson and Cherry). Water Resources Res., v. 25, p. 1097–1109, 1989. Copyright by American Geophysical Union.

variation and latitude dependence of ^3H levels in precipitation are presented by Gat (1980) and Fontes (1980).

Ideally, by knowing the concentration of ^3H in precipitation (the source) and its distribution in ground water, one should be able to date the water. In most cases, however, it is not possible to work with tritium in such a quantitative way. The main problems stem from the uncertainty and complexity of atmospheric loading. For example at most places, there are usually insufficient data to establish the historical pattern of ^3H loading. Sometimes this limitation can be overcome to some extent by correlating partial local records to stations like Ottawa, Canada, with long-term records. The complex loading function also provides a problem in interpreting ^3H data. Without a great deal of information about tritium distributions in the ground water, it is difficult to determine whether a sample with 30 TU is a late 1950s water that has decayed through three half lives or a 1970s water originally with 75 TU that has decayed through one half life.

The main application of tritium is to differentiate pre-1952 water from younger water. The logic is that, assuming pre-1952 water to have had an original ^3H concentration of 5 TU, the concentration in 1988 would be at maximum 0.6 TU, which is close to the detection limit even using enrichment techniques of analysis. Thus, any detectable tritium in a sample implies that the water contains some component of more recent or post-1952 water (Fontes, 1980).

Carbon-14

Measurements of ^{14}C are reported as percent modern ^{14}C (pmc) determined as the ratio of the sample activity to that of the international standard expressed as a percentage. ^{14}C originates naturally in the upper atmosphere through a reaction involving nitrogen and neutrons. Like ^3H, weapons testing in the atmosphere has affected its concentration in recent years. However, except for young waters, this increase does not affect the interpretation.

^{14}C in groundwater comes from the solution of $CO_2(g)$ in the soil zone. The activity of ^{14}C in CO_2 gas is approximately 100 pmc (Fritz and Fontes, 1980) and slightly higher in the ions coming from the dissolution of CO_2. The dating method works because, once carbonate species move below the water table, ^{14}C begins to decay and there are no additional sources.

The main problem in applying this method is that some reactive minerals contain carbon, and carbon transfers in and out of the ground water. These interactions can reduce the ^{14}C activity in the water and thus need to be fully accounted for in estimating the age. In terms of Eq. 14.11, the value of A_0 (the ^{14}C activity assuming no decay) would be lower than 100 pmc, reflecting the fact that other processes besides radioactive decay influence the ^{14}C activity of the sample. Any age calculation is meaningful as long as A_0 and A_{obs} differ only due to the effects of radioactive decay.

Mook (1980), Reardon and Fritz (1978), and Wigley and others (1978) list the following processes that can alter the ^{14}C activity of ground water:

1.   The congruent dissolution of carbonate minerals, which adds "dead carbon" or carbon without ^{14}C activity to the ground water. Overall, this process lowers the ^{14}C activity measured for the sample.

2.   The incongruent dissolution of carbonate or other Ca containing minerals accompanied by the precipitation of calcite. This process will

remove ^{14}C as calcite precipitates and if dolomite is the mineral dissolving add dead carbon through 1. This process could occur in the zone of saturation following the rapid solution of calcite to equilibrium with subsequent precipitation as dolomite slowly dissolved.

3. The addition of dead carbon from other sources such as the oxidation of old organic matter, sulfate reduction, and methanogenesis—these again reduce the ^{14}C activity of the sample.

4. Possibly isotopic exchange involving CO_3^{2-} and carbonate minerals, which could lower the ^{14}C activity. This process is generally considered to have a negligible effect at normal ground water temperatures.

We normally follow one of two approaches in estimating a ground water age (Kimball, 1984). One way is to interpret ages on the basis of the ion and isotopic data for a single sample without information from other samples. The second approach involves using the ion or isotopic data from many samples in an integrated way and mass balance modeling to sort out all the major inputs and outputs of carbon.

The simplest way to establish A_0 is to account only for the most important process affecting ^{14}C activity, which is the congruent dissolution of calcite. The reaction between water containing CO_2 and calcite is

$$CO_2 + H_2O + CaCO_3(s) = Ca^{2+} + 2HCO_3^- \qquad (14.12)$$

At equilibrium, according to this reaction, half the bicarbonate would be generated from a source containing ^{14}C (CO_2), and the other half would be generated from a dead source (calcite). Assuming that the activity of the CO_2 is 100 pmc and the calcite is 0 pmc, A_0 would be 50 pmc, reflecting the equal contribution of carbon from both sources.

In many cases, it is unlikely that the reaction would be at equilibrium due to the lack of carbonate minerals or kinetic effects (Pearson and Hanshaw, 1970). The following form of the reaction describes this more realistic situation (Mook, 1980):

$$(a + 0.5b)\, CO_2 + 0.5b\, CaCO_3 + H_2O = 0.5b\, Ca^{2+} + b\, HCO_3^- + a\, CO_2 \quad (14.13)$$

Excess CO_2 in this reaction results in a ^{14}C activity greater than 50 pmc and perhaps close to 85 pmc.

Two of the simpler techniques for correcting A_0 values are based on these ideas. For example, one empirical approach simply assumes A_0 has a value of 85 ± 5 pmc. Measurements of dissolved carbon in ground water of northwest Europe suggested that this value was in fact representative of soil water and shallow ground water in temperate climates (Vogel, 1967, 1970). Another less empirical approach, based on Eq. 14.13, provides an A_0 value that is the weighted contribution of CO_2 and calcite. Mathematically, this correlation is a simple mixing equation or

$$(a + b)A_0 = (a + 0.5b)A_{CO_2} + 0.5b\, A_c \qquad (14.14)$$

where A_{CO_2} is the estimated activity of CO_2 in the soil zone (usually 100 pmc), A_c is the estimated activity of calcite (usually 0 pmc), and $(a + b)$ is the total moles of carbon in the water (C_T). Rearranging Eq. 14.14 and assuming A_c is 0 provides the desired correction equation for A_0

$$A_0 = \frac{(a + 0.5b)A_{CO_2}}{(a + b)} \tag{14.15}$$

or, written in terms of C_T,

$$A_0 = \frac{(C_T - 0.5b)A_{CO_2}}{C_T} \tag{14.16}$$

Equation 14.16 is known as the Tamers equation (Tamers, 1967, 1975). The term on the right side of Eq. 14.16 assumes that for water with a normal pH, the total carbon in the water is found as CO_2 (or $H_2CO_3^*$ depending upon conventions) and HCO_3^-. C_T is determined approximately using equilibrium relations or the major ion data together with a speciation program like WATEQF. The molar concentration of HCO_3^- (that is, b) is known from the water analysis and A_{CO_2} is taken as 100 pmc.

Another way of establishing how much dead carbon has been contributed to the chemical system from carbonate dissolution is to use $\delta^{13}C$ as a measure of the extent of carbonate reactions (Ingerson and Pearson, 1964). Again, a carbon mixing equation can be written (Mook, 1980)

$$(a + b)\delta^{13}C_T = (a + 0.5b)\delta^{13}C_{CO_2} + 0.5b\delta^{13}C_c \tag{14.17}$$

where $\delta^{13}C_T$, $\delta^{13}C_{CO_2}$, and $\delta^{13}C_c$ are the measured values of δ^{13} in the ground water (as total carbon), soil gas, and calcite, respectively, and the other terms are molar concentrations. With some algebraic manipulation, we can express Eq. 14.17 in terms of $(a + 0.5b)/(a + b)$ or

$$\frac{(a + 0.5b)}{(a + b)} = \frac{(\delta^{13}C_T - \delta^{13}C_c)}{(\delta^{13}C_{CO_2} - \delta^{13}C_c)}$$

where the left side of the equation is the unknown factor in Eq. 14.15 correcting the A_0 value (Mook, 1980). The $\delta^{13}C$ content of CO_2 gas in the soil zone is known reasonably well (that is, $-27 \pm 5\permil$ PDB in temperate regions and $-13 \pm 4\permil$ in tropical regions; see Mook, 1980) as is the value for marine carbonate (0 to $+2\permil$).

All these approaches are useful to some extent. However, they do not account for many of the processes affecting the ^{14}C content of the water (Pearson and Hanshaw, 1970). More comprehensive schemes do exist, which we will briefly mention but not discuss in detail. Fontes and Garnier (1979), for example, account for mineral dissolution and isotopic exchange reactions. The most comprehensive model to date is that of Wigley and others (1978). Their scheme, which is complicated enough to require a computer code, takes into account an arbitrary number of carbon sources (for example, dissolution of carbonate minerals and oxidation of organic matter) and sinks (for example, mineral precipitation CO_2 degassing, and methane production), and the equilibrium fractionation between phases.

One way in which the Wigley and others (1978) model can be used is in conjunction with the code BALANCE (Kimball, 1984; Plummer, 1984). As a first step, BALANCE estimates the transfer of carbon into and out of solution. These estimates are input data for the computer model of Wigley and others (1978) to calculate fractionation factors, the predicted $\delta^{13}C$ and the corrected value of A_0.

Kimball (1984) examines several methods for interpreting ^{14}C ages of ground water in the Piceance Creek Basin in Colorado. He compares four different techniques: the mass balance model coupled with the procedure of Wigley and others (1978), the Tamers equation, the Ingerson and Pearson equation, and the Fontes and Garnier equation. The results in particular for waters from the lower aquifer illustrate some important features about ^{14}C dating. First, there is a clear requirement to account for the chemical processes in producing a date. In all cases, the corrected ages are less than the analytical ages and in a few cases more than an order of magnitude less. Second, there is a marked variability in the dates, depending upon which correction technique is used. For example, the dates derived from mass balance modeling sometimes were an order of magnitude less than those obtained from the other methods (that is, 1600 versus 18,000 yr). It appears from Kimball's work that the Tamers equation, which only accounts for mineral dissolution gives the oldest dates. The other approaches yield younger dates. Beyond this, there is little other consistency among the results.

These and other results raise the question as to what interpretive method yields the best date. Can we say that because the mass balance procedure accounts for more processes the date is better? Probably not. Maybe the reaction model could be wrong, or perhaps important processes were left out. In spite of a long history of development, the ^{14}C method is at best a semiquantitative tool. Confident predictions can be made only when the processes affecting the carbon chemistry are absolutely defined—the exception rather than the rule. The hope of collecting a single sample of water and extracting a date seems to have faded in light of the effort required for process identification.

Chlorine-36

There is strong evidence (Bentley and others, 1986; Phillips and others, 1986) that ^{36}Cl will emerge as a useful radioisotope for dating water up to 2 million years old. Again, the most important source of this isotope is fallout from the atmosphere, although small quantities are produced in the subsurface. The results of analyses are reported as the ratio of $^{36}Cl/Cl$ with a typical value for meteoric water lying in a range from 100 to 500×10^{-15}. Because there appear to be comparatively few processes that affect the $^{36}Cl/Cl$ ratio, the dating equations are straightforward modifications of Eq. 14.11. The simple geochemistry of Cl in ground water avoids the complexity inherent with ^{14}C. About the only processes contributing dead Cl to ground water are the dissolution of chloride salts and mixing of older water containing Cl^-. To date, ^{36}Cl has been used to date ground water in the Great Artesian Basin of Australia (Bentley and others, 1986) and the Milk River aquifer in Alberta, Canada (Phillips and others, 1986).

Suggested Readings

Bethke, C. M., Harrison, W. J., Upton, C., and Altaner, S. P., 1988, Supercomputer analysis of sedimentary basins, Science, v. 239, p. 261–267.

Hendry, M. J., and Schwartz, F. W., 1990, The chemical evolution of ground water in the Milk River aquifer, Canada, Ground Water, v. 28, no. 2, p. 253–261.

Mook, W. G., 1980, Carbon-14 in hydrogeological studies, Chapter 2, Handbook of Environmental Isotope Geochemistry, Fritz, P., and Fontes, J., eds., Amsterdam, Elsevier, v. 1, p. 49–74.

Kimball, B. A., 1984, Ground water age determinations, Piceance Creek Basin, Colorado, First Canadian/American Conference on Hydrogeology, Hichon, B., and Wallick, E. I., eds., National Water Well Assoc., Dublin, Ohio, p. 267–283.

Palmer, C. D., and Cherry, J. A., 1984, Geochemical evolution of groundwater in sequences of sedimentary rocks, J. Hydrol., v. 75, p. 27–65.

Plummer, L. N., 1984, Geochemical modeling: A comparison of forward and inverse methods, First Canadian/American Conference on Hydrogeology, Hitchon, B., and Wallick, E. I., eds., National Water Well Assoc., Dublin, Ohio, p. 149–177.

Problems

1. Explain why the isotopic composition of ground water found in glacial till in Canada and parts of the United States may be depleted in $\delta^{18}O$ and δD relative to precipitation sampled at the same locations.

2. Over the plains region of the northern United States and Canada, carbonate-rich till overlies marine shale and sandstone. In nonarid areas, recharge from snowmelt might take on the following chemical composition as moves it downward through till and into shale bedrock:

Unit	Concentration (milligrams per liter)						
	Ca^{2+}	Mg^{2+}	Na^+	HCO_3^-	SO_4^{2-}	Cl^-	pH
till	79.0	50.0	210.0	436.0	61.0	14.0	7.80
shale	5.0	0.5	450.0	1044.0	6.0	53.0	8.10

Interpret the chemical evolution of the water in terms of the most likely mass-transport processes.

3. Explain why the order in which ground water encounters minerals can be important in determining how the major ion chemistry evolves.

4. In carbonate rocks subject to recharge, the most dramatic changes in major ion chemistry occur over a relatively short distance as infiltration first enters the unit. Later changes are often almost insignificant by comparison. Explain why, using arguments related to kinetics and mineral equilibrium.

5. Using the concept of redox zones, explain why H_2S gas is rarely found in ground water close to the water table in recharge areas.

6. Explain how tritium can be used in hydrogeological investigations.

7. Along a single pathline, the concentration of Ca, Mg, Na, K, and Cl changes by -1.0 mM, 2.0 mM, 10.2 mM, 0.1 mM, and 0.3 mM between two sampling points. With the dissolution of calcite, dolomite, halite, sylvite, and ion exchange (that is, $Ca^{2+} + 2Na\text{-clay} = 2Na^+ + Ca\text{-clay}$) as the plausible reactions, determine the net mass transfer reaction.

8. Isotopic measurements on a sample yield a measured carbon-14 activity of 6.9 percent modern carbon, and a measured $\delta^{13}C_T$ as total carbon of $4.6‰$. Assuming a $\delta^{13}C$ of $8‰$ for calcite and a $\delta^{13}C$ of $-27‰$ for soil gas, estimate the age of the ground water.

9. Use the Tamers equation to determine the age of a water sample having a total inorganic carbon content (C_T) of 16.7 mM and a HCO_3^- concentration of 997 mg/L. The measured carbon-14 activity of the sample is 2.7 percent modern carbon.

Chapter Fifteen

Mass Transport in Ground Water Flow: Geologic Systems

15.1 Mass Transport in Carbonate Rocks

15.2 Economic Mineralization

15.3 Migration and Entrapment of Hydrocarbons

15.4 Self-Organization in Hydrogeologic Systems

15.5 Coupled Phenomena

The ability of ground water to dissolve rocks and minerals and to redistribute large quantities of dissolved mass has broad geologic implications. The so-called "geologic work" of ground water that is related to mass transport thus includes classical problems such as chemical diagenesis, the formation of some types of ore deposits, soil salinity, and evaporite formation. Our objective in treating these topics here is to develop a sense of how mass transport processes in ground water work at various scales and to describe specific physical and chemical factors that relate to these problems. Another topic treated in this chapter is the migration and entrapment of hydrocarbons. While more strictly a problem of multicomponent flow, this subject naturally falls in this geologically oriented discussion.

The chapter concludes with a discussion of coupling in mainly chemical systems and its geological implications. Here, the term coupling describes a situation where the operation of a flow or transport process is related directly to another process. For example, the property of self-organization in some geological systems is a condition where chemical reactions and flow are coupled. Similarly, flow may be coupled to gradients in temperature imposed by energy transport, or gradients in concentration imposed by mass transport. All these effects are sufficiently complex that we will do little else here other than to make readers aware of some fundamental principles.

15.1   Mass Transport in Carbonate Rocks

Hydrogeologists for years have had a fascination with carbonate rocks. Not only are they often productive aquifers, but with karst give rise to spectacular surface landforms and cave systems. However, karst is but one manifestation of chemical

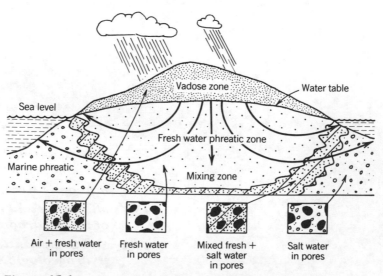

Figure 15.1
Diagrammatic cross section showing mixing zone in an emerging carbonate rock (after Longman, 1982, as modified from Vernon, 1969). Reprinted by permission.

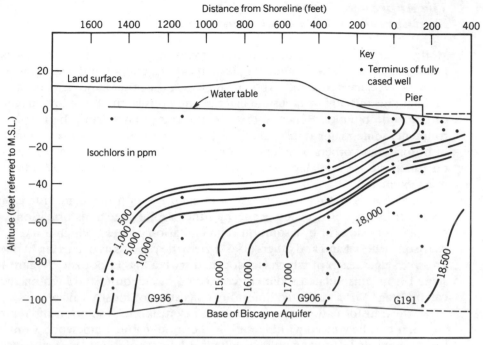

Figure 15.2
Cross section of a mixing zone near Miami, Florida (from Kohout and Klein, 1967).
Reprinted with permission of Intl. Assoc. of Hydrol. Sci.

diagenesis. The term chemical diagenesis refers to any chemical change that occurs in a sediment following deposition. Now, let us examine in more detail some of diagenetic changes that affect carbonate rocks. The process responsible for most of these changes is mixing—the displacement of connate water by recharging meteoric water, the same process we discussed in Chapter 14.

The environment of diagenesis for emerging carbonates is demonstrated in Figure 15.1. This idea was advanced by Vernon (1969) and modified by Hanshaw and others (1971). This type of environment develops everywhere on an emerging carbonate platform where fresh water drives out the marine fluids to some depth below sea level. Three diagenetic settings are evident: (1) a fresh water zone, replenished by meteoric water; (2) a mixing zone where fresh water mixes with sea water; and (3) a deep marine zone, dominated by sea water. Several modern-day mixing zones have been documented, one of which occurs in carbonate rocks near Miami, Florida (Figure 15.2). Note the length of this zone exceeds 2000 feet, with sea water and fresh water interfusing about this boundary. There is ample evidence to indicate that the mixing zone moves seaward during periods of heavy recharge and landward when recharge is limited. At the seaward side, incoming seawater migrates to the bottom of the mixed zone due to its greater density, displacing the lighter mixed water upward. Environments such as those depicted on Figure 15.1 do not represent small-scale phenomena as this type of environment must, at some point in time, exist everywhere on an emerging carbonate bank. The slower the rate of emergence, the longer each part of the bank becomes subjected to meteoric water. Diagenesis progressively advances in the direction of the retreat of the sea from the area.

The Approach Toward Chemical Equilibrium in Carbonate Sediments

Modern carbonate sediment in marine environments consists of the metastable phase aragonite ($CaCO_3$), high-magnesium calcite, and a limited amount of low-magnesium calcite. Ancient carbonate rocks, on the other hand, are virtually void of aragonite and high-magnesium calcites and consist chiefly of low-magnesium calcite and dolomite. Hence, sometime after emergence from the marine environment, some major transformations take place, some accompanied by a gain in porosity and others by a loss in pore space. Virtually hundreds of references on this topic are available, with the broad review in Bathurst (1971) being the most extensive.

Inversion is the replacement of a mineral by its polymorph, for example, aragonite by calcite. Inversion occurs by either a solid-state transformation, which makes it very slow, or by a solution-reprecipitation process, which is likely faster (Carlson, 1983). Land and others (1967) view the process as occurring in five progressive stages, each of which is associated with a loss in porosity (Figure 15.3): stage I is unconsolidated sediment consisting of aragonite and high-magnesium calcite; stage V is a stabilized limestone consisting of mostly calcite.

If the transformation from aragonite to calcite is one of solution and reprecipitation, it may be viewed within a simple thermodynamic framework. Comparing the phases calcite and aragonite, calcite is the least soluble. However, we cannot stipulate a solubility for calcite without further information for some of the environmental parameters. For example, calcite has a solubility of about 100 mg/L if the partial pressure of CO_2 is about 10^{-3} bars and 500 mg/L if the partial pressure of CO_2 is 10^{-1} bars, again at a pH of 7. Whatever the solubility of calcite, the solubility of aragonite is 1.38 times higher (D. Langmuir, 1987, personal communica-

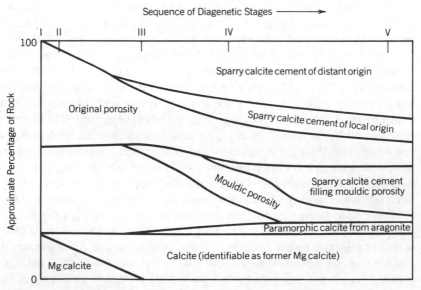

Figure 15.3
Sequential stages in the calcitization of a marine carbonate sand consisting of aragonite and Mg calcite (from Land and others, 1967. Geol. Soc. Amer. Pub., from Geol. Soc. Amer. Bull., v. 78, p. 993–1006).

tion). Thus, fresh water entering the emerged aragonite is undersaturated with respect to both calcite and aragonite and starts to dissolve the aragonite. When a sufficient amount of $CaCO_3$ is in solution, the solubility of calcite is achieved so that the solution becomes saturated with respect to calcite but remains undersaturated with respect to aragonite. With continued dissolution of aragonite, the waters become supersaturated with respect to calcite, initiating its precipitation. With only aragonite as our initial starting material, and with calcite having a lower solubility than aragonite, it is patently impossible to achieve saturation with respect to the high-solubility phase, and aragonite is always in a state of dissolution. Hence, in the absence of an infinite supply, it disappears from the rocks, with all the $CaCO_3$ going to calcite. With high-magnesium calcites and aragonite in the starting material, the dissolution-reprecipitation process continues down the solubility gradient until the transformation to the lowest-solubility phase (calcite) is complete. Thus, with both aragonite and high-magnesium calcites in the initial starting material, the waters can be supersaturated with respect to the two lowest-solubility phases but never supersaturated with respect to all three. When the sediment becomes depleted of the high-solubility phase, the saturation concentration shifts downward, becoming lower than the saturation concentration of the remaining high-solubility phase. Thus, the ultimate transformation to calcite.

The foregoing arguments are based purely on thermodynamic grounds. What cannot be ascertained from thermodynamic reasoning is the distance over which the flow system must persist until saturation with respect to the lowest-solubility phase (calcite) is achieved. If this state of saturation and, ultimately, supersaturation cannot be achieved, the whole aragonite deposit will be completely dissolved and removed from the geologic record with no coprecipitation of calcite. Because calcitic limestone is ubiquitous in the geologic record and absent in modern sediments, supersaturation with respect to calcite is undoubtedly achieved over distances that are short relative to size of the emerging carbonate body. The determination of "how short" is a problem for transport theory.

Consider Figure 15.4a, which shows the spatial distribution of $CaCO_3$ in solution for the problem being addressed. Meteoric water enters the emerged aragonite deposit undersaturated with respect to aragonite and starts to dissolve the $CaCO_3$ deposit. The concentration eventually builds up to the saturation concentration of calcite. Continued dissolution of aragonite gives rise to a state of supersaturation with respect to calcite, and calcite starts to precipitate. As the concentration continually rises, aragonite is dissolving faster than calcite is precipitating. Finally, where the concentration no longer changes with distance in the system, the net dissolution is zero, that is, as much calcite is precipitating as aragonite is dissolving. Note that this occurs at a solubility less than the solubility of aragonite. Eventually, aragonite will be totally removed from the system with precipitation of calcite. The distance to saturation with respect to calcite shown on Figure 15.4a is described as (Palciauskas and Domenico, 1976)

$$x_s = \frac{-2\alpha_x}{1 - \left(1 + \dfrac{4S^*k\alpha_x}{v}\right)^{1/2}} \ \log \frac{C_{ss} - C_o}{C_{ss} - C_{eq}} \qquad (15.1)$$

where x_s is the saturation distance; α_x is the longitudinal dispersivity; v is the ground water velocity; S^* is the specific surface of the original material, defined as the surface area of the pores per unit bulk volume of sediment; k is a reaction

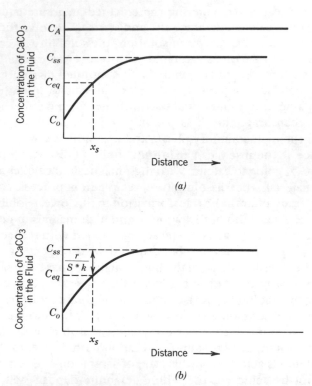

Figure 15.4
Spatial variation of CaCO$_3$ in solution where the entering fluid has a concentration of C_0, the equilibrium concentration for calcite is C_{eq}, the concentration C_{ss} denotes the position where the net dissolution is zero, and C_A indicates the saturation concentration of aragonite (from Palciauskas and Domenico, 1976. Geol. Soc. Amer. Pub., from Geol. Soc. Amer. Bull., v. 87, p. 207–214).

rate coefficient, defined as the volume of the original deposit dissolved per unit surface area of rock per unit time; and the concentrations C are as defined on Figure 15.4.

The numerator on the right-hand side of Eq. 15.1 suggests that the saturation distance increases with increasing values of longitudinal dispersivity. Within the denominator, the important dimensionless quantity is the Damköhler number

$$\frac{S^*k}{v}\, \alpha_x \qquad\qquad (15.2)$$

The larger this quantity, the smaller the saturation distance. Hence, this interesting interplay of variables suggests that if the dissolution rate of the original material is fast with respect to the velocity at which the dissolution products can be carried away by a moving ground water, the saturation distance is rapidly achieved, and the transformation to calcite takes place within a relatively short distance. If the rate of dissolution is slow compared to the velocity at which the products can be carried away, more of the sediment is dissolved with no coprecipitation of calcite;

that is, the solution remains undersaturated with respect to calcite. If the reaction rate coefficient is close to zero, the saturation distance occurs at infinity; that is, the fluid remains undersaturated throughout the flow domain. Thus, a highly dispersive system thins the mass in solution, whereas a strongly advective one reduces the contact time between the dissolving agent (water) and the solid surfaces making up the porous medium, and both intrude upon the kinetics.

Equation 15.1 also applies to the approach to calcite equilibrium in terrestrial limestones where meteoric water enters the formations in an undersaturated state. In limestones, supersaturation may be achieved because of the dissolution of Ca- and CO_3-bearing minerals other than calcite (for example, gypsum), which are more soluble than calcite. The solution to this problem is given in graphical form in Figure 15.4b. As noted, the concentration once more increases exponentially with distance from C_0 to C_{ss}, passing through the saturation concentration at some intermediate point. In this diagram r is the production rate of the high-solubility phase, which produces ions in solution contributing to the supersaturation, and S^*k is the dissolution rate constant for the low-solubility phase. The quantity r has the units $\dot{M}/L^3T$, or mass produced per unit volume per unit time, and S^*k has the units T^{-1}, or volume dissolved per unit volume per unit time. As the production rate r becomes small with respect to S^*k, C_{ss} approaches C_{eq}, and the saturation distance x_s becomes larger. Conversely, if r is large relative to S^*k, the saturation distance decreases. In a monomineralic terrain, that is $r = 0$, the terminal concentration is the saturation concentration and supersaturation is not possible. It thus follows that

$$C_{ss} = C_{eq} + \frac{r}{S^*k} \tag{15.3}$$

so that if r is greater than zero, C_{ss} is greater than C_{eq}.

The competition between kinetic and physical processes has been observed at the laboratory scale for dissolution-precipitation reactions at large Peclet numbers (advective systems). This is demonstrated in Figure 15.5, which shows the departure from saturation in two calcite packs that differ only in particle diameter. In these experiments, water that is undersaturated with respect to calcite is sent through a calcite pack at different velocities, and the concentration at the end of the pack is determined for each test. Given the exit concentration, the degree of saturation with respect to calcite is then determined. The departure from saturation is plotted against $1/v$ so that the slope of the line equals S^*kL. As shown in the figure, the departure from saturation increases (approaches one) with increasing velocity. For a given velocity, the degree of saturation increases with increasing values of S^*k. The dimensionless Damköhler group S^*kL/v for experiment I is 0.713 and for the second experiment is only 0.206. The most likely reason for this is that experiment I had finer particles and, consequently, a higher specific surface.

The Problem of Undersaturation

Up to now our main interest has been in the rate at which states of saturation and supersaturation are achieved in carbonate terrain. However, we recognize the existence of karst is proof of the ability of water to remain in some undersaturated state over relatively large flow distances. Explaining how this can occur can be difficult. Thrailkill (1968) proposes three possible mechanisms: (1) a flow rate effect, (2) a mixing effect, and (3) a temperature effect.

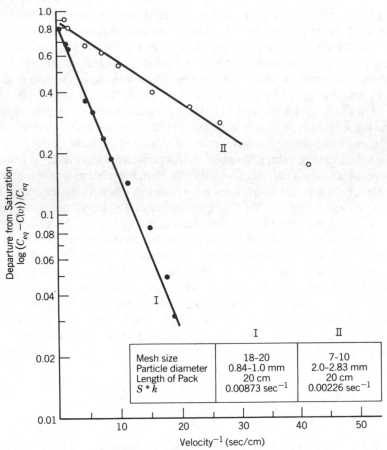

Figure 15.5
Departure from saturation versus velocity⁻¹ from calcite-pack experiments (from data presented by Weyl, 1958, pp. 173–174) (from Palciauskas and Domenico, 1976. Geol. Soc. Amer. Pub., from Geol. Soc. Amer. Bull., v. 87, p. 207–214).

The flow rate effect explains the undersaturation by recharge from two different sources. The first is as diffuse infiltration through the unsaturated zone in small fractures, and the second is as surface water inflow through discrete features such as sinkholes. The first source likely provides water that is saturated with respect to calcite, while the second with shorter contact times and smaller contact areas with the rock, is characterized by water that is undersaturated. This latter source is typically most active during storm runoff periods. Thus, the flow rate effect is in reality a mixing effect.

Long-term monitoring of one of the karst springs at Mammoth Cave, Kentucky showed this variability in composition (Hess and White, 1988). Specific conductance and hardness in spring discharge decline markedly during storm runoff periods. Hess and White (1988) interpret this response as the dilution of ground water by dilute surface water. Once a storm passes, the conduits drain and diffuse ground water inflow to the conduits becomes the main control on the spring chemistry. The conduit system effectively transmits water on a time scale

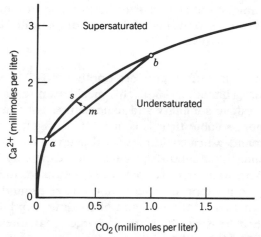

Figure 15.6
Solubility of calcite in pure water at 25°C and
1 atm total pressure as a function of Ca^{2+} and
CO_2 concentrations. The line *a-b* is the locus
of compositions of mixtures of waters *a* and
b (from Langmuir, 1984). Figure from Guide
to the Hydrology of Carbonate Rocks, Studies
and Reports of Hydrogeology No. 41. ©
Unesco 1984. Reproduced with the permission
of Unesco.

that is short with respect to the time required for calcite equilibrium to be
achieved (Hess and White, 1988).

Development of conduits could also be enhanced by a mixing effect. Two
carbonate waters of differing composition, both in equilibrium with respect to
calcite, can mix to produce a water that is undersaturated. This behavior is illustrated in Figure 15.6. The curved line defines equilibrium with respect to calcite
in terms of Ca^{2+} and P_{CO_2}. Mixtures of waters *a* and *b* will have a composition that
lies somewhere along the mixing line joining these points. A mixture at point *m*
is undersaturated with respect to calcite and will tend to return to equilibrium at
point *s* by dissolving more calcite. This phenomenon may be common in carbonate terrane where fractures and conduits form an integrated drainage network.

Thrailkill (1968) and Langmuir (1984) list some reasons why this mechanism
might be less effective in certain circumstances:

1. The shape of the calcite saturation curve is such that significant undersaturation only develops when the water exhibits major compositional
   differences. In other words, points *a* and *b* (Figure 15.6) must be widely
   separated on the saturation curve.

2. Supersaturation of either of the end member waters (that is, *a* or *b*) will
   counteract the mixing effect. Supersaturation could develop when
   ground water with an elevated P_{CO_2} encounters an atmospheric P_{CO_2} in a
   conduit and is slow to return to equilibrium.

3. If other sources or sinks for Ca^{2+}, H^+, CO_2, HCO_3^- and CO_3^{2-} exist other
   than calcite dissolution or precipitation, the effect depicted on Figure

15.6 might not materialize. In these more complex cases, Langmuir (1984) points to the need for more sophisticated computer models of the processes involved.

The temperature effect provides another possible way for enhancing calcite dissolution (Thrailkill, 1968). This mechanism relies on the inverse relationship between calcite solubility and temperature. For example, at 10°C calcite is 1.34 times more soluble than it is at 25°C (Langmuir, 1984). During summer, recharging ground water could be several degrees warmer than the ground temperature. Cooling in the subsurface will create a capability for dissolving additional calcite even with water that was previously saturated. In addition, from what we have learned about temperature redistributions in ground water flow (Chapter 9), temperature gradients increase with increasing depth in recharge areas and decrease with increasing depth in discharge areas, thereby affecting solubility. Sipple and Glover (1964) have attempted to examine solution alteration from this perspective.

There are a few other different ways for creating conditions of undersaturation. Lawrence and Upchurch (1982) and Upchurch and Lawrence (1984) focus on the Floridian aquifer, which contains upland active zones of karst formation in spite of the fact that water is saturated or supersaturated with respect to calcite. Indications from data on color and iron content are that Ca^{2+} is complexing with a humic substance, which was not accounted for in the saturation calculations. Reducing the activity of Ca^{2+} through complexation with organics would make the water more aggressive in dissolving calcite.

On the other hand, karst is well developed in lowland areas where ground water is generally undersaturated with respect to calcite. Upchurch and Lawrence (1984) point to the recharge of dilute water through sinks during scarp retreat, short residence times, and perhaps a mixing effect with water flowing from beneath the highlands as the main factors.

Dolomitization

The mineral dolomite, $CaMg(CO_3)_2$, frequently forms by replacement of calcium carbonate, $CaCO_3$. This obviously requires some input of Mg, normally considered to be provided by a moving fluid. The transformation ultimately results in a 13% reduction of the space occupied by the minerals. Land (1973) performed some calculations that focus on the number of pore volumes of fluid required to dolomitize completely 1 m^3 of sediment with 40% porosity without any loss of porosity. His calculations are as follows:

44 pore volumes of hypersaline brine that has precipitated gypsum

807 pore volumes of normal sea water

8070 pore volumes of normal sea water diluted by a factor of 10 by fresh water

From these calculations, dolomites may form in a variety of environments where the conditions just described hold. Folk and Land (1975) focus on these environments with a diagram that plots salinity against Mg/Ca ratios of the aqueous environment in which the transformation takes place (Figure 15.7). Thus, for

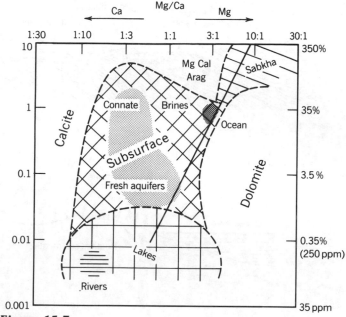

Figure 15.7
Fields of occurrence of common natural waters plotted on a
graph of salinity versus Mg/Ca ratios. As salinity rises, higher
Mg/Ca ratios are required to promote dolomitization (from Folk
and Land, 1975. Reprinted by permission).

dolomites to form in high-salinity *sabkhas* (the Arab word for salt flats), the Mg/Ca
ratio must be on the order of 5:1 to 10:1, whereas for low-salinity lakes, the ratio
can be less than 1:1. With fresh water aquifers with mixing zones similar to that
shown in Figure 15.1, the ratio must be on the order of 1:1 or 2:1, depending on
the salinity.

From the preceding discussion, the lower the salinity, the lower the required
Mg/Ca ratio required for dolomitization. Three models have been proposed
where ground water is the carrier fluid for the magnesium. Figure 15.8*a* is
the reflux model proposed by Zenger (1972) where hypersaline waters in the
sabkha environment have a density greater than that of the lagoonal sea water,
so that waters with a high Mg/Ca ratio seep downward, dolomitizing the carbonates beneath the lagoon. A variation of this model, also proposed by Zenger
(1972), is the evaporative pumping model, where marine waters are pulled
landward toward the *sabkha* due to evaporation on the sabkha during dry spells
(Figure 15.8*b*). Yet a third conceptual model has been proposed for the mixing
zone concept of Figure 15.1 by Hanshaw and others (1971). In these environments, dolomite is able to form at low Mg/Ca ratios at progressively reduced
salinities. At salinities on the order of 90 to 95 percent fresh water and 10 to
5 percent sea water, dolomite forms at a Mg/Ca ratio of about 1:1 (Scoffin, 1987).
Figure 15.8*c* is taken from Land (1973) to account for the dolomitization of
Pleistocene limestones of Jamaica.

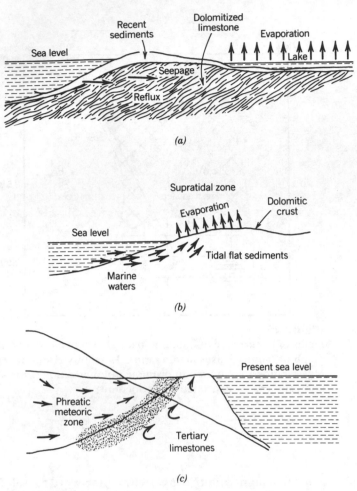

Figure 15.8
Seepage reflux (*a*) and evaporative pumping (*b*) models for
dolomitization (after Zenger, 1972). Figure (*c*) shows mixing
zone dolomite on the north coast of Jamaica (from Land, 1973).
Panel (*a*) and panel (*b*) reproduced with permission of J. of Geological Education. Panel (*c*) reprinted by permission of Elsevier
Science Publishers from Sedimentology, v. 20.

15.2 Economic Mineralization

White (1968) has identified four factors responsible for the formation of ore
deposits by groundwater:

1. A source for the mineral constituents
2. Dissolution of the minerals in water
3. Migration of the fluids
4. Precipitation of the minerals in response to physical and chemical
   changes in the fluid and/or the porous medium.

With some rewording, a similar set of conditions can be applied to the migration and entrapment of hydrocarbons.

Origin of Ore Deposits

Mass precipitating from ground water or left as a product of weathering can in some cases form valuable mineral deposits. Examples of deposits formed in this way are listed in Table 15.1. These deposits form in a variety of different ground water settings that range from local water table conditions in the case of nickel laterites or supergene sulfides, to regional scale convection deep in the crust in the case of porphyry copper or lode gold deposits. Space is not sufficient here to examine all these deposits in detail. However, a glance at the factors contributing to precipitation clearly demonstrates the role of the mixing zone in the precipitation process. The dominant contributing factors include abrupt chemical changes (changing E_H-pH conditions, decomplexation), mixing of oxidizing and reducing waters with precipitation at the redox front, the mixing of meteoric and magmatic fluids, and declining temperatures.

Roll Front Uranium Deposits

A classic study on the relationship between the occurrence of uranium and ground water flow is the work of Galloway (1978) on the uranium deposits of the Catahoula Formation of the Texas Coastal Plain. This Oligocene to Miocene–aged formation ranges in thickness from 60 to 300 m and was deposited in a complex fluvial environment. The uranium deposits in cross section are commonly crescent shaped and are formed by mineral precipitation at a redox boundary. An important aspect of this study involved mapping the distribution of fine- and coarse-grained facies because the distribution of uranium coincided with the distribution and orientation of permeable sand belts.

Galloway's conceptual model (Figure 15.9) proposed that leaching of volcanic ash layers provided the source of uranium. This process began shortly after the fluvial sediments were deposited. Flow through semiconfined sands transported uranium through the most permeable parts of the flow system. The flow pattern was controlled by the geometry of the channel sands and the presence of down-dip faults (Figure 15.9). The transport continued until uranium and other trace metals like iron precipitated somewhere close to the sulfate-sulfide redox boundary. The redox front itself would have continued to move downgradient with existing minerals continually solubilized and precipitated. Effectively, the redox front provides a place for initially dispersed trace metals to be concentrated. The migration of the front must eventually stop because of postdepositional changes to the flow system that reduce the overall permeability and the flux of water. These changes could have included (1) the compaction and sealing of boundary aquitards, (2) the continued displacements along faults, and (3) the diagenetic modification of porous, permeable units (Galloway, 1978).

In oxidizing environments, the uranyl (+VI) species (for example, UO_2^{2+}, $UO_2CO_3^0$, $UO_2SO_4^0$, and UO_2OH^+) are mobile. Various uranyl complexes contribute to this mobility in an important way (Langmuir, 1978). Thus, uranium would be transported at early stages when the system is oxidizing. Mobility would be maintained as E_H declined through oxygen and iron manganese redox zones. Galloway (1978) suggested that reductants might include organic debris and sulfides in the sandstone or gases migrating vertically. The highly insoluble solids such as coffinite ($USiO_4$) or uraninite (UO_2) would be the stable phase as redox fell to a

Table 15.1
Some of the different kinds of ore deposits whose origin depends in part on flowing ground water

Type of deposit	Example	Type of flow system	Factors contributing to precipitation
Nickel laterite	New Caledonia	Shallow, water table	Weathering and changing E_H-pH at the water table
Laterite bauxite	Jamaica	Shallow, water table drainage helped by karst	Accumulation as residual deposit accompanying weathering
Supergene sulfide	Chuquicamate, Chile	Shallow, water table	Weathering and changing E_H-pH at the water table
Calcrete uranium	Yeelirrie, Australia	Discharge end of shallow ground water flow system	Dissolution from source rock, transport, and precipitation due to evaporation and decomplexation
Roll-front uranium	Texas coastal plain	Shallow ground water	Leaching of ash, transport, and precipitation at redox front
Unconformity-related uranium	Athabasca district Saskatchewan, Canada	Deep ground water flow related to faulting	Mixing of oxidizing uraniferous and reducing waters
Mississippi Valley type Lead Zinc deposits	Pine Point, Northwest Territories, Canada	Gravity or compaction flow of brines from deep sedimentary basins	Leaching from sedimentary source rocks, transport, and deposition due to declining temperatures and possibly changing E_H-pH
Porphyry copper	San Manuel, Kalamazoo, Arizona	Convection in response to intrusion of a stock or dike	Mixing of meteoric and magmatic fluids and cooling
Lode gold deposits	Carlin, Nevada	Fluid convection of meteoric water deep in the crust	Leaching of source rocks, transport and deposition in fractured rocks due to declining temperature

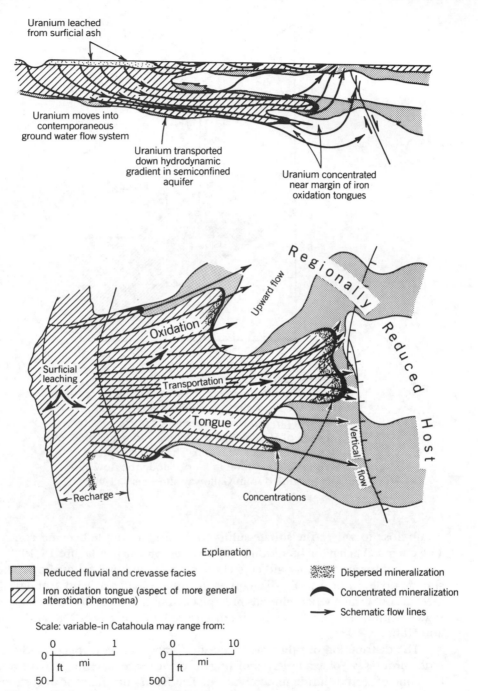

Figure 15.9
Diagrammatic representation of the origin of roll-front-type uranium deposits in the
Catahoula fluvial systems of Texas (modified from Galloway, 1978). Reproduced from
Econ. Geol., 1978, v. 73, p. 1656–1676.

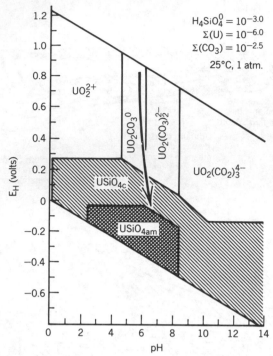

Figure 15.10
E_H-pH diagram showing the stability fields of crystalline and amorphous coffinite ($USiO_4$) in carbonate-rich ground water containing approximately
60 mg/L of H_4SiO_2, 200 mg/L total carbonate species, and 240 µg/L uranium. The arrow depicts
the possible variation in E_H-pH along the flow system (modified from Galloway and Kaiser, 1980).

point close to where the sulfate-sulfide or a similar couple became controlling.
The chemical behavior of uranium is shown graphically in Figure 15.10. The stability field for amorphous and crystalline forms of uraninite (not shown) lies in a
similar position on the E_H diagram as coffinite (Galloway and Hobday, 1983).
Details concerning what minerals precipitate and the origin of the strongly reducing conditions in the sandstones are discussed by Galloway (1978) and Galloway
and Hobday (1983).

The distribution of other metals besides uranium is controlled by these same
redox processes. For example, paralleling the occurrence of uranium in the roll front
is a zone of iron disulfide mineralization. The E_H-pH diagram for iron shows that
the stability field for pyrite lies in a similar position to that of coffinite or uraninite.

Mississippi Valley–Type Lead-Zinc Deposits
Carbonate-hosted lead-zinc deposits are found in the midcontinent region of the
United States and in Canada and are often referred to as Mississippi Valley–type
deposits. Ohle (1959), and later Anderson and Macqueen (1982), summarized the
similarities between these deposits. All occur primarily, though not exclusively, in
preferred horizons in carbonate rocks and are sufficiently conformable to the bedding so as to be described as stratiform or stratabound. All contain fluid inclusions

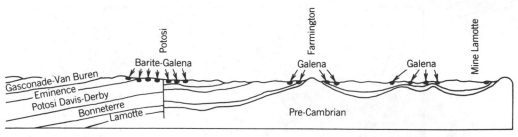

Southeast Missouri

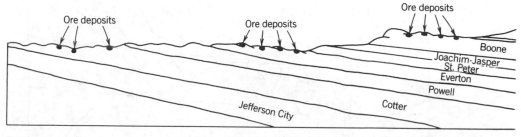

Northern Arkansas

Figure 15.11
Generalized sections of two lead-zinc mining districts showing structural and strati-
graphic relations of the ore deposits (from Bastin, 1939. Geol. Soc. Amer. Pub., from
Geol. Soc. Amer. Special Paper, v. 24, 16 p).

filled with brines that indicate a temperature of formation of 100°C to 150°C.
Most have similar structural associations, being located in regional highs such as
the Ozarks, Cincinnati Arch, and so on, where they are found near the top or
along the flanks of the domal structures (Figures 15.11 and 15.12).

A postulated ground water origin for these deposits was stated as early as 1854
by Whitney. By the late 1800s, most investigators agreed on the meteoric origin.
Their differences, however, had to do with the source of the meteoric water. One
group favored a downward percolation theory from overlying source beds whereas
the other group favored lateral- or upward-moving waters. Some difficulties with
both these theories came about with the later introduction of fluid inclusion data,
which suggested a formation temperature of 100°C to 150°C. With the current
geothermal gradient, the lowest temperature cited requires a depth in excess of
5000 ft whereas the deposits known to date are above 2000 feet in depth.

Water from Compaction   Stoiber (1941) was one of the first to propose that
the fluids that formed the lead-zinc deposits in the Tri State region were closely
related to oil field brines and were derived from compaction. Nobel (1963)
advanced this idea in general to account for all Mississippi Valley–type deposits.
This idea was accepted and amplified by Jackson and Beals (1967), Dunham
(1970), and Dozy (1970). Dozy (1970) also noted that both oil and mineral
deposits were common around the Ozark uplift area, indicating that the processes
that control the migration of oil and ore should be in agreement. The conceptual
model of Nobel (1963) is shown in Figure 15.13. As noted, the ore forming fluids
move out as hot brine through permeable zones to precipitate eventually on basin
margins that currently act as ground water recharge areas.

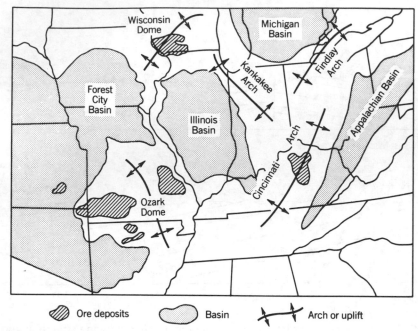

Figure 15.12
Map showing the relationship of the ore deposits to sedimentary basins and arches and uplifts (modified from Heyl and others, 1970).

The hypothesis of Nobel (1963) and others has been examined from the perspective of mathematical (numerical) models by Sharp (1978) and Cathels and Smith (1983). Sharp (1978) used a one-dimensional compactional model coupled to heat transport with provisions for faulting for the overpressured fluids of the Ouachita basin to account for lead-zinc deposits near the Ozark Dome. Faulting was presumed to rupture the overpressured basin during the Late Pennsylvanian–Permian period, providing hot pore fluids to the northern Arkansas–southeastern Missouri area where lead-zinc deposits are known to occur. The main emphasis of

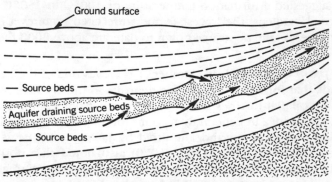

Figure 15.13
Idealized section showing aquifer transmitting water of compaction from source beds (from Noble, 1963).
Reproduced from Econ. Geol., 1963, v. 58, p. 1145–1156.

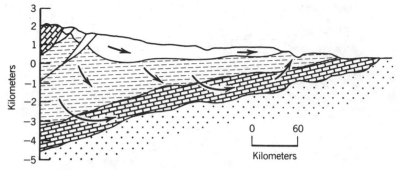

Figure 15.14
Conceptual model of fluid flow in a carbonate unit that could be the locus of a stratabound orebody (from Amer. J. Sci., Garven and Freeze, 1984a. Reprinted by permission of Amer. J. Sci.).

Sharp's (1978) work was to determine if critical temperatures could be maintained along with peak flow rates required in the mineralization. Cathels and Smith (1983) suggested a pulselike release of compactional fluids based on the modeling of a basin similar to the Illinois Basin. The consensus of these studies indicates that the postulated compactional origin of Mississippi Valley–type lead-zinc deposits is possible.

Gravity Flow Origins   The Pine Points district in the Northwest Territories of Canada contains carbonate-hosted lead-zinc deposits similar in many respects to those of midcontinental United States. In this region, Jackson and Beals (1967) postulated that the metals were derived from an argillaceous sediment or from metal sulfides in associated black shales. A compaction model was proposed to account for the Pine Point ore deposit. Garven and Freeze (1984a, b) examined a gravity-driven flow system as a potential origin for these stratabound ore deposits. Their conceptual model is shown in Figure 15.14, where fluid flow is directed across a thick shale unit and focused into a basal carbonate unit. The ore forming constituents are leached from the shales and are deposited at the discharge end of the basin. The mathematical models included not only fluid and energy transport, but mass transport as well. Their simulation results indicate that gravity-driven flow can provide favorable flow rates, temperatures, and metal concentrations for the formation of an ore deposit in a relatively short geologic time. Bethke (1986) also considered a topographic drive system to account for Mississippi Valley–type deposits from Illinois basin brines.

The Expulsion of Fluids from Orogenic Belts and Continental Collisions
In 1986, Oliver proposed a mineralization hypothesis of continental proportions, the essence of which calls for the expulsion of fluids from buried and deformed continental margin sediments to foreland basins and the continental interior during overthrusting associated with plate collisions. A simplified history of the Appalachian orogen in response to the closing of the ancestral Atlantic Ocean and collision of North America and Africa is shown on Figure 15.15. As noted in the figure, the fluid in the margin sediments may be forced into the adjacent continent, that is, according to Oliver (1986), the thrust sheet acts like a squeegee driving fluids ahead of it.

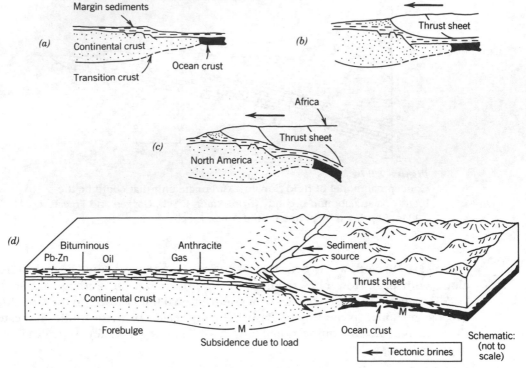

Figure 15.15
Simplified history of Appalachian orogen. (*a*) Passive continental margin with margin sediments. (*b*) Thrust sheet overrides margin sediments; load causes subsidence. (*c*) Continent-continent collision advances thrust sheet and terminates orogeny. (*d*) Block diagram of orogen at time corresponding to Figure 15.15(*b*). Heavy arrows schematically illustrate flow of the brines expelled to buried sediments. Gas and anthracite deposits are closer to orogen than oil and bituminous coal, respectively. Continental crust is 35 km thick; horizontal dimension in diagram is 500 km. Geol. Soc. Amer. Pub., from Geology, v. 14, p. 99–102.

As noted in Figure 15.15, metamorphism of coal decreases with distance from the orogenic belt and gas fields are closer to the orogenic belt than oil fields, both presumably due to higher temperatures. The lead-zinc deposits of the type we have been discussing are farthest from the belt and presumably were carried by the same fluids.

It has been mentioned that Dozy (1970) noticed the close association between lead-zinc deposits of the Mississippi Valley–type and liquid hydrocarbons. This association is shown in Figure 15.16. The hypothesis of Oliver (1986) suggests that both oil and minerals are transported by the tectonic brines from the main orogenic belts on the continent. Such a system may be seen to be compatible with the Garven-Freeze (1984) gravity flow model. Here the tectonic brines originating from the precollision marginal sediments may mix with meteoric water driven by the hydraulic head established by tectonically created topographic relief (Figure 15.17).

In recent years we have seen parts of the hypothesis by Oliver (1986) embraced by members of the modeling community—at least those parts that can be readily programmed for a digital computer. Thus, in 1988, Garven proposed

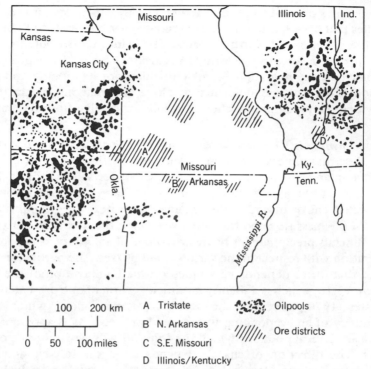

Figure 15.16
Distribution of oil and ore occurrences around Ozark uplift (from
Dozy, 1970. Reprinted by permission of the Institution of Mining
and Metallurgy).

that oil migration in the Alberta Basin throughout Tertiary time was the result of
a foreland basin uplift. The similarity between the Appalachian orogen and the
Cordilleran orogen in Alberta has been reasonably well established. Garven (1988)
states that the transport of oil waned in the Late Tertiary as the regional flow dissipated due to further erosion of the landscape. Koons (1988) discusses the fluid
flow regime due to a continental collision in New Zealand and the occurrence of

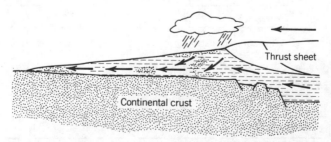

Figure 15.17
Simplified illustration showing injection of tectonic brines
into hydrologic flow induced by tectonically created
topography (from Oliver, 1986. Geol. Soc. Amer. Pub.,
from Geology, v. 14, p. 99–102).

active zones of mineralization. Bethke (1988) concludes on the basis of model studies that deep basin brines have migrated for hundreds of kilometers through the interior basin of North America. The brines redistributed hydrocarbons in their present reservoirs and formed Mississippi Valley–type deposits. He mentions further that the migration events coincide both temporally and spatially with episodes of orogenic deformation along the margins of the craton, which provided the topographic relief to establish the paleo flow systems.

Noncommercial Mineralization: Saline Soils and Evaporites

Any time that mineralized ground water gets within a few meters of the ground surface in an arid climate, the potential exists for saline soils or thick evaporite deposits to form. Evaporation concentrates whatever salts were initially present in the water. The quantity of salts generated at or near the ground surface is a function of the mass flux and the time over which the flux is operative.

Minerals precipitate in the inverse order of solubility as the solution reaches saturation with respect to the various solid phases. The particular mineral assemblage that forms depends very much on the composition of the starting water. Figure 15.18, which is a slightly modified version of one presented by Hardie and Eugster (1970), describes the possible pathways along which the major ion chemistry of an evaporating water could proceed. As is apparent on the figure, there are critical points at which the solution will evolve in one of two possible ways. The direction of change depends upon the relative abundance of key cations and anions. Drever (1982) discusses this model in detail along with a few more recent refinements.

Here we will consider two examples. The first is a case of saline soil development in east-central Saskatchewan, Canada (Meyboom, 1966b), and the second is an evaporite deposit in the northern Salt Basin in west Texas and New Mexico

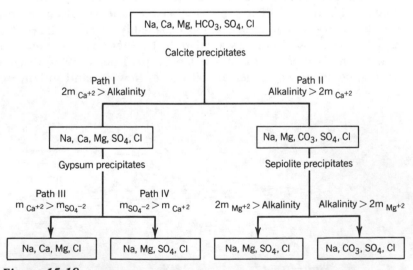

Figure 15.18
Conceptual model of brine evolution and the sequence of mineral precipitation (from Boyd and Kreitler, 1986; modified from Hardie and Eugster, 1970. Mineral. Soc. Am. Spec. Pub., v. 3, p. 273–290). Reprinted with permission.

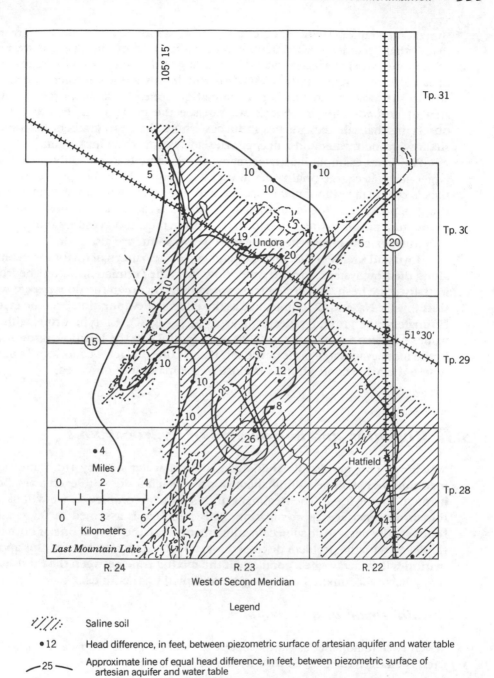

Legend

⌇⁄⁄⁄⁄⁄⁞:   Saline soil

•12   Head difference, in feet, between piezometric surface of artesian aquifer and water table

—25—   Approximate line of equal head difference, in feet, between piezometric surface of
         artesian aquifer and water table

Figure 15.19
Saline soils on the Great Salt Plain, Saskatchewan, and their relationship to the
hydrogeology. (Reprinted from Meyboom, P., 1966b. Groundwater Studies in the
Assiniboine River Drainage basin, Part 1: The Evaluation of a Flow System in South-
central Saskatchewan. Geological Survey of Canada, Bull. v. 139, 65 p. Reproduced
with the permission of the Minister of Supply and Services Canada, 1990).

(Boyd and Kreitler, 1986). Shown in Figure 15.19 are features of "the Great Salt Plain" that occupies a broad depression between hills on the east and west sides of the study area in Saskatchewan. The shading outlines an area with large patches of salt efflorescence (mainly Na_2SO_4) on which only a few salt tolerant plants are able to survive. The area of salt accumulation coincides almost exactly with an area of ground water discharge. We can see the evidence in Figure 15.19 by observing that salt accumulates in places where the head gradient between the shallow ground water and a deeper artesian system is at a maximum.

The Texas example illustrates many of the same features, only on a slightly larger scale. Recharge originating in rocks and alluvial fans peripheral to the salt flats is eventually evaporated in areas where the water table is less than 1 m from the ground surface (Boyd and Kreitler, 1986). Gypsum is the most commonly found mineral, occurring in the form of gypsum mud and gypsum sand. However, dolomite, calcite, magnesite, halite, and native sulfur are also found.

Boyd and Kreitler (1986) have documented the pattern of major ion evolution along the flow system. Ground water flow from the Permian limestone bedrock is initially quite fresh. Salinity increases progressively down the flow system with a shift toward Na^+ and Cl^- dominance as the effects of evaporation become evident. Ultimately, the ground water evolves to a Na-Mg-SO_4-Cl–type brine with TDS values between 50,000 and 300,000 mg/L (path IV in the Hardie-Eugster model) in the most evaporated parts of the salt flats. Halite, in fact, occurs locally in areas where the brine concentration is close to the maximum observed.

15.3   *Migration and Entrapment of Hydrocarbons*

Whenever two immiscible fluids like oil and water occupy the same porous medium, there will be a simultaneous flow of each fluid, with each propelled by its own driving force. The treatment of this topic is complex so we will use here the abrupt interface approximation. This interface is assumed to segregate the fluids rigidly but will be subject to adjustments with the flow of one or both of the fluids. As with the problem discussed for the intrusion of salt water, this approximation will generally be a good one if the mixing zone between the fluids is relatively narrow compared to the region occupied by the fluids.

Displacement and Entrapment

When two homogeneous fluids occupy adjacent regions in space (Figure 15.20), each is characterized by its own hydraulic head (Hubbert, 1953)

$$b_1 = z + \frac{P}{\rho_1 g} \tag{15.4a}$$

$$b_2 = z + \frac{P}{\rho_2 g} \tag{15.4b}$$

where ρ_2 is greater than ρ_1. Because the pressure is continuous across the interface, the interface is shared by both fluids. This makes it possible to solve for the pressure in either of Eqs. 15.4 and substitute the result in the other equation. Solving for the less dense fluid

Figure 15.20
Slope of the interface between two immiscible fluids (from Hubbert, 1940. Reprinted by permission of the J. of Geol. Univ. of Chicago Press. Copyright 1940).

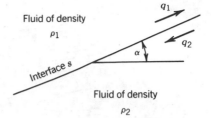

$$h_1 = \left(\frac{\rho_2}{\rho_1}\right) h_2 - \left[\frac{\rho_2 - \rho_1}{\rho_1}\right] z \qquad (15.5)$$

which is an expression for the head of a lighter fluid h_1 in terms of the head of the more dense fluid. Solving this expression for z gives

$$z = \left(\frac{\rho_2}{\rho_2 - \rho_1}\right) h_2 - \left(\frac{\rho_1}{\rho_2 - \rho_1}\right) h_1 \qquad (15.6)$$

where z is the elevation of the interface given in Figure 15.20.

A question now arises as to the slope of the interface if one or both of the fluids are in movement. The slope of the interface is given as the change in its elevation z along the interface s

$$\sin \alpha = \frac{\partial z}{\partial s} = \left(\frac{\rho_2}{\rho_2 - \rho_1}\right) \frac{\partial h_2}{\partial s} - \left(\frac{\rho_1}{\rho_2 - \rho_1}\right) \frac{\partial h_1}{\partial s} \qquad (15.7)$$

Substituting Darcy's law, where the gradient is taken as $\partial h/\partial s$, gives

$$\sin \alpha = \frac{\partial z}{\partial s} = -\left(\frac{\rho_2}{\rho_2 - \rho_1}\right) \frac{q_2}{K_2} + \left(\frac{\rho_1}{\rho_2 - \rho_1}\right) \frac{q_1}{K_1} \qquad (15.8)$$

If the dense fluid q_2 is not in motion

$$\sin \alpha = \frac{\partial z}{\partial s} = \left(\frac{\rho_1}{\rho_2 - \rho_1}\right) \frac{q_1}{K_1} \qquad (15.9)$$

As $\rho_2 > \rho_1$, $\sin \alpha > 0$ and the slope of the interface will increase upward in the direction of flow. The greater the flow rate q_1 of the lighter fluid, the greater the slope. It follows that a light fluid will always displace a more dense fluid if the latter is not in motion, provided the displacement is done immiscibly. Consider, for example, that the heavy fluid is taken as a liquid hydrocarbon immobilized within an anticlinal structure and the light fluid is taken as a gas. If the anticlinal structure is filled with the liquid hydrocarbon and a later generated gas phase enters the reservoir, the liquid hydrocarbon can be totally displaced by the gas. A similar type of displacement is shown in Figure 15.21 for salt water and fresh water in a synclinal structure where the salt water is not in motion and the displacement is assumed to be immiscible. Normally, highly saline water would tend to lie at the bottom of such synclines and less dense moving fluids would tend to rise and pass over or around the more dense mass, giving the pattern shown on Figure 15.21.

If the more dense fluid is in motion and the light fluid static, $\sin \alpha$ is negative, and the slope of the interface will increase downward in the direction of flow. The

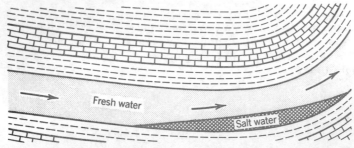

Figure 15.21
Position of salt water in syncline under conditions of fresh water movement (from Hubbert, 1953. Reprinted by permission).

greater the flow rate of the more dense fluid q_2, the greater the negative tilt. This is demonstrated in Figure 15.22 for ground water movement in an anticlinal structure containing a nonmoving liquid hydrocarbon. The relationship between the slope of the oil-water interface and the hydraulic gradient for the fluid can be expressed (Hubbert, 1953)

$$\tan \alpha = \frac{dz}{dx} = \left(\frac{\rho_w}{\rho_w - \rho_o} \right) \frac{db}{dx} \qquad (15.10)$$

where the subscript w is for water and o is for oil. As the hydraulic gradient decreases and approaches zero, the slope of the oil-water interface approaches zero (horizontal). According to Hubbert (1953), structures with leeward closing dips less than the tilt of the oil-water interface cannot hold oil. Hence, in the presence of low-velocity ground water (flat hydraulic gradients), even minor structures become more efficient in their capability to hold the hydrocarbon. On the other hand, for steep hydraulic gradients, the hydrocarbon may be flushed out of the trap.

Basin Migration Models

Tóth (1988) has summarized the various forces that give rise to hydrocarbon migration. These include sediment compaction and rebound, buoyancy, gravity flow, confined flow, gas expansion, thermal expansion of liquids, molecular diffusion, and osmosis. Most of these driving forces have been previously discussed. Tóth (1988) then arranges basin migration models into four classes according to the dominant force driving the fluids. These include (1) compaction, (2) compaction-heat, (3) compaction-gravitational, and (4) gravitational.

The compaction models selected for review are those of Jacquin and Poulet (1970, 1973), Bonham (1980), and Bethke (1985). Jacquin and Poulet (1970, 1973) set out to determine the flux, the pressure and the flow direction as a function of time in a hypothetical basin. They conclude that the potential for migration is greatest in reservoir rocks during the active sedimentation period. Bonham (1980) considers progressive compaction wherein during the early years of deposition, water is expelled upward from the older beds to the younger ones. With

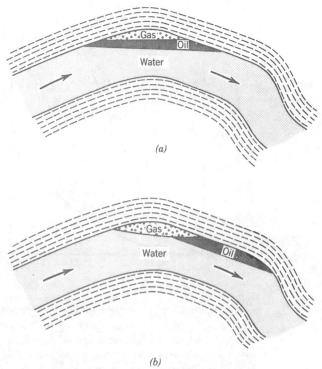

Figure 15.22
Relationship between the tilt of the oil-water interface
and the intensity of ground water flow (from Hubbert,
1953. Reprinted by permission).

increased time, the older sediments achieve their minimum porosity with no further expulsion of the pore waters. In the later stages of compaction, no additional water for migration is provided from the already compacted units. Thus the driving force diminishes. Bethke (1985) examines these conditions with a numerical model (see Chapter 9).

The compaction-heat drive model merely recognizes that the specific volume of water increases with increasing depth (temperature) (Figure 15.23) and the direction of movement of this water coincides with the direction of movement caused by isothermal compaction.

The compaction-gravity drive system has been discussed by Coustau and others (1975) along with several other investigators and is based essentially on the evolution of abnormal pressure basins to mature gravity flow systems as demonstrated on Figure 8.2. The compaction system is thought to be the cause of hydrocarbon accumulation in traps. The gravity flow superimposed on this system can flush these hydrocarbons, except in the centrally located parts of the basin. Tóth (1988) presents an interesting modification of this system based on the hydrogeologic regimes shown in Figure 8.1. Given a suitable cap rock, no vertical or lateral escape of hydrocarbons is possible in the vicinity of converging flow fields. This has already been demonstrated in Figure 14.4 for the same geometry of flow exhibited in Figure 8.1.

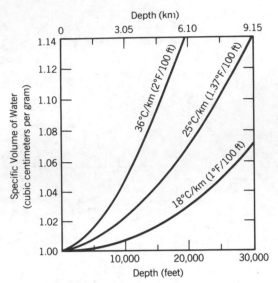

Figure 15.23
Specific volume of water as a function of depth
for specified geothermal gradients (from Magara,
1978). Reprinted with permission of Elsevier
Science Publishers from Compaction and Fluid
Migration: Practical Petroleum Geology.

Figure 15.24 demonstrates the various hydraulic, hydrodynamic, and structural
traps that can develop within the gravity-driven flow systems. Structural traps occur
where faulting results in offset of permeable horizons, causing a potential reservoir
rock to be bounded by an impermeable one. Hydrodynamic traps can occur in anticlinal structures whereas hydraulic traps occur where two flow systems converge.

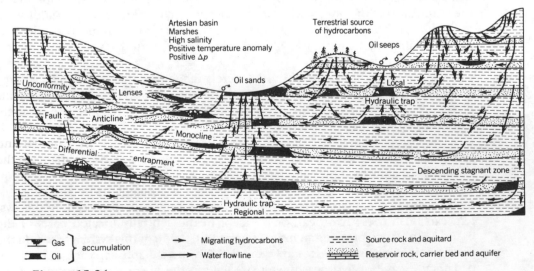

Figure 15.24
Hydraulic and hydrodynamically aided geologic traps in regionally unconfined fields of
gravity flow of formation water (from Tóth, 1980. Reprinted by permission).

15.4 Self-Organization in Hydrogeologic Systems

As illustrated in the previous sections on mixing zone phenomena and economic mineralization, the interactions between fluid flow and chemical reactions in disequilibrium can result in changes in the types, amounts, and locations of minerals and solutes within a flow system. These changes fall into the general category of self-organization, defined as the patterning of one or more descriptive variables that results from reaction-transport feedbacks (Ortoleva and others, 1987a,b). Self-organization in ground water flow systems may commonly involve patterning associated with dissolution, with precipitation, and with interactions between dissolution and precipitation and other processes.

Patterning Associated with Dissolution

Dissolution-induced patterning is the simplest of the self-organizational mechanisms noted and, as such, has been most widely modeled, both numerically and in the laboratory. Consider the most basic case, that of a medium containing a single reactive component into which undersaturated ground water is flowing. Zones of initially higher porosity and hydraulic conductivity become preferentially more porous and permeable as a result of dissolution. In turn, flow is increasingly directed toward those zones, and dissolution can be enhanced there (Figure 15.25). The feedback loop thus established is integral to self-organization. Numerical modeling of a generic, single-component system by Steefel and Lasaga (1988)

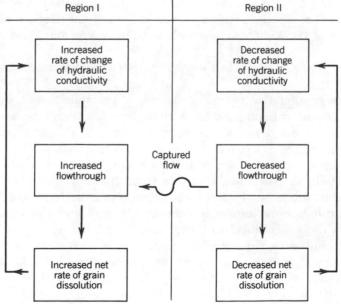

Figure 15.25
Dissolution-transport feedback loop for a region (I) of initially higher porosity and hydraulic conductivity adjacent to a region (II) (from Amer. J. Sci., Ortoleva and others, 1987a. Reprinted by permission of Amer. J. Sci.).

indicates that the propagation of preferential channels in zones of dissolution, or patterning of porosity and hydraulic conductivity, is a function of the rate of reaction relative to the rate of fluid flow. When the reaction is sufficiently rapid relative to flow, the length of such channels is proportional to the ratio of advective to dispersive transport, which is embodied in the Peclet number. That is, when there is no competition between flow and reaction, the Damköhler number plays no role in describing the process. Where the reaction is slow relative to flow, channel length directly depends on the ratio of the reaction rate constant to the flow rate, embodied in the Damkohler number. Not surprisingly, in both situations, the relative spatial extent of patterning is dependent on the scale of transport being considered.

An obvious natural analog to the single-component system is a carbonate medium. (It should be remembered that dissolution of the single-phase aragonite creates a multicomponent problem involving precipitation; this more complicated case will be touched on later.) For a calcitic limestone without a relatively inert matrix (as would be provided by quartz grains), so that dispersion is minimal, significant channel lengths should be possible. The formation of karst conduits is a logical consequence. In a related vein, petroleum engineers have conducted experimental studies of flow enhancement by acid injection into carbonate reservoir rocks. Hoefner and Fogler (1988) introduced dilute hydrochloric acid into limestone and dolomite cores, then, after evacuating the cores, injected an alloy into each to make a casting of the channel network resulting from dissolution. Each core was subsequently dissolved in concentrated acid, leaving only the casting. The relatively rapid dissolution of calcite at moderately acidic pH was reflected in the limestone castings by the persistence of a single, long channel. For dolomite, whose dissolution was significantly slower, pronounced channelization was only evident at low flow rates. These results, while forced by lower values of pH than those typical of natural systems, confirm the relations of patterning to the Peclet and Damkohler numbers obtained by Steefel and Lasaga (1988).

Patterning Associated with Precipitation and Mixed Phenomena

It is conceivable that repetitive patterns of precipitates and, in turn, of reduced porosity and hydraulic conductivity could result from ground water mixing in a porous medium without concurrent dissolution. This situation might exist, for example, where different flow systems meet at a common discharge zone. However, given the prevalence of multimineralic media, it is possible that precipitation-induced patterning linked to local dissolution is more widespread. The systematic inversion of aragonite to calcite is one such water-rock interaction. The supersaturation-nucleation-depletion cycle (Ortoleva and others, 1987a) represents a theoretical mechanism for repetitive banding of precipitates. Consider water containing an aqueous species A as infiltrating a reactive medium, dissolving a species B from disseminated grains of a mineral X, so that supersaturation results with respect to a solid AB. Precipitation of AB at a site within the medium can repress nucleation of the solid at another site downgradient in favor of continued growth of the solid at the first site. Eventually, though, upgradient diffusion of B will become negligible, permitting nucleation and growth at a site even further downgradient, and so forth.

The localization of precipitation by upgradient diffusion may yet be overwhelmed by considerations of hydraulic efficiency. If precipitation of a relatively insoluble solid significantly reduces the available pore space, initial dissolution channels can be abandoned in favor of forming new, adjacent channels. Rege and Fogler (1989) have demonstrated this process by injecting an acidic solution of ferric chloride into limestone cores, inducing calcite dissolution and ferric hydroxide precipitation. Rege and Fogler (1989) observed fluctuations in hydraulic conductivity during runs by monitoring changes in the pressure gradient along each core. After each run, following Hoefner and Fogler (1988), a casting of the reaction-induced channel network was created. At low initial flow rates (high Damkohler numbers), precipitation caused eventual refocusing of flow into a few channels in which flow rates increased sufficiently to prevent further plugging. For high initial flow rates (low Damkohler numbers), plugging and diversion of initial channels were minimized.

Beyond the simple cases of reaction-transport feedbacks discussed thus far, self-organizational phenomena may extend to multiple, interactive precipitate bands and dissolution channels in "impure" media. Mechanical feedbacks can additionally come into play, as dissolution may induce compaction, thereby potentially reducing porosity. Other phenomena operative with fluid flow may include buoyancy-driven convective cells resulting from density differences associated with dissolution and precipitation. Dewers and Ortoleva (1988) have considered the role of mechanical feedbacks and convective cells in the generation of relatively impermeable seals in sedimentary basins.

Self-organizational phenomenon may also extend to reactions that are strongly dependent upon temperature. For heterogeneous chemical processes, it can be argued that it is not possible for the overall rate of reaction to increase without limit if heat is added. If surface reactions are normally small compared to diffusion rates at a given temperature, an increase in temperature can cause an exponential rise in the rate of surface reaction, but will have a small effect on the mass transport controlled by diffusion. Surface kinetics will then speed up and eventually overhaul and become fast with respect to transport, thereby forcing the reaction process from the kinetic to the diffusion-controlled regime (Frank-Kamenetskii, 1969). Homogeneous liquid-liquid reactions, on the other hand, may increase without limit in high-temperature environments. These same ideas apply to exothermic reactions taking place in the absence of outside energy. At least in theory, chemically controlled exothermic reactions can pass from the kinetic to the diffusion regime, whereas endothermic reactions, which are self regulating in a thermal sense, cannot. An overall view of self regulatory or cybernetic chemical systems is given by Frank-Kamenetskii (1969)

15.5 *Coupled Phenomena*

In earlier chapters we have dealt with the flow of fluid in response to a single gradient in hydraulic head. In some cases, fluid can move in response to other gradients, for example, gradients in concentration, or temperature, or in voltage. Additionally, mass can move in response to a gradient in head or temperature. These processes are part of a larger family of similar processes that are referred to as "coupled" in the sense that the movement may not be described as a simple relationship between one flux and one driving force. Such coupled flow processes may be of importance in certain geologic environments.

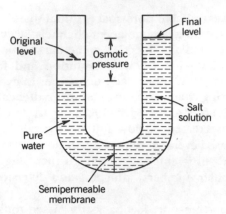

Figure 15.26
U tube demonstration of osmosis.

Chemical Osmosis

Chemical osmosis is the movement of fluid in response to a concentration gradient. A simple experiment demonstrating the phenomenon of chemical osmosis is shown in Figure 15.26, where a column of relatively pure water is hydrostatically balanced by a column of saline water through a semipermeable membrane. Because water moves from the pure water side to the saline side, the pure water must somehow possess more energy than the saline solution. The reasons for this lie in the concept of free energy, which is a thermodynamic potential not unlike potential energy except that free energy encompasses more than energy of position. In this regard,

1. Free energy increases with increasing hydrostatic pressure.
2. Free energy decreases with increasing concentration of dissolved material.
3. Free energy decreases with increasing temperature.

Thus we see that the saline solution was initially at a lower free energy. A semipermeable membrane can pass fluid but not solute so that the saline solution becomes diluted and its concentration decreases. As the system is presumed closed, the transport of water causes an increase in hydrostatic pressure for the saline solution, and a decrease in hydrostatic pressure for the fresh water. In terms of energy, the free energy of the saline solution increases because of decreased salinity and increased hydrostatic pressure, and the free energy of the pure water decreases due to a decrease in hydrostatic pressure. Movement will cease when the free energies are the same on both sides. In this case the osmotic pressure may be described as

$$\pi = 2 \, \Delta P \qquad (15.11)$$

where ΔP is the loss of pressure on the pure water side of Figure 15.26 (or the gain of pressure on the salt solution side) and is defined as the hydrostatic pressure that must be placed on a concentrated solution to establish equilibrium between it and the pure solvent, that is, to raise the free energy of the concentrated solution to that of pure water. For example, from the initial conditions of Figure 15.26 and the developments cited earlier, no net movement could have taken place if the pressure of the saline solution was raised by $2\Delta P$.

The field counterpart of the apparatus in Figure 15.26 does not involve a pure solvent and a concentrated solution, but salinities of different concentration. The general example is as follows: if the salinities of the water in two aquifers separated by a thin layer of clay or shale are unequal, a net movement of water may occur from the aquifer of low salinity to the aquifer of high salinity. During this process, the ions contained in water in the more diluted aquifer cannot move freely across the shale membrane. As a consequence, the dissolved constituents of the originally diluted aquifer become more concentrated and its salinity increases. The dissolved constituents of the originally salty aquifer become diluted. This movement can be opposed if pressure can be applied to the fluid of high salinity. This balancing pressure may be provided by the hydrostatic head that is developed. According to Hill and others (1961), highly compacted shale membranes with an abundance of clay minerals may develop osmotic pressures that can be 12 to 14 lb/in.^{2} (say, 1 bar) for each 1000 mg/L salinity difference. The upper limit of about one bar suggests a 10 m head differential per 1000 mg/L concentration difference to promote equilibrium.

The phenomenon of chemical osmosis is called upon to explain certain pressure and salinity anomalies in sedimentary basins. In Figure 15.27, two aquifers are separated by a semipermeable membrane. The aquifers have characteristic heads Δh_I and Δh_{II} and the activities of their waters are denoted by $\alpha_{H_2O}^I$ and $\alpha_{H_2O}^{II}$. Activities of water are inversely related to salinities, so that if $\alpha_{H_2O}^{II} > \alpha_{H_2O}^I$, aquifer I is the saltier of the two. If $\Delta h_{II} > \Delta h_I$, the water will flow from aquifer II to aquifer I. As ions cannot move freely across the membrane, the water in aquifer II gets saltier and the water in aquifer I gets fresher. As free energy increases with decreasing salinity, a counterforce is set up, and the flow may cease when the increased free energy of the water in aquifer I balances the head difference between the aquifers.

The process described here is called salt filtering, or ultrafiltration (Back and Hanshaw, 1965), a mechanism often called upon to explain the occurrence of brines in deep sedimentary basins. Bredehoeft and others (1963) recognized that brines are usually found (1) in deeper parts of formations that contain fresh water at shallow levels near the outcrop areas, (2) in close proximity to evaporites or other soluble strata, (3) in formations near saline surface-water bodies with hydraulic conditions that are favorable for brine encroachment, and

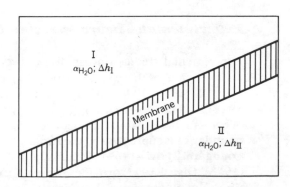

Figure 15.27
Shale membrane separating two aquifers.

(4) in formations that have been subject to migration of brines from one or more of these sources. The mechanisms for the formation of brines in categories 2 through 4 are obvious. Category 1 is more difficult to explain. Because many brines are as much as six times more concentrated than normal sea water, the suggestion that original marine water was trapped in the sediment is generally ruled out. The osmotic mechanism described was proposed by DeSitter (1947) to account for the occurrence of brines in basins that are not subject to migration of brines from other sources and that do not have extensive saline deposits.

Returning now to Figure 15.27, if there is no appreciable difference in Δb_I and Δb_{II}, and if $\alpha_{H_2O}^{II} > \alpha_{H_2O}^{I}$, water may flow from aquifer II to I. If such flow takes place in a poorly transmissive environment, the net effect may approximate that of the closed system of Figure 15.26, where no flow is contributed from the outside. Under these conditions, the pressure of the fluid in aquifer II may decrease in response to osmotic withdrawal, and the pressure in aquifer I may increase. Areas of high or low fluid pressures often constitute anomalies in certain environments and have been typically explained by chemical osmosis (Berry, 1959; Berry and Hanshaw, 1960; Back and Hanshaw, 1965).

Electro- and Thermo-osmosis

The movement of water across a semipermeable membrane may also take place in response to an electrical potential across the membrane, or to a temperature gradient. In the experimental example of electro-osmosis, two electrodes are installed in a saturated mass and a direct current applied. There results a movement of water from the positive electrode to the negative electrode. This phenomenon is associated with exchangeable ions that, under the influence of an electrical current, are positively charged and move toward the negative electrode. They are replaced by positive hydrogen ions produced by electrolysis of water. In nature, a membrane may develop a potential difference when it separates two electrolyte solutions containing cation species at different concentrations (Back and Hanshaw, 1965). Casagrande (1953) and Zaslavsky and Ravina (1965) provide comprehensive reviews of electro-osmosis.

Thermo-osmosis is the movement of water in response to a temperature gradient. Studies by Rastoga and others (1964) demonstrated a linear relationship between the temperature gradient and the osmotic flow. The membrane in these experiments was cellophane with very small pore sizes. The authors also noted that the conductance of the membrane varied with its temperature, generally decreasing with increasing temperature over the range $40°C-52°C$ and increasing over the range $52°C-60°C$.

Experimental, Theoretical, and Field Studies

Experimental studies on osmotic effects abound in the literature. Berry (1969) reviews some of these studies, including the demonstration of ultrafiltration with clays (McKelvey and Milne, 1962), the electrical properties of membranes (Wyllie, 1948), and the observation of osmotic pressure in bentonite and molded material (Kemper, 1960). Ionic impedance of salt solutions across low-permeability boundaries is supported by the experimental work of McKelvey and Milne (1962), Young and Low (1965), Kharaka and Berry (1973), and Hanshaw and Coplin (1973). Olsen (1972) provides some key experiments that may have direct application to geologic materials at depth. Experimental activity in this area in general appears to have diminished considerably in the last 15 years, possibly because the field verification of such phenomena is difficult even when all the symptoms (low pressure, high salinity) are present. In more recent years, laboratory studies of electromigration demonstrate a possible field technique for the removal of certain contaminants from ground water (Runnells and Larson, 1986).

Relatively few theoretical studies have been performed on osmotic effects. Greenberg (1971), and later Mitchell and others (1973), provided flow equations and some solutions for specified conditions.

Field studies regarding osmotic effects are more appropriately referred to as hydraulic hypotheses. That is, field studies are not normally conducted to find or verify osmotic effects; instead, such effects are normally proposed to account for a certain set of field observations that were collected from studies designed to serve other purposes. Hence, Hill and others (1961) argue for osmotic withdrawals on the basis of a high salinity association with the closure of an equipotential low in a permeable formation overlain by shale; Hanshaw and Hill (1969) present similar arguments; Graf (1982), on the basis of extensive studies of the Illinois Basin and membrane phenomena in general, strongly argues for ultrafiltration; and Marine and Fritz (1981) propose an osmotic model to account for anomalous head data. On the other hand, on the basis of experience in the Gulf Coast, Hanor (1983) is not a proponent of osmotic effects as an explanation for the occurrence of brines. Hence, there appears to be little question that the phenomenon operates, but there are arguments for and against the fact that it is responsible for the geologic work attributed to it.

Generalized Treatment

To move from the specifics of ground water flow from the Darcy perspective to a broader view based on coupled phenomena, it is useful to introduce the concept of a general rate equation of the form

$$\mathbf{F} = -B \operatorname{grad} \phi \tag{15.12}$$

where **F** is a flux, or the flow of something per unit area per unit time, B is a conductance, or a material property, and grad ϕ is the gradient of some potential ϕ. The potential is some scalar quantity and the gradient of the potential represents the driving force. It is immediately recognized that Darcy's law is of the form given by Eq. 15.12, where the flux is the specific discharge $\mathbf{q}$, the conductance is the hydraulic conductivity K, and the potential is hydraulic head h. Other common rate equations of the form given by Eq. 15.12 include

$$\mathbf{J} = -D_d \operatorname{grad} C \tag{15.13a}$$

$$\mathbf{H} = -\kappa \operatorname{grad} T \tag{15.13b}$$

$$\mathbf{I} = -\left(\frac{1}{R}\right) \operatorname{grad} V \tag{15.13c}$$

which are known, respectively, as Fick's law, Fourier's law, and Ohm's law. In these formulations, **J** is a chemical mass flux in response to a gradient in concentration C, **H** is a heat flux in response to a gradient in temperature T, and **I** is a current flow in response to a gradient in voltage V. It is noted that concentration, temperature, and voltage are all scalar quantities. The coefficients D_d, κ, and R

Flux	Head gradient	Temperature gradient	Voltage gradient	Concentration gradient
Fluid	Darcy's law	Thermo-osmosis	Electro-osmosis	Chemical osmosis
Heat		Fourier's law		DuFour effect
Current			Ohm's law	
Mass		Soret effect		Fick's law

Figure 15.28
Matrix diagram relating forces and fluxes.

represent conductivities of some sort. They are, respectively, a diffusion coefficient, a coefficient of thermal conductivity, and an electrical resistance. All these laws are phenomenological; that is, they were derived by experimentation without regard for the microscopic details of the process. In addition, the laws apply to stationary single-phase mediums (nonmoving fluids, gases, or solids) as well as to nonmoving fluids in porous solids.

The coupling effects between these various rate equations are shown schematically in Figure 15.28. The driving forces are listed across the top, and the flows or the fluxes are cited vertically down the side. The diagonal terms from the top left to the lower right are the conventional rate equations. The off diagonal terms are the cross phenomena. From this matrix, it is possible to establish the following set of equations

$$q = L_{11}X_1 + L_{12}X_2 + L_{13}X_3 + L_{14}X_4$$
$$H = L_{21}X_1 + L_{22}X_2 + L_{23}X_3 + L_{24}X_4$$
$$I = L_{31}X_1 + L_{32}X_2 + L_{33}X_3 + L_{34}X_4$$
$$J = L_{41}X_1 + L_{42}X_2 + L_{43}X_3 + L_{44}X_4$$

(15.14)

In this set of equations, the Xs represent the driving forces and the Ls represent the coefficients. The equations are set up according to Figure 15.28 so that the diagonals are the conventional rate equations. Hence, L_{11} is the hydraulic conductivity and X_1 is the hydraulic gradient, L_{22} is the thermal conductivity and X_2 is the temperature gradient, L_{33} is the electrical conductivity and X_3 is the voltage gradient, and L_{44} is the diffusion coefficient and X_4 is the concentration gradient. So, for fluid flow, it is noted that fluid will flow in response to a hydraulic gradient, but can also have components of flow in response to temperature (X_2) voltage (X_3), and concentration (X_4) gradients. Note in Eq. 15.14 that the coupling coefficient L_{12} is for fluid flow in response to a temperature gradient and L_{21} for heat flow in response to a hydraulic gradient. The Onsager reciprocal relations state that

$$L_{12} = L_{21}$$

or, in general,

$$L_{ij} = L_{ji} \tag{15.15}$$

Thus, the Soret effect is mass flow in response to a temperature gradient, whereas the Dofour effect is heat flow in response to a concentration gradient, that is, $L_{24} = L_{42}$.

Much of the difficulty in research treating cross phenomena is in the determination of the coupling coefficients. It is not yet fully clear how important cross phenomena are in hydrologic problems. They may certainly be important for some geologic problems where long periods of time are involved, such as the movement of salts out of shales under the influence of voltage gradients, but such processes are not yet part of the mainstream of research.

Suggested Readings

Garven, G., and Freeze, R.A., 1984, Theoretical analysis of the role of ground-water in the genesis of stratabound ore deposits, Quantitative results, v. 2: Amer. J. Sci., v. 248, p. 1125–1174.

Galloway, W.E., 1978, Uranium mineralization in a coastal-plain fluvial aquifer system; Catahoula Formation, Texas: Economic Geology, v. 73, p. 1655–1676.

Hanshaw, B., Back, W., and Deike, R., 1971, A geochemical hypothesis for dolomitization by groundwater: Economic Geology, v. 66, p. 710–724.

Oliver, J., 1986, Fluids expelled tectonically from orogenic belts: Their role in hydrocarbon migration and other geologic phenomena: Geology, v. 14, p. 99–102.

Tóth, J., 1988, Ground water and hydrocarbon migration. The Geology of North America, v. 0-2, Hydrogeology, Back, W., Rosenshein, J., and Seaber, P., eds.: Geol. Soc. Amer. Pub., p. 485–502.

Problems

1. Consider the following reactions:

$$A \text{ (solid)} \; \overset{kC_b}{\underset{r_2}{\rightleftarrows}} \; B \text{ (solution)} \; \overset{}{\underset{r_1}{\leftarrow}} \; D \text{ (solid)}$$

a. In the absence of the solid D, write an equation for the equilibrium concentration C_{eq} if there is equilibrium between A and B.

b. Provide an expression for the concentration when the net dissolution rate is zero.

2. For oil and gas in an anticline consider the following:

$$\frac{\rho_w}{\rho_w - \rho_o} = 10 \qquad \frac{\rho_w}{\rho_w - \rho_g} = 2$$

The hydraulic gradient is 10 ft/mile and the leeward closing-dip of the anticline is 80 ft/mile. Can the structure hold oil? Can the structure hold gas?

3. From the following figure, h_w may be taken as a point on a piezometric surface and z as a point on a structure contour map. After multiplying these by appropriate density amplification factors, the difference gives the head at this point in an oil body (h_o). Explain (hint: see Eq. 15.5).

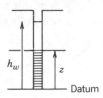

4. Develop a hydrodynamic approach for petroleum exploration that would take advantage of the typical kinds of reservoir data that become available once a sedimentary basin becomes extensively drilled.

5. Consider aquifer I characterized by head h_1 and concentration C_1 underlying and separated by a membrane from aquifer II characterized by h_2 and C_2. Select the appropriate answer from the choices given here that describes the following four conditions:

_____ $h_1 > h_2, C_1 = C_2$

_____ $h_1 > h_2, C_1 > C_2$

_____ $h_1 > h_2, C_2 > C_1$

_____ $h_1 = h_2, C_2 > C_1$

   a. Ground water moves from aquifer I to aquifer II in strict accordance with Darcy's law.

   b. Ground water moves from aquifer I to aquifer II but not in accordance with Darcy's law.

   c. It is not possible to determine the direction of movement.

   d. Ground water moves from aquifer II to aquifer I in strict accordance with Darcy's law.

   e. Ground water moves from aquifer II to aquifer I but not in accordance with Darcy's law.

6. Assume an interface between two fluids of different density in a homogeneous, isotropic medium where $\rho_2 > \rho_1$. Comment on whether or not the following statements are true. Include the reasoning you apply in arriving at the conclusion.

   a. A sloping interface always indicates movement of one or both of the fluids.

b. A horizontal interface is only possible if $q_1 > q_2$.

c. If one of the fluids is a liquid hydrocarbon and the other is a dense salt water in motion, $\sin \alpha < 0$ when the hydrocarbon is trapped in an anticlinal structure and the interface slopes downward in the direction of flow. The slope will be steeper with increasing q_2.

d. If one of the fluids is a gaseous hydrocarbon and the other is a moving salt water, $\sin \alpha < 0$ when the hydrocarbon is trapped in an anticlinal structure and the interface slopes downward in the direction of flow. The slope, however, will be less steep than in (c).

e. A light fluid will always displace a heavier fluid provided the heavy fluid is not in motion and the displacement is done immiscibly.

Chapter Sixteen

Introduction to Contaminant Hydrogeology

16.1 Sources of Ground Water Contamination

16.2 Solute Plumes as a Manifestation of Processes

16.3 Multifluid Contaminant Problems

16.4 Design and Quality Assurance Issues in Water Sampling

16.5 Sampling Methods

16.6 Indirect Methods for Detecting Contamination

This chapter introduces theoretical and practical concepts related to the occurrence of contaminants in the subsurface. A contaminant is any dissolved solute or nonaqueous liquid that enters ground water as a consequence of people's activities. Contaminants can originate from an amazingly diverse array of sources, so many, in fact, that attributing them to specific activities is just about impossible. As this chapter unfolds, we will show how the distribution of contaminants in the subsurface is a manifestation of mass transport and mass transfer processes. Establishing this link is important because from it flows understanding of how a particular plume has developed, how it can be expected to behave in the future, and how effective a remedial strategy might be. Once this conceptual framework is in place, we will examine the more practical problem of how one actually goes about characterizing contaminant distributions in the field.

16.1 Sources of Ground Water Contamination

Three important attributes distinguish sources of ground water contamination: (1) their degree of localization, (2) their loading history, and (3) the kinds of contaminants emanating from them. Given the large number of ways of contaminating ground water, there is a spectrum of source sizes ranging from an individual well to areas of 100s km² or more. In practice, the terms "point" or "nonpoint" describe the degree of localization of the source. A point source is characterized by the presence of an identifiable, small-scale source, such as a leaking storage tank, one or more disposal ponds, or a sanitary landfill. Usually, this source produces a reasonably well-defined plume. A nonpoint problem refers to larger-scale, relatively diffuse contamination originating from many smaller sources, whose locations are often poorly defined. Examples of nonpoint contaminants could include herbicides or pesticides that are used in farming, nitrates that originate as effluents from household disposal systems, salt derived from salting highways in winter, and acid rain. Typically, there are no well-defined plumes in these cases but a large enclave of contamination with extremely variable concentrations.

The loading history describes how the concentration of a contaminant or its rate of production varies as a function of time at the source. A spill is an example of pulse loading, where the source produces contaminants at a fixed concentration for a relatively short time (Figure 16.1a). This loading could occur from a one-time release of contaminants from a storage tank or storage pond. Long-term leakage from a source is termed continuous source loading. Figure 16.1b is one type of continuous loading where the concentration remains constant with time. This loading might occur, for example, when small quantities of contaminants are leached from a volumetrically large source over a long time.

Most sources of long-term leakage cannot be described in terms of a constant loading function. For example, the concentration of chemical wastes added to a storage pond at an industrial site can vary with time due to changes in a manufacturing process, seasonal or economic factors, or the addition of other reactive wastes (for example, Figure 16.1c). Leaching rates for solid wastes at a sanitary landfill site could be controlled by seasonal factors related to recharge, or by a decline in source strength as components of the waste (for example, organics) biodegrade. This latter source behavior could result in the loading history shown

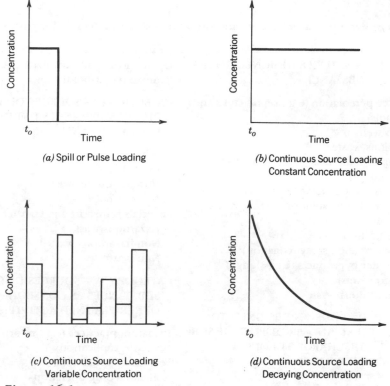

Figure 16.1

Examples of functions used to characterize contaminant loading from a spill (*a*) or long-term leakage (*b*, *c*, and *d*).

on Figure 16.1*d*. Typically, increasing complexity in the loading function translates directly into increasingly variable concentration distributions.

An amazingly large list of potential activities can produce ground water contamination (Table 16.1). As a consequence of these many different industrial, agricultural, and domestic activities, the list of potential contaminants can number in the thousands or tens of thousands of compounds. The question of how to organize this list for study is a difficult one. One approach has been to concentrate on a subset of this list and to pay special attention to contaminants that commonly occur in effluents and drinking water, produce adverse health effects, or persist within the food chain. An example is the U.S. Environmental Protection Agency's list of 129 priority pollutants, containing 114 organic compounds and 15 inorganic species, mainly trace metals (Table 16.2).

The organic priority pollutants are subdivided into four groups: volatiles, base-neutral extractables, acid extractables, and pesticides (Table 16.2). The method of analysis is different for each group, but all compounds are finally examined using combined gas chromatography–mass spectrometry (GC-MS). The inorganic group is analyzed with other techniques. One advantage of working with a standard list (for example, Table 16.2) is that highly automated procedures screen water samples for subsets of priority pollutants at a reasonable cost. Custom analyses for other species are often more costly.

Table 16.1
Sources of ground water contamination (from Office of Technology Assessment, 1984)

CATEGORY I—SOURCES DESIGNED TO DIS-
  CHARGE SUBSTANCES

Subsurface percolation (e.g., septic tanks and
  cesspools)
Injection wells
  Hazardous waste
  Non-hazardous waste (e.g., brine disposal
  and drainage)
  Non-waste (e.g., enhanced recovery, arti-
    ficial recharge, solution mining, and in-
    situ mining)
Land application
  Wastewater (e.g., spray irrigation)
  Wastewater byproducts (e.g., sludge)
  Hazardous waste
  Non-hazardous waste

CATEGORY II—SOURCES DESIGNED TO
  STORE, TREAT, AND/OR DISPOSE OF SUB-
  STANCES; DISCHARGE THROUGH
  UNPLANNED RELEASE

Landfills
  Industrial hazardous waste
  Industrial non-hazardous waste
  Municipal sanitary
Open dumps, including illegal dumping
  (waste)
Residential (or local) disposal (waste)
Surface impoundments
  Hazardous waste
  Non-hazardous waste
Waste tailings
Waste piles
  Hazardous waste
  Non-hazardous waste
Materials stockpiles (non-waste)
Graveyards
Animal burial
Aboveground storage tanks
  Hazardous waste
  Non-hazardous waste
  Non-waste
Underground storage tanks
  Hazardous waste
  Non-hazardous waste
  Non-waste
Containers
  Hazardous waste
  Non-hazardous waste
  Non-waste

Open burning and detonation sites
Radioactive disposal sites

CATEGORY III—SOURCES DESIGNED TO
  RETAIN SUBSTANCES DURING TRANS-
  PORT OR TRANSMISSION

Pipelines
  Hazardous waste
  Non-hazardous waste
  Non-waste
Materials transport and transfer operations
  Hazardous waste
  Non-hazardous waste
  Non-waste

CATEGORY IV—SOURCES DISCHARGING
  SUBSTANCES AS CONSEQUENCE OF
  OTHER PLANNED ACTIVITIES

Irrigation practices (e.g., return flow)
Pesticide applications
Fertilizer applications
Animal feeding operations
De-icing salts applications
Urban runoff
Percolation of atmospheric pollutants
Mining and mine drainage
  Surface mine-related
  Underground mine-related

CATEGORY V—SOURCES PROVIDING CON-
  DUIT OR INDUCING DISCHARGE
  THROUGH ALTERED FLOW PATTERNS

Production wells
  Oil (and gas) wells
  Geothermal and heat recovery wells
  Water supply wells
Other wells (non-waste)
  Monitoring wells
  Exploration wells
Construction excavation

CATEGORY VI—NATURALLY OCCURRING
  SOURCES WHOSE DISCHARGE IS
  CREATED AND/OR EXACERBATED BY
  HUMAN ACTIVITY

Ground water—surface water interactions
Natural leaching
Salt-water intrusion/brachish water upconing
  (or intrusion of other poor-quality natural
  water)

Table 16.2

Environmental Protection Agency list of priority pollutants. Organic compounds are subdivided into four categories according to the method of analysis

BASE-NEUTRAL EXTRACTABLES

Acenaphthene	Diethyl phthalate
Acenaphthylene	Dimethyl phthalate
Anthracene	2,4-Dinitrotoluene
Benzidine	2,6-Dinitrotoluene
Benzo[a]anthracene	Di-n-octyl phthalate
Benzo[b]fluoranthene	1,2-Diphenylhydrazine
Benzo[k]fluoranthene	Fluoranthene
Benzo[ghi]perylene	Fluorene
Benzo[a]pyrene	Hexachlorobenzene
Bis(2-chloroethoxy)methane	Hexachlorobutadiene
Bis(2-chloroethyl) ether	Hexachlorocyclopentadiene
Bis(2-chloroisopropyl) ether	Hexachloroethane
Bis(2-ethylhexyl) phthalate	Indeno[1,2,3-cd] pyrene
4-Bromophenyl phenyl ether	Isophorone
Butyl benzyl phthalate	Naphthalene
2-Chloronaphthalene	Nitrobenzene
4-Chlorophenyl phenyl ether	N-Nitrosodimethylamine
Chrysene	N-Nitrosodiphenylamine
Dibenzo[a,b]anthracene	N-Nitrosodi-n-propylamine
Di-n-butyl phthalate	Phenanthrene
1,2-Dichlorobenzene	Pyrene
1,3-Dichlorobenzene	2,3,7,8-Tetrachlorodibenzo-p-dioxin
1,4-Dichlorobenzene	1,2,4-Trichlorobenzene
3,3'-Dichlorobenzidine	

ACID EXTRACTABLES

p-Chloro-m-cresol	2-Nitrophenol
2-Chlorophenol	4-Nitrophenol
2,4-Dichlorophenol	Pentachlorophenol
2,4-Dimethylphenol	Phenol
4,6-Dinitro-o-cresol	2,4,6-Trichlorophenol
2-4-Dinitrophenol	Total phenols

VOLATILES

Acrolein	1,2-Dichloroethane
Acrylonitrile	1,1-Dichloroethylene
Benzene	trans-1,2-Dichloroethylene
Bis(chloromethyl) ether	1,2-Dichloropropane
Bromodichloromethane	cis-1,3-Dichloropropene
Bromoform	trans-1,3-Dichloropropene
Bromomethane	Ethylbenzene
Carbon tetrachloride	Methylene chloride
Chlorobenzene	1,1,2,2-Tetrachloroethane
Chloroethane	1,1,2,2-Tetrachloroethene
2-Chloroethyl vinyl ether	Toluene
Chloroform	1,1,1-Trichloroethane
Chloromethane	1,1,2-Trichloroethane
Dibromochloromethane	Trichloroethylene
Dichlorodifluoromethane	Trichlorofluoromethane
1,1-Dichloroethane	Vinyl chloride

PESTICIDES

Aldrin	Dieldrin	PCB-1016[a]
α-BHC	α-Endosulfan	PCB-1221[a]
β-BHC	β-Endosulfan	PCB-1232[a]
γ-BHC	Endosulfan sulfate	PCB-1242[a]
δ-BHC	Endrin	PCB-1248[a]
Chlordane	Endrin aldehyde	PCB-1254[a]
4,4'-DDD	Heptachlor	PCB-1260[a]
4,4'-DDE	Heptachlor epoxide	Toxaphene
4,4'-DDT		

[a]not pesticides

INORGANICS

Antimony	Chromium
Arsenic	Copper
Asbestos	Cyanide
Beryllium	Lead
Cadmium	Mercury

Nickel
Selenium
Silver
Thallium
Zinc

The way that we organize the contaminants is by grouping according to reaction type and mode of occurrence. Thus, dissolved compounds that are affected similarly by chemical, nuclear, or biological processes are grouped together. Contaminants that occur as a nonaqueous phase liquid (NAPL) are a separate subgroup. The major groups of contaminants include (1) radionuclides, (2) trace elements, (3) nutrients, (4) other inorganic species, (5) organic contaminants, and (6) microbial contaminants. The NAPLs that are included in (5) naturally subdivide into light nonaqueous phase liquids (LNAPLs) and dense nonaqueous phase liquids (DNAPLs), depending upon whether they are less or more dense than water.

All these contaminants have the potential to produce health problems. In moving through the groups, we will point out the most serious ones. As a broad generalization, too much of anything in water can produce health problems in humans. For some contaminants, particularly the radionuclides, increasing exposure results in increasing health consequences. Thus, any exposure above background levels can be of concern. For other contaminants, such as the major ions Na^+ or Cl^-, there is often a threshold below which no serious health effects will occur.

Radioactive Contaminants

The nuclear industry is the main generator of radioactive contaminants. Potential sources occur throughout the nuclear fuel cycle, which involves the mining and milling of uranium, uranium enrichment and fuel fabrication, power plant operation, fuel reprocessing, and waste disposal. The kinds of contaminants depend upon the type of reactor and the extent of spent-fuel reprocessing. For example in the United States, Japan, France, and Germany, there are large numbers of light-water reactors (LWR) that utilize enriched uranium (^{235}U) as the predominant fuel source, and possibly ^{239}Pu and ^{233}U. In Canada, heavy-water reactors (HWRs) use natural or enriched uranium as a fuel, and heavy water (D_2O) as a coolant and moderator.

During mining when the raw ore is processed, ^{238}U, ^{230}Th, ^{226}Ra, and ^{222}Rn(gas) are potential contaminants along with nonradioactive contaminants that include trace constituents, and major ions such as SO_4^{2-} or Cl^-. Areas where this contamination occurs are Colorado, New Mexico, Texas, Utah, Wyoming, northern Saskatchewan, and Ontario. The enrichment and fuel fabrication step treats raw uranium concentrate to increase the concentration of ^{235}U in the fuel relative to the more abundant ^{238}U, and produces UO_2, which is the actual fuel. The most common contaminants are ^{238}U, ^{235}U, ^{90}Sr, and ^{137}Cs.

Many radionuclides are generated as fission products from the decay of ^{235}U or ^{239}Pu and from neutron activation of stable elements in the coolant or metallic components of the reactor. Fission is the power-generating part of a nuclear reaction whereby the heavy nucleus is split into nuclei of lighter elements and neutrons. Neutron activation is a process wherein neutrons are added to the nucleus of a stable isotope to produce a radioactive one. These processes produce radionuclides like ^{137}Cs, ^{134}Cs, ^{58}Co, ^{51}Cr, ^{54}Mn, ^{55}Fe, ^{3}H, and ^{131}I. Overall in excess of 75 radionuclides could potentially be produced. Fortunately, most of these radioisotopes, except for ^{3}H, remain in the spent fuel or the reactor and are not a serious source of contamination at this stage in the fuel cycle. Reprocessing treats spent fuel to remove ^{235}U and ^{239}Pu. However, the other radionuclides remain to produce the high-level waste problems common to many countries of the world.

The health hazards associated with ionizing radiation are well known. An exposed individual can be affected by cancer, or genetic defects that could affect his or her offspring. These risks are difficult to assess at low levels of exposure. It is still not known with certainty whether a threshold exists below which some exposure is not hazardous, or whether the probability of inducing cancer or genetic damage increases as a function of the dose without a threshold. Most conservative assessments of health risk assume the latter. Other health problems associated with exposure to radiation include, for example, cataracts, nonmalignant skin damage, depletion of bone marrow, and infertility.

Trace Metals

The next major group of contaminants is the trace metals (Table 16.3). As a group, trace metals contain the largest proportion of elements found in the periodic table. The most common sources of contamination include (1) effluents from mining; (2) industrial waste water; (3) runoff, solid wastes, or waste water contributed from urban areas; (4) agricultural wastes and fertilizers; and (5) fossil fuels. Readers can refer to an excellent survey of sources and background concentrations in a book by Forstner and Wittman (1981).

Trace metals can be toxic and even lethal to humans even at relatively low concentrations because of their tendency to accumulate in the body. Some studies have found positive correlations between the concentration of trace metals in water (for example, Be, Cd, Pb, and Ni) and death rates from some cancers. Bioaccumulation of trace metals in the food chain has produced the most well-known cases of metal poisonings (for example, Minamata, Japan). Organisms higher up the food chain progressively accumulate metals. Eventually, humans at the top of the chain can experience severe health problems.

Nutrients

This group of potential contaminants includes those ions or organic compounds containing nitrogen or phosphorus. By far, the dominant nitrogen species in

Table 16.3
Examples of trace metals occurring in ground water

Aluminum	Gold	Silver
Antimony	Iron	Strontium
Arsenic	Lead	Thallium
Barium	Lithium	Tin
Beryllium		
Boron	Manganese	Titanium
Cadmium	Mercury	Uranium
	Molybdenum	Vanadium
Chromium	Nickel	Zinc
Cobalt	Selenium	
Copper		

ground water is nitrate (NO_3^-), then to a lesser extent ammonium ion (NH_4^+). Agricultural practices including the use of fertilizers containing nitrogen, cattle feeding operations, and the cultivation of virgin soils (leading to the oxidation of large quantities of nitrogen existing in organic matter in the soil) are important sources of contamination. Other sources are sewage that could enter ground water from septic tank systems or irrigated waste water. According to Bouwer (1978), effluent from sewage treatment plants in the United States contains 30 mg/L of nitrogen, mainly as NH_4^+, and organic nitrogen in the form of nitrobenzenes or nitrotoluenes.

The main health effects related to contamination by nitrogen compounds are (1) methemoglobinemia, a type of blood disorder in which oxygen transport in young babies or unborn fetuses is impaired (Craun, 1984) or (2) the possibility of forming cancer-causing compounds (for example, nitrosamines) after drinking contaminated water. Typically, phosphorus contamination is considered together with nitrogen. However, it is less important because of the low solubility of phosphorus compounds in ground water, the limited mobility of phosphorus due to its tendency to sorb on solids, and the lack of proven health problems. The major sources of phosphorus are again soil-applied fertilizers and waste water.

Other Inorganic Species

This miscellaneous group includes metals present in nontrace quantities such as Ca, Mg, and Na plus nonmetals such as ions containing carbon and sulfur (for example, HCO_3^-, HS^-, CO_3^{2-}, SO_4^{2-}, and H_2CO_3) or other species such as Cl^- and F^-. Many of these ions are major contributors to the overall salinity of ground water. Extremely high concentrations of these species makes water unfit for human consumption and for many industrial uses. The health-related problems are not as serious as those caused by the other contaminant groups. However, high concentrations of even relatively nontoxic salts, particularly Na^+, can disrupt cell or blood chemistry with serious consequences. At lower concentrations, an excessive intake of Na^+ may cause less serious health effects such as hypertension (Craun, 1984).

The potential sources of major ion salinity include (1) saline brine that is produced with oil, (2) leachate from mine tailings, mine spoil, or sanitary landfills, and (3) industrial waste water that often has large concentrations of common ions in addition to heavy metals or organic compounds. Fluoride is probably the best example of a trace nonmetal occurring as a contaminant. Like the metals, trace quantities of these contaminants can produce health problems at relatively low concentrations. With fluoride, an increase in concentration to as little as 7 or 8 times the levels for combating tooth decay can cause skeletal fluorosis (Craun, 1984).

Organic Contaminants

Contamination by organic compounds either dissolved in water or present as a separate liquid is a serious problem faced by hydrogeologists today. Extremely large quantities of organic compounds are manufactured and used, in addition to unrefined petroleum products that contain many soluble organic compounds. Of the list of sources we considered previously, almost every one either is known to contribute or has the potential to contribute organic contaminants to ground water (Table 16.4). A few types of organic contaminants occur much more frequently than the others. One such group is the soluble aromatic hydrocarbons that

Table 16.4
Occurrence of organic contaminants in relation to potential sources (modified from Office of Technology Assessment, 1984)

	Organic chemicals			
	Aromatic hydrocarbons	Oxygenated hydrocarbons	Hydrocarbons with specific elements	Other hydrocarbons
CATEGORY I				
Subsurface percolation	◧	◧	◧	◧
Injection wells	☐	☐	◧	☐
Land application[b]				
a. Wastewater				◧
b. Wastewater byproducts				◧
c. Hazardous waste	◧		◧	
CATEGORY II				
Landfills	◧	◧	◧	◧
Open dumps	◧	◧	◧	◧
Residential disposal	◧	◧	◧	◧
Surface impoundments	◧	◧	◧	◧
Waste tailings				
Waste piles				
Materials stockpiles				
Graveyards				
Animal burial				
Aboveground storage tanks	☐	☐	☐	◧
Underground storage tanks	◧	◧	◧	◧
Containers	☐	☐	☐	☐
Open burning and detonation sites				
Radioactive disposal sites				
CATEGORY III				
Pipelines	☐	☐	☐	◧
Materials transport and transfer operations	☐	◧	◧	◧
CATEGORY IV				
Irrigation practices				
Pesticide applications			◧	
Fertilizer applications			◧	
Animal feeding operations				
Deicing salts applications				
Urban runoff	☐	☐	◧	☐
Percolation of atmospheric pollutants				
Mining and mine drainage				

Key:

■ Contaminant in class has been found in ground water associated with source.

☐ Potential exists for contaminant in class to be found in ground water associated with source.

Table 16.5
The 20 most abundant organic compounds found at 183 waste disposal sites in the United States (from Plumb and Pitchford, 1985)*

Rank	Ground water contaminant	Sites
1	Trichloroethene	63
2	Methylene chloride	57
3	Tetrachloroethene	57
4	Toluene	57
5	1,1-Dichloroethane	52
6	bis-(2-ethylhexyl)phthalate	52
7	Benzene	50
8	1,2-trans-Dichloroethane	50
9	1,1,1-Trichloroethane	49
10	Chloroform	46
11	Ethyl benzene	46
12	1,2-Dichloroethane	39
13	1,1-Dichloroethane	37
14	Phenol	35
15	Vinyl chloride	30
16	Chlorobenzene	30
17	Di-n-butyl phthalate	28
18	Naphthalene	23
19	Chloroethane	23
20	Acetone	22

*Reprinted by permission of Second Canadian/American Conference on Hydrogeology. Copyright © 1985. All rights reserved.

enter ground water following a spill of petroleum fuels or lubricants. The group of compounds includes benzene, toluene, ethylbenzene, para-xylene, meta-xylene, and ortho-xylene. They are called the BTX or BTEX compounds, and comprise a package of contaminants often investigated in relation to a petroleum spill.

Other compounds occur frequently beneath waste disposal sites. Plumb and Pitchford (1985) tabulated the 20 most abundant organic compounds found at 183 waste disposal sites located across the United States (Table 16.5). The list is probably biased toward the priority pollutants because they are the most frequently analyzed.

There are important health effects related to drinking water contaminated by organic compounds. However, as Craun (1984) points out, it is difficult to establish which compounds are most toxic because not all have been tested, and health risks are inferred from studies of laboratory animals, poisonings or accidental ingestion, and occupational exposures. Furthermore, there is a serious lack of information on the health effects related to the combined effect of several compounds, and on the epidemiology of populations consuming contaminated water. Organic contamination may cause cancer in humans and animals and a host of other problems, including liver damage, impairment of cardiovascular function, depression of the nervous system, brain disorders, and various kinds of lesions. More detailed information on health effects related to organic materials can be obtained from Craun (1984).

Biological Contaminants

The important biological contaminants include pathogenic bacteria, viruses, or parasites. It does not take a degree in medicine to be aware of the serious health problems from typhoid fever, cholera, polio, and hepatitis. Other less serious abdominal disorders are often too well known by travelers to countries with poor sanitation. As a group, these health effects are some of the most significant related to the contamination of ground water.

The main source of biological contamination is from human and animal sewage or waste water. Ground waters become contaminated due to (1) land disposal of sewage from centralized treatment facilities or septic tank systems, (2) leachates from sanitary landfills, and (3) various agricultural practices such as the improper disposal of wastes from feedlots.

Particulate contaminants are less mobile than are those dissolved in water. For this reason, most reported problems are of a local nature, related, for example, to (1) poor well construction, which enables surface runoff or sewage to enter the well, and (2) sewer line breaks or septic tank fields located close to a well.

16.2 *Solute Plumes as a Manifestation of Processes*

The important processes and parameters that influence the transport of mass should be familiar. Now, we will examine how these process work together to control the spread of contaminants. The natural starting place is with the mass transport processes because they determine the maximum extent of plume spread and the geometric character of the concentration distribution. The chemical, nuclear, and biological processes mainly attenuate the spread of contaminants, reducing the size of the contaminated region to a fraction of that attributable to mass transport alone. Advection is by far the most dominant mass transport process in shaping the plume. Hydrodynamic dispersion is usually a second-order process, except in some cases involving fractured rocks.

The magnitude and direction of advective transport is controlled by (1) the hydraulic conductivity distribution within the flow field, (2) the configuration of the water table or potentiometric surface, (3) the presence of sources or sinks (for example, wells), and (4) the shape of the flow domain. All these parameters are important in controlling the ground water velocity, which drives advective transport. We can illustrate this concept with the simulation results on Figure 16.2. In these examples, the water table configuration and the other boundary conditions for the steady-state flow system remain fixed. Contaminants are added at the same place in the recharge area by fixing concentration along part of the inflow boundary.

Because there is no dispersion or reactions, the plumes have a uniform concentration equal to the source concentration. Their shape depends upon the hydraulic conductivity distribution. A reduction in hydraulic conductivity reduces the extent of plume spread by simply lowering the ground water velocity (compare Figures 16.2*a* and *b*). Adding two or three layers also influences plume shape (Figures 16.2*c*, *d*, and *e*) because in these cases both the magnitude and direction of ground water flow change. Again, there is a link between advective transport and the pattern of flow.

We will not take time to examine the other three parameters in the list that influence advection. Earlier parts of the book, we hope, have developed a sense

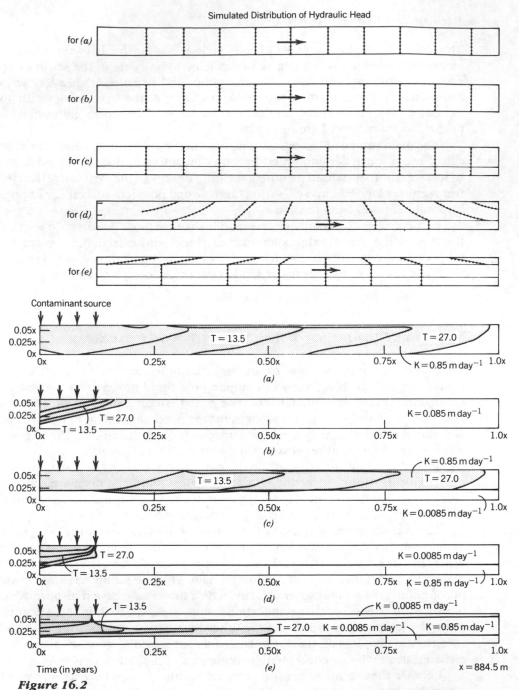

Figure 16.2
Changing plume shapes with time (T) in years as a function of the different patterns of layering. In all cases, advection is the only process operating (modified from Schwartz, 1975). Reprinted with permission of Elsevier Science Publishers from J. Hydrol., v. 27.

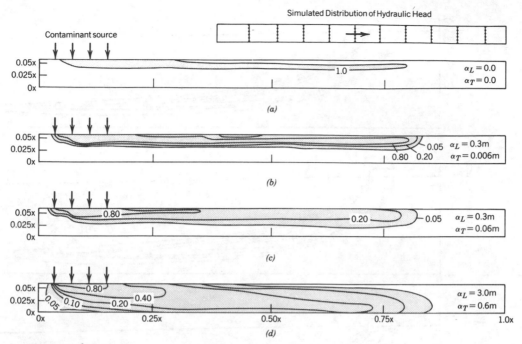

Figure 16.3
Changing plume shapes as a function of the longitudinal and transverse dispersivities:
(*a*) $\alpha_L = 0.0$; $\alpha_T = 0.0$, (*b*) $\alpha_L = 0.03$ m; $\alpha_T = 0.006$ m, (*c*) $\alpha_L = 0.3$ m; $\alpha_T = 0.06$ m, and (*d*) $\alpha_L = 3.0$ m; $\alpha_T = 0.6$ m (modified from Schwartz, 1975). Reprinted with permission of Elsevier Science Publishers from J. Hydrol., v. 27.

of how patterns of flow respond to water table configuration, pumping, and region shape. The extension of these basic concepts to advective transport simply involves visualizing the movement of a contaminant along stream tubes—at least for steady-state flow.

Adding dispersion to advective transport can cause important changes in the shape of a plume. Consider the nonlayered system shown on Figure 16.3. All of the transport parameters except the longitudinal (α_L) and transverse (α_T) dispersivities are held constant. As the magnitude of the dispersivities (α_L, α_T) increases from (0.0, 0.0) m to (0.03, 0.006) m to (0.3, 0.06) m and, finally, to (3.0, 0.6) m (Figures 16.3*a*, *b*, *c*, and *d*), the size of the plume increases markedly. However as the plume size increases, the maximum concentration decreases. Increased dispersion mixes the contaminant with an increasing proportion of the uncontaminated water.

Because both α_L and α_T increase proportionately, we cannot separate their overall effects on concentration distributions. A series of model trials published by Frind and Germain (1986) shows how the shape of a continuous plume changes only as a function of the transverse dispersivity. Flow is downward and laterally away from a source at the left end of the system (Figure 16.4). The transport simulations on Figure 16.4 document the change in plume shape as values of α_T are increased from 0.001 m to 0.01 m to 0.1 m. For comparative purposes, the position of the front due only to advection is shown by the dashed lines. As transverse dispersion increases, the plume changes from one that is compact and not much larger than the stream tube initially containing the contaminant (Figure

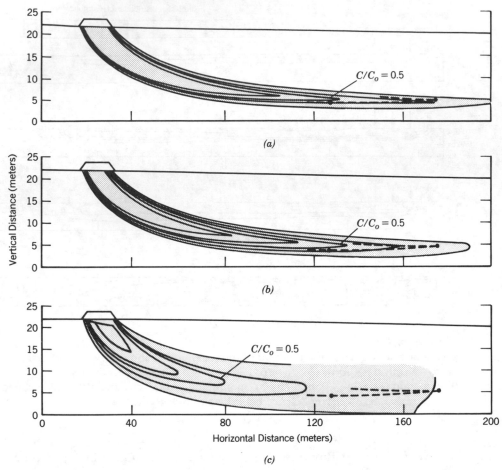

Figure 16.4

Changing plume shapes as a function of the transverse dispersivity: (*a*) $\alpha_T = 0.001$ m; (*b*) $\alpha_T = 0.01$ m; and (*c*) $\alpha_T = 0.1$ m. Dashed curves mark the position of the front due to advection alone (modified from Frind and Germain, Water Resources Res., v. 22, p. 1857–1873, 1986. Copyright by American Geophysical Union).

16.4*a*) to one that has spread laterally into adjacent stream tubes (Figure 16.4*c*). Accompanying this transverse spreading is a reduction in the overall length of the plume, exemplified by the position of the 0.5 contour line relative to the position of the advective front.

Other important groups of processes include (1) radioactive decay, biodegradation, or hydrolysis that may be represented by a simple, first-order kinetic reaction, and (2) sorption that may be characterized by an equilibrium, binary exchange reaction. Let us compare a series of plumes in which the half life of the kinetic reaction decreases from 30 yr, to 3.0 yr, to 0.33 yr (Figures 16.5*a*, *b*, and *c*). In general, the more rapidly a reaction removes a contaminant the smaller the plume will be at a given time. Thus, the smallest plume is associated with the smallest half life (Figure 16.5*c*).

Ion exchange can also produce the same dramatic attenuation in concentration. The changing plume shapes on Figure 16.6 are due to a changing selectivity coefficient (K_s). When $K_s = 0$, there is no exchange, and the contaminant moves

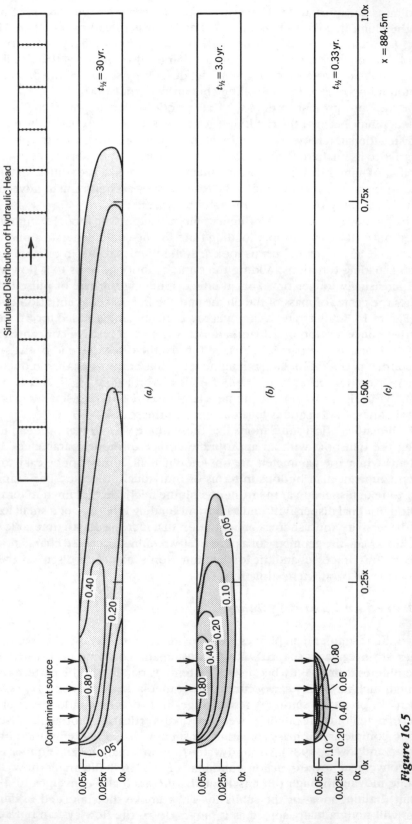

Figure 16.5

Changing plume shapes as a function of the half-life of a kinetic reaction: (*a*) $t_{1/2} = 30$ yr, (*b*) $t_{1/2} = 30$ yr, (*b*) $t_{1/2} = 3.0$ yr, (*c*) $t_{1/2} = 0.33$ yr (modified from Schwartz, 1975). Reprinted with permission of Elsevier Science Publishers from J. Hydrol., v. 27.

due to mass transport alone. However, as K_s increases over four orders of magnitude, the plume becomes smaller as exchange retards spreading (compare Figures 16.6*a*, *b*, *c*, and *d*). Other parameters that control ion exchange reactions (for example, cation exchange capacity or ion charge) also influence the distribution of contaminants. Irrespective of the model describing sorption, the process is of paramount importance in controlling contaminant transport.

We will not present examples of all reactions that might occur in a contaminant plume because the results are generally similar. The greater the tendency for any reaction to remove a contaminant from solution the smaller the plume will be relative to its unaltered size. In some cases, the geochemical processes have the capability of immobilizing particular contaminants right at the source. In terms of the transport processes as a whole, the reactions are as important as advection in finally determining the disposition of contaminants.

The loading function also influences plume shape. One modeler humorously pointed out that given a proper loading function he could generate a plume that "looked like an elephant." Let us look at some plume shapes in relation to nonconstant loading functions. Adding the same quantity of mass to a flow system over increasingly longer times or, in effect, reducing the rate of pulse loading, changes the center of mass of the plume and the internal concentration distribution. Figure 16.7*a* illustrates a case where a contaminant is added for 5.1 years at a relative concentration of 1.0 and after this time at a relative concentration of 0.0. This plume is compact and located toward the discharge end of the system. As the initial period of loading is lengthened while at the same time reducing the concentration (for example, $C_0 = 0.5$ for 10.2 yr in Figure 16.7*b* and $C_0 = 0.25$ for 20.4 yr in Figure 16.7*c*), the plume is larger in size but has a smaller maximum internal concentration and is located closer to the source.

With mathematical simulations like these, there was no real problem in controlling the transport with an appropriate choice of model parameters. In real problems where the parameters are not known at all, it may not be easy to interpret contaminant distributions in terms of individual processes. For example, a broad zone of dispersion at the front of a plume might be explained in terms of a large longitudinal dispersivity and a constant loading function, or a small longitudinal dispersivity and a source concentration that increased with time, at least initially. To explain the geometry of a plume fully requires a detailed characterization of the various processes and the loading function. Chapter 18 discusses the issue of process and parameter estimation.

Fractured and Karst Systems

Life would be simpler if all plumes were as "ideal" as those just discussed. Often, plumes are irregularly shaped with nonsystematic concentration distributions. These plumes form mainly because of extreme variability in the pattern of flow. Although such variability can occur in porous media, karst and some types of fractured rocks provide a setting where changes in hydraulic conductivity of many orders of magnitude are possible over very small vertical and horizontal distances.

The complexity of mass spreading in such a system is exemplified by karst (Quinlan and Ewers, 1985). Ground water flow in a mature karst aquifer is typically convergent toward major conduits. Figure 16.8 illustrates how diffuse recharge moving through the unsaturated zone gradually converges to a major conduit draining most of the subbasin. Thus unlike the simulated examples, a plume will not gradually spread as it moves down the flow system but actually

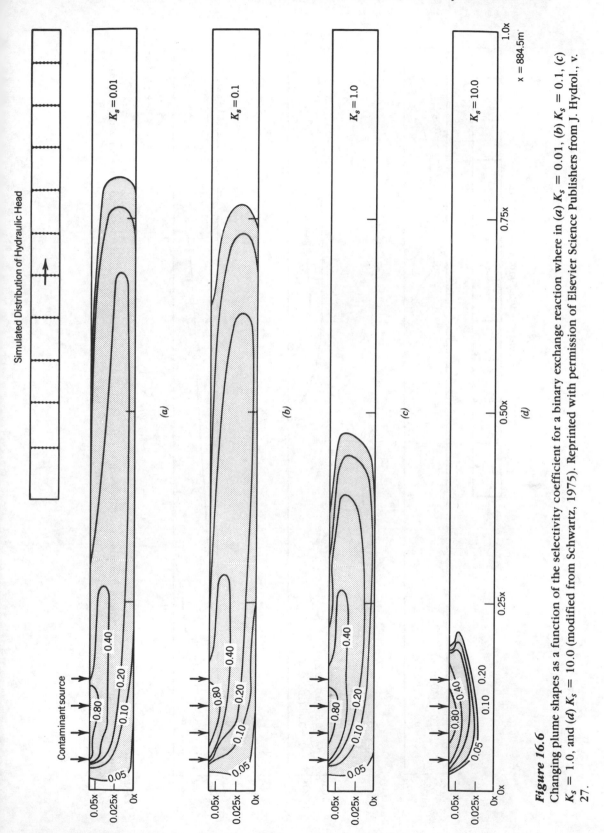

Figure 16.6

Changing plume shapes as a function of the selectivity coefficient for a binary exchange reaction where in (*a*) $K_s = 0.01$, (*b*) $K_s = 0.1$, (*c*) $K_s = 1.0$, and (*d*) $K_s = 10.0$ (modified from Schwartz, 1975). Reprinted with permission of Elsevier Science Publishers from J. Hydrol., v. 27.

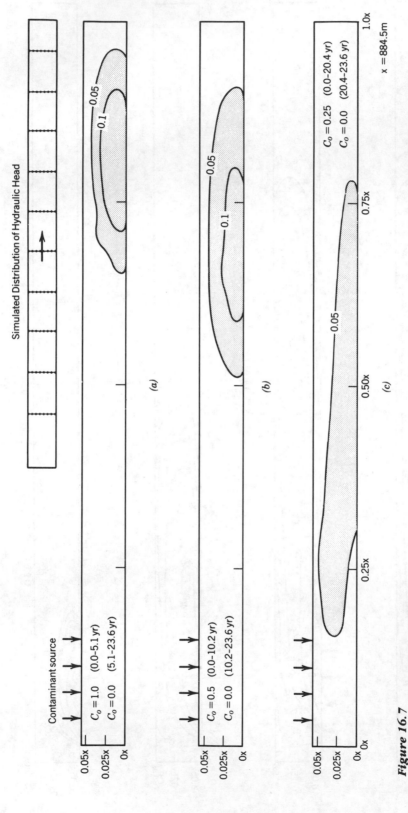

Figure 16.7

Changing plume shapes as a function of the non-constant loading rates indicated on each cross section (modified from Schwartz, 1975). Reprinted with permission of Elsevier Science Publishers from J. Hydrol., v. 27.

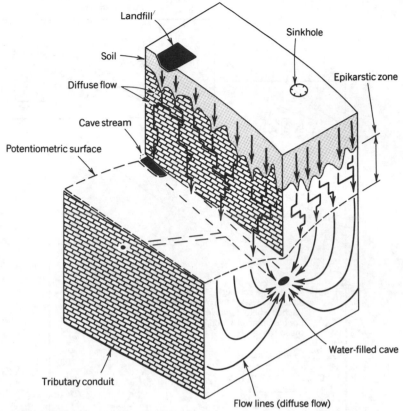

Figure 16.8
The complex pattern of flow and contaminant migration in a maturely
karsted terrane. Flow converges to a cave stream which is the ultimate
pathway for contamimant migration. Diffuse flow in the ground water
system is occurring along bedding planes and other joints (from Quinlan and Ewers, 1985). Reprinted by permission of Fifth National Symposium and Exposition on Aquirer Restoration and Ground Water
Monitoring. Copyright © 1985. All rights reserved.

converges toward a cave stream and move rapidly to the spring or springs at the
downstream end of the basin (Quinlan and Ewers, 1985). The geometry of the
network of small fractures and conduits determines the plume geometry. This
problem is in many respects more analogous to those commonly encountered in
surface water rather than ground water with rapid turbulent flow in a network of
channels and tributaries.

Babylon, New York Case Study

Now let us consider some real examples of how concentration distributions
depend on processes. At Babylon, New York (see Figure 16.9), Cl⁻, nitrogen compounds, trace metals, and various organic compounds originating from a landfill
have contaminated shallow ground water (Kimmel and Braids, 1980). Landfilling
at the site began in 1947 with urban refuse, incinerated garbage, cesspool waste,
and industrial refuse. The refuse is 18 to 24 m thick and often placed below the

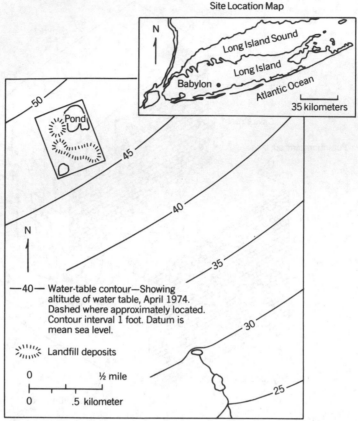

Figure 16.9
Maps showing the location of the Babylon landfill and the water-table contours in April 1974 (modified from Kimmel and Braids, 1980).

water table. The cesspool wastes were treated to some extent before being discharged into lagoons on the north and west parts of the property.

The surficial sand aquifer is approximately 27.5 m thick and has a maximum hydraulic conductivity of 1.7×10^{-3} m/s. Flow in the aquifer is generally to the south-southeast. Contaminants include major ions Ca^{2+}, Mg^{2+}, Na^+, K^+, HCO_3^-, SO_4^{2-}, and Cl^-; nitrogen species such as NH_4^+ and NO_3^-; heavy metals, particularly iron and manganese; and organic compounds.

Shown in Figure 16.10 is the Cl^- plume between 9.1 to 12.1 m below the water table. Because Cl^- does not react, the plume is a manifestation of only the mass transport processes. Of these, the most important is advection. Longitudinal dispersion is relatively significant but transverse dispersion is negligible. A three-dimensional analysis of the main portion of the Cl^- plume by Kelly (1985) determined an advective velocity of 2.9×10^{-4} cm/s, which is similar to the velocity reported by Kimmel and Braids (1980). Values of dispersivity in the longitudinal and two transverse directions (y and z) are estimated to be 18.6 m, 3.1 m, and 0.6 m, respectively. The tendency for α_x to be about six times larger than α_y and for α_z to be much smaller than either of the others is in line with expectations from carefully run tracer tests. The relatively smooth decline in Cl^- concentration away

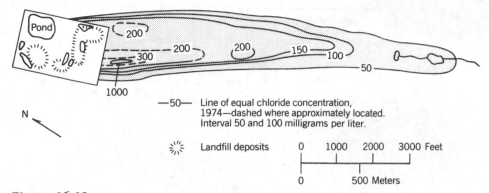

Figure 16.10
Map of the Cl^- plume in 1974 at depths below the water table ranging between 9.1 and 12.1 m (modified from Kimmel and Braids, 1980).

from the source suggests a relatively constant rate of loading. However, there is no way to establish with the given information whether the estimate of α_x contains a component due to variable source loading that might inflate the dispersivity value.

The ratio of NO_3^--N to total N (Figure 16.11) indicates that at the source most of the N is present as NH_4^+, which is indicative of reducing conditions near the landfill. The source of NH_4^+ is the microbial decomposition of organic wastes containing nitrogen. The increase in the proportion of NO_3^--N as a function of the travel distance away from the source is probably due to the oxidation of NH_4^+ to NO_3^- as mixing brings oxygen into the plume. Mapping the distribution of various nitrogen species often can be helpful in assessing redox conditions.

This reduced zone also explains why metals, particularly iron and manganese, are so mobile. The E_H-pH diagram for Fe (Figure 12.9) illustrates that with moderate reducing conditions in a pH range of 6.0 to 6.5 (as observed near the source), Fe^{2+} is the stable form of Fe. Similarly, Mn^{2+} is the stable form of manganese. The gradual transition to more oxidizing conditions down the plume creates a situation where solids (for example, $Fe(OH)_3$ and $MnOOH$) are the most stable form. The mobility of the metal ions is dramatically reduced.

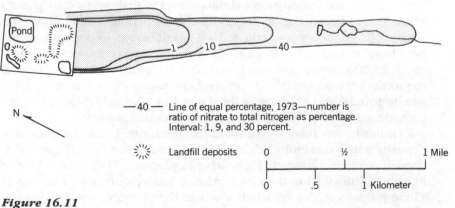

Figure 16.11
Map showing the ratio of NO_3^--N to total N expressed as a percentage. Low values close to the source reflect the presence of NH_4^+-N as the dominant N species (modified from Kimmel and Braids, 1980).

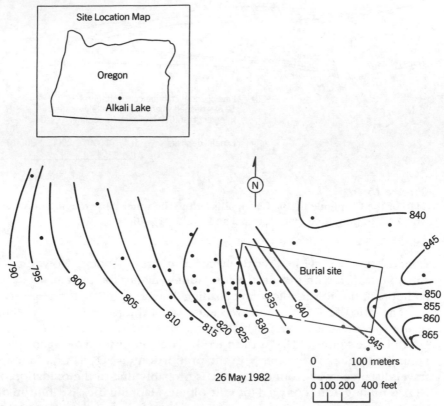

Figure 16.12

Maps showing the site location (inset) and the configuration of the water table in May 1982 (modified from Pankow and others, 1984). Reprinted by permission of Ground Water. Copyright © 1984. All rights reserved.

Alkali Lake, Oregon Case Study

The second case examines contamination of a shallow ground water system at Alkali Lake, Oregon (Figure 16.12) by chlorophenolic compounds (Pankow and others, 1984; Johnson and others, 1985). It illustrates important features of transport involving dissolved organic compounds subject to sorption. In November 1976, 25,000 barrels of wastes from the manufacture of 2,4-D (2,4-dichlorophenoxyacetic acid) and MCPA (4-methyl-2-chlorophenoxyacetic acid) were crushed and disposed of in 12 shallow unlined trenches. The eight different chlorophenolic compounds listed on Table 16.6 were of particular interest.

Ground water flows westward in fractured, fine-grained, eolian, or lacustrine deposits with a maximum thickness of approximately 30 m (Figure 16.12). These deposits contain 2.4% ± 0.3% solid organic matter. Hydraulic conductivities fall in a range from 0.01 to 0.10 cm/s, while effective porosities are from 1% to 5%. These porosity values are much less than the measured total porosities of 0.6 to 0.7. With the given hydraulic gradient, the mean ground water velocity ranges from 3.9 to 1040 cm/d.

Table 16.6

List of contaminants investigated by Johnson and others (1985) at Alkalai Lake, Oregon*

2,4-DCP	2,4-Dichlorophenol	CPP's including CL2D2, CL3D3, CL4D2
2,6-DCP	2,6-Dichlorophenol	chlorophenoxyphenol dimers with two,
2,4,6-TCP	2,4,6-Trichlorophenol	three, and four chlorines
TeCP	2,3,4,6-Tetrachlorophenol	
PCP	Pentachlorophenol	

All eight contaminants have spread to some extent. The plumes for 2,4-DCP (Figure 16.13*a*), 2,6-DCP, and 2,4,6-TCP are the largest. Spreading is less with TeCP (Figure 16.13*b*) and PCP. The CPP plumes are the smallest especially CL4D2 shown in Figure 16.13*c*. Contaminants are moving approximately in the direction of the hydraulic gradient. However, because of anisotropism, the direction of ground water flow and the gradient do not exactly coincide. The plume is elongated in the direction of the hydraulic gradient implying that advection is the dominant mechanism of spreading. The 2,4-DCP plume is unaffected by sorption or other retardation processes and thus is useful for evaluating the mass transport processes. Qualitatively, the shape of the plume indicates that longitudinal dispersion is more significant than transverse dispersion. An unpublished analysis of this plume has determined a mean flow velocity of 1.1×10^{-4} cm/s and values of α_x and α_y of approximately 7.0 m, and 1.0 m respectively.

Of the eight contaminants studied at the site, 2,4-DCP, 2,6-DCP, and 2,4,6-DCP were shown through batch experiments to have a K_d of zero. There is thus no tendency for sorption on solid organic carbon in the aquifer. In ground water of pH 10 at this site, these compounds ionize to negatively charged species. They are therefore much more mobile than their published K_{ow} values would suggest. This example points out the need to understand fully the limitations in predicting K_{oc} values from regression relationships. The other contaminants TeCP, PCP, and the three CPPs are also largely ionized in this ground water. However, these compounds do sorb to some extent on organic carbon. Even so, the retardation in these plumes is much less than predicted from K_d values determined from batch experiments. This behavior may be due in part to fractures in sediments at the site. The data from the batch experiments do predict correctly the relative order of retardation for all eight compounds. This study is instructive because it is one of the first to document compound-dependent retardation for organic compounds.

16.3 Multifluid Contaminant Problems

Some contamination problems involve two or more fluids. Examples include air, water, and organic liquids in the unsaturated zone, or organic liquids and water in an aquifer. These problems are in general more complex than are the single-fluid systems and require us to develop several basic concepts related to multifluid flow.

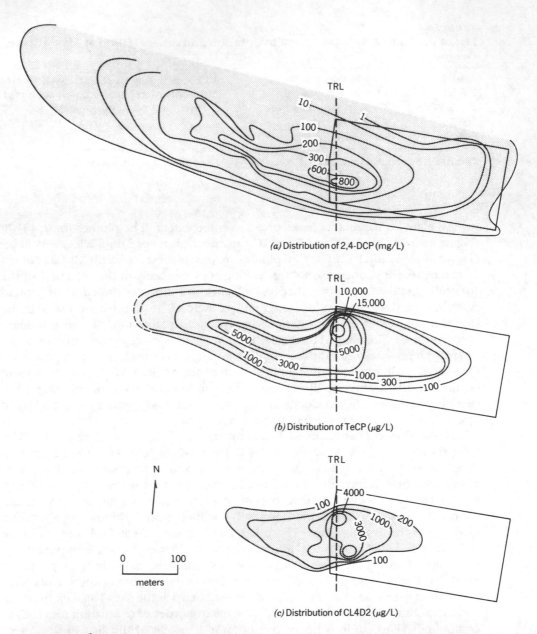

(a) Distribution of 2,4-DCP (mg/L)

(b) Distribution of TeCP (μg/L)

(c) Distribution of CL4D2 (μg/L)

Figure 16.13
Maps illustrating the pattern of spreading for (a) 2,4-DCP, (b) TeCP, and (c) CL4D2 at Alkalai Lake (modified from Johnson and others, 1985). Reprinted by permission of Ground Water. Copyright © 1985. All rights reserved.

Saturation and Wettability

Saturation describes the relative abundance of fluid in a porous medium as the volume of the *i*th fluid per unit void volume for a representative elemental volume

$$S_i = \frac{V_i}{V_{\text{voids}}} \tag{16.1}$$

where V_i is the volume of the *i*th fluid. In a multicomponent system, the sum of all the component saturations is equal to one.

Wettability is the tendency for one fluid to be attracted to a surface in preference to another. According to Demond and Roberts (1987), the only direct measure of wetting is the contact angle. The idea of the contact angle can be illustrated by an experiment in which a surface such as a piece of glass is immersed in a bath of some reference fluid such as water and a drop of the test fluid is placed on the surface. The contact angle is established by measuring the tangent to the droplet from a point where all three components are in contact (that is, surface, reference liquid, and test liquid; Figure 16.14). A contact angle of less than 90° (Figure 16.14) implies that the test liquid is wetting. A contact angle >90° implies that the reference liquid is wetting and the test liquid is nonwetting.

Wettability is unique for given types of solids and fluids. However, a few generalizations hold:

- Water is always the wetting fluid with respect to oil or air on rock-forming minerals (Albertsen and others, 1986).

- Oil is a wetting fluid when combined with air, but a nonwetting fluid when combined with water (Albertsen and others, 1986).

- Oil is the wetting fluid on organic matter (for example, peat or humus) in relation to either water or air (Albertsen and others, 1986).

- The wetting character of organic contaminants is often similar to oil.

Imbibition and Drainage

Imbibition and drainage are the dynamic processes by which fluids displace one another. Imbibition is the displacement of the nonwetting fluid by the wetting

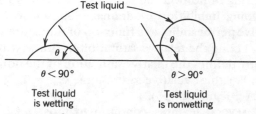

Figure 16.14
Definition of wetting on the basis of the contact angle (modified from Demond and Roberts, Water Resources Bulletin, v. 23, p. 617–628, 1987. Reproduced by permission of Amer. Water Resources Association).

fluid, and drainage is just the opposite (Albertsen and others, 1986). Thus, water being added to a dry soil or oil being gradually displaced by water from a water-wet oil reservoir are examples of imbibition. The entry of a nonwetting organic liquid into a water-wet aquifer is drainage.

It is necessary to distinguish the two processes because the physical parameters do depend on which process is operating. One parameter exhibiting hysteretic behavior is relative permeability, which we consider in detail in the next section.

Relative Permeability

The concept of relative permeability accounts for the tendency for fluids to interfere with one another as they flow (Demond and Roberts, 1987). The idea is best understood in relation to a Darcy's law expression for multifluid flow. For one-dimensional flow in a homogeneous medium, Darcy's equation written in terms of pressure gradients is (Muskat and Meres, 1936)

$$q_i = - \frac{k_i}{\mu_i} (\nabla P_i - \rho_i g \nabla h) \tag{16.2}$$

where q_i is the flow of the ith fluid per unit area of the medium, k_i is the effective permeability of the medium to the ith fluid, μ is viscosity, P is pressure, ρ is density, g is the gravitational acceleration constant, and h is elevation. The relative permeability to the ith fluid is

$$k_{ri} = \frac{k_i}{k} \tag{16.3}$$

where k_i as before is the effective permeability and k is the intrinsic permeability. Substitution of Eq. 16.3 into Eq. 16.2 yields the general form of Darcy's equation for multifluid flow (Demond and Roberts, 1987)

$$q_i = - \frac{k \, k_{ri}}{\mu_i} (\nabla P_i - \rho_i g \nabla h) \tag{16.4}$$

In multifluid systems, k_{ri} ranges between zero and one. The product $k \, k_{ri}$ represents the reduction in the intrinsic permeability because two or more liquids are present in the system. With a single-fluid system, k_{ri} is one and Eq. 16.4 reduces to the more familiar form of Darcy's equation. Exactly how k_{ri} varies between zero and one is a complex function of the relative saturation, whether the fluid is wetting or nonwetting with respect to the solids and whether the system is undergoing imbibition or drainage. Shown in Figure 16.15 are typical curves of relative permeability as a function of the saturation for wetting and nonwetting fluids. Fixing the relative saturation of the wetting fluid (for example, $S_w = 0.8$) implicitly defines the relative saturation of the nonwetting fluid (for example, $S_n = 0.2$). For these relative saturations, $k_{rw} = 0.47$ and $k_{rn} = 0.04$ (for the imbibition curve, Figure 16.15).

There are several features common to most relative permeability curves (Demond and Roberts, 1987). First, when both fluids are present, the relative permeabilities rarely sum to one. For our example saturations, the sum is about 0.5. Second, as shown on the figure, the relative permeabilities of both the wetting and nonwetting fluids approach zero at finite saturations. In other words,

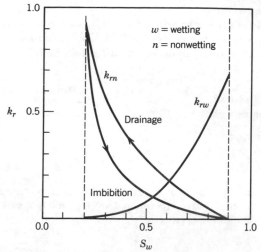

Figure 16.15
Typical relative permeability of curves (from
Demond and Roberts, Water Resources Bulletin,
v. 23, p. 617–628, 1987. Reproduced by permission of Amer. Water Resources Association).

some quantity of either wetting or nonwetting fluid in the pore system cannot
move below some saturation threshold. In Figure 16.15, the relative permeability
of the wetting fluid becomes zero at $S_w = 0.2$, and $k_{rn} = 0$ at $S_{nw} = 0.1$. These
saturations are characteristic parameters known as residual saturations. The pattern of residual saturation shown by the example (that is, $S_{rw} > S_{rn}$) is observed
frequently. A fluid at residual saturation is not capable of flow because at the low
levels of saturation the fluid is not connected across the network of pores.

Residual saturation of the wetting fluid is sometimes called pendular saturation (Albertsen and others, 1986). The fluid is held by capillary forces in the nar-
rowest parts of the pore space. Figure 16.16a provides an example of pendular
saturation with water as the wetting fluid at residual saturation and air as the nonwetting fluid. Residual saturation of the nonwetting fluid is sometimes called

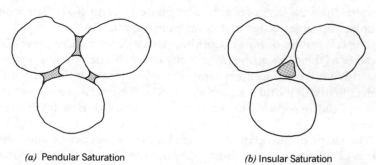

(a) Pendular Saturation   (b) Insular Saturation

Figure 16.16
Examples of different types of fluid saturation. In (a) the wetting
fluid is at residual saturation; in (b) the nonwetting fluid is at
residual saturation.

Table 16.7
Residual saturation data for a hydrocarbon in a uniform glass bead column for three fluid (vadose zone) and two fluid (saturated zone) systems (modified from Wilson and Conrad, 1984)*

Mean bead diameter (mm)	k (Darcy)	Residual saturation (%)	
		Vadose zone	Saturated zone
0.655	147	3	14
0.327	85	5	14
0.167	22	8	14

*Reprinted by permission of Petroleum Hydrocarbons and Organic Chemicals in Ground Water—Prevention, Detection and Restoration. Copyright © 1984. All rights reserved.

insular saturation. This saturation is illustrated in Figure 16.16b, where a small quantity of organic liquid is at residual saturation in a water-wet aquifer. The fluid occurs as an isolated blob in the center of the pore.

A common situation is for water, air, and a contaminant such as oil to be found together in the unsaturated zone. Water is the wetting fluid, and air is the nonwetting fluid. Oil has intermediate wetting properties. It is nonwetting with respect to water but wetting with respect to air (Wilson and Conrad, 1984). The oil at residual saturation is caught as blobs and pendular rings, trapped between the water in the small parts of pores (Wilson and Conrad, 1984).

Because a contaminant such as oil behaves differently in the unsaturated versus the saturated zone, the residual saturations differ. For uniform glass beads packed in a column, the residual saturation is two to five times larger in the saturated zone depending upon bead size (Table 16.7).

At saturations at or below the residual value, the fluid cannot flow, and will remain captured in the pore. As Chapter 19 illustrates, this persistence has important implications for spill cleanups.

Another feature of relative permeability curves is that for comparable saturations $k_{rn} > k_{rw}$. For the curves we show, at $S_w = 0.4$ and $S_n = 0.4$, the value of k_{rn} is about 0.1 (for imbibition) with $k_{rw} = 0.04$. The reason is that the wetting fluid occupies the smaller pores that contribute least to flow. The nonwetting fluid occupies the larger pores that contribute most to flow.

The last feature of note about the relative permeability curves is the hysteretic character of the nonwetting fluid. Many times this behavior is not evident because in petroleum studies authors typically only present imbibition curves for water displacing oil (Demond and Roberts, 1987). Hysteresis is evident because the nonwetting fluid occupies a different pore network during imbibition than during drainage.

The shape of the relative permeability curves depends upon several variables such as (1) intrinsic permeability, (2) pore-size distribution, (3) viscosity ratio, (4) interfacial tension, and (5) wettability. Space is not sufficient here for us to discuss these factors. Readers can refer to an excellent article by Demond and Roberts (1987) and the text by Dullien (1979).

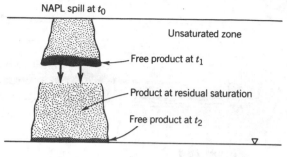

Figure 16.17
The downward percolation of an NAPL in the
unsaturated zone. As the contaminant moves
downward the quantity of mobile fluid decreases
as an increasing quantity is trapped as residual
saturation.

Features of Nonaqueous Contaminant Spreading

Consider the problem of a spill near the ground surface. With time, the contaminant percolates downward through the unsaturated zone toward the water table.
The most important process influencing the downward migration of the liquid is
flow due to a potential gradient. In the unsaturated zone, water on the solid grains
of the soil is the wetting phase and the nonaqueous phase liquid (NAPL) is the wetting phase with respect to air on the water film. Because water wets the solids, the
NAPL does not displace the water from the surface but moves from pore to pore
once saturation exceeds the residual saturation. The NAPL displaces pore water
that is not strongly held and soil gas.

Several important factors control the way in which the contaminant spreads. In
the case of a noncontinuous source (Figure 16.17), the quantity of mobile contaminant gradually decreases because some of the liquid is trapped in each pore at
residual saturation. Thus the quantity of free product reaching the water table is less
than the spill volume. If the spill is relatively small, downward percolation in the
unsaturated zone will stop once the total volume is at residual saturation. Another
pulse of NAPLs or more recharge is necessary to move the contaminants down to
the capillary fringe. The main threat to ground water from small volume spills is the
opportunity for continuing dissolution or volatilization as the spill remains.

The plume in Figure 16.17 also exhibits a tendency to spread horizontally as
it moves downward. This spreading is due to capillary forces, which operate
together with gravity forces to control migration. The presence of layers of varying hydraulic conductivity also influences NAPL spread (Farmer, 1983). Even a
relatively thin, low permeability unit (Figure 16.18) will inhibit downward percolation and force the contaminant to spread laterally. If such a layer is continuous,
the contaminant will spread only within the unsaturated zone. If the layer is discontinuous, the NAPL will eventually spill over and continue to move downward
toward the water table.

Spreading also depends on the way the spill occurs (Farmer, 1983). Releasing
a relatively large volume of contaminants over a relatively short time causes rapid
downward and lateral migration (Figure 16.19*a*). Spreading is maximized and a

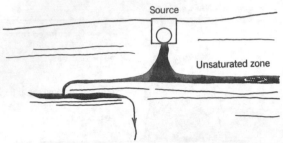

Figure 16.18
The presence of zones of low hydraulic conductivity within the unsaturated zone can cause
NAPLs to mound and spread laterally. In this case,
some of the contaminant is able to move around
the ends of discontinuous units (modified from
Farmer, 1983).

relatively large volume of residual contamination remains in what Farmer (1983)
refers to as the descent cone (Figure 16.19*a*). With slow leakage over a long time,
the contaminant moves along the most permeable pathways (Figure 16.19*b*).
These pathways can be a single channel or a more complex array of smaller channels, arranged in a dendritic pattern. Both the extent of lateral spreading and the
volume of product held at residual saturation are considerably less than with a
large volume spill. Overall, more of the mobile liquid will reach the water table
from slow leakage.

When the NAPL finally reaches the capillary fringe, the pattern of spreading
becomes more complex. Because water saturates a large proportion of the pores,
the relative permeability to the immiscible liquid declines and consequently there
is a tendency for NAPLs to spread somewhat at the top of the capillary fringe. With
LNAPLs such as gasoline, kerosine, and some oils, free product accumulates near
the top of the capillary fringe. The contaminant will flow near the top capillary
fringe once a critical thickness is achieved. The direction of flow coincides with
the gradient of the water table.

The way LNAPLs interact with the capillary fringe depends on the rate at
which product is supplied (Farmer, 1983). A large volume of fluid, reaching the
capillary fringe over a relatively short period of time, collapses the fringe and
depresses the water table (Figure 16.19*a*). The extent of depression depends upon
the quantity of product and its density. Even with this loading, the fluid spreads
as a relatively thin layer in the upper part of the capillary fringe (Figure 16.19*a*).
Alternatively, a slow rate of supply has little effect on the capillary fringe or the
configuration of the water table (Figure 16.19*b*). For spills that reach the capillary
fringe, spreading continues until the total spill is at residual saturation.

Not all contaminants are less dense than water. When DNAPLs reach the capillary fringe, they slow down (Figure 16.20). However, as the quantity of contaminant within a pore exceeds the residual saturation, downward percolation
continues. This process is similar to what happens in the unsaturated zone. However now, the water and not soil gas is being displaced. The dense fluid moves to
the base of the permeable unit because of driving forces related to the difference

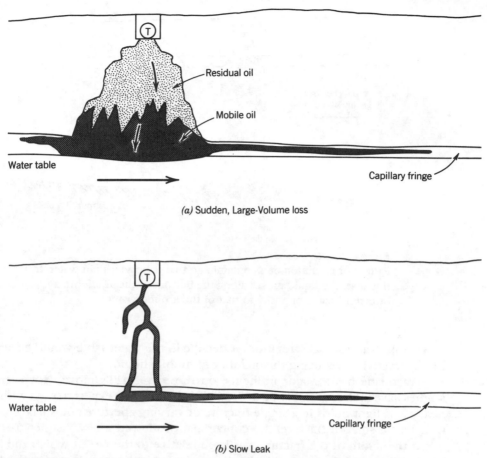

(a) Sudden, Large-Volume loss

(b) Slow Leak

Figure 16.19
Pattern of migration is determined by the nature of the spill. In (*a*), a sudden, large
volume loss results in maximum spreading, large residual saturation, and collapse of the
capillary fringe. In (*b*) a slow leak causes the product to follow a set of channels, a large
volume of product to reach the water table, and minimal disturbance of the capillary
fringe (modified from Farmer, 1983).

in density between the DNAPL and ground water. DNAPL accumulating on low-
permeability units will move downhill following the topography of the boundary.
This flow in many cases will be in a direction that is different than the ground
water. Spreading continues until the spill is at residual saturation.

Heterogeneities within the saturated zone also may cause the DNAPLs to
spread laterally. The direction as before depends upon the attitude of the unit and
the extent depends upon how continuous the unit is. Thus, in complex geologic
settings, predicting migration pathways for DNAPLs can be relatively difficult.

Before leaving this overview on NAPLs, we need to mention the influence of
water table fluctuations and fracturing on their behavior. A fluctuating water table
has the greatest influence on less dense fluids. Just above the water table, there are
both a zone of residual saturation and a lens of mobile liquid. Any movement of
the water table up or down reduces the quantity of mobile liquid and increases

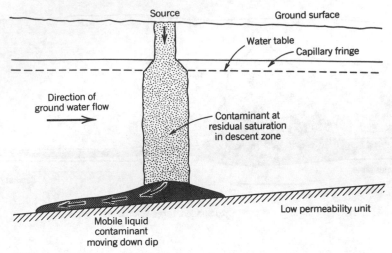

Figure 16.20
Pattern of migration of a contaminant that is heavier than water. In this case, the liquid moves along the bottom of the aquifer in a direction that is opposite to that of the ground water.

that trapped at residual saturation. A decline in the water table would be slightly more effective in reducing the quantity of mobile liquid.

Fracturing has a major influence on the way NAPLs move. However, few studies have looked at this issue in detail. Schwille (1984) experimented with both heavy and light fluids in a single fracture of varying aperture and surface roughness. For a smooth fracture, the contaminant followed a few distinct pathways when the width of the fracture was large relative to the rate at which the liquid was being loaded. Increasing the loading rate produced spreading through a larger number of pathways. This behavior is qualitatively similar to what we described earlier for porous media. A similar set of experiments carried out for rough-walled fractures resulted in a much more branched set of pathways.

Secondary Contamination Due to NAPLs

NAPLs in ground water can serve as important sources of secondary contamination. Problems develop when organic contaminants present as the free or residually saturated products partition into the soil gas through volatilization or into the ground water through dissolution. Figure 16.21*a* illustrates how even a small volume spill of a volatile organic liquid in the unsaturated zone may produce a

▶

Figure 16.21
NAPLs and DNAPLs as sources of secondary ground water contamination; (*a*) dissolution of volatile compounds from the soil gas (modified from Mendoza and McAlary, 1990), and (*b,c*) dissolution and volatilization of residually saturated fluids (from Walther and others, 1986). Panel (*a*) is reprinted by permission of Ground Water. Copyright © 1990. Panels (*b*) and (*c*) reprinted by permission of Conference on Petroleum Hydrocarbons and Organic Chemicals in Ground Water—Prevention, Detection and Restoration. Copyright © 1986. All rights reserved.

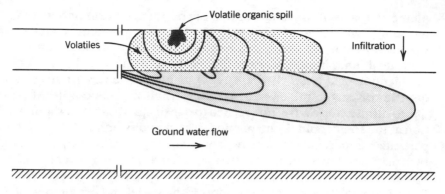

(a)

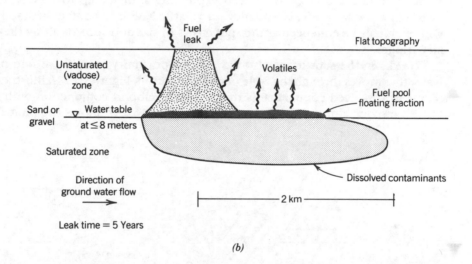

(b)

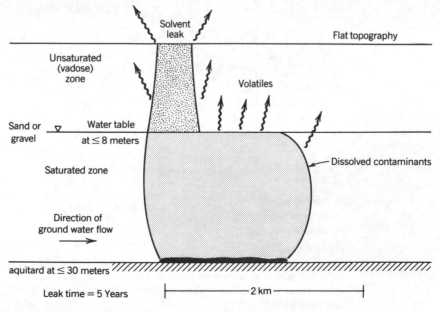

(c)

plume of dissolved contaminants that can be much more important than the original spill (Mendoza and McAlary, 1990). Volatiles spreading laterally and downward away from the source can partition into available soil moisture. Through continued infiltration, this water ultimately contaminates the aquifer.

Mendoza and Frind (1990a, 1990b) simulated the vapor transport of volatile organic solvents in the unsaturated zone. Their model considered (1) transport due to diffusion and advection related to density gradients and vapor mass generation at the source, and (2) attenuation related to dissolution into soil moisture. Of particular importance is density-driven advection caused by density gradients that may develop with dense (that is, relative to soil gas) chlorinated solvent vapors moving downward and away from a spill (Mendoza and Frind, 1990a). What is surprising about this transport is the speed and extent to which volatiles are able to spread through the soil gas system. Simulation results presented by Mendoza and Frind (1990a) showed vapor migration of up to several 10's of meters in a few weeks. This tendency for volatile NAPLs to contaminate soil gases will obviously be enhanced as the spill grows in size and spreads along the capillary fringe.

NAPLs in the subsurface also have the opportunity to dissolve into mobile water that moves through the zone of contamination. Figure 16.21*b* illustrates the plume of dissolved contaminants that could develop in conjunction with a free LNAPL moving along the capillary fringe. Figure 16.21*c* shows the plume from

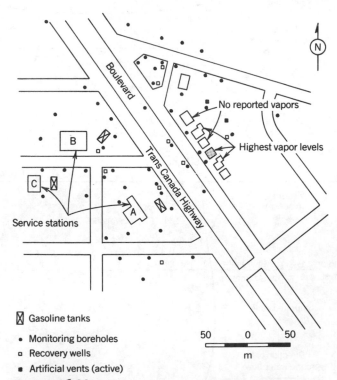

Figure 16.22
Site plan showing the location of buried tanks, basements in which gasoline fumes were detected, the location of monitoring and recovery wells, and vents (from O'Connor and Bouckhaut, 1983).

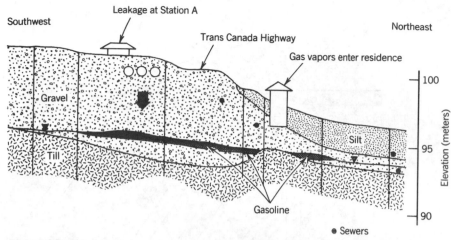

Figure 16.23
Cross section in the direction of spreading showing the general geology of the site and the pattern of gasoline spreading. The gasoline is moving downgradient along the water table, moving around the till highs (modified from O'Connor and Bouckhout, 1983).

DNAPLs at residual saturation within the cone of descent and from the free product. Experimental studies by Anderson and others (1987) and work they review show that water moving through DNAPLs at residual saturation requires only 10 centimeters or so of contact distance before saturation is reached in the ground water. Although the solubilities of volatile organic compounds (for example, benzene, carbon tetrachloride, trichloroethane) in water can be relatively low, the quantity of dissolved contaminant in water is often several orders of magnitude greater than the current standards permit (Anderson and others, 1987). Thus, these secondary sources of contamination are important in a regulatory sense.

A Case Study of Gasoline Leakage

Case histories provide a more realistic view of how NAPLs behave. The first example, summarized from work by O'Connor and Bouckhout (1983), concerns a spill of gasoline from underground storage tanks in an urban area. This case is typical of the large number of spills involving petroleum products. More than 75,000 liters of gasoline leaked from buried underground storage tanks at Station A, and possibly others nearby (Figure 16.22). The spill was first discovered when fumes leaked into sanitary sewers and the basements of houses. In the western half of the study area, from 4 to 8 m of gravel and sand overlie till (Figure 16.23). To the east, the gravel is thinner and capped by a thin silt or fine sand layer.

Gasoline occurs at the water table beneath and east of Station A, but apparently bypasses areas where the till is at elevations above the water table (Figure 16.23). Thus, the spill while generally moving down the gradient of the water table is strongly influenced by the irregular topography of the till surface. The shaded areas (Figure 16.24) illustrate places where the surface of the till is generally above the water table. The arrows are pathways of migration where the water table is located in gravel. This complex pattern of flow explains why gasoline vapors were only detected in some of the buildings and why it was difficult to

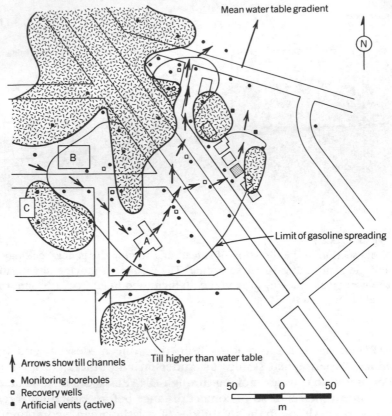

Figure 16.24
The pattern of gasoline migration in relation to the gradient on the water table. In general, the gasoline moves through the gravel, avoiding areas (hatched) where the till is at a higher elevation than the water table (from O'Connor and Bouckhout, 1983).

establish whether other stations were leaking. It also added to the difficulty in designing the collection system.

Hyde Park Landfill Case Study

The second case study, taken from Faust (1985) and Cohen and others (1987), examines the complex migration of both DNAPLs and associated aqueous phase liquids (APLs) from the Hyde Park Landfill at Niagara Falls, New York (Figure 16.25). Disposal of chemical wastes began at the site in 1954 by the then Hooker Chemical and Plastics Corporation. Ultimately, Hyde Park Landfill received approximately 80,000 tons of chemical wastes in both solid and liquid form. The wastes consist generally of chlorinated benzenes (21%), chlorinated toluenes (16%), chlorofluorotoluenes (11%), hexachlorocyclopenadiene ("C-56") (75%), trichlorophenol (4%), chlorinated organic acids, pesticides, metal chlorides, and various other chemicals. A partial list of contaminants is presented in Table 16.8. Some of the contaminants, particularly chloroform, polychlorinated biphenyls, and mirex, are of particular concern because of their toxicity and persistence. The

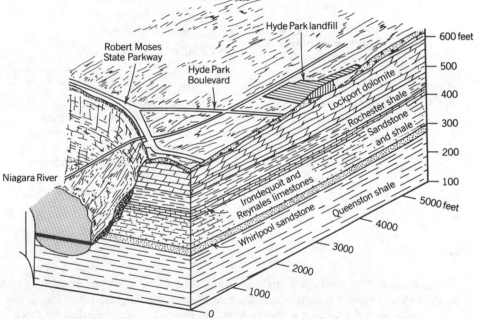

Figure 16.25
Location and geologic setting of the Hyde Park Landfill (modified from Faust, 1985).

landfill also contains an estimated 0.6 to 1.6 tons of 2,3,7,8-tetrachlorodibenzo-p-dioxin ("TCDD"), which is one of the most potent carcinogens known.

At the site less than approximately 10 m of overburden overlies bedrock. Overburden consists of silt, clay, silty clay till, and a zone of sand. The uppermost bedrock units (Figure 16.25) include the Lockport dolomite, Rochester shale, and Irondequoit/Renales limestones. The Lockport dolomite ranges in thickness from approximately 40 m immediately southeast of the landfill to approximately 20 m at the Niagara Gorge. Bedrock dips southward.

Ground water flows in a northwesterly direction toward the Niagara River in both overburden and bedrock units. The hydraulic head is higher in the overburden than the Lockport dolomite, providing a downward component of flow. The fractured upper three to 10 m of the Lockport dolomite provide a highly permeable pathway for ground water flow. Deeper parts of the Lockport are fractured but less

Table 16.8
List of contaminants identified at the Hyde Park Landfill
(summarized from Faust, 1985)

HALOGENATED ALIPHATICS	PHENOLS/CHLOROPHENOLS
chloroform	2,4,5-Trichlorophenol
tetrachloroethylene	phenol
trichloroethylene	
	BENZOHALIDES
AROMATICS	chlorobenzotrifluorides
Toluene	
Benzene	**BENZOIC ACID DERIVATIVES**
Xylenes	benzoic acid
	m-chlorobenzoic acid
CHLOROBENZENES	o-chlorobenzoic acid
Monochlorobenzene	p-chlorobenzoic acid
Tetrachlorobenzenes	
Trichlorobenzenes	**MISCELLANEOUS COMPOUNDS**
CHLOROTOLUENES	2,3,7,8-TCDD
chlorotoluenes	hexachlorocyclohexanes
dichlorotoluenes	alpha-Hexachlorocyclohexane
	beta-Hexachlorocyclohexane
	chlorendic acid
	gamma-Hexachlorocyclohexane
	(Lindane)
	polychlorinated biphenyls
	(as 1248)

permeable. The Rochester shale is considered to be a low-permeability barrier to downward flow. Ground water discharges into the Niagara River along the gorge.

Figures 16.26a and b illustrate the extent of contamination by DNAPLs and dissolved compounds in the overburden and Lockport dolomite. By far the greatest mass of contaminants is represented by the DNAPLs in the Lockport dolomite. Dense liquids from the landfill are localized within the Lockport dolomite to the top of the Rochester shale. Lateral spreading of both dissolved contaminants and DNAPLs within the overburden has been relatively limited. Much of the contamination in the overburden (Figure 16.26a) is thought to be related to surface runoff more than lateral subsurface transport.

NAPLs in the landfill and those in the Lockport dolomite are the source for dissolved contaminants that have been transported more than 750 m in the Lockport dolomite. Contaminants discharged into the Niagara River northwest of the site. However, the DNAPLs in the Lockport dolomite are not moving in the same direction as the ground water and the dissolved contaminants. These dense liquids are moving southward following the regional dip of the bedrock. This complex pattern of spreading helps to explain the differences in the DNAPL distribution and plumes of dissolved contaminants. Heterogeneity due to fracture development within the upper and lower parts of the Lockport dolomite also results in localized variability in the contaminant distributions.

16.4 Design and Quality Assurance Issues in Water Sampling

The problem of how to characterize the distribution of contaminants is of great practical concern because plume definition is such an integral part of any field study. A lesson too often learned the "hard way" is that all the basic theory and background knowledge in the world cannot correct problems with poor measurements. This section deals with network design and issues of quality assurance in sampling water.

Design of Sampling Networks

A major concern in sampling is the design of the network. Important considerations in design include the need for close interval point sampling and sample locations that take into account the character and complexity of flow. With piezometers used for measuring water levels, often little attention is paid to screen length. Consider the aquifer in Figure 16.27. As far as hydraulic head is concerned, there is little difference in whether we measure head in a small volume of the aquifer (point sample) or in a larger volume (nonpoint sample). The hydraulic head is about the same in either case (Figure 16.27a) because in a permeable unit the gradients in hydraulic head are usually small. Using the piezometers for chemical sampling, however, can produce dramatically different results (Figure 16.27b). For this example, only the point samples provide concentrations suitable for interpreting the contaminant distribution. Because of mixing within the sampling device, estimates of the plume geometry based on the nonpoint samples would be in error. Thus, all measurements not concerned simply with presence/absence indications should involve point sampling.

Because point sampling usually involves removing a small volume of water, systems should be designed with a minimum internal volume. This consideration requires a small sampling chamber or small-diameter casing (tubing) connecting the intake system to the surface. Keeping the internal volume small avoids mixing a sample with a large preexisting volume of water of different composition. Such mixing in large-diameter monitoring wells could represent a major source of error. Unfortunately, a system that is ideal for chemical sampling is usually not very good for measuring hydraulic heads with an electric tape.

The location of sampling points should account for the character and complexity of flow. In many respects, defining the vertical extent of the plume is most difficult because often plumes are not very thick. For example, Smith and others (1987) compare the vertical variation in specific conductance and nitrate concentration as deduced from four standpipe piezometers with results from 15 multiport samplers installed at the same site over an interval of about 25 m (Figure 16.28a). The loss in resolution in defining the plume with a small number of vertical samples is obvious (Figures 16.28b and c). With NO_3^-, four piezometers failed even to detect the plume (Figure 16.28c).

For determining plume boundaries, wells need be installed within and closely adjacent to the plume. One line of samplers installed along the midline of the plume will determine its overall length and thickness. Lines of sampler installed normal to the midline serve to define the plume's width and thickness. This simple design is not universally applicable. Care must be taken to design a system that is appropriate for the overall complexity of flow conditions. A good illustration of this point

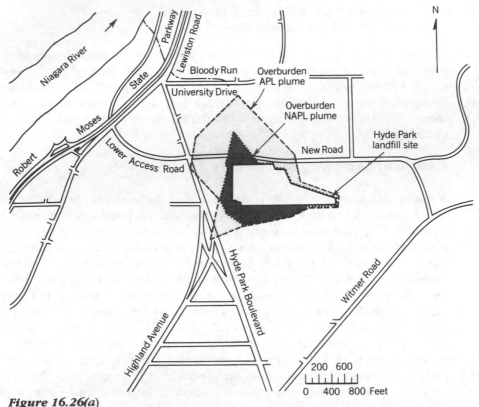

Figure 16.26(a)
NAPL and dissolved contaminant plumes in overburden (modified from Faust, 1985).

comes in the paper by Quinlan and Ewers (1985) who looked at the problem of how to define plumes in karst terranes. As we showed with Figure 16.13, contaminants from a surface source could migrate through a succession of larger and larger conduits to one large conduit possibly 3 to 10 m wide. With a simple network design as we just described, there is almost no chance of intersecting this single conduit and collecting samples along the critical pathway for contaminant migration. Quinlan and Ewers (1985) propose a site-specific monitoring scheme that requires that major elements of the karst system be defined. Spring mapping and dye tracing are some of the tools that can be used to establish connections between the disposal site, springs, and the conduit systems. Ultimately, a few critical springs and wells on the major conduits provide the proper monitoring sites.

Assuring the Quality of Chemical Data

Another concern in sampling is with the overall quality of the chemical data. Reliable chemical data are not the result of chance but the care and precautions taken in study programs. There are four potential problems that can have a serious impact on chemical data, including (1) contamination of samples with fluids that were used to drill the hole, (2) changes in water quality caused by the presence of the well, (3) sample deterioration, and (4) sloppy field and laboratory practices.

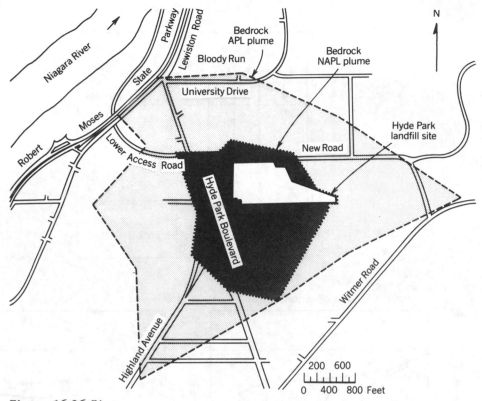

Figure 16.26 (b)
NAPL and dissolved contaminant plumes in the Lockport dolomite (modified from
Faust, 1985).

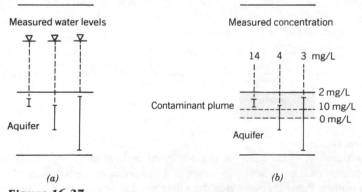

Figure 16.27
(*a*) In many cases there may not be a substantial difference in
hydraulic head values measured in an aquifer using point and
non-point sampling procedures. (*b*) Concentrations of contaminants can change markedly over relatively small vertical distances. In this example, nonpoint sampling will yield
concentrations that are much less than the true concentration.

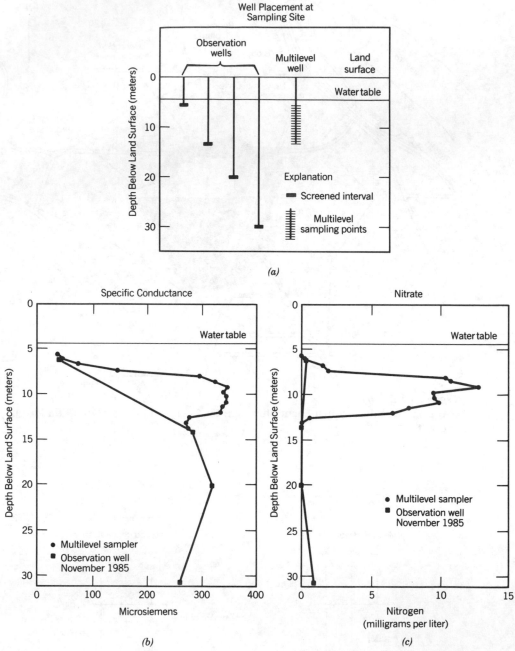

Figure 16.28
Demonstration of the need for close-interval vertical sampling. Panel (*a*) shows the
instrumentation at the test site—a set of 4 observation wells versus 15 multilevel sampling points. Panel (*b*) compares the interpreted distributions for specific conductance
and (*c*) for nitrate (from Smith and others, 1987).

The first problem has to do with the question of exactly what fluid is being sampled. Occasionally, samples will contain some of the gas, water, or oil in combination with foams, emulsions, and muds that were required to drill the hole (Hull and others, 1984). Once these exogenous fluids are present in the subsurface it is often difficult to remove them completely. One way to avoid sampling drilling water is to develop the well. A tracer, added to the drilling fluid, can be monitored as the well is developed to check that development is complete.

Changes in the chemistry of the ground water may also be related to the presence of the well. The most serious potential problem arises from the use of cement (or other soluble materials) as a seal above the intake of a well or piezometer. High-sample pHs (for example, above 9) mostly reflect a continuing interaction between the cement seal and the water being sampled. This problem can persist for a long time and is difficult to fix. Samples can also be affected when water inside or outside of the well casing dissolves small quantities of the casing or glues used at joints. A possible solution to these problems is to use threaded plastic casing or at least remove several standpipe volumes of water before sampling. A proper selection of materials for the sampling device is the only sure fix for these problems.

Another problem is sample deterioration. Once the temperature, pressure, and gas content of a water sample change following collection, the chemistry can change as well. For example, the addition of oxygen to an anoxic sample results in a rapid change in E_H that can cause dissolved trace metals to precipitate as solids. Loss of CO_2 can raise the pH and decrease the HCO_3^- concentration and in some cases cause $CaCO_3$ to precipitate. The extent of the change in composition depends in many respects on the chemistry of the particular sample. These problems have been addressed by specialized sampling equipment for high-pressure–high-temperature environments (Hull and others, 1984), the use of flow cells at the well head (Jackson and Inch, 1980), and methods of sample treatment to avoid deterioration. In designing a sampling program, it is a good idea to refer to information on standard practice, for example, Rainwater and Thatcher (1960), Ball and others (1976), Wood (1976), Gibb and others (1981), Scalf and others (1981), Claasen (1982), and Lico and others (1982). Methods for preserving samples containing organic compounds are described by Goerlitz and Brown (1972), Wershaw and others (1983), and Minear and Keith (1984).

Problems can also be related to the handling of samples in the field or laboratory analyses. In many studies, these problems go unrecognized and have an important impact on the quality of the chemical data. Sample contamination caused by improper bottle washing, filtering, or the use of impure preservatives is the main concern in handling samples in the field. It is prudent to check for this problem on an ongoing basis by running blanks of ultrapure distilled water through the various field treatments and submitting these samples to a laboratory. In the case of organic compounds, adsorption onto containers or filtering devices and loss of volatiles from the sample can also pose problems (Gibb and Barcelona, 1984; Jackson and others, 1985).

Likewise, the quality of the laboratory analyses should be checked on an ongoing basis. Some common approaches are (1) submitting ''spiked'' samples with known concentrations, (2) submitting duplicate samples to different laboratories, and (3) submitting replicate samples. The first two of these checks generally assess the analytical accuracy of the laboratory, while the third tests analytical repeatability. Guidance for developing quality assurance programs is

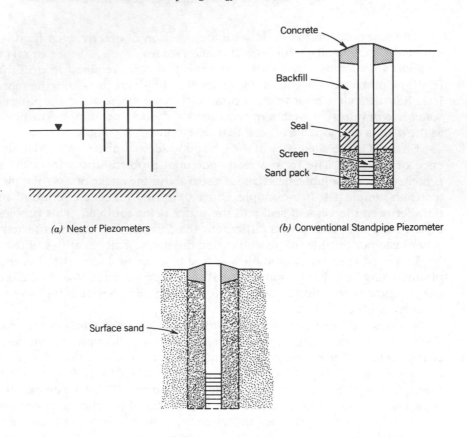

(a) Nest of Piezometers

(b) Conventional Standpipe Piezometer

(c) Piezometer Installed in Caving Materials

Figure 16.29
Basic designs for standpipe piezometers. (*a*) illustrates the concept of how a nest of piezometers provides spatially distributed concentration. Panels (*b*) and (*c*) are examples of typical piezometers for noncaving and caving materials (modified from Cherry, 1983).

provided by Wilson (1952), the American Chemical Society (1980), and Friedman and Erdman (1982).

16.5 Sampling Methods

Solutes in the Zone of Saturation

Sampling systems capable of providing point samples of fluid from the zone of saturation include (1) nests of conventional standpipe piezometers, (2) various multilevel devices installed in a single borehole, and (3) a packer arrangement that can be moved to various positions in an uncased borehole in rock or cohesive sediments (Cherry, 1983).

A nest of conventional piezometers can provide vertically spaced water samples (Figure 16.29*a*). The advantage of completing each piezometer in a separate borehole is that there is less difficulty in installing seals as compared to multilevel samplers installed in one borehole. Another important advantage of this system is the ease in measuring water levels. These installations are also extremely durable with little possibility of failure once the standpipes have been emplaced. The main disadvantage of piezometers for sampling is the higher cost of drilling additional boreholes versus the installation of many multilevel samplers in a single borehole.

Figure 16.29*b* illustrates a conventional standpipe piezometer. The design shown in the figure with filter sand placed around the intake is most appropriate for shallower depths. When these samplers are installed in relatively deep boreholes, the seal is set above a metal-petal basket (or similar device) rather than a sand pack. In deep boreholes, placing sand around the intake is much more difficult than at shallow depths. Standpipes placed in deposits that cave are often installed without a filter sand or seal (Figure 16.29*c*). They are usually emplaced using hollow stem augers, which hold the borehole open until the casing is run in to the proper depth. A variation on this design involves placing a bentonite seal above the intake by passing material through the hollow stem auger. While a good idea, this installation procedure is tedious and may not work.

The concept of multilevel sampling involves placing several samplers at various depths in a single borehole (Cherry, 1983). The advantage of this method is the relatively large number of discrete sampling points that can be installed at a relatively low cost. The method is most economical for near-surface investigations in noncohesive sand or silts, where filter sands and seals are not required. For cohesive deposits, it is necessary to place seals between intakes. Whether or not filter sand is required depends upon the type of seal. Extra difficulties related to placing the seals and filter sands correctly can affect the cost and reliability of these installations. The small size of casing typically used in these samplers makes it difficult to develop the screen and sandpack and may promote plugging. Further, the design of an appropriate sounder for measuring water levels in these samplers can challenge one's ingenuity. However, the small internal volume of these devices is a significant benefit. Because small quantities of water are required for flushing the system and the actual sample, sampling will probably not perturb the contaminant plume.

A variety of multilevel devices (from Cherry, 1983) are shown in Figures 16.30*a*, *b*, *c*, *d*, and *e*. These can be installed in cohesive or noncohesive units, but the examples show either one or the other type. The first system (Figure 16.30*a*) consists of a large number of small-diameter standpipes placed in a single borehole. The main requirement for installing this system is patience in carefully placing the filter sand and seals.

Boreholes that cave usually require a hollow stem auger to install the multilevel device. To facilitate the installation, the standpipes (sometimes polyethylene tubing) are bundled around a more rigid casing (Figure 16.30*b*). Another variation on this design places the tubing inside a solid casing to protect the tubing against damage (Figure 16.30*c*). When there is some possibility of units not caving to prevent vertical communication in the borehole, seals need to be provided between the intakes. Cherry (1983) discusses ways of installing these seals.

Another innovative procedure involves sampling through specially designed valved couplings (Figure 16.30*d*) that join sections of PVC or stainless steel casing. The sampling device passes through a valve or port to sample fluids outside of the casing. Another tool containing a downhole transducer measures hydraulic head through these same ports. Packers prevent communication in the borehole between the sampling ports. To reduce the complexity of the installation, a

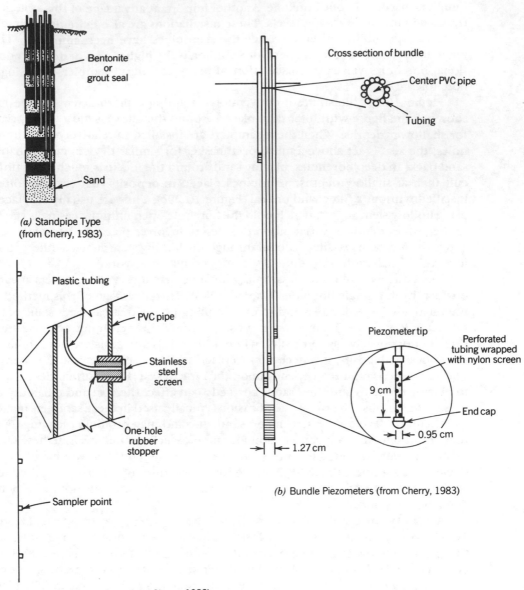

(a) Standpipe Type
(from Cherry, 1983)

(b) Bundle Piezometers (from Cherry, 1983)

(c) Suction Type (from Cherry, 1983)

Figure 16.30
Examples of multilevel sampling devices. Portions of panel *(e)* are reprinted by permission of Ground Water Monitoring Review. Copyright © 1981. All rights reserved.

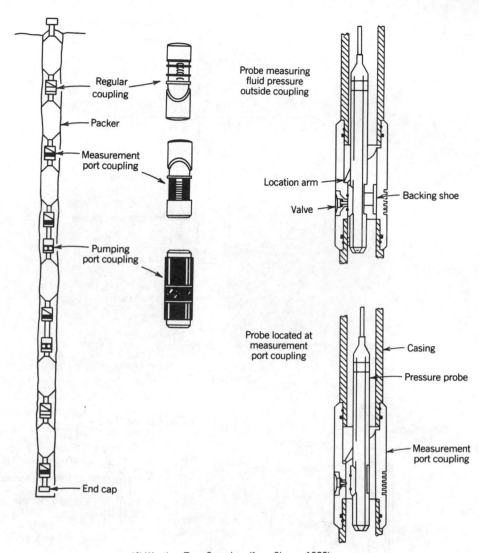

(d) Westbay-Type Samplers (from Cherry, 1983)

Figure 16.30d

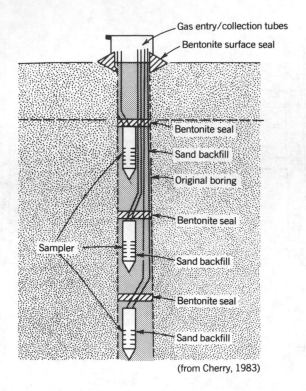

(from Cherry, 1983)

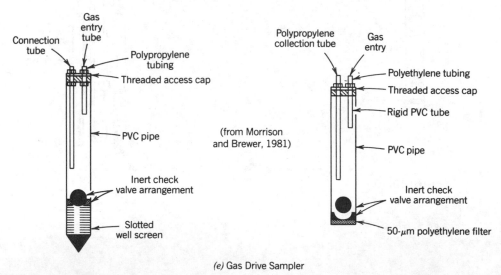

(from Morrison and Brewer, 1981)

(e) Gas Drive Sampler

Figure 16.30e

Figure 16.31
A technique for collecting water samples in a cased or uncased borehole (from Cherry, 1983).

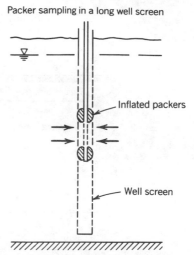

special tool inflates the packers through the casing. A stainless steel version of this sampling system is particularly well suited for deep boreholes in bedrock.

The final type of multilevel device is what Cherry (1983) refers to as gas-drive samplers. They consist of a sample chamber with narrow-diameter tubes connecting the device to the surface (Figure 16.30*e*). Compressed gas forced down one of the tubes will empty the sample chamber for flushing or sampling. The figure illustrates the completion details for these devices.

It is possible to sample continuously in an uncased borehole following a procedure that is the ground water equivalent of a drill-stem test (Figure 16.31). The method is attractive for reasons of cost. Unfortunately, experience has found the method to be unreliable, especially if some time passes between drilling and sampling. The main problem is the possibility for fluid communication within the borehole. After a short time, fluid migration can significantly alter the pore water chemistry. Attempts have been made to remedy this problem by installing removable plugs in the borehole between sampling trips. However, some mixing will still occur. We would recommend that this sampling scheme be avoided and be replaced by one of the multilevel approaches.

Nonaqueous Phase Liquids in the Zone of Saturation

Characterizing the distribution of LNAPLs requires a different sampling approach. Instead of collecting samples for analysis, a specially designed sounder measures the thickness of LNAPL floating on the water column in a water table observation well. This measurement provides the ''apparent thickness'' of spilled product, which is often larger than the ''true'' product thickness.

The difference between apparent and true thicknesses can be explained in the following way. An LNAPL floats near the top of the capillary fringe above the water table (Figure 16.32*a*). The product flows into the well above the water table and fills the casing between the top of the layer of product and the water table (Figure 16.32*a*). Thus, the apparent thickness reflects the thickness of the capillary fringe in addition to the true product thickness. The weight of the product also forces down the water level in the well, increasing the apparent product

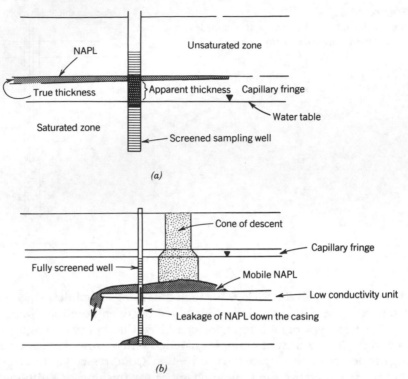

Figure 16.32
Problems associated with sampling NAPLs. In panel (*a*) product near the top of the capillary fringe moves down to the free surface in the well, making the apparent thickness much greater than the true thickness. Non point sampling with the well in panel (*b*) shows the heavy NAPL at the bottom of the samples rather than its actual position within the aquifer.

thickness even more. With large accumulations of LNAPL, collapse of the capillary fringe reduces the discrepancy between the apparent and true product thicknesses. The ratio of apparent product thickness to true product thickness could range from two to four. However, this ratio will be unique to each site, a function of the product and the porous medium.

Because of the difficulty in establishing the true product thickness, most investigators never really worry about it. Apparent product thickness is used as a relative measure. If estimates of true thicknesses are required, a variety of testing methods are available including well tests (Gruszczenski, 1987; Hughes and others, 1988; and Sullivan and others, 1988) and dielectric well logging (Keech, 1988).

A variation of the same problem manifests itself with DNAPLs. The combination of a long screened interval in a well and the physical characteristics of the contaminant also can result in a significant error in establishing where the contaminant is located in the aquifer. DNAPLs entering a well from zones in the upper part of an aquifer will flow down to the bottom of the well and out (Figure 16.32*b*). The well acts as a permeable conduit, much like a large fracture, to help the fluid move to the bottom of the aquifer. Any hope of establishing the distribution of heavy NAPLs with certainty requires properly designed, point sampling systems.

Figure 16.33
A typical device for collecting soil-water samples from the unsaturated zone (from Soil-moisture Equipment Corp). Reprinted with permission.

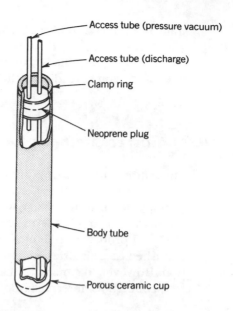

Access tube (pressure vacuum)

Access tube (discharge)

Clamp ring

Neoprene plug

Body tube

Porous ceramic cup

Dissolved Contaminants and NAPLs in the Unsaturated Zone

Nearly all sampling techniques in the unsaturated zone involve pulling the fluid under vacuum through a porous ceramic cup into a container (Figure 16.33). Once a sufficient volume of fluid has been collected, the sample is removed by suction or by gas displacement depending upon how deep the sampler is installed. Signor (1985) describes varieties of these samplers.

These devices function in both unsaturated and saturated zones. This characteristic makes them useful for sampling when the state of saturation changes frequently. An example might be a facility for artificial recharge that is operated with only periodic flooding of the infiltration basins. The main problem with this method is the possibility for the fluid composition to change from reaction with the ceramic cup or from the loss of gases as the sample is collected under vacuum. These effects can be examined in the laboratory and, it is hoped, accounted for in the interpretation of results.

Solid and Fluid Sampling

Sometimes it is worthwhile to collect samples of both the porous medium and the fluids by drilling and coring. Fluids squeezed out of the sample can be analyzed in the normal manner. Another approach is to determine contaminant concentrations in the dried solid and report contaminant concentrations as mass of contaminant per unit mass of solid.

The main advantages of this sampling technique are, first, the spatial control in being able to relate specific concentrations of contaminant to specific lithologies, and, second, the ability to minimize contamination due to the sampling method. In all the direct sampling approaches we discussed earlier, the possibility exists for a "memory effect." Some of the contaminants brought into the sampling device at an earlier time may remain sorbed on the internal surfaces. These compounds can desorb during later sampling and result in errors. The approach of solid and fluid sampling can eliminate this possible problem. The main limitation with this approach is the expense of bringing a rig back onto the site if more than one round of sampling is required. In addition, too much drilling may contribute to contaminant spreading if the boreholes are not carefully plugged.

16.6 Indirect Methods for Detecting Contamination

The indirect methods for plume delineation can provide a rapid and inexpensive alternative to conventional sampling. These approaches are successful when contamination is manifested in a secondary way, for example, by the presence of volatile organic compounds in the soil gas found above a spill or the change in electrical properties of a unit caused by the presence of the plume. A variety of published case histories point to the usefulness of these approaches in site evaluation. However, like many of the indirect approaches, they do not always provide an unambiguous interpretation.

Soil-Gas Characterization

Characterizing the composition of soil gases has emerged as an excellent technique for tracing volatile organic compounds in ground water. The approach involves defining zones of ground water contamination based on the presence of volatile components in the soil gas. With time, any volatiles at the capillary fringe partition into the soil gas and gradually diffuse upward to the ground surface. The presence of volatiles is established commonly by collecting soil gas from some fixed depth and analyzing the sample with a gas chromatograph. However, this approach does not detect all organic contaminants because not all are volatile. In addition, the organic compound should not be too soluble. Highly soluble volatiles moving through the unsaturated zone will dissolve into any water present. Reisinger and others (1987) illustrate the typical components of gasoline, which might be successfully detected in a soil-gas survey (Figure 16.34).

A soil-gas survey is conducted by driving a hollow metal probe with a perforated tip from 2 to 4 m into the unsaturated zone and extracting small samples of soil gas by pumping. Portable equipment analyzes samples on-site and provides an estimate of the mass of a particular contaminant per volume of soil gas (for example, µg/L). When plotted on a map, these data may delineate the zones of contamination. By relating contaminant concentrations in the soil gas to measured concentrations in the ground water, the soil-gas data can be transformed to provide a quantitative estimate of concentrations in ground water.

Malley and Bath (1985) describe a variation on this approach, which involves sampling soil gases with a static collector. This device is constructed by applying a small quantity of activated charcoal to the tip of a ferromagnetic wire. The charcoal is cleaned by heating and kept clean by storage in an inert atmosphere. Once buried in a shallow auger hole and left for several weeks, the collector equilibrates with the soil atmosphere. A grid of samplers provides an estimate of the spatial variability. Once the collectors are returned to the laboratory, the volatiles are desorbed and analyzed using mass spectrometry. This analytical procedure provides for the detection and identification of compounds up to mass 240 (Malley and Bath, 1985). Maps of mass fluxes (or relative ion counts) provide an estimate of the plume position.

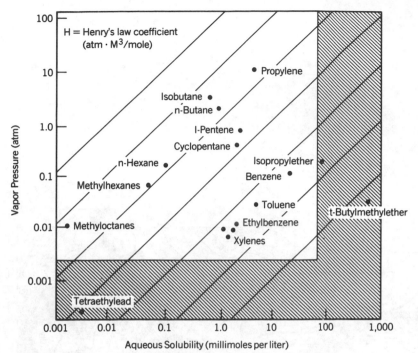

Figure 16.34
The unshaded area shows the range in vapor pressure and aqueous solubility for the common constituents in gasoline for which soil-gas surveying will be successful (from Reisinger and others, 1987). Reprinted by permission of First Outdoor Action Conference on Aquifer Restoration, Ground Water Monitoring and Geophysical Methods. Copyright © 1987. All rights reserved.

    Both these approaches provide the possibility of sampling at a much larger number of locations than is economically feasible with direct ground water sampling. This feature, coupled with the relatively short time necessary to run a sampling program, makes the technique attractive for reconnaissance studies. Knowing what volatile contaminants are present and their approximate location is extremely useful in designing a conventional sampling program. In some studies, the information provided by a soil-gas survey may be all that is required to establish whether or not a contamination problem exists, or to identify a source.

    There are some circumstances when these approaches may not work well. A low-permeability layer in the unsaturated zone can inhibit the upward diffusion of vapor and promote extensive horizontal spreading. This situation will produce an estimated distribution of contaminants that is much larger than the actual one. Other problems relate to how the contaminant occurs. If the contaminant occurs as an LNAPL, vapor concentrations in the unsaturated zone will be larger and easier to detect than if the contaminant is dissolved in the ground water (Reisinger and others, 1987). Furthermore when contaminants are present in localized fracture zones (for example, in sandstone or limestone), the rates of diffusion away from the fractures may be so slow that vapor phase transport is limited in extent (Reisinger and others, 1987). Thus, as in all sampling, care must be exercised in interpreting the results, and where necessary conclusions should be confirmed using an independent approach.

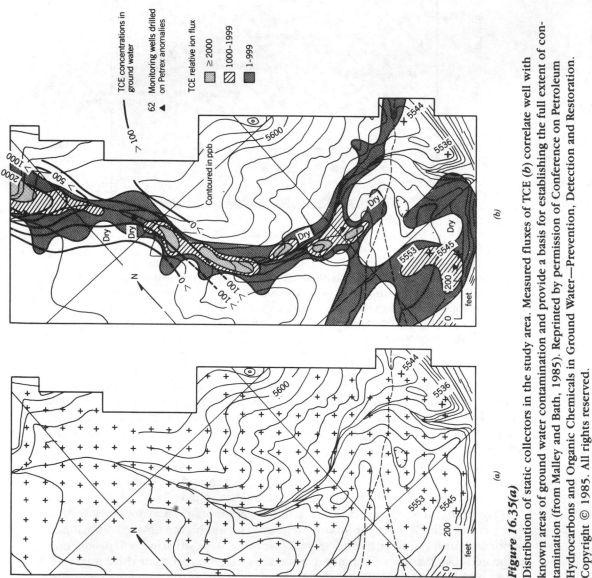

Figure 16.35(a)

Distribution of static collectors in the study area. Measured fluxes of TCE (*b*) correlate well with known areas of ground water contamination and provide a basis for establishing the full extent of contamination (from Malley and Bath, 1985). Reprinted by permission of Conference on Petroleum Hydrocarbons and Organic Chemicals in Ground Water—Prevention, Detection and Restoration. Copyright © 1985. All rights reserved.

Malley and Bath (1985) present a case study that demonstrates the potential of soil-gas sampling for delineating contaminated ground water. They reexamine a problem of contamination due to trichloroethylene (TCE) using a static sampling approach. Shown on Figure 16.35*a* is the dense array of collectors installed at the site. As expected, the survey confirmed the presence of TCE previously found in the ground water (near top Figure 16.35*b*). In addition, the study inferred the southeasterly spread of TCE beyond the area previously studied. The installation of several monitoring wells subsequently confirmed the existence of this larger zone of contamination.

Geophysical Methods

Geophysical methods have been used for years as a standard investigative tool for plume delineation. Usually surface resistivity or electromagnetic conductivity methods are used to delineate boundaries between waters of different chemical composition. The greatest success has been in mapping saline plumes associated, for example, with the leakage of brine from storage ponds and the migration of leachates from sanitary landfills.

One way of measuring resistivity is from the surface by adding electrical current to the ground with a pair of electrodes and measuring the resulting drop in voltage with a second set of electrodes (Figure 16.36). Apparent resistivity values

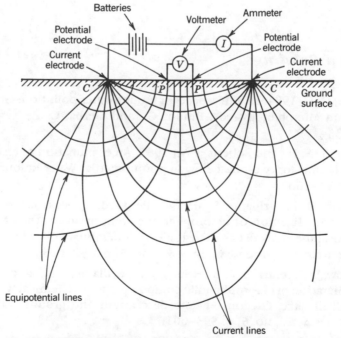

Figure 16.36
A surface technique for measuring the resistivity and outlining some types of contaminant plumes (from Todd, D.K., 1980, Groundwater Hydrology). Copyright © 1980 by John Wiley & Sons, Inc. Reprinted by permission of John Wiley & Sons, Inc.

are either read directly from the instrument or calculated for a given arrangement of electrodes. Lateral profiling is the most common survey. It is run by maintaining a fixed electrode spacing (producing a constant depth of current penetration) and moving from station to station following a grid. Anomalies on the resistivity map may coincide with contaminant distributions.

Electromagnetic conductivity methods are also designed to identify anomalous zones in the electrical properties of an area. However, the approach is very different from a conventional resistivity survey. An electrical magnetic field is directed into the ground to induce a secondary field, which is measured by a receiver. This technique provides an estimate of the bulk electrical conductivity. Because the field is induced in the subsurface, it is not necessary to hammer electrodes into the ground as before. This feature speeds up the surveys and makes it possible to survey some sites where resistivity surveying may not be possible (sites with very dry surface soils, frozen ground, and near-surface conductors such as railway tracks or pipes).

These electrical approaches often fail to delineate zones of contamination. The commonest problems are (1) an insufficient contrast in electrical conductivity or resistivity between the plume and the native ground water; (2) a plume that is small relative to the volume of rock or sediment being tested (the plume is lost by weighted averaging); (3) natural variability in geology or water chemistry, which makes it difficult to identify the plume; and (4) problems on the surface that directly interfere with running the survey. Because of these potential problems, data from electrical surveys should never be used by themselves to define a plume without confirmation from direct sampling procedures.

Suggested Readings

Craun, G. F., 1984, Health aspects of groundwater pollution: Groundwater Pollution Microbiology, eds. Bitton, G., and Gerba, C. P., New York, Wiley, p. 135–179.

Demond, A. H., and Roberts, P. V., 1987, An examination of relative permeability relations for two-phase flow in porous media: Water Resources Bull., v. 82, p. 1055–1062.

Johnson, R. L., Brillante, S. M., Isabelle, L. M., Houck, J. E., and Pankow, J. F., 1985, Migration of chlorophenolic compounds at the chemical waste disposal site at Alkali Lake, Oregon –2. Contaminant distributions, transport and retardation: Ground Water, v. 23, no. 5, p. 652–665.

Pankow, J. F., Johnson, R. L., Houck, J. E., Brillante, S. M., and Bryan, W. J., 1984, Migration of chlorophenolic compounds at the chemical waste disposal site at Alkali Lake, Oregon –1. Site description and ground-water flow: Ground Water, v. 22, no. 5, p. 593–601.

Quinlan, J. F., and Ewers, R. O., 1985, Ground water flow in limestone terranes: Strategy, rationale and procedure for reliable, efficient monitoring of ground water quality in Karst areas: Proc. 5th National Symposium and Exposition on Aquifer Restoration and Ground Water Monitoring, National Water Well Assoc., Dublin, Ohio, p. 197–234.

Schwartz, F. W., 1975, On radioactive waste management: An analysis of the parameters controlling subsurface contaminant transfer: J. Hydrol., v. 27, p. 55–71.

Problems

1. Match the breakthrough curves given on the diagram with the proper description given here.

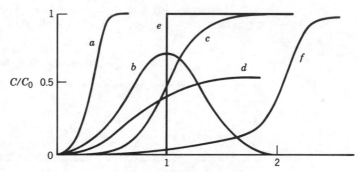

Number of pore volumes[1]

[1] A pore volume in a column is the volume of water that will completely fill all of the voids in the column.

_____ This curve was obtained for continuous injection in a one-dimensional laboratory column where the constituents in the displacing fluid were being sorbed by the medium.

_____ This curve was obtained for continuous injection in a one-dimensional column with such a low velocity that plug flow (no dispersion) was observed.

_____ This curve was observed for a pulse (instantaneous) type source with both longitudinal and transverse dispersion.

_____ This curve was obtained for continuous injection in a one-dimensional (no transverse dispersion) laboratory column.

_____ This curve was obtained for continuous injection with both longitudinal and transverse dispersion.

2. On diagrams (*a*) and (*b*), illustrate what the pattern of contaminant distribution might look like for the organic compounds with the specified chemical properties. Consider all important spreading mechanisms.

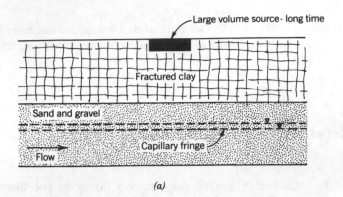

(a)

2(a) Gasoline Leak over Fissured Clay

Two soluble and volatile components of concern in gasoline

benzene— log K_{ow} = 2.04: solubility 1780 mg/L
ethylbenzene— log K_{ow} = 3.15: solubility 140 mg/L

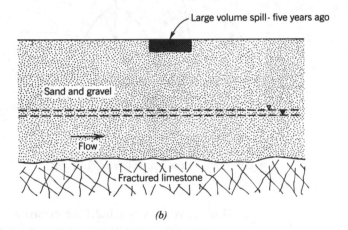

(b)

2(b) PCB Spill over Sand and Gravel Aquifer

PCB's: slightly soluble in water; log K_{ow} = 4 to 6; not volatile;
specific gravity = 1.5

3. For the spill in (*a*), develop a chemical sampling network that could be used in the field to describe a complex problem of this kind. Add notes to a copy of the figure to document your answer.

4. Shown on the accompanying figure is a series of plumes from a sanitary landfill. Examine these plumes in detail and answer the following questions.

    a. Evaluate in a qualitative manner the extent to which advection and dispersion are important in controlling contaminant spread at the site.

    b. Given the type of source and the resulting plume shapes, what can you say about the type of source loading?

c. Suggest what processes could be operating to cause the pH to increase away from the source.

d. Metals, for example Fe, tend to be relatively abundant in landfill leachates. However, Fe^{2+} is often strongly attenuated relative to mobile species like Cl^-. Explain why iron behaves in this way.

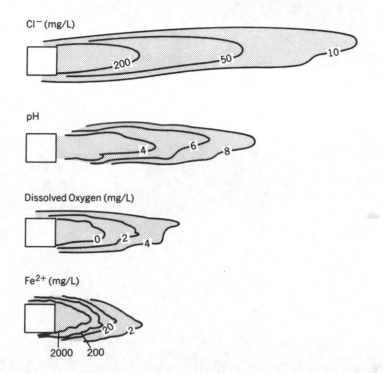

5. Some preliminary studies had indicated that the contamination problem identified in problem 4 had developed in a surficial sand aquifer about 15 m thick. Develop a plan for a detailed site investigation that involves both direct and indirect methods of detecting contamination. The plan should provide specific details about the types of instrumentation/surveys required and plans for water-quality analyses.

Chapter Seventeen

Modeling
Contaminant Transport

17.1 *Analytical Approaches*

17.2 *Programming the
Analytical Solutions
for Computers*

17.3 *Semianalytical Approaches*

17.4 *Numerical Approaches*

17.5 *Case Study in the
Application of a Numerical
Model*

This chapter is concerned exclusively with modeling the spread of contaminants dissolved in water. The starting point in modeling is with the various mass transport equations from Chapter 13. These equations describe in mathematical form the mass transport processes. Models are available to consider multiphase problems of contamination involving, say, gasoline and water or DNAPLs and water. However, the topic is sufficiently complicated to leave to more advanced texts.

17.1   *Analytical Approaches*

Chapter 16 explained the nature of contaminant plumes as a manifestation of physical and chemical processes. The approach was a conceptual one, simply illustrating what effect processes could have on contaminant transport. An alternative way is to study processes quantitatively with analytical models. Because analytical solutions abound in the literature, the discussion here will focus on the more practical ones that can be used in the field. Normally, this would preclude all one-dimensional solutions. However, in some cases, the pertinent features of a one-dimensional solution may be embedded in a more complex three-dimensional equation. Take, for example, three one-dimensional solutions of the form

$$\frac{C(x,\,t)}{C_0} = F_1(\alpha_x,\,x,\,t); \qquad \frac{C(y)}{C_0} = F_2(\alpha_y,\,y); \qquad \frac{C(z)}{C_0} = F_3(\alpha_z,\,z)$$

where C is the concentration, C_0 is the source concentration, α_x, α_y, and α_z are the dispersivities, x, y, and z are spatial coordinates, and t is time. If the three one-dimensional solutions can be found, an approximate solution to a three-dimensional problem in some cases will be

$$\frac{C(x,\,y,\,z,\,t)}{C_0} = F_1(\alpha_x,\,x,\,t)\,F_2(\alpha_y,\,y)\,F_3(\alpha_z,\,z)$$

In words, the three-dimensional solution is the product of the one-dimensional solutions. Thus, understanding the three-dimensional result follows from knowledge about the one-dimensional results. This is exactly the approach we will follow in this section.

Advection and Longitudinal Dispersion

Advective transport is frequently demonstrated with a plug flow model that neglects all longitudinal and lateral mixing. The contaminant moves with the velocity as stated by Darcy's law corrected for flow through the pores. In Figure 17.1a the source material enters the flow tube with a concentration C_0 and displaces the original fluid. The advective front is located at the position $x = vt$. The concentrations are at steady state and everywhere equal to the source concentration C_0. Such behavior is unrealistic in the field or the laboratory. But the concept is useful in describing other processes.

Figure 17.1b shows advection with longitudinal dispersion. In the absence of lateral or transverse dispersion, the displacing fluid mixes with and displaces the original fluid in the x direction strictly within the flow tube. The concentration at the advective front $x = vt$ in this case is now less than the original concentration

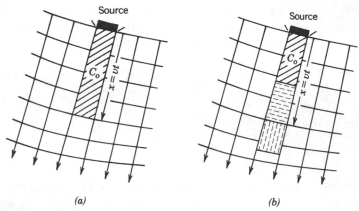

(a) (b)

Figure 17.1
Schematic diagram showing the essential features of (*a*) plug
flow, and (*b*) plug flow accompanied by longitudinal dispersion.

C_0. Longitudinal dispersion causes some of the dissolved mass to move ahead of
the advective front at the expense of material behind the advective front. However, at some point $x << vt$, the concentration is steady and is the same as the
source C_0. This steady-state part of the plume possesses all the properties of the
plug flow model of Figure 17.1*a*.

The situation just described can be reproduced in a laboratory column (Figure
10.5). The resulting breakthrough curve (Figure 17.2*a*) can be thought of as a
window fixed at some point *x* along the column. For some time after the experiment is begun, the concentration ratio at this point (that is, C/C_0) will be zero as
only water initially present in the column passes by. Prior to the breakthrough of
the advective front, the concentration ratio C/C_0 becomes finite and equals 0.5 at
the instant the advective front breaks through. Eventually, with increasing time,
$C/C_0 = 1$, indicating that the original fluid has been totally displaced and only the
tracer of concentration C_0 is passing through the column.

An alternative way to demonstrate the column experiment is shown in Figure
17.2*b*. This concentration distribution is the spatial distribution of the ratio C/C_0
throughout the column at one point in time. Here we see that $C/C_0 = 1, 0.5$, and
0 at some $x << vt$, $x = vt$, and $x >> vt$, respectively.

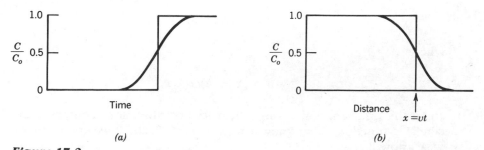

(a) (b)

Figure 17.2
(*a*) Breakthrough curve at the end of a column describing the variation in relative
concentration as a function of time and (*b*) the spatial variation in relative concentration along the column at one time.

The concentration distribution in a column is completely described by the Ogata-Banks equation (1961), which is a solution to the one-dimensional dispersion advection equation introduced in Chapter 13

$$D_x \frac{\partial^2 C}{\partial x^2} - v_x \frac{\partial C}{\partial x} = \frac{\partial C}{\partial t} \tag{17.1}$$

for the conditions

$$C(0, t) = C_0 \quad \text{and} \quad C(x, 0) = 0$$

The first boundary condition states that at $x = 0$, for all time t, the concentration is C_0 (that is, a continuous source). The second is an initial condition, and states that at all points x at time $t = 0$, the concentration is zero. That is, at the start of the test, the concentration of tracer in the water in the column is zero. An abbreviated solution of Eq. 17.1 for these boundary conditions is

$$C(x, t) = \left(\frac{C_0}{2} \right) \text{erfc} \left[\frac{(x - vt)}{2(\alpha_x vt)^{1/2}} \right] \tag{17.2}$$

where α_x is the longitudinal dispersivity, erfc is the complementary error function, and v is the linear velocity of the water, which in this case is identical to the velocity of the tracer.

We have taken certain liberties with this solution. First, a second term that is added to this expression has been ignored. Its value is so small as to be negligible in virtually all cases. Second, the quantity $(\alpha_x vt)^{1/2}$ is frequently written as $(Dt)^{1/2}$, where D is the coefficient of hydrodynamic dispersion. However, if diffusion is small compared to mechanical dispersion, the dispersion coefficient is simply $\alpha_x v$. In this expanded form, the solution becomes

$$C(x, t) = \left(\frac{C_0}{2} \right) \left\{ \text{erfc} \left[\frac{(x - vt)}{2(Dt)^{1/2}} \right] + \exp(vx/D)\text{erfc} \left[\frac{(x + vt)}{2(Dt)^{1/2}} \right] \right\} \tag{17.3}$$

Further discussions will focus only on the abbreviated form of Eq. 17.3.

The complementary error function occurs frequently in solutions to the advection-dispersion equation. Like the various well functions in hydraulic testing, it is a well-tabulated function so that graphs or tables can provide numerical values. It is convenient to express this complementary error function as

$$\text{erfc} (\beta)$$

where

$$\beta = (x - vt)/2(\alpha_x vt)^{1/2}$$

and is the argument of the complementary error function. Thus, as with all well-tabulated functions, if we know β, we know erfc β, and if we know erfc β, we likewise know β.

Values for erfc β and erf β (or error function of β) for various values of the argument β are shown on Table 17.1 and are given graphically on Figure 17.3. Note on the figure that erf β ranges from -1 to $+1$, whereas erfc β ranges from 0

Table 17.1
Values of erf (β) and erfc (β) for positive values of β

β	erf (β)	erfc (β)
0	0	1.0
0.05	0.056372	0.943628
0.1	0.112463	0.887537
0.15	0.167996	0.832004
0.2	0.222703	0.777297
0.25	0.276326	0.723674
0.3	0.328627	0.671373
0.35	0.379382	0.620618
0.4	0.428392	0.571608
0.45	0.475482	0.524518
0.5	0.520500	0.479500
0.55	0.563323	0.436677
0.6	0.603856	0.396144
0.65	0.642029	0.357971
0.7	0.677801	0.322199
0.75	0.711156	0.288844
0.8	0.742101	0.257899
0.85	0.770668	0.229332
0.9	0.796908	0.203092
0.95	0.820891	0.179109
1.0	0.842701	0.157299
1.1	0.880205	0.119795
1.2	0.910314	0.089686
1.3	0.934008	0.065992
1.4	0.952285	0.047715
1.5	0.966105	0.033895
1.6	0.976348	0.023652
1.7	0.983790	0.016210
1.8	0.989091	0.010909
1.9	0.992790	0.007210
2.0	0.995322	0.004678
2.1	0.997021	0.002979
2.2	0.998137	0.001863
2.3	0.998857	0.001143
2.4	0.999311	0.000689
2.5	0.999593	0.000407
2.6	0.999764	0.000236
2.7	0.999866	0.000134
2.8	0.999925	0.000075
2.9	0.999959	0.000041
3.0	0.999978	0.000022

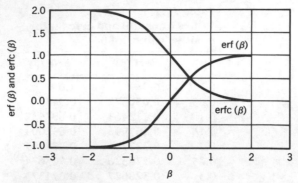

Figure 17.3
erf(β) and erfc(β) plotted versus β.

to +2. Further, erfc(β) takes on numbers greater than one only for negative values of the argument. This feature is not evident in Table 17.1. For negative values of the argument β, erfc(−β) may be determined as

$$\text{erfc}(-\beta) = 1 + \text{erf }\beta$$

Other required or useful relationships include

$$\text{erf}(-\beta) = -\text{erf }\beta \quad \text{and} \quad \text{erfc }\beta = 1 - \text{erf }\beta$$

Now let us examine the physical meaning of the argument of erfc starting first with the numerator $x - vt$, where both x and vt are lengths. This expression simply identifies the point of observation x with respect to the position of the advective front vt. The point of observation can be in one of three possible places: at the advective front, in front of the advective front, or behind the advective front. At the advective front where $x = vt$, the argument is zero and erfc(0) = 1 so that from Eq. 17.2, $C = 0.5C_0$. This result is observed on the breakthrough curves of Figure 17.2. Beyond the advective front where $x >> vt$, the argument is positive, and as it approaches infinity, erfc(∞) = 0 so that from Eq. 17.2, $C = 0$. In other words, the observation point x is far in front of the zone of dispersion so that mixing has not yet occurred, and we are observing the concentration of the original fluid. However in a practical sense, a positive argument equal to 2 or 3 suggests that erfc β is on the order of 10^{-3} to 10^{-5} (Table 17.1). Hence for all practical purposes, $C = 0$ at positive values of β greater than 2.

If the point of observation is behind the advective front, the argument β is negative, and as it approaches negative infinity, erfc(−∞) equals 2 (Figure 17.2). This means that $C = C_0$, Eq. 17.2. However, a negative argument on the order of (−2) suggests that erfc (−2) is about 1.999. Thus, for all practical purposes, $C = C_0$ at small negative values of β. The denominator of the argument has the units of length and may be regarded as the longitudinal spreading length or a measure of the spread of the mass around the advective front. The larger the denominator, the greater the spread about $C/C_0 = 0.5$.

Thus, the simple expression Eq. 17.2 tells us something about longitudinal dispersion. First, from a modeling perspective, longitudinal dispersion merely moves mass ahead of the advective front in a given flow tube. It contributes nothing to the lateral movement of material out of the tube. Second, as the distance of the advective front from the source increases with increasing time, the spread becomes larger around the advective front. Last, at some position $x << vt$, the concentration is maintained at the original source concentration C_0. The Ogata-Banks solution is not a field equation but a laboratory one because no matter how far the point of observation is from a continuous source, the equation predicts that the maximum concentration will eventually be equal to the source concentration. This condition is unrealistic, but such is the nature of models with only advection and longitudinal dispersion as the operative processes.

The example that follows illustrates one of the ways that analytical solutions can be used in this case to estimate a longitudinal dispersivity from breakthrough data.

Example 17.1

A nonsorbing species is sent through a column 30 cm in length at a velocity of 1×10^{-2} cm/s. C/C_0 ratios of 0.42 and 0.573 are noted at 46.6 and 53.3 minutes, respectively, after the test started. What is the longitudinal dispersivity?

Using the first breakthrough concentration

$$C/C_0 = (1/2) \, \text{erfc} \left[\frac{(x - vt)}{2(\alpha_x vt)^{1/2}} \right]$$

$$0.42 = (1/2) \, \text{erfc} \left[\frac{30 - 28}{2(\alpha_x 28)^{1/2}} \right]$$

$$0.84 = \text{erfc} \, (\beta)$$

Now solve for α_x

$$\beta = 0.14$$

$$0.14 = \frac{(30 - 28)}{2(\alpha_x 28)^{1/2}}$$

$$\alpha_x = 1.8 \text{ cm}$$

The second calculation is essentially the same

$$0.573 = (1/2) \, \text{erfc} \left[\frac{(30 - 32)}{2(\alpha_x 32)^{1/2}} \right]$$

$$1.146 = \text{erfc}(-\beta) = 1 + \text{erf} \, (\beta)$$

$$\beta = 0.13 \quad \text{and} \quad \alpha_x = 1.8 \text{ cm}$$

The Retardation Equation

For the case of mass transport accompanied by linear sorption, the governing equation is, as we discussed earlier

$$\frac{D_x \partial^2 C}{R_f \partial x^2} - \frac{v_x \partial C}{R_f \partial x} = \frac{\partial C}{\partial t} \tag{17.4}$$

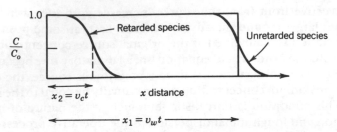

Figure 17.4
A comparison of concentration distributions for retarded and unretarded tracers.

where

$$R_f = 1 + \left(\frac{1-n}{n}\right)\rho_s K_d \qquad (17.5)$$

The effect of sorption is to decrease the value of the transport parameters D and v. The solution to this equation for the same boundary conditions used with the Ogata-Banks (1961) equation is

$$C = \left(\frac{C_0}{2}\right)\text{erfc}\left[\frac{(R_f x - v_w t)}{2(\alpha_x v_w t R_f)^{1/2}}\right] \qquad (17.6)$$

If $R_f = 1$, the Ogata-Banks result is recovered exactly. Note that v_w is the velocity of the water, which in this case is not the same as velocity of the species undergoing sorption.

The relationship between an unretarded species, as described by the Ogata-Banks (1961) equation, and a retarded one, as described by Eq. 17.6 is shown in Figure 17.4. Essentially, the velocity of the contaminant becomes less than the velocity of the ground water. Mathematically, the following equation relates these two velocities:

$$v_c = \frac{v_w}{R_f} = \frac{v_w}{\{1 + [(1-n)/n]\rho_s K_d\}} \qquad (17.7)$$

Equation 17.7 is known as the retardation equation. It predicts the position of the front of a plume due to advective transport with sorption described by a simple linear isotherm. The ratio v_w/v_c describes how many times faster the ground water (or a nonsorbing tracer) is moving relative to the contaminant being sorbed. Note that when K_d is zero (no sorption), the velocities are the same. In normal application of Eq. 17.7, v_c and v_w can be measured in any convenient units. However, K_d is normally reported in milliliters per gram and ρ_s in grams per cubic centimeter so that their product is milliliters per cubic centimeter. No further conversions are required because of the units chosen where there are 1000 cm³ per liter (or per 10^3 ml). In applications, ρ_s may be taken as 2.65 and $(1-n)/n$ is the inverse of the void ratio so that $(1-n)\rho_s/n = \rho_s/e$, where e is the void ratio introduced in Chapter 2.

The retardation equation takes on a slightly different form if instead of defining a K_d value one uses

$$\frac{K_s CEC}{\tau} = K_d \tag{17.8}$$

where K_s is the selectivity coefficient, CEC is the cation exchange capacity (meq/mass), and τ is the total competing cation concentration in solution (meq/mass). Thus for a problem involving binary exchange, the retardation equation is

$$\frac{v_w}{v_c} = 1 + \frac{\rho_s(1 - n)K_s CEC}{n\tau} \tag{17.9}$$

The retardation equation has proven useful in evaluating problems of contamination from organic compounds and to a lesser extent radionuclides and trace metals. With rudimentary information about a site, one can begin to explain differences in the observed distributions of various contaminants and predict how contaminants might spread in the future. To illustrate how the equation could be applied in practice, consider the following example based on an analysis presented by Grisak and Jackson (1978) for the Whiteshell Nuclear Research Establishment (WNRE) in Canada.

Example 17.2

At WNRE, a sequence of glacial drift overlies Precambrian bedrock. The unit of interest in this problem is the basal sand, which overlies the bedrock and is confined on top by clay-loam till. The question to be answered is how fast a hypothetical contaminant such as Sr^{90} might migrate in the basal sand should the primary containment fail. The seepage velocity through the sand is estimated to be 1.85 cm/d. Values of other necessary parameters are $\rho_s(1 - n)$ is 1.9 g/cm³, CEC is 1.4 meq/100g, τ is 4.8 meq/L, n is 0.35, and K_s is 1.3.

The velocity of Sr^{90} can be determined by substituting the known parameters in consistent units in the retardation equation

$$
\begin{aligned}
V_{Sr-90} &= \frac{v_w}{[1 + \rho_s(1 - n)\mathrm{CEC} \cdot K_s/(n\tau)]} \\[2mm]
&= \frac{1.85 \text{ cm/d}}{[1 + 1.9 \text{ g/cm}^3 \cdot 0.014 \text{ meq/g} \cdot 1.3/[0.35 \cdot 0.0048 \text{ meq/ml}]} \\[2mm]
&= 0.086 \text{ cm/d}
\end{aligned}
$$

Thus, the velocity of Sr^{90} is about 5% of that of the ground water.

The simplicity of this calculation makes it useful for back-of-the-envelope-type analyses of contamination problems. The method obviously breaks down when the physical setting is complex, or processes other than advection or sorption/exchange operate to control contaminant distributions.

Radioactive Decay, Biodegradation, and Hydrolysis

Let us now consider mass transport involving a first-order kinetic reaction. Mathematically, this problem is described by Eq. 13.14. For the same initial and boundary conditions of the Ogata-Banks (1961) equation, the solution in one dimension is (Bear, 1979)

$$C = \left(\frac{C_0}{2}\right)\exp\left\{\left(\frac{x}{2\alpha_x}\right)\left[1 - \left(\frac{1 + 4\lambda\alpha_x}{v}\right)^{1/2}\right]\right\}$$

$$\text{erfc}\left[\frac{x - vt(1 + 4\lambda\alpha_x/v)^{1/2}}{2(\alpha_x vt)^{1/2}}\right] \tag{17.10}$$

where λ is the decay constant, or 0.693 divided by the half life. Although this equation is not sufficiently complete for describing a field problem, it provides insight that does apply to field situations. First if λ is zero, Eq. 17.10 reduces to the Ogata-Banks (1961) solution. For this condition, the exponential term in the solution is one because $e^0 = 1$. Further for a finite λ, the exponent will be raised to a negative number so that we immediately establish the appropriate limits for the exponential term

$$e^0 = 1 \qquad e^{-\infty} = 0$$

The important dimensionless group in the exponential term is $4\lambda\alpha_x/v$. As λ gets large with respect to the other variables in this dimensionless group, the exponential term approaches zero, and the concentration approaches zero. The material is reacting or decaying faster than it can be transported through the system. As the velocity becomes large with respect to the other variables, the dimensionless group approaches zero, and we approach the Ogata-Banks (1961) solution. Transporting the mass faster than it can degrade or decay makes the decay ineffective. Competition thus exists between the transport and the kinetics, in particular the decay of a species and its velocity. As the longitudinal dispersivity α_x gets large, the dimensionless group of Eq. 17.10 gets large and decay again dominates. However, the dispersivity term occurs in two places in the exponential term of Eq. 17.10 and these apparently offset each other. In other words, any reasonable value of α_x can be used in the exponential term, and the value of this term will generally change by less than 1% for a few orders in magnitude range in longitudinal dispersivity. For example, for a nuclide with a half life of 1000 years, a velocity of 4 m/yr, a distance of 5000 m, and $\alpha_x =$ to 30 m, 100 m, and 300 m, the exponential term equals 0.5, 0.505, and 0.51, respectively.

Equation 17.10 reflects a steady state when the argument of erfc approaches -2, that is, the point of observation is behind the modified advective front $vt(1 + 4\lambda\alpha_x/v)^{1/2}$. This steady-state solution is

$$C = C_0 \exp\left\{\left(\frac{x}{2\alpha_x}\right)\left[1 - \left(\frac{1 + 4\lambda\alpha_x}{v}\right)^{1/2}\right]\right\} \tag{17.11}$$

Although there appears to be a contradiction of terms when a steady state can be achieved for a decaying species, the condition can be realized if the source is continually renewed. As with the Ogata-Banks equation, the steady-state concentration is independent of the longitudinal dispersivity and in this case depends only on

Figure 17.5
Advection accompanied by transverse dispersion.

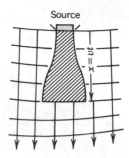

advection and decay. For all practical purposes, Eq. 17.11 describes the concentration of a species undergoing decay or degradation in a purely advective system.

There is one more thing to be learned from Eq. 17.11. The term v is velocity, but what velocity? The answer is the velocity of the contaminant. For an unretarded contaminant, the velocity of the contaminant is simply equal to the linear velocity of the water. For a retarded contaminant we can use the retarded velocity directly, or we may use v_w/R_f from Eq. 17.7. Thus we note the importance of retardation in problems of decay or degradation in that retarded velocities favor kinetics over transport. Slow-moving nuclides or organics have a greater opportunity to decay or degrade.

Transverse Dispersion

Figure 17.5 illustrates an advective model with transverse spreading (lateral dispersion). In the absence of longitudinal dispersion, no mass travels beyond the advective front. However, the mass is not restricted to a single-flow tube but can spread in a direction transverse to the flow both laterally and vertically. Concentrations are everywhere less than the source concentration C_0, except at $x = 0$. Concentrations less than C_0 occur because the mass is equally distributed throughout the plume, but the volume which it occupies increases with increasing distance from the source.

In all developments thus far, we specified a source concentration but no information concerning the geometry of the source. With transverse dispersion, something must be stated about the source geometry. Possible geometrical configurations include a point, a vertical line, or a planar area (Figure 17.6). The equations describing the parts of a plume controlled exclusively by transverse spreading for the point or line source are complicated. Thus, the equations that follow are limited to describing the maximum concentrations along the plane of symmetry, that is, along the x axis for y and $z = 0$ or

$$C_{\text{max}} = C_0 \text{ (plug flow)}$$

$$C_{\text{max}} = \frac{C_0 Q}{2v(\pi \alpha_y x)^{1/2}} \text{ (line source)}$$

$$C_{\text{max}} = \frac{C_0 Q}{4x\pi v(\alpha_y \alpha_z)^{1/2}} \text{ (point source)} \tag{17.12}$$

where Q is a volumetric flow rate with units of $L^2 T^{-1}$ for the line source and $L^3 T^{-1}$ for the point source so that $C_0 Q$ has the units of $ML^{-1}T^{-1}$ and MT^{-1}, respectively.

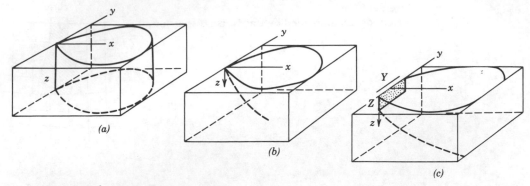

Figure 17.6
Geometrical considerations for a contaminant source. Shown in (*a*) is a vertical line
source, in (*b*) a point source, and in (*c*) a finite plane source (from Huyakorn and others,
1987). Reprinted by permission of Ground Water. Copyright © 1987. All rights reserved.

The line source result was developed by Wilson and Miller (1978), and the point
source result by Hunt (1978).

Equations 17.12 clearly demonstrate the role of the transverse dispersivity in
transport problems. For the line source, the larger α_y the greater the lateral extent
of the plume and the smaller the concentrations. The point source result incorporates lateral (α_y) as well as vertical (α_z) spreading. The source flow rate C_0Q
reflects the geometry of the source, expressed as mass per length per time for the
line source and mass per time for the point source. Further, the concentrations are
independent of time and they approach infinity as x approaches zero. Thus, these
equations are only useful for far field conditions where x is large.

The planar source is obviously the more practical one with lateral spreading
in two directions and vertical spreading downward (Figure 17.6). Solutions for
this source geometry have been developed by Domenico and Palciauskas (1982).
For the lateral (y) component of spreading

$$C = \left(\frac{C_0}{2}\right)\left\{ \text{erf}\left[\frac{(y + Y/2)}{2(\alpha_y x)^{1/2}}\right] - \text{erf}\left[\frac{(y - Y/2)}{2(\alpha_y x)^{1/2}}\right]\right\} \tag{17.13}$$

where the half source size $Y/2$ is part of the solution. The z component of spreading is

$$C = \left(\frac{C_0}{2}\right)\left\{ \text{erf}\left[\frac{(z + Z)}{2(\alpha_z x)^{1/2}}\right] - \text{erf}\left[\frac{(z - Z)}{2(\alpha_z x)^{1/2}}\right]\right\} \tag{17.14}$$

where the full source size Z becomes part of the solution. For spreading directions
in both y and z, we adhere to the product rule discussed earlier (Domenico and
Palciauskas, 1982)

$$C = \left(\frac{C_0}{4}\right)\left\{ \text{erf}\left[\frac{(y + Y/2)}{2(\alpha_y x)^{1/2}}\right] - \text{erf}\left[\frac{(y - Y/2)}{2(\alpha_y x)^{1/2}}\right]\right\}$$
$$\left\{ \text{erf}\left[\frac{(z + Z)}{2(\alpha_z x)^{1/2}}\right] - \text{erf}\left[\frac{(z - Z)}{2(\alpha_z x)^{1/2}}\right]\right\} \tag{17.15}$$

For the plane of symmetry ($y = z = 0$),

$$C_{\max} = C_0 \, \text{erf}\left[\frac{Y}{4(\alpha_y x)^{1/2}}\right] \text{erf}\left[\frac{Z}{2(\alpha_z x)^{1/2}}\right] \qquad (17.16)$$

The source dimensions and transverse dispersivities control the maximum concentrations that are encountered in a steady-state plume. For example, as Y or Z is increased, or α_y and α_z decreased, the arguments of the error function approach $+2$, and the value of the error functions approach one; that is, spreading is restricted and the concentration is maintained near C_0 over small values of x.

Models for Multidimensional Transport

Multidimensional transport involves both longitudinal and transverse dispersion in addition to advection. The most complex form of the dispersion-advection equation that is amenable to an analytical solution includes three dispersive components, a constant advective velocity, and one kinetic term, as follows

$$D_x\frac{\partial^2 C}{\partial x^2} + D_y\frac{\partial^2 C}{\partial y^2} + D_z\frac{\partial^2 C}{\partial z^2} - v_x\frac{\partial C}{\partial x} - \frac{r}{n} = \frac{\partial C}{\partial t} \qquad (17.17)$$

where r is defined by some mathematical rate law.

Solutions to Eq. 17.17 provide the concentration distribution resulting from continuous or instantaneous sources. The following sections will examine these different source conditions in detail.

Continuous Sources

Models that include transverse spreading must incorporate information on the source geometry. Wilson and Miller (1978) provide a complete solution for a nonreacting species emanating from a line source, and Hunt (1978) presents a three-dimensional solution for a nonreacting species associated with a point source. However because the finite planar source is the most realistic geometry for field problems, it has received the most attention. Two types of models are available. First, there are finite source models that require numerical integration (for example, Cleary, 1978; Huyakorn and others, 1987). A second type of solution is given in a closed form format, where no numerical integrations are necessary. Two models are available, one for advection and dispersion alone (Domenico and Robbins, 1985b) and another for mass transport together with a first-order reaction (Domenico, 1987).

The essence of the Domenico and Robins model is apparent in Figure 17.7. Because of an assumed one-dimensional velocity, the plume shown in Figure 17.7*d* can be referred to as a plug flow model with both longitudinal and transverse spreading (dispersion). The distribution of concentration (Figure 17.7*d*) has features common to both the longitudinal (Figure 17.7*b*) and the transverse spreading model (Figure 17.7*c*). There is a frontal zone of dispersion beyond the advective front caused exclusively by longitudinal dispersion, and a zone of mass depletion zone behind the advective front. At some distance behind the advective front, depending largely on the value of the longitudinal dispersivity, the plume is at steady state. Because of transverse spreading, the steady-state concentrations

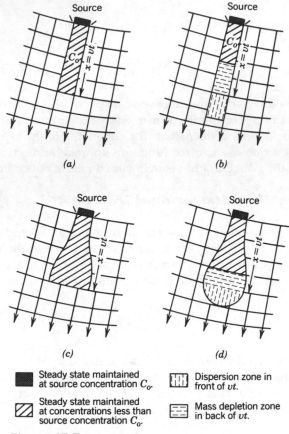

Steady state maintained at source concentration C_o.

Steady state maintained at concentrations less than source concentration C_o.

Dispersion zone in front of vt.

Mass depletion zone in back of vt.

Figure 17.7
Idealization of the dispersion process showing (*a*) plug flow, (*b*) longitudinal dispersion, (*c*) transverse dispersion, and (*d*) longitudinal and transverse dispersion (after Domenico, 1987). Reprinted with permission of Elsevier Science Publishers from J. Hydrol., v. 91.

are everywhere less than the source concentration C_0, except at $x = 0$. Results of the product rule discussed earlier capture the essential features of this plume for the source geometry of Figure 17.6*c* (Domenico and Robbins, 1985b)

$$C(x, y, z, t) = \left(\frac{C_0}{8}\right) \text{erfc} \left[\frac{(x - vt)}{2(\alpha_x vt)^{1/2}}\right]$$
$$\left\{ \text{erf} \left[\frac{(y + Y/2)}{2(\alpha_y x)^{1/2}}\right] - \text{erf} \left[\frac{(y - Y/2)}{2(\alpha_y x)^{1/2}}\right] \right\} \qquad (17.18)$$
$$\left\{ \text{erf} \left[\frac{(z + Z)}{2(\alpha_z x)^{1/2}}\right] - \text{erf} \left[\frac{(z - Z)}{2(\alpha_z x)^{1/2}}\right] \right\}$$

For $x \ll vt$, the argument of erfc approaches -2, and we recover the Domenico-Palciauskas (1982) steady-state model of Eq. 17.15. For the plane of symmetry $y = z = 0$

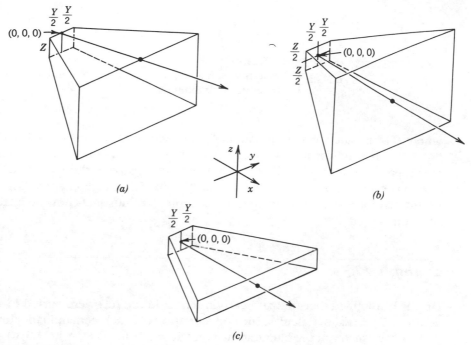

Figure 17.8
Idealized contaminant migration geometries for various transverse spreading directions (from Domenico and Robbins, 1985b). Reprinted by permission of Ground Water. Copyright © 1985. All rights reserved.

$$C(x, 0, 0, t) = \left(\frac{C_0}{2}\right) \operatorname{erfc}\left[\frac{(x - vt)}{2(\alpha_x vt)^{1/2}}\right]\left\{\operatorname{erf}\left[\frac{Y}{4(\alpha_y x)^{1/2}}\right] \operatorname{erf}\left[\frac{Z}{2(\alpha_z x)^{1/2}}\right]\right\}$$

$$(17.19)$$

This form also shows that as α_y and α_z approach zero, we recover the Ogata-Banks (1961) Eq. (17.2).

As with all the transport equations, the velocity v is the contaminant velocity. For a retarded species, the contaminant velocity is used directly, or v_w/R_f is substituted directly for v in Eq. 17.19. There is one other feature about this result that helps to accommodate various field conditions. As developed Eq. 17.18 applies to a source geometry with two transverse and one vertical spreading direction (Figure 17.8a). For two spreading directions in z (Figure 17.8b), the source dimension Z terms become $Z/2$. If z spreading is eliminated altogether (Figure 17.8), the error functions containing the z terms in Eq. 17.18 are ignored, and $C_0/8$ becomes $C_0/4$. If the aquifer thickness is small, the plume may spread vertically and occupy the entire thickness (Figure 17.9). The length x' over which this spreading occurs can be approximated as

$$x' = \frac{(H - Z)^2}{\alpha_z}$$

$$(17.20)$$

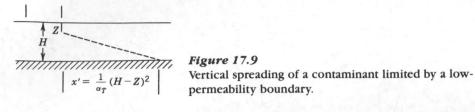

Figure 17.9
Vertical spreading of a contaminant limited by a low-permeability boundary.

where H is the thickness of the aquifer (Domenico and Palciauskas, 1982). Hence, Eq. 17.18 is valid for all distances x equal to or less than $(H - Z)^2/\alpha_z$. For distances greater than x', the distance x in the denominator of the error function of z term is replaced by x' (for example, Eq. 17.20). Thus, further spreading in z is prohibited for $x > x'$. Following is an example that illustrates how an analytical solution is used.

Example 17.3

Drums of diethyl ether (de) and carbon tetrachloride (ct) were buried in a sand aquifer 15 years ago. Calculate the concentrations of each contaminant along the plane of symmetry of the plume at the point ($x = 225$ m, $y = 0$ m, $z = 0$ m) at time 15 years (4.73×10^8 sec). The velocity of the ground water is 1×10^{-6} m/sec. The retardation factor for de is 1.5 and for ct 27.4. The source concentration for de is 1×10^4 μg/L and for ct is 5×10^2 μg/L. The source size in Y is 25 m and in Z is 5 m. The estimated dispersivities are $\alpha_x = 1.0$ m, $\alpha_y = 0.1$ m, and $\alpha_z = 0.01$ m. Assume that the plume can spread in two z directions and two y directions so that

$$C(x, 0, 0, t) = \left(\frac{C_0}{2}\right) \text{erfc}\left[\frac{(x - vt)}{2(\alpha_x vt)^{1/2}}\right] \text{erf}\left[\frac{Y}{4(\alpha_y x)^{1/2}}\right] \text{erf}\left[\frac{Z}{4(\alpha_z x)^{1/2}}\right]$$

For de,

$$C = \left(\frac{1 \times 10^4}{2}\right) \text{erfc}\left[\frac{(225 - 312)}{2(1 \times 312)^{1/2}}\right] \text{erf}\left[\frac{25}{4(0.1 \times 225)^{1/2}}\right]$$

$$\text{erf}\left[\frac{5}{4(0.01 \times 225)^{1/2}}\right]$$

$$= 5 \times 10^3 \text{ erfc}(-2.46) \text{ erf}(1.3) \text{ erf}(0.83)$$

$$= 5 \times 10^3 (2)(0.93)(0.759) = 7 \times 10^3 \text{ μg/L}$$

Because the argument of the complementary error function has a negative value greater than -2, the plume is at steady state at this point, and the calculated concentration is a maximum value.

For ct,

$$C = \left(\frac{5 \times 10^2}{2}\right) \text{erfc}\left[\frac{(225 - 17.2)}{2(1 \times 17.2)^{1/2}}\right] \text{erf}\left[\frac{25}{4(0.1 \times 225)^{1/2}}\right]$$

$$\text{erf}\left[\frac{5}{4(0.01 \times 225)^{1/2}}\right]$$

Because erfc[$(225 - 17.2)/2(1 \times 17.2)^{1/2}$] = erfc(25) $\approx$ 0, it follows that carbon tetrachloride has not yet reached this point because of greater retardation. The maximum concentration that ct will eventually attain is

$$C_{max} = (5 \times 10^2)(0.93)(0.759) = 3.5 \times 10^2 \text{ μg/L}$$

Domenico (1987) uses similar arguments to incorporate decay or degradation in the following solution for multidimensional transport

$$
\begin{aligned}
C(x, y, z, t) = {} & \left(\frac{C_0}{8}\right) \exp\left\{\left(\frac{x}{2\alpha_x}\right)\left[1 - \left(\frac{1 + 4\lambda\alpha_x}{v}\right)^{1/2}\right]\right\} \\
& \text{erfc}\left[\frac{x - vt(1 + 4\lambda\alpha_x/v)^{1/2}}{2(\alpha_x vt)^{1/2}}\right] \\
& \left\{\text{erf}\left[\frac{(y + Y/2)}{2(\alpha_y x)^{1/2}}\right] - \text{erf}\left[\frac{(y - Y/2)}{2(\alpha_y x)^{1/2}}\right]\right\} \\
& \left\{\text{erf}\left[\frac{(z + Z)}{2(\alpha_z x)^{1/2}}\right] - \text{erf}\left[\frac{(z - Z)}{2(\alpha_z x)^{1/2}}\right]\right\}
\end{aligned}
\tag{17.21}
$$

which reduces to the Domenico-Robbins (1985b) result when λ = 0. The concentration along the plane of symmetry is

$$
\begin{aligned}
C(x, y, z, t) = {} & \left(\frac{C_0}{2}\right) \exp\left\{\left(\frac{x}{2\alpha_x}\right)\left[1 - \left(\frac{1 + 4\lambda\alpha_x}{v}\right)^{1/2}\right]\right\} \\
& \text{erfc}\left[\frac{x - vt(1 + 4\lambda\alpha_x/v)^{1/2}}{2(\alpha_x vt)^{1/2}}\right] \\
& \left\{\text{erf}\left[\frac{Y}{4(\alpha_y x)^{1/2}}\right]\text{erf}\left[\frac{Z}{2(\alpha_y x)^{1/2}}\right]\right\}
\end{aligned}
\tag{17.22}
$$

As α_y and α_z approach zero, the result of Bear (1979) is recovered (Eq. 17.10). Further a steady-state form may be obtained when the argument of the complementary error function approaches -2.

Virtually everything that has been stated for the Domenico-Robbins (1985b) result applies here also. Thus, Eq. 17.21 can be modified to incorporate different spreading geometries as well as to account for limitations in z spreading. Further, the velocity v in these equations is the contaminant velocity so that retardation is already incorporated or can be easily accommodated by substituting $v = v_w/R_f$.

The models discussed have been applied to field problems by LaVenue and Domenico (1986) and Fryar and Domenico (1989). Their main use is in determining the transport parameters from concentration distributions observed in the field (Chapter 18).

The Instantaneous Point Source Model

The accidental spill, frequently referred to as an instantaneous or pulse-type problem, represents another potential contamination problem. Models for two types are available: the parallelpiped finite source model of Hunt (1978) and the point source model of Baetsle (1969). The point source of Baetsle (1969) is the

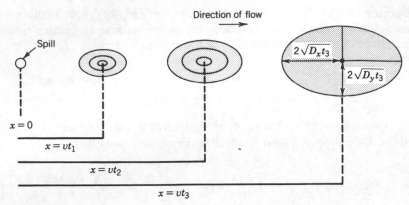

Figure 17.10
Plan view of the plume developed from an instantaneous point source at three different times.

more useful one because the geometry of a spill more closely resembles a point, at least when viewed from the far field, than a parallelepiped. Baetsle's model is

$$C(x, y, z, t) = \left[\frac{C_0 V_0}{8(\pi t)^{3/2}(D_x D_y D_z)^{1/2}} \right] \exp\left[-\frac{(x - vt)^2}{4D_x t} - \frac{y^2}{4D_y t} - \frac{z^2}{4D_z t} - \lambda t \right]$$

$$(17.23)$$

where C_0 is the original concentration; V_0 is the original volume so that the produce $C_0 V_0$ is the mass involved in the spill; D_x, D_y, D_z are the coefficients of hydrodynamic dispersion; v is the velocity of the contaminant; x, y, and z are space coordinates; t is time, and λ is the disintegration constant for a radioactive substance. For a nonradioactive substance, the term λt is ignored. Again, this solution can be obtained readily as the product of three one-dimensional solutions.

With an idealized three-dimensional point source spill, spreading occurs in the direction of flow, and the peak or maximum concentration occurs at the center of the "cloud," that is, where $y = z = 0$ and $x = vt$ (Figure 17.10)

$$C_{\max} = \frac{C_0 V_0 e^{-\lambda t}}{[8(\pi t)^{3/2}(D_x D_y D_z)^{1/2}]}$$

$$(17.24)$$

The dimensions of the cloud, assuming it actually started as a point are

$$3\sigma_x = (2D_x t)^{1/2}; \qquad 3\sigma_y = (2D_y t)^{1/2}; \qquad 3\sigma_z = (2D_z t)^{1/2}$$

where σ is the standard deviation so that $3\sigma_x$, $3\sigma_y$, and $3\sigma_z$ represent three spreading lengths within which about 99.7% of the mass is contained.

Example 17.4 (after Baetsle, 1969)

A leak in a storage tank for radioactive waste results in an accidental release of 1000 ci of 10 yr cooled fission products, along with tritium. The waste contains 400 ci Cs-137 ($t_{1/2}$ 33 yr), 400 ci Sr-90 ($t_{1/2}$ 28 yr), 100 ci Pm-147 ($t_{1/2}$ 2.7 yr), and

100 ci T-3 ($t_{1/2}$ 12.26 yr). Assume that a river exists 100 m from the spill. Determine how much time is required for each constituent to reach the river and what the maximum concentrations will be. The velocity of the ground water is 10 cm/day, or 1.16×10^{-4} cm/sec. Other data include

Nuclide	K_d(ml/mg)	R_f	D_x(cm²/sec)	$D_y = D_z$
Cs-137	10	47.6	10^{-4}	10^{-3}
Sr-90	0.6	2.885	10^{-3}	10^{-5}
Pm-147	100	476.	10^{-5}	10^{-5}
H-3	0	1.	10^{-5}	10^{-5}

(a) Time to Reach a Point at 100 m

From Eq. 17.7 where $R_f x = v_w t$, let x be the distance traveled by the nuclide. For tritium,

$$t = \frac{R_f x}{v_w} = \frac{(1) \cdot 1 \times 10^4}{1.16 \times 10^4} = 0.864 \times 10^8 \text{ sec} = 2.75 \text{ yr}$$

Similar calculations show that Cs-147 will take 112 yr, Sr-90 6.9 yr, and Pm-147 1120 yr.

(b) Peak Concentration at 100 m

The half life for tritium is 12.26 yr so that $e^{-\lambda t} = e^{-0.693 t / (12.26 \times 365 \times 86,400)}$ because $\lambda = \ln 2/\text{half life}$ and $\ln 2 = 0.693$. The value of time for the calculation comes from part (a).

$$C_{max} = \frac{100 \, e^{(-0.693 \times 0.864 \times 10^8)/(12.26 \times 365 \times 86,400)}}{8(3.14 \times 0.864 \times 10^8)^{3/2} \times 10^{-7.5}}$$

$$C_{max} = 7.6 \times 10^{-5} \text{ ci/ml}$$

The rest of the results are Cs-137 = 1.08×10^{-7}; Sr-90 = 0.608×10^{-5}; and Pm-147 negligibly small, 10^{-125} ci/ml.

17.2 Programming the Analytical Solutions for Computers

The analytical solutions can be applied directly to contamination problems. However an alternative and more efficient approach is to construct codes to perform these calculations. Almost any type of computer can be used, and the necessary codes can be constructed with little difficulty.

Most codes of this type carry out three operations: reading values of the transport parameters, solving for concentration at specified times and locations, and writing the final results. We can illustrate these steps in a practical way by developing a code for evaluating the analytical solution to Eq. 17.1 for one-dimensional mass transport subject to simple sorption. This example is a demonstration that is not to be confused with a real approach to a problem. According to the solution (17.2), suitably modified to account for sorption, three transport parameters v, α_x, and R_f and the concentration at the upstream boundary (C_0) need to be

Table 17.2
FORTRAN code for evaluating the analytical solution for one-dimensional transport with retardation

```
C       CODE FOR EVALUTING ONE-DIMENSIONAL DISPERSION
C          WITH RETARDATION
C
        DIMENSION C(50)
C
C       READ NECESSARY INPUT DATA
C
        OPEN(5,FILE='IN.DAT')
        OPEN(6,FILE='OUT.DAT')
        READ(5,100) NX
        WRITE(6,100)NX
        READ(5,110) CO,ALFX,VX,RF,TYR,DELX
        WRITE(6,110)CO,ALFX,VX,RF,TYR,DELX
C
        TSEC=TYR*365.*86400.
        X=0.0
C
C       CALCULATE CONCENTRATIONS
C
        DO 200 I=1,NX
        X=X+DELX
    200 C(I)=(CO/2)*ERFC((RF*X-VX*TSEC)/(2*SQRT(ALFX*VX*TSEC*RF)))
C
C       WRITE RESULTS
C
        WRITE(6,120)TYR
        WRITE(6,110)(C(I),I=1,NX)
C
    100 FORMAT(I5)
    110 FORMAT(6F10.2)
    120 FORMAT(F5.0,'  YEARS')
        STOP
        END
C
C       CALCULATE ERFC
C
        FUNCTION ERFC(Z)
        DATA B,C,D,E,F,G/.0705230784,.0422820123,.0092705272,
       * 1.520143E-04,2.765672E-04,4.30638E-05/
        IF(Z.GT.-3.AND.Z.LT.3.)GO TO 100
        IF(Z.LE.-3.)ERFC=2.0
        IF(Z.GE.3.)ERFC=0.0
        RETURN
    100 X=ABS(Z)
        IF(X.GT.3.0)ERFC=0.56419*EXP(0.0-X*X)/
       * (X+.5/(X+1/(X+1.5/(X+2/(X+2.5/(X+1))))))
        IF(X.LE.3.0)ERFC=1./(1.+B*X+C*X*X+D*X**3+E*X**4
       * +F*X**5+G*X**6)**16.
        IF(Z.LT.0.0)ERFC=2.0-ERFC
        RETURN
        END
```

specified. Coding begins (Table 17.2) by reading and writing these values and defining the time and positions where concentrations are to be calculated. Input data are read from the file IN.DAT and output data are written to the file OUT.DAT. The analytic solution is evaluated within a DO loop for various distances from the source. The function subroutine ERFC evaluates the error functions. Finally, we write the calculated concentrations to the specified output file. The following hypothetical example illustrates how this code is used.

Example 17.5

An organic contaminant is disposed of continuously in a long narrow trench that fully penetrates a shallow, semi-infinite and aquifer (see Figure 13.1a). For this aquifer $v = 2.31 \times 10^{-6}$ m/s, $\alpha_x = 4.3$ m, and $R_f = 3.0$. The contaminant concentration at the source (C_0) remains constant with time at 100 mg/L. Calculate the contaminant concentration after three years at 10-m intervals from the source.

   This problem is solved with the code listed in Table 17.2. The data set looks like this

```
 18
      100      4.3 .00000231       3.        3.        10.
```

For consistency in units, years are converted to seconds in the code. Following is the calculated concentration of contaminants at the end of three years down the system at 10-m intervals beginning with $x = 10$ m.

```
3. YEARS
          0.99     0.98     0.96     0.91     0.82     0.70
          0.55     0.39     0.25     0.14     0.07     0.03
          0.01     0.00     0.00     0.00     0.00     0.00
```

   Modeling is one of those activities that must be practiced to develop necessary coding skills, the ability to formulate a problem in terms of boundary conditions and data, and the ability to use a computer. Let us consider another more complex problem.

Example 17.6

For the problem in Example 17.3, develop a computer code to predict the concentration of diethyl ether (de) along a horizontal plane through the middle of the source (that is, $z = 0$ m).

   There are five steps necessary to solve this problem: (1) formulate the problem mathematically and select an appropriate analytical solution, (2) establish what input parameters are required, (3) construct the computer code, (4) run the code using known or estimated information about the site, and (5) interpret the results.

Step 1
The information provided for this problem points to three-dimensional spreading of a plume in a one-dimensional velocity field. Diethyl ether with fixed initial

concentration of 10,000 μg/L enters the ground water from a source of finite size. This problem can be solved using Eq. 17.18. This form of the transport equation accounts for dispersion in three directions, advection in one direction, and retardation due to sorption. The downstream side of the disposal site acts as the finite source with an area YZ and specified nonzero concentrations of each contaminant. The initial condition is that the concentration of diethyl ether (C_{de}) is zero within the semi-infinite domain.

Step 2

The analytical solution requires values for v_x, α_x, α_y, α_z, C_0, $Z/2$, $Y/2$, and the location of points where we intend to calculate concentrations. This code has a similar scheme for data entry as the previous one except for requiring additional dispersivities and the dimensions of the source. Concentrations are calculated for a grid of points on a horizontal plane passing through the middle of the source at $z = 0$.

Step 3

Construct the code as follows:

```
C       CODE FOR EVALUATING THREE-DIMENSIONAL TRANSPORT
C           WITH RETARDATION
C
        DIMENSION C(30,30)
C
C       READ NECESSARY INPUT DATA
C
        OPEN(5,FILE='IN3.DAT')
        OPEN(6,FILE='OUT3.DAT')
        READ(5,100) NX, NY
        WRITE(6,100)NX,NY
C
C       TRANSPORT PARAMETERS
C
        READ(5,110) CO,ALFX,ALFY,ALFZ,VX,RF
        WRITE(6,110)CO,ALFX,ALFY,ALFZ,VX,RF
C
C       TIME/SPACE PARAMETERS
C
        READ(5,110) TYR,DELX,DELY,Z
        WRITE(6,110)TYR,DELX,DELY,Z
C
C       SOURCE SIZE
C
        READ(5,110)BZ,BY
        WRITE(6,110)BZ,BY
C
        TSEC=TYR*365.*86400.
        VX=VX/RF
        X=0.0
C
C       CALCULATE CONCENTRATIONS
C
```

```
      DO 200 I=1,NX
      X=X+DELX
      Y=-DELY*FLOAT((NY+1)/2)
      DO 200 J=1,NY
      Y=Y+DELY
      AA=(CO/8)ERFC((X-VX*TSEC)/(2*ALFX*VX*TSEC)**0.5))
      BB=(ERF((Y+BY/2)/(2*(ALFY*X)**0.5))-ERF((Y-BY/2)/(2*
     1   (ALFY*X)**0.5)))
      CC=(ERF((Z+BZ/2)/(2*(ALFZ*X)**0 .5))-ERF((Z-BZ/2)/(2*
     1   (ALFZ*X)**0.5)))
      C(I,J)=AA*BB*CC
  200 CONTINUE
C
C     WRITE RESULTS
C
      WRITE(6,120)TYR
      DO 210 I=1,NX
  210 WRITE(6,130)(C(I,J),J=1,NY)
C
  100 FORMAT(2I5)
  110 FORMAT(6F10.2)
  120 FORMAT(F5.0,'     YEARS')
  130 FORMAT(7F9.2)
      STOP
      END
C
C     CALCULATE ERFC
C
      FUNCTION ERFC(Z)
      DATA B,C,D,E,F,G/.0705230784,.0422820123,.0092705272,
     *  1.520143E-04,2.765672E-04,4.30638E-05/
      IF(Z.GT.-3.AND.Z.LT.3.)GO TO 100
      IF(Z.LE.-3.)ERFC=2.0
      IF(Z.GE.3.)ERFC=0.0
      RETURN
  100 X=ABS(Z)
      IF(X.GT.3.0)ERFC=0.56419*EXP(0.0-X*X)/
     *   (X+.5/(X+1./(X+1.5/(X+2/ (X+2.5/(X+1))))))
      IF(X.LE.3.0)ERFC=1./(1.+B*X+C*X*X+D*X**3+E*X**4
     *   +F*X**5+G*X**6)**16.
      IF(Z.LT.0.0)ERFC=2.0-ERFC
      RETURN
      END
C
C     CALCULATE ERF
C
      FUNCTION ERF(Z)
      IF(Z.LE.0.0)ERF=-ERFC(Z)+1.
      IF(Z.GT.0.0)ERF=1.-ERFC(Z)
      RETURN
      END
```

Step 4

A grid provides a convenient way to define the concentration of points on the plane through the center of the plume. NX and NY are the number of points in the x and y directions. DELX and DELY are the spacings between points.

x dir: NX = 10 y dir: NY = 7
      DELX = 50.0 m   DELY = 15.0 m

The resulting set of input data is

```
 10    7
   10000.        1.      0.1      .01    1.0E-06      1.5
      15.       75.      15.      0.0
       5.       25.
```

All units are consistent in meters and seconds. Because R is dimensionless we can use another set of internally consistent units. The simulated distribution of *de* is

```
  15. YEARS
  0.00      0.00    2486.12    9575.76    2486.12      0.00      0.00
  0.00      5.95    2757.84    8319.54    2757.84      5.95      0.00
  0.00     34.59    2700.07    7137.75    2700.07     34.59      0.00
  0.07     60.30    1886.02    4514.41    1886.02     60.30      0.07
  0.00      1.21      21.63      47.73      21.63      1.21      0.00
  0.00      0.00       0.00       0.00       0.00      0.00      0.00
  0.00      0.00       0.00       0.00       0.00      0.00      0.00
  0.00      0.00       0.00       0.00       0.00      0.00      0.00
  0.00      0.00       0.00       0.00       0.00      0.00      0.00
  0.00      0.00       0.00       0.00       0.00      0.00      0.00
```

17.3 Semianalytical Approaches

The semianalytical method describes mass transport due to advection or advection modified by sorption. Unlike the analytical approaches, this procedure applies to complex, two-dimensional flow problems involving an arbitrary number of sources and sinks. Its most severe limitation is the inability to account for dispersion and other geochemical processes besides sorption. Nevertheless, this approach is potentially applicable to an array of practical problems.

Details of the theoretical development for steady-state systems are given by Nelson (1978) and Javandel and others (1984). The essence of the technique is first to develop analytical expressions characterizing the hydraulic potential and components of velocity v_x and v_y that result from various combinations of what are termed "simple flow components." These simple flow components might include uniform regional flow, recharge wells, discharge wells, and finite circular sources of contamination (for example, a pond). For each of these components there are elemental solutions describing the fluid potential and velocities. Combining these solutions in various ways lets us model even a relatively complicated flow system. For example in Figure 17.11, the flow components include a pond,

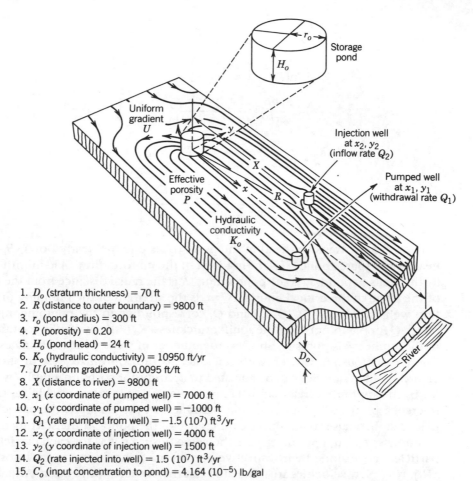

1. D_o (stratum thickness) = 70 ft
2. R (distance to outer boundary) = 9800 ft
3. r_o (pond radius) = 300 ft
4. P (porosity) = 0.20
5. H_o (pond head) = 24 ft
6. K_o (hydraulic conductivity) = 10950 ft/yr
7. U (uniform gradient) = 0.0095 ft/ft
8. X (distance to river) = 9800 ft
9. x_1 (x coordinate of pumped well) = 7000 ft
10. y_1 (y coordinate of pumped well) = −1000 ft
11. Q_1 (rate pumped from well) = −1.5 (10^7) ft^3/yr
12. x_2 (x coordinate of injection well) = 4000 ft
13. y_2 (y coordinate of injection well) = 1500 ft
14. Q_2 (rate injected into well) = 1.5 (10^7) ft^3/yr
15. C_o (input concentration to pond) = 4.164 (10^{-5}) lb/gal

Figure 17.11
An example of how simple flow components including a uniform, regional flow field, a recharging well, a discharging well, and a pond (source) can be represented in a semianalytical model. The list of 15 parameters are the data required as input to the calculation (from Nelson, R.W., Water Resources Res., v. 14, p. 416–428, 1978. Copyright by American Geophysical Union).

which is the contaminant source, an injection well, and a pumped well. All these sources and sinks are superimposed on a regional flow system. The following equations describe the potential distribution (ϕ) and the components of the ground water velocity (dx/dt, dy/dt) at any point (x, y) within the domain (Nelson, 1978)

$$\phi = H'H_o - Ux + \frac{U(r_o)^2 x}{x^2 + y^2} - \frac{H'H_o}{\ln(R/r_o)} \ln\left[\frac{(x^2 + y^2)^{1/2}}{r_o}\right]$$
$$- \sum_{j=1}^{N} \frac{m_j}{2\pi} \ln\left[\frac{(x - x_j)^2 + (y - y_j)^2}{(x_j^2 + y_j^2)}\right]^{1/2} \quad N=2 \qquad (17.25)$$

$$\frac{dx}{dt} = \frac{K_0 U}{P}\left[1 - \frac{r_0^2}{x^2 + y^2}\left(1 - \frac{2x^2}{x^2 + y^2}\right)\right]$$

$$+ \frac{K_0 H' H_0}{P \ln(R/r_0)}\left(\frac{x}{x^2 + y^2}\right) \tag{17.26}$$

$$+ \sum_{j=1}^{N} \frac{K_0 m_j}{2\pi P}\left[\frac{x - x_j}{(x - x_j)^2 + (y - y_j)^2}\right] \qquad N=2$$

$$\frac{dy}{dt} = \frac{2K_0 U}{P}\left(\frac{(r_0)^2 xy}{[x^2 + y^2]^2}\right) + \frac{K_0 H' H_0}{P \ln(R/r_0)}\left(\frac{y}{x^2 + y^2}\right)$$

$$+ \sum_{j=1}^{N} \frac{K_0 m_j}{2\pi P}\left[\frac{y - y_j}{(x - x_j)^2 + (y - y_j)^2}\right] \tag{17.27}$$

where H' is the transient head in the pond (H' is one for steady flow), H_0 is the head in the source pond, U is the gradient of the uniform flow field uninfluenced by the wells or pond, r_0 is the pond radius, R is the radial distance from the center of the pond to the critical boundary, m_j is the strength of the pumping or injection wells with $m_j = Q_j/(D_0 K_0)$ and Q_j corresponding to the pumping/injection rate ($-Q$ for pumping), D_0 is the aquifer thickness, K_0 is the hydraulic conductivity of the aquifer, P is porosity, and N is the number of wells with coordinates x_j, y_j. These equations are written with reference to the coordinate system shown in Figure 17.11. Velocities can be adjusted to account for sorption simply by dividing by the appropriate retardation factor. A simulation involves tracking a series of reference particles as they trace out pathlines. In the example, the particles are placed at intervals around the circumference of the pond. At the starting particle positions $x(t_0), y(t_0)$, we determine the velocities $v(x), v(y)$ for each particle. The particles move along their respective vectors for a short time to new positions, $x(t_1), y(t_1)$. New velocities are calculated and the particles move again. After a large number of time steps, each particle will have traced out a pathline. Particular times such as one year, or five years and so on can be flagged on each pathline. When all the points of equal time are joined, these lines define the position of the contaminant front due to advection, or advection and sorption.

The approach provides other useful information. For example, with the code RESSQ, Javandel and others (1984) show how concentration versus time data for a single producing well can be mapped to other observation points to estimate spatial distributions in concentration. Nelson (1978) shows how the semianalytical approach provides (1) location/arrival time distributions, and (2) location/outflow quantity distributions at critical outflow boundaries of the system. These distributions provide a convenient way of establishing the environmental consequences of ground water contamination.

The problem in Figure 17.11 is useful in illustrating the semianalytical approach. Note that water produced at the pumped well is reinjected in the other well. The simulation results (Figure 17.12) include the numbered pathlines traced out by the reference particles and time contours illustrating the progress of the advective fronts. Contaminants arrive at the pumped well after approximately nine years. Arrival times at the river range from approximately 15 years to greater than 21 years. Nelson (1978) summarizes arrival times at the river in terms of a location/arrival time distribution (Figure 17.13). This figure shows where the contaminants enter the river at a given time. For example at 16 years, the entry zone is between about 950 and 2400 m (Figure 17.13).

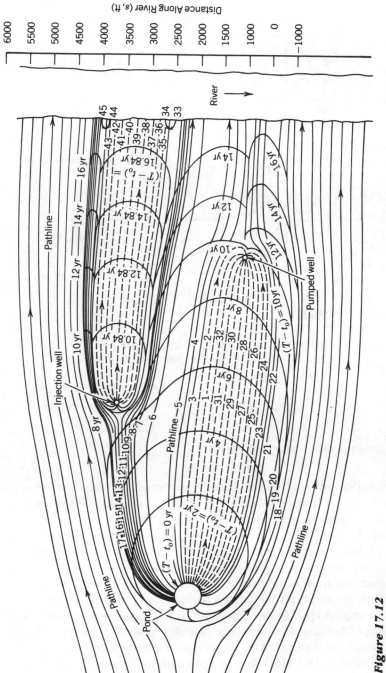

Figure 17.12

The calculated distribution of pathlines and time contours for the example data on the previous figure (from Nelson, R.W., Water Resources Res., v. 14, p. 416–428, 1978. Copyright by American Geophysical Union).

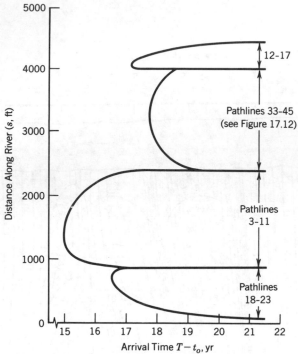

Figure 17.13
The location/arrival time distribution for the example problem (from Nelson, R.W., Water Resources Res., v. 14, p. 416–428, 1978. Copyright by American Geophysical Union).

17.4   *Numerical Approaches*

The numerical approaches are a family of computer-based techniques for solving the contaminant transport equations. They approximate forms of the advection-dispersion equation as a system of algebraic equations or alternatively simulating transport through the spread of a large number of moving reference particles. Whatever the procedure used, it invariably has to be coded for solution on a high-speed computer.

Numerical approaches easily deal with variability in the flow and transport parameters (for example, hydraulic conductivity, porosity, dispersivity, and cation exchange capacity). It is this flexibility in representing parameters that facilitates modeling of layering or other more complex geometries in two and three dimensions. Thus, one can simulate the complex plume shapes that often develop in natural systems. The analytic and semianalytic approaches cannot account for variability in transport parameters. Thus, numerical approaches are readily adapted to site-specific problems, which makes them particularly useful in practice.

Other important features of the numerical approaches are the flexibility in implementing complex boundary conditions, and in accounting for a variety of important mass transport processes. Codes are now available to model (1) radionuclide chains in which radioisotopes and their daughters decay, each with a variable half life; (2) equilibrium reactions such as precipitation, dissolution, and complexation; and (3) competitive sorption or ion exchange. Developing the increased sophistication to deal with reactions is necessary to model accurately contaminant groups such as trace metals or radionuclides.

A Generalized Modeling Approach

The modeling process can be examined independently of a particular solution technique. The starting place in nearly all cases is with one of the various forms of the advection-dispersion equation. Because of the power of the numerical approaches, the transport equation is commonly formulated in two or three dimensions. Although our emphasis here is with mass transport involving a single contaminant, modeling approaches have evolved to the point where a coupled set of differential equations can be formulated for problems involving several reacting species.

One of the first steps in developing a computer model is to subdivide the region in terms of cells or elements. This process makes it possible to account for varying parameter values within the domain. Further, subdividing the region also serves to define the nodes at which concentrations are calculated. The way in which nodes are defined and how the domain is subdivided (for example, squares, rectangles, triangles, and stream tube segments) depends upon the specific numerical technique. The Otis Air Base case study presented later shows how one goes about subdividing a region and assigning boundary conditions.

There is really no difference in the boundary conditions and transport parameters required in the numerical approaches as compared to the analytical approaches. Complicating the situation with numerical models is the need to provide boundary conditions at a large number of node points or nodal blocks (areas around the nodes). For example, values of concentration or loading rates defining various boundary conditions now need to be specified for all nodes located along the boundary of the domain. Initial conditions and transport parameters are specified for all nodes except in some cases for nodes with constant concentrations.

It is no real problem to assign transport parameters to nodal areas, other than perhaps not knowing what the values are. Velocity values are an important exception because they are more difficult to specify. In most real systems with pumping and injection wells, and hydraulic conductivity values that vary, the velocity field can be extremely variable. Both the magnitude and the direction of flow can change in space and time. In terms of modeling, measured or guessed velocity values are not adequate. Continuity considerations in all numerical solutions of transport equations require a smooth and accurate representation of the velocity field.

This kind of field can only be obtained by simulation with a flow model. Approaches to calculating velocity values include applying the Darcy equation with calculated values of hydraulic and known parameters, or directly calculating velocity values at the nodes. Figure 17.14 illustrates how flow and mass transport models are used together to predict contaminant distributions. Input necessary for the flow model could include the hydraulic conductivity distribution, the water table configuration, and other boundary conditions. The transport model is coupled to the flow model by the velocity terms as shown in Figure 17.14.

Flow in many cases is assumed to be independent of the mass transport. In other words, the concentration of contaminants does not influence the flow by changing the fluid density. In this situation, there is no requirement to solve the flow and transport equations simultaneously. For steady flow, the ground water

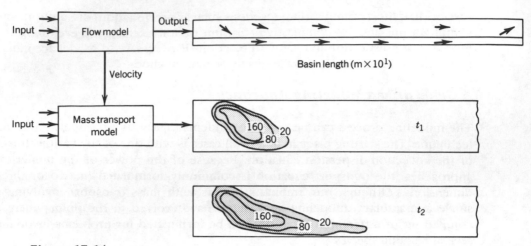

Figure 17.14
The relationship between ground water flow and mass transport models in simulating the distribution of contaminants (from Schwartz and others, 1985). Reprinted with permission of Second Canadian/American Conference on Hydrogeology. Copyright © 1985. All rights reserved.

flow equation is solved once and a single velocity field applies for all time. If flow is transient, the velocity field must be calculated each time that the contaminant concentration is required. However, again, the flow and transport equations are solved separately.

Contaminant concentrations are sometimes so large that the presence of the spreading plume causes the flow pattern to change. In these cases, the ground water flow depends on the mass transport just as the mass transport depends on the ground water flow. In Figure 17.14, the arrows joining the flow and transport models now will point in both directions. For this fully coupled situation, the flow and transport equations need to be solved simultaneously.

The Common Solution Techniques

The techniques for solving the advection-dispersion equation are extremely varied and would require almost a complete textbook to discuss them all in detail. Here, we present a brief summary of the most common methods mainly to introduce the pertinent literature and to create an awareness of the basic terminology. There are two general approaches available for solving mass transport problems. The direct solution techniques are the most common and involve a numerical solution of the advection-dispersion equation. Included in this group are the well-known finite difference and finite element techniques, and the method of characteristics.

Research is continuing to develop new and improved approaches for solving mass transport equations. The continuing effort to develop numerical approaches is attempting to reduce the computational effort thereby facilitating the analysis of complex three-dimensional problems. Examples of recent modeling approaches or significant refinements of existing methods include the principal direction method (Frind, 1982), the alternating direction Galerkin method (Daus and Frind, 1985), and the Laplace transform Galerkin technique (Sudicky, 1989).

The finite difference methods have traditionally been applied to solve flow and transport equations. One of the most important implementations of the finite difference approach is in the powerful code SWIFT (Dillon and others, 1978) and its succeeding versions. This code has become a standard for use in studying the most complicated mass transport problems.

The essence of the finite difference method is to replace the governing differential equation by a set of difference equations applicable to the system of nodes. The difference equations approximate the first- and second-order derivatives in the transport equation (that is, $\partial C/\partial x$ or $\partial^2 C/\partial x^2$) by concentration differences between node points. When each node in the network is considered, the result is a system of algebraic equations. This system can be solved with matrix methods. The interested reader will readily discover a voluminous literature on the subject. An important feature of the finite difference method is the relative ease in formulating the difference equations. Of all the numerical techniques available to solve differential equations, finite difference techniques are easiest to understand from a conceptual point of view.

Finite element approaches are common in solving mass transport equations. Although the approach was first developed in the petroleum industry, much of the subsequent refinements and applications were by hydrogeologists. The most important early studies included work by Pinder (1973) and Duguid and Reeves (1976) on the modeling of contaminant transport, and by Rubin and James (1973) on transport with multicomponent ion exchange. Various extensions of the technique examined density-dependent problems (see Huyakorn and Pinder, 1983).

Unlike the finite difference methods, which involve solving the mass transport equation directly, the finite element techniques deal with a mathematically equivalent integral form of the mass transport equation. This integral form can be developed in using the method of weighted residuals or the variational method. Huyakorn and Pinder (1983) present a detailed discussion of these approaches and in particular the Galerkin procedure (a special case of the method of weighted residuals).

The method of characteristics is another useful approach for solving mass transport equations. It was first applied by Garder and others (1964). Subsequently, the method was used in hydrogeological applications to study salt water intrusion (Pinder and Cooper, 1970), ground water contamination (Reddell and Sunada, 1970; Pinder, 1973), and the role of various parameters in controlling the spread of contaminants (Schwartz, 1975). Konikow and Bredehoeft (1978) implemented this approach in a versatile two-dimensional computer code, which today is one of the most commonly used codes.

The method takes the advection-dispersion equation and breaks it down into a set of simpler differential equations. This formulation in effect provides a frame of reference that is moving with the mean ground water velocity. Thus, the advection-dispersion equation, which is relatively difficult to solve, becomes a diffusion equation, which is simpler to solve.

The moving particle method is different from the previous three numerical approaches. It does not solve the mass transport equation directly but simulates the spread of mass dissolved in water. This approach is formulated as a classical random walk problem involving the motion of a swarm of reference particles. It has been applied in modeling atomic particle motion in nuclear reactors, the spread of air pollutants, and oil spills in oceans. Ahlstrom and others (1977) introduced the method which subsequently implemented in several computer codes (Schwartz and Crowe, 1980; Prickett and others, 1981).

The transport of contaminants is simulated by adding reference particles and moving them in a prescribed manner. By varying the number of particles added at the source during any one time step, it is possible to simulate complex loading functions. To account for advection, each particle moves in the direction of flow a distance that is determined by the product of the magnitude of the velocity and the size of the time step. With a small time step, this particle motion traces a pathline through the system. Dispersion is accounted for in the particle motion by adding to the deterministic motion a random component, which is a function of the dispersivities.

The mean concentration for each grid block is calculated as the sum of the mass carried by all the particles located in a given block divided by the total volume of water in the block. By appropriate adjustments to the quantity of mass carried by each particle, first-order kinetic reactions or simple linear sorption can be simulated with little difficulty.

Adding Chemical Reactions

Going beyond problems involving simple reactions like linear sorption, or first-order kinetic decay to those with several interacting solutes is much more difficult than any numerical approach considered so far. With a few exceptions, the possible approaches fall into two categories. With the "one-step" approach, a complete mass transport equation including all the appropriate reactions is written for each solute species. The entire coupled set of algebraic equations is then solved simultaneously using either a finite element or finite difference method. This procedure is numerically complicated but is the most rigorous way to handle reactions in a transport framework.

Most of the more general implementations of this scheme have been for one-dimensional transport. Extending this approach to two-dimensional problems is a formidable task. One of the most sophisticated examples of the one-step procedure is a code developed by Miller and Benson (1983). Their model and the planned extensions account for one-dimensional advection and dispersion along with a host of equilibrium chemical reactions. Included in the suite of reactions are complexation in the aqueous phase, exchange of both ions and complexes, and mineral precipitation. Other examples of these models include Willis and Rubin (1987) and Ortoleva and others (1987a, b).

The "two-step" procedure separates the physical and chemical processes. A solution to the advection-dispersion equation provides an initial estimate of concentration. A second step corrects these concentrations to account for the partitioning of mass due to a suite of chemical reactions. Operationally, the two-step procedures require iteration between the steps until some specified convergence criterion is met. Iteration assures that the calculated concentrations are solutions to both the advection-dispersion equation and the material balance equations describing the chemical system. This method is simpler computationally although less rigorous than the one-step procedure. For this reason, the approach is amenable to two- and three-dimensional problems. It also provides a way of grafting sophisticated geochemical codes like PHREEQE, MINTEQ, or MINEQL to transport models.

An example in the development of this approach is described by Cederberg (1985). The transport equations are solved in two dimensions using a Galerkin–finite element procedure and the algebraic equation set for the chemical reactions using a Newton-Raphson method. Chemical processes considered in this model include complexation and complex sorption, which considered the effects of surface ionization and complexation at the solid-solution interface. Other examples of two-step models come from work by Narasimhan and others (1986) and Liu and Narasimhan (1989).

17.5   Case Study in the Application of a Numerical Model

A model study of contamination at Otis Air Base on Cape Cod, Massachusetts (LeBlanc, 1984b), illustrates some of the steps in using a mass transport model. At Otis Air Base, approximately 1740 m³/d of treated waste water has been disposed in the subsurface since 1936. The disposal unit is an unconfined sand and gravel aquifer approximately 35 m thick. A zone of contamination has developed that is approximately 915 m long, 23 m thick, and 3.35 km long. This site was studied by the U.S. Geological Survey as part of the Toxic-Waste Ground-Water Contamination Program (LeBlanc, 1984a; Franks, 1987; Ragone, 1988).

The treated sewage effluent contains above-background concentrations of Na^+, Cl^-, ammonium, nitrate, phosphate, detergents, and several different volatile organic compounds. Boron, one of the trace metals in the effluent, was selected by LeBlanc (1984b) for a detailed model study because (1) its concentration has remained relatively constant in the effluent at 500 µg/L, (2) it has a relatively low background level in the native ground water (30 µg/L), and (3) it tends not to react chemically during transport. The objective in modeling was to guide the collection of data from the site and to test hypotheses concerning the character of contaminant migration.

The first step in developing the flow and boron transport models was to define the region of interest and establish boundary conditions for flow. LeBlanc (1984b) used the following guidelines, which should apply generally to most studies. The domain had to be large enough to include all the existing plume (Figure 17.15) as well as providing room for further spreading in the future. The side and bottom boundaries downgradient of the disposal area are ponds, rivers, and salt water bays. These features are a good choice for boundaries because in the absence of detailed hydraulic head data, they provide the best places to estimate boundary conditions. In effect by assuming ground water discharges at these locations, boundaries for ground water flow can be estimated in terms of constant head nodes or leakage fluxes. North of Coonamessett and Johns ponds, there are no natural hydrogeologic boundaries. In this area, flow lines formed the side boundaries (Figure 17.15). The flow lines are imaginary boundaries, located so as not to intersect the contaminant plume. The top or northern boundary was arbitrarily defined by the 60-ft equipotential line. This boundary could have been placed anywhere north of the site with the proviso again that contaminants from the site could not intersect this boundary.

The Konikow and Bredehoeft (1978) model requires that the site be subdivided into a regular finite difference grid. In this example, the grid consisted of 40 rows and 36 columns. The nodal blocks are square with dimensions chosen so that there is not an unmanageably large number of cells (an upper limit might be 50 or so cells in the row/column directions). However, the blocks are sufficiently small to ensure that the plume is not localized in just a few cells. The plume in this case has a width of about seven cells. Figure 17.16 shows how the model replicates the region shape and how the boundary conditions for flow are included.

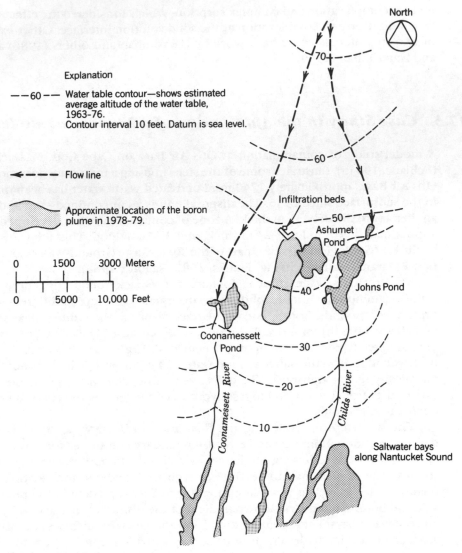

Figure 17.15
Map of important hydrologic features at the site in relation to the boron plume in 1978–1979 (modified from LeBlanc, 1984b).

LeBlanc modeled the pattern of ground water flow first. Because the density and viscosity of the contaminated ground water are nearly the same as the uncontaminated water, the distribution of hydraulic head and hence the velocity field are unaffected by the migration of the plume. The flow equation therefore can be solved first independently of the mass transport equation. Observations at the site indicate further that the gradients in hydraulic head do not change significantly with time. Thus ground water flow is steady, which means that the ground water velocities need only to be calculated once.

Figure 17.17 compares the observed configuration of the water table with the "best fit" from a series of simulations. Such trials are designed to calibrate the

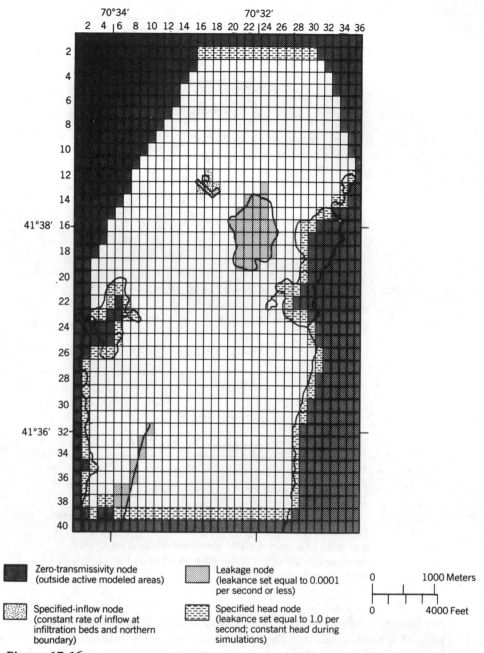

Figure 17.16
Grid system used to model both ground water flow and mass transport. For ground water flow modeling, the combination of zero transmissivity cells, leakance nodes, and specified head nodes define the region shape and boundary conditions (from LeBlanc, 1984b).

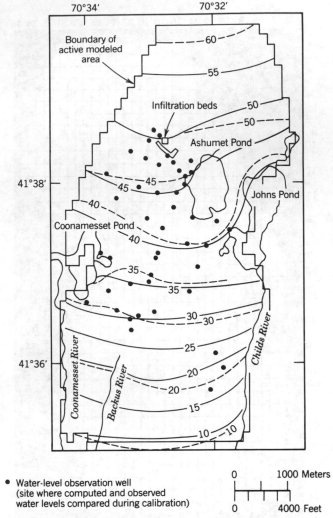

Figure 17.17
A comparison between the observed (dashed lines) and simulated (solid lines) water table elevations for November 1979 (from LeBlanc, 1984b).

model—a process of selecting the set of parameters that produce the best simulation of a known history. The calibration here took initial estimates of model parameters (for example, hydraulic conductivity, recharge rates) and adjusting them by trial and error until the model successfully reproduced the observed configuration of the water table. Because of the lack of available data, no attempt was made to reproduce spatial variability in hydraulic conductivity or recharge rates. Nevertheless, the simulated and observed results compare well (Figure 17.17). The model as represented by the parameters and boundary conditions is not unique. Another set of different parameters and boundary conditions could give the same or a better fit. By keeping the hydraulic conductivity close to the measured values, the simulated hydraulic heads and resulting velocity field should be a reasonable representation of the flow system.

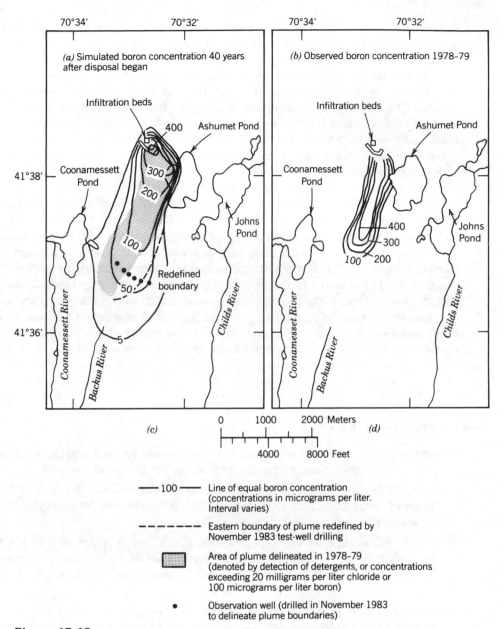

Figure 17.18
A comparison between the simulated distribution of boron in the aquifer (*a*) with that observed in 1978–1979 (*b*). Also shown on (*a*) are indications of the plume size provided by Cl⁻ and detergents (from LeBlanc, 1984b).

   The only information required for the transport simulation with a conservative species like boron is the longitudinal and transverse dispersivities and the loading function at the source. In the absence of actual dispersivity data, values were selected from the literature for similar types of geologic materials. Here, this uncertainty in choosing a value is not a problem because analysis shows that with active recharge simulated concentrations are not sensitive to the particular choice of dispersivity value. A lack of data may not be so easily dealt with in every model study. Contaminant loading was approximated by fixing the boron concentration at 500 µg/L at the four cells representing the infiltration beds.

Tests with the model attempted to reproduce the historical spread of contaminants over a 40-yr period between 1940 and 1978–1979. Compared in Figures 17.18*a* and *b* are the simulated and the observed distributions of boron after 40 yr. As LeBlanc (1984b) points out, the simulated path of the plume agrees reasonably well with that observed in 1978–1979, particularly at concentrations above 50 µg/L. However, at lower concentrations, the center line of the simulated plume was located east of the observed center line. The simulated plume is also somewhat longer and wider than the observed plume.

LeBlanc (1984b) explained the difference between observed and simulated plumes as follows. The observed plume could in fact be larger than is represented in Figure 17.18*b* because of uncertainty in establishing the edges of the plume at concentrations below 100 µg/L. Inaccuracies in the simulated flow field could have existed, which produced somewhat more divergent flow than actually occurs. This problem could be related to the complex interaction between ground waters and surface waters. Finally, the model could contain undetected numerical dispersion.

This study went on to evaluate the suitability of management options for the contamination. With reasonable confidence in the ability of this model to predict boron distributions, LeBlanc (1984b) examined the consequences of operating the site in the future as it is now or eliminating the boron source.

Suggested Readings

Domenico, P. A., 1987, An analytical model for multidimensional transport of a decaying contaminant species: J. Hydrol., v. 91, p. 49–58.

Javandel, I., Doughty, C., and Tsang, C. F., 1984, Groundwater Transport: Handbook of mathematical models: American Geophys. Union, Washington, Water Resources Monograph 10, 228 p.

Nelson, R. W., 1978, Evaluating the environmental consequences of groundwater contamination 2. Obtaining location/arrival time and location/outflow quality distributions for steady flow systems: Water Resources Res., v. 14, no. 3, p. 416–428.

LeBlanc, D. R., 1984b, Digital modeling of solute transport in a plume of sewage-contaminated ground water. Movement and Fate of Solutes in a Plume of Sewage-Contaminated Ground Water, Cape Cod, Massachusetts: U.S. Geological Survey Toxic Waste Ground-Water Contamination Program, U.S. Geol. Surv. Open-File Rept., p. 11–45.

Problems

1. In a one dimensional column, a retarded species has moved a certain distance when $v_w t$ of a nonretarded species is taken as 28 cm and α_x is = 2 cm. The retardation factor is two.

How far *behind the advective front* of the nonretarded species is the advective front of the retarded species?

2. In a one-dimensional column test, C/C_0 is noted to be on the order of 0.65. This means:

    _____ *a.* the advective front has already broken through

    _____ *b.* the advective front has not yet broken through

    _____ *c.* the argument of erfc is negative

    _____ *d.* the argument of erfc is positive

    _____ *e.* a and c are correct

    _____ *f.* b and d are correct

3. The erfc of negative beta is 1.146. What is the value of beta?

4. An experiment with a step function input of nonreactive tracer is conducted. The velocity of the tracer through the column is about 36 cm/hr. The length of the flow tube is 30 cm. The $C/C_0 = 0.25$ point is arrived at the end of the column in approximately 0.7 hours. At time equal to 0.992 hr., $C/C_0 = 0.75$, and at time 0.833 hr., $C/C_0 = 0.5$.

    a. Estimate the dispersion coefficient in cm²/hr.

    b. Estimate the dispersivity in cm.

    c. Can you use the $C/C_0 = 0.5$ at $t = 0.833$ hr. in the preceding calculations? Why?

5. In a plan view of a contaminant plume you notice that chloride has moved approximately 1000 m whereas cadmium has moved about 100 m. Assuming both constituents were released from the source at the same time, find the distribution coefficient for cadmium if the porosity is 0.2 and the density ρ_s is 2.65.

6. Answer the following with regard to the Ogata Banks equation.

$$C = \frac{C_0}{2} \, \text{erfc} \, \frac{x - vt}{2(D_x t)^{1/2}} = \frac{C_0}{2} \, \text{erfc} \, \frac{x - vt}{2(\alpha_x vt)^{1/2}}$$

a. The maximum observed concentration will be equal to _____ .

b. This maximum will occur at some point x where x is _____ .

c. At $x = vt$, the concentration C is equal to _____ .

d. The concentration approaches zero at some point x where x

_____

Prove or disprove each of the following.

e. The larger α_x, the more extensive the spread ahead of the center of mass.

f. The larger α_x, the further behind the center of mass the displacing fluid takes on its maximum concentration.

g. The larger α_x, the longer the time required to reach steady state at the column exit.

h. When determining α_x in the laboratory, it is advisable to use the C/C_0 data at the column exit when $x = vt$.

i. The larger α_x, the lower the concentration at some point just ahead of the center of mass.

7. A spill from a high-level storage tank releases the following radionuclides.

Nuclide	Activity ci	Half life years	K_d ml/gm	D_x cm²/sec	D_y	D_z
A	400	33	8	10^{-4}	10^{-5}	10^{-6}
B	400	28	52	10^{-3}	10^{-5}	10^{-7}
C	400	2.7	40	10^{-5}	10^{-5}	10^{-5}
D	400	20	0.1	10^{-6}	10^{-5}	10^{-6}

_____ a. What constituent will be first to arrive at the property boundary?

_____ b. What constituent will be last to arrive at the property boundary?

_____ c. What constituent will have the smallest maximum concentration at the property boundary?

_____ d. What constituent will have the largest maximum concentration at the property boundary?

_____ **e.** What constituent will have the smallest spread around the center of mass in the direction of flow?

_____ **f.** What constituent will have the largest spread around the center of mass in the direction of flow?

_____ **g.** What constituent will have the same spread around the center of mass in all directions, that is, in the flow direction, transverse to the flow direction, and vertically as well.

8. A hazardous waste facility is to be constructed in a given area. The state requires a buffer zone between the waste trenches and the property boundary. Over 500 different constituents are to be included in the waste, some with reasonably large distribution coefficients, others with a likely distribution coefficient of zero. A few years after construction, a plume is first noted by observation wells at the waste boundary. The maximum concentration in the plume occurs in a zone of about 6 m thick and extends the entire length of the waste boundary, or about 60 m. Because of low permeability bounding materials, the plume is only capable of lateral spreading in Y but *not* vertical spreading in Z.

Calculate the maximum concentration that might be expected (in terms of C_0, the plume concentration at the waste boundary) once the plume arrives at the edge of the buffer zone, which is 150 m from the waste boundary. The transverse dispersion coefficient is 1×10^{-3} m²/sec and the linear velocity (flow through the pores) is 1×10^{-3} m/sec.

9. Ground water travels with a linear velocity of 15 cm/d. What will the transport velocity be for an organic contaminant having a K_d of 6.6 ml/g in a medium with a porosity of 0.35 and a solids density (ρ_s) of 2.65 g/cm³?

10. Following is a site map and hypothetical plume. At the F-area and H-area seepage basins marked on the figure, waste water containing a variety of contaminants was disposed of over a 13-yr period into an unconfined aquifer. Your objective is to model the transport of a contaminant at the F-area that is moderately sorbed, and is lost via a first-order kinetic reaction. Formulate this problem for a numerical solution by (1) defining a local region of interest and appropriate boundary conditions for flow and transport, (2) selecting appropriate forms of the flow and transport equations, and (3) indicating what parameter values are required for a simulation. (Figures reprinted by permission of Solving Ground Water Problems with Models. Copyright © 1987. All rights reserved.)

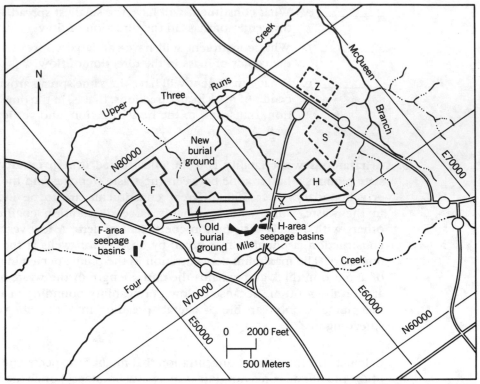

(from Duffield and others, 1987)

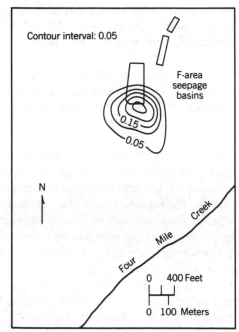

(from Duffield and others, 1987)

11. A plume of bromoform dissolved in ground water originates from two storage ponds with a total length of 150 m. The ponds penetrate the unconfined aquifer to a depth of 5 m. The total saturated thickness of the aquifer is 15 m. Use the computer code that was developed in Example 17.6 to solve this problem. Note that is will be necessary to modify the code to consider only one direction of z spreading (that is, Eq. 17.18). Predict the concentration distribution along the water table after 5110 days assuming a linear ground water velocity of 1.6×10^{-3} cm/sec, dispersivities in x, y, and z of 5.0 m, 0.5 m, and 0.1 m respectively, $C_0 = 4000$ mg/L, $f_{oc} = 0.0045$, log $K_{ow} = 2.38$, porosity 0.30, and a solids density of 2.65 g/cm^3.

Chapter Eighteen

Process and Parameter Identification

18.1 **Tracers and Tracer Tests**

18.2 **The Diffusional Model of Dispersion**

18.3 **Dispersivities Estimated Using the Advection-Dispersion Equation**

18.4 **New Approaches in Estimating Dispersivity**

18.5 **Identification of Geochemical Processes**

18.6 **Quantification of Geochemical Processes**

Any attempt to interpret mass or contaminant distributions or to analyze a problem quantitatively requires estimates of the important transport parameters. In this chapter, we will examine the variety of field and laboratory techniques available for this purpose and discuss their problems and limitations. This knowledge provides the parameter-based framework that is essential for designing field studies. The variety of different chemical processes often means that any one of several reactions might affect dissolved contaminants in surprisingly similar ways (for example, mineral precipitation and cation exchange, or sorption and bacterial reduction). Thus, identifying processes or combination of processes can be a difficult and not insignificant problem. At sites of ground water contamination, predictions of contaminant behavior or choices for remedial strategies depend on understanding the geochemical processes.

There is often the need to go one step beyond identifying a process to characterizing it in terms of equilibrium constants or kinetic terms. For example as we saw in Chapter 17, such information is absolutely necessary for contaminant transport models. In many respects, this investigation of geochemical processes is a new area for most practitioners. Parameters are required to characterize processes like hydrophobic sorption and organic biodegradation, which a few years ago many people really never even knew existed.

18.1   Tracers and Tracer Tests

Because tracers reflect the outcome of mass transport process, they are useful for characterizing mass transport and mass transfer processes. The most important tracers are (1) ions that occur naturally in a ground water system such as Br^- or Cl^-; (2) environmental isotopes such as ^2H, ^3H, or ^{18}O; (3) contaminants of all kinds that enter a flow system; and (4) chemicals added to a flow system as part of an experiment. This last group could include radioisotopes such as ^3H, ^{131}I, ^{82}Br; ionic species such as Cl^-, Br^-, I^-; and organic compounds such as rhodamine WT, lissamine FF, and amino G acid. Many of these ions or compounds do not react to any appreciable extent with other ions in solution and the porous medium. These are what classically have been referred to as "ideal tracers." Others do react and for this reason are particularly useful in defining the nature of reactions. Excellent reviews of ground water tracers are given by Davis and others (1985) and Smart and Laidlaw (1977).

Given this choice of tracers confusion can arise as to which to use. In most cases, the choice is linked inexorably to the scale of the study or the presence of a contaminant plume. For example, tracing flow and dispersion in a unit of regional extent (10s to 100s of km) will involve either naturally occurring ions or environmental isotopes. Conducting a tracer experiment on such a large scale is simply not feasible because of the long times required for tracers to spread regionally. On a more localized or site-specific basis (several kilometers), the presence of a contaminant plume that has spread over a long time automatically makes it the tracer of choice. Again, time is insufficient (except perhaps in karst) to run an experiment at the scale of interest. Only for small systems (for example, some fraction of a kilometer) is there any real justification for running a field tracer experiment. It is only for these experiments when a selection has to be made among the various tracers. For reactive tests, the best tracer is one that participates fully in the reaction of interest.

Tracers are transported in two kinds of flow environments. In a natural gradient system, tracers move due to the natural flow of ground water. In a system stressed by injection and/or pumping, transport occurs in response to gradients typically much larger than those in natural systems. A small-scale tracer experiment can involve either of these flow conditions.

Success in all these tests depends on adequately characterizing concentration distributions in space and time. Typically, a large three-dimensional network of monitoring points is necessary to define the tracer distribution accurately. Point sampling is essential to avoid concentration averaging within the well bore in cases where concentrations vary over relatively small vertical distances. For example, estimated values of dispersivity tend to be larger when the number of individual sampling points is small or samples are collected over relatively large vertical intervals. This excess dispersion represents error due to inadequacies of the sampling network.

Field Tracer Experiments

Field tracer experiments can be run with a natural flow system or one modified by pumping or withdrawal. It should be possible in theory to estimate mass transport parameters from almost any test where a tracer is added in a controlled way. However, a few more or less standard tests are preferred because in most cases there are simple procedures for interpreting the results. Unfortunately, many of these tests are subject to error because run in their basic configuration there is often an insufficient number of monitoring points to provide a three-dimensional representation of the tracer distribution.

Natural Gradient Test

The natural gradient test involves monitoring a small volume of tracer as it moves down the flow system. Keeping the quantity of tracer small minimizes the initial disturbance of the natural potential field. The resulting concentration distributions are the data necessary to determine advective velocities, dispersivities, and occasionally equilibrium and kinetic parameters.

This experiment typically requires a dense network of sampling points. In one research study at Canadian Forces Base Borden, Canada (Mackay and others, 1986), the network contained more than 5000 separate sampling points. Not every experiment requires this kind of instrumentation. However, serious attempts at estimating transport parameters will require many tens to several hundred monitoring points.

Single-Well Pulse Test

The single-well pulse test (Figure 18.1a) involves first injecting a tracer followed by water into an aquifer at some constant rate. After an appropriate period of injection, the aquifer is pumped at the same rate. The concentration of tracer in water being withdrawn is monitored as a function of time or total volume of water pumped. These concentration/time data provide a basis for estimating the longitudinal dispersivity and chemical parameters like K_d within a few meters of the well (Pickens and others, 1981). Fried (1975) describes a more refined version of this test where concentration/time data are collected at different positions in the well with a downhole probe.

While historically of some interest, this test has limited applicability in estimating values of dispersivity. It is generally not possible to scale up dispersivity

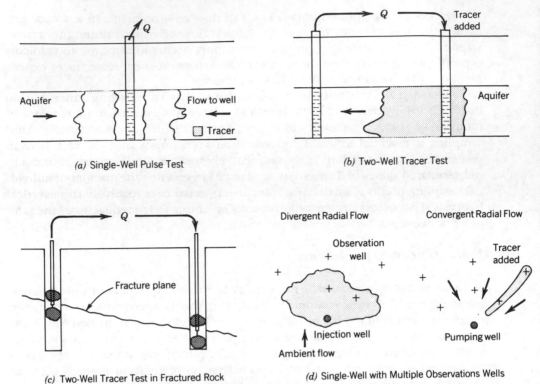

Figure 18.1
Examples of field tracer tests.

measurements made at such a small scale to the larger scales of interest in most problems. In addition, the lack of detailed observations on how the tracer is spreading in the vicinity of the well makes any dispersivity estimate quite crude. This test holds more promise for evaluating geochemical processes. Because reactions operate on a local scale, there is no real scaling problem.

Two-Well Tracer Test

The two-well test involves pumping water from one well and injecting it into another at the same rate to create a steady-state flow regime (Figure 18.1b). The tracer is added continuously at a constant concentration at the injection well and monitored in the withdrawal well. The simplest way of running this test is by recirculating the pumped water back to the injection well as is shown on the figure. However, the tracer concentration will begin to increase at the injection well once breakthrough occurs at the pumped well. The test can also be run without recirculation by providing the water for the injection well from an alternative source. As before, the resulting concentration versus time data at the pumped well are interpreted in terms of processes and parameters. These tests can be conducted over several hundred meters in highly permeable systems. However with only a single monitoring point, the test provides at best only a crude estimate of dispersivity. Estimates can be improved by adding more observation wells between the pumping/injection doublet.

Modified versions of a two-well test have been conducted in fractured rocks (Raven and Novakowski, 1984). Packing off a small section of the borehole (Figure

18.1*c*) isolates a single fracture plane between pumping and injection wells. While this test is technically more demanding because it requires working with packers, it involves the same kind of interpretive technique and is subject to the same limitations we just discussed. In all cases, there is the inherent assumption of radial flow.

Single-Well Injection or Withdrawal with Multiple-Observation Wells

These tests create a transient radial flow field by injection or withdrawal. The radially divergent test (Figure 18.1*d*) involves monitoring the tracer as it moves away from the well. The radially convergent test involves adding the tracer at one of the observation wells and monitoring as it moves toward the pumped well. Parameters can be estimated at scales of practical interest with reasonable accuracy provided a reasonable number of observation wells is provided. If a choice exists as to which of these two tests to run, the divergent flow test is preferable (Gelhar and others, 1985). The converging flow field (Figure 18.1*d*) counteracts spreading due to dispersion and is thought to be less useful.

18.2 The Diffusional Model of Dispersion

This section illustrates how the diffusional model of dispersion provides a basis for estimating dispersivities from concentration data. Concentration distributions are assumed to be normally distributed with the dispersion coefficient related to the variance of the tracer distribution and time and dispersivity related through velocity to the dispersion coefficient. A significant advantage of working with plume statistics alone is that dispersivity estimates are independent of the assumptions implicit in the advection-dispersion equation. As Freyberg's (1986) work at the Borden site showed, this method provides information on how dispersivities vary as a function of travel distance. In the following section, we will explore other approaches that involve fitting various forms of the advection-dispersion equation to concentration data. In most cases, interpretations based on these approaches require that dispersivities remain constant as a function of travel time (Freyberg, 1986).

Consider the two-dimensional plume shown on Figure 18.2*a* produced by a continuous, point source. At any point along the middle of the plume (for example, *x*,0) a breakthrough curve (Figure 18.2*b*) can be constructed by plotting the relative concentration as a function of time (Robbins, 1983). The relative concentration is C/C_{max}, where C_{max} is the highest concentration that will be observed at $(x,0)$. C_{max} is always less than the source concentration C_0.

This breakthrough curve is a cumulative normal distribution with a 2σ value (σ is the standard deviation) that can be derived graphically from Figure 18.2*b* as

$$2\sigma_t = (t_{84} - t_{16}) \tag{18.1}$$

where t_{84} and t_{16} are the breakthrough times corresponding to relative concentrations of 0.84, and 0.16, respectively. Having calculated σ, the dispersion coefficient is

$$D_L = \frac{v^2 \sigma_t^2}{2t} \tag{18.2}$$

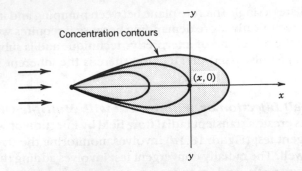

Concentration contours

(a) Spreading of a Tracer from a point Source

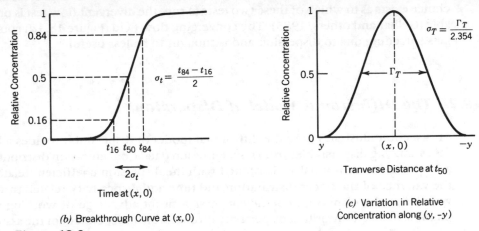

$$\sigma_t = \frac{t_{84} - t_{16}}{2}$$

$2\sigma_t$

Time at $(x, 0)$

(b) Breakthrough Curve at $(x, 0)$

$$\sigma_T = \frac{\Gamma_T}{2.354}$$

Tranverse Distance at t_{50}

(c) Variation in Relative
Concentration along $(y, -y)$

Figure 18.2
Pattern of spreading of a tracer from a continuous point source in a one-dimensional flow field. Two graphical procedures for estimating the standard deviation of a normal distribution are shown in (b) and (c) (from Robbins, 1983).

The following equation relates the dispersion coefficient to the dispersivity (α) and the linear flow velocity

$$D_L = \alpha_L v \tag{18.3}$$

The same approach can be followed to establish values of dispersivity in the transverse direction. This time the relative concentration is plotted as a function of distance traveled transverse to the direction of flow at t_{50} (Robbins, 1983). The resulting concentration distribution is again normal (Figure 18.2c). Once the variance is estimated, the dispersivity value is

$$D_T = \frac{\sigma_T^2}{2t} \tag{18.4}$$

The standard deviation in this case (Figure 18.2c) is related to the half-width of the distribution (Γ) at a relative concentration of $0.5C_{max}$ (Robbins, 1983)

$$\sigma_T = \frac{\Gamma}{2.354} \tag{18.5}$$

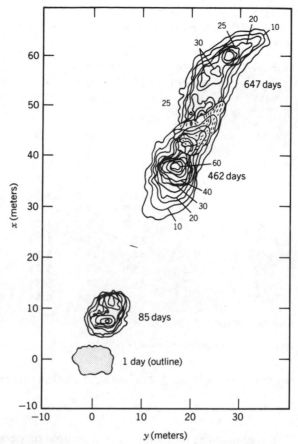

Figure 18.3
Vertically averaged concentration distribution of Cl⁻
at various times after injection (from Mackay and
others, Water Resources Res., v. 22, p. 2017–2029,
1986. Copyright by American Geophysical Union).

This same graphical approach can be used with any normal distribution. As
before, space/time statistics can be interchanged in a constant velocity field with
this equation

$$\Gamma_T = \Gamma_t v \tag{18.6}$$

Borden Tracer Experiment

A massively instrumented tracer test at Canadian Forces Base Borden (Mackay and
others, 1986; Freyberg, 1986) provides an excellent illustration of how the spreading pattern of a plume can be interpreted in terms of transport parameters. The
concentration of two nonreactive tracers and five volatile organic compounds
was monitored for slightly more than 1000 days at over 5000 observation points.
The experiment was conducted as a natural gradient test. Our immediate interest
here is with dispersion so first we consider the behavior of Cl⁻. The addition of
10.7 kg of Cl⁻ at the source provided a mean concentration of 892 mg/L (Freyberg,
1986). The test was conducted in a shallow sand aquifer with a mean ground
water velocity measured to be 0.091 m/d. Figure 18.3 illustrates the vertically

Table 18.1

Estimates of mass in solution, location of the center of mass, and spatial covariance for the chloride plume*

Date	Elapsed time, days	Mass in solution, kg	Center of mass[1] x_c, m	Center of mass[1] y_c, m	Center of mass[1] z_c, m	Spatial covariance[2] $\sigma_{x'x'}$, m²	Spatial covariance[2] $\sigma_{y'y'}$, m²	Spatial covariance[2] $\sigma_{x'y'}$, m²
			Chloride					
Aug. 24, 1982	1	6.7	0.2	0.1	2.78	2.1	2.4	0.5
Sept. 1, 1982	9	9.2	0.7	0.4	3.02	1.7	2.4	0.7
Sept. 8, 1982	16	9.2	1.6	0.7	3.06	2.3	2.8	0.8
Sept. 21, 1982	29	11.5	2.9	0.9	3.27	2.5	2.6	0.9
Oct. 5, 1982	43	11.3	4.1	1.6	3.34	4.4	2.7	1.2
Oct. 25, 1982	63	9.0	5.7	2.0	3.50	4.4	2.4	1.1
Nov. 16, 1982	85	11.2	7.7	3.2	3.75	5.7	3.3	0.8
May 9, 1983	259	11.5	22.7	11.6	4.52	17.8	4.4	3.7
Sept. 8, 1983	381	9.6	32.3	15.3	5.18	20.6	4.4	3.9
Oct. 26, 1983	429	9.2	35.9	17.2	5.25	24.3	6.0	3.2
Nov. 28, 1983	462	8.2	38.2	17.4	5.33	27.8	5.5	2.1
May 31, 1984	647	9.1	53.1	23.9	5.55	51.5	5.5	3.0

[1]Given in the field coordinate system.

[2]Given in rotated coordinates: x' is parallel to linear horizontal trajectory, y' is perpendicular to linear horizontal trajectory.

*From Freyberg, Water Resources Res., v. 22, p. 2031–2046, 1986. Copyright by American Geophysical Union.

averaged Cl⁻ distribution at four times. Longitudinal dispersion is obvious as the plume gradually elongates in the direction of flow. Transverse spreading normal to the mean flow direction is less marked (Figure 18.3).

Freyberg (1986) calculates the center of mass of the Cl⁻ plume in three dimensions (Table 18.1) as well as the variance in concentration distributions in the longitudinal and horizontal transverse directions. Transverse vertical spreading is negligible. Details of the calculation are discussed by Freyberg (1986).

Several different techniques were used to interpret the data. The simplest is based on the relationships among variance, time, and dispersivity (Eq. 10.3), assuming that dispersivities are constants. Fitting a line through σ_L^2 and σ_T^2 versus time data (Figure 18.4) provides an estimate of the dispersion coefficients D_L and D_T. From these values, Freyberg (1986) calculates α_L and α_T to be 0.36 and 0.039 m, respectively. At later times, the straight-line approximation does not fit the data too well (Figure 18.4), indicating that dispersivities may not be constant with time or travel distance.

Freyberg (1986) used another interpretive technique that is based on the time rate of change in the variance. Starting with Eq. 10.3 and substituting for the dispersion coefficient with Eq. 18.3 provides the following equation

$$\alpha_L v = \frac{\sigma_L^2}{2t} \tag{18.7}$$

or

$$2\alpha_L x = \sigma_L^2 \tag{18.8}$$

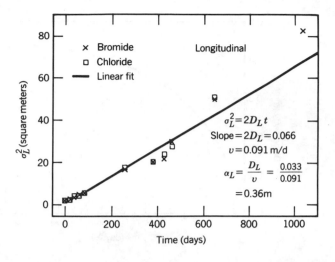

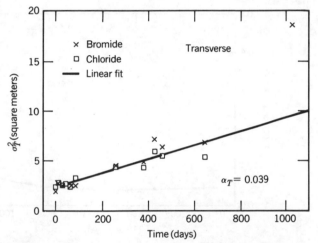

Figure 18.4
Fit of the Borden data to a constant dispersivity model
(from Freyberg, Water Resources Res., v. 22, p. 2031–
2046, 1986. Copyright by American Geophysical Union).

Differentiation with respect to time gives

$$2\alpha_L \left(\frac{dx}{dt} \right) = \frac{d\sigma_L^2}{dt} \tag{18.9}$$

The rate at which the variance ($d\sigma_L^2/dt$) grows is a function of velocity dx/dt and dispersivity α_L. This equation is used by expressing the differentials in Eq. 18.9 in difference form

$$\alpha_L = \frac{1}{2v} \left[\frac{\sigma_L^2(t_i) - \sigma_L^2(t_{i-1})}{t_i - t_{i-1}} \right] \tag{18.10}$$

where t_i and t_{i-1} refer to the present and previous times at which the concentration distribution is measured. Taking successive pairs of variances from Table 18.1, Freyberg (1986) calculated how dispersivities varied as the plume moved

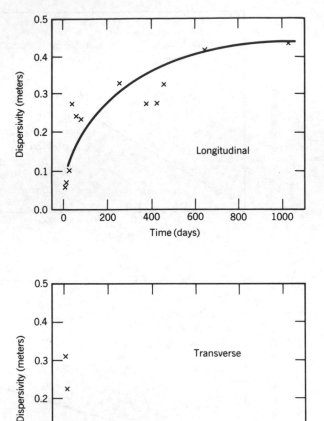

Figure 18.5
Variation in the longitudinal and transverse dispersivity as a function of travel time (modified from
Freyberg, Water Resources Res., v. 22, p. 2031–2046,
1986. Copyright by American Geophysical Union).

down the aquifer. The longitudinal dispersivity appears to be scale dependent
(Figure 18.5). Values increase from 0.06 m near the source to about 0.43 m after
about 1000 days. This behavior is in line with concepts of an asymptotic dispersivity discussed in Chapter 10. The horizontal transverse dispersivity remained
more or less constant throughout the test.

18.3 Dispersivities Estimated Using the Advection-Dispersion Equation

Most other approaches for evaluating concentration distributions involve fitting the
data to some form of the advection-dispersion equation. This fitting process can
involve curve matching or other graphical methods, or more complex modeling

Figure 18.6
Patterns of ground water flow in a two-well tracer test (from Güven and others, 1986). Reprinted by permission of Ground Water. Copyright © 1986. All rights reserved.

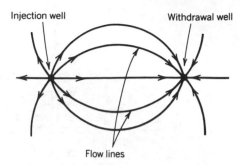

techniques. These latter techniques match observed and simulated concentrations through a trial-and-error choice of transport parameters. However, as we will see shortly, work is going on to estimate parameters using formal inverse methods.

Interpreting Data from a Two-Well Tracer Test

The two-well test is commonly used for estimating longitudinal dispersivities. Several different interpretive approaches are available, but one of the simplest is Grove and Beetem (1971). An injection/withdrawal doublet operating in a homogeneous and isotropic unit produces the idealized pattern of flow shown in Figure 18.6. A series of crescent-shaped flow tubes connects the two wells. The Grove and Beetem method subdivides the flow domain into 179 of these flow tubes. The individual flow tubes are treated as a column of length L for which a one-dimensional transport equation is assumed to hold. For a given time, the series of one-dimensional equations is solved at the withdrawal well with calculated travel times, and assumed values of dispersivity and porosity. The sum of relative concentrations from all flow tubes is an estimate of the overall composition of discharge water. Repeating this calculation for different times yields a composite breakthrough curve, which can be compared to the measured breakthrough curve. Repeated simulation trials with different dispersivity and porosity values should produce an acceptable match between the calculated and observed breakthrough curve. Grove (1971) discusses the procedure for interpreting two-well data and provides a computer code.

When the tracer experiment is of the recycling type, the shape of the composite breakthrough curve is affected by the eventual change in tracer concentration at the source. Grove (1971) discusses a procedure to account for recycling by superimposing segments of the original breakthrough curve. More complexity arises when ambient regional flow is not negligible. Readers again can refer to Grove (1971) who provides a theoretical approach for handling this situation.

Heterogeneity in hydraulic conductivity influences the outcome of two-well tests. Of the available approaches for dealing with heterogeneous systems, one of the simplest is that of Güven and others (1986). Their model accounts for nonuniform tracer advection due to layering, but neglects local hydrodynamic dispersion. In other words, the variability in the breakthrough curve at the pumped well is simply due to the variability in advective transport caused by layering.

The aquifer is assumed to be horizontal and confined with both the overall porosity and total aquifer thickness constant (Figure 18.7). The aquifer is divided into n horizontal layers each having a constant thickness and hydraulic conductivity. The pattern of layering is defined for example by hydraulic testing using

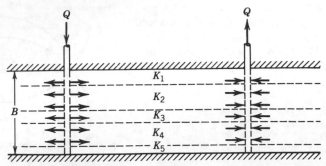

Figure 18.7

Idealization of aquifer heterogeneity for a two-well tracer experiment (modified from Güven and others, 1986). Reprinted by permission of Ground Water. Copyright © 1986. All rights reserved.

multilevel piezometers located between the two wells or by flow meter surveys in the wells.

The flow field is subdivided into 180 crescent-shaped flow tubes. Travel times are calculated for each. Now, however, the calculation is carried out on a layer by layer basis. The concentration at the withdrawal well is determined as the flow-weighted contribution of $180 \times n$ flow tubes, the length of the tracer injection period (for pulse tests), and corrections for recirculation. The fitting exercise involves varying porosity and the hydraulic conductivity of the different units to match the simulated and observed breakthrough curves.

Huyakorn and others (1986a) present a finite element approach for layered systems that includes dispersion. The method involves the formulation of three-dimensional, curvilinear flow net elements. Each layer in an aquifer is subdivided by an identical set of elements that are defined along and normal to the pathlines.

The application of the layered models for interpreting tracer tests (Huyakorn and others, 1986b) shows why it is important to account for variations in hydraulic conductivity. Shown in Figure 18.8 are breakthrough data for a non-recirculating test at a site near Mobile, Alabama. Plotted on the same figure are three breakthrough curves. One is simulated by a two-dimensional model that ignores variability in hydraulic conductivity. The remaining two are runs from a three-dimensional model with slightly different hydraulic conductivity distributions that account for the stratification. When the interpretation ignores layering, the aquifer dispersivity (α_L) is 4.0 m. When layering is considered, α_L is 0.15 m or more than 25 times smaller. In effect, representing the system with a large dispersivity and no variability in hydraulic conductivity, or a small dispersivity and layering in hydraulic conductivity, produces much the same breakthrough curve. As a general rule, the more faithfully the actual hydraulic conductivity field is represented, the smaller the dispersivity becomes.

Deciding which of these dispersivity values is correct may not be possible. A more detailed representation of the hydraulic conductivity field could produce an even smaller dispersivity value. The state of practice in modeling is choosing a dispersivity value that is appropriate for the detail at which the hydraulic conductivity field is described.

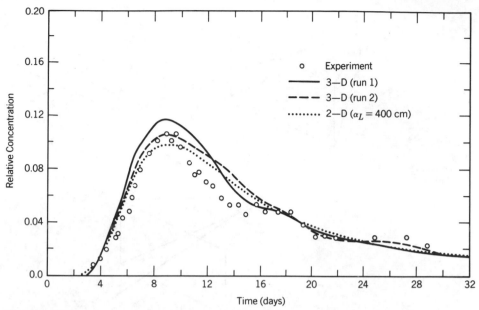

Figure 18.8
A comparison of three different model fits to breakthrough data collected in the withdrawal well for a two-well test at the Mobile site (from Huyakorn and others, Water Resources Res., v. 22, p. 1016–1030, 1986. Copyright by American Geophysical Union).

Domenico and Robbins (1985b) Graphical Procedure

Unknown transport parameters can be estimated directly from concentration data using a graphical/mathematical procedure developed by Domenico and Robbins (1985b). The theory of the procedure resides in the finite source multidimensional models for mass transport discussed in Chapter 17. This approach provides estimates for the longitudinal and two transverse dispersivities, the source concentration, the size of the source, and the position of the advective front. The source is assumed to be a semi-infinite parcel with dimensions Y and Z normal to the direction of ground water flow (Figure 18.9a). The advective front is defined by the product of the contaminant velocity and the time since the contaminant first entered the ground water system.

Now if n unknown parameters are to be determined, we require n different expressions. If the analysis is restricted to a single horizontal plane (for example, the water table Figure 18.9a) in the near field or steady-state part of the plume, the governing equation has three unknowns α_y, Y, and C_o. By taking a ratio of this form of the equation for two points in the near field, the source concentration is eliminated, providing one expression with two unknowns. It is not possible to solve for the unknowns. However, one can assume a variety of different values for one of the unknowns and solve for the other. On a plot of Y versus α_y, these pairs of values form a line (Figure 18.10). Carrying out the same procedure at two or more points at different distances from the source provides a family of curves. Ideally, these curves all pass through a single point giving unique values of Y and α_y. The value of C_o is determined next from the steady-state form of the equation. A similar procedure can be applied for Z and α_z.

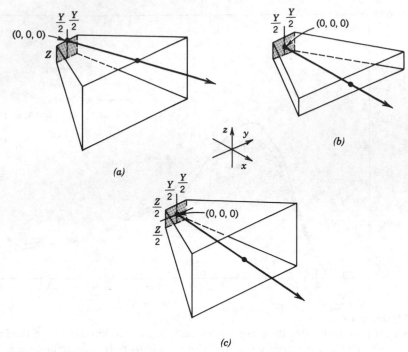

Figure 18.9
Idealized patterns of spreading for the Domenico and Robbins (1985b) model (from Domenico and Robbins, 1985b). Reprinted by permission of Ground Water.

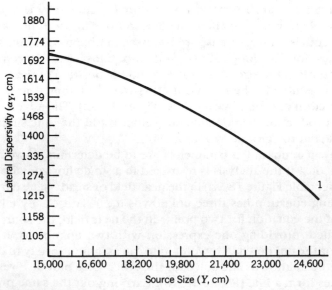

Figure 18.10
Iterative diagram for a single well pair (from Mirsky, 1989).

Figure 18.11
Chloride distribution in the Brunswick aquifer (modified from Bredehoeft and Pinder, Water Resources Res.,
v. 9, p. 194–210, 1973. Copyright by American
Geophysical Union).

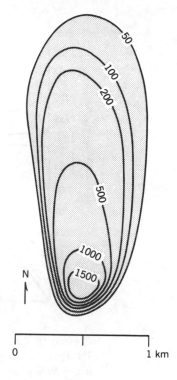

The two-well iteration is limited because it requires that wells be located at the
same x coordinate. To this end, an equation that relates data from three wells (x_1,
y_1, C_1; x_2, y_2, C_2; x_3, y_3, C_3) with α_y and source dimension Y has been developed
by Edwards (1988) and a four well iteration by Fryar and Domenico (1987).

This procedure has been applied to the Brunswick chloride plume. The plume
was studied by Bredehoeft and Pinder (1973) and is shown in Figure 18.11. The
iteration procedure yields the common point $Y = 133$ m and $\alpha_y = 11.5$ m (Figure
18.12), which are well-defined and unique parameters. Each line on this diagram
was computed by the ratio method discussed earlier. Given values for Y, α_y, and
C_0, the steady-state concentration can be calculated for every part of the plume.
In the near field, the actual values will be at their steady (maximum) values. In the
far field, the actual concentrations will be less than their steady-state values.
Given far-field concentrations, manipulation of the governing equation will permit calculations of concentrations along the plane of symmetry of the plume
independent of information on the longitudinal dispersivity and the position of
the advective front. This calculation is shown in Figure 18.13 for the Brunswick
plume where curve B gives the observed concentrations, curve A gives the calculated steady-state concentrations, and curve C gives the ratio of curve B to A, the
so-called relative concentration profile, which is of the form

$$\frac{C(x, 0, 0, t)}{C'(x, 0, 0)} = \left(\frac{1}{2} \right) \text{erfc} \left[\frac{(x - vt)}{2(\alpha_x vt)^{1/2}} \right] \tag{18.11}$$

where C' is the steady-state concentration along the plane of symmetry. Note
that the plume is at its steady state to a distance of about 700 m from the source

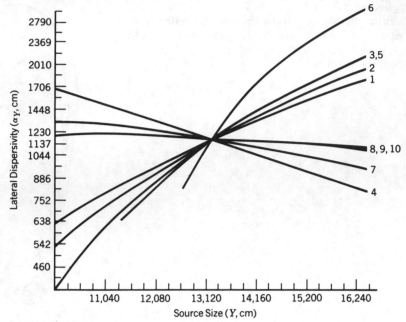

Figure 18.12
Ten iterative diagrams for five wells within the steady-state part of the
Brunswick plume (from Mirsky, 1989).

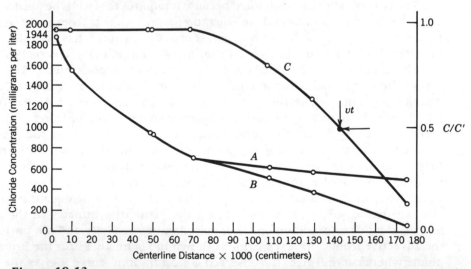

Figure 18.13
Diagram showing the plane of symmetry actual concentrations (*B*), actual and
projected steady state concentrations (*A*), and the ratio of curve *B* to curve *A* (*C*) for
the Brunswick plume (from Mirsky, 1989).

($C/C' = 1$; Figure 18.13). For $x = vt$, the value of the complementary error function goes to one. Thus, the advective front is located at the distance where $C/C' = 0.5$ or approximately 1400 m from the source. Because v is the contaminant velocity, this procedure can be applied to a retarded species with no information required on the degree of retardation.

The last unknown is the longitudinal dispersivity, which can be readily calculated by using an unsteady concentration in Eq. 18.11 and solving for the remaining unknown.

Applications of the method include the work of Domenico and Robbins (1985b) and LaVenue and Domenico (1986). Similar procedures can be used to analyze plumes with a first-order reaction term (Fryar and Domenico, 1989). In all these applications, the equations and methods given were readily programmed for a personal computer.

Trial-and-Error Fit of a Numerical Model

Another way of working with concentration data for a plume is to use a numerical model to try and simulate the system. A satisfactory match between the observed and simulated concentration distribution implies that the parameters used in the model describe the system. One weakness with this approach, unfortunately, is that the "best-fit" parameters cannot be shown to be unique. A different, untested combination of parameters could provide an equally good or better fit. In spite of this limitation, the approach is common in practice with constraints by field data to help justify the final choice of parameters.

The technique usually involves the sequential fitting of flow and transport models to various subsets of the available data. The first step is estimating the velocity field. Except in rare instances when the magnitude and direction of flow are constant (that is, unidirectional flow), the velocity field has to be calculated with a ground water flow model. A series of trial-and-error runs attempts to match calculated values of hydraulic head with observed data.

The next step is the calibration of the mass transport model through a series of trial-and-error runs. With a conservative species like Cl⁻, the only adjustable parameters are the transverse and longitudinal dispersivities. If the velocity field is correct, a few runs could be all that is necessary to match the plume. The fit may not be too good initially because of the sensitivity of the plume geometry to the velocity field. It is often necessary to go back to the ground water flow model to adjust the velocity field. Finally after a number of trials, the simulated plume can be coaxed to advect and disperse like the observed plume. An example of this procedure is provided by the Otis Air Base case study in Section 17.6.

18.4 New Approaches in Estimating Dispersivity

The problem of estimating values of dispersivity from various types of field data continues to be an area of active research in contaminant hydrogeology. To conclude the discussion on dispersivity, we would like to mention two approaches that could help with this task: (1) formal techniques for parameter estimation and (2) techniques based on stochastic transport theory.

The formal inverse approaches use measured values of concentration as a basis for calculating a set of optimal dispersivity values. Examples of work in this

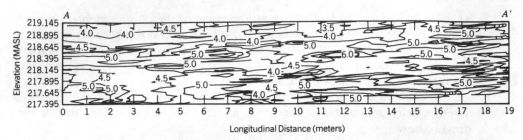

Figure 18.14
Variation in hydraulic conductivity (as -ln *K*) along the cross section *A-A'* at Canadian Forces Base Borden (modified from Sudicky, Water Resources Res., v. 22, p. 2069–2082, 1986. Copyright by American Geophysical Union).

area are Murty and Scott (1977), Umari and others (1979), Jury and Sposito (1985), Strecker and Chu (1986), Wagner and Gorelich (1986), and Yeh and Wang (1987). Yeh (1986) reviews the general topic of inverse techniques. The goal of the continuing work in this area is to develop easy to use computer codes for mathematically extracting a set of transport parameters from information about the plume.

The stochastic approach involves using the geostatistical concepts developed in Chapter 10. Thus, information about hydraulic conductivity and not concentration provides the basic data for the analysis. An example in the estimation of longitudinal asymptotic dispersivity is provided by Frind and other (1987) using data collected by Sudicky (1986). Shown in Figure 18.14 is the contoured hydraulic conductivity field based on 1279 measurements along a single cross section. This section is oriented parallel to the direction of tracer migration at Canadian Forces Base Borden (Freyberg, 1986; Mackay and others, 1986). The distribution in hydraulic conductivity when expressed as ln(K) is normally distributed with $\sigma_Y^2 = 0.38$. Sudicky (1986) estimated on the horizontal and vertical correlation length scales (λ_x and λ_z) to be 2.8 m and 0.12 m, respectively. These values confirm what is apparent from looking at the contoured field, namely, that the correlation structure is anisotropic.

The contribution of heterogeneity to the asymptotic longitudinal dispersivity can be calculated from Eq. 10.23 as

$$A_L = \frac{\sigma_Y^2 \lambda_x}{\gamma^2} = \frac{0.38 \times 2.8}{1^2} = 1.1 \text{ m}$$

where the flow factor γ is taken to be one. With the contribution from diffusion essentially negligible and the local-scale dispersivity (α_L) estimated to be 0.05 m, most of the longitudinal dispersion arises due to heterogeneity in the hydraulic conductivity field (Frind and others, 1987). The overall value (that is, 1.15 m) is of a similar magnitude to that reported by Freyberg (1986). The reasonable agreement in two independent methods in this test points to the potential of statistical methods in estimating dispersivity values.

An attractive feature of this method is that it does not require concentration distributions. Thus, it has the potential to be applied to large-scale systems, too big to be tested using more conventional methods. One of the obvious limitations of this approach is the relatively large number of hydraulic conductivity values that are required to defines the spatial structure of the medium accurately. The only practical way to address this limitation is through indirect determination of hydraulic conductivity using various kinds of borehole logs or a borehole flow meter.

18.5 Identification of Geochemical Processes

A problem peculiar to geochemical processes is knowing whether a particular reaction is operative. This issue was not important for mass transport processes because both advection and dispersion usually operate to some extent. Approaches to deal with this problem of process identification can involve the interpretation of field or laboratory data and the evaluation of observed contaminant distributions in light of various theoretical models. Following are a few of the strategies, illustrated by examples, that one can follow.

Discovering Systematics in Plume Configurations

With organic contaminants, it is often difficult to determine whether hydrophobic sorption is responsible for apparent retardation in contaminant spread. However, with site investigations and tracer tests, one can test whether the systematics predicted by a hydrophobic sorption model are reflected in plume geometries. In other words, a compound with a large octanol/water partition coefficient should be strongly sorbed and form a small plume, whereas a compound with a small K_{ow} should form a large plume.

The plumes of organic contaminants at the Gloucester site near Ottawa, Canada, are a case in point (Jackson and others, 1985). A variety of organic contaminants were disposed of in shallow trenches in glacial drift. Identified as contaminants were diethyl ether, tetrahydrofuran, 1,4-dioxane, carbon tetrachloride, benzene, and 1,2-dichloroethane. Plumes for three of these illustrate the variable extent of spreading (Figure 18.15). However, the size of the plume for all compounds is inversely related to the hydrophobicity of the compound as measured by the K_{ow}. This relationship strongly suggests that hydrophobic sorption is a dominant process. However, the f_{oc} is relatively low for units at the site (Table 12.8) and the possibility exists for sorption onto mineral surfaces.

Complete Characterization of Possible Reactions

Another useful strategy for unraveling complex chemical data is to characterize possible reactions in terms of reactants, products, and biological facilitators such as bacteria. Thus while it may not be possible to say why the concentration of a contaminant is declining, the presence or absence of degradation products or elevated numbers of bacteria could provide valuable clues.

The first example to illustrate this idea describes how an increase in the number of bacteria in water is evidence for microbial degradation of organic contaminants (Ceazan and others, 1984). The example of contamination due to the disposal of treated sewage at Otis Air Base has been introduced already in Chapter 17. One other aspect of the study involved quantifying bacterial levels in water by counting the number of colony-forming units on both nutrient agar and dilute soil-extract media (Ceazan and others, 1984). Both methods of plating give

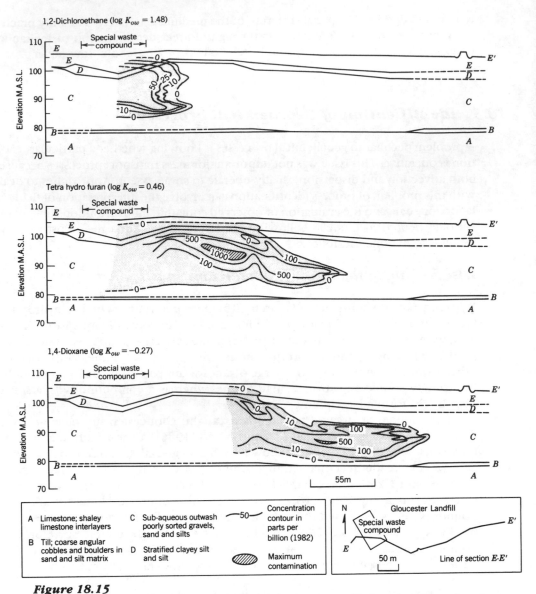

Figure 18.15
Sections illustrating the occurrence of three of the organic contaminants found at the Gloucester sanitary landfill near Ottawa, Ontario. Reprinted from Jackson and others, 1985. Contaminant Hydrogeology of Toxic Organic Chemicals at a Disposal Site, Gloucester, Ontario, 1. Chemical Concepts and Site Assessment. NHRI Paper No. 23, Environment Canada, 114 p. Reproduced with the permission of the Minister of Supply and Services Canada, 1990.

approximately the same results (Figures 18.16*a* and *b*). The areas with elevated bacterial counts coincide with the position of the plume estimated using measured contaminant concentrations. The strong correlation between bacterial counts and the distribution of dissolved organic carbon (Figure 18.16*c*) is evidence that microbial processes are removing organic compounds from the plume (Ceazan and others, 1984). Other work by Harvey and others (1984) showed that

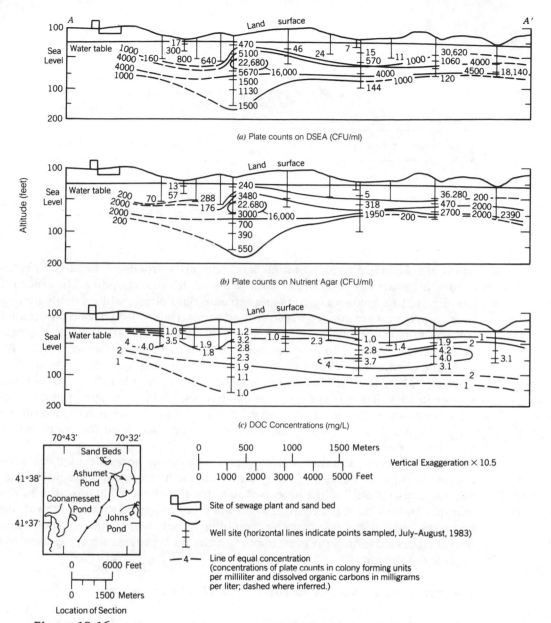

Figure 18.16
Bacterial distributions as reflected by plate counts and their relationship to dissolved organic carbon content (from Ceazan and others, 1984).

these free-living bacteria in the ground water represent a small proportion of the total bacterial population. Most of the bacteria are bound in biofilms on solid surfaces. The relatively large organic carbon content in downstream parts of the plume (that is, 2750 m from the source) is not matched by large bacterial populations. A significant proportion of the organics, however, are nonbiodegradable detergents used before 1964.

The second example, a problem of ground water contamination from a pipeline break at Bemidji, Minnesota, in 1979, illustrates how the careful documen-

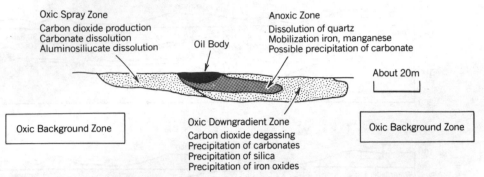

Figure 18.17
Redox environments related to a crude oil spill at Bemidji, Minnesota (from Siegel, 1987).

tation of reactions provides a useful way to identify processes. Crude oil percolated to the water table and began to move in a shallow sand aquifer. The crude oil has migrated approximately 20 m as a separate fluid phase, while soluble derivatives from the oil are moving in the ground water (Hult, 1987). Based on detailed studies of the major ions, some trace metals, pH, E_H, and dissolved oxygen, Siegel (1987) defined four redox environments related to the spill (Figure 18.17). By using the chemical data together with an extremely detailed study of sediments (Berndt, 1987), the processes shown in Figure 18.17 were identified. Work on the sediments included bulk chemistry using direct current plasma spectrometry and selective extraction techniques to establish how the metals are partitioned on the solids. In addition, a scanning electron microscope was used for grain-surface analyses (Bennett, 1987).

At the same site, Eganhouse and others (1987) show how a detailed examination of the organic chemistry of the monoaromatic hydrocarbons provides a useful "molecular probe" of the processes affecting the dissolved hydrocarbons. For example, by looking at the changes in the apparent disappearance rates in various benzenes, and the n-alkybenzene homology, they were able to show that microbial degradation is the most important chemical process affecting dissolved organic compounds.

Interpreting Reaction Pathways in the Laboratory

Laboratory experiments can identify processes particularly in complex chemical systems. The goal of such experiments is often to define reaction pathways without necessarily trying to obtain precise estimates of transport parameters. The reason for using laboratory experiments is the control that is afforded in running the experiment. An example is in the use of microcosms to explore the transport and fate of organic compounds and trace metals in biologically mediated systems. Unlike most inorganic chemical reactions, which are reasonably systematic in thermodynamic terms, biological reactions are inherently much more complex. The main problem is determining whether a given contaminant will react at a meaningfully rapid rate. For example, in Section 12.5, we discussed a variety of factors influencing the biotransformation of organic compounds. Experiments with microcosms provide the window to the world of biological processes. The term microcosm refers generally to experimental systems in which various components of the subsurface environment are isolated and studied

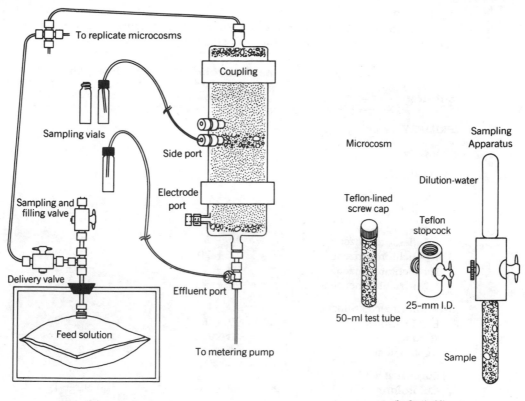

(a) Flow-through Microcosm (from Bengtsson, 1985)

(b) Static Microcosm
(from Wilson and others, 1985)

Figure 18.18
Examples of (a) flow-through and (b) static microcosms (in Ground Water Quality, eds. C.H. Ward, W. Giger, and P.L. McCarty). Copyright © 1985 by John Wiley & Sons, Inc. Reprinted by permission of John Wiley & Sons, Inc.

(Wilson and Noonan, 1984). More commonly, however, the term is used specifically to describe experimental systems that are biologically active. These systems consist of a porous medium, water, and a functioning population of microorganisms isolated in a container. The particular design is a trade-off between the convenience of operation in maintaining a given environment, cost, and the faithfulness with which given processes are simulated (Wilson and Noonan, 1984). With the flowthrough systems (Figure 18.18a), fluids are moved continuously through the microcosm in a column-style experiment. Figure 18.18b shows a static system that provides a much simpler design. Wilson and Noonan (1984) discuss the main features and limitations of these and other similar systems.

An example in the use of microcosms is work by Bouwer and McCarty (1984) to evaluate the secondary utilization of organic contaminants. The experiments involved flowthrough studies in columns under both aerobic and methanogenic conditions. Acetate was the primary substrate to study the secondary utilization of halogenated organic compounds. Glass beads in the columns supported the biofilms and minimized potential complications due to sorption of the organic compounds. With aerobic conditions, the chlorinated benzenes and nonchlorinated aromatic compounds were effectively removed by biooxidation (Table 18.2). However, the halogenated aliphatic compounds were not degraded significantly (Table

Table 18.2
Average utilization of substrates in aerobic and methanogenic biofilm reactors (modified from Bouwer and McCarty, 1984)**

Substrate	Aerobic column[1] (% removal)	Methanogenic column[1] (% removal)
PRIMARY		
Acetate	95 ± 1	99 ± 1
SECONDARY		
Chlorinated benzenes		
Chlorobenzene	91 ± 4	7 ± 26
	85 ± 6	—
1,2-dichlorobenzene	96 ± 2	4 ± 27
1,3-dichlorobenzene	22 ± 20	-2 ± 29
1,4-dichlorobenzene	98 ± 1	-4 ± 29
1,2,4-trichlorobenzene	90 ± 5	11 ± 25
Nonchlorinated aromatics		
Ethylbenzene	99 ± 1	7 ± 26
Styrene	>99	8 ± 26
Naphthalene	99 ± 1	-2 ± 29
Halogenated aliphatics		
Chloroform	-2 ± 20	99 ± 1
Carbon tetrachloride	—	>99
1,2-dichloroethane	—	-1 ± 20
1,1,1-trichloroethane	3 ± 27	97 ± 3
1,1,2,2-tetrachloroethane	—	97 ± 3
Tetrachloroethylene	-1 ± 50	76 ± 10
Bromodichloromethane	—	>99
Dibromochloromethane	—	>99
Bromoform	—	>99
1,2-dibromoethane	—	>99

[1]One standard deviation of the mean values is given.

18.2). Under methanogenic conditions, the results were just the opposite, negligible loss of the aromatic compounds and utilization of the aliphatic compounds. In summary, these experiments showed that the secondary utilization of organic compounds was possible. Furthermore, the type of electron acceptor is critical to the overall process.

18.6 *Quantification of Geochemical Processes*

We have already discussed some aspects of the problem of quantifying geochemical processes. Classical equilibrium theory provides a starting point for describing acid-base, solution, volatilization, precipitation, and complexation reactions.

Data are tabulated for many of the chemical systems of interest to hydrogeologists. However, experiments will be necessary occasionally to refine available data. Similarly, certain families of kinetic reactions like radioactive decay and hydrolysis are well characterized in terms of rate laws and parameters.

Of the important reactions, sorption and oxidation reduction are the most uncertain. Following here is a discussion of how parameters describing these reactions can be estimated. There are as before opportunities for studying these reactions in the field or the laboratory. However, most of the work in quantifying sorption and redox reactions has gone on in the laboratory.

Field Approaches

Many of the same strategies for estimating dispersivity apply to characterizing chemical processes. This statement assumes of course that the chemical processes are correctly identified. All the approaches generally involve fitting a form of the advection-dispersion reaction equation to a set of observed concentration data. As before, the measurements can be from an actual plume, or from one or more tracers added during a field experiment. To make the methods work, it is necessary to know the concentration distribution of one of the unreactive species, like Cl^-. This information provides a basis for estimating the mass transport parameters independently of the reactive species.

The most common field techniques involve application of (1) the retardation equation, (2) an extended form of the Domenico-Robbins graphical procedure, or (3) a trial-and-error fit with a numerical model. Application of the retardation equation simply interprets concentration distributions in terms of advection and linear sorption. By calculating the retardation factor for the plume, one can deduce the distribution coefficient because the retardation factor is related to the distribution coefficient (K_d)

$$R_f = 1 + \frac{\rho_s K_d (1 - n)}{n}$$
(18.12)

Thus, either the retardation factor (R_f) or K_d serves as the characteristic parameter describing the sorption process.

Let us examine another aspect of the large-scale tracer experiment at Canadian Forces Base Borden to illustrate this approach further. Five halogenated organic tracers—carbon tetrachloride (CTET), bromoform (BROM), tetrachloroethylene (PCE), 1,2-dichlorobenzene (DCB), and hexachloroethane (HCE)—were also part of the experiment. These compounds sorb on solid organic carbon (Roberts and others, 1986). Retardation in the spread of these compounds as compared to Cl^- was evident with the more strongly sorbed (HCE and DCB) being most retarded. Figure 18.19 illustrates the distribution of two of the lesser retarded compounds (CTET and PCE) in relation to Cl^-. Roberts and others (1986) characterized the sorptive behavior of these compounds in terms of retardation factors estimated from the velocity ratios, v_{Cl}/v_i (Table 18.3). In line with sorption being a dominant process, the more hydrophobic compounds exhibit the most marked retardation. Interestingly, all the calculated retardation factors increased as a function of time. This behavior is explained as a deviation from local equilibrium (Roberts and others, 1986). In other words, the assumption that the rate of sorption is fast relative to physical transport is not completely valid.

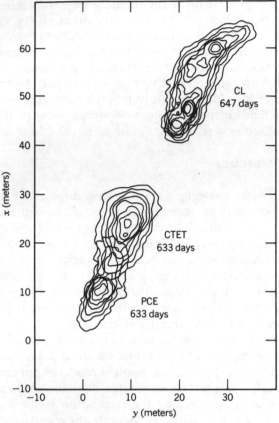

Figure 18.19
Retardation of carbon tetrachloride (CTET) and
tetrachloroethylene (PCE) as compared to chloride (CL). Contour interval depicted for CL is 5
mg/L beginning at an outer contour of 10 mg/L.
Contour intervals depicted for CTET and PCE are
0.1 µg/L beginning at an outer contour of 0.1
µg/L (from Roberts and others, Water Resources
Res., v. 22, p. 2047–2058, 1986. Copyright by
American Geophysical Union).

   The Domenico-Robbins procedure provides another way of working with
field data to quantify a given reaction. The method we discussed in Section 18.3
can be extended to look at linear sorption defined in terms of a distribution coefficient or retardation factor. Having identified sorption as the cause of retardation,
one evaluates the plume exactly as before. However, the velocity that is determined now is that of the contaminant or expressed another way v_w/R_f. The
retardation factor for the plume can be determined if the velocity of the water or
an unretarded species like Cl^- is known.
   The simplicity of this approach stems from the fact that the advection-dispersion equation for sorbing and nonsorbing compounds differs only by the retardation factor, which divides the dispersion coefficients and the velocity term. Adding
a process like oxidation or radioactive decay that is described by a first-order

Table 18.3
Calculated retardation factors for five chlorinated
hydrocarabons as a function of time*

Time	Retardation factors for				
(days)	CTET	BROM	PCE	DCB	HCE
15	1.8	1.9	2.7	3.9	5.1
30	1.9	2.0	3.1	4.6	6.1
85	2.1	2.2	3.9	5.7	7.9
250	2.3	2.5	4.8	7.3	Not quantifiable[1]
400	2.4	2.6	5.3	8.0	Not quantifiable[1]
650	2.5	2.8	5.9	9.0	Not quantifiable[1]

[1]Not quantifiable owing to disappearance of HCE.

*From Roberts and others, Water Resources Res., v. 22, p.
2047–2058, 1986. Copyright by American Geophysical Union.

kinetic reaction requires another term in the transport equation. The analytical solution will change to reflect this additional process. Domenico (1987) shows how to work with this solution in the same manner as before to derive information about the source size, initial concentration, position of the mean advective front, and now the decay-rate constant (or half-life) of the reaction.

The trial-and-error modeling approach begins as before to determine the velocity field and porous medium dispersivities in a set of calibration trials. Once the values of the mass transport parameters are fixed, the parameters describing reactions (for example, half life for decay or distribution coefficients) are varied to match the simulated distribution of a reactive species with observed data. Proceeding through the fitting exercise in a stepwise fashion minimizes the number of parameters that need to be adjusted in any given set of simulations. While this approach is certainly feasible, there are few studies that have used it.

Laboratory Techniques

Another approach to quantifying sorption and redox reactions is with batch and column experiments. Batch sorption experiments provide isotherms that model the partitioning of a species between the aqueous and solid phases. These laboratory-measured parameters are useful in understanding how systems behave at the field scale. For example, a series of batch sorption studies on samples from Canadian Forces Borden concentrated on the use of laboratory methods in quantifying the sorptive behavior of the five organic tracers (Curtis and others, 1986). The first tests determined the length of time required for equilibrium to be achieved. This experiment is typically run with a series of identical soil/water/tracer mixtures where each batch is stopped at a specified time. The calculated quantity of tracer sorbed on the solid for each test plotted versus time provides the reaction rate. The study by Curtis and others (1986) showed that two or three days were required for equilibrium to be achieved. Thus while not instantaneous, the reactions are relatively fast.

Other experiments defined the equilibrium isotherms. The results in general fit a linear sorption model (see examples, Figure 18.20). As expected, calculated

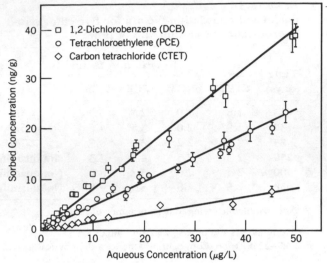

Figure 18.20
Sorption isotherms for CTET, PCE, and DCB (from Curtis
and others, Water Resources Res., v. 22, p. 2059–2067,
1986. Copyright by American Geophysical Union).

distribution coefficients increased with increasing hydrophobicity as measured
by the octanol/water partition coefficient (Table 18.4).

These laboratory values compare well with those computed by Roberts and
others (1986) from observed distributions in the field (Table 18.5). What is not so
good are estimates of retardation factors from K_d values determined as the
product of the fraction organic carbon and the K_{oc} that comes from the Schwarzenbach and Westall (1981) equation. Apparently, this approach underestimates
laboratory values, because sorption on inorganic surfaces is not considered (Curtis and others, 1986). The f_{oc} is sufficiently small for this aquifer that this contribution to sorption probably cannot be ignored.

Column-type experiments are also helpful in studying sorption reactions. As
compared to batch tests, the sediment/water ratio in a column is more representative of natural conditions as is the reaction setting. Moving pore fluids in the

Table 18.4
Measured distribution coefficients[1] from batch sorption experiments*

	CTET	BROM	PCE	DCB	HCE
Number of data	11	10	25	24	52
C_{max}, ng/cm³	47	13	49	50	34
K_d, 95% CI	0.17 ± 0.03	0.17 ± 0.02	0.48 ± 0.02	0.81 ± 0.03	0.81 ± 0.03
R^2	0.972	0.992	0.994	0.993	0.981

[1](cm³/g)

*From Curtis and others, Water Resources Res., v. 22, p. 2059–2067, 1986. Copyright by American
Geophysical Union.

Table 18.5

Comparison of retardation factors determined three different ways*

	Predicted[1]	Batch experiments	Field experiments[2]
CTET	1.3	1.9 ± 0.1	1.8–2.5
BROM	1.2	2.0 ± 0.2	1.9–2.8
PCE	1.3	3.6 ± 0.3	2.7–5.9
DCB	2.3	6.9 ± 0.7	3.9–9.0
HCE	2.3	5.4 ± 0.5	5.1–7.9

[1]Calculated from the regression by Schwarzenbach and Westall (1981) with $f_{OC} = 0.02\%$.

[2]From Roberts and others (1986).

*From Curtis and others, Water Resources Res., v. 22, p. 2059–2067, 1986. Copyright by American Geophysical Union.

column actively redistribute the products of reaction unlike batch experiments. On the negative side, column experiments are technically more challenging to run.

Working with reactive contaminants requires no change in the column design except that the apparatus should be constructed from materials that minimize sorption of the tracer. The interpretation of observed breakthrough curves in terms of parameters, however, is more complicated than is the case for nonreactive transport. If linear equilibrium sorption is the only chemical process, retardation factors again provide a simple way to characterize the process. If the reactions are kinetic or involve multiple sites for sorption, the preferred approach is to fit the breakthrough data to various models of the exchange process. Because of the simple flow conditions in the column, analytical approaches are employed rather than more complex numerical ones. These fits can be accomplished by trial and error or using a least-squares approach that automatically calculates the best set of exchange parameters fitting a given breakthrough curve (van Genuchten, 1981).

Kan and Tomson (1986) use column experiments to study sorption phenomena. Their problem was to explain why toluene and napthalene although apparently strongly sorbed on solid organic carbon could be much more mobile than predicted from hydrophobic sorption theory. The working hypothesis was that transport is being enhanced or facilitated by cosolution with dissolved organic compounds moving at velocities close to that of the ground water. Kan and Tomson tested this idea with a small column (0.9 cm in diameter and 15 cm long) packed with sieved soil and test solutions containing both an organic contaminant and molecules of bovine serum albumin (BSA) or Triton X-100 micelles. These latter pair of compounds have been shown to have sorptive properties similar to naturally occurring organic compounds. Breakthrough curves from these experiments for tritiated water, benzene ($C_0 = 100$ µg/ml), BSA ($C_0 = 250$ µg/ml), napthaline/BSA, and napthaline ($C_0 = 2$ µg/ml) are shown in Figure 18.21. Retardation factors are used to interpret the breakthrough curves and to quantify the combined sorption processes. The breakthrough curves and calculated retardation factors for benzene and BSA (1.3), napthaline/BSA (4.8) and naphthalene

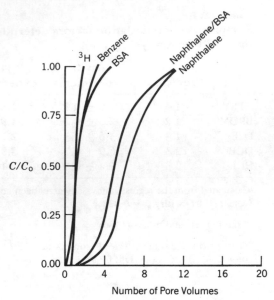

Figure 18.21
Normalized breakthrough curves for tritiated water, benzene, BSA, napthalene/BSA, and napthaline (from Kan and Tomson, 1986). Reprinted by permission of Conference on Petroleum Hydrocarbons and Organic Chemicals in Ground Water—Prevention, Detection and Restoration. Copyright © 1986. All rights reserved.

(6.1) point to facilitated transport. The following modified form of the retardation equation describes the processes

$$R_f = 1 + \frac{K_{oc} \cdot f_{oc} \cdot \rho_s(1 - n)/n}{1 + K_{DOC} \cdot DOC \cdot 10^{-6}} \tag{18.13}$$

where DOC is the dissolved organic carbon content (micrograms per milliliter) and K_{DOC} is the distribution coefficient describing the partitioning of organic solute between dissolved organic carbon and water. This relationship provides a basis for estimating the effect of this process on mass transport.

Redox reactions, particularly those involving the biotransformation of organic compounds, are beginning to receive attention in laboratory studies. Most emphasis is in the quantification of rates. As with sorption, both batch and column techniques are useful. These have been summarized in our earlier discussion of microcosms. Techniques available to interpret the data are essentially the same as for sorption experiments. With the batch techniques, the concentration of the contaminant in solution is monitored as a function of time with the resulting data interpreted in terms of a kinetic model of the process. An example of this approach is work by White and others (1985) looking at the biodegradation rates of methanol and tertiary butyl alcohol, compounds used commercially as gasoline additives. They interpreted their experimental results in terms of the order of the degradation reaction and the utilization rate (milligrams per liter per day) and suggested that published biokinetic parameters for a Monod-type model provide a reasonable description of the process.

Suggested Readings

Curtis, G. P., Roberts, P. V., and Reinhard, M., 1986, A natural gradient experiment on solute transport in a sand aquifer 4. sorption of organic solutes and its influence on mobility: Water Resources Res., v. 22, n. 13, p. 2059–2067.

Freyberg, D. L., 1986, A natural gradient experiment on solute transport in a sand aquifer 2. spatial moments and the advection and dispersion of nonreactive tracers: Water Resources Res., v. 22, n. 13, p. 2031–2046.

Fryar, A. E., and Domenico, P. A., 1989, Analytical inverse modeling of regional-scale tritium waste migration: J. Contam. Hydrol., v. 4, p. 113–125.

Güven, O., Falta, R. W., Molz, F. J., and Melville, J. G., 1986, A simplified analysis of two-well tracer tests in stratified aquifers: Ground Water, v. 24, n. 1, p. 63–71.

Jackson, R. E., Patterson, R. J., Graham, B. W., Bahr, J., Belanger, D., Lockwood, J., and Priddle, M., 1985, Contaminant hydrogeology of toxic organic chemicals at a disposal site, Gloucester, Ontario: 1. Chemical Concepts and Site Assessment. Environment Canada, National Hydrol. Research Instit., Paper No. 23, Ottawa, 114 p.

Roberts, P. V., Goltz, M. N., and Mackay, D. M., 1986, A natural gradient experiment on solute transport in a sand aquifer 3. retardation estimates and mass balances for organic solutes: Water Resources Res., v. 22, n. 13, p. 2047–2058.

Problems

1. The accompanying diagram is an enlargement of the Cl⁻ plume measured after 462 days during the tracer experiment at Canadian Forces Base Borden and shown previously in Figure 18.3.

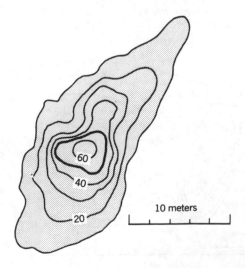

    a. Determine σ_L^2 from the field data using the graphical procedure and a conventional statistical calculation.

    b. Assuming that the longitudinal dispersivity and velocity are constant and that the plume originates as a point source at the coordinates (0.0, 0.0 in Figure 18.3), estimate α_L.

    c. Explain in this case why estimates of σ_T will not yield estimates of the transverse dispersivity. *Hint*: Examine Figure 18.3.

    d. Examine the character of transverse dispersion in Figure 18.3. What qualitative assessment can you make about the importance of transverse dispersion?

2. A series of hydraulic conductivity measurements for an unconfined aquifer provide a mean hydraulic conductivity (Y) of 0.004 m/s, where $Y = \ln K$, a variance in the log transformed hydraulic conductivity (σ_Y^2) of 1.0, and a correlation length in the direction of the mean flow (λ_x) of 10.0 m. Estimate the asymptotic macroscale dispersivity for the aquifer.

3. Shown here is the Cl⁻ plume in the confined outwash aquifer at the Special Waste Compound at Gloucester, after about 13 yr. Make some simplifying assumptions about the nature of dispersion and the source loading and determine the retardation factors for the three organic compounds whose plumes are shown in Figure 18.13. (Figure is reprinted from Jackson and others, 1985. Contaminant Hydrogeology of Toxic Organic Chemical at a Disposal Site, Gloucester, Ontario, 1. Chemical Concepts and Site Assessment. NHRI Paper No. 23, Environment Canada, 114 p. Reproduced with the permission of the Minister of Supply and Services Canada, 1990.)

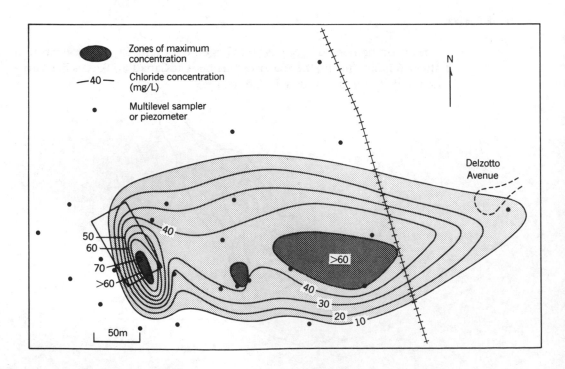

4. **a.** Take the estimated retardation factors in problem 3 and calculate K_d values for the three compounds. Assume a porosity of 0.32 and a solids density of 2.65 g/cm³.

   b. The measured f_{oc} for the Gloucester site is 0.0006. Calculate the K_d value using the Hassett and others (1983) equation. The K_{ow} values for the compounds are given in Figure 18.3.

   c. Compare the two sets of results in (a) and (b). Explain why one set of distribution coefficients is preferable for further modeling studies.

5. Use the analytic solution coded for Problem 16.11, and model the Cl⁻ distribution shown previously for Problem 3. List the best set of transport parameters determined from the trial-and-error fitting exercise. The Z dimension of the source is 10 m. To simulate the nonconstant source loading, it is necessary to use superposition in time of several analytical solutions. For example, adding the concentrations due to a constant 100 mg/L source that begins at t_0 and a constant -50 mg/L source that begins at 2 yr simulates the effect of the source concentration changing from 100 mg/L to 50 mg/L after 2 yr.

6. Explain how the modeling approach in Problem 5 could be extended to determine whether a contaminant like diethyl ether that is obviously being retarded through sorption is also being degradated biologically.

7. You are given the information shown in the accompanying figure for the plane of symmetry of a continuous source plume. Curve B gives the actual concentrations along the centerline versus distance from the source. Curve A gives the maximum (steady state) concentrations that will ultimately be achieved. Curve C is the ratio of curve B to A, that is, $C_{actual}/C_{maximum}$. Note that values for $C_{actual}/C_{maximum}$ are plotted along the vertical right-hand side of the figure. As an example, at $x = 7000$ cm, $C_{actual}/C_{maximum}$ is about 0.27.

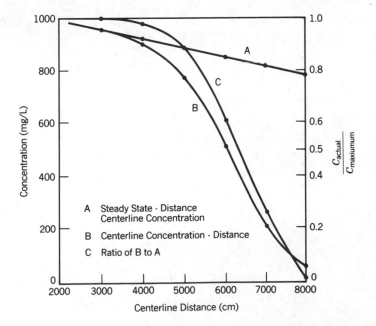

    a. Over what distance from the source is this plume in a steady state?

    b. At what distance x is the advective front vt located?

    c. What is the value for the dispersivity α_x?

8. In a laboratory experiment, the concentration of an organic compound is found to decline over a period of 15 days.

Elapsed Time (days)	Concentration ($\mu g/L$)
0	123.0
1	111.7
2	102.2
4	84.5
8	58.9
15	30.5

Estimate the decay rate constant and the half-life for this particular compound.

Chapter Nineteen

Remediation

19.1   Overview of Corrective Action Alternatives

19.2   Technical Considerations with Injection Withdrawal Systems

19.3   Methods for Designing Injection/Withdrawal Systems

19.4   Interceptor Systems for NAPL Recovery

19.5   Bioremediation

19.6   Steps for Dealing with Problems of Contamination

19.7   Case Studies in Site Remediation

So far, we have examined the causes of ground water contamination, how contaminants are transported, and how to characterize transport in terms of key parameters. What remains to be considered is how to correct a contamination problem once it has developed. As this chapter will show, alternatives for control and remediation exist that can be combined to provide an overall corrective strategy. The particular choice of what to do depends on the conditions existing at a site and regulatory constraints.

19.1 Overview of Corrective Action Alternatives

There are four main alternatives for dealing with problems of contamination (1) containing the contaminants in place, (2) removing contaminants from the ground altogether, (3) treating the contaminants in situ, and (4) attenuating the possible hazard by institutional controls (OTA, 1984). These methods are surveyed in the following sections.

Containment

The containment options are a series of control measures that keep the contaminants in the ground but prevent further spread through the use of physical or hydrodynamic barriers. The alternatives listed in this discussion are standard geotechnical approaches that have been used historically to control ground water seepage but have been adapted to problems of contaminant control.

1. *Slurry walls* are low-permeability barriers emplaced in the subsurface. They confine contaminants by either surrounding the entire spill (Figure 19.1) or by removing the potential for flow through the source with an upstream barrier. Slurry walls can be constructed in several different ways. For example, with the trench method, a trench is dug through a bentonite slurry to the desired depth (Knox and others, 1984). In most cases, the slurry wall is solidified either by incorporating excavated material with the bentonite, or by adding cement in the original slurry.

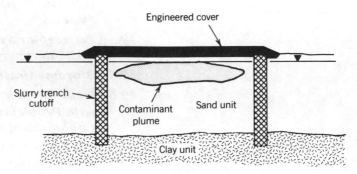

Figure 19.1
Example of a dissolved contaminant plume contained by slurry walls (modified from Knox and others, 1984).

With the vibrating beam method a steel plate is forced into the ground. As this plate is gradually removed, bentonite is injected to fill the space created by the beam.

A typical slurry wall emplaced by trenching ranges in width from about 0.5 to 2 m and can be installed to depths of up to about 50 m. The slurry walls created with the vibrating beam method are much narrower and typically emplaced at a shallower depth (Knox and others, 1984). Because the walls are thin and placed in segments, the continuity of the barrier is often of concern.

2. *Sheet pile* cutoff walls are constructed by driving interlocking steel piles into the ground (Rogoshewski and others, 1983). Because of the joints between piles, pile walls typically leak. Leakage can be reduced through newly developed sealing techniques or in some cases as fines fill the joints. However, the ability of sheet pile walls to control contaminant spreading effectively is a concern. Given this problem and the relatively large costs associated with constructing a pile wall, there are not many examples of their use in remedial work.

3. *Grouting* is another direct way to control the migration of contaminants. A grout wall or curtain is constructed by injecting fluids under pressure into the ground. The grouting material moves away from the zone of injection eventually to gel or solidify thereby reducing the hydraulic conductivity (Rogonshewski and others, 1983). The distance of grout penetration varies from site to site but can be relatively small requiring closely spaced injection holes (for example, 1.5 m). In practice, holes are drilled in two or three staggered rows to ensure that a more or less continuous barrier is constructed. Typical grouting fluids include cement, bentonite, or specialty fluids like silicate or lignochrome grouts (Rogoshewski and others, 1983).

Grouting techniques are most useful for sealing fractured rocks, where it is not possible to construct a low-permeability barrier in any other way. The high cost and potential for contaminant/grout interactions limit the use of this technique for contamination problems. In addition, most grouts can only be injected in materials with sand-sized and larger grain sizes (Knox and others, 1984).

4. *Geomembranes* are synthetic sheets installed in open trenches to control contaminant spread. They have attributes of both a slurry trench and a pile wall. This technology is generally considered to be in the research and development stage with concerns about long term performance and the compatability of geomembranes with organic solvents (OTA, 1984).

Surface seals and surface drainage are often used together to control infiltration moving downward through the contaminant source or into the plume. If the source is isolated from infiltration, the rate of generation of the contaminants could be substantially reduced together with the ground water flow velocity. In addition, covering and controlling the drainage will reduce the possibility of exposure and erosion of contaminated materials, and the rate of transfer of volatiles or other hazardous materials (for example, asbestos) into the atmosphere.

5. *Surface seals* are a common element in many different types of waste management problems. Given that the construction of a surface seal is

FMC Vault, Minneapolis, Minnesota

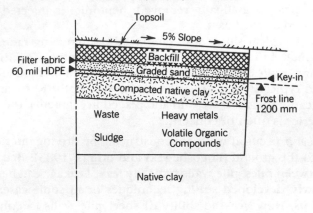

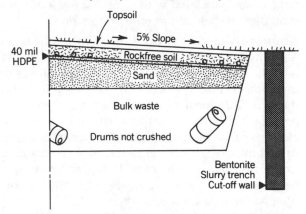

Figure 19.2
Examples of engineered covers (from Mclelwain and Reades, 1985). Reprinted by permission of Second Canadian/American Conference on Hydrogeology. Copyright © 1985. All rights reserved.

not as difficult as a vertical cutoff wall, a much greater variety of materials can be used. Examples of cover materials include compacted clay and silty clay, mixtures of natural soils, and stabilizers such as cement, bitumen or fly ash; bentonite layered with other natural materials; sprayed bituminous membranes; synthetic membranes [for example, high-density polyethylene (HDPE) or polyvinyl chloride (PVC)] again used in combination with other materials; and waste materials such as furnace slag, incinerator residue, fly ash, and clinkers (Rogoshewski and others, 1983). Figure 19.2 shows examples of these surface liners used at hazaradous waste sites in North America.

The technology for sealing sites is proven. The main source of uncertainty in seal performance is the quality of ongoing maintenance. Possibilities exist for damage by erosion from runoff, cracking due to desiccation, frost effects or subsidence, punctures created by burrowing animals or plant roots, and deterioration of synthetic membranes by tearing or exposure to sunlight. Thus, there is a continuing need to inspect and repair surface seals on an ongoing basis.

6. *Surface drainage* involves grading the surface of the seal to enhance drainage to ditches or other drainageways. The goal is to move water off of the surface seal as rapidly as possible to minimize infiltration, while avoiding excessive erosion. General specifications for the design of a landscape in terms of slope lengths and gradients are provided by Rogoshewski and others (1983). Also included in their discussion are considerations in designing and constructing the different elements of a drainage network including drainage ditches, diversions (a combined ditch and dike constructed across a slope), grassed waterways, and drainage benches (terraces built across long slopes to reduce the steepness). The technology for these surface controls is well established. However like surface seals, a system needs to be inspected and maintained on an ongoing basis.

The last alternative for waste containment is pumping or injection to control the spread of contaminants with hydraulic or pressure gradients. These approaches are used both for contaminated ground water and soil gases contaminated by volatiles.

7. *Hydrodynamic control* of ground water contamination can have as its objective to lower the water table to prevent discharge to rivers or lakes, to reduce the rate of contaminant generation by desaturating the waste, or to confine the plume in a potentiometric low created by an appropriate combination of pumping/injection wells (Rogoshewski and others, 1983). Design is a straightforward extension of the well hydraulics concepts developed in Chapter 5. The assessment of this technology (OTA, 1984) considers it to be unconventional but viable given sufficient care in design and operational monitoring. Of all the techniques considered so far this one requires the greatest ongoing care. Considerations include keeping the wells and pumps properly serviced, and responding to changed ground water conditions. For example, above-average recharge may raise water levels sufficiently to destroy confinement, while below-average recharge may result in widespread lowering of levels due to excess withdrawals. In addition, the possibility that the wells could begin to pump contaminated water could require that onsite treatment facilities be constructed.

    Positive or negative differential pressure systems can be used to control vapor migration (O'Connor and others, 1984a). In spills involving hydrocarbons, the most immediate problem is often the rapid movement of vapors. Even when the free liquid has been removed, the vapors may continue to be a problem at a site (O'Connor and others, 1984a). By manipulating pressure gradients in the soil by adding air or removing soil gases, contaminant spread in the vapor phase can be controlled. Figure 19.3*a* shows two ways of keeping volatiles out of a structure using positive pressure displacement systems. The same result can be achieved also with vapor extraction systems (Figure 19.3*b*).

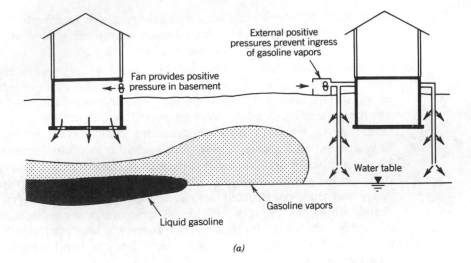

Fan provides positive
pressure in basement

External positive
pressures prevent ingress
of gasoline vapors

Water table

Gasoline vapors

Liquid gasoline

(a)

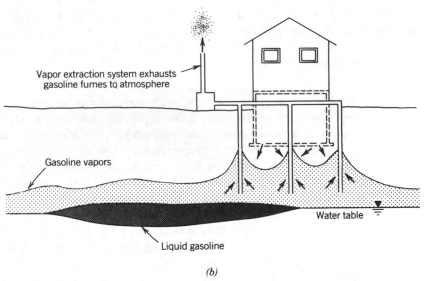

Vapor extraction system exhausts
gasoline fumes to atmosphere

Gasoline vapors

Water table

Liquid gasoline

(b)

Figure 19.3
Control of subsurface gasoline vapors using (*a*) positive and (*b*) negative
differential pressure systems (from O'Connor and others, 1984a). Reprinted by
permission of Conference on Petroleum Hydrocarbons and Organic Chemicals
in Ground Water - Prevention, Detection and Restoration. Copyright © 1984.
All rights reserved.

Contaminant Withdrawal

The alternatives for removing contaminants from the ground are often an integral
part of an overall strategy for site remediation. Here, we will consider four different approaches: pumping, interceptor systems, soil venting, and excavation.

1. *Pumping* uses wells to remove contaminants from the ground. This technique has proven to be successful for a variety of different site conditions.
   Because of the overall importance of this approach, we will discuss issues

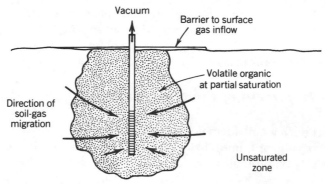

Figure 19.4

Use of a vacuum system for removing volatile organic compounds from the unsaturated zone (from Schwartz, 1988). Reprinted with permission of Elsevier Science Publishers from J. Hydrol., v. 100.

of design and operation in detail in Sections 19.2 and 19.3. With pumping, there is always the problem of what to do with the contaminated water removed from the ground. Of necessity, onsite treatment is required before injecting the water to the subsurface or releasing it to surface water bodies. Knox and others (1984) extensively discuss the physical and chemical/biological treatment methods.

2. *Interceptor systems* use drains, trenches, and lined trenches to collect contaminants close to the water table. A drain is a line of buried perforated pipe or drainage tiles (Knox and others, 1984). A trench is an excavation that is either open or backfilled with gravel to promote wall stability. Typically, interceptor systems are installed close to the ground surface and one or two meters below the water table. They function as an infinite line of wells and are thus efficient at removing shallow contamination. As we will discuss in more detail later in this chapter, trenches are used most often for collecting LNAPLs like crude oil and gasoline that move in the capillary fringe above the water table.

3. *Soil venting* removes volatile organic compounds from the unsaturated zone. Soil gas is removed from a well by vacuum pumping, which causes large volumes of air to circulate through the spill (Figure 19.4). The presence of the circulating air promotes volatilization and possibly biodegradation. Surface treatment is required in most cases to remove contaminants from the airstream before release to the atmosphere (Rogonshewski and others, 1983). Data from Batchelder and others (1986) and Crow and others (1985) indicate removal rates of approximately 0.5 to 40 kg of product (for example, hydrocarbon vapor) per day might be expected from a single well. The radius of influence of a single well is about 13 m for airflow rates of greater than 15 ft³/min through a well point in coarse to fine sand (Batchelder and others, 1986).

The method is one of the few alternatives to soil removal when the quantity of volatile contaminants in the ground is below residual saturation. Work to date has shown the method to be effective. However, some of the constituents in complex mixtures like gasoline are not particularly volatile, and become difficult to remove.

4. *Excavation* is an obvious alternative for dealing with contamination at a site. Not much needs to be said about the method. It is an exercise in digging and trucking. The greatest problems are the costs involved and finding an appropriate place to dispose of the contaminated soils.

In Situ Treatment of Contaminants

There has been a considerable research effort to develop techniques for degrading, detoxifying, or immobilizing contaminants in situ. The most promising technologies rely on biological and chemical processes.

1. *Biological degradation* involves using organic contaminants as an energy source for bacteria and producing simple compounds like water and CO_2 from complex organic molecules. The most commonly used techniques are aimed at enhancing the growth of existing populations of aerobic bacteria through the addition of nutrients like nitrogen and phosphorous, and electron acceptors like oxygen. A less common procedure is to use specially engineered bacteria that have a predisposition to metabolize particular contaminants. Because of the complexity and importance of these techniques, we will discuss them in more detail in a later section.

    Methods for in situ biodegradation are at a relatively early stage of development with most existing applications related to the cleanup of petroleum spills. Issues of efficiency and predictability of performance were cited as the main sources of uncertainty in the method by the OTA (1984) assessment.

2. *Chemical degradation* attempts to treat contaminants in situ by injecting an appropriate chemical or treatment agent. These chemicals are usually added through a designed network of well points. As discussed by Knox and others (1984), the treatment agent has to be specific to particular classes of contaminants. For example, the addition of alkalies or sulfides can cause heavy metals to precipitate as insoluble minerals, the addition of oxygen can destroy cyanide, and the addition of particular anions can cause cations to be removed as mineral precipitates. Overall, this technology is much less developed than are the biological approaches with few commercial applications (OTA, 1984). The greatest problem is the uncertainty in performance when compared to other more reliable and proven techniques. If a chemical degradation scheme fails for example, one still has to clean up the original contaminant as well as the chemicals that were injected.

Management Options

The management options (Table 19.1) are institutional actions that can be taken to avoid health problems or shut down a source of contamination without specifically dealing with the contamination in the ground. Because these alternatives are straightforward, we will not consider them individually. Included in Table 19.2 is the OTA (1984) assessment of the state of development and principal sources of uncertainty with these measures.

Table 19.1
Management alternatives for dealing with ground water contamination (modified from Office of Technology Assessment, 1984)

Management options: Management options are usually applied either to prevent further contamination or to protect potential exposure points from contaminated ground water. These methods thus focus on sources and exposure points rather than on the contaminants per se. The methods also tend to be institutionally based rather than technology based.

1. *Limit/terminate aquifer use:* Limits access or exposure of receptors to contaminated ground water.

2. *Develop alternative water supply:* Involves the substitution of contaminated ground water with alternative supplies (for example, surface water diversions and/or storage, desalination, and new wells).

3. *Purchase alternative water supply:* Includes bottled water and water imports.

4. *Source removal:* Involves the physical removal of the source of contamination and includes measures to eliminate, remove, or otherwise terminate source activities; could also include modification of a source's features (for example, operations, location, or product) to reduce, eliminate, or otherwise prevent contamination.

5. *Monitoring:* Involves an active evaluation program with a "wait and see" orientation.

6. *Health advisories:* Involves the issuance of notifications about ground water contamination to potential receptors.

7. *Accept increased risk:* Involves the decision to accept increased risk; is usually a "no action" alternative.

19.2   Technical Considerations with Injection Withdrawal Systems

Of the three main contamination problems—contaminants dissolved in water, light nonaqueous phase liquids (LNAPLs), and dense nonaqueous phase liquids (DNAPLs)—the first is the easiest to deal with in terms of well design. Pumping water containing dissolved contaminants presents no special problems so that water well technology is usually appropriate. Some care is required in selecting casing materials and pumps that will not fail due to chemical interaction with the contaminant.

The removal of LNAPLs at the water table presents a greater technical challenge. The collection device needs to accomplish two things: (1) develop a cone of depression on the water table that results in product flow toward the collection well and (2) remove free product moving into the collection well. Wells designed for this purpose are usually one of three types. The single pump system with one recovery well (Figure 19.5) is the simplest and least expensive (Blake and Freyberger, 1983). The pump is positioned in the well so that as it cycles on and off, both water and product are removed. This pumping creates the necessary depression of the water table and removes the floating product. The greatest problem is that a high-speed pump emulsifies the LNAPL in water. This mixture requires treatment at the surface to separate the LNAPL and water. However, even after separation, the waste water may contain soluble compounds (for example,

Table 19.2
Assessment of management alternatives (modified from Office of Technology Assessment, 1984)

Technique	Principal components of uncertainty affecting performance	Development status Summary[a]	Remarks
1. Limit/ terminate aquifer use	• The ability to shut down domestic wells due to possible public resistance. • The ability to enforce usage patterns in cases of environmental exposure (for example, to sport fish or streams).	6	Historically this is a common response to aquifer contamination.
2. Develop alternative water supply	• Availability of water supply alternatives, especially in water-short areas which may, in turn, limit the long-term growth of an area.	6	In conjunction with limiting/terminating aquifer use, alternative water supply development is a frequently implemented response.
3. Purchase alternative water supply	• Reliance on imports, especially in water-short areas where the supply may be terminated or depleted. • Potential opposition to interbasin transfers.	6	This is a frequently implemented response although generally considered a short-term solution.
4. Source removal	• Increased contaminant migration (for example, via breaking of drums or additional infiltration during precipitation). • Availability of secure disposal options. • Extent of contamination and resulting costs. (See Excavation, above.)	1	Conventional construction techniques are used for source removal although substantial increases in health and safety precautions are required for ground water contamination applications. Current activity already involves significant health and safety measures.
5. Monitoring	• Undetected plume migration because of improper placement or sampling of wells. • Mistakes are difficult to detect until a problem occurs or backup wells around key exposure points are installed.	1,6	Conventional technology is used for monitoring ground water contamination problems and conducting hydrogeologic investigations. If methods are used properly, reliable plume delineation and migration data can be generated.
6. Health advisories	• The ability to enforce usage patterns in cases of environmental exposure (for example, to sport fish or streams). • The ability to shut down domestic wells due to possible public resistance.	6	This option is a conventional practice of state and local health departments.
7. Accept increased risk	• The ability to predict contaminant migration. • Corrective action alternatives can be more expensive as the contaminant spreads out (in other words, a larger plume).	6	Historically this option is the response to many contamination incidences. Impacts on population are unclear.

[a]Key: 1—Technology is proven; performance data are available from applications to ground water contamination problems.

2—Technology is proven in applications other than ground water contamination; long-term performance data are unavailable for ground water contamination.

3—Technology is in R&D stage with respect to ground water contamination applications, although proven for other applications; performance is generally unknown for ground water contamination problems.

4—Application of technology has been limited to specific, narrowly defined site conditions.

5—Technology is generally in R&D stage; results are unreliable.

6—Technology has been applied historically, for example, before the development of regulatory programs and consideration of potential long-term impacts.

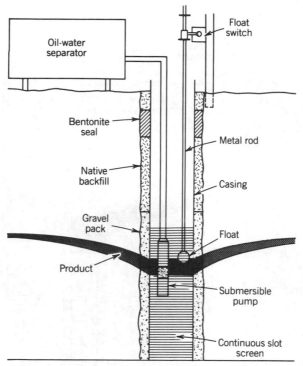

Figure 19.5

Example of a one-well, one-pump system for recovering NAPLs (from Blake and Lewis, 1982). Reprinted by permission of Second National Symposium on Aquifer Restoration and Ground Water Monitoring. Copyright © 1982. All rights reserved.

hydrocarbons) that may have to be removed (Blake and Freyberger, 1983). The treatment problem can be reduced to some extent by using piston pumps or other noncentrifugal pumps, which produce less mixing.

The other pumping schemes avoid the problem of surface treatment by preventing the LNAPL and water from mixing. With the two-pump, two-well systems (Figure 19.6), the deep water well screened below the spill lowers the water table and the shallow well collects the free product (Blake and Freyberger, 1983). Two-well, two-pump systems are more expensive than are the one-well, one-pump systems. The two-pump, one-well design (Figure 19.7) is a compromise design that is less expensive and avoids treating large volumes of water. Because two pumps and operational monitoring equipment are required in the same well, casing diameters are larger than either of the other two types of systems.

The operation of product recovery systems has to be carefully monitored. For example, with the one-pump, one-well system, the pump has to be placed at the proper depth and the float switch carefully adjusted so that product is being pumped toward the end of the pumping cycle. With the two-pump systems, the water well should not be overpumped to the extent that product is pumped through the water pump (Blake and Freyberger, 1983). Failure of the product pump in a two-pump, one-well system may also cause the product to accumulate in the well and eventually to reach the intake of the water pump. Part of the job of the control system is to shut off the water pump when free product is detected by sensors above the pump. Blake and Freyberger (1983) provide additional

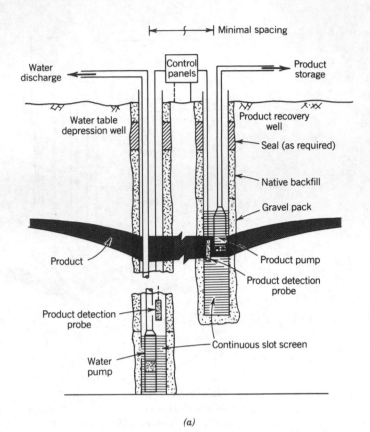

(a)

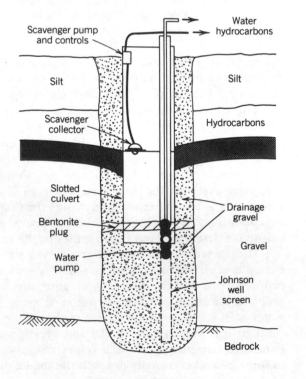

Figure 19.6
Examples of a two-well, two-pump system for the recovery of NAPLs. Panel (*a*) is from Blake and Lewis (1982) and is reprinted by permission of Second National Symposium on Aquifer Restoration and Ground Water Monitoring. Copyright © 1982. Panel (*b*) is modified from O'Connor and others (1984b) and is reprinted by permission of Conference on Petroleum Hydrocarbons and Organic Chemicals in Ground Water - Prevention, Detection, and Restoration. Copyright © 1984. All rights reserved.

(b)

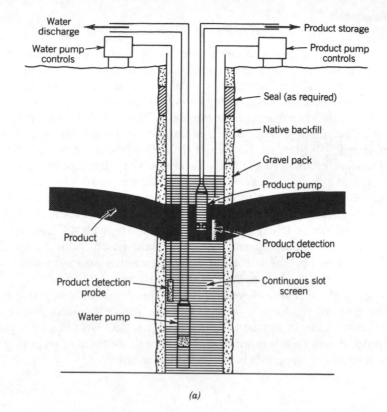

(a)

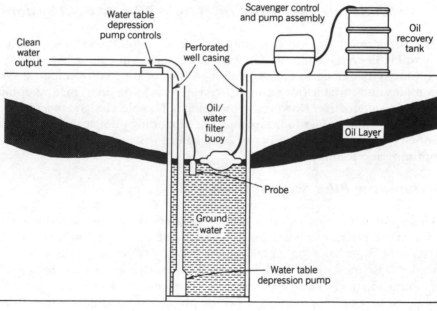

(b)

Figure 19.7

Examples of a one-well, two-pump system for the recovery of NAPLs. Panel (*a*) is from Blake and Lewis (1982) and is reprinted by permission of Second National Symposium on Aquifer Restoration and Ground Water Monitoring. Copyright © 1982. Panel (*b*) is modified from Yaniga and Mulry (1984) and is reprinted by permission of Conference on Petroleum Hydrocarbons and Organic Chemicals in Ground Water - Prevention, Detection and Restoration. Copyright © 1984. All rights reserved.

guidance on operational and safety details that need to be considered, especially in selecting the proper equipment and working with potentially explosive liquids.

Research on the technical aspects of emplacing wells for the collection of DNAPLs is extremely limited. Ferry and others (1986) studied how the efficiency of removing DNAPLs from low-permeability bedrock depended on the pumping system. The first scheme they examined involved simply removing the accumulated DNAPL at the bottom of the well with an air-activated purge pump. This approach of collecting DNAPL through gravity drainage into the well was not successful in this test. Once the fluid in the well was removed, there was no further inflow up to a day afterward. In their second test, a submersible pump was installed higher in the well to produce a cone of depression. This scheme promoted the inflow of DNAPL into the bottom of the well. In one case after about 1140 minutes of pumping and a final drawdown of about 15 m, the level of DNAPL increased by about 1 meter. Creating a situation for a natural water drive appears to be helpful in moving DNAPL into a well.

In general, we cannot overemphasize the difficulties in removing DNAPLs occurring as a pumpable free product or as an immobile phase. Freeze and Cherry (1989) state ''there is now little doubt at sites where DNAPLs are a problem, the local ground-water has terminal cancer. A cure in the form of returning the aquifer to drinking-water standards is unachievable at any cost.''

19.3 Methods for Designing Injection/Withdrawal Systems

The first critical issue in designing an injection/withdrawal system is whether pumping is even technically feasible. Conditions will exist where the hydraulic conductivity of a contaminated unit is so low, or the hydrogeological system so complex and variable (for example, fractured rocks or karst) that pumping is not a reliable alternative. However, if pumping is feasible, design strategies do exist. Here, we will consider four possibilities: expanding pilot scale systems, capture-zone-type curves, trial-and-error simulations using models, and simulation-optimization techniques.

Expanding Pilot Scale Systems

This approach simply involves starting recovery operations at a site with a pilot system and carefully monitoring how the system performs. Evaluation of the field data provides a basis for expanding the pilot by selecting places where the next well will do most good. In the case of LNAPLs, the next well could be placed at a point where (1) large apparent product thicknesses exist and are not being reduced by existing pumping or (2) the product is moving in a direction that is inappropriate (for example, toward wells, rivers, and buildings). In the case of dissolved contaminants, again the pumping wells could be placed where the plume is not under control. Injection wells are located to provide barriers to flow or in areas of low gradient to increase the velocity of flow to collection wells.

These systems rely on experience and a cautious, incremental expansion of the network to ensure the remedy is effective and economical. For spills involving LNAPLs, there are as yet no practical quantitative approaches to optimize the elements of the design. The situation for dissolved contaminants is much better with several useful analytical and numerical procedures.

Figure 19.8
The effects of pumping superimposed on a regional flow system creates a capture zone (modified from Gorelick, 1987). Reprinted by permission of Solving Ground Water Problems with Models. Copyright © 1987. All rights reserved.

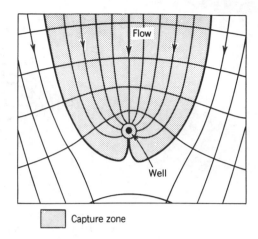

Capture-Zone-Type Curves

Javandel and Tsang (1986) describe an analytical approach for the design of recovery well systems. The number of injection or withdrawal wells and the pumping/injection rates are minimized through a proper choice of well location and the distance between wells. The approach is based on the concept of a capture zone (Figure 19.8). When the drawdown caused by a well in a homogeneous aquifer is superimposed upon a steady-state, hydraulic head field, some flow tubes (see shading) will terminate in the pumping well. This shaded area is the capture zone for the well, marking that part of the plume controlled by the collection well. For a given aquifer, the size of the capture zone depends upon the pumping rate. Pump the well harder and the capture zone becomes larger. The optimal pumping rate for a well is the one that provides a capture zone that is slightly larger than the plume.

Well hydraulics theory tells us that a well can only be pumped to some maximum rate determined by the available drawdown. However, pumping one well at this maximum rate may not provide a capture zone that is large enough.

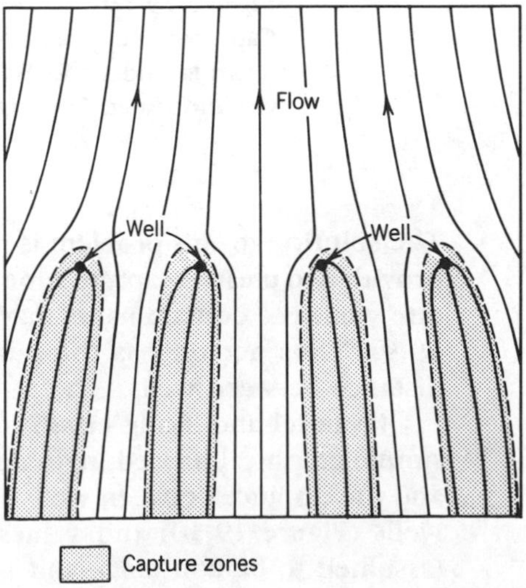

Figure 19.9
An illustration of leakage between wells that can occur when the spacing between the wells is too large (modified from Javandel, 1986).

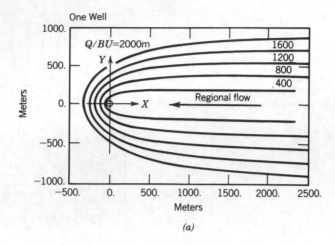

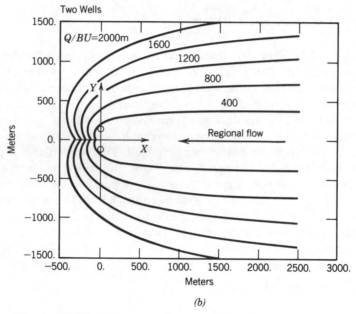

Figure 19.10
Capture-zone type curves for one, two, three and four wells
(from Javandel and Tsang, 1986). Reprinted by permission of
Ground Water. Copyright © 1986. All rights reserved.

The solution to this problem is to add wells until sufficient pumping capacity is
provided to create a capture zone of the appropriate size. However with more than
one well, the contaminants may be able to pass between the two wells (Figure
19.9). Thus not only is the pumping rate of concern, but so is the optimum
distance between wells.

Javandel and Tsang (1986) use complex potential theory as the basis for a
simple graphical procedure to determine the pumping rate, the number of wells,
and the distance between wells. The method requires type curves for one to four
wells (Figure 19.10) and values for two parameters, B, the aquifer thickness
(assumed to be constant), and U, the specific discharge or Darcy velocity (also

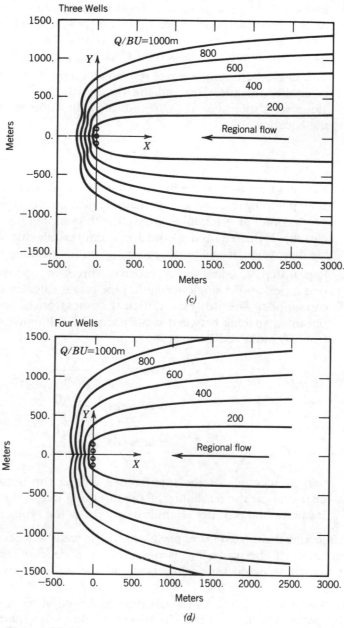

Figure 19.10
continued

assumed constant) for the regional flow system. The method involves the following five steps (Javandel and Tsang, 1986).

1. A map of the contaminant plume is constructed at the same scale as the type curves. The edge or perimeter of the plume should be clearly indicated together with the direction of regional ground water flow.

2. Superimpose the type curve for one well on the plume, keeping the x axis parallel to the direction of regional ground water flow and along the

midline of the plume so that approximately equal proportions of the plume lie on each side of the x axis. The pumped well on the type curve will be at the downstream end of the plume. The type curve is adjusted so that the plume is enclosed by a single Q/BU curve.

3. The single-well pumping rate (Q) is calculated using the known values of aquifer thickness (B) and the Darcy velocity for regional flow (U) along with the value of Q/BU indicated on the type curve (TCV) with the equation

$$Q = B \cdot U \cdot TCV \tag{19.1}$$

4. Next it is necessary to determine whether the well can support the calculated pumping rate over the estimated time required for the cleanup. If the production rate is feasible, one well with pumping rate Q is required for cleanup. If the required production is not feasible due to a lack of available drawdown, it will be necessary to continue adding wells (see step 5).

5. Repeat steps 2, 3, and 4 using the two-, three-, or four-well-type curves in that order, until a single-well pumping rate is calculated that the aquifer can support. The only extra difficulty comes from having to calculate the optimum spacing between wells using the following simple equations from Javandel and Tsang (1986)

$$\text{Two wells:} \quad \frac{Q}{\pi BU}$$

$$\text{Three wells:} \quad \frac{1.26Q}{\pi BU}$$

$$\text{Four wells:} \quad \frac{1.2Q}{\pi BU}$$

and to account for the interference among the pumped wells when checking on the feasibility of the pumping rates. The wells are always located symmetrically around the x axis, as the type curves show.

Reinjecting the treated water produced by the wells will accelerate the rate of aquifer cleanup. The procedure is essentially the same as that just discussed, except that the type curves are reversed and the wells are injecting instead of pumping.

Javandel and Tsang (1986) suggest that the injection wells should be moved slightly upstream of the calculated location to avoid causing parts of plume to follow a long flow path. Their rule of thumb is to place wells half the distance between the theoretical location and the tail of the plume. The following example taken from Javandel and Tsang (1986) illustrates how the technique is used.

Example 19.1

Shown in Figure 19.11 is a plume of trichloroethylene (TCE) present in a shallow confined aquifer having a thickness of 10 m, a hydraulic conductivity of 10^{-4} m/sec, an effective porosity of 0.2, and a storativity of 3×10^{-5}. The hydraulic gradient for the regional flow system is 0.002 and the available drawdown for wells in the aquifer is 7 m. Given this information design an optimum collection system.

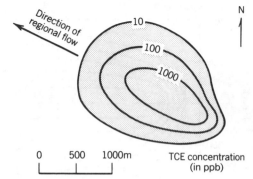

Figure 19.11
Hypothetical plume of TCE (from Javandel
and Tsang, 1986). Reprinted by permission
of Ground Water. Copyright © 1986. All
rights reserved.

Values of B and U are required for the calculation. B is given as 10 m, but U
needs to be calculated from the Darcy equation

$$U = K \operatorname{grad}(h) \quad \text{or} \quad U = 10^{-4} \times (0.002) = 2 \times 10^{-7} \text{ m/sec}$$

Now we are ready to work with the type curves following the steps just outlined.

Superposition of the type curve for one well on the plume (Figure 19.12*a*)
provides a Q/BU curve of about 2500. Using this number and the values of B and
U, the single-well pumping rate is

$$Q = B \cdot U \cdot TCV \quad \text{or} \quad 10 \times (2 \times 10^{-7}) \times 2500 = 5 \times 10^{-3} \text{ m}^3/\text{sec}$$

A check is required to determine whether this pumping rate can be supported for
the aquifer. The Cooper-Jacob (1946) equation provides the drawdown at the
well, assuming $r = 0.2$ m and the pumping period is one year

$$s = \frac{2.3Q}{4\pi T} \log \frac{2.25Tt}{r^2 S} \tag{19.2}$$

where $Q = 5 \times 10^{-3}$ m³/sec, $T = KB = 10^{-3}$ m²/sec, $t = 1$ yr or 3.15×10^7 sec, r
$= 0.2$ m, and $S = 3 \times 10^{-5}$.

The pumping period represents some preselected planning horizon for
cleanup. Substitution of the known values into Eq. 19.2 gives a drawdown of 9.85
m. Even without accounting for well loss, the calculated drawdown exceeds the
7 m available. Thus a multiwell system is necessary.

Superimposing the plume on the two-well-type curve provides a Q/BU value
of 1200, which in turn gives a Q for each of the two wells of $10 \times (2 \times 10^{-7})$
$\times 1200$ or 2.4×10^{-3} m³/sec. The optimum distance between wells is $Q/(\pi BU)$
or $2.4 \times 10^{-3}/[\pi \times 10 \times (2 \times 10^{-7})] = 382$ m. Again we check the predicted
drawdown at each well after one year against the available 7 m. Because of the
symmetry, the drawdown in each well is the same. The total drawdown at one

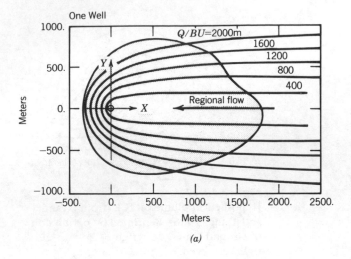

(a)

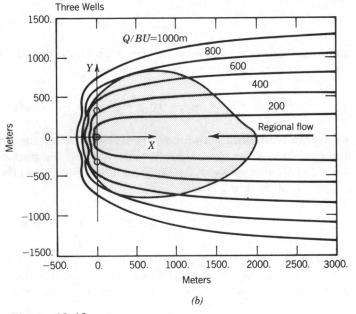

(b)

Figure 19.12
The TCE plume at the matching position with one-well and
three-well type curves (modified from Javandel and Tsang,
1986). Reprinted by permission of Ground Water. Copyright
© 1986. All rights reserved.

of the wells includes the contribution of that well pumping plus the second one
382 m away, or

$$s = \frac{2.3Q}{4\pi T}\left[\log\frac{2.25Tt}{r_w^2 S} + \log\frac{2.25Tt}{r^2 S}\right]$$

The calculated drawdown s is 6.57 m, which is less than the available drawdown.
However, well loss should be considered, which makes the two-well scheme
unacceptable. Moving to a three-well scheme, Q/BU is 800 (Figure 19.12b), which

translates to a pumping rate of 1.6×10^{-3} m³/sec for each well. Carrying out the drawdown calculation for three wells located $1.26Q/(\pi BU)$ or 320 m apart provides an estimate of 5.7 m for the center well, which is comfortably less than the available drawdown. Thus we have been able to ascertain the need for three wells, located 320 m apart, and each pumped at 1.6×10^{-3} m³/sec.

This method is extremely useful but it cannot be applied to every design problem encountered in practice. There are assumptions built into the formulation, such as constant aquifer transmissivity, fully penetrating wells, no recharge, and isotropic hydraulic conductivity, that have to be satisfied by the field problem or the method will not necessarily yield a correct result. Gorelick (1987) and Larson and others (1987) made this point in developing designs using alternative methods and comparing the results with those obtained with capture-zone-type curves.

Trial-and-Error Simulations With Models

Numerical models are also useful in designing well systems for contaminant recovery. These approaches rely on the modeler's experience and a series of model runs to test various designs. One design should be better than the rest in terms of minimizing the number of wells and pumping rates, while providing a capture zone of an appropriate size and shape. Thus designs are not optimized in the formal mathematical sense, but the modeling procedures are straightforward.

Larson and others (1987) use a three-dimensional ground water flow model in conjunction with a second program that calculates ground water pathlines from any starting point. A finite difference model solves for the steady-state distribution of hydraulic head on a three-dimensional nodal network. These values of hydraulic head and known values of porosity and hydraulic conductivity are input to the second program. A particle tracking procedure provides the pathlines where the components of velocity for a series of hypothetical particles are determined with the Darcy equation and the various input data. The pathlines help to define the capture zone for the well. The following example from Larson and others (1987) will illustrate this final step.

Assume a partially penetrating well is being pumped at 21.5 gpm in an aquifer having a $K_h = 100$ ft/day, $K_v = 1.00$ ft/day, and a recharge rate of 4 in./yr. By examining the pathlines in plane view and cross sections, the capture zone can be readily defined (Figure 19.13). Interestingly, the lateral extent of the capture zone is larger than predicted by the Javandel and Tsang analysis, while the vertical size is somewhat less (Figure 19.13). The partially penetrating well, the recharge, and anisotropism are complexities that violate the basic assumptions in the Javandel-Tsang procedure.

Simulation-Optimization Techniques

A promising approach to the design of contaminant recovery systems combines ground water simulation with mathematical optimization (Gorelick and others, 1984; Atwood and Gorelick, 1985; Lefkoff and Gorelick 1986; Wagner and Gorelick, 1989). The power of numerical flow modeling accounts for the complexities of the hydrogeological setting, while formal optimization techniques provide the best design subject to imposed constraints. Discussion of this sophisticated

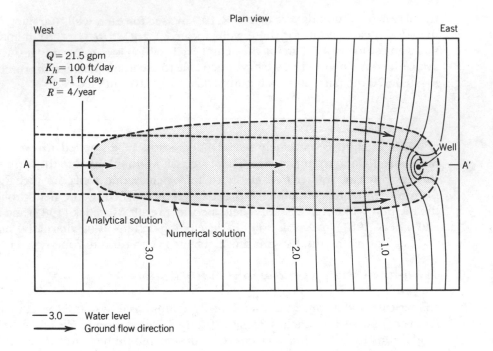

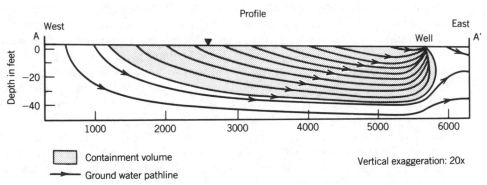

Figure 19.13
Capture zone calculated using a three-dimensional ground water model (from Larson and others, 1987). Reprinted by permission of Solving Ground Water Problems with Models. Copyright © 1987. All rights reserved.

modeling approach is beyond the scope of this book. This and similar approaches have promise of being able to overcome the inherent limitations of the analytical approaches in dealing with complex systems and the nonuniqueness of the trial-and-error procedures.

19.4   *Interceptor Systems for LNAPL Recovery*

Interceptor systems are useful in recovering LNAPLs. Figures 19.14*a* and *b* illustrate two different designs for open trench systems. In the first, a skimmer pump selectively removes product entering the trench. Because little water is removed,

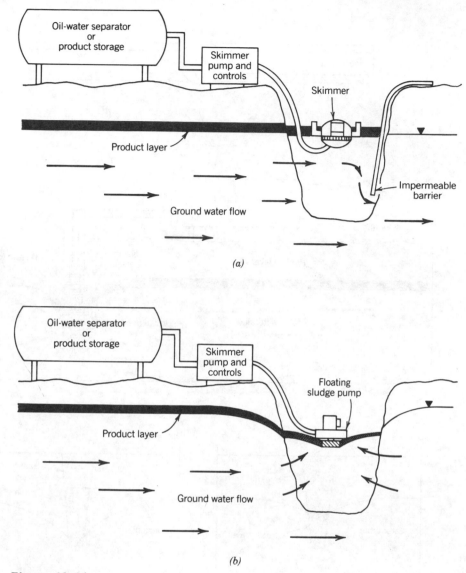

Figure 19.14
Examples of interceptor trenches used to control the spread of NAPLs. In (*a*) product is recovered using a skimmer pump with an impermeable barrier to control product leakage. In (*b*) product and water are recovered with a floating sludge pump. Potential product leakage is controlled hydrodynamically (from Blake and Fryberger, 1983).

a barrier needs to be installed to prevent the product from reentering the subsurface on the down gradient side of the trench (Figure 19.14*a*). The second design is similar except that a greater volume of water is removed to contain the product hydraulically in the trench (Figure 19.14*b*). This design avoids the impermeable barrier and speeds up product recovery because the flow gradient to the trench is increased (Blake and Freyberger, 1983). However, a greater volume of contaminated water has to be treated at the surface.

An alternative design involves closed trenches or drains (Figure 19.15). Product moves through gravel or drain pipes in the trench laterally to collection

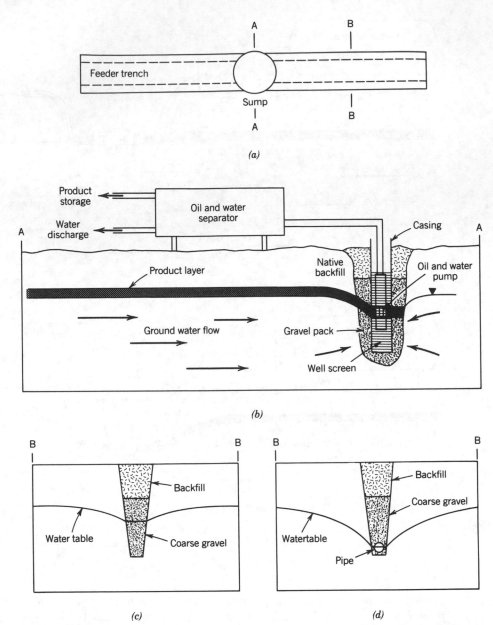

Figure 19.15
Schematic representation of two interceptor systems. Water and contaminants move along the feeder trench to a sump (*a*). A pumping system moves the fluids to the surface for treatment (*b*). The feeder trenches can be filled with gravel (*c*) or actual drains (*d*) (from Blake and Freyberger, 1983).

sumps and is removed by pumping. Again, hydraulic control keeps the product from simply passing through the trench (Figures 19.15*c*, *d*). Surface treatment of water pumped from the sumps again may be required. The main attractiveness of the interceptor systems is their overall simplicity, availability of construction materials, and speed of installation (Blake and Freyberger, 1983).

Unlike drainage systems for farmland, interceptor systems are not commonly installed as a large areal network to collect product everywhere the spill is evident. More typically, they are installed across the nose of a migrating spill or at critical points along its perimeter to prevent further spreading or inflow to surface water. Multicomponent flow is sufficiently complex that practical quantitative approaches for designing a system are still in their infancy. The main considerations in design are simply to ensure that the trench is long enough to avoid product migration around the end and deep enough both to provide the necessary hydraulic control and to avoid situations where the water table might fall below the bottom of the trench. For contaminants dissolved in water, a variety of more conventional design strategies are available.

19.5 Bioremediation

Bioremediation uses natural populations or seeded bacteria to transform organic contaminants into less hazardous compounds. Overall, this remedial strategy is important because it provides the opportunity to reduce concentrations of contaminants to low levels. Most of the other alternatives for site remediation simply do not have the potential to achieve this kind of removal. For example, in the case of a gasoline spill, a significant proportion of the original product can remain even after all the free product has been removed. If the gasoline held at residual saturation is left in place, it remains a source for dissolved compounds for a long time. Of the remedial alternatives for organic compounds, soil venting in the unsaturated zone or biodegradation in both the saturated and unsaturated zones provides the best hope for in situ treatment. Because most applications to date involve stimulating an existing population of bacteria, our discussion will be concerned with these approaches. Knox and others (1984) discuss the research and the few field studies that involved the seeding of acclimated or mutant microbial populations.

Bioremediation is not without its problems. The most important is a lack of well-documented field demonstrations to show just how effective the technique is and what if any are the long-term effects of this treatment on ground water systems. The need to learn about this technology is also being hindered by its significant commercial potential, which at the date of this writing means that even basic information is proprietary.

Biochemical Reactions

The essence of a biotransformation reaction is the oxidation of an organic compound by an electron acceptor such as O_2, NO_3^-, SO_4^{2-}, or CO_2 in a reaction whose rate is controlled by bacteria. In most applications, the contaminants are petroleum hydrocarbons such as gasoline, crude oil, heating oil, fuel oil, refinery waste, lube oil waste, and mineral oil (Lee and others, 1987). The monoaromatic and aliphatic constituents in these hydrocarbons are completely degraded in aerobic reactions similar to the following one

$$C_6H_6 + \tfrac{15}{2} O_2 \rightarrow 6CO_2 + 3H_2O \tag{19.3}$$

Lee and others (1987) discuss other types of organic compounds that have the potential to degrade. Examples include alcohols (isopropanol, methanol, ethanol), ketones (acetone, methyl ethyl ketone), and glycols (ethylene glycol).

Many of these compounds are degraded under anaerobic conditions with electron acceptors other than oxygen. This feature of the reaction and problems with the generation of undesirable intermediate compounds means that bioremediation techniques are much less developed for these compounds than petroleum products (Lee and others, 1987).

Another class of potentially treatable organic contaminants is halogenated aliphatic compounds (Lee and others, 1987). Because of their abundance and relative mobility, these compounds have received particular attention. The problem again with these compounds is that they degrade only to a limited extent in aerobic environments. Thus, many of the conventional approaches for treating petroleum spills are ineffective for these compounds. To biodegrade these contaminants, a mixed culture of bacteria is required that grows by oxidizing methane (usually added) and that can co-oxidize the halogenated aliphatic compounds (Wilson and Wilson, 1985; Fogel and others, 1987). Because the development of methods for dealing with compounds degraded in an anaerobic environment is still at a conceptual stage, the remainder of the discussion in this section will be concerned with the remediation of petroleum spills.

Oxygen and Nutrient Delivery

Enhanced bioreclamation for a petroleum spill is essentially a series of engineering activities that create favorable conditions for aerobic biodegradation (Hinchee and others, 1987). Most of the effort involves delivering oxygen and nutrients to the right place and in the right amounts. However actually delivering these materials, particularly oxygen, is a problem. The main constraint is the low solubility of oxygen in water. Pure oxygen in the form of a gas or liquid has a solubility of about 40 mg/L in water (Lee and others, 1987). If air is used to supply oxygen, the water will contain from 8 to 12 mg/L of dissolved oxygen (Lee and others, 1987). For a hypothetical spill involving 1000 gallons of fuel (Table 19.3) bioremediation using water with an oxygen content of 8 mg/L would require that 290,000,000 gallons of water be moved through the spill. With an oxygen content of 40 mg/L, 58,000,000 gallons would be required. Assuming that the water moves under a hydraulic gradient of 0.2, 0.4 to 4,000,000 yr would be required to move the necessary quantity of water through the spill depending upon hydraulic conductivity (Table 19.3). With 40 mg/L oxygen, this range becomes 0.08 to 800,000 years.

One way to overcome this problem in oxygen delivery is to use hydrogen peroxide. In water, hydrogen peroxide decomposes according to the following reaction (Hinchee and others, 1987)

$$H_2O_2 \rightarrow H_2O + \tfrac{1}{2}O_2 \tag{19.4}$$

The trick to making hydrogen peroxide work is to add a relatively large quantity to water and have oxygen released in a controlled manner as it advances through the aquifer (Hinchee and others, 1987). The techniques necessary to stabilize hydrogen peroxide are in general proprietary. However, it appears that the maximum upper limit of stabilized peroxide delivery is about 500 mg/L (Hinchee and others, 1987). With hydrogen peroxide, cleanup requires much less time because smaller volumes of water need to be moved through the spill (Table 19.3). The table, however, does show that bioremediation is probably only feasible in sand or

Table 19.3

Minimum pumping required for the enhanced biodegradation of a petroleum spill of 1000 gallons*

Condition	Form of Oxident Injection				
	Air Saturated Water	Oxygen Saturated Water	100 mg/L Peroxide	300 mg/L Peroxide	500 mg/L Peroxide
Available oxygen	8 mg/L	40 mg/L	50 mg/L	150 mg/L	250 mg/L
Minimum volume of water (gallons) required to remediate 1000 gallons of fuel	290,000,000	58,000,000	46,000,000	15,000,000	10,000,000

Theoretical minimum pumping time (years) required[1] to treat 1,000 gallons

Gravel ($K = 10^4$ gal/day/ft^2)	0.4	0.08	0.063	0.02	0.014
Sand ($K = 10^2$ gal/day/ft^2)	40	8.0	6.3	2.0	1.4
Silt ($K = 1$ gal/day/ft^2)	4,000	800	630	200	140
Clay ($K = 10^{-3}$ gal/day/ft^2)	4,000,000	800,000	630,000	200,000	140,000

[1]Assumes treatment area 100' × 100' with an average induced gradient of 0.2 (20' of differential across the site).

*Modified from Hinchee and others, 1987. Reprinted by permission of Conference on Petroleum Hydrocarbons and Organic Chemicals in Ground Water—Prevention, Detection and Restoration. Copyright © 1987. All rights reserved.

gravel irrespective of how oxygen is delivered. Fortunately, petroleum products themselves have difficulty spreading through the finer-grained units.

Several problems are associated with the use of hydrogen peroxide. The stability issue is one that we have talked about. If the hydrogen peroxide destabilizes, oxygen will come out of solution as a gas. This loss of available oxygen means that the process becomes much less efficient than otherwise might be expected. In severe cases, gas production (both O_2 and CO_2) can lead to a reduction in hydraulic conductivity (Brown and others, 1984). One undesirable side effect of stimulating the microbial oxidation of organic compounds is that other redox reactions may also be enhanced. For example, in a fuel spill, ferrous iron concentrations are often elevated because of the lack of oxygen. Aeration or the addition of hydrogen peroxide oxidizes Fe^{2+} to $Fe_2O_3(s)$. This solid precipitates and plugs up the pore system at places where oxygen is being added to the system (Hinchee and others, 1987).

Another possible problem is that large concentrations of hydrogen peroxide are toxic to microorganisms. Exact levels have not been established, but Brown and others (1984) cite concentrations as low as 200 mg/L as a possibility. One final issue of hydrogen peroxide use is cost. Hinchee and others (1987) indicate that the delivered 1987 cost of 35% hydrogen peroxide is $U.S. 4.20 per gallon. Using a simple assumed stoichiometry for the biodegradation reaction, the hydrogen peroxide necessary to degrade 1 gallon of fuel will cost $U.S. 50, assuming

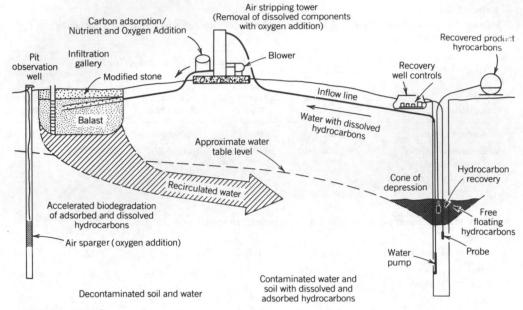

Figure 19.16
Example of a system for bioreclamation coupled with a two-pump, one-well collection system (from Yaniga and Mulry, 1984). Reprinted by permission of Conference on Petroleum Hydrocarbons and Organic Chemicals in Ground Water - Prevention, Detection and Restoration. Copyright © 1984. All rights reserved.

complete oxygen utilization. Thus, the cost of materials alone make bioremediation an expensive alternative for aquifer remediation.

We have not discussed the nutrients required for bacterial growth. Again details of the exact mixtures are proprietary. From the biochemistry of bacterial growth, commercial mixtures will contain nitrogen, phosphorous, and other nutrients in forms usable by bacteria. The most common inorganic sources of nitrogen and phosphorous would be nitrate, ammonia, and phosphate. Other elements required for growth would include potassium, magnesium, calcium, and sodium along with various trace metals depending upon the bacteria involved.

There are a variety of ways to deliver oxygen and nutrients to the spill. With air sparging, a small air compressor at the surface forces air down into one or more wells. This approach works best for small spills, but is not capable of delivering sufficient oxygen to treat large spills. Using compressed oxygen instead of air will help somewhat. Hydrogen peroxide and/or nutrients can be added to an aquifer by means of injection wells, infiltration galleries, or surface irrigation (Hinchee and others, 1987). Yaniga and Mulry (1984) describe a system to clean up gasoline. In the initial stages of remediation, free product was recovered using a one-well, two-pump system (Figure 19.16) with water treated and released. Once the recovery of free product dwindled, an enhanced bioreclamation scheme was implemented to deal with the residual contamination. Nutrients and hydrogen peroxide were added to the treated waste water, which was then recirculated through the spill by means of an infiltration gallery (Figure 19.16). Additional oxygen was provided by air sparging.

A similar system was used to assist in removing acetone and n-butyl alcohol from a shallow drift aquifer (Jhaveri and Mazzacca, 1986). Water rich in oxygen and nutrients was moved through the spill using injection trenches on the upstream end and a collection trench on the downstream end. A network of air injection wells was also installed within the spill to help maintain oxygen levels. Contaminated water removed from the ground was treated using a biological treatment system at the surface. Common elements in both of these systems include a regulated potential gradient so that nutrient and oxygen enriched water move in a controlled way and facilities at the surface to treat contaminated ground water and mix the oxygen and nutrients. Although we have largely ignored the problem of treating contaminated ground water at a site, this element of a contaminant recovery system is an expensive issue that needs to be considered.

Comments on System Design

The expense of chemicals, especially hydrogen peroxide, the facilities required on the surface, and the operating problems that can develop (for example, formation plugging or peroxide instability) demand that serious attention be given to system design. Some guidance on what needs to be done in a feasibility study is provided by Lee and others (1987). In addition to a somewhat more rigorous than usual field study (Table 19.4); significant efforts are required to characterize the contaminant and the microbial population thoroughly. With perhaps the exception of petroleum hydrocarbons, whose behavior is known, this work could involve detailed laboratory studies. For example, the efficacy of the system discussed by Jhaveri and Mazzacca (1986) was tested in laboratory experiments and a small field pilot. In many respects, this approach of scaling up the technology from laboratory, to pilot, to full implementation is not unlike one of the strategies for the design of injection/withdrawal systems.

19.6 Steps for Dealing with Problems of Contamination

A comprehensive plan for dealing with problems of ground water contamination is shown in Figure 19.17 as a general flowchart of the most important steps and critical decisions. In many cases, the first action will be an emergency response. Often, awareness of a problem is triggered by an emergency, for example gasoline fumes are detected in a building. Action is required immediately before anything is known about the site or the full extent of the problem. Success under emergency conditions depends on individuals with practical experience in dealing with emergency situations, the particular contaminants and the remedial action alternatives.

Eventually, attention shifts to an investigation of the problem. The level of effort required depends upon several factors. The most important is obviously the magnitude of the problem. A spill involving a large volume of contaminants and forming a large plume will of necessity require considerable fieldwork simply to document the problem. A small spill involving an exotic contaminant could require extensive laboratory studies to characterize fully the properties of the contaminant. Serious legal or public health considerations could require a much more thorough study to resolve completely issues of liability, potential health risks, and regulatory concerns. By the time the problem has been investigated, one should be able to answer these questions: (1) What is the source of contamination? (2) What contaminants are involved and in what forms do they occur (for example,

Table 19.4
Components of a study to establish the feasibility of bioremediation (from Lee and others, 1987)*

SITE CHARACTERISTICS

A. Aquifer properties
   1. Permeability
   2. Specific yield
   3. Clay content of soil
   4. Heterogeneity of formation
   5. Depth
   6. Thickness
   7. Direction of ground water flow
   8. Recharge and discharge areas
   9. Seasonal fluctuations in water table
B. Ground water quality
   1. Inorganic nutrient levels
   2. Precipitation of inorganic nutrients
   3. Dissolved oxygen content
C. Location of physical structures
D. Ground water pumpage controls
E. Potential for contamination of drinking water or agricultural wells

CONTAMINANT CHARACTERISTICS

A. Type
B. Concentration
C. Areal and vertical extent of contamination
D. Location of released material in aquifer—dissolved, floating, trapped, or sinking
E. Heterogeneity of contamination
F. Biodegradability of contaminants
G. Presence and quantity of toxic agents
H. Nature of release—acute, chronic, or periodic
I. Time since release
J. Effect of phyical weathering or abiotic reaction on contaminants

MICROBIAL CHARACTERISTICS

A. Presence of active microbial population
B. Acclimation to contaminants
C. Nutrient requirements for optimal growth
D. Extent of biodegradation that can be achieved
E. Rate of biodegradation

*Reprinted by permission of Conference on Petroleum Hydrocarbons and Organic Chemicals in Ground Water—Prevention, Detection and Restoration. Copyright © 1987. All rights reserved.

dissolved in water, NAPLs)? (3) What are the characteristics of these wastes (for example, volumes, chemical properties, toxicity)? (4) What is the site geology in terms of major units, their distribution, and mineralogical/chemical characteristics? (5) What is the hydrogeological setting in terms of patterns of flow, water chemistry, and key flow and transport parameters? (6) How have the characteristics of the contaminants, the geological setting, and the hydrogeology controlled the pattern of contaminant spreading? For problems involving dissolved contaminants, knowledge should be quantitative to the extent that it should be possible to model the mass transport processes. For spills involving LNAPLs or DNAPLs, there is much less effort to define the hydrogeologic setting in a quantitative way because so few

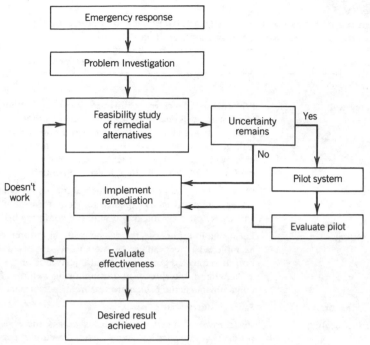

Figure 19.17
Flowchart of the steps involved in remediating a problem of
ground water contamination.

practical mathematical tools exist. However, before proceeding with remediation,
a clear conceptual model of the problem is needed.

The next step in the process (Figure 19.17) is the assessment of action alternatives. Essentially, the feasibility of different overall strategies is evaluated in terms
of their ability to meet specific cleanup goals, their cost, and regulatory acceptability. Candidate strategies are developed by piecing together the various alternatives for remedial action and where possible checking the design elements with
mathematical and/or experimental approaches.

This process usually leads to an overall remedial strategy that should have a
reasonable probability of success. Nevertheless, a possibility of failure can exist
because of uncertainty inherent in these problems. Even after detailed studies, one
may not be completely sure of what contaminants are present or where they are distributed. Aspects of the geological and hydrogeological settings may not be well
defined, especially in complex settings. There are as well uncertainties in scientific
knowledge about the way contaminants might behave and the performance of
some remedial alternatives. For example, given the particular conditions at a site, a
slurry cutoff wall unexpectedly might perform badly. However, in more general
terms, there is not a great deal of information in the literature of how well various
types of slurry walls are likely to perform even under optimum conditions.

Uncertainty creates problems in the implementation of an overall strategy.
Some proposed remedies might be ineffective, certain equipment may be inadequate, or additional pieces may need to be added to a recovery system. Examination of case studies shows that it is relatively rare to design an overall strategy
for remedial action that works perfectly when implemented. The flowchart

Table 19.5

Summary of ways for measuring performance of remedial alternatives (modified from Office of Technology Assessment, 1984)

Technique	Measuring performance
CONTAINMENT	
1. Slurry wall	Performance of a slurry wall is determined by various methods. Monitoring well data outside of the wall can indicate the degree of leakage. Hydraulic head differences determine leakage potential; actual leakage can be calculated. Head measurements in underlying aquifers determine potential for vertical leakage. Permeability measurements of confining bed also determine leakage potential.
2. Sheet pile	Same as slurry wall except that measurements are taken at specific places where leakage is expected to occur, for example, at pile joints and where piles are integrated with the confining bed.
3. Grouting	Encapsulation processes cannot be easily monitored or controlled (for example, a barrier wall can be more easily inspected during construction than injected grout; and grout injection is not as easily controlled as trenching). Interpretation of monitoring well data for downgradient contaminants is the principal measure of performance.
4. Geomembrane	Same as slurry wall.
5. Surface sealing	Visual inspection can locate holes or cracks. Increased leachate production indicates leakage. Also, increased pumpage requirements in head management system may indicate leakage.
6. Diversion ditches	Visual inspection is used to measure performance—for example, during precipitation events.
7. Hydrodynamic control	Water levels can be monitored in surrounding wells to observe gradients.
WITHDRAWAL	
1. Pumping	Contaminant concentration levels can be measured in produced water to determine removal rates; and effects of pumping can be verified by monitoring water levels in surrounding wells. Underlying aquifers must be monitored to detect downward migration. Concentration levels after pumping is terminated must be monitored to determine increases in concentrations due to desorption. Geochemical interactions between contaminants and the aquifer matrix affect the partitioning of the contaminant between solid and water phases; the potential effectiveness and length of operations are dependent on these interactions.
2. Ineceptor systems	Same as pumping.
3. Soil venting	Measurements can be taken using gas collection probes. Mass balance calculations can be made on the quantity of contaminants removed. Soil sampling.
4. Excavation	Contaminant concentration levels can be measured in surrounding soil and aquifer materials and in surrounding waters to verify total removal. Measurements are most accurate if contaminants are highly concentrated and limited in depth and volume.
IN-SITU REHABILITATION	
1. Biological degradation	Contaminant levels can be monitored in soil and water.
2. Chemical degradation	Same as biological degradation.

Table 19.5
Continued

Technique	Measuring performance
MANAGEMENT OPTIONS	
1. Limit/terminate aquifer use	Exposure levels can be monitored; actual use patterns over time can be determined. Performance is also economic, for example, it may be cheaper to terminate use and import water or develop alternative supplies than to treat supplies or otherwise correct contamination.
2. Develop alternative water supply	Performance is mainly economic, for example, it may be cheaper to terminate use and develop alternative supplies or import water than to treat supplies or otherwise correct contamination.
3. Purchase alternative water supply	Performance is mainly economic, for example, it may be cheaper to terminate use and develop alternative supplies or import water than to treat supplies or otherwise correct contamination.
4. Source removal	Contaminant concentration levels can be measured in surrounding soils, aquifer materials, and waters to verify total removal.
5. Monitoring	Performance can be measured by duplicating samples and analyses. Use of qualified personnel are essential for proper well placement and for the overall ground water quality investigation.
6. Health advisories	Exposure levels can be monitored; actual use pattern over time can be determined. Performance is also economic, for example, it may be cheaper to terminate use and develop alternative supplies or import water than to treat supplies or otherwise correct contamination.
7. Accept increased risk	Performance is often measured in economic terms.

(Figure 19.17) acknowledges that in spite of detailed site and feasibility studies uncertainties about the performance of the system could exist. A step to reduce the uncertainty especially with large and expensive systems would be to implement a pilot-scale system. If successful, the pilot can be expanded to an operational system. If unsuccessful, alternative measures can be developed without a great deal of expense. Where little uncertainty is evident, there is no need to test a pilot system.

Once remedial activities are underway, monitoring of critical parameters provides evidence that the expected performance is being achieved (Figure 19.17). There are tests that can be used with each of the remedial alternatives to verify performance (Table 19.5). Failure of elements in the overall strategy should prompt corrective action. If the system is in fact performing adequately, remedial action and monitoring will continue until the desired result is achieved.

19.7 Case Studies in Site Remediation

The following four case studies illustrate different remedial alternatives and the basic steps in conducting a study. The first two cases are small-scale problems involving spills of gasoline and oil. They are typical of situations that occur commonly because of the large number of petroleum storage tanks in North America. The last two case studies describe larger and much more serious problems resulting from the disposal of toxic organic chemicals at uncontrolled sites.

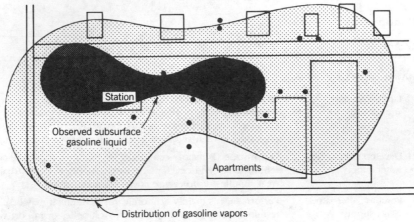

Figure 19.18

The original distribution of gasoline and gasoline vapor (from O'Connor and others, 1984a). Reprinted by permission of Conference on Petroleum Hydrocarbons and Organic Chemicals in Ground Water - Prevention, Detection and Restoration. Copyright © 1984. All rights reserved.

Gasoline Spill

O'Connor and others (1984a) describe studies and remedial action in relation to a small gasoline spill in alluvial gravels at an undisclosed location. The presence of gasoline fumes in an occupied apartment building necessitated a rapid emergency response. By sealing cracks in the basement floor of the building and applying positive air pressure inside the basement, the vapor problem was eliminated in a matter of hours. Subsequent investigations discovered free gasoline at the water table as well as a large area affected by gasoline vapor (Figure 19.18). Initial remedial action at the site involved installation of a vapor extraction system (VES) to protect the building. Gasoline vapor was extracted from the unsaturated zone using six wells located close to the apartment building and in an area where the thickness of the gasoline was the greatest. Monitoring showed that the VES system not only was effective in controlling the vapor but also removed free gasoline by volatilization reducing the quantity of free product (Figure 19.19). After continued monitoring and fine tuning of the system, a major portion of the spill was cleaned up.

Oil Spill: Calgary, Alberta

Another larger and more serious spill of petroleum products in floodplain sediment is related to an asphalt plant and terminal at Calgary, Alberta (O'Connor and others, 1984b). The spill first came to light when petroleum products discharged at the surface in the areas marked by "Early oil seeps" (Figure 19.20). After plant personnel constructed an interceptor ditch along the southern plant boundary (Figure 19.20), it was thought that the problem was solved. However, at this early stage the full extent of the problem was not known. Oil subsequently discharged into lagoons of the Inglewood Bird Sanctuary.

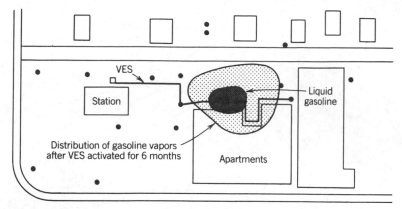

Figure 19.19
Map showing the location of the vapor extraction system and the distribution of gasoline vapor and liquid gasoline following six months
of operation (from O'Connor and others, 1984a). Reprinted by permission of Conference on Petroleum Hydrocarbons and Organic
Chemicals in Ground Water - Prevention, Detection and Restoration.
Copyright © 1984. All rights reserved.

The need to control seepage into the bird sanctuary in advance of returning
geese required emergency action. An interceptor drain was installed across the
zone of seepage (Figure 19.20). A horizontal box culvert and gravel buried in the
trench provided lateral drainage of the oil to a series of sumps. Migration of oil
past the interceptor system was eliminated by adding an impermeable barrier to

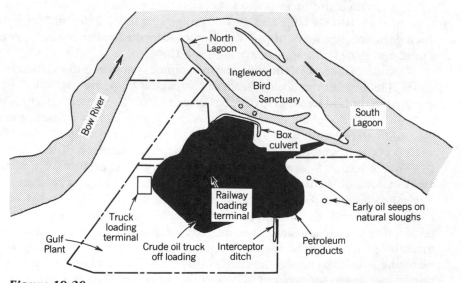

Figure 19.20
Site of an oil spill at Calgary, Alberta (from O'Connor and others, 1984b).
Reprinted by permission of Conference on Petroleum Hydrocarbons and Organic
Chemicals in Ground Water - Prevention, Detection and Restoration. Copyright ©
1984. All rights reserved.

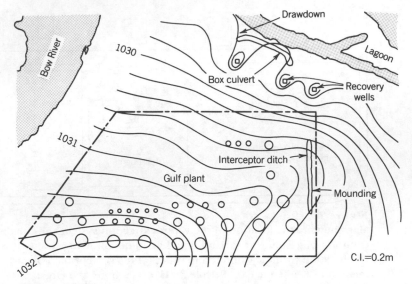

Figure 19.21
Elevation of the water table (meters) following start-up of the recovery wells autumn 1979 (from O'Connor and others, 1984b). Reprinted by permission of Conference on Petroleum Hydrocarbons and Organic Chemicals in Ground Water -Prevention, Detection and Restoration. Copyright © 1984. All rights reserved.

the downstream wall of the culvert. Cutoff walls added to both ends of the interceptor system deflected oil into the box culvert system. This action eliminated oil seepage into the north lagoon.

The detailed site investigation revealed that oil had collected on the water table over a relatively large area and was migrating toward the bird sanctuary following the local gradient of the water table. The contaminants were spreading in a sand and gravel unit approximately 10 m thick.

Because of the size of the spill, the interceptor system was not sufficient to contain and recover all the oil. A system of two-well, two-pump extraction wells was installed in the bird sanctuary (refer to Figure 19.6b). The water wells produced up to 2.5×10^{-2} m³/sec and created the necessary potential low to move the oil into the product collection wells (Figure 19.21). Clean water pumped from the water wells was discharged into the interceptor ditches constructed earlier at the site.

This collection system was run for five years and its performance reviewed in 1983. A lack of well maintenance did contribute to reduced product recovery. Nevertheless during this period of operation approximately 1,300,000 L of oil were collected. A significant quantity of product remained in the subsurface especially under the plant site and away from the area of most intense remedial activity. The increasing thickness of product at the plant site pointed to the continued leakage of oil at the source. Based on the performance of the system up to 1983, about six more years would be required to collect the remaining free product.

Gilson Road: Nashua, New Hampshire

The Gilson Road site in New Hampshire is important historically because it was the subject of the first cooperative agreement signed under the Superfund law in

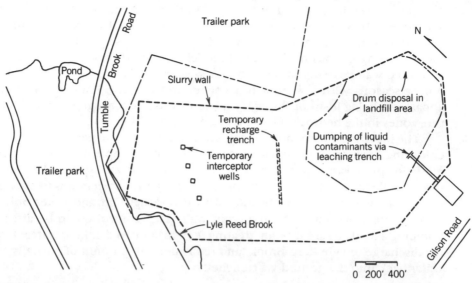

Figure 19.22
Layout of the Gilson Road hazardous waste-disposal site at Nashua, New Hampshire
showing the landfill and liquid waste-disposal areas as well as facilities for controlling
contaminant migration (modified from Ayres and others, 1983). Reprinted by permission
of Third National Symposium on Aquifer Restoration, and Ground Water Monitoring.
Copyright © 1983.

the United States. Chemical sludge and organic wastes in drums were placed
together with refuse in a sand and gravel pit. In addition, some 4,000,000 L of
liquid chemical wastes were discharged into the ground at a nearby trench (Figure
19.22; Ayres and others, 1983). The most hazardous contaminants included heavy
metals and volatile and extractable organic compounds. Spreading of contaminants was facilitated by the presence of 6–35 m of permeable sand and gravel,
which overlie till and fractured bedrock. Initial studies in 1980 defined a contaminant plume more than 450 m long and up to 33 m thick (Ayres and others,
1983). By December 1980, contaminants from the front of the plume had begun
to discharge into Lyle Reed Brook, located at the western edge of the study area.
Following completion of the site investigation in 1981, a feasibility study was
conducted to establish a strategy for remedial action. The possible alternatives
included combined hydrologic containment and ground water treatment as well
as total removal (Ayres and others, 1983). The overall strategy that was adopted
involved isolating the most contaminated parts of the plume using a cutoff wall
and surface seals and a withdrawal-treatment system to remove the contaminants
from the ground water. Before the remedy could be implemented in 1982,
monitoring showed that a highly contaminated portion of the plume was about to
discharge into the brook. To delay the arrival of this front an emergency recirculation system was designed to remove contaminated ground water from four wells
at the nose of the plume (Figure 19.22) and to return it without treatment using
a temporary recharge trench. A two-dimensional computer model was developed
and calibrated against field data to assist in the design of this system (Ozbilgin and
Powers, 1984). Subsequent monitoring indicated that pumping the wells at a total
of 0.25 m³/min was successful in temporarily retarding the advance of the plume
(Ozbiligin and Powers, 1984).

The conventional cutoff wall ranged in thickness from 0.5 to 1.25 m (Ayres and others, 1983; Schulz and others, 1984). It surrounded an area of about 80,000 m² to depths ranging from 10 to 33 m. Extensive analyses of the performance of the wall using various mathematical models indicated that the quantity of contaminated water passing through the site was reduced from 284 m³/d to 114 m³/d during seasons with low rainfall (Ozbilgin and Powers, 1984). Containment is apparently being lost not through the grout wall but through fractures and fracture zones in the bedrock (Ozbilgin and Powers, 1984).

The literature we examined discussed the design of the permanent system for collecting, treating, and recirculating water but not its implementation. Computer modeling played an important role in the design of the well system. In addition to simply collecting water for treatment at the surface, the system was to provide further containment for contaminants. With this design, ground water only flows into the area surrounded by the grout wall. This control could be affected by pumping a total of 1.14 m³/min, treating 0.19 m³/min to drinking water standards for discharge to Lyle Reed Brook, and returning 0.95 m³/min of partially treated water back into the ground via trenches.

Hyde Park Landfill: Niagara Falls, New York

Coming out of the Settlement Agreement for Hyde Park Landfill in 1982 was a provision to evaluate the available data and to develop a plan for remedial action. The remedial measures finally agreed upon were reviewed in detail by Faust (1985) and are described here as an illustration the variety of factors that must be considered when dealing with a complex site and a variety of different contaminants.

The overall strategy for remediation involves three parts: (1) source control, (2) control and collection of dissolved and aqueous phase contaminants from the overburden, and (3) control and collection of contaminants from the Lockport dolomite. The complexity of the geologic setting and uncertainty about how well certain remedial strategies might work necessitated a phased approach that involved prototype systems along with extensive data collection. Proceeding in this way avoided "the situation where remedial measures are installed and later to be found ineffective without any simple recourse to correct the deficiencies" (Faust, 1985).

The main reason for source control is to reduce the quantity of contaminants moving out of the landfill. A surface seal will reduce infiltration from precipitation and the flux of water and contaminants. Collection wells will be used to remove fluids from the actual landfill itself. Initial designs are for this to be a prototype system involving two wells.

Remedial action in the overburden will (1) contain the lateral spread of contaminants and (2) provide a way of collecting them. The measure that is proposed is a subsurface drain around the area where free organic liquids are known to occur. Contaminants not collected by the drain will move down into bedrock where other collection systems are in place. These bedrock systems include an injection/withdrawal scheme for DNAPLs (Figure 19.23) and a withdrawal scheme for collecting dissolved contaminants close to the Niagara Gorge. The system proposed for DNAPLs is analogous to a secondary recovery scheme for oil (Faust, 1985). The advantages of the injection/withdrawal scheme as compared to pumping alone include (1) it avoids dewatering of the upper part of the Lockport

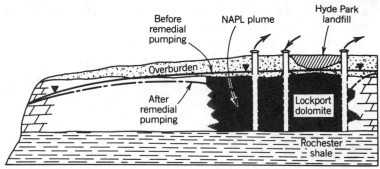

Figure 19.23
Conceptual diagram of the injection/withdrawal system for removing DNAPLs from the Lockport dolomite at the Hyde Park Landfill (from Faust, 1985).

dolomite that is contaminated and requires ground water to circulate for cleanup to occur, and (2) larger volumes of water can be circulated through contaminated parts of the plume. Both two- and three-dimensional flow modeling was conducted to evaluate the effectiveness of the prototype system, which involves four collection and two injection wells. Pumping rates of 20 gallons per minute for each withdrawal well and 10 gallons per minute for each injection well should provide a capture zone larger than the DNAPL plume. After further study, an operational system will be put in place. The second withdrawal system will collect dissolved contaminants in the Lockport dolomite. These contaminants are sufficiently far from the site that they will be in general unaffected by the cone of depression created by the first DNAPL removal wells. Flow, however, will eventually cause these contaminants to discharge along the Niagara Gorge. The plan is to install a line of interceptor wells across the plume west of the site.

In concluding this discussion about plans for the remedial system at Hyde Park landfill, it is worth pointing out how monitoring is such an integral part of ongoing operations. Monitoring will establish whether a particular remedial alternative has met specific performance goals, provide criteria for logically expanding the prototype systems, and safeguard the public against accidental exposure to the contaminants.

Suggested Readings

Javandel, I., and Tsang, C., 1986, Capture-zone type curves: a tool for aquifer cleanup: Ground Water, v. 24, n. 5, p. 616–625.

Hinchee, R. E., Downey, D. C., and Coleman, E. J., 1987, Enhanced bioreclamation soil venting and ground water extraction; a cost-effectiveness and feasibility comparison: Proc. of the NWWA/API Conference on Petroleum Hydrocarbons and Organic Chemicals in Ground Water-Prevention, Detection and Restoration, National Water Well Assoc., Dublin, Ohio, p. 147–164.

Knox, R. D., Canter, L. W., Kincannon, D. F., Stover, E. L., and Ward, C. H., 1984, State-of-the-art of aquifer restoration: National Center for Groundwater

Research, University of Oklahoma, Oklahoma State University and Rice University, Norman, Oklahoma 73019, National Technical Information Service, 371 p.

OTA, 1984, Protecting the Nation's Groundwater from Contamination: Office of Technology Assessment, OTA-0-233, Washington, D.C., 244 p.

Problems

1. The capture-zone type curve for a single well is superimposed on an outline of a dissolved contaminant plume as follows

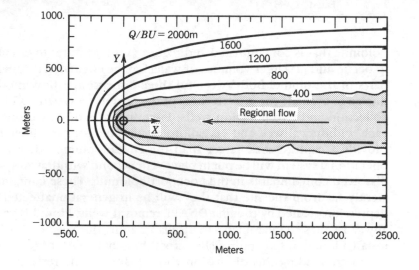

a. Given that the thickness of the aquifer is 10 m, and the Darcy flow velocity is 1×10^{-6} m/sec, calculate the single-well pumping rate.

b. For an aquifer transmissivity of 0.005 m²/sec, storativity of 1×10^{-4}, and available drawdown of 10 m, determine whether the calculated production rate is feasible.

2. In Chapter 18, we described the case of contamination of Gloucester, Ontario, which came about due to a variety of dissolved organic compounds. Develop three feasible strategies in qualitative terms for remediating the site. Be sure to explain the positive and negative factors that would influence the choice of a particular strategy.

3. An underground storage tank at an industrial facility leaked approximately 90,000 liters of gasoline over a five-year period into a shallow sand aquifer. An aggressive program of product recovery collected approximately 35,000 liters of free product.

a. Explain what potential problems could occur by simply leaving the residually-saturated fraction of the spill in the ground.

b. What are the feasible alternatives for removing this remaining volume of gasoline?

Answers to Problems

Chapter 3

1. $q = 1.89 \times 10^{-8}$ m/sec

2. effective porosity = 0.1, $v = 1.89 \times 10^{-7}$ m/sec
   effective porosity = 0.0001, $v = 1.89 \times 10^{-4}$ m/sec

7. Total head: 900 ft, 800 ft, 700 ft, 600 ft
   Pressure head: 600 ft, 575 ft, 150 ft, 25 ft
   Elevation head: 300 ft, 225 ft, 550 ft, 575 ft

8. $K = 3.28 \times 10^{-6}$ ft/sec, $K = 2.12$ gal/day ft; $k = 1 \times 10^{-9}$ cm^2,
   $k = 1.04 \times 10^{-1}$ darcys

11. $K_x = 3.7 \times 10^{-7}$ m/sec, $K_z = 2.7 \times 10^{-8}$ m/sec

12. $q_x = 1 \times 10^{-4}$ cm/sec

13. Section B, 1×10^{-6} m/sec, Section C, 2×10^{-6} m/sec,
    Section D, 5×10^{-7} m/sec

16. $q_z = 5 \times 10^{-8}$ cm/sec, downward flow

17. ***a.*** Inflow both aquifers, $Q = 1 \times 10^{-4}$ m^3/sec

    b. Outflow Aquifer A, $Q = 1.5 \times 10^{-4}$ m^3/sec
    Outflow Aquifer B, $Q = 5 \times 10^{-5}$ m^3/sec

    c. Outflow area of A, 1.5×10^{-1}
    Outflow area of B, 0.5×10^{-1}

Chapter 4

2. *a.* 300 ft³

   b. $S = 8.8 \times 10^{-5}$

   c. 1.7×10^{-8} ft²/lb

4. *a.* 6250 lb/ft², 4368 lb/ft², 1882 lb/ft²

   b. 4998 lb/ft², 4368 lb/ft², 630 lb/ft²

   d. 30 ft

5. $T = 4380$ ft²/day

7. $S = 2.2 \times 10^{-4}$

Chapter 5 (Answers subject to interpolation of tabulted values)

1. $T = 19.3 \times 10^3$ ft²/day

2. $T = 5 \times 10^3$ ft²/day, $S = 2.7 \times 10^{-4}$

3. 1 year, 162 ft; 3 year, 168 ft

4. $Q = 15.5 \times 10^3$ ft³/day

5. $Q = 54 \times 10^4$ gal/day

6. zero drawdown at stream, 12.4 ft at midpoint

7. approximately 40 percent

11. *a.* approximately 75 ft

12. *a.* $T = 10,000$ gal/day ft

    c. Specific capacity = 10 gpm/ft

    d. $T = 20,000$ gal/day ft (approximately)

13. *a.* 145,800 ft

    b. 8.2 ft

14. in order of the blanks and selections: 100; 6; 40; 4; add; 30; 2; add.

15. 2000 ft

16. approximately 89 ft

Chapter 6

2. *a.* elastic compression = 0.042 ft

   b. inelastic compression = 2 ft

   c. bulk modulus = 3×10^7 lb/ft²

4. $Q = 0.247$ ft³/sec

5. inland distance = 4125 ft; inland distance with seaward flow reduced = 8250 ft

Chapter 7 (Answers subject to interpolation of tabulated values)

6. *a.* at the canal, 25.9 ft, 17.5 ft, 14 ft, 9.1 ft
      at the divide, 70 ft, 70 ft, 66.5 ft, 61 ft

   b. $q' = 15,529$ ft²/yr at 10 yrs; $q' = 914$ ft²/yr at 100 yrs

8. *a.* $x = 0.6$ ft, $h = 23$ ft, $x = 3.1$ ft, $h = 29$ ft, $x = 10.1$ ft, $h = 35$ ft, $x = 27.6$ ft, $h = 48$ ft, $x = 70.1$ ft, $h = 50$ ft

   b. $Q = 94 \times 10^{-3}$ ft³/day

Chapter 8

1. 19% at 2000m, 60% at 10,000m

2. *a.* 150 m/my

   b. 300 m/my

   d. 15

5. about 40°C above ambient

7. Change in total stress 450 kbars, change in fluid pressure 54 kbars, change in effective stress 396 kbars

Chapter 9

1. 1.9 and 0.8 cal m^{-1} sec^{-1} $°C^{-1}$ for the saturated sandstone

3. $q = 2.5 \times 10^{-3}$ cm/sec

4. *a.* $k = 7 \times 10^{-9}$ cm²

Chapter 10

1. fractured = 2.85 yr; unfractured = 2850 yr

2. approximately 180 yrs

3. 4×10^{-6} cm²/sec

7. advection = 0.1 mg/m²sec, diffusion = 0.4×10^{-5} mg/m²sec

8. *b.* 2×10^{-2} cm

9. 28.2m, 8.9m, and 2.8m in x, y, z, respectively

10. approximately 0.25m

Chapter 11

1. *a.* 2.3×10^{-3} M; 0.83×10^{-3} M; 5.5×10^{-3} M; 0.25×10^{-3} M
   4.7 meq/L; 0.83 meq/L; 5.47 meq/L; 0.25 meq/L

   b. 0.012 M

   c. 0.66 and 0.91

3. SI = 0.03 slightly oversaturated

4. cation/anion balance = 0.99; good analysis

Chapter 12

2. *a.* 1.7×10^{-2} atm

   b. strongly oversaturated — IAP/K = 9.2

3. *a.* 2×10^{-5} atm

   b. 0.1 atm

4. *a.* $n = 0.75$; $K = 30$

   b. $K_d = 0.60$ ml/gm

5. *a.* (i) 3480 (ii) 355 (iii) 5.2×10^5

   b. (i) 34.8 (ii) 3.6 (iii) 5180

6. 4.16, 3.36, 4.57

7. *a.* 5, 1, 6, 2, 3, 4

   b. 4, 3, 2, 6, 1, 5

8. $pe = 4.65$; $E_H = 0.27$ V

9. $Fe^{3+} = 9.1 \times 10^{-7}$ M; $Fe^{2+} = 9.1 \times 10^{-6}$ M

10. 13.9 days

11. 13.8µg/L

13. $\delta^{18}O = -18‰$; $\delta D = -135‰$ SMOW

Chapter 14

8. water sample 1 + 2 calcite + 2 dolomite + 0.2 halite + 0.2 sylvite + 10 Na (from exchange) − 5 Ca (from exchange) = water sample 2

9. 2850 years old

10. 24,300 years old

Chapter 17

1. 14 cm

3. −0.13

4. *a.* 36 cm²/hr

   b. 1.0 cm

5. 0.85 ml/g

8. $C = 0.91\ C_o$

9. 0.45 cm/d

11. (input data)

```
10     7
4000.00     5.00   0.05  0.10  0.00  5.97
   14.00   200.00  50.00  0.00
    5.00   150.00
```

(output data)

14 Years

0.00	88.02	2195.20	2283.22	2195.20	88.02	0.00
0.15	179.12	1516.28	1695.10	1516.28	179.12	0.15
1.55	216.39	1191.31	1404.60	1191.31	216.39	1.55
4.92	231.56	997.65	1219.36	997.65	231.56	4.92
9.33	226.28	828.12	1035.77	828.12	226.28	9.33
6.75	104.52	339.73	430.82	339.73	104.52	6.75
0.49	5.48	16.27	20.78	16.27	5.48	0.49
0.00	0.01	0.04	0.05	0.04	0.01	0.00
0.00	0.00	0.00	0.00	0.00	0.00	0.00
0.00	0.00	0.00	0.00	0.00	0.00	0.00

Chapter 18

1. *a.* 15 m²

   b. 0.18 m

2. 10 m

3. 1,2-dichloroethane: $R_f = 9.2$
   tetra hydrofuran: $R_f = 2.1$
   1,4-dioxane: $R_f = 1.5$

4. **a.** 1,2-dichloroethane: K_d = 1.45 ml/g
   tetra hydrofuran: K_d = 0.19
   1,4-dioxane: K_d = 0.088

   b. K_d = 0.016 ml/g
   K_d = 0.0019
   K_d = 0.0004

5. C_0 = 50, 100, 125, 80, 70, and 60 mg/L at 0.0, 2.75, 5.5, 8.25, 9.5, and 11.0 yr respectively.
   $v = 1 \times 10^{-4}$ cm/sec, Y = 100m, Z = 10m, R_f = 1.0 and $\alpha_{x,y,z}$ = 4.0m, 1.5m, and 0.15m respectively.

7. **a.** 3500 cm

   b. 6200 cm

   c. 111 cm

8. 0.092 d^{-1}, 7.5 d

Chapter 19

1. **a.** 6×10^{-3} m³/s

   b. feasible — drawdown about 2.5m after 1 yr

References

Abelin, H., Gidlund, J., and Neretnieks, I., 1982, Migration in a single fissure. Scientific Basis for Nuclear Waste Management, v. 5. New York, Elsevier, p. 529–538.

Ahlstrom, S. W., Foote, H. P., Arnett, R. C., Cole, C. R., and Serne, R. J., 1977, Multicomponent mass transport model: Theory and numerical implementation (Discrete-Parcel-Random-Walk Version): Richland, Wash., Battelle Northwest Laboratories, BNWL no. 2127.

Albertsen, M., and an ad hoc task force, 1986, Beurteilung und behandlung von Mineralölschadensfallen im Hinblick auf den Grundwasserschutz, Jeil 1, Die wissenschaftlichen Grundlagen zum Verständnis des Verhaltens von Mineralöl im Untergrund: Federal Office of the Environment, LTwS No. 20, Berlin, 178 p.

Alpay, O. A., 1972, A practical approach to defining reservoir heterogeneity: J. Petrol. Technol., v. 20 no. 7, p. 841–848.

American Chemical Society, 1980, Guidelines for data acquisition and data quality evaluation in environmental chemistry: Anal. Chem., v. 52, p. 2242–2249.

Amundson, N. R., 1950, Mathematics of adsorption in beds, II: J. Phys. Colloid Chem., v. 54, p. 812–820.

Anderson, G. M., and Macqueen, R. W., 1982, Ore deposit models—Mississippi Valley type lead zinc deposits: Geoscience Canada, v. 9, p. 108–117.

Anderson, M. P., 1984, Movement of contaminants in groundwater: Groundwater transport—advection and dispersion. Groundwater Contamination, NRC Studies in Geophysics: Washington, D.C., National Academy Press, p. 37–45.

————, and Munter, J. A., 1981, Seasonal reversals of groundwater flow around lakes and the relevance to stagnation points and lake budgets: Water Resources Res., v. 17, p. 1139–1150.

Anderson, R. E., and Laney, R. L., 1975, The influence of late Cenozoic stratigraphy on distribution of impoundment related seismicity at Lake Mead, Nevada, Arizona: U.S. Geol. Survey J. Res., v. 3, p. 337–343.

Aris, R., and Amundson, N. R., 1973, Mathematical Methods in Chemical Engineering, II: Englewood Cliffs, N.J., Prentice-Hall, 369 p.

Athy, L. F., 1930, Density, porosity, and compaction of sedimentary rocks: Amer. Assoc. Petrol. Geol. Bull., v. 14, p. 1–24.

Atwood, D. F., and Gorelick, S. M., 1985, Sensitivity analysis in aquifer studies. Water Resources Res., v. 13, no. 4, p. 733–737.

Ayres, J. E., Lager, D. C., and Barvenik, M. J., 1983, The first EPA superfund cutoff wall: Design and specifications. Third National Symposium on Aquifer Restoration and Ground Water Monitoring, Proc., ed., D. M. Nielsen: Dublin, Ohio, National Water Well Assoc., p. 13–22.

Back, W., 1960, Origin of hydrochemical facies of groundwater in the Atlantic Coastal Plain: Internat. Geol. Cong., Copenhagen, Proc., p. 87–95.

————, 1961, Techniques for mapping of hydrochemical facies: U.S. Geol. Surv. Prof. Paper 424-D, p. 380–382.

————, Cherry, R., and Hanshaw, B., 1966, Chemical equilibrium between water and the minerals of a carbonate aquifer: Nat'l Speleo. Soc. Bull., v. 28, p. 119–126.

————, and Freeze, R. A., 1983, Chemical Hydrogeology. Benchmark Papers in Geology: Stroudsburg, Pa., Hutchinson Ross.

————, and Hanshaw, B., 1965, Chemical Geohydrology. Advances in Hydroscience, ed. V. T. Chow: New York, Academic Press, Vol. 2, p. 49–109.

————, Rosenshein, J. S., and Seaber, P. R., eds., 1988, The Geology of North America, Hydrogeology, 0–2: Boulder, Colorado: Geol. Soc. Amer. Pub., p. 15–263.

Baetslé, L. H., 1969, Migration of radionuclides in porous media. Progress in Nuclear Energy Series XII, Health Physics, ed. A. M. F. Duhamel: Elmsford, N.Y., Pergamon Press, p. 707–730.

Bagley, E. S., 1961, Water rights law and public policies relating to groundwater mining in the southwestern states: J. Law and Econ., v. 4, p. 144–174.

Bahr, J. M., and Rubin, J., 1987, Direct comparison of kinetic and local equilibrium formulations for solute transport affected by surface reactions. Water Resources Res., v. 23, no. 3, p. 438–452.

Bailey, J. E., and Ollis, D. F., 1987, Biochemical Engineering: New York, McGraw-Hill, 753 p.

Bair, E. S., 1987, Analysis of hydraulic gradients across the host rock at the proposed Texas Panhandle Nuclear Waste Reposition Site: Ground Water, v. 25, p. 440–447.

Ball, J. W., Jenne, E. A., Burchard, J. M., and Truesdell, A. H., 1976, Sampling and preservation techniques for water with a section on gas collection: First Workshop on Sampling Geothermal Effluents, Proc., EPA-600/9-76-0111, p. 219–234.

————, Jenne, E. A., and D. K. Nordstrom, 1979, WATEQ 2-A computerized chemical model for trace and major element speciation and mineral equilibria of natural waters. Chemical Modeling in Aqueous Systems, ed. E. A. Jenne: American Chemical Society, Symposium Series 93, p. 815–835.

Banks, H. O., 1953, Utilization of underground storage reservoirs: Trans. Amer. Soc. Civil Engrs., v. 118, p. 220–234.

_____, and Richter, R. C., 1953, Sea water intrusion into groundwater basins bordering the California coast and inland bays: Trans. Amer. Geophys. Union., v. 34, p. 575–582.

Barari, A., and Hedges, L., 1985, Movement of water in glacial till. Memoirs, Hydrogeology of Rocks of Low Permeability: Tucson, Ariz., International Association of Hydrogeologists, p. 129–134.

Barker, C., 1972, Aquathermal pressuring-role of temperature in development of abnormal pressure zones: Amer. Assoc. Petrol. Geol. Bull. v. 56, p. 2068–2071.

Barker, J. A., and Black, J. H., 1983, Slug tests in fissured aquifers: Water Resources Res., v. 19, p. 1558–1564.

Barnes, I., and Clark, F. E., 1969, Chemical properties of groundwater and their corrosion and encrustation effect on wells: U.S. Geol. Survey, Prof. Paper 498-D.

Bastin, E. S. ed., 1939, Contributions to the knowledge of the lead-zinc deposits of the Mississippi Valley Region: Geol. Soc. Amer. Paper 24, 16 p.

Batchelder, G. V., Panzeri, W. A., and Phillips, H. T., 1986, Soil ventilation for the removal of adsorbed liquid hydrocarbons in the subsurface. NWWA/API Conference on Petroleum Hydrocarbons and Organic Chemicals in Ground Water—Prevention, Detection, and Restoration, Proc.: Dublin, Ohio, National Water Well Assoc., p. 672–688.

Bathhurst, R. G. C., 1971, Carbonate sediments and their diagenesis. Developments in sedimentology, no. 12, New York, Elsevier.

Baumann, P., 1963, Theoretical and practical aspects of well recharge: Trans. Amer. Soc. Civil Engrs., v. 128, p. 739–764.

_____, 1965, Technical development in groundwater recharge. Advances in Hydroscience, v. 2, ed. V. T. Chow: New York: Academic Press, p. 209–279.

Bear, J., 1972, Dynamics of Fluids in Porous Media: New York, American Elsevier, 764 p.

_____, 1979, Hydraulics of Groundwater: New York, McGraw-Hill.

Behnke J., and Bianchi, W., 1965, Pressure distributions in layered sand columns during transient and steady state flows: Water Resources Res., v. 1 p. 557–562.

Bell, M. L., and Nur, A., 1978, Strength changes due to reservoir-induced pore pressure and stresses and application to Lake Oroville: J. Geophys. Res., v. 83, p. 4469–4483.

Bennett, P. C., 1987, Solid phase studies of aquifer sediments, Bemidji, Minnesota, research site. U.S. Geol. Surv. Program on Toxic Waste—Ground-Water Contamination: Proc. Third Technical Meeting, Pensacola, Florida, March 23–27, 1987, U.S. Geol. Surv. Open-File Rept. 87–109, p. C-21–C-22.

Bennett, R. R., and Meyer, R. R., 1952, Geology and groundwater resources of the Baltimore area: Maryland Board Nat. Resources, Dept. Geology, Mines, Water Resources Bull. 4.

Bennett, S. S., and Hanor, J. S., 1987, Dynamics of subsurface salt dissolution at the Welsh Dome, Louisiana, Gulf coast. Dynamical Geology of Salt and Related Structures, eds. I. Lerche and J. J. O'Brien: Orlando, Fla., Academic Press, p. 653–677.

Bennion, D. W., and Griffiths, J. C., 1966, A stochastic model for predicting variations in reservoir rock properties: Trans. Amer. Inst. Mining Met. Engrs., v. 237, no. 2, p. 9–16.

Bentley, H. W., Phillips, F. M., Davis, S. N., Habermehl, M. A., Airey, P. L., Calf, G. E., Elmore, D., Grove H. E., and Torgersen, T., 1986, Chlorine 36 dating of very old groundwater 1. The great artesian basin, Australia: Water Resources Res., v. 22, no. 13, p. 1991–2001.

Berndt, M. P., 1987, Metal partitioning in aquifer sediments, Bemidji, Minnesota, research site. U.S. Geol. Surv. Program on Toxic Waste—Ground-Water Contamination, Proc. Third Technical Meeting, Pensacola, Florida, March 23–27, 1987,: U.S. Geol Surv. Open-File Rept. 87–109, p. C-17–C-19.

Berner, R. A., 1974. In the Sea, ed. E. D. Goldberg, v. 5: New York: John Wiley, p. 427–450.

Berner, R. A., 1980, Early Diagenesis: Princeton, N.J., Princeton University Press, 241 p.

Berry, F., 1959, Hydrodynamics and geochemistry of the Jurassic and Cretaceous systems in the San Juan basin, northwestern New Mexico and southwestern Colorado. Ph.D. dissertation, School of Mineral Sciences. Stanford, Calif.

———, 1969, Relative factors influencing membrane filtration effects in geologic environments: Chem. Geol., v. 4, p. 295–301.

———, and Hanshaw, B., 1960, Geologic evidence suggesting membrane properties of shales [abs.]: Internat. Geol. Congr. 21st, Copenhagen, Proc.

Betcher, R. N., 1977, Temperature distribution in deep groundwater flow systems—a finite element model. Master's thesis, University of Waterloo, Ontario, Canada.

Bethke, C. M., 1985, A numerical model of compaction driven groundwater flow and heat transfer and its application to the paleohydrology of intracratonic sedimentary basins: J. Geophys. Res., v. 90, p. 6817–6828.

———, 1986, Hydrologic constraints on the genesis of the upper Mississippi Valley Mineral District from Illinois Basin Brines: Econ. Geol., v. 81, p. 233–249.

———, 1988, Origin of long range brine migration across the North American craton [abs.]: EOS. Trans. Amer. Geophys. Union, v. 69, p. 470.

———, Harrison, W. J., Upton, C., and Altaner, S. P., 1988, Supercomputer analysis of sedimentary basins: Science, v. 239, p. 261–267.

Beven, K., and Germann, P., 1982, Macropores and water in soils: Water Resources Res., v. 18, p. 1311–1325.

Bianchi, W. C., and Muckel, D. C., 1970, Groundwater recharge hydrology, ARS 41–161: Washington, D.C., Agri. Res. Service, U.S. Dept. Agriculture.

Biot, M. A., 1941, General theory of three dimensional consolidation: Appl. Physics, v. 12, p. 155–164.

———, and Willis, D. G., 1957, The elastic coefficients of the theory of consolidation: J. App. Mech., v. 24, p. 594–601.

Bittinger, M. W., Duke, H. R., and Longenbaugh, R. A., 1967, Mathematical simulations for better aquifer management: Internat. Assoc. Sci. Hydro. Symposium Haifa, Publ. 72, p. 509–519.

———, and Trelease, F. J., 1965, The development and dissipation of a groundwater mound: Trans. Amer. Soc. Agric. Engrs., v. 8, p. 103–104, 106.

Blake S. B., and Fryberger, J. S., 1983, Containment and recovery of refined hydrocarbons from groundwater. Seminar on Groundwater and Petroleum Hydrocarbons Protection, Detection, Restoration, Ottawa, Proc.: Petroleum Assoc. for Conservation of the Canadian Environment, p. 4-1-4-47.

———, and Lewis, R. W., 1982, Underground oil recovery. Second National Symposium on Aquifer Restoration and Ground Water Monitoring, Proc., ed. D. M. Nielsen; Dublin, Ohio, National Water Well Assoc., p. 69–75.

Blanchard, F. G., and Byerly, P., 1935, A study of a well gage as a seismograph: Bull. Seismol. Soc. Amer., v. 25, p. 313–321.

Blanchard, P. E., and Sharp, J., 1985, Possible free convection in thick Gulf Coast sandstone sequences: Southwest Section of the Amer. Assoc. Petro. Geol. Trans., p. 6–12.

Blatt, H., 1979, Diagenetic processes in sandstone. Aspects of Diagenesis, eds., P. A. Scholle and P. R. Schluger: Soc. Econ. Palentol. and Mineral Special Pub. 26, p. 141–157.

Bodvarsson, G., 1970, Confined fluids as strain meters: J. Geophys. Res., v. 75, p. 2711–2718.

Boggs, S., 1987, Principles of Sedimentology and Stratigraphy: Columbus, Ohio, Charles E. Merrill.

Bogomolov, V., and others, 1978, The principles of paleohydrological reconstruction of groundwater formation: Internat. Geol. Congress, 24th Sec. II, p. 205–226.

Bolt, G. H., and Groenvelt, P. H., 1969, Coupling phenomena as a possible cause for non-Daracian behavior of water in soil: Bull. Intern. Assoc. Sci. Hydrol., v. 14, p. 17–26.

Bonham, L. C., 1980, Migration of hydrocarbons in compacting basins. Problems of Petroleum Migration, eds., W. H. Roberts, and R. J. Cordell: Amer. Assoc. Petrol. Geol. Studies in Geology, Vol. 10, p. 69–88.

Borden, R. C., and Bedient, P. B., 1986, Transport of dissolved hydrocarbons influenced by oxygen-limited biodegradation. 1, Theoretical development: Water Resources Res., v. 22, no. 13, p. 1973–1982.

Born, S. M., Smith, S. A., and Stephenson, D. A., 1974, The hydrologic regime of glacial terrain lakes with management and planning applications. An inland lake renewal and management demonstration project report: Madison, Wisconsin, Upper Great Lakes Regional Comm.

_____, Smith, S. A., and Stephenson, D. A., 1979, The hydrologic regime of glacial terrain lakes: J. Hydrol., v. 43, p. 7–44.

Bosworth, W., 1981, Strain induced partial dissolution of halite. Tectonphysics, v. 78, p. 509–525.

Boulton, N. S., 1954, The drawdown of the water table under nonsteady conditions near a pumped well in an unconfined formation. Inst. Civil Engrs., Proc., v. 3, p. 564–579.

_____, 1955, Unsteady radial flow to a pumped well allowing for delayed yield from storage. Gen. Assembly, Rome Internat. Assoc. Sci. Hydrol. Publ., Proc., v. 37, p. 472–477.

_____, 1963, Analysis of data from nonequilibrium pumping test allowing for delayed yield from storage: Inst. Civil Engrs., Proc., v. 26, p. 469–482.

_____, and Streltsova, S. V., 1977, Unsteady flow to a pumped well in a fissured water bearing formation: J. Hydrol., v. 35, p. 257–269.

Bouma, J., and Dekker, I. W., 1978, A case study on infiltration into dry clay. I, Morphological Observations: Geoderma, p. 27–40.

Bouwer, E. J., and McCarty, P. H., 1984, Modeling of trace organics biotransformation in the subsurface: Ground Water, v. 22, no. 4, p. 433–440.

Bouwer, H., 1978, Groundwater Hydrology: New York, McGraw-Hill, 480 p.

_____, 1985, Renovating waste water with groundwater recharge. Issues in Groundwater Management, eds. E. T. Smerdon and W. R. Jordan: College Station, Texas A&M University, p. 331–346.

Box, G. E. P., and Jenkins, G. W., 1976, Time Series Analysis: Forecasting and Control: San Francisco, Holden-Day, 575 p.

Boyd, F. M., and Kreitler, C. W., 1986, Hydrogeology of a gypsum playa, Northern Salt Basin, Texas: Bureau of Economic Geology, The University of Texas at Austin, Report of Investigations No. 158, 37 p.

Brace, W. F., 1980, Permeability of crystalline rocks and argellaceous rocks: Internat. J. Rock Mech. Min. Sci. Geomechanics., v. 17, p. 241–251.

_____, Paulding, B. W., and Sholz, C., 1966, Dilatency in the fracture of crystalline rocks: J. Geophy. Res., v. 71, p. 3939–3953.

_____, Walsh, J. B., and Fangos, W. T., 1968, Permeability of granite under high pressure: J. Geophys. Res., v. 63, p. 2225–2236.

Bradbury, K. R., 1984, Major ion and isotope geochemistry of groundwater in clayey till, northwestern Wisconsin, U.S.A. First Canadian/American Conference on Hydrogeology, eds. B. Hitchon and E. I. Wallick: Dublin, Ohio, National Water Well Assoc., p. 284–289.

Bradley, J. S., 1975, Abnormal formation pressure: Amer. Assoc. Petrol. Geol. Bull., v. 59, p. 957–973.

Brashears, M. L., 1946, Artificial recharge on Long Island, New York: Econ. Geol., v. 41, p. 503–516.

Bredehoeft, J. D., 1965, The drill stem test: The petroleum industry's deep-well pumping test: Ground Water, v. 3, p. 31–36.

_____, 1967, Response of well-aquifer systems to earth tides: J. Geophys. Res., v. 72, p. 3075–3087.

_____, and Hanshaw, B., 1968, On the maintenance of anomalous fluid pressures. I, Thick sedimentary sequences: Geol. Soc. Amer. Bull., v. 79, p. 1097–1106.

_____, and Papadopulos, I. S., 1965, Rates of vertical groundwater movement estimated from the earth's thermal profile. Water Resources Res., v. 1, p. 325–328.

_____, and Papadopulos, I. S., 1980, A method for determining the hydraulic properties of tight formations: Water Resources Res., v. 16, p. 233–238.

_____, and Pinder, G. F., 1973, Mass transport in flowing groundwater. Water Resources Res., v. 9, no. 1, p. 194–210.

_____, and Young, R., 1970, The temporal allocation of groundwater—A simulation approach: Water Resources Res., v. 16, p. 3–21.

_____, and Young, R., 1983, Conjunctive use of groundwater and surface water for irrigated agriculture: Risk aversion: Water Resources Res., v. 19, p. 1111–1121.

_____, Blyth, C. R., White, W. A., and Maxey, G. B., 1963, Possible mechanism for concentration of brines in subsurface formations: Amer. Assoc. Petrol. Geol. Bull., v. 47, p. 257–269.

Brown, J. A., 1937, Discussion on "Effect of a sea level canal on the groundwater level of Florida": Econ. Geol., v. 32, p. 589–599.

Brown, R. A., Norris, R. D., and Raymond, R. L., 1984, Oxygen transport in contaminated aquifers with hydrogen peroxide. NWWA/API Conference on Petroleum Hydrocarbons and Organic Chemicals in Ground Water—Prevention, Detection, and Restoration, Proc.: Dublin, Ohio, National Water Well Assoc., p. 441–450.

Brown, R. H., and Parker, G. G., 1945, Salt water encroachment in limestone of Silver Bluff, Miami, Florida: Econ. Geol., v. 40, p. 235–262.

Browne, P. R. L., 1978, Hydrothermal alteration in active geothermal fields: Ann. Rev. Earth, Planet Sci., eds. F. A. Donath, F. G. Stehli, and G. Wetherill, v. 6, p. 229–250.

Bryan, K., 1925, Geology of reservoir and dam sites: U.S. Geol. Survey Water Supply Papers, 597A.

Buras, N., 1966, Dynamic programming in water resources development. Advances in Hydroscience, ed. V. T. Chow, Vol. 3: New York, Academic Press, p. 367–412.

Bureau of Reclamation, 1960, Design of Small Dams: Washington D.C., U.S. Government Printing Office.

Burns, W. A., 1969, New single well test for determining vertical permeability: Trans. Amer. Inst. Mining Engrs., v. 246, p. 743–752.

Burst, J. F., 1969, Diagenesis of Gulf Coast clayey sediments and its possible relation to petroleum migration: Amer. Assoc. Petrol. Geol. Bull. v. 53, p. 73–93.

Butler, R. D., 1984, Hydrogeology of the Dakota aquifer system, Williston Basin, North Dakota. D. G. Jorgensen, and D. C. Signor: Geohydrology Dakota Aquifer Symposium, Proc., Worthington, Ohio, Water Well Journal Pub. Co., p. 12–23.

Byerlee, J. D., 1967, Frictional characteristics of granite under high confining pressure: J. Geophys. Res., v. 72, p. 2629–2648.

Cady, R. C., 1941, Effect upon groundwater levels of proposed surface water storage in Flatland Lake, Montana: U.S. Geol. Survey Water Supply Paper 849 b, p. 51–81.

California Dept. Water Resources, 1958, Sea water intrusion in California: Sacramento, California Dept. Water Res. Bull. 63.

———, 1980, Groundwater basins in California: Sacramento, Calif. Dept. Water Res. Bull. 118–80.

Cant, D. J., 1982, Fluvial facies models and their applications. Sandstone Depositional Environments, eds. P. A. Scholle and D. Spearing: Amer. Assoc. Petrol. Geol. Mem. 31, p. 115–138.

Carder, D. S., 1970, Reservoir loading and local earthquakes. Engineering Seismology, ed. W. M. Adams: Geol. Soc. Amer., Eng. Geol. Case Histories #8.

Cardwell, W. T., and Parsons, R. L., 1945, Average permeabilities of heterogenous oil sands. Trans. Amer. Inst. Min. Met. Engrs., v. 169, p. 34–42.

Carlson, W. D., 1983, The polymorphs of $CaCO_3$ and the aragonite-calcite transformation, Carbonates: Mineralogy and Chemistry, ed., R. J. Reader: Mineral. Soc. Amer., Reviews in Mineralogy, v. 11, p. 191–225.

Carlston, C. W., Thatcher, L. L., and Rhodehamel, E. C., 1960, Tritium as a hydrologic tool, the Wharton Tract Study: Internat. Assoc. Sci. Hydrol, Publ. 52, p. 503–512.

Carothers, W. W., and Kharaka, Y. K., 1978, Aliphatic acid anions in oil field water— implications for origin of natural gas: Amer. Assoc. Petroleum Geol. Bull., v. 62, p. 2441–2453.

Carr, P. A., and Van Der Kamp, G. S., 1969, Determining aquifer characteristics by the tidal method: Water Resources Res., v. 5, p. 1023–1031.

Carrillo, N., 1948, Influence of artesian wells in the sinking of Mexico City. Proc. 2nd Internat. Conf. Soil Mech. and Found. Eng., v. 7, p. 156–159.

Carslaw, H., and Yeager, J., 1959, Conduction of Heat in Solids: Oxford, Clarendon Press.

Cartwright, K., 1970, Groundwater discharge in the Illinois Basin as suggested by temperature anomalies: Water Resources Res., v. 6, p. 912–918.

———, 1973, The effect of shallow groundwater flow systems on rock and soil temperatures. Ph.D. thesis, Geology Dept., University of Illinois, Urbana.

Casagrande, L., 1953, Review of past and current work on electro osmotic stabilization of soils. Harvard Soil Mech. Series No. 45. Harvard University, Cambridge, Mass.

Case, H. L., 1984, Hydrology of Inyan Kara and Dakota-NewCastle aquifer system, North Dakota. Geohydrology Dakota Aquifer Systems, North Dakota. eds., D. G. Jorgensen, and D. C. Signor: Worthington, Ohio, Nat. Water Well. Assoc., p. 24–42.

Cathles, L. M., 1977, An analysis of cooling of intrusives by groundwater convection which includes boiling: Econ. Geol., v. 72, p. 804–826.

_____, and A. T. Smith, 1983, Thermal constraints on the formation of Mississippi Valley type lead-zinc deposits and their implications for episodic basin dewatering and deposit genesis: Econ. Geol., v. 78, p. 983–1002.

Ceazan, M. L., Updegraff, D. M., and Thurman, E. M., 1984, Evidence of microbial processes in sewage-contaminated ground water. Movement and Fate of Solutes in a Plume of Sewage-Contaminated Ground Water: Cape Cod, Mass., U.S. Geol. Surv. Toxic Waste Ground-Water Contamination Program, U.S. Geol. Surv. Open-File Rept. 84-475, p. 115–138.

Cederberg, G. A., 1985, TRANQL: A ground-water mass-transport and equilibrium chemistry model for multicomponent systems. Ph.D. dissertation, Stanford University, Stanford, Calif., 117 p.

Cedergren H. R., 1967, Seepage, Drainage, and Flow Nets: New York: John Wiley.

Chamberlin, T. C., 1885, The requisite and qualifying conditions of artesian wells: U.S. Geol. Survey 5th Ann. Rept., p. 131–173.

Champ, D. R., Gulens, J., and Jackson, R. E., 1979, Oxidation-reduction sequences in ground-water flow systems: Can. J. Earth Sci., v. 16, no. 1, p. 12–23.

Chapelle, F. H., Zelibor, J. L., Jr., Grimes, D. J., and Knobel, L. L., 1987, Bacteria in deep coastal plain sediments of Maryland: A possible source of CO_2 to groundwater: Water Resources Res., v. 23, no. 8, p. 1625–1632.

Chapman, R. E., 1980, Mechanical versus thermal causes of abnormally high pore pressure in shales: Amer. Assoc. Petrol. Geol. Bull., v. 64, p. 2179–2183.

_____, 1981, Geology and Water: Boston, Martinus Nijhoff/Dr. W. Junk.

Charlax, E., Guyon, E., and Rivier, N., 1984, A criterion for percolation threshold in a random array of plates: Solid State Communications, v. 20, p. 999–1002.

Chebotarev, I. I., 1955, Metamorphism of natural water in the crust of weathering: Geochim. Cosmochim. Acta, v. 8, p. 22–48, 137–170, 198–212.

Cherry, J. A., 1983, Piezometers and other permanently installed devices for groundwater quality monitoring. Seminar on Groundwater and Petroleum Hydrocarbons Protection, Detection, Restoration, Ottawa, Proc.: Petroleum Assoc. for Conservation of the Canadian Environment, p. IV-1–IV-39.

Chia, Y. P., 1979, Digital simulation of compaction of sedimentary sequences. Ph.D. thesis, Geology Dept., University of Illinois, Urbana.

Chiou, C. T., Schmedding, D. W., and Manes, M., 1982, Partitioning of organic compounds in octanol-water systems: Environ. Sci. Technol., v. 16, p. 4–10.

Chun, R. Y. D., Mitchell, L. R., and Mido, K. W., 1964, Groundwater management for the Nation's future-optimum conjunctive operation of groundwater basins: J. Hydrauls. Div. Amer. Soc. Civil Engrs. HY4. v. 90, p. 79–105.

Claasen, H. C., 1982, Guidelines and techniques for obtaining water samples that accurately represent the water chemistry of an aquifer: U.S. Geol. Surv. Open-File Report 82-1024, 49 p.

Cleary, R. W., 1978, Report on 208 Long Island pollution study: Water Resources Program, Princeton University, Princeton, N.J.

Cochran, G., 1968, Optimization of conjunctive use of groundwater and surface water for urban supply. M.S. thesis in Civil Engineering, University of Nevada, Reno.

Cohen, P., Frank, O., and Foxworthy, B., 1968, An atlas of Long Island's water resources: N.Y. State Water Res. Comm. Bull. GW62.

Cohen, R. M., Rambold, R. R., Faust, C. F., Rumbaugh, J. O., III, and Bridge, J. R., 1987, Investigation and hydraulic containment of chemical migration: Four landfills in Niagara Falls. Civ. Eng. Pract., Spring, p. 33–58.

Coleman, J. M., and Prior, D. B., 1982, Deltaic environments of deposition. Sandstone Depositional Environments, eds. P. A. Scholle and D. Spearing: Amer. Assoc. Petrol. Geol. Mem., v. 31, p. 139–178.

Collins, R. E., 1961, Flow of Fluids Through Porous Materials. New York: Reinhold.

Collins, W. D., 1923, Graphical representation of water analysis: Ind. Eng. Chem., v. 15, p. 394.

———, 1925, Temperature of water available for individual use in the United States: U.S. Geol. Surv. Water Supply Papers 520-F, p. 97–104.

Combarnous, M. A., and Bories, S. A., 1975, Hydrothermal convection in saturated porous media. Advances in Hydroscience, ed. V. T. Chow: New York, Academic Press, Vol. 10, p. 231–307.

Conkling, H., 1946, Utilization of groundwater storage in stream system development: Trans. Amer. Soc. Civil Engrs., v. 111, p. 275–305.

Cooper, H. H., 1966, The equation of groundwater flow in fixed and deforming coordinates: J. Geophy. Res., v. 71, p. 4785–4790.

———, and Rorabaugh, M. I., 1963, Groundwater movements and bank storage due to flood stages in surface streams: U.S. Geol. Survey Water Supply Paper 1536-J, p. 343–366.

———, Bredehoeft, J. D., and Papadopulos, I. S., 1967, Response of a finite diameter well to an instantaneous charge of water: Water Resources Res., v. 3, p. 263–269.

———, and Jacob, C. E., 1946, A generalized graphical method for evaluating formation constants and summarizing well field history. Trans. Amer. Geophys. Union., v. 27, p. 526–534.

———, and others, 1965, The response of well aquifer systems to seismic waves. J. Geophys. Res., v. 70, p. 3915–3926.

Corey, J. D., Nielsen, D. R., and Biggar, J. W., 1963, Miscible displacement in saturated and unsaturated sandstone: Soil Sci. Soc. Am., Proc., v. 27, p. 258–262.

Coustau, H., and others, 1975, Classification hydrodynamique des basins sedimentaires utilisation combinée avec d'autres méthodes pour rationaliser l'exploration dans des bassins non produotifs: 9th World Petrol. Congr., Tokyo, Proc., Section II, p. 105–119.

Craig, H., 1961, Isotopic variations in meteoric water: Science, v. 133, p. 1702–1703.

Craun, G. F., 1984, Health aspects of groundwater pollution. Groundwater Pollution Microbiology, eds. G. Bitton and C. P. Gerba. New York, John Wiley, p. 135–179.

Croff, A. G., Lomenick, T. F., Lowrie, R. S., and Stow, S. H., 1985, Evaluation of five sedimentary rocks other than salt for high level waste repository sitting purposes: Oak Ridge, Tenn., Oak Ridge Nat. Lab. ORNL/CF-85/2/V2.

Crow, W. L., Anderson, E. P., and Minugh, E. M., 1985, Surface venting of vapors emanating from hydrocarbon product on ground water. NWWA/API Conference Petroleum Hydrocarbons and Organic Chemical, Protection, Detection and Restoration, Proc.: Dublin, Ohio, National Water Well Assoc., p. 536–554.

Csallany, S., and Walton, W. C., 1963, Yields of shallow dolomite wells in Northern Illinois. Ill. State Water Survey Rept. Invest. 46.

Cuevas, J. A., 1936, Foundation conditions in Mexico City. Internat. Conf. Soil Mech., Proc., v. 3, Cambridge, Mass.

Curtis, G. P., Roberts, P. V., and Reinhard, M., 1986, A natural gradient experiment on solute transport in a sand aquifer. 4, sorption of organic solutes and its influence on mobility: Water Resources Res., v.22, no. 13, p. 2059–2067.

Dagan, G., 1967, A method of determining the permeability and effective porosity of unconfined anisotropic aquifers: Water Resources Res. v. 3, p. 1059–1071.

———, 1982, Stochastic modeling of groundwater flow by unconditional and conditional probabilities. 2, the solute transport: Water Resources Res., v. 18, no. 4, p. 835–848.

———, and Bear, J., 1968, Solving the problem of local interface upconing in a coastal aquifer by the method of small perturbations: J. Hydr. Research, v. 6, p. 15–44.

Daines, S. R., 1982, Aquathermal pressuring and geopressure evaluation: Amer. Assoc. Petrol. Geol. Bull., v. 66, p. 931–939.

Darcy, H. P. G., 1856, Les fontaines publiques de la Ville de Dijon: Paris, Victor Dalmont.

Darton, N. H., 1909, Geology and underground waters of South Dakota: U.S. Geol. Survey Water Supply Papers 227.

Daubert, J. T., and Young, R., 1982, Groundwater development in western river basins: Large economic gains with unseen costs: Ground Water, v. 20, p. 80–85.

Daus, A. D., and Frind, E. O., 1985, An alternating direction Galerkin technique for simulation of contaminant transport in complex groundwater systems. Water Resources Res., v. 21, no. 5, p. 653–664.

Davidson, M. R., 1984, A green-ampt model of infiltration in a cracked soil. Water Resources Res., v. 20, p. 1685–1690.

———, 1985, Numerical calculations of saturated-unsaturated infiltration in a cracked soil: Water Resources Res., v. 21, p. 709–714.

Davis, G. H., and others, 1959, Groundwater conditions and storage capacity in the San Joaquin Valley, California: U.S. Geol. Survey Water Supply Papers 1469.

Davis, G. H., Small, J. B., and Counts, H. B., 1963, Land subsidence related to decline of artesian pressure in the Ocala limestone at Savannah, Georgia. Eng. Geol. Case Histories, v. 4, Geol. Soc. Amer., p. 1–8.

Davis, J. C., 1986, Statistics and Data Analysis in Geology: New York, John Wiley, 646 p.

Davis, S. N., 1969, Porosity and permeability of natural materials. Flow Through Porous Materials, ed. R. J. M. DeWiest: New York, Academic Press, p. 54–89.

———, Campbell, D. J., Bentley, H. W. and Flynn, T. J., 1985, Ground-water tracers. Cooperative agreement CR-810036, Robert S. Kerr Environmental Research Laboratory, Ada, Oklahoma, National Water Well Assoc., 200 p.

———, and De Wiest, R. J. M., 1966, Hydrogeology: New York, John Wiley, 463 p.

de Swaan, A. O., 1976, Analytical solutions for determining naturally fractured reservoir properties by well testing. Soc. Pet. Eng. J., v. 16, p. 117–122.

de Marsily, G., 1985, Flow and transport in fracture rock. Memoirs. Hydrogeology of Rocks of Low Permeability: Tucson, Ariz., International Association of Hydrogeologists, p. 267–277.

DeBoer, R. B., 1977, On the thermodynamics of pressure solution-interaction between chemical and mechanical forces: Geochim. Cosmochim. Acta, v. 41, p. 249–256.

Deere, D. U., and Patton, F. D., 1967, Effect of pore pressure on stability of slopes: Geol. Soc. Amer. Soc. Civil Engrs. Symp., New Orleans, La.

Deju, R. A., 1973, A worldwide look at the occurrence of high fluid pressures in petroliferous basins. Special publication of Gulf Research and Development Co, Pittsburgh, Pa.

Demond, A. H., and Roberts, P. V., 1987, An examination of relative permeability relations for two-phase flow in porous media: Water Resources Bull., v. 23, no. 4, p. 617–628.

Department of Energy, 1980, Statement of position of the United States Department of Energy, proposed rule making on the storage and disposal of nuclear waste: Washington, D.C.

———, 1986, Environmental assessment, reference repository location, Hanford Site, Washington: Washington, D.C., DOE/RW-0070.

———, 1988, Consultation draft site characterization plan, Yucca Mountain site, Nevada. Washington, D.C., US DOE OCRWM, I-III.

Desai, C. T., 1979, Elementary Finite Element Method: Englewood Cliffs, N.J., Prentice Hall.

Desaulniers, D. E., Cherry, J. A., and Fritz, P., 1981, Origin, age, and movement of pore water in argillareous quaternary deposits at four sites in southwestern Ontario: J. Hydrol., v. 50, p. 231–257.

DeSitter, L. U., 1947, Diagenesis of oil field brines: Amer. Assoc. Petrol. Geol. Bull., v. 31, no. 11, p. 2030–2040.

———, 1956, Structural Geology: New York, McGraw-Hill.

Dewers, T., and Ortoleva, P., 1988, The role of geochemical self-organization in the migration and trapping of hydrocarbons: Appl. Geochem., v. 3, p. 287–316.

DeWiest, R. J. M., 1965, Geohydrology: New York: John Wiley, 1965.

———, 1966, On the storage coefficient and the equations of groundwater flow: J. Geophys. Res., v. 71, p. 1117–1122.

Dickinson, G., 1953, Geologic aspects of abnormal reservoir pressures in Gulf Coast Louisiana: Amer. Assoc. Petrol. Geol. Bull., v. 37, p. 410–432.

Dienes, J. K., 1982, Permeability, percolation, and statistical crack mechanisms. Issues in Rock Mechanics: 22nd Symp. on Rock Mechanics, Proc., Berkeley, Calif.

Dillon, R. T., Lantz, R. B., and Pahwa, S. B., 1978, Risk Methodology for Geologic Disposal of Radioactive Waste. The Sandia Waste Isolation and Flow (SWIFT) Model, NUREG/CR-0424: Washington, D.C., Nuclear Regulatory Commission.

Drever, J. I., 1982, The Geochemistry of Natural Waters: Englewood Cliffs, N.J., Prentice Hall, 388 p.

Domenico, P. A., 1972, Concepts and Models in Groundwater Hydrology: New York, McGraw-Hill.

———, 1977, Transport phenomena in chemical rate processes in sediments: Ann. Rev. Earth Planet Sci., v. 5 eds., F. A. Donath, F. G. Stehli, and G. Wetherill, p. 287–317.

———, 1983, Determination of bulk rock properties from Groundwater level fluctuations. Bull. Assoc. Eng. Geol., v. 20, p. 283–287.

———, 1987, An analytical model for multidimensional transport of a decaying contaminant species: J. Hydrol., v. 91, p. 49–58.

———, and M. D. Mifflin, 1965, Water from low permeability sediments and land subsidence: Water Resources Res., v. 4, p. 563–576.

———, Mifflin, M. D., and Mindling, A. L., 1966, Geologic controls on land subsidence in Las Vegas Valley: Ann. Eng. Geol, Soils Eng. Symp., Proc., Moscow, Idaho, p. 113–121.

———, and Palciauskas, V. V., 1973, Theoretical analysis of forced convective heat transfer in regional groundwater flow. Geol. Soc. Amer. Bull., v. 84, p. 3803–3814.

———, and Palciauskas, V. V., 1979, Thermal expansion of fluids and fracture initiation in compacting sediments: Geol. Soc. Amer. Bull., Part II, v. 90, p. 953–979.

———, and Palciauskas, V. V., 1982, Alternative boundaries in solid waste management. Ground Water, v. 20, p. 303–311.

———, and Palciauskas, V. V., 1988, The generation and dissipation of abnormal fluid pressures in active deposition environments. The Geology of North America, Hydrogeology, Vols. 0–2, eds. W. Back, J. S. Rosenshein, and P. R. Seaber: p. 435–445.

———, and Robbins, G. A., 1984, A dispersion scale effect in model calibrations and field tracer experiments: J. Hydrol., v. 7, p. 121–132.

———, and Robbins, G. A., 1985a, The displacement of connate water from aquifers: Geol. Soc. Am. Bull., v. 96, p. 328–335.

———, and Robbins, G. A., 1985b, A new method of contaminant plume analysis: Ground Water, v. 23, no. 4, p. 476–485.

———, Tapia Garcia, F. J., and Uribe, L., 1974, Alternative policies for water management in the Costa De Hermosillo, a report to the World Bank, Washington, D.C.

Donaldson, I. G., 1962, Temperature gradients in the upper layers of the earth's crust due to convective water flows: J. Geophy. Res., v. 67, p. 3449–3459.

Douglas, J., and Peaceman, D., 1955, Numerical solution of two dimensional heat flow problems: J. Amer. Inst. Chem. Eng., v. 1, p. 505–512.

Downey, J. S., 1984a, Hydrodynamics of the Williston Basin in the northern Great Plains. Geohydrology Dakota Aquifer Symposium, eds., Jorgensen and D. C. Signor: Worthington, Ohio, Nat. Water Well Assoc., p. 55–61.

———, 1984b, Geology and hydrology of the Madison limestone and associated rocks in parts of Montana, Nebraska, North Dakota, South Dakota, and Wyoming: U.S. Geol. Survey Prof. Paper 1273-G, 152 p.

Dozy, J. J., 1970, A geologic model for genesis of lead-zinc ores of the Mississippi Valley, U.S.A. Inst. Min. Metal. Trans., v. 79, p. B163–B170.

Driscoll, F. G., 1986, Groundwater and Wells: St. Paul, Minn., Johnson Division.

Drost, W. D., and others, 1968, Point dilution methods of investigating groundwater flow by means of radioisotopes: Water Resources Res., v. 4, p. 125–146.

Duffield, G. M., Buss, D. R., Stephenson, D. E., and Mercer, J. W., 1987, A grid refinement approach to flow and transport modeling of a proposed groundwater corrective action at the Savannah River Plant Aiken South Carolina. Solving Ground Water Problems with Models, Proc., Vol. 2: Dublin, Ohio, National Water Well Assoc., p. 1087–1120.

Dullien, F. A. L., 1979, Porous Media: Fluid Transport and Pore Structure: New York, Academic Press, 396 p.

Dunbar, C. O., and Waage, K. M., 1969, Historical Geology: New York: John Wiley.

Dunham, K. C., 1970, Mineralization by deep formation waters: A review. Inst. Min. Metall. Trans. 79, p. B127–B136.

Eakin, T. A., 1964, Groundwater appraisal of Coyote Spring and Kane Spring Valleys and Muddy River Springs area, Lincoln and Clark Counties, Nevada. Groundwater Resources Reconnaissance Series Rept. 25: Carson City, Nev., U.S. Geol. Survey.

————, 1966, A regional interbasin groundwater system in the White River area, southeastern Nevada. Water Resources Res., v. 2, no. 2, p. 251–271.

Eakin, T. A., and others, 1976, Summary appraisals of the Nations groundwater resources: Great Basin: U.S. Geol. Survey Prof. Papers 813-G.

Earlougher, R. C., 1977, Pressure buildup testing: Advances in well test analysis, Monograph 5. New York, Soc. Petrol. Engrs. Amer. Inst. Mec. Engrs.

Eaton, J. P., and Takasaki, K. J., 1959, Seismological interpretation of earthquake induced water level fluctuations in wells. Bull. Seismol. Soc. Amer., v. 49, p. 227–245.

Edwards, D. A., 1988, A solute transport model calibration procedure as applied to a tritium plume in the Savannah River Plant F-area, South Carolina. M.Sc. dissertation, Texas A&M University, College Station.

Eganhouse, T. F., Dorsey, T. F., and Phinney, C. S., 1987, Transport and fate of monoaromatic hydrocarbons in the subsurface, Bemidgi, Minnesota, research site. U.S. Geol. Surv. Program on Toxic Waste—Ground-Water Contamination: Third Technical Meeting, Pensacola Florida, March 23–27, 1987, Proc. U.S. Geol. Surv. Open-File Rept. 87-109, p. C-29.

Eliott, T., 1978, Deltas. Sedimentary Environments and Facies, ed., H. G. Reading: Oxford, Blackwell, p. 97–142.

Elzeftawy, A., Mansell, R. S., and Selim, H. M., 1976, Distribution of water and herbicide in lakeland sand during initial stages of infiltration: Soil Sci., v. 122, p. 297–307.

Endoe, H. K., 1984, Mechanical transport in two dimensional networks of fractures. Ph.D. dissertation, University of California, Berkeley.

Energy Research and Development Administration, 1976, A Bibliography of Geothermal Resources—Exploration and Exploitation: Washington, D.C., ERDA.

Eng. News Record, 1960, Flowing sand plugs two miles of tunnel—and it won't come out, v. 165, p. 38–40.

Ernest, L. F., 1969, Groundwater flow in the Netherlands delta area and its influence on the salt balance of the future Lake Zeeland: J. Hydrol., v. 8, p. 137–172.

Eshett, A., and Longenbaugh, R., 1965, Mathematical model for transient flow in porous media: Civil Eng. Dept. Colorado State University, Fort Collins.

Evans, D., 1966, The Denver area earthquakes and the Rocky Mountain Arsenal Disposal Well: Mtn. Geol., v. 3, p. 23–36.

Farmer, V. E., 1983, Behaviour of petroleum contaminants in an underground environment. Seminar on Groundwater and Petroleum Hydrocarbons, Protection, Detection, Restoration, Ottawa, Proc.: Petroleum Assoc. for Conservation of the Canadian Environment, p. II-1–II-16.

Farvolden, R. N., 1964, Geologic controls on groundwater storage and base flow: J. Hydrol., v. 1, p. 219–250.

Faust, C. R., 1985, Affidavits in Civil Action No. 79–989: U.S. District Court for the Western District of New York, 92 p.

Faust, C. R., and Mercer, J. W., 1984, Evaluation of slug tests in wells containing a finite thickness skin: Water Resources Res., v. 20, p. 504–506.

Felmy, A. R., Girvin, D., and Jenne, E. A., 1983, MINTEQ: A Computer Program for Calculating Aqueous Geochemical Equilibria: Athens, Georgia, U.S. Environmental Protection Agency, EPA-600/3/84-032, PB84-157148.

Ferris, J. G., 1951, Cyclic fluctuations of water levels as a basis for determining aquifer transmissibility: Int. Assoc. Sci. Hydrol. Pub. 33, p. 148–155.

_____, 1959, Groundwater. Hydrology, eds., C. O. Wisler and E. F. Brater: New York, John Wiley.

_____, and others, 1962, Theory of aquifer tests. U.S. Geol. Survey Water Supply Papers 1536-E, p. 69–174.

Ferry, J. P., Dougherty, P. J., Moser, J. B., and Schuller, R. M., 1986, Occurrence and recovery of a DNAPL in a low-yielding bedrock aquifer. Proc. of the NWWA/API Conference on Petroleum Hydrocarbons and Organic Chemicals in Ground Water—Prevention, Detection, and Restoration: Dublin, Ohio, National Water Well Assoc., p. 722–735.

Fischel, V. C., 1956, Long term trends of groundwater levels in the United States: Trans. Amer. Geophys. Union, v. 37, p. 429–435.

Fletcher, J. B., and Sykes, L. R., 1977, Earthquakes related to hydraulic mining and natural seismic activity in western New York State: J. Geophy. Res., v. 82, p. 3767–3780.

Fogel, S., Findlay, M., Moore, A., and Leahy, M., 1987, Biodegradation of chlorinated chemicals in groundwater by methane oxidizing bacteria. NWWA/API Conference Petroleum Hydrocarbons and Organic Chemicals in Ground Water—Prevention, Detection, and Restoration, Proc.: Dublin, Ohio, National Water Well Assoc., p. 167–185.

Fogg, G. E., and Kreitler, C. W., 1982, Ground-water hydraulics and hydrochemical facies in Eocene aquifers of the East Texas Basin. Bureau of Economic Geology, The University of Texas at Austin, Report of Investigations No. 127, 75 p.

Folk, R. L., and Land, L. S., 1975, Mg/Ca ratio and salinity: two controls over crystallization of dolomite: Amer. Assoc. Petrol. Geol. Bull., v. 59, p. 60–68.

Fontes, J. C., 1980, Environmental isotopes in groundwater hydrology. Chapter 3 in Handbook of Environmental Isotope Geochemistry, Vol. 1, eds. P. Fritz and J. C. Fontes: New York, Elsevier, p. 75–140.

_____, and Garnier, J. M., 1979, Determination of the initial ^{14}C activity of the total dissolved carbon: A review of the existing models and a new approach: Water Resources Res., v. 15, no. 2, p. 399–413.

Forster, C., and Smith, L., 1988, Groundwater flow systems in mountainous terrain. 2, Controlling factors: Water Resources Res., v. 24, p. 1011–1023.

_____, and Smith, L., 1989, The influence of groundwater flow on thermal regimes in mountainous terrain. J. Geophys. Res., v. 94, p. 9439–9451.

Forstner, U., and Wittman, T. W., 1981, Metal Pollution in the Aquatic Environment: Berlin, Heidelberg, Springer-Verlag 486 p.

Forsythe, G. E., and Wasow, W. R., 1960, Finite Difference Methods for Partial Differential Equations: New York, John Wiley.

Foster, M. D., 1950, The origin of high sodium bicarbonate waters in the Atlantic and Gulf coastal plains: Geochim. Cosmochim. Acta, v. 1, p. 33–48.

Franks, B. J., ed., 1987, U.S. Geological Survey Program on Toxic Waste—Ground-Water Contamination: Third Technical Meeting, Pensacola, Florida, March 23–27, 1987, Proc.: U.S. Geol. Surv. Open-File Report 87–109.

Frank-Kamenetskii, P. A., 1955, Diffusion and Heat Transfer in Chemical kinetics, trans. by J. P. Appleton (1969): New York, Plenum Press.

Freeze, R. A., 1969, The mechanism of natural groundwater recharge and discharge. I, One dimensional vertical, unsteady unsaturated flow above a recharging or discharging groundwater flow system: Water Resources Res., v. 5, p. 153–171.

————, 1971, Three dimensional transient saturated-unsaturated flow in a groundwater basin: Water Resources Res., v. 7, p. 347–366.

————, and Cherry, J. A., 1979, Groundwater: Englewood Cliffs, N.J., Prentice Hall, 604 p.

————, and Cherry, J. A., 1989, Guest editorial—what has gone wrong: Ground Water, v. 27, no. 4, p. 458–464.

————, Masmann, J., Smith, L., Sperling, T., and James, B., 1989, Hydrogeologic decision analysis. 1, A framework. in press, Ground Water.

————, and Witherspoon, P. A., 1966, Theoretical analysis of regional groundwater flow. I, Analytical and numerical solutions to the mathematical model: Water Resources Res., v. 2, p. 641–656.

————, and Witherspoon, P. A., 1967, Theoretical analysis of regional groundwater flow. II, Effect of water table configuration and subsurface permeability variations: Water Resources Res., v. 3, p. 623–634.

Freyberg, D. L., 1986, A natural gradient experiment on solute transport in a sand aquifer. 2, Spatial moments and the advection and dispersion of nonreactive tracers: Water Resources Res., v. 22, no. 13, p. 2031–2046.

Fried, J. J., 1975, Groundwater Pollution: Amsterdam, Elsevier, 330 p.

————, and Combarnous, M. A., 1971, Dispersion in porous media. Advances in Hydroscience, V. T. Chow, ed.: New York, Academic Press, v. 7, p. 169–282.

Friedman, L. C., and Erdman, D. E., 1982, Quality assurance practices for the chemical and biological analyses of water and fluvial sediments: U.S. Geol. Surv. TWRI, Book 5, Chapter A6, 181 p.

Frind, E. O., and Germain, D., 1986, Simulation of contaminant plumes with large dispersive contrast: Evaluation of alternating direction Galerkin models: Water Resources Res., v. 22, no. 13, p. 1857–1873.

————, Sudicky, E. A., and Schellenberg, S. L., 1987, Micro-scale modelling in the study of plume evolution in heterogeneous media: Stochastic Hydrol. Hydraul., v. 1, p. 263–279.

————, 1982, The principal direction technique a new approach to groundwater contaminant transport modeling. Fourth International Conference on Finite Elements in Water Resources, Proc.: New York, Springer-Verlag, p. 13-25–13-42.

Fritz, P., and Fontes, J. C., 1980, Introduction. Handbook of Environmental Isotope Geochemistry, Vol. 1, eds. P. Fritz and J. C. Fontes: New York, Elsevier, p. 1–19.

Fryar, A. E., and Domenico, P. A., 1989, Analytical inverse modeling of regional scale tritium waste migration: J. Contam. Hydrol., v. 4, p. 113–125.

Gabrysch, R. K., and Bonnet, C. W., 1975, Land surface subsidence in the Houston-Galveston region, Texas: Texas Water Dev. Board Rept. 188.

Gale, J. E., 1975, A numerical field and laboratory study of flow in rocks with deformable fractures. Ph.D. dissertation, University of California, Berkeley.

————, Rouleau, A., and Atkinson, L. C., 1985, Hydraulic properties of fractures. Memoires. Hydrogeology of Rocks of Low Permeability: International Association of Hydrogeologists (Tucson, Ariz.), p. 1–11.

Galloway, W. E., 1978, Uranium mineralization in a coastal-plain fluvial aquifer system, Catahoula formation, Texas: Econ. Geol., v. 73, p. 1655–1676.

———, and Hobday, D. K., 1983, Terrigenous Clastic Depositional Systems: Application to Petroleum, Coal and Uranium Exploration: New York, Springer-Verlag, 423 p.

———, and Kaiser, W. R., 1980, Catahoula Formation of the Texas coastal plain: Origin, geochemical evolution, and characteristics of uranium deposits: Bureau of Economic Geology, The University Texas, Austin, Report Invest., No. 100.

Gambolati, G., and Freeze, R. A., 1973, Mathematical simulation of the subsidence of Venice. I, Theory: Water Resources Res., v. 9, p. 721–733.

———, Gatto, P., and Freeze, R. A., 1974a, Mathematical simulation of the subsidence of Venice. 2, Results: Water Resources Res., v. 10, p. 563–577.

———, Gatto, P., and Freeze, R. A., 1974b, Predictive simulation of the subsidence of Venice: Science, v. 183, p. 849–851.

Gangi, A. F., 1978, Variation of the whole and fractional porous rock permeability with confining pressure. Internat. J. Rock Mech., v. 15, p. 249–257.

Garder, A. O., Peaceman, A. L., and Pozzi, R., 1964, Numerical calculation of multi-dimensional displacement by the method of characteristics. Soc. Pet. Eng. J., v. 4, no. 1, p. 26–38.

Garrels, R. M., 1960, Mineral Equilibrium at Low Temperature and Pressure: New York, Harper.

———, and Christ, C. L., 1965, Solutions, Minerals and Equilibria: New York, Harper & Row, 450 p.

———, and MacKenzie, F. T., 1967, Origin of the chemical compositions of some springs and lakes. Equilibrium Concepts in Natural Water Systems, ed. R. F. Gould: Washington, D. C., American Chemical Society, Adv. Chem. Ser. 67, p. 222–242.

———, and Thompson, M. E., 1962, A chemical model for seawater at 25°C and one atmosphere total pressure: Am. J. Sci., v. 260, p. 57–66.

Garrison, A. W., Keith, L. H., and Shackelford, W. H., 1977, Occurrence, registry, and classification of organic pollutants in water, with development of a master scheme for their analysis. Aquatic Pollutants: Transformation and Biological Effects, eds. O. Hutzinger, I. H. Van Lelyveld, and B. C. J. Zoeteman: Oxford, Pergamon Press, p. 39–68.

Garvin, G., 1988, Paleohydrology of oil migration in the Alberta Basin (abst). EOS, Trans. Amer. Geophy Union, v. 69, p. 360.

———, and Freeze, R. A., 1984a, Theoretical analysis of the role of groundwater flow in the genesis of stratabound ore deposits. I, Mathematical and numerical model: Amer. J. Sci., v. 284, p. 1085–1124.

———, and Freeze, R. A., 1984b, Theoretical analysis of the role of groundwater flow in the genesis of stratabound ore deposits. II, Quantitative results: Amer. J. Sci., v. 284, p. 1125–1174.

Gary, S. K., and Kassoy, D. R., 1981, Convective heat and mass transfer in hydrothermal systems. Geothermal Systems, Principles and Case Histories, eds., L. Rybach and L. J. P. Muffler. New York, John Wiley, p. 37–76.

Gat, J. R., 1980, The isotopes of hydrogen and oxygen in precipitation. Chapter 1 in Handbook of Environmental Isotope Geochemistry, Vol. 1, eds. P. Fritz and J. C. Fontes: New York, Elsevier, p. 21–47.

Gelhar, L. W., and Axness, C. L., 1983, Three-dimensional stochastic analysis of macrodispersion in aquifers: Water Resources Res., v. 19, no. 1, p. 161–180.

———, Gutjahr, A. L., and Naff, R. L., 1979, Stochastic analysis of macrodispersion in a stratified aquifer: Water Resources Res., v. 15, no. 6, p. 1387–1397.

_____, Mantoglou, A., Welty, C., and Rehfeldt, K. R., 1985, A review of field-scale physical solute transport processes in saturated and unsaturated porous media: Electric Power Research Institute EPRI EA-4190 Project 2485-5, 116 p.

Ghyben, W. B., 1899, Notes in verband met Voorgenomen Put boring Nabji Amsterdam, Tijdschr: The Hague, Koninhitk, Inst. Ingrs.

Gibson, R. E., 1958, The progress of consolidation in a clay layer increasing in thickness with time: Geotechnique, v. 8, p. 171–182.

Gibb, J. P., and Barcelona, M. J., 1984, Sampling for organic contaminants in ground water: J. AWWA, v. 76, no. 5, p. 48–51.

_____, Schuller, R. M., and Griffin, R. A., 1981, Procedures for the Collection of Representative Water Quality Data from Monitoring Wells. Cooperative Ground Water Report: Illinois State Water Survey and U.S. Environ. Protection Agency, 60 p.

Glasstone, S., and Lewis, D., 1960, Elements of Physical Chemistry: Princeton, N.J., D. Van Nostrand Company, 758 p.

Gilluly, J., and Grant, U. S., 1949, Subsidence in Long Beach Harbor area, California. Geol. Soc. Amer. Bull., v. 60, p. 461–521.

Glover, R. E., 1964, The pattern of freshwater flowing in a coastal aquifer, in sea water in coastal aquifers: U.S. Geol. Survey Water Supply Paper 1613-C, p. 32–35.

Goerlitz, D. F., and Brown, E., 1972, Methods for analysis of organic substances in water: U.S. Geol. Surv. TWRI, Book 5, Chapter A3, 40 p.

Goguel, J., 1976, Geothermics, trans. by A. Rite. New York, McGraw-Hill.

Goodman, R. E., and others, 1965, Groundwater inflows during tunnel driving: Engineering Geol., v. 2, p. 39–56.

Gordon, F. R., 1970, Water level changes preceding the Neckering Western Australia earthquake of October 14, 1968: Bull. Seismol. Soc. Amer., v. 60, p. 1739–1740.

Gorelick, S., 1983, A review of distributed parameter groundwater management modeling methods: Water Resources Res., v. 19, p. 305–319.

_____, 1987, Sensitivity analysis of optimal ground water contaminant capture curves. Solving Ground Water Problems with Models: Dublin, Ohio, National Water Well Assoc., p. 133–146.

_____, Voss, C. I., Gill, P., Murray, W., Saunders, M., and Wright, M., 1984, Aquifer reclamation design: The use of contaminant transport simulation combined with nonlinear programming: Water Resources Res., v. 20, no. 4, p. 415–427.

Graf, D. L., 1982, Chemical osmosis, reverse chemical osmosis, and the origin of subsurface brines: Geochim. Cosmochim. Acta, v. 48, p. 1431–1448.

Graton, L. C., and Fraser, H. J., 1935, Systematic packing of spheres with particular relation to porosity and permeability and experimental study of the porosity and permeability of clastic sediments. Geol., v. 43, p. 785–909.

Green, D. H., and Wang, H. F., 1986, Fluid pressure response to undrained compression in saturated sedimentary rock: Geophysics, v. 51, p. 948–956.

Green, D. W., and Perry, R. H., 1961, Heat transfer with a fluid flowing through porous media: Chem. Eng. Prog. Symp. Ser., v. 57, p. 61–68.

Green, H. W., 1984, "Pressure solution" creep: Some causes and mechanisms: J. Geophys. Res., v. 89, p. 4313–4318.

Greenberg, J. A., 1971, Diffusional flow of salt and water in soils. Ph.D. dissertation, University of California, Berkeley.

Greenkorn, R. A., 1983, Flow Phenomena in Porous Media: Fundamentals and Applications in Petroleum, Water and Food Production: New York, Marcel Dekker, 550 p.

_____, and Kessler, D. P., 1972, Transfer Operations: New York, McGraw-Hill, 548 p.

Gretener, P. E., 1972, Thoughts on overthrust faulting in a layered sequence: Bull. Canadian Petrol. Geol., v. 20, p. 583–607.

_____, 1981, Pore pressure: Fundamentals, general ramification, and implications for structural geology (revised): Tulsa, Okla., Education Course Note Series 4, Amer. Assoc. Petrol. Geol.

Griffin, R. A., and Roy, W. R., 1985, Interaction of Organic Solvents with Saturated Soil-Water Systems: Environmental Institute for Waste Management Studies, The University of Alabama, Open-File Report No. 3, 86 p.

Grisak, G. E., and Jackson, R. E., 1978, An Appraisal of the Hydrogeological Processes Involved in Shallow Subsurface Radioactive Waste Management in Canadian Terrain, Scientific Series No. 84: Ottawa, Inland Waters Directorate, Environment Canada, 194 p.

_____, Merritt, W. F., and Williams, D. W., 1977, A fluoride borehole dilution apparatus for groundwater velocity measurements: Can. Geotech. J., v. 14, p. 554–561.

_____, and Pickens, J. F., 1980, Solute transport through fractured media, 1, the effect of matrix diffusion: Water Resources Res., v. 16, no. 4, p. 719–730.

Grove, D. B., 1971, U.S. Geological Survey tracer study, Armargosa Desert, Nye County, Nevada. II, An analysis of the flow field of a discharging-recharging pair of wells: U.S. Geol. Surv. Rep. USGS, 474-99, 56 p.

Grove, D. B., and Beetem, W. A., 1971, Porosity and dispersion constant calculations for a fractured carbonate aquifer using the two well tracer method: Water Resources Res., v. 7, p. 128–134.

Gruszczenski, T. S., 1987, Determination of a realistic estimate of the actual formation product thickness using monitor wells: A field bailout test. NWWA/API Conference on Petroleum Hydrocarbons and Organic Chemicals in Ground Water—Prevention, Detection, and Restoration, Proc.: Dublin, Ohio, National Water Well Assoc., p. 235–253.

Gupta, M. L., and Rastogi, B. K., 1970, Dams and earthquakes: Devel. in Engr. Geol., v. 11.

_____, Rastogi, B. K., and Narain, H., 1972, Common features of the reservoir associated seismic activities: Bull. Seismol. Soc. Amer., v. 6, p. 481–492.

_____, and Suknija, B. S., 1974, Preliminary studies of some geothermal areas in India: Geothermics, v. 3, p. 105–112.

Guth, P. L., Hodges, K. V., and Willemin, J. H., 1982, Limitations on the role of pore pressure in gravity sliding: Geol. Soc. Amer. Bull. v. 93, p. 606–612.

Güven, O., Falta, R. W., Molz, F. J., and Melville, J. G., 1986, A simplified analysis of two-well tracer tests in stratified aquifers: Ground Water, v. 24, no. 1, p. 63–71.

Haga, K., 1961, The fractured zone of the Karobe transportation Tunnel: Geologie and Bauwesen, v. 26, p. 60–78.

Handin, J., and others, 1963, Experimental deformation of sedimentary rocks under confining pressure: Pore pressure tests: Amer. Assoc. Petrol. Geol. Bull., v. 47, p. 717–755.

Hanor, J. S., 1983, Fifty years of development of thought on the origin and evolution of subsurface sedimentary brines. Revolution in the Earth Sciences, Advances in the Past Half Century, ed. S. J. Boardman: Dubuque, Iowa, Kendall/Hunt, p. 99–111.

Hansch, C., and Leo, A., 1979, Substituent Constants for Correlation Analysis in Chemistry and Biology: New York, John Wiley, 339 p.

Hanshaw, B., Back, W., and Deike, R., 1971, A geochemical hypothesis for dolomitization by groundwater: Econ. Geol., v. 66, p. 710–724.

———, and Coplan, T. B., 1973, Ultrafiltration by a compacted clay membrane. II, Sodium ion exclusion at various ioninc strengths. Geochim. Cosmochim. Acta., v. 37, p. 2311–2327.

———, and Hill, G., 1969, Geochemistry and hydrodynamics of the Paradox Basin region, Utah, Colorado and New Mexico: Chem. Geol., v. 4 p. 263–294.

———, and Zen, E., 1965, Osmotic equilibrium and overthrust faulting: Geol. Soc. Amer. Bull., v. 76, p. 1379–1386.

Hantush, M. S., 1956, Analysis of data from pumping tests in leaky aquifers: Trans. Amer. Geophys. Union, v. 37, p. 702–714.

———, 1957, Nonsteady radial flow to a well partially penetrating an infinite leaky aquifer: Proc. Iraqi. Sci. Soc., v. 1, p. 10–19.

———, 1960, Modification of the theory of leaky aquifers: J. Geophys. Res., v. 65, p. 3713–3725.

———, 1961, Aquifer tests on partially penetrating wells: Proc. Amer. Soc. Civ. Engrs., v. 87, p. 171–195.

———, 1964, Hydraulics of wells. Advances in Hydroscience, ed., V. Chow: New York, Academic Press, Vol. 1, p. 281–432.

———, 1966a, Wells in homogenesis anisotropic aquifers: Water Resources Res., v. 2, p. 273–279.

———, 1966b, Analysis of data from pumping test in anisotropic aquifers: J. Geophys. Res., v. 71, p. 421–426.

———, 1967, Growth and decay of groundwater mounds in response to uniform percolation: Water Resources Res., v. 3, p. 227–234.

———, and Jacob, C. E., 1955, Nonsteady radial flow in an infinite leaky aquifer: Trans. Amer. Geophy. Union, v. 36, p. 95–100.

———, and Thomas, R. G., 1966, A method for analyzing a drawdown test in anisotropic aquifers: Water Resources Res., v. 2, p. 281–285.

Hardie, L. A., and Eugster, H. P., 1970, The evolution of closed-basin brines: Mineral. Soc. Am. Spec. Publ., v. 3, p. 273–290.

Harleman, D. R. E., Melhorn, P. F., and Rumer, R. R., 1963, Dispersion-permeability correlation in porous media: J. Hydraul. Div., Amer. Soc., Civil Engrs., v. 89, p. 67–85.

Harris, J. C., 1982, Rate of hydrolysis. Handbook of Chemical Property Estimation Methods, eds. W. J. Lyman, W. F. Reehl, and D. H. Rosenblatt: New York, McGraw-Hill, 977 p.

Harris, J. F., Taylor, G. L., and Walper, J. L., 1960, Relation of deformational fracture in sedimentary rocks to regional and local structure: Amer. Assoc. Petrol. Geol. Bull., v. 44, p. 1853–1873.

Harriss, D. K., Ingle, S. E., Magnuson, V. R., and Taylor, D. K., 1982, Programmer's Manual for REDEQL.UMD: Duluth, University of Minnesota (unreleased preliminary document).

Harvey, R. W., Smith, R. L., and George, L., 1984, Microbial distribution and heterotrophic uptake in a sewage plume. Movement and Fate of Solutes in a Plume of Sewage-Contaminated Ground Water: Cape Cod, Mass., U.S. Geol. Surv. Toxic Waste Ground-Water Contamination Program, U.S. Geol. Surv. Open-File Rept. 84–475, p. 139–152.

Hashin, Z., and Shtrikman, S., 1962, A variation approach to the theory of effective magnetic permeability of multiphase materials: J. Appl. Phys., v. 33, p. 3125–3131.

Hassett, J. J., Banwart, W. L., and Griffin, R. A., 1983, Correlation of compound properties with sorption characteristics of nonpolar compounds by soils and sediments: Concepts and limitations. Chapter 15 in Environment and Solid Wastes: Characterization, Treatment and Disposal, eds. C. W. Francis and S. I. Auerbach: Stoneham, Mass., Butterworth, p. 161–178.

Hastings, T., 1986, A theoretical approach for assessing the role of rock and fluid properties in the development of abnormal pressure. M.S. thesis, Texas A&M University.

Hayes, J. B., 1977, Sandstone diagenesis, the hole truth. Aspects of Diagenesis, eds., P. A. Scholle and R. R. Schluger: Soc. Econ. Paleontol. and Mineral Spec Pub. 26, p. 127–139.

Hazen, A., 1911, Discussion of "Dams on Soil Foundations": Trans. Amer. Soc. Civil Engrs., v. 73, p. 199.

Healy, J. W., and others, 1968, The Denver earthquakes: Science, v. 161, p. 1301–1310.

Heard, H. C., and Rubey, W. W., 1966, Tectonic implications of gypsum dehydration: Geol. Soc. Amer. Bull., v. 77, p. 741–760.

Heath, R. C., 1982, Classification of groundwater systems of the United States: Ground Water, v. 20, p. 393–401.

_____, 1984, Groundwater regions of the United States: U.S. Geol. Survey Water Supply Paper 2242.

Helfferich, F., 1966, In Ion Exchange, A Series of Advances, ed. J. A. Marinsky: New York, Marcel Dekker, p. 65–100, 424 p.

Helgeson, H. C., 1968, Evaluation of irreversible reactions in geochemical processes involving minerals and aqueous solutions. 1, Thermodynamic relations: Geochim. Cosmochim. Acta, v. 32, p. 853–877.

Helgeson, H. C., Brown, T. H., Nigrini, A., and Jones, T. A., 1970, Calculation of mass transfer in geochemical processes involving aqueous solutions: Geochim. Cosmochim. Acta, v. 34, p. 569–592.

Helm, D. C., 1975, One dimensional simulation of aquifer compaction system near Pixley, California: Water Resources Res., v. 11, p. 465–478.

Hem, J. D., 1959, Study and interpretation of the chemical characteristics of natural water: U.S. Geol. Survey Water Supply Papers, 1473.

Henderson, L. H., 1939, Detailed geologic mapping and fault studies of the San Jacinto tunnel line and vicinity: J. Geol., v. 47, p. 314–324.

Hendry, M. J., Cherry, J. A., and Wallick, E. I., 1986, Origin and distribution of sulfate in a fractured till in southern Alberta, Canada: Water Resources Res., v. 22, no. 1, p. 45–61.

_____, and Schwartz, F. W., 1988, An alternative view on the origin of chemical and isotopic patterns in groundwater from the Milk River aquifer: Water Resources Res., v. 24, no. 10, p. 1747–1764.

_____, and Schwartz, F. W., 1990, The chemical evolution of ground water in the Milk River aquifer, Canada: Ground Water, v. 28, no. 2, p. 253–261.

Herrick, C. C., 1976, Los Almos Scientific Laboratory, Los Alamos, New Mexico (private communication).

Herrin, H., and Goforth, T., 1975, Environmental problems associated with power production from geopressured reservoirs: Proc. 1st Geopress. Geotherm. Conf., p. 311–320.

Herzberg, B., 1901, Die Wasserversovgung einiger Nordseebaser: J. Gasbeleucht and wasserversov, v. 44, p. 815–819.

Hess, J. W., and White, W. B., 1988, Storm response of the karstic carbonate aquifer of south central Kentucky: J. Hydrol., v. 99, p. 235–252.

Heyl, A. V., and others, 1970, Guidebook to the Upper Mississippi Valley basic-metal district: Madison, University of Wisconsin Geol. and Nat. History Survey, Inf. Circ. No. 16, 49 p.

Hill, G. A., Colburn, W. A., and Knight, J. W., 1961, Reducing oil finding costs by use of hydrodynamic evaluations. Economics of Petroleum Exploration, Development, and Property Evaluation: Englewood Cliffs, N.J., Prentice Hall, p. 38–69.

Hillel, D., 1971, Soil and Water: Physical Principles and Processes: New York, Academic Press.

Hinchee, R. E., Downey, D. C., and Coleman, E. J., 1987, Enhanced bioreclamation soil venting and ground water extraction; a cost-effectiveness and feasibility comparison. NWWA/API Conference on Petroleum Hydrocarbons and Organic Chemicals in Ground Water—Prevention, Detection, and Restoration, Dublin, Ohio, Proc.: National Water Well Assoc., 147–164.

Hirasaka, G. J., 1974, Pulse tests and other early transient pressure analysis for in-situ estimation of vertical permeability: Trans. Amer. Inst. Mining Engr., v. 257, p. 75–90.

Hoefner, M. L., and Fogler, H. S., 1988, Pore evolution and channel formation during flow and reaction in porous media: AIChE J., v. 34, p. 45–54.

Hoek, E., and J. W., Bray, 1981, Rock Slope Engineering. London, Inst. of Min. and Met.

Hornberger, G. M., Ebert, J., and Remson, I., 1970, Numerical solution of the Boussinesq equation for aquifer-stream interaction: Water Resources Res., v. 6, p. 601–608.

Hornberger, G. M., Remson, I., and Fungaroli, A. A., 1969, Numeric studies of a composite soil moisture groundwater system: Water Resources Res., v. 5, p. 797–802.

Horner, D. R., 1951, Pressure buildup in wells. Third World Petrol. Congress, Proc., Sect. II: Leiden, Holland. E. J. Brill, p. 503–521.

Horton, C. W., and Rogers, F. T., 1945, Convection currents in a porous medium: J. Appl. Phys., v. 16, p. 367–370.

Horton, R. E., 1933, The role of infiltration in the hydrologic cycle: Trans. Amer. Geophy. Union, v. 14, p. 446–460.

———, 1940, Approach toward a physical interpretation of infiltration capacity: Soil Sci. Soc. Amer. Proc., v. 5, p. 339–417.

Hsieh, P. A., and Neuman, S. P., 1985, Field determination of the three dimensional hydraulic conductivity tensor of anisotropic media. I, Theory: Water Resources Res., v. 21, p. 1655–1665.

———, and Bredehoeft, J. D., 1981, A reservoir analysis of the Denver earthquakes. J. Geophys. Res., v. 86, p. 903–920.

———, and others, 1985, Field determination of the three dimensional hydraulic conductivity tensor of anisotropic media. II, Methodology and application to fractured rocks: Water Resources Res., v. 21, p. 1667–1676.

Hsu, K. J., 1977, Studies of Ventura field, California. II, Lithology, compaction, and permeability of sands: Amer. Assoc. Petrol. Geol. Bull., v. 61, p. 169–191.

Hubbert, M. K., 1940, The theory of groundwater motion: J. Geol., v. 48, p. 785–944.

_____, 1953, Entrapment of petroleum under hydrodynamic conditions. Amer. Assoc. Petrol. Geol. Bull., v. 37, p. 1944–2026.

_____, 1956, Darcy's law and the field equations of the flow of underground fluids: Trans. Amer. Inst. Min. Met. Engrs., v. 207, p. 222–239.

_____, and Rubey, W. W., 1959, Role of fluid pressure in mechanics of overthrust faulting. Part I, Mechanics of fluid filled porous solids and its application to overthrust faulting: Geol. Soc. Amer. Bull., v. 70, p. 115–166.

_____, and Willis, D. G., 1957, Mechanics of hydraulic fracture: Amer. Inst. Min. Engr. Trans., v. 210, p. 153–168.

Hufschmidt, M. M., and M. B. Fiering, 1966, Simulation Techniques for Design of Water Resource Systems: Cambridge, Mass., Harvard University Press.

Hughes, J. P., Sullivan, C. R., and Zinner, R. E., 1988, Two techniques for determining the true hydrocarbon thickness in an unconfined sandy aquifer. NWWA/API Conference on Petroleum Hydrocarbons and Organic Chemicals in Ground Water—Prevention, Detection, and Restoration, Proc.: Dublin, Ohio, National Water Well Assoc., p. 291–314.

Huitt, J. L., 1956, Fluid flow in simulated fractures: Amer. Inst. Chem. Eng., v. 2, p. 259.

Hull, R. W., Kharaka, Y. K., Maest, A. S., and Fries, T. L., 1984, Sampling and analysis of subsurface water. First Canadian/American Conference on Hydrogeology, eds. B. Hitchon and E. I. Wallick: Dublin, Ohio, National Water Well Assoc., p. 117–126.

Hult, M. F., 1987, Introduction. U.S. Geol. Surv. Program on Toxic Waste— Ground-Water contamination. Third Technical Meeting, Pensacola Florida, March 23–27, 1987, Proc.: U.S. Geol. Surv. Open-File Rept. 87-109, p. C-3–C-5.

Hunt, B. W., 1978a, Dispersive sources in uniform groundwater flow: J. Hydraulics. Div., ASCE, v. 99, no. HYi, p. 13–21.

_____, 1978b, Dispersive source in uniform groundwater flow. J. Hydrol. Div., Proc. Amer. Soc. Civil Eng., v. 104, p. 75–85.

Hurr, R. T., and Richards, D. B., 1966, Groundwater engineering of the Straight Creek Tunnel (Pilot Bore), Colorado: Eng. Geol., v. 3, p. 80–90.

Huyakorn, P., and others, 1987, A three dimensional analytical method for predicting leachate migration: Ground Water, v. 25, p. 588–598.

_____, and Pinder, G. F., 1983, Computational Methods in Subsurface Flow: New York, Academic Press, 473 p.

_____, Andersen, P. F., Güven, O., and Molz, F. J., 1986a, A curvilinear finite element model for simulating two-well tracer tests and transport in stratified aquifers: Water Resources Res., v. 22, no. 5, p. 663–678.

_____, Andersen, P. F., Molz, F. J., Güven, O., and Melville, J. G., 1986b, Simulation of two-well tracer tests in stratified aquifers at the Chalk River and the Mobile sites: Water Resources Res., v. 22, no. 7, p. 1016–1030.

Hvorslev, M. J., 1951, Time lag and soil permeability in groundwater observations: U.S. Army Corps of Engr. Waterway Exp. Stat. Bull. 36 (Vicksburg, Miss.).

Ibrahim, H. A., and Brutsaert, W., 1965, Inflow hydrographs from large unconfined aquifers: J. Irr. Drain. Div., Proc. Amer. Soc. Civil Engrs, v. 91, p. 21–38.

Ingerson, E., and Pearson, J. F., Jr., 1964, Estimation of age and rate of motion of ground-water by the ^{14}C method. Recent research in the Fields of Hydrosphere, Atmosphere and Nuclear Chemistry: Tokyo, Maruzen, p. 263–283.

Jackson, R. E., and Patterson, R. J., 1982, Interpretation of pH and Eh trends in a fluvial-sand aquifer system: Water Resources Res., v. 18, no. 4,5 p. 1255–1268.

———, Patterson, R. J., Graham, B. W., Bahr, J., Belanger, D., Lockwood, J., and Priddle, M., 1985, Contaminant Hydrogeology of Toxic Organic Chemicals at a Disposal Site, Gloucester, Ontario. 1, Chemical Concepts and Site Assessment: Ottawa, Environment Canada, National Hydrol. Research Instit., Paper No. 23, 114 p.

———, and Inch, J., 1980, Hydrogeochemical processes affecting the migration of radionuclides in a fluvial sand aquifer at the Chalk River Nuclear Laboratories: Ottawa, Environment Canada, National Hydrol. Research Instit., Paper No. 7, Science Series, 58 p.

Jackson, S. A., and Beales, F. W., 1967, An aspect of sedimentary basin evolution: The concentration of Mississippi Valley type ores during late stages of diagenesis: Canadian Petrol. Geol. Bull., v. 15, p. 383–437.

Jacob, C. E., 1939, Fluctuations in artesian pressure produced by passing railroad trains as shown in a well on Long Island, New York: Trans. Amer. Geophys. Union, v. 20, p. 666–674.

———, 1940, On the flow of water in an elastic artesian aquifer: Trans. Amer. Geophys. Union, v. 22, p. 574–586.

———, 1941, Notes on the elasticity of the Lloyd Sand on Long Island, New York: Trans. Amer. Geophys. Union, v. 22, p. 783–787.

———, 1950, Flow of Groundwater. Engineering Hydraulics, ed. H. Rouse: New York, John Wiley, p. 321–386.

Jacquin, C., and Poulet, M., 1970, Study of hydrodynamic pattern in a sedimentary basin subject to subsidence: Soc. Petrol. Engrs. Amer. Inst. Mining, Metal. Petrol. Engs. Paper no SPE2988, 10 p.

———, and Poulet, M., 1973, Essas de Restitution des conditions hydrodynamiques regnant dans un bassin sedimentarire au coors de son evolution: Revue de L'Institut Français du Petrole., v. 28, p. 269–297.

Jaeger, J. C., and Cook, N. G. W., 1969, Fundamentals of Rock Mechanics: London, Methuen Press.

Jamieson, G. R., and Freeze, R. A., 1983, Determining hydraulic conductivity distribution in a mountainous area using mathematical modeling: Ground Water, v. 21, p. 168–177.

Javandel, I., 1986, Application of capture-zone type curves for aquifer cleanup. Groundwater Hydrology, Contamination, and Remediation, eds. R. Khanbilvard and J. Fillos: Washington, D.C., Scientific Publications, p. 249–279.

———, Doughty, C., and Tsang, C. F., 1984, Groundwater Transport: Handbook of Mathematical Models: American Geophys. Union (Washington), Water Res. Monograph 10, 228 p.

———, and Tsang, C., 1986, Capture-zone type curves: a tool for aquifer cleanup: Ground Water, v. 24, no. 5, p. 616–625.

Jhaveri, V., and A. J. Mazzacca, 1986, Bio-reclamation of ground and groundwater by in-situ biodegradation. Groundwater Hydrology, Contamination, and Remediation, eds. R. Khanbilvardi and J. Fillos: Washington, D.C., Scientific Publications, p. 281–321.

Johnson, A. I., 1948, Groundwater recharge on Long Island: J. Amer. Water Works Assoc., v. 49, p. 1159–1166.

_____, 1967, Specific yield-compilation of specific yields for various materials. U.S. Geol. Survey Water Supply Paper 1662-D, 74 p.

_____, and D. A. Morris, 1962, Physical and hydrologic properties of water bearing deposits from core holes in the Las Banos–Kettleman City area, California. Denver, Colo., U.S. Geol. Survey Open. File Rept.

Johnson, K. S., 1981, Dissolution of salt on the east flank of the Permian Basin in the southwestern U.S.A. J. Hydrol., Special Issue, Symposium on Geochemistry of Groundwater, eds. W. Back and R. Létolle, v. 54, p. 75–94.

Johnson, R. L., Brillante, S. M., Isabelle, L. M., Houck, J. E., and Pankow, J. F., 1985, Migration of chlorophenolic compounds at the chemical waste disposal site at Alkali Lake, Oregon, 2, Contaminant distributions, transport and retardation: Ground Water, v. 23, no. 5, p. 652–665.

Jones, J. W., and others, 1985, Field and theoretical investigation of fractured crystalline rock near Oracle, Southern Arizona, NUREG/CR-3736. Washington, D.C.: Nuclear Regulatory Commission.

Jouanna, P., 1972, In situ permeability tests under applied stresses. Proc. Symp. Percolation Through Fissured Rock, Stuttgart, Germany: International Association of Engineering Geologists.

Jury, W. A., and Sposito, G., 1985, Field calibration and validation of solute transport models for the unsaturated zone: Soil Sci. Soc. Am. J., v. 49, p. 1331–1341.

Kan, A. T., and Tomson, M. B., 1986, Facilitated transport of naphthalene and phenanthane in a sandy soil column with dissolved organic matter—macromolecules and micelles. NWWA/API Conference on Petroleum Hydrocarbons and Organic Chemicals in Ground Water—Prevention, Detection, and Restoration, Proc.: Dublin, Ohio, National Water Well Assoc., p. 393–406.

Karickhoff, S. W., 1981, Semi-empirical estimation of sorption of hydrophobic pollutants on natural sediments and soils: Chemosphere, v. 10, no. 8, p. 833–846.

_____, 1984, Organic pollutant sorption in aquatic systems: J. Hydraulic Engr. (ASCE), v. 110, no. 6, pp. 707–735.

_____, 1985, Pollutant sorption in environmental systems. Environmental Exposure Form Chemicals, Vol. I, eds. W. B. Neely and G. E. Blau: Fla.: Boca Raton, CRC Press, p. 49–62.

_____, Brown, D. S., and Scott, T. A., 1979, Sorption of hydrophobic pollutants on natural sediments: Water Res., v. 13, p. 241–248.

Kasten, P. R., Lapidus, L., and Amundson, N. R., 1952, Mathematics of adsorption in beds. V, Effect of intraparticle diffusion in flow systems in fixed beds: J. Phys. Chem., v. 56, p. 683–688.

Kazmann, R. G., 1956, Safe yield in groundwater development, reality or illusion. Proc. Amer. Soc. Civil Engrs., v. 82, 1956.

Keech, D. A., 1988, Hydrocarbon thickness on groundwater by dielectric well logging. NWWA/API Conference on Petroleum Hydrocarbons and Organic Chemicals in Ground Water—Prevention, Detection, and Restoration, Proc.: Dublin, Ohio, National Water Well Assoc., p. 275–289.

Kemper, W. D., 1960, Water and ion movement in thin films as influenced by the electrostatic layer of cations associated with clay mineral surfaces: Soil Sci. Soc. Amer. Proc., v. 24, p. 10–16.

Kenaga, E. E., 1980, Predicted bioconcentration factors and soil sorption coefficients of pesticides and other chemicals. Ecotoxicology and Environ. Safety, v. 4, p. 26–38.

Kharaka, Y. K., and Barnes, I., 1973, SOLMNEQ: Solution-Mineral Equilibrium Computations. National Tech. Infor. Serv. Tech. Rept. PB214-899, 82 p.

_____, and Berry, F. A., 1973, Simultaneous flow of water and solutes through geologic membranes. I. Experimental investigations. Geochem. Cosmochim. Acta, v. 37, p. 2577–2603.

Kiersch, G., 1973, The Vaiont Reservoir disaster. Focus on Environmental Geology, ed. R. Tank: New York, Oxford Press.

Killey, R. W. D., McHugh, J. O., Champ, D. R., Cooper, E. L., and Young, J. L., 1984, Subsurface cobalt-60 migration from a low-level waste disposal site: Environ. Sci. Technol., v. 18, p. 148–157.

Kilty, K., and Chapman, D. S., 1980, Convective heat transfer in selected geological situations. Ground Water, v. 18, p. 386–394.

Kimball, B. A., 1984, Ground water age determinations, Piceance Creek Basin, Colorado. First Canadian/American Conference on Hydrogeology, eds. B. Hitchon and E. I. Wallick: Dublin, Ohio, National Water Well Assoc., p. 267–283.

Kimmel, G. E., Braids, O. C., 1980, Leachate Plumes in Groundwater from Babylon and Islip Landfills, Long Island, N. Y.: U.S. Geol. Surv., Prof. Paper 1085, 38 p.

Kincaid, C. T., and Morrey, J. R., 1984, Geochemical Models for Solute Migration. Vol. 2, Preliminary Evaluation of Selected Computer Codes: Palo Alto, Calif., Electric Power Research Instit., EA-3417.

_____, Morrey, J. R., and Rogers, J. E., 1984, Geohydrochemical Models for Solute Migration. Vol. 1, Process Description and Computer Code Selection: Palo Alto, Calif., Electric Power Research Instit., EA-3417.

King, F. H., 1899, Principles and conditions of the movements of groundwater: U.S. Geol. Survey 19th Ann. Rept., pt 2, p. 59–294.

Kirkham, D., and Powers, W. L., 1972, Advanced Soil Physics: New York, Wiley-Interscience.

Kissen, I. G., 1978, The principal distinctive features of the hydrodynamic regime of intensive earth crust downwarping areas. Hydrogeology of Great Sedimentary Basins: Inter. Assoc. Sci. Hydrol, no. 120, p. 178–185.

Klotz, I. M., 1946, The adsorption wave: Chem. Rev., v. 39, p. 241–268.

_____, 1950, Chemical Thermodynamics. Englewood Cliffs, N.J., Prentice Hall, 369 p.

Knapp, R. B., and Knight, J. E., 1977, Differential thermal expansion of pore fluids: fracture propagation and microearthquake production in hot pluton environments: Jour. Geophys. Res., v. 82, p. 2515–2522.

Knox, R. D., Canter, L. W., Kincannon, D. F., Stover, E. L., and Ward, C. H., 1984, State-of-the-art of aquifer restoration: National Center for Groundwater Research, University of Oklahoma, Oklahoma State University, and Rice University, National Technical Information Service, 371 p.

Kohout, F. A., 1961, Case history of salt water encroachment caused by a storm sewer in Miami: J. Amer. Water Works Assoc., v. 53, p. 1406–1416.

_____, and Klein, H., 1967, Effect of pulse recharge on the zone of diffusion in the Biscayne aquifer: Inter. Assoc. Sci. Hydrol. Symp. Haifa, Pub. 72, p. 252–270.

Konikow, L. F., and Bredehoeft, J. D., 1978, Computer model of two-dimensional solute transport and dispersion in ground water: Reston, Va., U.S. Geol. Surv. TWRI, Book 7,1 Chap. C2, 40 p.

Koons, P. O., 1988, The evolution of the fluid flow regime during continental collision [abst.]: EOS. Trans. Amer. Geophys Union, v. 69, p. 361.

Kozeny, J., 1927, Uber Kapillare leitung des Wassers in boden. Sitzungsber Akad. Wiss. Wien., v. 136, p. 271–306.

Kreitler, C. W., 1977, Fault control of subsidence, Houston, Texas: Ground Water, v. 3, p. 203–214.

Krumbein, W. C., and Monk, G. D., 1943, Permeability as a function of the size parameters of unconsolidated sand: Trans. Amer. Inst. Min. Met. Engrs., v. 151, p. 153–163.

Krupp, H. K., Biggar, S. W., and Nielsen, D. R., 1972, Relative flow rates of salt in water in soil: Soil Sci. Soc. Am. Proc., v. 36, p. 412–417.

Krynine, D. P., and W. R. Judd, 1957, Principles of Engineering Geology and Geotechnics: New York, McGraw-Hill.

Kunii, D., and Smith, J. M., 1961, Heat transfer characteristics of porous rocks. II, Thermal conductivities of unconsolidated particles with flowing fluids: J. Amer. Inst. Chem. Eng. v. 7, p. 29–31.

Lachenbruch, A. H., and Sass, J. H., 1977, Heat flow in the United States. The Earth's Crust, Monogr. Ser., ed. J. Heacock: Washington, D.C., Amer. Geophys. Union, p. 626–675.

Lai, S. H., and Jurinak, J. J., 1971, Numerical approximation of cation exchange in miscible displacement through soil columns: Soil Sci. Soc. Am. Proc., v. 35, p. 894–899.

Lallemand-Barrès, A., and Peaudecerf, P., 1978, Recherchedes relations entre la valeur de la dispersivité macroscopique d'un milieu aquifère, ses autres caracteristiques et les conditions de mesure: Bull. BRGM (2)III, Urbana, 4, p. 277.

Lamarck, J. B., 1984, Hydrogeology, trans. by A. V. Carozzi: Urbana, University of Illinois Press.

Land, L. S., 1973, Holocene meteoric dolomitization of Pleistocene limestone in North Jamaica: Sedimental, v. 20, p. 411–422.

———, MacKenzie, F. T., and Gould, S. J., 1967, Pleistocene history of Bermuda: Geol. Soc. Amer. Bull., v. 78, p. 933–1006.

———, and Prezbindowski, D., 1981, The origin and evolution of saline formation waters, Lower Cretaceous carbonates, South-Central Texas, U.S.A.: J. Hydrol., v. 54, p. 51–74.

Landes, K. K., and others, 1960, Petroleum resources in basement rock: Amer. Assoc. Petrol. Geol. Bull., v. 44, p. 1682–1691.

Landström, O., Klockars, C. E., Holmberg, K. E., and Westerberg, S., 1978, In situ experiments on nuclide migration in fractured crystalline rocks, KBS Technical Rept., Stockholm Sweden, TR110.

———, Klockars, C. E., Persson, O., Andersson, K., Torslenfelt, B., Allard, B., Larson, S. A., and Tullborg, E. L., 1982, A comparison of in situ radionuclide migration studies in the Studsvik area and laboratory measurements: New York, Elsevier, Scientific Basis for Nuclear Waste Management v. 5, p. 697–706.

Langmuir, D., 1971, The geochemistry of some carbonate ground waters in central Pennsylvania: Geochim. Cosmochim. Acta, v. 35, p. 1023–1045.

———, 1978, Uranium solution-mineral equilibria at low temperatures with applications to sedimentary ore deposits. Geochim. Cosmochim. Acta, v. 42, p. 547–569.

———, 1984, Physical and chemical characteristics of carbonate water: UNESCO Studies and Reports in Hydrology, v. 41, p. 69–130.

———, and Mahoney, J., 1984, Chemical equilibrium and kinetics of geochemical processes in ground water studies. First Canadian/American Conference on Hydrogeology, eds. B. Hitchon and E. I. Wallick: Dublin, Ohio, National Water Well Assoc., p. 69–95.

Lapidus, L., and Amundson, N. R., 1952, Mathematics of adsorption in beds. VI, The effect of longitudinal diffusion in ion exchange and chromatographic columns: J. Phys. Chem., v. 56, p. 984–988.

Lapwood, E. R., 1948, Convection of a fluid in a porous medium: Cambridge Phil. Soc. Proc., v. 44, p. 508–521.

Larson, S. P., McBride, M. S., and Wolf, R. J., 1976, Digital models of a glacial outwash aquifer in the Pearl-Sallie Lakes area, West-Central Minnesota: Water Res. Invest., U.S. Geol. Survey, p. 40–75.

———, Andrews, C. B., Howland, M. D., and Feinstein, D. T., 1987, A three-dimensional modeling analysis of ground water pumping schemes for containment of shallow ground water contamination. Solving Ground Water Problems with Models: Dublin, Ohio, National Water Well Assoc., p. 517–531.

Lattman, L. A., and Parizek, R. R., 1964, Relationship between fracture traces and the occurrence of groundwater in carbonate rocks: J. Hydrol., v. 2, p. 73–91.

Lau, L. K., Kaufman, W. J., and Todd, D. K., 1957, Studies of Dispersion in a Radial Flow System. Progress report no. 3 of Canal Seepage Research: Dispersion Phenomena in Flow Through Porous Media, I.E.R. Series No. 93, Issue No. 3: Berkeley, Sanitary Eng. Res. Lab., Dept. of Eng. and School of Pub. Health, University of California.

LaVenue, A. M., and Domenico, P. A., 1986, A preliminary assessment of the regional dispersivity of selected basalt flows at the Hanford Site, Washington, U.S.A.: J. Hydrol., v. 85, p. 151–167.

Law, J., 1944, A statistical approach to the interstitial heterogeneity of sand reservoirs: Trans. Amer. Inst. Min. Met. Engr., v. 155, p. 202–222.

Lawrence, F. W., and Upchurch, S. B., 1982, Identification of recharge areas using geochemical factor analysis: Ground Water, v. 20, p. 680–687.

Lawrence Berkeley Laboratory, 1984, Panel report on coupled thermomechanical-hydro chemical process associated with a nuclear waste repository, prepared for office of Nuclear Regulator Commission, Rept. LBL-18250.

LeBlanc, D. R., ed., 1984a, Movement and Fate of Solutes in a Plume of Sewage-Contaminated Ground Water: Cape Cod, Mass., U.S. Geol. Survey Toxic Waste Ground-Water Contamination Program, U.S. Geol. Surv. Open-File Rept., 84-475, 175 p.

———, 1984b, Digital modeling of solute transport in a plume of sewage-contaminated ground water. Movement and Fate of Solutes in a Plume of Sewage-Contaminated Ground Water: Cape Cod, Mass., U.S. Geol. Survey Toxic Waste Ground-Water Contamination Program, U.S. Geol. Surv. Open-File Rept., p. 11–45.

Lee, C. H., 1915, The determination of safe yield of underground reservoirs of the closed basin type: Trans. Amer. Soc. Civil Engrs., v. 78, p. 148–151.

Lee, M. D., Jamison, V. W., and Raymond, R. L., 1987, Applicability of in-situ bioreclamation as a remedial action alternative. NWWA/API Conference on Petroleum Hydrocarbons and Organic Chemicals in Ground Water—Prevention, Detection, and Restoration, Dublin, Ohio, Proc.: National Water Well Assoc., p. 167–185.

Lefkoff, L. J., and Gorelick, S. M., 1986, Design and cost analysis of rapid aquifer restoration systems using flow simulation and quadratic programming: Ground Water, v. 24, no. 6, p. 777–790.

LeGrand, H. E., 1954, Geology and groundwater in the Statesville area, North Carolina: North Carolina, Raleigh, North Carolina Dept. Conser. Dev., Div. Mineral Resources, Bull. 68.

Lerman, A., 1971, Time to chemical steady states in lakes and oceans. Advances in Chemistry Series 106, Nonequilibrium Systems in Natural Water Chemistry: American Chemical Society Washington, D.C., p. 30–76.

Leonards, G. A., 1962, Engineering properties of soils. Foundation Engineering, ed. G. A. Leonards: New York, McGraw-Hill, p. 66–240.

Li, Y.-H., and Gregory, S., 1974, Diffusion of ions in sea water and in deep-sea sediments: Geochim. Cosmochim. Acta., v. 38, p. 703–714.

Lico, M. S., Kharaka, Y. K., Carothers, W. W., and Wright, V. A., 1982, Methods for collection and analysis of geopressured geothermal and oil field water: U.S. Geol. Surv., Water Supply Pap. 2194, 21 p.

Lindberg, R. D., and Runnells, D. D., 1984, Ground water redox reactions: An analysis of equilbrium state applied to Eh measurements and geochemical modeling: Science, v. 225, p. 925–927.

Linsley, R. K., Kohler, M. A., and Paulhus, J. L. H., 1958, Hydrology for Engineers: New York, McGraw-Hill.

Liu, C. W., and Narasimhan, T. N., 1989, Redox-controlled multiple-species reactive chemical transport. 1, Model development: Water Resources Res., v. 25, no. 5, p. 869–882.

Lohman, S. W., 1961, Compression of elastic aquifers. U.S. Geol. Survey Prof. Paper 424-B, p. 47–48.

Long, J. C. S., and others, 1982, Porous media equivalents for networks of discontinuous fractures: Water Resources Res., v. 18, p. 645–658.

Longman, M. W., 1982, Carbonate diagenesis as a control on stratigraphic traps: Amer. Assoc. Petrol. Geol. Education Course Notes 21.

Loo, W. W., Frantz, K., and Holzhausen, G. R., 1984, The application of telemetry to large scale horizontal anisotropic permeability determination by surface tiltmeter survey: Ground Water Monit. Rev., v. 4, p. 124–130.

Lovering, T. S., and Goode, H. D., 1963, Measuring geothermal gradients in drill holes less than 60 feet deep, East Tintic District, Utah: U.S. Geol. Survey Bull. 1172.

Lusczynski, N. J., and Swarzenski, W. V., 1966, Salt water encrouchment in southern Nassau and southeastern Queens counties, Long Island, New York: U.S. Geol. Survey Water Paper 1613-F.

Maass, A., and others, 1962, Design of Water Resources Systems: Cambridge, Mass., Harvard University Press.

Mackay, D. M., Freyberg, D. L., and Roberts, P. V., 1986, A natural gradient experiment on solute transport in a sand aquifer. 1. Approach and overview of plume movement: Water Resources Res., v. 22, no. 13, p. 2017–2029.

———, and Leinonen, P. J., 1975, Rate of evaporation of low-solubility contaminants from water bodies to atmosphere. Environ. Sci. Technol., v. 9, no. 8, p. 1178–1180.

———, Roberts, P. V., and Cherry, J. A., 1985, Transport of organic contaminants in groundwater: Environ. Sci. Technol., v. 19, no. 5, p. 384–392.

———, and Vogel, T. M., 1985, Ground water contamination by organic chemicals: Uncertainties in assessing impact. Second Canadian/American Conference on Hydrogeology, eds. B. Hitchon and M. Trudell: Dublin, Ohio, National Water Well Assoc., p. 50–59.

Macumber, P. G., 1984, Hydrochemical processes in the regional ground water discharge zones of the Murray Basin, Southeastern Australia. First Canadian/American Conference on Hydrogeology, eds. B. Hitchon and E. I. Wallick: Dublin, Ohio, National Water Well Assoc., p. 47–63.

Magara, K., 1975, Importance of aquathermal pressuring effect in Gulf Coast: Amer. Assoc. Petrol. Geol. Bull., v. 59, p. 2037–2045.

———, 1978, Compaction and Fluid Migration Practical Petroleum Geology. Amsterdam, Elsevier.

Malley, M. J., Bath, W. W., and Bongers, L. H., 1985, A case history: Surface static collection and analysis of chlorinated hydrocarbons from contaminated ground water. NWWA/API Conference on Petroleum Hydrocarbons and Organic Chemicals in Ground Water—Prevention, Detection, and Restoration, Proc.: Dublin, Ohio, National Water Well Assoc., p. 276–290.

Malmberg, G. T., 1960, An analysis of the hydrology of the Las Vegas groundwater basin Nevada: U.S. Geol. Surv. Open-File Rept.

Manheim, F. T., and C. K. Paull, 1981, Patterns of groundwater salinity changes in a deep continental-oceanic transect off the southeastern Atlantic coast of the U.S.A.: J. Hydrol., Special Issue, Symposium on Geochemistry of Groundwater, eds. W. Back and R. Létolle, 54, p. 95–106.

———, and J. F. Bischoff, 1969, Geochemistry of pore waters from Shell Oil Company drill holes on the continental slope of the northern Gulf of Mexico. Chem. Geol., Special Issue, Geochemistry of Subsurface Brines, eds. E. E. Angino and G. K. Billings, v. 4, p. 63–82.

Manov, G. C., Bates, R. C., Hamer, W. J., and Acree, S. F., 1943, Values of the constants in the Debye-Huckel equation for activity coefficients: Jour. Amer. Chem. Soc., v. 65, p. 1765–1767.

Marine, I. W., and Fritz, S. J., 1981, Osmotic model to explain anomalous hydraulic heads: Water Resources Res., v. 17, p. 73–82.

Martin, W. E., Burdak, T. G., and Young, R. W., 1969, Projecting hydrologic and economic interrelationships in groundwater basin management, paper presented at internat. conf. on Arid Lands in a Changing World, Tucson, Arizona.

Masch, F. D., and Denny, K. J., 1966, Grain size distribution and its effect on the permeability of unconsolidated sands: Water Resources Res., v. 2, p. 665–677.

Matthess, G., 1982, The Properties of Groundwater: New York, John Wiley, 406 p.

———, and Pekdeger, A., 1985, Survival and transport of pathogenic bacteria and viruses in ground water. Ground Water Quality, eds. C. H. Ward, W. Giger, and P. L. McCarty: New York, John Wiley, p. 472–482.

Matthews, C. S., and Russel, D. G., 1967, Pressure buildup analysis. Pressure Buildup and Flow Test in Wells. Monograph 1: New York, Soc. Petrol. Engrs., Amer. Inst. Mech. Engrs.

Maxwell, J. C., 1964, Influence of depth, temperature and geologic age on porosity of quartzose sandstone: Amer. Assoc. Petrol. Geol. Bull., v. 48, p. 697–709.

McBride, M. S., and H. O. Pfannkuch, 1975, The distribution of seepage within lake beds: J. Res., U.S. Geol. Survey, v. 3, p. 505–512.

McCarty, P. L., M. Reinhard, and B. E. Rittmann, 1981, Trace organics in groundwater. Environ. Sci. Technol., v. 15, no. 1, p. 40–51.

_____, Rittman, B. E., and Bouwer, E. J., 1984, Microbiological processes affecting chemical transformations in groundwater. Groundwater Pollution Microbiology, eds. G. Bitton and C. P. Gerba: New York, John Wiley, p. 89–115.

McCracken, D. D., and Dorn, W. S., 1964, Numerical Methods and Fortran Programming with Applications in Engineering and Science: New York, John Wiley.

McDonald, M. G., and Harbaugh, A. W., 1984, A modular three dimensional finite difference groundwater flow model: U.S. Geol. Surv., 528 p.

McDowell-Boyer, L. M., Hunt, J. R., and Sitar, N., 1986, Particle transport through porous media: Water Resources Res., v. 22, no. 13, p. 1901–1921.

McDuff, R. E., and Morel, F. M., 1973, REDEQL2: Description and use of the chemical equilibrium program: Pasadena, EQ 73-02, California Inst. of Tech.

McGarr A., and Gay, N. C., 1978, State of stress in the earth's crust: Ann. Reviews of Earth and Planetary Sci., v. 6, p. 405–436.

McGowen, J. H., and Groat, C. G., 1971, Van Horn sandstone, West Texas: An alluvial fan model for mineral exploration: Texas Bur. Econ. Geol. Rept. Inv., v. 72, Austin.

McGuinness, C. L., 1963, The role of groundwater in the national water situation: U.S. Geol. Survey Water Supply Papers, 1800.

McKelvey, J. S., and Milne, I. H., 1962, The flow of salt solutions through compacted clay: Clays and Clay Min., v. 9, p. 248–259.

Mclelwain, T. A., and Reades, D. W., 1985, Status Report: The use of engineered covers at waste disposal sites. Second Annual Canadian/American Conference on Hydrology, Proc.: Dublin, Ohio, National Water Well Assoc., p. 176–182.

Mead, D. W., 1919, Hydrology: New York, McGraw-Hill.

Meinzer, O. E., 1917, Geology and water resources of Big Smokey, Clayton, and Alkali Springs Valleys, Nevada: U.S. Geol. Survey Water Supply Papers 423.

_____, 1922, Map of pleistocene lakes of the Basin and Range Province and its significance: Geol. Soc. Amer. Bull., v. 33, p. 541–552.

_____, 1923, Outline of groundwater in hydrology with definitions. U.S. Geol. Surv. Water Supply Papers, 494.

_____, 1923, The occurrence of groundwater in the United States with a discussion of principles: U.S. Geol. Surv. Water Supply Papers, 489.

_____, 1927, Plants as indicators of groundwater: U.S. Geol. Surv. Water Supply Papers, 577.

_____, 1928, Compressibility and elasticity of artesian aquifers: Econ. Geol., v. 23, p. 263–291.

_____, 1932, Outline of methods for estimating groundwater supplies. U.S. Geol. Surv. Water Supply Papers, 638-C, p. 94–144.

_____, 1939, Groundwater in the United States: U.S. Geol. Surv. Water Supply Papers 836-D, p. 157–232.

_____, 1942 ed., Hydrology: New York, McGraw Hill.

————, and Hard, H. H., 1925, The artesian water supply of the Dakota sandstone in North Dakota, with special reference to the Edgeley Quadrangle: U.S. Geol. Surv. Water Supply Papers, 520-E, p. 73–95.

————, and Stearns, N. D., 1928, A study of groundwater in the Pomperaug Basin, Conn., with special reference to intake and discharge: U.S. Geol. Surv. Water Supply Papers 597, p. 73–146.

Melnyk, T. W., Walton, F. B., and Johnson, H. L., 1983, High-level Waste Glass Field Burial Tests and CRNL: The Effect of Geochemical Kinetics on the Release and Migration of Fission Products in a Sandy Aquifer. Rep. AECL-6836: Chalk River, Ontario, Atom. Energy Can., Ltd.

Mendoza, C. A., and Frind, E. O., 1990a, Advective-dispersive transport of dense organic vapors in the unsaturated zone. 2, Sensitivity analysis: Water Resources Res., v. 26, no. 3, p. 388–398.

————, and Frind, E. O., 1990b, Advective-dispersive transport of dense organic vapors in the unsaturated zone. 1, Model development: Water Resources Res., v. 26, no. 3, p. 379–387.

————, and McAlary, T. A., 1990, Modeling of ground-water contamination caused by organic solvent vapors: Ground Water, v. 28, no. 2, p. 199–206.

Mercado, A., 1966, Recharge and mixing tests at Yaune 20 well field. Underground Water Storage Study Tech. Report 12, Pub. 611: Tel Aviv, TAHAL—Water Planning for Israel Ltd.

Mercer, J. W., and Faust, C. R., 1981, Groundwater Modeling: Dublin, Ohio, National Water Well Assoc. Pub.

————, Pinder, G. F., and Donaldson, I. G., 1975, A Galerkin finite element analysis of the hydrothermal system at Wairakei: New Zealand J. Geophys. Res. v. 80, p. 2608–2621.

Meyboom, P., 1960, Geology and Groundwater Resources of the Milk River Sandstone in Southern Alberta, Mem. 2: Edmonton, Alberta, Alta. Res. Counc.

————, 1961, Estimating groundwater recharge from stream hydrographs: J. Geophys. Res., v. 66, p. 1203–1214.

————, 1962, Patterns of groundwater flow in the Prairie environment: Third Can. Hydrol. Symp., p. 5–33.

————, 1966a, Unsteady groundwater flow near a willow ring in hummocky moraine: J. Hydrol., v. 4, p. 38–62.

————, 1966b, Groundwater studies in the Assiniboine River drainage basin. Part 1: The evaluation of a flow system in south-central Saskatchewan: Geological Survey of Canada, Bull. 139, 65 p.

————, 1967, Mass transfer studies to determine the groundwater regime of permanent lakes in hummocky moraine of western Canada: J. Hydrol., v. 5, p. 117–142.

Meyer, R. R., 1963, A chart relating well diameter, specific capacity, and the coefficients of transmissibility and storage. Methods of Determining Permeability, Transmissibility, and Drawdown: U.S. Geol. Surv. Water Supply Papers, 1536-I.

Mifflin, M. D., 1968, Delineation of groundwater flow systems in Nevada: Reno Desert Research Inst. Tech. Rept. Ser. H-W, no. 4.

Miller, C. W., and Benson, L. V., 1983, Simulation of solute transport in a chemically reactive heterogenous system: Model development and application: Water Resources Res., v. 19, no. 2, p. 381–391.

Minear, R. A., and Keith, L. H., eds., 1984, Water Analysis. V. III, Organic Species: Orlando, Fla., Academic Press, 456 p.

Mitchell, J. K., Greenberg, J. A., and Witherspoon, P. A., 1973, Chemico-osmotic effects in fine grained soils: Amer. Soc. Civi. Eng. J., Soil Mech. Found. Div. 99, p. 307–322.

Mitsdarffer, A. R., 1985, Hydrodynamics of the Mission Canyon formation in the Billings Nose area, North Dakota, Master of Science thesis. College Station, Geology Dept, Texas A&M University.

Molinari, J., and Peaudecerf, P., 1977, Essais conjoints en laboratoire et sur le terrain en vue d'une approche simplifiee de la prevision des propagations de substances miscibles dans les aquiferes reels. Symposium on Hydrodynamic Diffusion and Dispersion in Porous Media, Proc. A.I.R.H., Italy, p. 89–102.

Molz, F. J., Parr, A. D., and Anderson, P. F., 1981, Thermal energy storage in a confined aquifer, second cycle: Water Resources Res., v. 17, p. 641–645.

_____, M. A. Widdowson, and Benefield, L. D., 1986, Simulation of microbial growth dynamics coupled to nutrient and oxygen transport in porous media: Water Resources Res., v. 22, no. 8, p. 1207–1216.

Montazer, P., and Wilson, W., 1984, Conceptual hydrologic model of flow in the unsaturated zone, Yucca Mountain, Nevada: U.S. Geol. Survey Water Resources Inv., Report 84-4345.

Moody, J. B., 1982, Radionuclide migration/retardation: Research and development technology status report: Office of Nuclear Waste Isolation, Battelle Memorial Inst., ONWI-321, 61 p.

Mook, W. G., 1980, Carbon-14 in hydrogeological studies. Chapter 2 in Handbook of Environmental Isotope Geochemistry, Vol. 1, eds. P. Fritz and J. C. Fontes: Amsterdam, Elsevier, pp. 49–74.

Moran, S. R., Cherry, J. A., Fritz, P., Peterson, W. M., Somerville, M. H., Stancel, S. A., and Ulmer, J. H., 1978b, Geology, Groundwater Hydrology, and Hydrochemistry of a Proposed Surface Mine and Liquite Gasification Plant Site Near Dunn Center, North Dakota: North Dakota Geological Survey, Report of Investigation 61, 263 p.

Moran, S. R., Groenwold, G. H., and Cherry, J. A., 1978a, Geologic, Hydrologic, and Geochemical. Concepts and Techniques in Overburden Characterization for Mined-Land Reclamation: North Dakota Geological Survey, Report of Investigation No. 63, 152 p.

Morel, F. M. M., 1983, Principles of Aquatic Chemistry: New York, John Wiley, 446 p.

_____, and Morgan, J. J., 1972, A numerical method for computing equilbria in aqueous systems: Environ. Sci. Technol., v. 6, p. 58–67.

Moreno, L., Neretnieks, I., and Klockars, C. E., 1983, Evaluation of some tracer tests in the granitic rock at Finnsjön, KBS Technical Report, Stockholm, Sweden, TR 83–83.

_____, Neretnieks, I., and Eriksen, T., 1984, Analysis of some laboratory tracer runs in natural fissures. Manuscript for publication, March 1984.

Morgan, P. V., and others, 1981, A groundwater convective model for Rio Grande Rift geothermal systems: Geotherm. Resour. Council Trans., v. 5, p. 193–196.

Morganstern, N. R., 1970, The influence of groundwater on Stability. 1st Intern. Conf. on Stability in Open Pit Mining, Proc., Vancouver, B.C.

Morrey, J. R. and Shannon, D. W., 1981, Operator's manual for EQUILIB—A computer code for predicting mineral formation in geothermal brines, Vols. 1 and 2. Palo Alto, Calif., Electric Power Research Instit., Project RP 653-1.

Morrison, R. D., and Brewer, P. E., 1981, Airlift samples for zone-of-saturation monitoring: Ground Water Monitoring Review, v. 1, no. 1, p. 52–55.

Muckel, D. C., 1959, Replenishment of groundwater supplies by artificial means: U.S. Dept. Agric., Agric. Res. Service Tech. Bull. 1195.

Munn, M. J., 1909, The anticlinal and hydraulic theory of oil and gas accumulation: Econ. Geol., v. 4, p. 509–529.

Murty, V. V. N., and Scott, V. H., 1977, Determination of transport model parameters in groundwater aquifers: Water Resources Res., v. 13, no. 6, p. 941–947.

Muskat, M., 1937, The Flow of Homogeneous Fluids Through Porous Media: New York, McGraw-Hill.

———, and Meres, M. W., 1936, The flow of heterogeneous fluids through porous media: Physics, v. 7, p. 346–363.

Narasimhan, T. N., White, A. F., and Tokunaga, T., 1986, Groundwater contamination from an inactive uranium mill tailings pile. 2, Application of a dynamic mixing model: Water Resources Res., v. 22, no. 13, p. 1820–1834.

Nathenson, M., and Muffler, L. J. P., 1975, Geothermal resources in hydrothermal convection systems and conduction dominated areas. Assessment of Geothermal Resources of the United States—1975, U.S. Geol. Surv. Circ., 726, p. 104–121.

National Academy of Science, 1957, The Disposal of Radioactive Wastes on Land: Washington, D.C., NAS.

Neely, W. B., 1985, Hydrolysis. Environmental Exposure from Chemicals, Vol. I, eds. W. B. Neely and G. E. Blau. Boca Raton, Fla., CRC Press, p. 157–173.

Nelson, R. W., 1978, Evaluating the environmental consequences of groundwater contamination. 2, Obtaining location/arrival time and location/outflow quality distributions for steady flow systems: Water Resources Res., v. 14, no. 3, p. 416–428.

Neretnieks, I., 1985, Transport in fractured rocks. Proc. of Hydrogeology of Rocks of Low Permeability. Tucson, Arizona, International Association of Hydrogeologists, p. 306.

———, Eriksen, T., and Tähtinen, P., 1982, Tracer movement is a single fissure in granitic rock: Some experimental results and their interpretation: Water Resources Res., v. 18, p. 849.

Neuman, S. P., 1972, Theory of flow in unconfined aquifers considering delayed response of the water table: Water Resources Res., v. 8, p. 1031–1045.

———, 1974, Effect of partial penetration on flow in unconfined aquifers considering delayed gravity response: Water Resources Res., v. 10, p. 303–312.

———, 1975, Analysis of pumping test data from anisotropic unconfined aquifers considering delayed gravity response: Water Resources Res., v. 11, p. 329–342.

———, and Witherspoon, P. A., 1969a, Theory of flow in confined two aquifer system: Water Resources Res., v. 5, p. 803–816.

———, and Witherspoon, P. A., 1969b, Applicability of current theories of flow in leaky aquifers: Water Resources Res., v. 5, p. 817–829.

———, and Witherspoon, P. A., 1972, Field determination of hydraulic properties of leaky multiple aquifer systems: Water Resources Res., v. 8, p. 1284–1298.

Neuzil, C. E., 1982, On conducting the modified slug test in tight formation: Water Resources Res., v. 18, p. 439–441.

———, 1986, Groundwater in low permeability environments: Water Resources Res., v. 22, p. 1163–1195.

———, and Pollock, D. W., 1983, Erosional unloading and fluid pressures in hydraulically tight rocks. J. Geol., v. 9, p. 179–193.

Nield, D. A., 1968, Onset of thermohaline convection in a porous medium: Water Resources Res., v. 4, p. 1553–1560.

Nobel, E. A., 1963, Formation of ore deposits by water of compaction: Econ. Geol., v. 58, p. 1145–1156.

Nordstrom, D. K., and Munoz, J. L., 1986, Geochemical Thermodynamics. Palo Alto, Calif., Blackwell Scientific Publications, 477 p.

_____, Plummer, L. N., Wigley, T. M. L., Wolery, T. J., Ball, J. W., Jenne, E. A., Bassett, R. L., Crerar, D. A., Florence, T. M., Fritz, B., Hoffman, M., Holdren, G. R., Lafon, G. M., Mattigod, S. V., McDuff, R. E., Morel, F., Reddy, M. M., Sposito, G., and Thrailkill, J., 1979, A comparison of computerized chemical models for equilibrium calculations in aqueous systems. Chemical Modeling in Aqueous Systems, ed. E. A. Jenne. Washington, D.C., Chemical Society, Symposium Series 93, p. 857–892.

Norton, D., 1978, Sourcelines, source regions, and pathlines in hydrothermal systems related to cooling plutons. Econ. Geol., v. 73, p. 21–28.

_____, and Knapp, R., 1977, Transport phenomena in hydrothermal systems: Nature of porosity: Amer. J. Sci., v. 27, p. 913–936.

Nur, A., 1972, Dilatancy, pore fluids and premonitory variations of ts/tp travel times. Bull. Seismol. Soc. Amer., v. 62, p. 1217–1222.

O'Connor, M. J., and Bouchout, L. W., 1983, Gasoline spills in urban areas: a comparison of two case histories. Seminar on Groundwater and Petroleum Hydrocarbons, Protection, Detection, Restoration, Ottawa, Proc.: Petroleum Assoc. for Conservation of the Canadian Environment, p. VIII-1–VIII-34.

_____, Agar, J. G., and King, R. D., 1984a, Practical experience in the management of hydrocarbon vapors in the subsurface. NWWA/API Conference on Petroleum Hydrocarbons and Organic Chemicals in Ground Water—Prevention, Detection, and Restoration, Proc.: Dublin, Ohio, National Water Well Association, p. 519–533.

_____, Wofford, A. M., and Ray, S. K., 1984b, Recovery of subsurface hydrocarbons at an asphalt plant: Results of a five year monitoring program. NWWA/API Conference on Petroleum Hydrocarbons and Organic Chemicals in Ground Water—Prevention, Detection, and Restoration, Proc.: Dublin, Ohio, National Water Well Assoc., p. 359–376.

Office of Technology Assessment, 1982, Summary: Managing Commercial High Level Radioactive Work: Washington, D.C., U.S. Congress.

_____, 1984, Protecting the Nation's Groundwater from Contamination. Office of Technology Assessment, OTA-0-233: Washington, D.C., OTA, 244 pp.

Ogata, A., and Banks, R. B., 1961, A solution of the differential equation of longitudinal dispersion in porous media: U.S. Geol. Surv. Prof. Papers, 411-A.

Ohle, E. L., 1959, Some consideration in determining the origin of ore deposits of the Mississippi Valley Type, Part I: Econ. Geol., v. 54, p. 769–789.

Oliver, J., 1986, Fluids expelled tectonically from orogenic belts: Their role in hydrocarbon migration and other geologic phenomena. Geology, v. 14, p. 99–102.

Olsen, H., 1972, Liquid movement through kaolinite under hydraulic, electric, and osmotic gradients: Amer. Assoc. Petrol. Geol. Bull., v. 56, 2022–2028.

Ortoleva, P., Merino, E., Moore, C., and Chadam, J., 1987a, Geochemical self-organization. I, Reaction-transport feedbacks and modeling approach: Am. J. Sci., v. 287, p. 979–1007.

———, Chadam, J., Marino, E., and Sen, A., 1987b, Geochemical self-organization. II, The reactive-infiltration instability: Am. J. Sci., v. 287, p. 1008–1040.

Ozbiligin, M. M., and Powers, M. A., 1984, Hydrodynamic isolation in hazardous waste containment. Fourth National Symposium and Exposition on Aquifer Restoration and Groundwater Monitoring, Proc.: Dublin, Ohio, National Water Well Assoc., p. 44–49.

Paige, S., 1938, Effect of a sea-level canal on the groundwater level of Florida—A reply: Econ. Geol., v. 33, p. 647–665.

———, 1936, Effect of a sea-level canal on the groundwater level of Florida: Econ. Geol. v. 31, p. 537–570.

Palciauskas, V. V., and Domenico, P. A., 1989, Fluid pressure in deforming porous rocks: Water Resources Res., v. 25, p. 203–213.

———, and Domenico, P. A., 1976, Solution chemistry, mass transport, and the approach to chemical equilibrium in porous carbonate rocks and sediments: Geol. Soc. Amer. Bull., v. 87, p. 207–214.

———, and Domenico, P. A., 1980, Microfracture development in compacting sediments: Relation to hydrocarbon maturation: Amer. Assoc. Petrol. Geol. Bull, v. 64, p. 927–937.

———, and Domenico, P. A., 1982, Characterization of drained and undrained response of thermally loaded repository rocks: Water Resources Res., v. 18, p. 281–290.

Palmer, C. D., and Cherry, J. A., 1984, Geochemical evolution of groundwater in sequences of sedimentary rocks: J. Hydrol., v. 75, p. 27–65.

Pankow, J. F., Johnson, R. L., Houck, J. E., Brillante, S. M., and Bryan, W. J., 1984, Migration of chlorophenolic compounds at the chemical waste disposal site at Alkali Lake, Oregon. 1, Site description and groundwater flow: Ground Water, v. 22, no. 5, p. 593–601.

Papadopulos, I. S., 1965, Nonsteady flow of a well in an infinite anisotropic aquifer. Dubrovnik Symp. on Hydrol. of Fractured Rock, Proc.: Internat. Assoc. Sci. Hydrol., p. 21–31.

———, Bredehoeft, J. D., and Cooper, H. H., 1973, On the analysis of slug test data: Water Resources Res., v. 9, p. 1087–1089.

———, and Cooper, H. H., 1967, Drawdown in a well of large diameter: Water Resources Res., v. 3, p. 241–244.

———, and others, 1975, Assessment of onshore geopressured-geothermal sources of the United States, in Assessment of Geothermal Resources in the United States—1975. U.S. Geol. Surv. Circ. 726, p. 125–142.

Parker, G. G., and Springfield, V. T., 1950, Effects of earthquakes, rains, tides, winds, and atmospheric pressure changes on the water in geologic formations of southern Florida. Econ. Geol., v. 45, p. 441–460.

Parkhurst, D. L., Plummer, L. M., and Thorstenson, D. C., 1982, BALANCE—A Computer Program for Calculation of Chemical Mass Balance, U.S. Geol. Surv. Water-Resour. Invest. 82-14: Springfield, Va., NTIS Tech. Report PB82-255902, 33 p.

———, Thorstenson, D. C., and Plummer, L. N., 1980, PHREEQE—A Computer Program for Geochemical Calculation: U.S. Geol. Surv. Water Resources Invest. Report 80-96: 210 p.

Parsons, M. L., 1970, Groundwater thermal regime in a glacial complex: Water Resources Res., v. 6, p. 1701–1720.

Patton, F. D., and Hendron, A. J., Jr., 1974, General report on mass movements. Second Internat. Congress, San Paulo, Brazil, Proc.: Internat. Assoc. Eng. Geol., v. 2, p. V-GR.1–V-GR.57.

Peaceman, D. W., and Rachford, H. H., 1955, The numerical solution of parabolic and elliptical difference equations: J. Soc. Ind. Appl. Math., v. 3, p. 28–41.

Pearson, F. J., Jr., and Hanshaw, B. B., 1970, Sources of dissolved carbonate species in groundwater and their effects on carbon-14 dating. Isotope Hydrology: Vienna, Internat. Atomic Energy Agency, p. 271–286.

Perdue, E. M., and Lytle, C. R., 1983, Distribution model for binding of protons and metal ions by humic substances: Environ. Sci. Technol., v. 17, p. 654–660.

Perkins, T. K., and Johnston, O. C., 1963, A review of diffusion and dispersion in porous media: J. Soc. Petrol. Eng., v. 3, p. 70–83.

Peter, K. L., 1984, Hydrochemistry of lower Cretaceous sandstone aquifers, Northern Great Plains. Geohydrology Dakota Aquifer Sympsoium, Proc., eds. D. G. Jorgenson, and D. C. Signor: Worthington, Ohio, Water Well Journal Publ., p. 163–174.

Peters, R. R., and Klavetter, E. A., 1988, A continuum model for water movement in an unsaturated fractured rock mass: Water Resources Res., v. 24, p. 416–430.

Pfannkuch, H. O., 1962, Contribution à l'étude des deplacement de fluides miscible dans un milieu poreux. Rev. Inst. Fr. Petrol., v. 18, no. 2, p. 215–270.

Phillips, F. M., Bentley, H. W., Davis, S. N., Elmore, D., and Swanick, G., 1986, Chlorine 36 dating of very old groundwater. 2, Milk River Aquifer, Alberta: Water Resources Res., v. 22, no. 13, p. 2003–2016.

Pickens, J. F., Jackson, R. E., Inch, K. J., and Merritt, W. F., 1981, Measurement of distribution coefficients using a radial injection dual-tracer test: Water Resources Res., v. 17, no. 3, p. 529–544.

Pinder, G. F., 1973, A Galerkin finite-element simulation of groundwater contamination on Long Island, New York: Water Resources Res., v. 9, no. 6, p. 1657–1664.

_____, 1979, State of the art review of geothermal reservoir engineering, Rept. LBL-9093. Berkeley, Calif., Lawrence Berkeley Lab.

_____, and Bredehoeft, J. D., 1968, Application of digital computer for aquifer evaluation: Water Resources Res., v. 4, p. 1069–1093.

_____, and Cooper, H. H., 1970, A numerical technique for calculating the transient position of the salt-water front. Water Resources Res., v. 6, no. 3, p. 875–882.

_____, and Frind, E. O., 1972, Application of Galerkin's procedure to aquifer analysis: Water Resources Res., v. 8, p. 108–120.

_____, and Gray, W. G., 1977, Finite Element Simulation in Surface and Subsurface Hydrology: New York, Academic Press.

Piper, A. M., 1944, A graphic procedure in the geochemical interpretation of water analysis. Trans. Amer. Geophys. Union, v. 25, p. 914–923.

Pitzer, K. S., and Kim, J. J., 1974, Thermodynamics of electrolytes. 4, activity and osmotic coefficients for mixed electrolytes: J. Am. Chem. Soc., v. 96, p. 5701–5707.

Plumb, R. H., and Pitchford, A. M., 1985, Volatile organic scans: Implications for ground water monitoring. NWWA/API Conference on Petroleum Hydrocarbons and Organic Chemicals in Ground Water—Prevention, Detection, and Restoration, Proc.: Dublin, Ohio, National Water Well Assoc., p. 207–223.

Plummer, L. N., 1984, Geochemical modeling: A comparison of forward and inverse methods. First Canadian/American Conference on Hydrogeology, eds. B. Hitchon and E. I. Wallick: Dublin, Ohio, National Water Well Assoc., p. 149–177.

———, Jones, B. F., and Truesdell, A. H., 1976, WATEQF—A Fortran IV version of WATEQ, a Computer Program for Calculating Chemical Equilibrium of Natural Waters: U.S. Geol. Surv. Water Resources Investigations, 76–13, 61 p.

———, Parkhurst, D. L., and Thorstenson, D. C., 1983, Development of reaction models for groundwater systems. Geochim. Cosmochim. Acta, v. 47, p. 665–685.

———, Parkhust, D. L., and Kosuir, D. R., 1975, MIX2— A computer program for modelling chemical reactions in natural waters: U.S. Geol. Surv. Water Resources Investigations, no. 75–76, p. 61–75.

Poland, J. F., 1961, The coefficient of storage in a region of major subsidence caused by compaction of an aquifer: U.S. Geol. Prof. Papers, 424-B p. 52–54.

———, and Davis, G. H., 1956, Subsidence of the land surface in Tulare-Wasco (Delano) and Los Banos-Kettleman City area, San Joaquin Valley, California: Trans. Amer. Geophys. Union, v. 37, p. 287–296.

———, and Davis, G. H., 1969, Land subsidence due to withdrawal of fluids: Geol. Soc. Amer, Rev. Eng. Geol., v. 2, p. 817–829.

———, and others, 1975, Land subsidence in the San Joaquin Valley, California, as of 1972: U.S. Geol. Survey Prof. Papers, 437-H.

Powell, J. D., 1984, Geochemical models of the relation between water quality and mineralogy in coal-producing strata of southwestern Virginia. First Canadian/American Conference on Hydrogeology, eds. B. Hitchon and E. I. Wallick: Dublin, Ohio, National Water Well Assoc., p. 103–112.

Powers, M. C., 1967, Fluid-release mechanisms in compacting marine mudrocks and their importance in oil exploration: Amer. Assoc. Petrol. Geol. Bull., v. 51, p. 1240–1254.

Pratts, M., 1966, The effect of horizontal flow on thermally induced convection currents in porous mediums: J. Geophys. Res., v. 71, p. 4835–4838.

———, 1970, A method for determining the net vertical permeability near a well from in-situ measurements: Trans. Amer. Inst. Mining Engrs., v. 249, p. 637–643.

Prichett, W. C., 1980, Physical properties of shales and possible origin of high pressures: Soc. Pet. Eng. J., v. 20, p. 341–348.

Prickett, T. A., 1965, Type curve solution to aquifer tests under water table conditions: Ground Water, v. 3, p. 5–14.

———, and Lonnquist, C. G., 1968, Comparison between analog and digital simulation techniques for aquifer evaluation. Use of Analog and Digital Computers in Hydrology Symp., Tucson, Ariz., p. 625–634.

———, Naymik, T. G., and Lonquist, C. G., 1981, A Random-Walk Solute Transport Model for Selected Groundwater Quality Evaluations. Illinois State Water Surv. Bull., v. 65, Champaign, 103 p.

Pytkowicz, R. P., 1983a, Equilibria, Nonequilibria & Natural Waters, Vol. 1: New York, John Wiley, 351 p.

———, 1983b, Equilibria, Nonequilibria & Natural Waters, Volume 2: New York, John Wiley, 353 p.

Quinlan, J. F., and Ewers, R. O., 1985, Ground water flow in limestone terranes: Strategy, rationale and procedure for reliable, efficient monitoring of ground water quality in Karst areas. Fifth National Symposium and Exposition on Aquifer Restoration and Ground Water Monitoring, Proc.: Fifth National, Dublin, Ohio, National Water Well Assoc., p. 197–234.

Quon, D., and others, 1965, A stable explicit computationally efficient method for solving two-dimensional mathematical models of petroleum reservoirs: J. Can. Petrol. Tech., v. 4, p. 53–58.

———, and others, 1966, Application of alternating direction explicit procedure to two-dimensional natural gas reservoirs: J. Petrol. Technol. Soc. Petrol. Engrs. Amer. Inst. Mining Engrs., p. 137–142.

Ragone, S. E., ed., 1988, U. S. Geological Survey Program on Toxic Waste—Ground Water Contamination. Second Technical Meeting, Cape Code, Mass., October 21–25, 1985, Proc., U.S. Geol. Surv. Open-File Rept., 86-481.

Rainwater, F. H., and Thatcher, L. L., 1960, Methods for Collection and Analysis of Water Samples: U.S. Geol. Surv. Water Supply Papers, 1454, 301 p.

Raleigh, C. B., 1971, Earthquakes and fluid injection: Amer. Assoc. Petrol. Geol. Memoirs, no. 18, p. 273–279.

———, Healy, J. H., and Bredehoeft, J. D., 1972, Faulting and crustal stress at Rangely, Colorado: Geophys. Union Monog. Series 16, p. 275–284.

———, Healy, J. H., and Bredehoeft, J. D., 1976, An experiment in earthquake control at Rangeley, Colorado: Science, v. 191, p. 1230–1236.

Ranganathan, V., and Hanor, J. S., 1987, A numerical model for the formation of saline waters due to diffusion of dissolved NaCl in subsiding sedimentary basins with evaporites: J. Hydrol., v. 92, p. 97–120.

Rantz, S. E., 1961, Effect of tunnel construction on flow of springs in the Tecolote tunnel area of Santa Barbara County, California: U.S. Geol. Survey. Prof. Papers, 424-C, Art. 227.

Rasmussen, T. C., Blanford, J. H., and Sheets, P. J., 1989, Characterization of unsaturated fractured tuff at the Apache Leap Site: Comparison of laboratory/field, matrix/fracture and water/air data. Paper presented at Nuclear Waste Isolation in the Unsaturated Zone, joint sponsorship by the American Nuclear Society and the Geological Society of America, Las Vegas, Nevada.

Rastogi, R. P., and others, 1964, Cross-phenomenalogic coefficients. Part I, Studies on thermo osmosis: Trans. Far. Soc., v. 60, p. 7–12, p. 1386–1390.

Raven, K., and Novakowski, K. S., 1984, Field investigation of the solute-transport properties of fractures in monzonific gneiss. Internat. Symposium on Groundwater Resources Utilization and Contaminant Hydrogeology, Vol. II: Pinawa, Manitoba, Atomic Energy of Canada Ltd., p. 507–516.

Reardon, E. J., and Fritz, P., 1978, Computer modelling of ground water ^{13}C and ^{14}C isotope compositions: J. Hydrol., v. 36, p. 201–224.

Reddell, D. L., and Sunada, D. K., 1970, Numerical simulation of dispersion in groundwater aquifers: Ft. Collins, Colorado State University, Hydrol. Paper, 41, 79 p.

Regan, L. J., and Hughes, A. W., 1949, Fractured reservoirs of Santa Maria District, California. Amer. Assoc. Petrol. Geol. Bull., v. 33, p. 32–51.

Rege, S. D., and Folger, H. S., 1989, Competition among flow, dissolution, and precipitation in porous media: AIChE J., v. 35, p. 1177–1185.

Reinson, C. E., 1984, Barrier island and associated strand plain systems. Facies Models, ed. R. G. Walker: GeoSciences Canada Reprint Ser. 1, p. 119–140.

Reisinger, H. J., Burris, D. R., Cessar, L. R., and McCleary, G. D., 1987, Factors affecting the utility of soil vapor assessment data. First Outdoor Action Conference on Aquifer Restoration, Ground Water Monitoring and Geophysical Methods, Proc.: Dublin, Ohio, National Water Well Assoc., p. 425–435.

Relyea, J. F., 1982, Theoretical and experimental considerations for the use of the column method for determining retardation factors. Radioactive Waste Manage. Nucl. Fuel Cycle, v. 3, p. 151–166.

Remson, I., Hornberger, G. M., and Molz, F. J., 1971, Numerical Methods in Subsurface Hydrology: New York, Wiley-Interscience.

Renner, J. L., White, D. E., and Williams, D. L., 1975, Hydrothermal convection systems, in Assessment of Geothermal Resources of the United States—1975: U.S. Geol. Survey Circ. 726, p. 47–56.

Richards, L. A., 1931, Capillary conduction of liquids through porous mediums: Physics, v. 1, p. 318–333.

Rittmann, B. E., and McCarty, P. L., 1980a, Model of steady-state biofilm kinetics: Biotech. Bioeng., v. 22, p. 2343–2357.

———, and McCarty, P. L., 1980b, Evaluation of steady-state biofilm kinetics: Biotech. Bioeng., v. 22, p. 2359–2373.

Robbins, G. A., 1983, Determining dispersion parameters to predict groundwater contamination, Ph.D. dissertation. College Station, Texas A&M University, 226 p.

Roberts, P. V., Goltz, M. N., and Mackay, D. M., 1986, A natural gradient experiment on solute transport in a sand aquifer. 3, Retardation estimates and mass balances for organic solutes: Water Resources Res., v. 22, no. 13, p. 2047–2058.

Robertson, W. D., and Cherry, J. A., 1989, Tritium as an indicator of recharge and dispersion in a groundwater system in central Ontario: Water Resources Res., v. 25, no. 6, p. 1097–1109.

Robinson, P. C., 1982, Connectivity of fracture systems: A percolation theory approach: Theor. Phys. Div., AERE Harewll, HL82/960.

Robinson, T. W., 1939, Earth tides shown by fluctuations of water levels in wells in New Mexico and Iowa: Trans. Amer. Geophy Union, v. 20, p. 656–666.

Rogoshewski, P., Bryson, H., and Wagner, K., 1983, Remedial Action Technology for Waste Disposal Sites. Pollution Technology Review No. 101: Park Ridge, N.J., Noyes Data Corp., 498 p.

Rojstaczer, S., and Bredehoeft, J. D., 1988, Groundwater and fault strength. Hydrogeology, The Geology of North America, ed. W. Back, J. J. Rosenshein, and P. R. Seaber: Geological Society of America, p. 447–460.

Romm, E. S., 1966, Flow characteristics of fractured rocks (in Russian): Moscow, Nedra.

Roth, J. P., 1969, Earthquakes and reservoir loading. Fourth World Conf. on Earthquake Engineering, Santiago, Chile, Proc.

Royster, D. L., 1979, Landslide remedial measures. Bull. Assoc. Eng. Geol., v. 16, p. 301–352.

Rubin, J., 1968, Theoretical analysis of two dimensional transient flow of water in unsaturated and partly saturated soils: Soil Sci. Amer. Proc., v. 32, p. 607–615.

———, and James, R. V., 1973, Dispersion affected transport of reacting solutes in saturated porous media. Galerkin method applied to equilibrium controlled exchange in unidirectional steady water flow: Water Resources Res., v. 9, no. 5, p. 1332–1356.

———, and Steinhardt, R., 1963, Soil water relations during rain infiltration. I: Theory: Soil Sci. Amer. Proc., v. 27, p. 246–251.

———, and Steinhardt, R., and Reinsiger, P., 1964, Soil-water relations during rain infiltration. II, Moisture content profiles during rains of low intensity. Soil Sci. Soc. Amer. Proc., v. 28, p. 1–5.

Runnells, D., and Larson, J., 1986, A laboratory study of electromigration as a possible field technique for the removal of contaminants from ground water: Ground Water Monitoring Rev., v. 6, no. 3, p. 85–91.

Russell, W. L., 1972, Pressure-depth relations in Appalachian region: Amer. Assoc. Petrol. Geol. Bull., v. 56, p. 528–536.

Rust, B. R., and Koster, E. H., 1984, Coarse alluvial deposits. Facies Models, ed. R. G. Walker: Geoscience Canada Reprint Series 1.

Rutter, E. H., 1983, Pressure solution in nature, theory and experiment: J. Geol. Soc., v. 140, p. 725–740.

Sadovsky, M. A., and others, 1969, The processes preceding strong earthquakes in some regions of Middle Asia: Tectonophysics, v. 23, p. 247–255.

Sass, J. W., and H. Lachenbrach, 1982, Preliminary interpretation of thermal data from the Nevada Test Site: U.S. Geol. Surv. Open-File Rept. 82-973, Denver.

———, and others, 1976, A new heat flow contour map of the conterminous United States: U.S. Geol. Surv. Open-File Rept. 76-756, Denver.

Sauty, J. P., and others, 1982a, Sensible energy storage in aquifers. I, Theoretical study: Water Resources Res., v. 18, p. 245–252.

———, and others, 1982b, Sensible energy storage in aquifers. II, Field experiments and comparison with theoretical results: Water Resources Res., v. 18, p. 253–265.

Savin, S. M., 1980, Oxygen and hydrogen isotope effects in low-temperature mineral-water interactions. Chapter 8 in Handbook of Environmental Isotope Geochemistry, Vol. 1, eds. P. Fritz and J. C. Fontes: New York, Elsevier, p. 283–327.

Scalf, M. R., McNabb, J. F., Dunlap, W. J., Cosby, R. L., and Fryberger, J., 1981, Manual of Ground Water Sampling Procedures: U.S. Environmental Protection Agency, Reports EPA 660/2-81-160, 93 p.

Schiff, L., 1955, The status of water spreading for groundwater replenishment: Trans. Amer. Geophys. Union, v. 36, p. 1009–1020.

Schleicher, D., 1975, A model for earthquakes near Palisades Reservoir, Southeast Idaho: U.S. Geol. Surv. J. Res., v. 3, p. 393–400.

Schmidt, V., and McDonald, D. A., 1977, The role of secondary porosity in the course of sandstone diagenesis. Aspects of Diagenesis, eds. P. A. Scholle, and P. R. Schluger: Soc. Econ. Paleontol. and Mineral. Spec. Pub. 26, p. 175–225.

Schmorak, S., 1967, Saltwater encroachment in the Coastal Plain of Israel. Internat. Assoc. Sci. Hydrol. Symp. Haifa, Pub. 72, p. 305–318.

———, and Mercado, A., 1969, Upconing of freshwater-seawater interface below pumping wells: Water Resources Res., v. 5, p. 1290–1311.

Schneider, R., 1964, Relation of temperature distribution to groundwater movement in carbonate rocks of Central Israel. Geol. Soc. Amer. Bull., v. 75, p. 209–216.

Scholz, C. H., Sykes, L. R., and Aggerwal, Y. P., 1973, Earthquake prediction. A physical basis: Science, v. 181, p. 803–809.

Schuler, R. W., Stallings, V. P., Smith, J. M., 1952, Heat and mass transfer in fixed bed reactors. Chem. Eng. Prog. Symp. Ser. 48, p. 19–30.

Schulz, D., Barvenik, M., and Ayres, J., 1984, Design of soil-bentonite backfill mix for the first Environmental Protection Agency Superfund Cutoff wall. Fourth

National Symposium and Exposition on Aquifer Restoration and Groundwater Monitoring, Proc.: Dublin, Ohio, National Water Well Assoc., p. 8–17.

Schwartz, F. W., 1975, On radioactive waste management: An analysis of the parameters controlling subsurface contaminant transfer. J. Hydrol., v. 27, p. 51–71.

———, 1988, Contaminant hydrogeology—dollars and sense: J. Hydrol., v. 100, p. 453–470.

———, and Crowe, A. S., 1980, A Deterministic-Probabilistic Model for Contaminant Transport, NUREG/CR-1609. Washington, D.C., Nuclear Regulatory Commission, 158 p.

———, and Domenico, P. A., 1973, Simulation of hydrochemical patterns in regional groundwater flow: Water Resources Res., v. 9, no. 3, p. 707–720.

———, and Gallup, D., 1978, Some factors controlling the major ion chemistry of small lakes: Examples from the Prairie Parkland of Canada: Hydrobiologia, v. 58, no. 1, p. 65–81.

———, McClymont, G. L., and Smith, L., 1985, On the role or mass transport modeling. Canadian/American Conference on Hydrogeology, Proc.: Dublin, Ohio, National Water Well Assoc., p. 2–12.

———, and Muehlenbachs, K., 1979, Isotope and ion geochemistry of groundwaters in the Milk River aquifer, Alberta: Water Resources Res., v. 15, no. 2, p. 259–268.

———, Muehlenbachs, K., and Chorley, D. W., 1982, Flow systems controls on the chemical evolution of groundwater. Developments in Water Science, eds. W. Back and R. Létolle, Symp on Geochem. of Groundwater. New York, Elsevier, p. 225–243.

Schwarzenbach, R. P., and Giger, W., 1985, Behavior and fate of halogenated hydrocarbons in ground water. Ground Water Quality, eds. C. H. Ward, W. Giger, and P. L. McCarty: New York, Wiley-Interscience, p. 446–471.

———, and J. Westall, 1981, Transport of nonpolar organic compounds from surface water to groundwater. Laboratory studies: Environ. Sci. Technol., v. 15, p. 1300–1367.

Schwille, F. W., 1984, Migration of organic fluids immiscible with water in the unsaturated zone. Pollutants in Porous Media, eds. B. Yaron, G. Dagan, and J. Goldschmid: New York, Springer-Verlag, p. 26–48.

———, 1985, Migration of organic fluids immiscible with water in the unsaturated and saturated zones. Second Canadian/American Conference on Hydrogeology, eds. B. Hitchon and M. Trudell: Dublin, Ohio, National Water Well Assoc., p. 31–35.

Scoffin, T., 1987, An Introduction to Carbonate Sediments and Rocks: New York, Chapman and Hall.

Scott, R. F., 1963, Principles of Soil Mechanics: Reading, Mass., Addison-Wesley.

Seaburn, G. E., 1970, Preliminary analysis of rate of movement of storm runoff through the zone of aeration beneath a recharge basin on Long Island, N.Y.: U.S. Geol. Surv. Prof. Papers, 700-B, p. 196–198.

Segol, G., and Pinder, G. F., 1976, Transient simulation of saltwater intrusion in southeastern Florida: Water Resources Res., v. 12, p. 65–70.

Senger, R. K., Fogg, G. E., and Kreitler, C., 1987, Effect of hydrostratigraphy and basin development on hydrodynamics of the Palo Duro Basin, Texas. Rept. of Invest. 165: Austin, Texas, Bureau Econ. Geol.

_____, and Fogg, G. E., 1987, Regional underpressuring in deep brine aquifers, Palo Duro Basin, Texas. 1, Effects of hydrostratigraphy and topography: Water Resources Res., v. 23, p. 1481–1493.

_____, Kreitler, C., and Fogg, G., 1987, Regional underpressuring in deep brine aquifers, Palo Duro Basin, Texas. 2, The effect of Cenozoic development: Water Resources Res., v. 23, p. 1494–1504.

Serafim, J. L., and del Campo, A., 1965, Interstitial pressure on rock foundations of dams. J. Soil Mech. and Found. Div., Amer. Assoc. Civil Engrs., v. 91, p. 65–85.

Shante, V. K. S., and Kirkpatrick, S., 1971, An introduction to percolation theory: Adv. Phys., v. 20, p. 352–357.

Sharp, J. C., and Maini, Y. N. T., 1972, Fundamental considerations on the hydraulic characteristics of joints in rock in percolation through fractured rock, ed. W. Wittke: Stuttgart, Intern. Soc. Rock Mech.

Sharp, J. M., and Domenico, P. A., 1976, Energy transport in thick sequences of compacting sediment: Geol. Soc. Amer. Bull., v. 87, p. 390–400.

_____, 1983, Permeability control on aquathermal pressure: Amer. Assoc. Petrol. Geol. Bull, v. 67, p. 2057–2061.

_____, 1978, Energy and momentum transport model of the Ouachita Basin and its possible impact on formation of economic mineral deposits: Econ. Geol., v. 73, p. 1057–1068.

_____, and Kyle, J. R., 1988, The role of groundwater processes in the formation of ore deposits. The Geology of North America, Hydrogeology eds. W. Back, J. Rosenshein, and P. Seaber, 0-2: Geol. Soc. Amer., p. 461–483.

Shi, Y., and Wang, C. Y., 1986, Pore pressure generation in sedimentary basins: Overloading versus aquathermal: J. Geophy. Res., v. 91, p. 2153–2162.

Siegel, D. I., 1987, Geochemical facies and mineral dissolution Bemidji, Minnesota, research site. U.S. Geol. Surv. Program on Toxic Waste—Ground-water contamination. Proceedings of the Third Technical Meeting, Pensacola, Florida, March 23–27, 1987: U.S. Geol. Surv. Open-File Rept. 87-109, p. C-13–C-15.

Signor, D. C., 1985, Groundwater sampling during artificial recharge: equipment, techniques, and data analyses. Artificial Recharge of Groundwater, ed. T. Asano: Stoneham, Mass., Butterworth, p. 151–202.

_____, and others, 1970, Annotated bibliography on artificial recharge on groundwater, 1955–1967: U.S. Geol. Surv. Water Supply Papers, 1990.

Simon, R. B., 1969, Seismicity of Colorado: Science, v. 165, p. 897–899.

Singer, E., and Wilhelm, R. H., 1950, Heat transfer in packed beds: Analytical solution and design methods: Chem. Eng. Prog., v. 46, p. 343–357.

Sipple, R. F., and Glover, E. D., 1964, The solution alteration of carbonate rocks, the effects of temperature and pressure: Geochim. Cosmochim. Acta, v. 28, p. 1401–1417.

Skibitzke, H. E., and da Costa, J. A., 1962, The groundwater flow system in the Snake River Plain, Idaho. An idealized analysis: U.S. Geol. Surv. Water Supply Papers, 1536-D.

Slattery, J. C., 1972, Momentum, Energy, and Mass Transfer in Continua: New York, McGraw-Hill.

Slichter, C. S., 1899, Theoretical investigation of the motion of groundwater: U.S. Geol. Surv. Ann. Rept. 19 (part 2), p. 295–384.

Smart, P. L., and Laidlaw, I. M. S., 1977, An evaluation of some fluorescent dyes for water tracing: Water Resources Res., v. 13, no. 1, p. 15–33.

Smith, L., and Chapman, D. S., 1983, On the thermal effects of groundwater flow. I, Regional scale systems: J. Geophys. Res., v. 88, p. 593–608.

Smith, R. E., and Woolhiser, D. A., 1971, Overland flow on an infiltrating surface: Water Resources Res., v. 7, p. 899–913.

Smith, R. L., Harvey, R. W., and Duff, J. H., and LeBlanc, D. R., 1987, Importance of close interval vertical sampling in delineating chemical and microbiological gradients in ground-water studies. U.S. Geol. Surv. Open-File Rept. 87-109, p. B33–B35.

Smoluchowski, M. S., 1909, Some remarks on the mechanics of overthrusting, Geol. Magazine, v. 6, p. 1133–1138.

Smyth, J. R., 1982, Zeolite stability constraints on radioactive waste isolation in zeolite bearing rocks: J. Geol., v. 90, p. 195–201.

Snow, D. T., 1968, Rock fracture spacings, openings, and porosities: J. Soil Mech., Found Div., Proc. Amer. Soc. Civil Engrs., v. 94, p. 73–91.

———, 1972, Geodynamics of seismic reservoirs: Symp. on Percolation Through Fissured Rock, Proc., Stuttgart, T2-J.

Solley, W. B., Chase, E. B., and Mann, W. B., 1983, Estimated use of water in the United States in 1980: U.S. Geol. Survey Circ. 1001.

Sopper, W. E., and Kardos, L. T., 1974, Conference on recycling treated municipal wastewater through forest and cropland, Report EPA-600/2-74-003: Washington, D.C., U.S. E.P.A.

Sorey, M. L., 1971, Measurement of vertical groundwater velocity from temperature profiles in wells: Water Resources Res., v. 7, p. 963–970.

———, 1978, Numerical modeling of liquid geothermal systems: U.S. Geol. Surv. Prof. Papers, 1044-D.

Southwell, R. V., 1946, Relaxation Methods in Theoretical Physics. London, Oxford University Press.

Sposito, G., 1984, The Surface Chemistry of Soils: New York, Oxford University Press, 234 p.

———, and Mattigod, S. V., 1980, GEOCHEM: A computer program for the calculation of chemical equilibria in soil solutions and other natural water systems. Riverside, Kearny Foundation, University of California.

Stallman, R. W., 1952, Nonequilibrium type curves for two well systems: U.S. Geol. Surv. Groundwater Notes, Open-File Rept. 3.

Stearns, D. W., 1967, Certain aspects of fracture in naturally deformed rock. Nat. Sci. Foundation Advanced Science Seminar in Rock Mechanics, ed. R. E. Riecker: Bedford, Mass., Airforce Cambridge Research Lab Spec. Rept., p. 97–118.

———, 1964, Macrofracture patterns on Tetron anticline, Northwest Montana [abs.]: Amer. Geophy. Union. Trans., v. 45, p. 107–108.

———, and Friedman, M., 1972, Reservoirs in fractured rock. Fracture Controlled Production. Amer. Assoc. Petrol. Geol. Reprint Series 21, p. 174–198.

Steefel, C. I., and Lasaga, A. C., 1988, The space-time evolution of dissolution patterns: Permeability change due to coupled flow and reaction. Submitted to Am. Chem. Soc. Symp. Set.

Stein, R. A., and Schwartz, F. W., 1989, On the origin of saline soils at Blackspring Ridge, Alberta, Canada: J. Hydrol. (in press).

Stiff, H. A., Jr., 1951, The interpretation of chemical water analysis by means of patterns: J. Petrol. Technol., v. 3, no. 10, p. 15–17.

Stoiber, R., 1941, Movement of mineralizing solutions in the Picher Field, Oklahoma-Kansas: Econ. Geol., v. 46, p. 800–812.

Strecker, E. W., and Chu, W., 1986, Parameter identification of a groundwater contaminant transport model: Ground Water, v. 24, no. 1, p. 56–62.

Streltsova, T. D., and Rushton, K., 1973, Water table drawdown due to a pumped well: Water Resources Res., v. 9, p. 236–242.

_____, 1972, Unsteady radial flow in an unconfined aquifer: Water Resources Res., v. 8, p. 1059–1066.

Stringfield, V. T., and LeGrande, H. E., 1966, Hydrology of limestone terranes: Geol. Soc. Amer. Spec. Paper 3.

Stults, J. M., 1966, Predicting farmer response to a falling water table–An Arizona Case Study. Proc. Econ. Water Res. Dev. Western Agri.: Las Vegas, Nevada, Econ. Res. Council, Report No. 5, p. 127–141.

Stumm, W., and Morgan, J. J., 1981, Aquatic Chemistry, 2nd ed.: New York: John Wiley, 780 p.

Sudicky, E. A., 1986, A natural gradient experiment on solute transport in a sand aquifer: Spatial variability of hydraulic conductivity and its role in the dispersion process: Water Resources Res., v. 22, no. 13, p. 2069–2082.

_____, 1989, The Laplace transform Galerkin technique: A time-continuous finite element theory and application to mass transport in groundwater: Water Resources Res., v. 25, no. 8, p. 1833–1846.

_____, Cherry, J. A., and Frind, E. O., 1983, Migration of contaminants in groundwater at a landfill: A case study: J. Hydrol., v. 63, p. 81–108.

Sullivan, C. R., Zinner, R. E., and Hughes, J. P., 1988, The occurrence of hydrocarbon on an unconfined aquifer and implications for the liquid recovery. NWWA/API Conference Petroleum Hydrocarbons and Organic Chemicals in Ground Water—Prevention, Detection, and Restoration, Proc.: Dublin, Ohio, National Water Well Assoc., p. 135–156.

Surdam, R. C., Boese, S. W., and Crossey, L. J., 1984, The chemistry of secondary porosity. Clastic Diagenesis, Amer. Assoc. Petrol. Geol. Memoir 37, eds. D. A. McDonald and R. C. Surdam: p. 127–149.

Swenson, F. A., 1968, New theory of recharge to the artesian basin of the Dakotas: Geol. Soc. Am. Bull., v. 79, p. 163–182.

Szymanski, J., 1989, Conceptual considerations of the Yucca Mountain groundwater system with special emphasis on the adequacy of this system to accomodate a high level nuclear waste repository: Las Vegas, Dept. of Energy Rept., Nevada operations office (unpublished).

Tamers, M. A., 1967, Surface-water infiltration and groundwater movement in arid zones of Venezuela. Isotopes in Hydrology. Vienna, Intl. Atomic Energy Agency, p. 339–351.

_____, 1975, Variability of radiocarbon dates on groundwater. Geophys. Surv., v. 2, p. 217–239.

Tang, D. H., Frind, E. O., and Sudicky, E. A., 1981, Contaminant transport in fractured porous media: Analytical solution for a single fracture: Water Resources Res., v. 17, no. 3, p. 555–564.

Terzaghi, K., 1950, Mechanism of landslides. Application of Geology to Engineering Practice, Berkeley Vol.: Geol. Soc. Amer. Publ. p. 83–123.

_____, and Peck, R. B., 1948, Soil Mechanics in Engineering Practice: New York, John Wiley.

Theis, C. V., 1935, The relation between the lowering of the piezometric surface and rate and duration of discharge of a well using groundwater storage. Trans. Amer. Geophys. Union, v. 2, p. 519–524.

————, 1938, The significance and nature of a cone of depression in groundwater bodies: Econ. Geol., v. 33, p. 889–902.

————, 1940, The source of water derived from wells—essential factors controlling the response of an aquifer to development: Civ. Eng., Amer. Soc. Civil Eng., p. 277–280.

————, Brown, R. H., Myers, R. R., 1963, Estimating the transmissibility of aquifers from the specific capacity of wells. Methods of determining permeability, transmissibility, and drawdown: U.S. Geol. Surv. Water Supply Papers, 1536-I.

Theim, G., 1906, Hydrologische Methode: Leipzig, Gebhardt.

Thomas, H. C., 1944, Heterogeneous ion exchange in a flowing system: J. Amer. Chem. Soc., v. 66, p. 1664–1666.

Thomas, H. E., 1951, The Conservation of Groundwater: New York, McGraw-Hill.

————, 1955, Water rights in areas of groundwater mining: U.S. Geol. Surv. Circ. 347.

Thompson, D. G., Meinzer, O. E., and Stringfield, V. T., 1938, Discussion on "Effect of a sea level canal on the groundwater level of Florida": Econ. Geol., v. 33, p. 87–107.

Thornthwaite, C. W., 1948, An approach toward a rational classification of climate: Geograph. Rev., v. 38, p. 55–94.

Thraikill, J., 1968, Chemical and hydrologic factors in the excavation of limestone caves: Geol. Soc. Amer. Bull., v. 79, p. 19–46.

Thurman, E. M., 1985, Organic Geochemistry of Natural Waters: Dordrecht, Martinus Nijhoff/Dr. W. Junk, 497 p.

Todd, D. K., 1959a, Annotated bibliography on artificial recharge of groundwater through 1954: U.S. Geol. Surv. Water Supply Papers, 1477.

————, 1959b, Groundwater Hydrology, 1st ed., New York, John Wiley.

————, 1980, Groundwater Hydrology, 2nd ed., New York, John Wiley.

Tokyo Inst. Civ. Eng., 1975, Subsidence as of 1974. Tokyo Metrop. Govt. Ann. Rept. (in Japanese).

Tolman, C. F., 1937, Groundwater. New York, McGraw-Hill.

Tóth, J., 1962, A theory of groundwater motion in small drainage basins in Central Alberta: J. Geophy. Res., v. 67, p. 4375–4387.

————, 1963, A theoretical analysis of groundwater flow in small drainage basins: J. Geophys. Res., v. 68, p. 4795–4812.

————, 1966, Mapping and interpretation of field phenomena for groundwater reconnaissance in a prairie environment, Alberta, Canada: Internat. Assoc. Sci. Hydro., 11 Année, no. 2, p. 1–49.

————, 1980, Cross-formational gravity flow of groundwater—a mechanism of the transport and accumulation of petroleum. Problems of Petroleum Migration, eds. W. H. Roberts and R. J. Cordell: Tulsa, Amer. Assoc. Petrol. Geol. Studies in Geol., no. 10, p. 121–167.

————, 1988, Groundwater and Hydrocarbon migration. The Geology of North America; Hydrogeology, eds. W. Back, J. Rosenshein, and P. Seaber: Boulder, Col., 0-2, Geol. Soc. Amer. Pub., p. 485–502.

————, and Corbett, T., 1986, Post-Paleocene evolution of regional groundwater flow systems and their relation to petroleum accumulations, Taber area, Southern Alberta: Canadian Petrol. Geol. Bull., v. 34, no. 3, p. 339–363.

Trescott, P. C., Pinder, G. F., and Larson, S. P., 1976, Finite difference model for aquifer simulation in two dimensions with results of numerical experiments: U.S. Geol. Surv. Tech. Water Res. Invest., Book 7, Chap. C1, 116 p.

Truesdell, A. H., and Hulston, J. R., 1980, Isotopic evidence on environments of geothermal systems. Chapter 5 in Handbook of Environmental Isotope Geochemistry, Vol. 1, eds. P. Fritz and J. C. Fontes: Amsterdam, Elsevier, p. 179–226.

_____, and Jones, B. F., 1974, WATEQ: A computer program for calculating chemical equilibria of natural water: Jour. Res. U.S. Geol. Surv., v. 2, p. 233–248.

Tsang, C. F., Buscheck, T., and Doughty, C., 1981, Aquifer thermal storage, A numerical simulation of Auburn University field experiments. Water Resources Res., v. 17, p. 647–658.

Tyson, N. H., and Weber, E. M., 1964, Groundwater management for the Nation's future—computer simulation of groundwater basins: J. Hydraulics Div., Amer. Soc. Civil Engrs. HY4, v. 90, p. 59–77.

Umari, A., Willis, R., and Liu, P. L.-F., 1979, Identification of aquifer dispersivities in two-dimensional transient groundwater contaminant transport: An optimization approach. Water Resources Res., v. 15, no. 4, p. 815–831.

United Nations, 1976, Second U.N. Symposium on Development and Use of Geothermal Resources, Proc., 3 vols.: San Francisco, United Nations.

United States Department of the Navy, Bureau of Yards and Docks, 1961, Design manual, Soil Mechanics, Foundations, and Earth Structures, DM-7, Chap. 4.

United States Geological Survey, 1970, The National Atlas of the United States of America: USGS.

Upchurch, S. B., and Lawrence, F. W., 1984, Impact of ground-water chemistry on sinkhole development along a retreating scarp. Sinkholes: Their Geology, Engineering and Environmental Impact, eds. B. F. Beck and W. L. Wilson: A. A. Balkema Publishers, Accord, Mass., p. 23–28.

Valocchi, A. J., 1985, Validity of the local equilibrium assumption for modeling sorbing solute transport through homogeneous soils: Water Resources Res., v. 21, p. 808–820.

_____, Roberts, P. V., Parks, G. A., and Street, R. L., 1981, Simulation of transport of ion-exchanging solutes using laboratory-determined chemical parameter values. Ground Water, v. 19, no. 6, p. 600–607.

Van Der Kamp, 1972, Tidal fluctuations in a confined aquifer extending under the sea: Ottawa, Env. Canada reprint 242, p. 101–106.

_____, and Gale, J. E., 1983, Theory of earth tide and barometric effects in porous formations with compressible grains. Water Resources Res., v. 19, p. 538–544.

Van Everdingen, R. O., 1972, Observed changes in groundwater regime caused by the creation of Lake Diefenbaker, Saskatchewan: Can. Dept. of the Envir., Inland Waters Branch Tech. Bull. 59.

Van Genuchten, M. Th., 1981, Non-equilibrium transport parameters from miscible displacement experiments: U.S. Dept. Agric. U.S. Salinity Lab., Res. Rept. No. 119, 88 p.

Van Golf-Racht, T. D., 1982, Fundamentals of Fractured Reservoir Engineering: New York, Elsevier.

Van Orstrand, C. E., 1934, Temperature gradients. Problems of Petroleum geology: Tulsa, Amer. Assoc. Petrol. Geol., p. 989–1021.

Verma, R. D., and Brutsaert, W., 1970, Unconfined aquifer seepage by capillary flow theory: J. Hydraul. Div., Proc. Amer. Assoc. Civil Engrs., v. 96, p. 1331–1344.

———, and Brutsaert, W., 1971, Unsteady free surface groundwater seepage. J. Hydraul. Div., Proc. Amer. Assoc. Civil Engrs., v. 97, p. 1213–1229.

Vernon, R. O., 1969, The geology and hydrology associated with a zone of high permeability (Boulder Zone) in Florida: Soc. Min. Engr. Preprint 69-AG12, 24 p.

Verschueren, K., 1977, Handbook of Environmental Data on Organic Chemicals: New York, Van Nostrand Reinhold.

Vogel, J. C., 1967, Investigation of groundwater flow with radiocarbon. Isotopes in Hydrology: Vienna, Internat. Atomic Energy Agency, p. 355–368.

———, 1970, Carbon-14 dating of groundwater. Isotope Hydrology: Vienna, Internat. Atomic Energy Agency, p. 235–237.

Vogel, T. M., and McCarty, P. L., 1987, Abiotic and biotic transformations of 1,1,1-trichloroethane under methanogenic conditions: Environ. Sci. Technol., v. 21, p. 1208–1213.

Vorhis, R. R., 1955, Interpretation of hydrologic data resulting from earthquakes: Geologischer Rundschar, v. 43, p. 47–52.

Wagner, B. J., and Gorelick, S. M., 1989, Reliable remediation in the presence of spatially variable hydraulic conductivity: From data to design: Water Resources Res., v. 25, no. 10, p. 2211–2225.

———, and Gorelick, S. M., 1986, A statistical methodology for estimating transport parameters: Theory and applications to one-dimensional advective-dispersive systems. Water Resources Res., v. 22, no. 8, p. 1303–1315.

Walder, J., and Nur, A., 1984, Porosity reduction and crustal pore pressure development: J. Geophysics Res., v. 89, p. 11539–11548.

Walker, R. G., and Cant, D. J., 1984, Sandy fluvial systems. Facies Models, ed. R. Walker, Geoscience Canada Reprint Series 1, p. 71–89.

Wallace, R. E., 1974, Goals, strategy, and tasks of the earthquake hazard reduction program: U.S. Geol. Survey Circular 701.

Wallick, E. I., Krouse, H. R., and Shakur, A., 1984, Environmental isotopes: Principles and applications in ground water geochemical studies in Alberta, Canada. First Canadian/American Conference on Hydrogeology, eds. B. Hitchon and E. J. Wallick: Dublin, Ohio, National Water Well Assoc., p. 249–266.

Wallis, P. M., Hynes, H. B. N., and Telang, S. A., 1981, The importance of groundwater in the transportation of allochthonous dissolved organic matter to the streams draining a small mountain basin: Hydrobiologia, v. 79, p. 77–90.

Walsh, J. B., 1978, The effect of fractures on compressibility, resistivity, and permeability: EOS, no. 59, p. 1193.

———, 1981, Effect of the pore pressure and confining pressure on fracture permeability: Internat. J. Rock Mech. Min. Soc. and Geomech., v. 18, p. 429–435.

Walther, E. G., Pitchford, A. M., and Olhoeft, G. R., 1986, A strategy for detecting subsurface organic contaminants. Proc. of the Petroleum Hydrocarbons and Organic Chemicals in Ground Water: Prevention, Detection, and Restoration: Dublin, Ohio, National Water Well Assoc., p. 357–381.

Walton, W. C., 1970, Groundwater Resource Evaluation: New York, McGraw-Hill.

Wang, H. F., and Anderson, M. P., 1982, Introduction to Groundwater Modeling—Finite Difference and Finite Element Methods. San Francisco, W. H. Freeman.

Wang, J. S. Y., and Narasimhan, T. N., 1985, Hydrologic mechanisms governing fluid flow in a partially saturated fractured porous medium: Water Resources Res., v. 21, p. 1861–1874.

Wang, J. S. Y., and others, 1978, Transient flow in tight fractures. Proc. Invitational Well Testing Symp.: University of California, Lawrence Berkeley Laboratory, p. 103–116.

Warren, J. E., and Price, H. S., 1961, Flow in heterogeneous porous media: Soc. Petrol. Eng. J., v. 1, p. 153–169.

Watts, E. V., 1948, Some aspects of high pressure in the D7 zone of the Ventura Avenue Field. Amer. Inst. Mining Engrs. Petrol. Div., v. 174, p. 191–200.

Way, S. C., and McKee, C. R., 1982, In-situ determination of three dimensional aquifer permeabilities: Ground Water, v. 20, p. 594–603.

Webster, D. S., Proctor, J. F., and Marine, I. W., 1970, Two well tracer test in fractured crystalline rock: U.S. Geol. Surv. Water Supply Papers, 1544–1.

Weeks, E. P., 1969, Determining the ratio of horizontal to vertical permeability by aquifer test analysis: Water Resources Res., v. 5, p. 196–214.

Wenzel, L. K., 1942, Methods for determining permeability of water bearing materials with special reference to discharging well methods: U.S. Geol. Surv. Water Supply Papers, 887.

Werner, P. W., and Noren, D., 1951, Progressive waves in non artesian aquifers: Trans. Amer. Geophys. Union, v. 32, p. 238–244.

Wershaw, R. L., Fishman, M. J., Gabbe, R. R., and Lowe, L. E., 1983, Methods for the determination of organic substances in water and fluvial sediments: U.S. Geol. Surv. TWRI, Book 5, Chapter A3, 173 p.

Wesson, R. L., 1981, Interpretation of changes in water level accompanying fault creep and implications for earthquake prediction: J. Geophys. Res., v. 86, p. 9259–9267.

Westall, J. C., Zachary, J. L., and Morel, F. M., 1976, MINEQL: A computer program for the calculation of chemical equilbrium composition of aqueous systems. Ralph M. Parsons Laboratory for Water Resources and Environ. Eng., Dept. of Civil Eng., M.I.T., Cambridge, Mass., Tech. Note 18.

Weyl, P. K., 1958, The solution kinetics of calcite: J. Geol., v. 66, p. 163–176.

———, 1959, Pressure solution and the force of crystalization. A phenomenological theory: J. Geophys. Res., v. 64, p. 2001–2025.

Whitaker, S., 1967, Diffusion and dispersion in porous media: Am. Inst. Chem. Eng. J., v. 13, p. 420–427.

White, D. E., 1968, Environments of generation of some base-metal ore deposits. Econ. Geol., v. 63, p. 301–335.

———, and Williams, D. L., 1975, Summary and conclusion. Assessment of Geothermal Resources of the United States—1975: U.S. Geol. Surv. Circ. 726, p. 147–155.

White, W. B., and E. L. White, 1987, Ordered and stochastic arrangements within regional sinkhole populations. Karst Hydrogeology: Engineering and Environmental Applications, eds. B. F. Beck and W. L. Wilson: Accord, Mass., A. A. Balkema Publishers, p. 85–90.

White, K. D., Novak, J. T., Goldsmith, C. D., and Bevan, S., 1985, Microbial degradation kinetics of alcohols in subsurface systems. NWWA/API Conference on Petroleum Hydrocarbons and Organic Chemicals in Ground Water—Prevention, Detection, and Restoration, Proc.: Dublin, Ohio, National Water Well Assoc., p. 140–159.

Whitney, J. D., 1854, Metallic Wealth of the United States: Philadelphia Lippincott, Grambo.

Wigley, T. M. L., Plummer, L. N., and Pearson, F. J., Jr., 1978, Mass transfer and carbon isotope evolution in natural water systems: Geochim. Cosmochim. Acta, v. 42, p. 1117–1139.

Willis, C., and Rubin, J., 1987, Transport of reacting solutes subject to a moving dissolution boundary: Numerical methods and solutions: Water Resources Res., v. 23, no. 8, p. 1561–1574.

Wilson, E. B., Jr., 1952, An Introduction to Scientific Research. New York, McGraw-Hill, 373 p.

Wilson, G., and Grace, H., 1942, The settlement of London due to underdrainage of the London Clay: J. Inst. Civil Engr. (London), v. 19, p. 100–127.

Wilson, J. L., and Conrad, S. H., 1984, Is physical displacement of residual hydrocarbons a realistic possibility in aquifer restoration? NWWA/API Conference on Petroleum Hydrocarbons and Organic Chemicals in Ground Water—Prevention, Detection, and Restoration, Proc.: Dublin, Ohio, National Water Well Assoc., p. 274–298.

———, and Miller, P. J., 1978, Two-dimensional plume in uniform groundwater flow: J. Hydraul. Div., Proc. Amer. Assoc. Civ. Eng., v. 104, p. 503–514.

Wilson, J. L., and Jordan, C., 1983, Middle shelf. Carbonate Depositional Environments, eds. P. A. Scholle, D. G. Bebout, and C. H. Moore: Amer. Assoc. Petrol. Geol. Mem. 33, p. 297–344.

Wilson, J., and Noonan, M. J., 1984, Microbial activity in model aquifer systems. Groundwater Pollution Microbiology, eds. G. Bitton and C. P. Gerba: New York: John Wiley, p. 117–133.

Wilson, J. T., and Wilson, B. H., 1985, Biotransformation of trichloroethylene in soil. Appl. Environ. Microbiol., v. 49, p. 242–243.

Winograd, I. J., 1962, Interbasin movement of groundwater at the Nevada Test Site, Nevada: U.S. Geol. Surv. Prof. Papers, 450C, p. 108–111.

———, and Thordarson, W., 1975, Hydrogeologic and hydrochemical framework, South Central Great Basin, with special reference to the nevada Test Site: U.S. Geol. Surv. Prof. Papers, 712-C.

Winslow, A. G., and Wood, L. A., 1959, Relation of land subsidence to groundwater withdrawals in the upper Gulf Coastal region of Texas: Amer. Inst. Mining Engrs. Mining Div., Mining Eng., p. 1030–1034.

Winter, T. C., 1976, Numerical simulation analysis of the interaction of lakes and groundwaters: U.S. Geol. Surv. Prof. Papers, 1001.

———, 1978, Numerical simulation of steady state three-dimensional groundwater flow near lakes: Water Resources Res., v. 14, p. 245–254.

Witherspoon, P. A., and Gale, J. E., 1977, Mechanical and hydraulic properties of rocks related to induced seismicity. Eng. Geol., v. 11, p. 23–55.

———, and others, 1980, Validity of cubic law for fluid flow in a deformable rock fracture: Water Resources Res., v. 16, p. 1016–1024.

Wolery, T. J., 1979, Calculation of Chemical Equilibrium Between Aqueous Solutions and Minerals: The EQ3/EQ6 Software Package, UCRL-52658: Livermore, University of California, Lawrence Livermore Laboratory, 41 p.

Wolff, R. G., 1970, Field and laboratory determination of the hydraulic diffusivity of a confining bed: Water Resources Res., v. 6, p. 194–203.

———, and Papadopulos, S. S., 1972, Determination of the hydraulic diffusivity of a heterogeneous confining bed: Water Resources Res., v. 8, p. 1051–1058.

Wood, J. R., and Hewett, T. A., 1982, Fluid convection and mass transfer in porous sandstone—a theoretical model. Geochim. Cosmochim. Acta, v. 46, p. 1707–1713.

Wood, W. W., 1976, Guidelines for the collection and field analysis of ground water samples for selected unstable constituents: U.S. Geol. Surv. TWRI, Book 1, Chapter D2, 24 p.

Woodbury, A. D., and Smith, L., 1985, On the thermal effects of three-dimensional groundwater flow: J. Geophy. Res., v. 90, p. 759–767.

Woodward and Clyde, Consultants, 1979, Study of reservoir induced seismicity, Reston, Va.: U.S. Geol. Survey, Technical Report Contract 14-08-0001-16809.

Wyllie, M. R. J., 1948, Some electrochemical properties of shales: Science, v. 108, p. 684–685.

Yagi, S., and Kunii, D., 1957, Studies on effective thermal conductivities in packed beds: Am. Inst. Chem. Eng. J., v. 3, p. 373–381.

Yaniga, P. M., and Mulry, J., 1984, Accelerated aquifer restoration: In-situ applied techniques for enhanced free product recovery/adsorbed hydrocarbon reduction via bioreclamation. NWWA/API Conference on Petroleum Hydrocarbons and Organic Chemicals in Ground Water—Prevention, Detection, and Restoration, Proc.: Dublin, Ohio, National Water Well Assoc., p. 421–440.

Yeh, W. W.-G., 1986, Review of parameter identification procedures in groundwater hydrology: The inverse problem: Water Resources Res., v. 22, no. 2, p. 95–108.

———, and Wang, C., 1987, Identification of aquifer dispersivities: Methods of analysis and parameter uncertainty: Water Resources Bull., v. 23, no. 4, p. 569–580.

Yoder, H. S., 1955, Role of water in metamorphism: Geol. Soc. Amer. Spec. Paper 62, p. 505–524.

Yong, R. N., 1985, Interaction of clay and industrial waste: A summary review. Second Canadian/American Conference on Hydrogeology, eds. B. Hitchon and M. Trudell: Dublin, Ohio, National Water Well Assoc., p. 13–25.

Young, A., and Low, P. F., 1965, Osmosis in argillaceous rocks: Amer. Assoc. Petrol. Geol. Bull., v. 49, p. 1004–1008.

Young, R., and Bredehoeft, J., 1972, Digital computer simulation for solving management problems of conjuctive groundwater and surface water systems: Water Resources Res., v. 8, p. 533–556.

Zaporozec, A., 1972, Graphical interpretation of water quality data: Ground Water, v. 10, p. 32–43.

Zaslavsky, D., and Ravina, I., 1965, Review of some studies of electrokinectic phenomena. Symposia on Moisture Equilibria and Moisture Changes in Soils. Australia, Sydney.

Zenger, D. H., 1972, Dolomitization and uniformitarianism: J. Geol. Educ., v. 20, p. 107–124.

Zimmerman, R. W., Somerton, W. H., and King, M. S., 1986, Compressibility of porous rocks: Jour. Geophys. Res., v. 91, p. 12,765–12,777.

Zoback, M. D., and Byerlee, J. D., 1976, A note on the deformation behavior and permeability of crushed granite: Internat. J. Rock Mech., Min. Sci., and Geomech., v. 13, p. 291–294.

Abdominal disorders, biological contaminants of ground water, 583

Abnormal fluid pressures, 284–307. *See also* Fluid pressure
compaction-driven flow system, 486
irreversible processes, 306–307
isothermal basin loading, 290–297
one-dimensional loading, 291–294
extensions of, 294–297
mathematical formulations of, 288–290
origin and distribution of, 284–288
phase transformations, 304–306
rock fracture and, 301–304
subnormal pressure, 306
tectonic strain as pressure-producing mechanism, 297–298
thermal expansion of fluids, 298–301

Acid-base reactions, 422–426. *See also* pH
alkalinity, 425–426

CO_2-water system, 424–425
natural weak acid-base systems, 423–424
overview of, 422–423
oxidation reactions and, 451–452
weak acid–strong base reactions:
inorganic reactions in the saturated zone, 500–504
inorganic reactions in the unsaturated zone, 494–497

Active depositional environments:
abnormal fluid pressures in, 284–307
forced convection in, 335–341

Activity models of reactions, described, 395–398

Advection:
chemical change and, 484
contaminant transport modeling, analytical approaches, 634–639
described, 358–360
plume manifestation, contaminant hydrogeology and, 583

Advection-diffusion equation, mass transport equations, 473–474. *See also* Diffusion equation

Advection-dispersion equation, 686–693
graphical procedure (Domenico & Robbins), 689–693
mass transport equations, 475
overview of, 686–687
trial-and-error fit of numerical model, 693
two-well tracer test data interpretation, 687–689

Advective flow regime, dispersion study errors and, 374

Advective front, zone of dispersion and, 364

Age dating of ground water, 525–530
carbon-14, 527–530
chlorine-36, 530
tritium, 526–527

Agriculture:
ground water resources, 236
nutrient contaminants of ground water, 580

Air, thermal conductivity of, 321. *See also* Atmospheric pressure

Alcohols:
described, 407
solubility of, 428

Aldehydes, described, 406

Alkali Lake, Oregon case study,
plume manifestation,
contaminant hydrogeology,
594–595
Alkalinity, described, 425–426
Alluvial apron, 34
Alluvial basin, described, 33–35
Alluvial fan, described, 33
Alluvial valley, described, 33, 34
Aluminum, organic reactions in
the unsaturated zone, 499
Aluminum silicates, dissolution
reactions of, 433
American practical hydrology unit,
transmissivity and, 149–150
Amines, described, 406
Amino acids:
described, 405
organic complexation, 439–440
Ammonia, acid-base reactions, 423
Ammonium, ground water
contaminants, 579–580
Ancestral seas:
Cenozoic rock group, 41–42
Mesozoic rock group, 41
Paleozoic rock group, 39–41
Anisotropicity:
Darcy's law and, 71–74
hydraulic conductivity and,
67–71
Aqueous geochemistry, 387–420.
See also Chemical reactions;
entries under Geochemical
aqueous solution phase,
389–390
chemical data description,
411–419
deviations from equilibrium,
398–399
equilibrium *vs.* kinetic
descriptions of reactions,
392–394
equilibrium models of reactions,
394–398
gas and solid phases, 391
ground water composition,
408–411
kinetic reactions, 399–401
organic compounds, 402–408
overview of, 387
water structure and occurrence
of mass in water, 391–392
Aquifer:
alluvial valleys and, 33
bounded aquifer testing,
175–181
classification by heterogeneity,
70–71
confined flow equation for,
115–116
conventional hydraulic test
procedures and methods,
144–154
early field studies of, 242

energy storage in, 348–349
flow nets and, 122
free convection in, 343
ground water basin management
and, 205–206
hydraulic testing prototype,
142–144
leaky aquifer method (hydraulic
testing), 155–159
nonequilibrium pumping test
method, 146, 150–151
partially penetrating wells, 172
permeability and, 27
potentiometric maps and, 78–80
pumping response of, 200–202
safe yield definition and,
202–203
sea water intrusion into,
230–236
specific yield of, 116–118
storage properties of porous
rock, 104–105
temperature profiles and, 329
water table aquifers (hydraulic
testing), 159–161
Aquifer simulation studies,
210–217. *See also* Simulation
studies
computational techniques:
steady flow, 214–216
unsteady flow, 216
described, 210–211
finite differences method,
211–214
finite elements method, 216
numerical model uses, 216–217
numerical simulation, 211
Aquitard, defined, 27–28
Aragonite, carbonate sediment
inversion, 536–537
Aromatic hydrocarbons:
described, 407
ground water contaminants,
580–582
vapor pressure data for, 431
Artesian well, early field studies
of, 242
Artificial recharge:
conjunctive use and, 210
ground water resource
development, 205–209
Atmospheric pressure:
abnormal fluid pressures and,
306
capillarity and, 11–12
flow in unsaturated zone,
88–89
water-level fluctuations and,
126, 128–129

Babylon, New York case study,
plume manifestation,
contaminant hydrogeology,
591–594

Bacteria:
biological contaminants of
ground water, 583
bioremediation, 735
chemical reactions
characterization, 695–698
laboratory interpretation of
chemical reaction pathways,
698–700
Barometric pressure, *see*
Atmospheric pressure
Barrier island system, described,
36–37
Basalt(s):
directional quality of, 67
elastic coefficients, 134
weathering of, 28
Base flow, hydrologic cycle and,
10, 14–17
Basin geology, ground water flow
and, 249–255
Basin geometry, ground water
flow and, 247–249
Basin loading:
abnormal fluid pressures and,
290–297
deformable porous media and,
131
Basin migration models,
hydrocarbons, 558–560
Beach system, described, 36–37
Benzene, ground water
contaminants, 582
Bernoulli equation, ground water
movement and, 58–59
Biocides, vapor pressure data for,
431
Biodegredation:
contaminant transport modeling,
analytical approaches,
642–643
mass transport mathematics,
476
plume manifestation,
contaminant hydrogeology
and, 586
Biological contaminants, ground
water contamination sources,
583
Biological degradation, in situ
contaminant treatment, 718
Bioremediation, 735–739
biochemical reactions,
735–736
feasibility study of, 740
oxygen and nutrient delivery,
736–739
system design comments, 739
Biotransformation, of organic
compounds, 457–460
Border tracer experiment,
diffusional model of
dispersion, contaminant
hydrogeology, 683–686

Borehole, water sampling methods, 617
Borehole-dilution tests, single-borehole tests, 171–172
Boundary:
  bounded aquifer testing, 175–181
  nonequilibrium pumping test method, 145–146
Boundary conditions:
  constant head boundary, 119–120
  contaminant transport modeling, analytical approaches, 636
  deformable porous media and, 126
  described, 118–124
  flux boundary, 120
  mass transport mathematics, 480–482
  no-flow boundary, 120
Bounded aquifers, hydraulic testing, 175–181
Braided river, fluvial deposits, 32
Braided stream, fluvial deposits, 32
Brine, *see* Sea water
BTEX compounds, ground water contaminants, 582
BTX compounds, ground water contaminants, 582
Bulk diffusion, dispersion coefficient and, 371

Calcite, carbonate sediment inversion, 536–537
Calcium:
  cation exchange, inorganic reactions in the saturated zone, 509
  ground water contaminants, 580
Calcium carbonate, dolomitization, 542
Cambrian-Ordovician aquifer, formation of, 41
Canals, landslides and, 279
Capillarity, described, 11
Capillary forces, flow in unsaturated zone, 89
Capillary fringe:
  nonaqueous contaminant spreading, multifluid contaminant problems, 602
  water sampling methods, 621
Capillary theory, flow in fractured rock, 92
Capillary water, hydrologic cycle and, 10–11
Capillary water zone, hydrologic cycle and, 11
Capture-zone-type curves, injection withdrawal systems, remediation, 725–731

Carbon-13, isotopic processes, 467
Carbon-14, age dating of ground water, 527–530
Carbonate rock(s), 534–544
  mass transport:
    chemical equilibrium in carbonate sediments, 536–539
    dolomitization, 542–544
    undersaturation problem, 539–542
  Paleozoic rock group, 41
  secondary porosity enhancement in, 49–50
  weathering of, 28
Carbonates, dissolution reactions of, 433
Carbonate system, pH and, 425–426
Carbonate terrain, ground water hydrologic cycle and, 262
Carbon dioxide:
  gas dissolution and redistribution, 492–493
  organic reactions in the unsaturated zone, 499
  weak acid–strong base reactions, inorganic reactions in the unsaturated zone, 494–497
Carbon dioxide-water system, described, 424–425
Carboxylic acids:
  described, 406
  organic complexation, 439–440
  solubility of, 428
Cation exchange, *see also* Ions
  inorganic reactions in the saturated zone, 509
  inorganic reactions in unsaturated zone, 497–498
Cenozoic rock group, ancestral seas, 41–42
CERCLA (Comprehensive Environmental Response, Compensation, and Liability Act), 6
Channeling, fractured media mixing, 382–383
Chemical change, mixing as agent for, 484–491
  deep sedimentary environments, 487–491
  meteoric and original formation waters, 485–487
Chemical data, quality assurance for, water sampling, 612–616
Chemical data description, 411–419
  abundance and patterns of change methods, 414–419
  abundance or relative abundance methods, 411–414

Chemical degradation, in situ contaminant treatment, 718
Chemical diagenesis, carbonate rock, 534–535
Chemical equilibrium, *see* Equilibrium models
Chemical hydrogeology:
  historical perspective on, 5
  marine environments, 44–46
Chemical osmosis, coupled phenomena, 564–566
Chemical reactions, 421–470. *See also* Aqueous geochemistry
  acid-base reactions, 422–426
  bioremediation, 735–736
  characterization of complete, 695–698
  complexation reactions, 434–440
  dissolution and precipitation of solids, 432
  gas solution and exsolution, 427–428
  hydrolysis, 460–462
  isotopic processes, 462–467
  laboratory interpretation of pathways in, 698–700
  numerical approaches, contaminant transport modeling, 664–665
  oxidation-reduction reactions, 450–460
  reactions on surfaces, 440–450
  solid solubility, 432–434
  solution of organic solutes in water, 428–430
  volatization, 430–432
Chemical thermodynamics, equilibrium approach to, 5
Chemical waste, hydraulic testing, 181
Chlorides, dissolution reactions of, 433
Chlorine-36, age dating of ground water, 530
Cholera, biological contaminants of ground water, 583
Clay(s):
  diagenesis of, 42
  elastic coefficients, 134
  land subsidence and, 225–226
  mineral properties of, 447
  mixing as agent for chemical change, 486–487
  particle transport, 360
  thermal conductivity of, 321
Clean Air Act, 6
Clean Drinking Water Act, 6
Clean Water Act, 6
Coal mining, sulfide oxidation, inorganic reactions in the unsaturated zone, 497

Coastal aquifer, sea water intrusion into, 230–236
Columbia Plateau, weathering of, 28
Compaction:
  lead-zinc deposits, 549–551
  marine environments, 42–44
Compaction-driven flow system, abnormal fluid pressure, 486
Compaction-gravity model, hydrocarbon migration, 559
Compaction-heat drive model, hydrocarbon migration, 559
Compaction models, hydrocarbon migration, 558–560
Compaction regime, abnormal fluid pressures, 285
Complexation reactions, 434–440
  enhancing mobility of metals, 438
  ion complexation and equilibrium calculations, 437–438
  organic complexation, 438–440
  overview of, 434
  stability of complexes and speciation modeling, 434–437
Comprehensive Environmental Response, Compensation, and Liability Act (CERCLA), 6
Compressibility of rock matrix, storage properties of porous rock, 107–115
Compressibility of water, storage properties of porous rock, 105–107
Compression, land subsidence and, 225–226
Computational techniques:
  steady flow, aquifer simulation studies, 214–216
  unsteady flow, aquifer simulation studies, 216
Computer:
  hydrogeology and, 6
  numerical approaches, contaminant transport modeling, 660–665
Computer programming, contaminant transport modeling, 651–656
Condensation, hydrologic cycle and, 9–10
Conduction:
  defined, 319
  effective conduction, 320
  equation for, 324–325
  Fourier's law, 319–321
Conduction-convection equation, described, 325–326
Conductivity, partial penetration, 172–173
Confined flow, equation for, in aquifer, 115–116

Conjunctive use, ground water resource development, 209–210
Connate water, *see* Original formation water
Connectivity, of fractures, 53–54
Conservation equation, mass transport equations, 472
Conservation of fluid mass:
  compressibility of rock matrix, 107
  mathematical statement of, 100–104
Conservation of mass, mass balance models, chemical pattern evaluation, 521
Conservation statement:
  energy transport equations, 323
  hydrologic equation and, 18–21
Constant head boundary, described, 119–120
Containment, corrective action alternatives, 712–716
Contaminant hydrogeology, 573–631
  ground water contamination sources, 574–583
    biological contaminants, 583
    listings of, 576–577
    miscellaneous inorganic species, 580
    nutrients, 579–580
    organic contaminants, 580–582
    overview of, 574–575, 578
    radioactive contaminants, 578–579
    trace metals, 579
  indirect detection methods, 624–628
    geophysical methods, 627–628
    soil-gas characterization, 624–627
  multifluid contaminant problems, 595–610
    gasoline leakage case study, 607–608
    Hyde Park landfill case study, 608–610
    imbibition and drainage, 597–598
    nonaqueous contaminant spreading features, 601–604
    overview of, 595–596
    relative permeability, 598–601
    saturation and wettability, 597
    secondary contamination due to NAPLs, 604–607
  process and parameter identification, *see* Process and parameter identification

  remediation, 711–749. *See also* Remediation
  sampling methods, 616–624
    saturation zone nonaqueous phase liquids, 621–622
    saturation zone solutes, 616–621
    solid and fluid sampling, 623–624
    unsaturated zone dissolved contaminants and nonaqueous phase liquids, 623
  solute plumes as manifestation of processes, 583–595
    Alkali Lake, Oregon case study, 594–595
    Babylon, New York case study, 591–594
    fractured and karst systems, 588–591
    generally, 583–588
  water sampling, 611–616
    chemical data quality assurance, 612–616
    design of networks for, 611–612
Contaminant migration:
  advection and, 359–360
  hydrogeology and, 6
  particle transport, 360–362
Contaminants:
  aqueous geochemistry, 388
  classification of, 578
  complexation reactions, 434
  corrective action alternatives:
    containment of, 712–716
    contaminant withdrawal, 716–718
    management options, 718–719
    in situ treatment of, 718
    withdrawal of, 716–718
  listings of, 576–577
  organic compounds, 402–405
  organic solutes in water, 428
  vapor pressure data for, 431
  volatilization, 430
Contaminant spreading features, nonaqueous, multifluid contaminant problems, 601–604
Contaminant transport, *see* Contaminant migration
Contaminant transport modeling, 633–675
  analytical approaches, 634–651
    advection and longitudinal dispersion, 634–639
    multidimensional transport models, 645–651
    overview of, 634
    radioactive decay, biodegradation and

hydrolysis, 642–643
retardation equation, 639–641
transverse dispersion,
643–645
case study in application of,
665–670
computer programming,
651–656
numerical approaches, 660–665
adding chemical reactions,
664–665
common solution techniques,
662–664
generalized modeling
approach, 661–662
overview of, 660–661
semianalytical approaches,
656–660
Contaminant withdrawal,
corrective action
alternatives, 716–718
Continental collisions, expulsion
of fluids and, 551–554
Continental environments, 28–36
marine environment boundary
with, 36–37
weathering, 28–31
Continuous source,
multidimensional transport,
contaminant transport
modeling, 645–649
Continuous source loading,
ground water contamination,
574
Continuum approach, ground
water movement, 83–87
Convection:
defined, 319
heat transport, 321–323
Conventional hydraulic testing,
see also Hydraulic testing
leaky aquifer method, 155–159
nonequilibrium pumping
method, 144–154
procedures and analysis,
144–161
steady-state behavior as
terminal case of transient
response, 155
water table aquifers, 159–161
Corrective action, *see* Remediation
Coupled phenomena, 563–569
chemical osmosis, 564–566
electro- and thermo-osmosis,
566
experimental, theoretical, and
field studies of, 566–567
generalized treatment of,
567–569
Crust, 283–316
abnormal fluid pressures,
284–307. *See also* Abnormal
fluid pressures
pore fluids in tectonic

processes, 307–315. *See also*
Tectonic processes
Crustal deformation, deformable
porous media and, 131
Crystalline rock:
hydraulic conductivity of, 65
porosity and, 25, 26
Cubic law, ground water
movement, 87–88
Curve-matching procedure:
leaky aquifer method (hydraulic
testing), 156
nonequilibrium pumping test
method, 148–150

Dakota sandstone, 27, 28
potentiometric map of, 77
response to pumping of, 202
storage properties of porous
rock, 105
Dam(s):
flow nets, 120–121
ground water hydrologic cycle
and, 268–269
Dangerous depletion, yield
anlaysis, 202–203
Darcy's equation:
advection and, 360
diffusion, 366
Darcy's law:
anisotropic material, 71–74
conservation of fluid mass, 102
contaminant transport
modeling, analytical
approaches, 634
convection and, 322
deformable porous media and,
124, 125
flow in unsaturated zone, 88, 90
flow nets and, 121
fractured rock flows, 83–84
ground water movement, 56–63
hydraulic conductivity
measurement and, 75
hydraulic head field mapping
and, 77–83
hydrocarbon displacement and
entrapment, 557
relative permeability, multifluid
contaminant problems, 598
Dating, *see* Age dating of ground
water
Davies equation, equilibrium
models of reaction, 397
DDT, solubility of, 428
Debye-Hückel equation,
equilibrium models of
reaction, 395–397
Decay, *see* Radioactive decay
Deep sedimentary waters, mixing
as agent for chemical change,
487–491
Deformable porous media,
124–136

drained response of water
levels to natural loading
events, 131–132
elastic properties, 132–135
flow equation development for,
124–126
flow equations for, 135–136
one-dimensional consolidation,
124–132
three-dimensional consolidation,
132–136
undrained response of water
to natural loading events,
126–130
Deltaic system, described, 36–37
Dense nonaqueous phase liquids,
see Nonaqueous phase liquids
(NAPLs)
Deposition, described, 31–36
Desert, evapotranspiration and,
12, 13
Deuterium, isotopic processes,
465–467
Dewatering operation, planning
for, 188–190
Diagenesis:
abnormal fluid pressures and,
304–306
free convection and, 346
marine environments, 42–46
uplift and erosion, 46–49
Differential equation:
boundary and initial
conditions, 480
finite differences method
(aquifer simulation studies),
211
hydrogeology and, 4
hydrologic equation and, 20–21
Diffusion:
deep sedimentary waters,
487–491
dispersion, 366–369
dispersion coefficient and, 371
fractured media mixing, 382
Diffusional model of dispersion,
process and parameter
identification, contaminant
hydrogeology, 681–686
Diffusion equation:
aquifer simulation studies, 211
confined flow in an aquifer,
115–116
conservation of fluid mass, 103
dimensional analysis, 136–137
effective stress concept and,
114–115
mass transport equations,
472–473
Dimensional analysis:
diffusion equations, 136–137
mass transport equations, 474
Dimensionless groups, heat
transport, 326–327

Diplacement, imbibition and
  drainage, multifluid
  contaminant problems,
  597-598
Discharge area, ground
  water-surface water
  interactions, 266
Discharge-recharge relations,
  ground water hydrologic
  cycle and, 260-264
Disequilibrium compaction,
  deformable porous media
  and, 124
Dispersion, 362-381
  basic concepts of, 362-370
  chemical change and, 484
  diffusion, 366-369
  diffusional model of,
    contaminant hydrogeology,
    681-686
  fractured media mixing,
    381-384
  geostatistical model of, 377-381
  longitudinal, contaminant
    transport modeling, analytical
    approaches, 634-639
  mechanical dispersion, 369-370
  plume manifestation,
    contaminant hydrogeology
    and, 585
  transverse, contaminant
    transport modeling, analytical
    approaches, 643-645
Dispersion coefficient, 371-377
  macroscopic and larger scale
    studies, 372-377
  as medium property, 372
  microscopic scale studies,
    371-372
Dispersivity, *see also* Process and
    parameter identification
  dispersion coefficient and,
    373-374
  estimation approaches of,
    693-695
  estimation of, 379-381
Displacement, hydrocarbons,
  556-558
Dissolution:
  diagenetic processes, 46
  patterning associated with,
    561-562
  of solids, 432
Dissolution of soluble salts,
  inorganic reactions in the
  saturated zone, 504
Dissolved contaminants, sampling
  methods, unsaturated zone,
  623
Distance-drawdown method,
  nonequilibrium pumping test
  method, 153-154
DNPLs, *see* Nonaqueous phase
  liquids (NAPLs)

Dolomite:
  thermal conductivity of, 321
  thermal diffusivities of, 326
Dolomitization, carbonate rock
  mass transport, 542-544
Dominant couples, oxidation-
  reduction reactions, 455
Double-porosity concept, hydraulic
  testing models, 144
Drainage:
  multifluid contaminant
    problems, 597-598
  stream hydrograph and, 15-17
Drained response, deformable
  porous media, 131-132
Drawdown:
  bounded aquifer testing,
    175-177
  hydraulic testing models, 143
  land subsidence and, 230
  leaky aquifer method (hydraulic
    testing), 155-156
  nonequilibrium pumping test
    method, 146-150
  response at pumped well test,
    169
  single-borehole tests, 162
Drill stem test, single-borehole
  tests, 163-164
Dunes, described, 35

Earthquake:
  deformable porous media and,
    131
  fluid injection and, 309-312
  landslides and, 279
  phreatic seismograph, 313-315
  pore fluids at midcrustal depths
    and, 312-313
  reservoir-induced, 312
Economic mineralization,
    544-556. *See also* Minerals
  expulsion of fluids from
    orogenic belts and
    continental collisions,
    551-554
  Mississippi Valley-type lead-zinc
    deposits, 548-551
  noncommercial mineralization
    (saline soils and evaporites),
    554-556
  roll front uranium deposits,
    545-548
Effective porosity, described,
  24-27
Effective stress:
  compressibility of rock matrix,
    107-115
  deformable porous media,
    124-125
Elasticity, land subsidence and,
  225-226
Elastic properties, deformable
  porous media, 132-135

Electolytes, aqueous geochemistry,
  388
Electro- and thermo-osmosis,
  coupled phenomena, 566
Electromagnetic conductivity,
  indirect detection of
  contamination methods,
  geophysical methods, 628
Electron acceptor availability,
  biotransformation of organic
  compounds, 459-460
Electron activity, described,
  450-455
Energy resources, 347-349
  aquifer storage and, 348-349
  geothermal energy, 347-348
Energy sources, geothermal
  energy and, 6
Entrapment, hydrocarbons,
  556-558
Environmental law, *see* Law
Environmental Protection Agency,
  see United States
  Environmental Protection
  Agency
Eolian deposit, described, 35
Equilibrium, chemical, in carbonate
  sediments, 536-539
Equilibrium models:
  complexation reactions,
    437-438
  described, 394-398
  deviations from, 398-399
  kinetic models *vs.*, 392-394
  quantitative approaches
    (chemical patterns
    evaluation), 515-520
Equilibrium reactions:
  equilibrium sorption reactions,
    mass transport mathematics,
    476-478
  sorption, mass transport
    mathematics, 476-478
Equipotential line:
  constant head boundary and,
    119
  ground water flow and,
    244-245
Erosion, 46-49. *See also*
    Weathering
  abnormal fluid pressures and,
    285
  described, 31-36
  subnormal pressures and, 306
Esters, described, 407
Estuarine system, described, 36-37
Ethylbenzene, ground water
  contaminants, 582
Ethylenediaminetetraacetic acid
  (EDTA), 439
Evaporation:
  hydrologic cycle and, 9-10
  hydrologic equation and, 18-21
  volatilization, 430

Evaporites, formation of, 554–556
Evapotranspiration, 9
  ground water hydrologic cycle
    and, 260–261
  ground water–surface water
    interactions, 266
  hydrologic cycle and, 9–10,
    12–14
  hydrologic equation and, 20–21
Excavation, contaminant
    withdrawal, 718
Excavations and tunnels, 269–278
  ground water inflow into,
    271–273, 274–278
  sea-level canals, 269–271
  stability of, in ground water
    discharge areas, 273–274

Facies, defined, 38
Faults, tunnel inflows and, 276
Fick's law:
  diffusion, 366–368
  diffusion equation, mass
    transport, 472
Field theory, ground water
    movement and, 60–61
Field tracer experiments, process
    and parameter identification,
    contaminant hydrogeology,
    679–681
Filter cakes, 361, 362
Finite differences method, aquifer
    simulation studies, 211–214
First-order kinetic reactions, mass
    transport mathematics, 476
Flooding, land subsidence and,
    217
Floridian aquifer, 28
Flow, *see also* Ground water flow
  continuum approach to, 83–87
  Darcy's law of, 56–63
  laws of, 56
Flow boundary, described, 120
Flow equation, deformable porous
    media, 124–126, 135–136
Flow lines, *see* No-flow boundary
Flow nets, described, 118–124
Flow theory, essentials of, 119
Fluid density, compressibility of
    water, 105–106
Fluid injection, seismicity induced
    by, 309–312
Fluid mass, *see* Conservation of
    fluid mass
Fluid pressure, *see also* Abnormal
    fluid pressures
  deformable porous media and,
    124, 127
  effective stress concept and, 110
  fluid injection and, 309–312
  geostatic pressure and, 285–286
  midcrustal depths, 312–313
  phreatic seismograph, 313–315
  porosity and, 53

reservoirs and, 312
  storage properties of porous
    rock, 105
  thrust faulting and, 308–309
Fluids, thermal expansion of,
    298–301
Fluid sampling, sampling methods,
    623–624
Fluid volume, compressibility of
    water, 105–106
Fluvial deposits, described, 32–33
Flux boundary, *see* Flow boundary
Forced convection, 327–343
  active depositional
    environments, 335–341
  classifications of, 327–328
  ground water velocity and,
    328–331
  heat transport, 322–323
  mountainous terrain, 341–343
  regional ground water flow,
    332–335
FORTRAN code, 652
Fourier's law:
  conduction, 324–325
  heat transport, 319–321
Fracture:
  defined, 50
  style of fracturing, 51–52
Fractured rock:
  abnormal fluid pressures and,
    301–304
  ground water movement in,
    83–88
  hydraulic testing, 181–185
  landslides and, 279–280
  mixing in, 381–384
  unsaturated flow in, 90–93
Fractured system, plume
    manifestation, contaminant
    hydrogeology, 588–591
Fracture formation, 50–54
Free convection, 343–347
  defined, 343
  geological implications,
    346–347
  heat transport, 322–323
  onset of, 344–345
  sloping layers, 345–346
Fresh water noses, development
    of, 485
Fresh water–salt water interface,
    sea water intrusion, 231–235
Frio deltaic aquifer, 37

Gas:
  aqueous geochemistry, 388, 391
  soil-gas characterization,
    indirect detection methods
    (of contamination), 624–627
  volatilization, 430
Gas dissolution and redistribution,
  inorganic reactions in
  unsaturated zone, 492–493

Gasoline, *see* Petroleum
Gasoline leakage case study,
    607–608
Gasoline spill case study, 744
Gas solution and exsolution,
    described, 427–428
Geochemical processes
    identification, 695–700
  complete characterization of
    possible reactions, 695–698
  laboratory interpretation of
    reaction pathways, 698–700
  plume configuration systematics,
    695
Geochemical processes
    quantification, 700–707
  field approaches to, 701–703
  laboratory techniques for,
    703–707
  overview of, 700–701
Geochemistry, *see* Aqueous
    geochemistry
Geohydrology, 2–3
  term of, 2
Geology, hydrogeology and, 7
Geomembranes, containment
    alternatives, 713
Geophysical methods, indirect
    detection methods (of
    contamination), 627–628
Geostatic pressure, fluid pressure
    and, 285–286
Geostatistical model, dispersion,
    377–381
Geothermal energy:
  energy resources, 347–348
  transport processes and, 6
Ghyben-Herzberg relation, sea
    water intrusion, 232–233
Gilson Road site, Nashua, New
    Hampshire, 746–748
Glacial deposit, described, 35–36
Gouge, tunnel inflows and, 276
Government:
  ground water resource
    management, 236–239
  nuclear waste storage and,
    349–350
Gradient, *see also* Potential
    gradient
  basin geology and, 249–255
  conservation of fluid mass, 103
  deformable porous media, 125
  flow boundary and, 120
Granite:
  recession curve and, 16
  thermal diffusivities of, 326
Graphical procedure (Domenico
    & Robbins), advection-
    dispersion equation, 689–693
Gravel pit, artificial recharge and,
    207
Gravitational loading, abnormal
    fluid pressure and, 288

Gravity:
  ground water flow and,
    242–243
  hydrocarbon migration, 559
Gravity flow origins, lead-zinc
    deposits, 551
Ground water:
  age dating of, 525–530
    carbon-14, 527–530
    chlorine-36, 530
    tritium, 526–527
  carbonate rock porosity and,
    49–50
  chemical composition of,
    408–411
  deviations from equilibrium, 398
  environmental law and, 6
  equilibrium models and, 392
  exploration for, 3, 4
  hydrogeology and, 2, 3
  hydrologic cycle and, 10–11
  hydrologic equation and, 18–21
  regional classification of, 33
  temperature of, 318
  water needs fulfilled by, 204
Ground water analysis, *see also*
    Water sampling; Water
    sampling methods
  routine, 408–410
  specialized, 410–411
Ground water basin management,
    defined, 205–206
Ground water contamination
    sources, 574–583
  biological contaminants, 583
  listings of, 576–577
  miscellaneous inorganic species,
    580
  nutrients, 579–580
  organic contaminants, 580–582
  overview of, 574–575, 578
  radioactive contaminants,
    578–579
  trace metals, 579
Ground water discharge area,
    excavation stability in,
    273–274
Ground water flow, 55–97. *See
    also* Ground water hydrologic
    cycle
  advection and, 358–360
  basin geology effects in, 249–255
  basin geometry effects in,
    247–249
  Darcy's law and field extensions
    of, 56–63
  early field studies of, 242–243
  fractured rock, 83–88
  geology and, 7
  heat transport and, 317–355.
    See also Heat transport
  hydraulic conductivity and,
    63–77
  hydraulic head field mapping

and, 77–83
  regional, 5
  research in, 4
  unsaturated zone flow, 88–93
Ground water hydrologic cycle,
    241–281
  surface features, 260–267
    ground water–lake
      interactions, 264–266
    ground water–surface water
      interactions, 266–267
    recharge-discharge relations,
      260–264
  topographic drive systems,
    268–280
    excavation stability in ground
      water discharge areas,
      273–274
    ground water inflows into
      excavations, 271–273
    ground water inflows into
      tunnels, 274–278
    landslides and slope stability,
      278–280
    large reservoir impoundments,
      268–269
    sea-level canal, 269–271
  topographic driving forces,
    242–259
    basin geology effects in,
      249–255
    basin geometry effects in,
      247–249
    early field studies, 242–243
    models of, 243–255
    mountainous terrain, 255–259
Groundwater hydrology, term of, 2
Ground water–lake interactions,
    ground water hydrologic
    cycle and, 264–266
Ground water mixing, *see* Mass
    transport
Ground water recession, defined,
    15
Ground water resource
    development:
  aquifer pumping response,
    200–202
  artificial recharge, 205–209
  conjunctive use, 209–210
  water law, 203–205
  yield analysis, 202–203
Ground water resource(s),
    199–240
  development of, 200–210
  land subsidence, 217–230
  overview of, 200
  sea water intrusion, 230–236
  simulation of aquifer response
    to pumping, 210–217
  simulation-optimization
    concepts, 236–239
Ground water–surface water
    interactions, ground

water hydrologic cycle and,
    266–267
Grouting, containment alternatives,
    713
Gypsum, organic reactions in the
    unsaturated zone, 499
Gypsum-anhydrite transformation,
    abnormal fluid pressures and,
    304
Gypsum precipitation and
    dissolution, inorganic
    reactions in unsaturated
    zone, 497

Hagen-Poiseuille equation, ground
    water movement and, 62
Half-reactions, described, 450–455
Halogenated hydrocarbons:
  contaminants, 402–405
  vapor pressure data for, 431
Hanford basalts, elastic
    coefficients, 134
Hawthorne formation, 28
Hazardous waste, *see also*
    Radioactive waste; Sewage
  flow in fractured rock, 90
  hydraulic testing and, 142
Health effects:
  biological contaminants of
    ground water, 583
  nutrient contaminants of ground
    water, 580
  organic contaminants of ground
    water, 582
Heat, *see also* Temperature
  abnormal fluid pressure and,
    288–289
  nonequilibrium pumping test
    method, 144–145
  thermal expansion of fluids,
    298–301
Heat flow:
  transport processes and, 6
  water flow and, 4
Heat transport, 317–355
  convection, 321–323
  energy resources, 347–349
  equations of, 323–327
  forced convection, 327–343
    active depositional
      environments, 335–341
    classifications of, 327–328
    ground water velocity and,
      328–331
    mountainous terrain, 341–343
    regional ground water flow,
      332–335
  Fourier's law, 319–321
  free convection, 343–347
    defined, 343–347
    geological implications,
      346–347
    onset of, 344–345
    sloping layers, 345–346

historical research in, 318-319
nuclear waste storage and, 349-354
Henry's law:
  gas solution and exsolution, 427
  volatilization, 430-431
Hepatitis, biological contaminants of ground water, 583
Heterogeneity:
  aquifer classification and, 70-71
  hydraulic conductivity and, 67-71
Heterogeneous equilibrium models, quantitative approaches (chemical patterns evaluation), 515-520
Heterogeneous kinetic reactions, mass transport mathematics, 478-480
Homogeneity, nonequilibrium pumping test method, 151
Homogeneous equilibrium models, quantitative approaches (chemical patterns evaluation), 515-520
Horizontal extension, abnormal fluid pressures and, 295-297
Hubbert's force potential, ground water movement and, 58-60
Hurricane, land subsidence and, 217
Hydaulic testing, superposition principle, 173-175
Hyde Park landfill, Niagara Falls, New York case study, 608-610, 748-749
Hydraulic conductivity:
  advection and, 360
  basin geology and, 251-252
  deformable porous media and, 124
  dispersivity estimation approaches, 693-695
  distribution character of, 66-67
  flow in fractured rock, 92
  flow in unsaturated zone, 89-90, 91
  flow nets and, 121
  fractured media mixing, 381
  geostatistical model of dispersion, 377
  ground water movement and, 62, 63-77
  heat transport and, 333
  leaky aquifer method (hydraulic testing), 156, 158
  measurement of, 74-77
  mechanical dispersion and, 369-370
  mountainous terrain, 255, 259
  particle transport and, 362
Hydraulic diffusivity:
  conservation of fluid mass, 103

deformable porous media and, 131
Hydraulic head:
  conservation of fluid mass, 102, 103
  field mapping of, 77-83
  ground water flow and, 244-245
  ground water movement and, 58-60, 62
  mountainous terrain and, 255
  recharge-discharge relations, 262, 264
  reservoir impoundments and, 268
  solute transport in vicinity of salt dome, 491
  water sampling network design, 611
Hydraulic pressure, mixing as agent for chemical change, 486-487
Hydraulic testing, 141-197
  applications to problems of, 185-192
  aquifer response to pumping, 200, 201
  bounded aquifers, 175-181
  conventional procedures and analysis, 144-161
  described, 142
  fractured or low-permeability rocks, 181-185
  leaky aquifer method, 155-159
  nonequilibrium pumping method, 144-154. *See also* Nonequilibrium pumping test method
  partial penetration, 172-173
  planning for (dewatering operation), 188-190
  planning for (pumping test), 185-188
  prototype geological models in, 142-144
  single-borehole tests, 161-172
  steady-state behavior as terminal case of transient response, 155
  water table aquifers, 159-161
Hydraulic transmissivity, *see* Transmissivity
Hydrocarbons:
  basin migration models, 558-560
  chlorinated, calculated retardation factors for, 703
  described, 407
  displacement and entrapment of, 556-558
  ground water contaminants, 580-582
  vapor pressure data for, 431

Hydrodynamic control, containment alternatives, 715
Hydrogen:
  acid-base reactions, 422
  half redox reactions, 453
  solubility of organic compounds, 428
Hydrogeology:
  defined, 7
  geologic studies and, 7
  history of field, 2-7
  hydrologic cycle and, 9-21
Hydrograph, *see* Stream hydrograph
Hydrologic cycle, 9-21. *See also* Ground water hydrologic cycle
  base flow in, 14-17
  components of, 9-12
  evapotranspiration and potential evapotranspiration, 12-14
  hydrologic equation and, 18-21
  infiltration and recharge in, 14
Hydrologic equation, hydrologic cycle and, 18-21
Hydrology, term of, 3
Hydrolysis:
  contaminant transport modeling, analytical approaches, 642-643
  described, 460-462
  mass transport mathematics, 476
  plume manifestation, contaminant hydrogeology and, 586
Hydrophobic sorption of organic compounds, surface reactions, 443-446
Hydrostatic balance, sea water intrusion, 232-233
Hydrostatic pressure, precipitation and, 243
Hydrostatics, capillarity and, 11
Hydrostratigraphic unit, defined, 28
Hydroxides, dissolution reactions of, 433

Igneous rock, weathering of, 28
Igneous silicate minerals, weathering of, 29
Image well theory, bounded aquifer testing, 175
Imbibition, multifluid contaminant problems, 597-598
Indirect detection methods (of contamination):
  geophysical methods, 627-628
  soil-gas characterization, 624-627
Infiltration:
  artificial recharge and, 206
  evapotranspiration and, 13

Infiltration (*Continued*)
  flow in fractured rock, 92
  flow in unsaturated zone, 89–90
  ground water–surface water
    interactions, 266, 267
  hydrologic cycle and, 10, 14
  mountainous terrain, 255
  particle transport, 361
Initial conditions, mass transport
    mathematics, 480–482
Injection withdrawal systems:
  design methods, remediation,
    724–732
  technical considerations,
    remediation, 719–724
Inorganic reactions in the
    saturated zone:
  cation exchange, 509
  dissolution of soluble salts, 504
  redox reactions, 504–509
  weak acid-strong base reactions,
    500–504
Inorganic reactions in unsaturated
    zone, 492–498
  cation exchange, 497–498
  gas dissolution and
    redistribution, 492–493
  gypsum precipitation and
    dissolution, 497
  sulfide oxidation, 497
  weak acid-strong base
    reactions, 494–497
Inorganic species (miscellaneous),
    ground water contamination
    sources, 580
In situ treatment, of contaminents,
    corrective action alternatives,
    718
Instantaneous point source
    model, multidimensional
    transport, contaminant
    transport modeling, 649–651
Interceptor systems:
  contaminant withdrawal, 717
  light nonaqueous phase liquids
    (LNAPLs), 732–735
Interflow, hydrologic cycle and,
    10
Intergranular porous rock, fluid
    flow in, 83–84
Intermediate zone, hydrologic
    cycle and, 10–11
Inversion:
  carbonate sediments, 536
  diagenetic processes, 46
Ions:
  acid-base reactions, 422
  advection and, 360
  alkalinity and, 425–426
  aqueous geochemistry, 388
  aqueous solution phase, 390
  cation exchange, inorganic
    reactions in the saturated
    zone, 509

chemical data description,
    411–419
complexation reactions, 434,
    437–438
deviations from equilibrium,
    398
diffusion coefficients for, 369
equilibrium models, 394, 397
ground water contaminants, 580
mobility of metals, 455–456
oxidation-reduction reactions,
    451
plume manifestation,
    contaminant hydrogeology
    and, 586, 588
solid solubility, 432–434
surface reactions, 447–450
weak acid-base systems,
    423–424
Iron, organic reactions in the
    unsaturated zone, 499
Irrigation:
  evapotranspiration and, 12, 13
  ground water demands of, 204
Isothermal basin loading,
    abnormal fluid pressures and,
    290–297
Isotopic processes, 462–467
  carbon-13 and sulfur-34, 467
  described, 464–465
  deuterium and oxygen-18,
    465–467
  isotopic reactions, 464–465
  radioactive decay, 462–463
Isotropicity, nonequilibrium
    pumping test method, 151

Karst:
  carbonate rock and, 534
  plume manifestation,
    contaminant hydrogeology,
    588–591
  undersaturation problem and,
    539–540
Karstification, described, 49
Kayenta sandstone, elastic
    coefficients, 134
Ketones, described, 406
Kinetic models, equilibrium
    models versus, 392–394
Kinetic reactions:
  described, 399–401
  first-order kinetic reactions,
    mass transport mathematics,
    476
  heterogeneous kinetic
    reactions, mass transport
    mathematics, 478–480
  oxidation-reduction reactions,
    455

Lacustrine deposit, described, 35
Lagoonal system, described,
    36–37

Lake-ground water interactions,
    ground water hydrologic
    cycle and, 264–266
Lakes, ground water hydrologic
    cycle and, 261
Landslides, ground water
    hydrologic cycle and,
    278–280
Land subsidence, 217–230
  causes of, 217
  consequences of, 217–218
  linearity and, 218–219
  mathematical treatment of,
    224–230
  physical properties of
    sediments, 219–224
  simulation studies of, 230
  time rates of, 228–230
  vertical compression, 224–227
Laplace's equation:
  conservation of fluid mass, 102,
    104
  flow nets and, 121
  ground water flow and,
    246–247
Law, *see also* Environmental law
  ground water resource
    development and, 203–205
  hydrogeology and, 6
Lead-zinc deposits, ground water
    formation of, 548–551
Leaky aquifer, temperature
    profiles and, 329
Leaky aquifer method:
  hydraulic testing, 155–159
  hydraulic testing
    (low-permeability rock), 184
Light nonaqueous phase liquids,
    see Nonaqueous phase liquids
    (NAPLs)
Limestone:
  elastic coefficients, 134
  recession curve and, 16
  thermal conductivity of, 321
  thermal diffusivities of, 326
  tunnel inflows and, 274
Lithology, fracture style and, 52
LNPLs, *see* Nonaqueous phase
    liquids (NAPLs)
Loading events:
  deformable porous media
    (drained response), 131–132
  deformable porous media
    (undrained response),
    126–130
Loading history, ground water
    contamination, 574–575
Loess, described, 35
Longitudinal dispersion:
  contaminant transport
    modeling, analytical
    approaches, 634–639
  dispersion coefficient and,
    373–374

transverse spreading and, 364
Long-term leakage, ground water contamination, 574
Low-permeability rock, hydraulic testing, 181–185

Magnesium:
cation exchange, inorganic reactions in the saturated zone, 509
ground water contaminants, 580
Management:
assessment of alternatives in, 720
corrective action alternatives, 718–719
remediation planning steps, 739–743
simulation-optimization concepts, 236–239
Maquoketa shale, 28
Marcasite, sulfide oxidation, inorganic reactions in the unsaturated zone, 497
Marginal marine environment, described, 36–37
Marine environment(s), 38–46
ancestral seas:
Cenozoic rock group, 41–42
Mesozoic rock group, 41
Paleozoic rock group, 39–41
carbonate sediments, 536
continental environment boundary with, 36–37
diagenesis in, 42–46
extent of, 38
stratigraphy of, 38–39
uplift from, 46–49
Mass balance models, quantitative approaches (chemical patterns evaluation), 520–523
Mass conservation, compressibility of water, 105
Mass transport, *see also* Contaminant transport modeling
advection, 358–360
dispersion, 362–381
basic concepts of, 362–370
diffusion, 366–369
mechanical dispersion, 369–370
dispersion coefficient, 371–377
geostatistical model of dispersion, 377–381
mixing in fractured media, 381–384
particle transport, 360–362
plume manifestation, contaminant hydrogeology and, 583
research in, 6
Mass transport (aqueous systems), 483–532

age dating of ground water, 525–530
carbon-14, 527–530
chlorine-36, 530
tritium, 526–527
inorganic reactions in the saturated zone, 500–509
cation exchange, 509
dissolution of soluble salts, 504
redox reactions, 504–509
weak acid–strong base reactions, 500–504
inorganic reactions in unsaturated zone, 492–498
cation exchange, 497–498
gas dissolution and redistribution, 492–493
gypsum precipitation and dissolution, 497
sulfide oxidation, 497
weak acid–strong base reactions, 494–497
Milk River aquifer case study, 510–515
mixing as chemical change agent, 484–491
deep sedimentary environments, 487–491
meteoric and original formation waters, 485–487
organic reactions in the unsaturated zone, 498–499
overview of, 484
quantitative approaches for evaluating chemical patterns, 515–525
homogeneous and heterogeneous equilibrium models, 515–520
mass balance models, 520–523
overview of, 515
reaction path models, 523–525
Mass transport (geologic systems), 533–571
carbonate rocks, 534–544
chemical equilibrium in carbonate sediments, 536–539
dolomitization, 542–544
undersaturation problem, 539–542
coupled phenomena, 563–569
chemical osmosis, 564–566
electro- and thermo-osmosis, 566
experimental, theoretical, and field studies of, 566–567
generalized treatment of, 567–569
economic mineralization, 544–556

expulsion of fluids from orogenic belts and continental collisions, 551–554
Mississippi Valley–type lead-zinc deposits, 548–551
noncommercial mineralization (saline soils and evaporites), 554–556
roll front uranium deposits, 545–548
hydrocarbon migration and entrapment, 556–560
overview of, 534
self-organization, 561–563
Mass transport mathematics, 471–482
boundary and initial conditions, 480–482
equations in, 472–475
advection-diffusion equation, 473–474
advection-dispersion equation, 475
diffusion equation, 472–473
mass transport with reaction, 475–480
equilibrium sorption reactions, 476–478
first-order kinetic reactions, 476
heterogeneous kinetic reactions, 478–480
Mathematical theory of electricity, bounded aquifer testing, 175
Mathematics (mass transport), *see* Mass transport mathematics
Matrix-fracture flow, flow in fractured rock, 92–93
Maximum stable basin yield, yield analysis, 203
Meandering river, fluvial deposits, 32
Mechanical dispersion:
described, 369–370
dispersion coefficient and, 371, 372
Meinzer unit, defined, 63
Mesozoic rock group, ancestral seas, 41
Metals:
complexation reactions, 438
control on mobility of, 455–456
ground water contaminants, 580
organic reactions in the unsaturated zone, 499
trace metal contaminants, 579
Metamorphic process(es), hydrogeology and, 7
Metamorphic rock:
directional quality of, 67
weathering of, 28

Meta-xylene, ground water
contaminants, 582
Meteoric water:
carbonate rock and, 535
isotopic processes, 466
mixing as agent for chemical
change, 485–487
Meteroic regime, abnormal fluid
pressures, 284–285
Methane zone, inorganic reactions
in the saturated zone, 505
Methanol, solubility of, 428
Methemoglobinemia, nutrient
contaminants of ground
water, 580
Microcosm, laboratory
interpretation of chemical
reaction pathways, 698–700
Microorganisms, particle transport,
360
Milk River aquifer case study,
mass transport (aqueous
systems), 510–515
Milk River sandstone,
heterogeneity, 68
Minerals, *see also* Economic
mineralization
alkalinity and, 425–426
aqueous geochemistry, 388
clay and, 447
complexation reactions,
437–438
dissolution reactions of, 433
homogeneous and heterogeneous
equilibrum models, chemical
pattern evaluation, 519
inorganic reactions in the
saturated zone, 504
mass balance models, chemical
pattern evaluation, 522–523
noncommercial mineralization,
554–556
precipitation of, 432
solid solubility, 432–434
sulfide oxidation, inorganic
reactions in the unsaturated
zone, 497
weak acid–strong base reactions,
inorganic reactions in the
unsaturated zone, 494–497
Miscellaneous inorganic species,
ground water contamination
sources, 580
Mississippi Valley–type lead-zinc
deposits, ore deposits,
548–551
Mixing, *see* Mass transport
Moisture content, defined, 11
Molal concentration, aqueous
solution phase, 389
Molar concentration, aqueous
solution phase, 389
Molecular diffusion, described,
367

Molecules, solubility of organic
compounds, 428
Montmorillonite-illite
transformation, abnormal fluid
pressures and, 304
Motion, laws of, 56
Mountainous terrain:
forced convection and, 341–343
ground water hydrologic cycle
in, 255–259
Multidimensional transport:
continous sources, contaminant
transport modeling, 645–649
instantaneous point source
model, contaminant transport
modeling, 649–651
Multifluid contaminant problems,
595–610
gasoline leakage case study,
607–608
Hyde Park landfill case study,
608–610
imbibition and drainage,
597–598
nonaqueous contaminant
spreading features, 601–604
overview of, 595–596
relative permeability, 598–601
saturation and wettability, 597
secondary contamination due
to NAPLs, 604–607
Multiparameter equilibrium
models, surface
reactions, 446–450
Multiple-borehead tests, hydraulic
testing (low-permeability
rock), 182–185
Multiple-lake systems, ground
water–lake interactions, 265
Multiple observation well test,
process and parameter
identification, contaminant
hydrogeology, 681

NAPL, *see* Nonaqueous phase
liquid (NAPL)
Natural gradient test, process
and parameter identification,
contaminant hydrogeology,
679
Newark Basin, 35
Nitrilotriacetic acid (NTA), 439
Nitrogen:
ground water contaminants,
579–580
half redox reactions, 453
No-flow boundary, described, 120
Nonaqueous contaminant
spreading features,
multifluid contaminant
problems, 601–604
Nonaqueous phase liquids
(NAPLs):
injection withdrawal systems,

contaminant remediation,
719–724
nonaqueous contaminant
spreading, multifluid
contaminant problems,
601–604
sampling methods:
saturation zone, 621–622
unsaturated zone, 623
secondary contamination due
to, multifluid contaminant
problems, 604–607
soil-gas characterization,
indirect contamination
detection, 625
Nonequilibrium pumping test
method, 144–154
aquifer response to pumping,
200, 201
assumptions and interpretations,
150–151
curve-matching procedure,
148–150
distance-drawdown method,
153–154
modifications of, 151–154
statement of, 145–148
temperature analogy in,
144–145
time-drawdown method,
152–153
Nonleaky response, hydraulic
testing models, 143
Nonpoint source, ground water
contamination, 574
Nuclear industry, radioactive
contaminant sources, 578
Nuclear waste storage, 349–354
burial sites for, hydraulic
testing and, 142
classification of materials,
349
hydraulic testing, 181
rock types and, 350–352
thermohydrochemical effects
and, 352–354
thermomechanical effects and,
354
Yucca Mountain site, 250
Numerical simulation studies,
aquifer simulation, 211
Nutrient delivery, bioremediation,
736–739
Nutrients, ground water
contamination sources,
579–580

Ogallala formation, 27
depletion of, 34
Ogata-Banks equation,
contaminant transport
modeling, analytical
approaches, 636, 640
Oil, *see* Petroleum

Oil spill, case study in, Calgary, Alberta, Canada, 744–746

One-dimensional basin loading, abnormal fluid pressures and, 290–297

One-dimensional consolidation, deformable porous media, 124–132

Ore deposits:
    expulsion of fluids from orogenic belts and continental collisions, 551–554
    ground water formation factors, 544
    Mississippi Valley-type lead-zinc deposits, 548–551
    noncommercial mineralization, 554–556
    roll front uranium deposits, 545–548

Organic complexation, described, 438–440

Organic compounds:
    aqueous geochemistry, 402–408
    biotransformation of, 457–460
    classification of, 402
    hydrolysis, 460–462
    hydrophobic sorption of, 443–446

Organic contaminants:
    bioremediation for, 735
    chemical reactions characterization, 695–698
    ground water contamination sources, 580–582
    laboratory interpretation of chemical reaction pathways, 698–700
    vapor pressure data for, 431

Organic reactions:
    rates of, 394
    in the unsaturated zone, mass transport (aqueous systems), 498–499

Organic solutes, chemical reactions, 428–430

Organic solvents, vapor pressure data for, 431

Organometallic compounds, described, 405

Original formation water:
    carbonate rock and, 535
    mixing as agent for chemical change, 485–487

Orogenic belts, expulsion of fluids from, 551–554

Ortho-xylene, ground water contaminants, 582

Osmosis:
    chemical osmosis, coupled phenomena, 564–566
    electro- and thermo-, couples phenomena, 566

Otis Air Base, Cape Cod, Massachusetts, contaminant transport modeling case study, 665–670

Overland flow, hydrologic cycle and, 10

Oxidation numbers, described, 450–455

Oxidation-reduction reactions, 450–460
    biotransformation of organic compounds, 457–460
    control on the mobility of metals, 455–456
    kinetics and dominant couples, 455
    overview of, 450–455

Oxygen, half redox reactions, 453

Oxygen 18, isotopic processes, 465–467

Oxygen delivery, bioremediation, 736–739

Paleozoic rock group, ancestral seas, 39–41

Parasites, biological contaminants of ground water, 583

Para-xylene, ground water contaminants, 582

Partial penetration, hydraulic testing, 172–173

Particle transport, described, 360–362

Patterning:
    dissolution and, 561–562
    precipitation and mixed phenomenon and, 562–563

Patuxent formation (Maryland), 27

PCBs, solubility of, 428

Percolation theory, connectivity of fractures and, 54

Permeability:
    abnormal fluid pressure and, 289
    described, 27–28
    fluid pressure and, 53
    ground water flow and, 251–255
    mountainous terrain, 255
    ground water movement and, 63–77
    relative, multifluid contaminant problems, 598–601

Pervious strata, artesian wells and, 242

Petroleum:
    bioremediation, 735, 736
    compaction and, 549
    gasoline leakage case study, multifluid contaminant problems, 607–608
    gasoline spill case study, 744
    ground water contaminants, 580–582
    hydrocarbon migration and displacement, 558–560

nonaqueous contaminant spreading, multifluid contaminant problems, 601–603

oil spill, case study, Calgary, Alberta, Canada, 744–746

relative permeability, multifluid contaminant problems, 600

Petroleum engineering, drill stem test, 163

Petroleum resources, ground water resources compared, 200

pH, *see also* Acid-base reactions
    acid-base reactions, 422
    carbonate system and, 425–426
    dissolution and precipitation of solids, 432
    metals mobility and, 438
    weak acid–strong base reactions, inorganic reactions in the unsaturated zone, 494–497

Phase transformation, abnormal fluid pressures and, 304–306

Phenols, described, 406

Phosphorous, ground water contaminants, 579–580

Phosphorous compounds, described, 405

Phreatic seismograph, earthquake, 313–315

Phreatic zone, hydrologic cycle and, 10–11

Piedmont region, weathering and, 31, 29

Pierre shale, 28

Piezometer:
    described, 58
    water sampling methods, 616–617

Pilot scale system, injection withdrawal systems, remediation, 724–725

Plume:
    configuration systematics, 695
    dispersion study errors and, 374
    indirect detection of contamination methods:
        geophysical methods, 627
        soil-gas characterization, 624
    low-pH contaminant plume, 432
    water sampling network design and, 611, 612

Plumes, manifestation of contaminant processes, 583–595

Point source, ground water contamination, 574

Polio, biological contaminants of ground water, 583

Polynuclear aromatic hydrocarbons, described, 407

Pore size, abnormal fluid pressure and, 288
Pore volume:
    abnormal fluid pressure and, 289
    compressibility of rock matrix, 110
Pore water density, solute transport in vicinity of salt dome, 491
Porosity:
    carbonate rocks, 49–50
    described, 24–27
    fluid pressure and, 53
    rock fracture and, 302
    storage properties of porous rock, 105–107
    uplift and, 46–49
Porosity reduction:
    chemical rock-water interactions, 44–46
    compaction and pressure solution forces in, 42–44
    sedimentation and, 284
Porous media, *see* Deformable porous media
Porous rock:
    elastic behavior of, 4
    storage properties of, 104–118
Potential evapotranspiration, hydrologic cycle and, 12–14
Potential gradient, ground water movement and, 88
Potentiometric map:
    aquifers and, 78
    hydraulic head field mapping, 77
Prairie environment, ground water hydrologic cycle and, 260–261
Precipitates:
    patterning associated with, 562–563
    solids, 432
Precipitation:
    evapotranspiration and, 13
    flow in fractured rock, 92
    gas dissolution and redistribution, 492–493
    ground water hydrologic cycle and, 262
    ground water-surface water interactions, 266
    gypsum precipitation and dissolution, inorganic reactions in the unsaturated zone, 497
    hydrologic cycle and, 9–10
    hydrologic equation and, 18–21
    hydrostatic pressure and, 243
    landslides and, 279
Pressure:
    compressibility of rock matrix, 107
    flow in unsaturated zone, 88–89
Pressure solution, marine environments, 43–44

Prickett-Lonnquist model, 217
Process and parameter identification, 677–710
    advection-dispersion equation, 686–693
        graphical procedure (Domenico & Robbins), 689–693
        overview of, 686–687
        trial-and-error fit of numerical model, 693
        two-well tracer test data interpretation, 687–689
    diffusional model of dispersion, 681–686
    dispersivity estimation approaches (new), 693–695
    geochemical processes identification, 695–700
    geochemical processes quantification, 700–707
    overview of, 678
    tracers and tracer tests, 678–681
Proportionality constant, ground water movement and, 61–63
Prototype geological models, hydraulic testing, 142–144
Pulse loading, ground water contamination, 574
Pump placement, aquifer response to pumping and, 202
Pyrite, sulfide oxidation, inorganic reactions in the unsaturated zone, 497

Quality, *see* Water quality
Quantitative approaches (chemical patterns evaluation), 515–525
    homogeneous and heterogeneous equilibrium models, 515–520
    mass balance models, 520–523
    overview of, 515
    reaction path models, 523–525
Quartz, thermal conductivity of, 321

Radiation, defined, 319
Radioactive contaminants, ground water contamination sources, 578–579
Radioactive decay:
    age dating of ground water, 525–530
    contaminant transport modeling, analytical approaches, 642–643
    described, 462–463
    mass transport mathematics, 476
    plume manifestation, contaminant hydrogeology and, 586
Radioactive waste, flow in fractured rock, 90

Rain, *see* Precipitation
Raoult's law, volatilization, 430–431
Rates of reaction, *see* Kinetic reactions; Reactions
RCRA, *see* Resource Conservation and Recovery Act (RCRA)
Reaction path models, quantitative approaches (chemical patterns evaluation), 523–525
Reactions, *see also* Redox reactions
    deviations from equilibrium, 398–399
    equilibrium models of, 394–398
    equilibrium *vs.* kinetic descriptions of, 392–394
    kinetic reactions, 399–401
    mass transport with, 475–480
    rates of, 393–394
Recession:
    defined, 14
    ground water and, 15
Recharge:
    abnormal fluid pressures and, 285
    aquifer response to pumping, 200–202
    artificial recharge, 205–209
    ground water-surface water interactions, 266
    hydrologic cycle and, 14
    leaky aquifer method (hydraulic testing), 155
Recharge area, ground water flow and, 245
Recharge-discharge area, sea-level canal and, 270
Recharge-discharge relations, ground water hydrologic cycle and, 260–264
Recharge wells, artificial recharge and, 207–208
Recrystallization, diagenetic processes, 46
Redistribution, flow in unsaturated zone, 90
Redox couples, homogeneous and heterogeneous equilibrium models, chemical pattern evaluation, 515
Redox potential, described, 450–455
Redox reactions, *see also* Reactions
    inorganic reactions in the saturated zone, 504–509
    rates of, 394
    sulfide oxidation, inorganic reactions in the unsaturated zone, 497
Regional ground water flow:
    heat transport in, 332–335
    research in, 5
Relative permeability:
    multifluid contaminant

problems, 598-601
nonaqueous contaminant spreading, multifluid contaminant problems, 602
Remediation, 711-749
    bioremediation, 735-739
        biochemical reactions, 735-736
        oxygen and nutrient delivery, 736-739
        system design comments, 739
    case studies in, 743-749
        gasoline spill, 744
        Gilson Road site, Nashua, New Hampshire, 746-748
        Hyde Park landfill, Niagara Falls, New York, 748-749
        oil spill, Calgary, Alberta, Canada, 744-746
    corrective action alternatives, 712-719
        containment, 712-716
        contaminant withdrawal, 716-718
        management options, 718-719
        in situ treatment of contaminants, 718
    injection withdrawal systems design methods, 724-732
    technical considerations, 719-724
    interceptors system for light nonaqueous phase liquids, 732-735
    planning steps for, 739-743
Replacement, diagenetic processes, 46
Replenishment, hydrostatic pressure and, 243
Representative elemental volume (REV), dispersion studies, 374-375
Reservoir impoundments, ground water hydrologic cycle and, 268-269
Reservoir(s):
    landslides and, 279
    seismicity induced by, 312
Residual drawdown, single-borehole tests, 162
Residual saturation, relative permeability, multifluid contaminant problems, 599-600
Resistivity, indirect detection of contamination methods, geophysical methods, 627-628
Resource, *see* Ground water resources
Resource Conservation and Recovery Act (RCRA), 6
Retardation equation, contaminant transport modeling, analytical approaches, 639-641

Retardation factors:
    chlorinated hydrocarbons, 703
    comparisons of, 705
Riparian doctrine, water laws and, 203-205
Rivers:
    erosion and, 31
    fluvial deposits, 32
River stage, deformable porous media and, 131
River water, homogeneous and heterogeneous equilibrium models, chemical pattern evaluation, 518
Rock, *see also entries under specific rock types*
    hydraulic conductivity and, 67
    nuclear waste storage, 350-352
    permeability and, 27-28
    porosity associated with, 24-26
    thermal conductivity of, 321
    thermal diffusivities of, 326
    tunnel inflows and, 274
Rock fracture, *see* Fractured rock
Rock matrix, compressibility of, 107-115
Roll front uranium deposits, ore deposits, 545-548
Routine water analysis, described, 408-410
Runoff, stream hydrograph and, 15-17

Safe yield:
    law and, 203, 204-205
    yield analysis, 202-203
Saline soils, formation of, 554-556
Salinity, dolomitization and, 542-544
Salt dome, solute transport in vicinity of, 491
Salt water-fresh water interface, sea water intrusion, 231-235
Sampling. *See* Water sampling
Sampling networks, design of, 611-612
Sandstone:
    beach and barrier island systems, 37
    Cenozoic rock group, 42
    diagenesis of, 42-43
    elastic coefficients, 134
    free convection and, 346
    heterogeneity, 68
    Paleozoic rock group, 41
    secondary porosity of, 44-46
    storage properties of porous rock, 105
    thermal conductivity of, 321
    thermal diffusivities of, 326
    tunnel inflows and, 274
Saturated zone, inorganic reactions in, 500-509
Saturation:
    defined, 10-11

multifluid contaminant problems, 597
Saturation zone, sampling methods:
    nonaqueous phase liquids, 621-622
    solutes in, 616-621
Screen diameter, pumping test planning, 186
Sea-level canal, ground water hydrologic cycle and, 269-271
Seasonal variation:
    ground water temperature variation, 318
    landslides and, 279
    reservoir impoundments and, 269
    safe yield definition and, 203
Sea water:
    compaction-driven flow system, 486
    diffusion in deep sedimentary environments, 487-491
    original formation waters, 485
Sea water intrusion, 230-236
    described, 230-231
    fresh water-salt water interface, 231-235
    Ghyben-Herzberg relation, 232-233
    sea-level canals and, 271
    submerged seepage surface interface, 233-235
    upconing of interface caused by pumping, 235-236
Secondary contamination due to nonaqueous phase liquid (NAPL), multifluid contaminant problems, 604-607
Secondary porosity, carbonate rocks, 49-50
Sedimentary materials:
    hydraulic conductivity of, 65
    porosity and, 26
Sedimentary rock:
    directional quality of, 67
    hydraulic conductivity of, 65
    porosity and, 25, 26
Sedimentary waters, mixing as agent for chemical change, 487-491
Sedimentation:
    forced convection and, 335-341
    porosity reduction and, 284
    weathering and, 28
Sediments:
    organic carbon content of, 444
    physical properties of, land subsidence, 219-224
Seepage velocity, borehole-dilution tests, 171-172
Seismicity:
    fluid-injection induced, 309-312
    pore fluids at midcrustal

Seismicity (*Continued*)
  depths and, 312–313
  reservoir-induced, 312
Self-organization, mass transport
  (geologic systems), 561–563
Semianalytical approaches,
  contaminant transport
  modeling, 656–660
Sewage, biological contaminants
  of ground water, 583
Sewage treatment, environmental
  law and, 6
Shale(s):
  aquitards, 28
  diagenesis of, 42–43, 45–46
  Paleozoic rock group, 41
  thermal diffusivities of, 326
  tunnel inflows and, 274
Sheet pile cutoff walls,
  containment alternatives, 713
Silicates, dissolution reactions of,
  433
Simulation-optimization
  techniques:
  ground water resource(s),
    236–239
  injection withdrawal systems,
    remediation, 731–732
Simulation studies, *see also*
  Aquifer simulation studies
  land subsidence, 230
  plume manifestation,
    contaminant hydrogeology,
    588
Single-borehole tests, 161–172
  borehole-dilution tests,
    171–172
  drill stem test, 163–164
  hydraulic testing
    (low-permeability rock), 182
  recovery in a pumped well,
    162–163
  response at pumped well
    (specific capacity), 168–171
  slug injection (withdrawal)
    tests, 164–168
Single-well injection or
  withdrawal test, process and
  parameter identification,
  contaminant hydrogeology,
  681
Single-well pulse test, process
  and parameter identification,
  contaminant hydrogeology,
  679–680
Sinkholes, classification of, 49
Slope stability, ground water
  hydrologic cycle and,
  278–280
Sloping layers, free convection in,
  345–346
Slug injection (withdrawal) tests:
  hydraulic testing (low-
    permeability rock), 182
  single-borehole tests, 164–168

Slurry walls, containment
  alternatives, 712–713
Socioeconomic factors, ground
  water management, 236–239
Sodium:
  cation exchange, inorganic
    reactions in the saturated
    zone, 509
  ground water contaminants, 580
Sodium bicarbonate water,
  chemical hydrogeology and, 5
Soil-gas characterization, indirect
  detection methods (of
  contamination), 624–627
Soil(s):
  flow in unsaturated zone, 89–90
  formation of, 28
  gas dissolution and
    redistribution, 492–493
  ground water hydrologic cycle
    and, 261
  infiltration and, 14
  organic reactions in the
    unsaturated zone, 499
  particle transport, 360
  saline soil formation, 554–556
  thermal diffusivities of, 326
Soil venting, contaminant
  withdrawal, 717
Soil water, hydrologic cycle and,
  10–11
Solids:
  aqueous geochemistry, 391
  dissolution and precipitation of,
    432
  solubility of, 432–434
Solid sampling, sampling
  methods, 623–624
Solubility, of solids, 432–434
Soluble salts, dissolution of,
  inorganic reactions in the
  saturated zone, 504
Solute plumes. *See* Plumes
Solutes:
  aqueous geochemistry, 388
  sampling methods, saturation
    zone, 616–621
Sorption isotherms, surface
  reactions, 440–443
Source geometry, continous
  sources, multidimensional
  transport, contaminant
  transport modeling, 645–649
Specialized water anaylsis,
  described, 410–411
Speciation calculation,
  homogeneous and
  heterogeneous equilibrium
  models, chemical pattern
  evaluation, 518
Speciation modeling, complexation
  reactions, 434–437
Specific capacity, response at
  pumped well test, 168–171
Specific storage:

compressibility of rock matrix
  and, 109–115
  porous rock, 104–105
Specific yield, geological materials,
  summarized, 118
Spills:
  gasoline spill case study, 744
  ground water contamination,
    574
  oil spill, case study in, Calgary,
    Alberta, Canada, 744–746
Stagnation zone, ground
  water-lake interactions,
  264–266
State law, *see* Law
Steady-state, finite differences
  study, aquifer simulation
  studies, 215
Steady-state behavior,
  conventional hydraulic
  testing, 155
Step-drawdown test, planning for,
  186–187
Storage properties (porous rock),
  104–118
  compressibility of rock matrix,
    107–115
  compressibility of water,
    105–107
  confined flow in an aquifer
    equation, 115–116
  research in, 104–105
  specific yield of aquifer, 116–118
Storativity:
  confined flow in an aquifer,
    115–116
  low-permeability rock hydraulic
    testing, 183
  nonequilibrium pumping test
    method, 146
Stream hydrograph, hydrologic
  cycle and, 15–17
Stream runoff, hydrologic cycle
  and, 10
Stress, *see* Effective stress
Submerged seepage surface
  interface, sea water intrusion,
  233–235
Subnormal pressure, abnormal
  fluid pressures, 306
Subsidence, abnormal fluid
  pressures and, 285
Substrate avilability,
  biotransformation of organic
  compounds, 458–459
Sulfates, dissolution reactions of,
  433
Sulfer-34, isotopic processes, 467
Sulfide oxidation, inorganic
  reactions in unsaturated
  zone, 497
Sulfides, dissolution reactions of,
  433
Sulfide zone, inorganic reactions
  in the saturated zone, 505

Sulfur, half redox reactions, 453
Superfund, *see* Comprehensive
   Environmental Response,
   Compensation, and Liability
   Act (CERCLA)
Superposition principle, hydaulic
   testing, 173–175
Surface drainage, containment
   alternatives, 715
Surface Mining Act, 6
Surface reactions, 440–450
   hydrophobic sorption of organic
     compounds, 443–446
   multiparameter equilibrium
     models, 446–450
   rates of, 394
   sorption isotherms, 440–443
Surface seals, containment
   alternatives, 713, 713–715
Surface tension, capillary water
   zone and, 11
Surface water-ground water
   interactions, ground water
   hydrologic cycle and, 266–267

Tectonic processes, 307–315
   abnormal fluid pressure and,
     288, 291, 297–298
   mixing as agent for chemical
     change, 486
   phreatic seismograph, 313–315
   seismicity and pore fluids at
     midcrustal depths, 312–313
   seismicity induced by fluid
     injection, 309–312
   seismicity induced by reservoirs,
     312
   thrust faulting, 308–309
Tectonism, 50–54
Temperature, *see also* Heat
   abnormal fluid pressure and,
     288–289
   evapotranspiration and, 13–14
   solute transport in vicinity of
     salt dome, 491
   thermal expansion of fluids,
     298–301
   weathering and, 29
Terzaghi consolidation equation,
   131
Testing, *see* Hydraulic testing
Theis nonequilibrium pumping
   test method, *see*
   Nonequilibrium pumping test
   method
Thermal expansion:
   abnormal fluid pressures and,
     298–301
   nuclear waste storage and, 354
Thermo- and electro-osmosis,
   coupled phenomena, 566
Thermobaric regime, abnormal
   fluid pressures and, 285
Thermodynamics, *see* Chemical
   thermodynamics

aqueous geochemistry, 391
equilibrium *vs.* kinetic
   descriptions of reactions,
   392–394
Thermohydrochemical effects,
   nuclear waste storage,
   352–354
Thermomechanical effects, nuclear
   waste storage, 354
Three-dimensional consolidation,
   deformable porous media,
   132–136
Thrust faulting, tectonic
   processes, 308–309
Tidal flat, described, 36–37
Tides, deformable porous media
   and, 127–129, 131
Time-drawdown:
   bounded aquifer testing, 177
   hydraulic testing models, 143
   leaky aquifer method (hydraulic
     testing), 156
   low-permeability rock hydraulic
     testing, 183
   nonequilibrium pumping test
     method, 152–153
   partial penetration, 173
   water table aquifers, 159–160
Toluene, ground water
   contaminants, 582
Topographic drive system,
   abnormal fluid pressures and,
   285
Topography:
   basin geology and, 249
   ground water flows and, 242–255
   landslides and, 279
   sea-level canal and, 269
Total porosity:
   described, 24–27
   storage properties of porous
     rock, 118
Toxic metals, complexation
   reactions, 434
Trace metals:
   ground water contamination
     sources, 579
   half redox reactions, 453–454
   laboratory interpretation of
     chemical reaction pathways,
     698–700
Tracer, dispersion study errors
   and, 374
Tracer plume, dispersion and,
   364–365
Tracers and tracer tests:
   Borden tracer experiment,
     683–686
   process and parameter
     identification, contaminant
     hydrogeology, 678–681
   two-well, data interpretation of,
     687–689
Transient flow, land subsidence
   and, 227

Transient response:
   aquifer simulation studies, 211
   steady-state behavior as
     terminal case of, 155
Transmissivity:
   American practical hydrology
     unit, 149–150
   confined flow in an aquifer, 116
   hydraulic testing models, 144
   low-permeability rock hydraulic
     testing, 183
   nonequilibrium pumping test
     method, 145
   water table aquifers, 159–161
Transpiration, *see*
   Evapotranspiration
Transportation, described, 31–36
Transport processes, *see* Mass
   transport
Transverse dispersion, contaminant
   transport modeling, analytical
   approaches, 643–645
Transverse spreading, dispersion
   and, 364
Trial-and-error fit of numerical
   model, advection-dispersion
   equation, 693
Trial-and-error simulations with
   models, injection withdrawal
   systems, remediation, 731
Trichloroethylene (TCE), soil-gas
   characterization, indirect
   contamination detection, 627
Tritium, age dating of ground
   water, 526–527
Tundra, evapotranspiration and,
   13
Tunnel construction, water table,
   188
Tunnels, *see* Excavations and
   tunnels
Two-well tracer test:
   data interpretation of, 687–689
   process and parameter
     identification, contaminant
     hydrogeology, 680–681
Typhoid fever, biological
   contaminants of ground
   water, 583

Undersaturation, carbonate rock,
   539–542
Undrained response, deformable
   porous media, 126–130
United States Environmental
   Protection Agency, pollutants
   listed by, 577
United States Geological Survey,
   4, 206, 217, 242
Unsaturated zone:
   ground water movement in,
     88–93
   ground water–surface water
     interactions, 267
   hydrologic cycle and, 10–11

Unsaturated zone (*Continued*)
inorganic reactions in, 492–498
organic reactions in, 498–499
Uplift, 46–49
abnormal fluid pressures and, 285
formational styles associated with, 47–49
ground water mixing caused by, 485
subnormal pressures and, 306
Uranium deposits, roll front, 545–548

Vadose zone, hydrologic cycle and, 10–11
Van Horn sandstone, 34–35, 35
Vapor pressure, volatilization, 431–432
Vegetation, evapotranspiration and, 13
Velocity:
advection and, 360
conservation of fluid mass, 102
Darcy's law of, 57–56
dispersion coefficient and, 371
mechanical dispersion and, 369–370
temperature profiles and, 328–331
Venting, soil, contaminant withdrawal, 717
Vertical compression:
abnormal fluid pressures and, 295
effective stress concept, 110–111
land subsidence, 224–227
Vertical flux, flow in fractured rock, 92–93
Viruses, biological contaminants of ground water, 583
Volatization, described, 430–432

Waste, *see* Sewage treatment
Waste disposal sites:
ground water contaminants, 582
single-borehole tests, 161
Waste water, biological contaminants of ground water, 583
Water:
atomic structure of, 391–392
compressibility of, 105–107
thermal conductivity of, 321

Water flow, heat flow and, 4
Water level, *see also* Water table
containment alternatives, 715
excavations and, 271–273
Water-level change, *see* Drawdown
Water level fluctuation, deformable porous media and, 126–130
Water quality, chemical hydrogeology and, 5
Water sampling, *see also* Ground water analysis
chemical data quality assurance, 612–616
design of networks for, 611–612
Water sampling methods, 616–624. *See also* Ground water analysis
indirect detection methods (of contamination):
geophysical methods, 627–628
soil-gas characterization, 624–627
saturation zone nonaqueous phase liquids, 621–622
saturation zone solutes, 616–621
solid and fluid sampling, 623–624
unsaturated zone dissolved contaminants and nonaqueous phase liquids, 623
Water spreading, artificial recharge and, 206–207
Water supply, hydraulic testing applications, 190–192
Water table, *see also* Water level
absolute definition of, 12
basin geology and, 249–255
basin geometry and, 247–249
excavation stability in ground water discharge areas, 273–274
flow in unsaturated zone, 88–89
heat conductivity and, 333
mountainous terrain and, 255, 259
reservoir impoundments and, 268
sea-level canals and, 269
topography and, 243, 244–246
tunnel construction, 188
tunnel inflows and, 277
Water table aquifers, conventional hydraulic testing, 159–161

Weak acid–base systems:
described, 423–424
homogeneous and heterogeneous equilibirum models, chemical pattern evaluation, 515
Weak acid–strong base reactions:
inorganic reactions in the saturated zone, 500–504
inorganic reactions in unsaturated zone, 494–497
Weathering, *see also* Erosion
continental environments, 28–31
Well capacity, response at pumped well test (specific capacity), 168–171
Wells:
connectivity of fractures and, 53–54
decrease in yields of, 31
early field studies of artesian wells, 242
fluid injection wells, 309–312
ground water flow and, 4–5
observation well placement, 187–188
partially penetrating wells, 172
pumping test planning, 185–188
recharge wells, 207–208
single-borehole tests, 161
step-drawdown test planning, 186–187
storage properties of porous rock, 105
water level fluctuations in, 126, 127
Wettability, multifluid contaminant problems, 597
Wetting, capillarity and, 11
Wilcox deltaic aquifer, 37
Willow rings, ground water hydrologic cycle and, 261
Wind, aeolian deposit, 35
Withdrawal (slug injection) tests, single-borehole tests, 164–168

Yield analysis, ground water resource development, 202–203

Zero-downdraw, bounded aquifer testing, 175–176
Zinc, *see* Lead-zinc deposits
Zone of dispersion, advective front and, 364

Printed in Singapore by
Chong Moh Offset Printing Pte. Ltd.